MW00907911

Instructor's Solution Manual & Test Item File

Calculus & Its Applications

SEVENTH EDITION

GOLDSTEIN • LAY • SCHNEIDER

PRENTICE HALL Upper Saddle River, NJ 07458

A Simon & Schuster/Viacom Company
Upper Saddle River, NJ 07458

10 9 8 7 6 5 4 3 2 1

ISBN 0-13-375171-6
Printed in the United States of America

CONTENTS

0. Functions

0.1.1: Graph the function $y = -\frac{1}{3}x + 2$ by plotting points.

Answer:

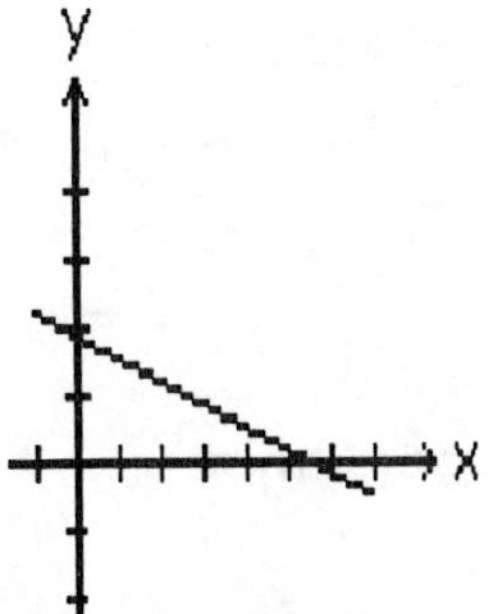

ES Moderate

0.1.2: Which of the following graphs are graphs of functions?

a)

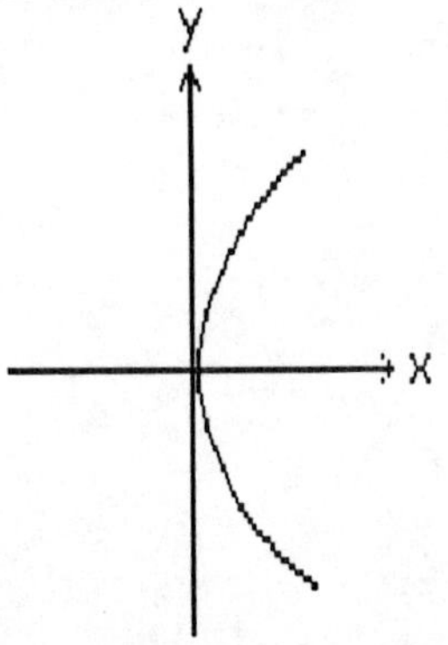

b)

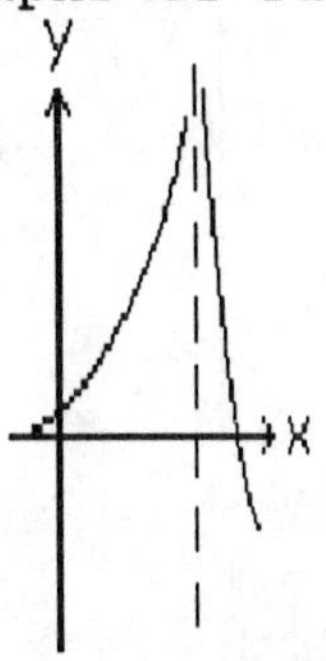

c)

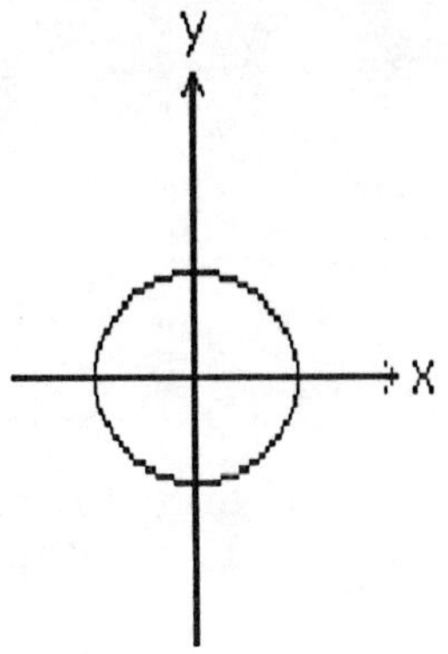

Answer: (b)
MC Moderate

0.1.3: If $f(x) = \frac{2x}{x^2 + 1}$, then $f(-2)$ equals

a) -2 b) $-\frac{4}{5}$ c) 0 d) $\frac{4}{3}$ e) $\frac{4}{5}$

Answer: (b)
MC Moderate

0.1.4: The function $f(x) = \dfrac{1}{(x - 1)^2 (x - 3)}$

a) has domain $x = 1, 3$
b) is not defined at $x = 1$, and $f(0) = \dfrac{1}{3}$
c) has domain $x \neq 1, 3$
d) is not defined for $x = -1, -3$, and $f(2) = 1$
e) none of the above

Answer: (c)
MC Moderate

0.1.5: Which of the graphs below are graphs of functions?

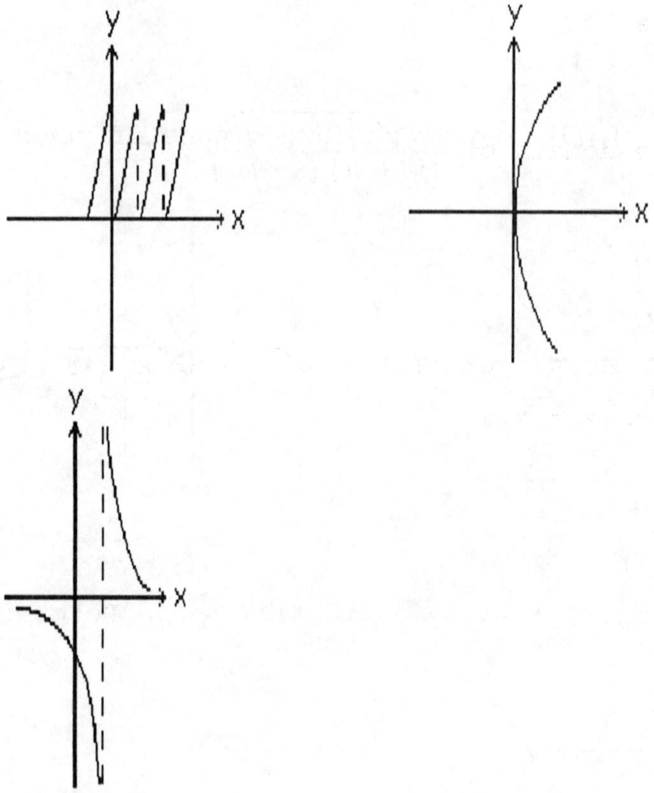

a) I only
b) I and II only
c) I and III only
d) all of the above

Answer: (c)
MC Moderate

0.1.6:

Let $f(x) = \sqrt{1 + \sqrt{x}}$. Then $f(9)$ equals:

a) 2
b) $\sqrt{3}$
c) 3
d) $\sqrt{10}$
e) none of the above

Answer: (a)
MC Moderate

0.1.7: The domain of the function $\frac{1}{\sqrt{1 - x}}$ is given by

a) $x \geq 0$
b) $x \leq 1$
c) $x \geq 1$
d) $x < 1$
e) none of the above

Answer: (d)
MC Moderate

0.1.8: Describe the domain of $f(x) = \frac{(x + 1)(x + 2)\sqrt{x - 3}}{(x^2 + 1)(x^2 + 5)}$.

Answer: $x \geq 3$
ES Moderate

0.1.9: Describe the domain of $\frac{\sqrt{|2 - x|}}{(x + 1)(x - 3)}$.

Answer: $x \neq -1, 3$
ES Moderate

0.1.10: Describe the domain of $f(x) = \sqrt{x} + \sqrt{x + 1} + \sqrt{x - 1} + \sqrt{x + 3} - \sqrt{x - 4}$.

Answer: $x \geq 4$
ES Moderate

0.1.11: Let $f(2x) = f(x) + 4x^2 + 5$ and $f(1) = 6$. Determine: $f(2)$

Answer: 15
ES Moderate

0.1.12: Let $f(x) = \frac{x + 4}{x^2 + 4x + 4}$. Determine: $f(1)$, $f(-1)$, $f(2)$

Answer: $f(1) = \frac{5}{9}$; $f(-1) = 3$; $f(2) = \frac{3}{8}$
ES Moderate

0.1.13: Which of the following could be the graph of $y = 2 + \sqrt{x}$?

a)

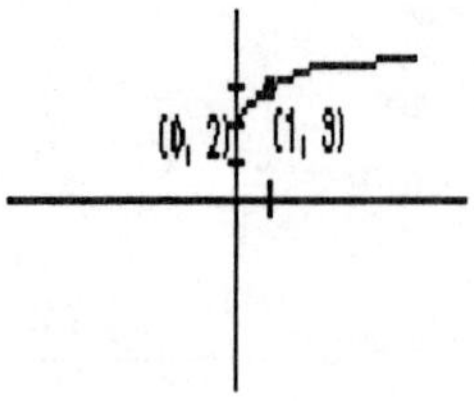

b)

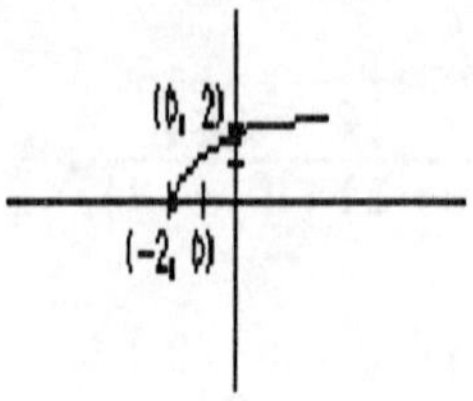

Answer: (a)
MC Moderate

0.1.14:

Describe the domain of the function $g(x) = \dfrac{\sqrt{x + 5}}{x + 3}$.

Answer: $x \geq -5$, $x \neq -3$
ES Moderate

0.1.15: If $f(x) = \dfrac{4x}{8(x + 2)(-x - 1)}$, which of the following are true?

(I) -1 is the domain.

(II) $f\left(\frac{1}{2}\right) = -\frac{1}{15}$.

(III) $f(0) = -\frac{1}{16}$.

(IV) f is not defined at $x = -2$ and $x = -1$

Answer: (II) and (IV)
ES Moderate

0.1.16: Consider the function $f(z) = \dfrac{5z}{(z + 3)\sqrt{z - 2}}$

(a) Describe the domain.
(b) Evaluate f(3).
(c) Find z when f(z) = 0.

Answer: (a) $z > 2$.
(b) $f(3) = \dfrac{5}{2}$.
(c) $z = 0$.
ES Moderate

0.1.17: Let $g(x) = \sqrt{\dfrac{1}{25 - x^2}}$. Determine g(-4).

Answer: $g(-4) = \dfrac{1}{3}$.

ES Moderate

0.1.18: If $f(t) = \dfrac{1}{\sqrt{-t}}$, then ...

a) f is not a function.
b) f has domain $t > 0$.
c) f has domain $t \leq 0$.
d) none of the above.

Answer: (d)
MC Moderate

0.1.19: If $h(t) = \dfrac{t^2}{t - 6}$, find an expression for h(a + 2)

Answer: $h(a + 2) = \dfrac{a^2 + 4a + 4}{a - 4}$.
ES Moderate

0.1.20: Is the point $\left(\dfrac{1}{2}, 2\right)$ on the graph of the function $f(x) = (x + 1)^2 - \dfrac{x}{2}$?

Answer: yes.
ES Moderate

0.2.1: Determine the domains of the following functions:

(a) $x^3 - 1$

(b) $\frac{x}{x - 2}$

(c) $\sqrt{x^2 + 1}$

(d) $\sqrt{(x - 1)(x - 2)}$

(e) $(x + 1)^{1/3}$

(f) $x^{3/4}$

Answer: (a) all x

(b) $x \neq 2$

(c) all x

(d) $x \geq 2$ or $x \leq 1$

(e) all x

(f) $x \geq 0$

ES Moderate

0.2.2: The y-intercept of the graph of the linear function $y = 3x + 8$ is

a) $(-\frac{8}{3}, 0)$

b) (0, 8)

c) (8, 0)

d) $(0, -\frac{8}{3})$

e) none of the above

Answer: (b)

MC Moderate

0.2.3: Find the x- and y-intercepts of the following functions:

(a) $f(x) = 5-2x$

(b) $f(x) = 3$

(c) $f(x) = x^2 -4$

Answer:

	x-int.	y-int.
(a)	5/2	5
(b)	none	3
(c)	2, -2	-4

ES Moderate

0.2.4: Evaluate $f(x) = \left| 3 - |x| \right|$ when $x = -6$.

Answer: $f(-6) = 3$.
ES Moderate

0.3.1: Let $f(x) = x^6 + 1$, $g(x) = x^3 + 1$. Calculate the following functions and express them in simplest terms:

(a) $f(x) + g(x)$

(b) $f(x) - g(x)$

(c) $f(x)g(x)$

(d) $f(x)/g(x)$

(e) $f(g(x))$

(f) $f(f(x))$

Answer:

(a) $x^6 + x^3 + 2$

(b) $x^6 - x^3$

(c) $x^9 + x^6 + x^3 + 1$

(d) $(x^6 + 1)/(x^3 + 1)$

(e) $(x^3 + 1)^6 + 1$

(f) $(x^6 + 1)^6 + 1$

ES Moderate

0.3.2: Let $f(x) = x^2 - 1$, $g(x) = \frac{1}{x}$. Determine the domains of $f(g(x))$ and $g(f(x))$ and calculate each of these functions.

Answer: $f(g(x)) = \frac{1}{x^2} - 1$, $x \neq 0$; $g(f(x)) = \frac{1}{x^2 - 1}$, $x \neq 1, -1$.
ES Moderate

0.3.3: Let $f(x) = x^3 + 2x^2$, $g(x) = x + 2$. Calculate $f(x)/g(x)$ and express in simplest terms.

Answer: $f(x)/g(x) = x^2$ $(x \neq -2)$
ES Moderate

0.3.4: Let $f(x) = \frac{x}{x - 1}$, $g(x) = \frac{2}{x + 1}$. Calculate the following functions and express them in simplest terms:

(a) $f(x) + g(x)$

(b) $f(x) - g(x)$

(c) $f(x)g(x)$

(d) $f(x)/g(x)$

(e) $f(g(x))$

(f) $f(f(x))$

Answer:

(a) $\frac{x^2 + 3x - 2}{x^2 - 1}$ $(x \neq \pm 1)$

(b) $\frac{x^2 - x + 2}{x^2 - 1}$ $(x \neq \pm 1)$

(c) $\frac{2x}{x^2 - 1}$ $(x \neq \pm 1)$

(d) $\frac{x^2 + x}{2x - 2}$ $(x \neq \pm 1)$

(e) $\frac{2}{1 - x}$ $(x \neq \pm 1)$

(f) x $(x \neq 1)$

ES Moderate

0.3.5: The graph of a function f(x) is sketched below. Which of the following statements is not true?

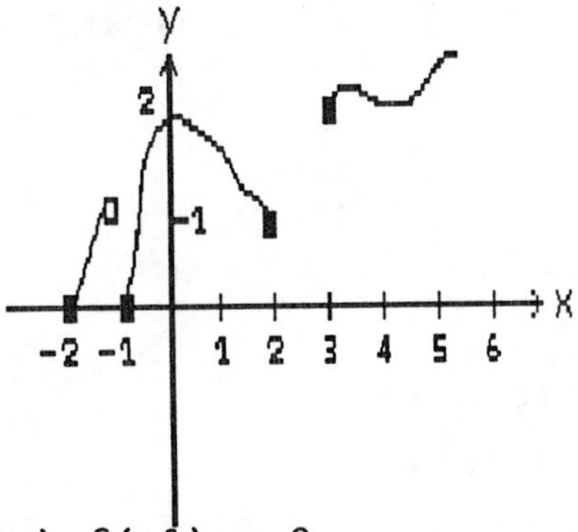

a) $f(-1) = 0$
b) $f(\frac{5}{2}) = 1$
c) f(x) is not defined for $2 < x \leq 3$
d) $f(0) = 2$
e) $f(-2) = 0$

Answer: (c)
MC Moderate

0.3.6: Let $g(x) = \sqrt{x}$, $h(x) = x^2 + 1$. Then g(h(-1)) is
a) 2
b) 0
c) $\sqrt{2}$
d) undefined
e) none of the above

Answer: (c)
MC Moderate

0.3.7: Let $f(x) = \frac{1}{x + 1}$, $g(x) = \frac{1}{x + 2}$. Then g(f(x)) equals
a) $\frac{x + 1}{x + 2}$
b) $\frac{x + 1}{3x + 2}$
c) $\frac{x + 1}{2x + 3}$, $x \neq -1$
d) $\frac{x}{3x + 2}$
e) none of the above

Answer: (c)
MC Moderate

0.3.8: Let $f(x) = x^{1/3}$, $g(x) = x^3$. Then $f(x)/g(x) - f(g(x))$ equals
a) $x^{-2/3} - x$
b) $x^{-8/3} - x$
c) $x^{-8/9} - x^3$
d) $x^{1/3} - x$
e) none of the above

Answer: (b)
MC Moderate

0.3.9: Let $f(x) = 2x^2 + 5x$. Then $\frac{f(x + h) - f(x)}{h}$ equals

a) $4x + 5xh + h^2 + 5h$

b) $4x + 5 + 4h^2 + 5h$

c) $4x + 5h + 4h^2$

d) $4x + 5 + 4h$

e) none of the above

Answer: (e)
MC Moderate

0.3.10: Determine the points of intersection:
$y = 2x^2 - 3x + 2$ and $y = x^2 + 2x - 4$

Answer: (3, 11) and (2, 4)
ES Moderate

0.3.11: Determine the points of intersection:
$y = -x^2 + 3x + 2$ and $y = 3x - 2$

Answer: (2, 4) and (-2, -8)
ES Moderate

0.3.12: Let $f(x) = x^2 - 3x + 2$.

Let $g(x) = 4x + 5$.

(a) Determine: $f(x) + g(x)$

(b) Determine: $f(x) \text{ x } g(x)$

(c) Determine: $f(g(x))$

Answer:

(a) $x^2 + x + 7$

(b) $4x^3 - 7x^2 - 7x + 10$

(c) $16x^2 + 28x + 12$

ES Moderate

0.3.13: Let $f(x) = 3x + 1$.

Let $g(x) = \sqrt{x + 3}$.

Let $h(x) = x^2$.

(a) Determine: $f(h(g(x)))$

(b) Determine: $g(h(f(x)))$

Answer: (a) $3x + 10,\ x \geq -3$

(b) $\sqrt{9x^2 + 6x + 4}$

ES Moderate

0.3.14:

If $f(x) = \dfrac{x}{2x - 1}$ and $g(x) = \dfrac{4\sqrt{x}}{x(x + 2)}$, what is the domain of $f(x) \cdot g(x)$?

a) $x \geq 0,\ x \neq \frac{1}{2}$ b) $x > \frac{1}{2}$ c) $x > 0,\ x \neq \frac{1}{2}$

Answer: (c)

MC Moderate

0.3.15: Let $f(t) = 5t - 4$. Find $\dfrac{f(t + h) - f(t)}{h}$.

Answer: 5

ES Moderate

0.3.16: Let $g(x) = 2x + 2$ and $h(x) = \dfrac{1}{x - 1}$. Calculate the function $h(g(x))$ and determine its domain.

Answer: $h(g(x)) = \dfrac{1}{2x + 1}$, domain: $x \neq \dfrac{-1}{2}$.

ES Moderate

0.3.17: Let $f(x) = -3 + 4x - x^2$ and $g(x) = x - 4$. When $x = 3$, $g(x)/f(x)$ is:

a) zero

b) undefined

c) -1

d) none of the above

Answer: (b)

MC Moderate

0.3.18: Let $f(x) = 2x + 3$ and $g(x) = \dfrac{1}{(2x + 3)(x - 1)}$.
(a) Express $f(x)/g(x)$ in simplest form.
(b) What is the domain of $f(x)/g(x)$?

Answer: (a) $\dfrac{f(x)}{g(x)} = \dfrac{1}{x - 1}$.
(b) $x \neq 1, -\dfrac{3}{2}$.
ES Moderate

0.3.19: Let $f(x) = \dfrac{5}{x + 1}$ and $g(x) = \dfrac{3x^2}{x - 1}$. Calculate:
(a) $f(x) \cdot g(x)$.
(b) $f(g(x))$

Answer: (a) $f(x)g(x) = \dfrac{15x^2}{x^2 - 1}$.
(b) $f(g(x)) = \dfrac{5(x - 1)}{3x^2 + x - 1}$.
ES Moderate

0.3.20: Find $\dfrac{g(a + h) - g(a)}{h}$ when $g(t) = -3t^2 + 2t - 3$.

Answer: $-6a + 2 - 3h$
ES Moderate

0.3.21: Let $g(t) = \dfrac{2t + 5}{t - 2}$ and $h(t) = \dfrac{t}{3t^2 + 1}$. Express $g(t) - h(t)$ as a rational function.

Answer: $g(t) - h(t) = \dfrac{6t^3 + 14t^2 + 4t + 5}{3t^3 - 6t^2 + t - 2}$.
ES Moderate

0.3.22: Let $f(x) = \dfrac{8}{3 - 5x}$ and $g(x) = x^2$. Calculate:
(a) $g(f(x))$
(b) $g(x) + f(x)$
(c) $g(x) \cdot f(x)$

Answer: (a) $g(f(x)) = \dfrac{64}{9 - 30x + 25x^2}$.
(b) $g(x) + f(x) = \dfrac{8 + 3x^2 - 5x^3}{3 - 5x}$.
(c) $g(x) \cdot f(x) = \dfrac{8x^2}{3 - 5x}$.
ES Moderate

0.4.1: Solve the following equations:

(a) $x^2 + 4x + 3 = 0$

(b) $x^2 - 3x - 4 = 0$

(c) $x - \frac{4}{x} = 0$

Answer: (a) $x = -3, -1$

(b) $x = 4, -1$

(c) $x = 2, -2$
ES Moderate

0.4.2: Solve the equation: $\frac{x^3 - 81x}{x^2 + 1} = 0$

Answer: $x = 0, 9, -9$
ES Moderate

0.4.3: Factor the following polynomials:

(a) $x^2 - 7x - 8$

(b) $3x^3 - 27x$

(c) $x^4 - 16$

Answer: (a) $(x - 8)(x + 1)$

(b) $3x(x - 3)(x + 3)$

(c) $(x^2 + 4)(x - 2)(x + 2)$
ES Moderate

0.4.4: Solve $2x^2 - 3x - 3 = 0$ using the quadratic formula.

Answer:
$x = \frac{3 \pm \sqrt{33}}{4}$
ES Moderate

0.4.5: Find the points of intersection of the graphs $y = 3x^2 - x - 2$ and $y = x^2 - 2x - 1$.

Answer: $(\frac{1}{2}, -\frac{7}{4})$, $(-1, 2)$
ES Moderate

0.4.6: The equation $-8x^2 - 3x + 6 = 0$ has as solution(s)

a) no value of x

b) $x = 0, -1$

c) $x = \dfrac{\pm 3 \pm \sqrt{39}}{16}$

d) $x = \dfrac{-3 \pm \sqrt{57}}{16}$

e) none of the above

Answer: (e)
MC Moderate

0.4.7: The expression $x^3 - 27x$ equals:

a) $x(x + \sqrt{27})(x - \sqrt{27})$

b) $(x + 3)^2(x - 3)$

c) $x(x + 3)(x - 3)$

d) $x^2(x + 3)$

e) none of the above

Answer: (a)
MC Moderate

0.4.8: The expression $x^2 - 5x + 4$ equals

a) $(x + 4)(x - 1)$

b) $(x - 4)(x - 1)$

c) $(x - 5)(x + 1)$

d) $(x + 4)(x + 1)$

e) none of the above

Answer: (b)
MC Moderate

0.4.9: The zeros of the function $f(x) = -x^2 - x - \frac{1}{4}$ are:

a) $x = \pm\frac{1}{2}$

b) $x = -\frac{1}{2}$

c) $x = 2$

d) $x = \dfrac{1 \pm \sqrt{2}}{2}$

e) none of the above

Answer: (b)
MC Moderate

0.4.10: The points of intersection of the graphs of $y = x^3 - x$ and $y = 3x$ are

a) (0, 0)

b) (0, 0) and (1, 0)

c) (0, 0), (2, 6) and (-2, -6)

d) (0, 0), (4, 0) and (-4, 0)

e) none of the above

Answer: (c)
MC Moderate

0.4.11: Determine all zeros of f for $f(x) = x^2 - 3x - 28$.

Answer: $x = -4, 7$
ES Moderate

0.4.12: Factor: $x^2 - 2x + 1$

Answer: $(x - 1)^2$
ES Moderate

0.4.13: Factor: $x^2 - x - 6$

Answer: $(x - 3)(x + 2)$
ES Moderate

0.4.14: Factor: $x^3 + 7x^2 - 8x$

Answer: $x(x + 8)(x - 1)$
ES Moderate

0.4.15: Factor: $6x^2 - 11x + 3$

Answer: $(3x - 1)(2x - 3)$
ES Moderate

0.4.16: Factor: $8x^3 - 14x^2 - 15x$

Answer: $x(4x + 3)(2x - 5)$
ES Moderate

0.4.17: Find the points of intersection of the graphs of the functions $y = \frac{1}{2}x + 4$ and $y = (x - 2)^2$ and sketch their graphs on the same coordinate axes.

Answer: (0, 4) and $\left(\frac{9}{2}, \frac{25}{4}\right)$

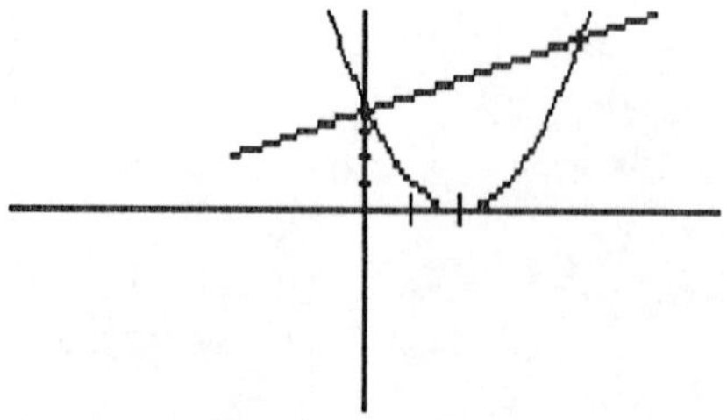

ES Moderate

0.4.18: Find the zeroes of the function $F(x) = x^5 - 9x^3 + 20x$.

Answer: $x = 0, \pm 2, \pm\sqrt{5}$
ES Moderate

0.4.19: The zeroes of the function $H(x) = x^4 + x^2 - 12$ are:
a) $x = \pm 2$ and $x = \pm 3$
b) $x = \pm\sqrt{3}$ and $x = -4$
c) $x = \pm 2$ and $x = \pm\sqrt{3}$
d) $x = \pm\sqrt{3}$

Answer: (d)
MC Moderate

0.4.20: The points of intersection of the graphs of $y = x^3 - 4x^2 - 16x$ and $y = 5$ are:
a) (5, -7) and (5, 3)
b) (0, 5), (7, 5) and (-3, 5)
c) (0, 0) and (7, -3)
d) (7, 5) and (-3, 5)

Answer: (b)
MC Moderate

0.4.21: The solutions of the equation $x^4 + 7x^2 + 10 = 0$ are:
a) $x = \pm 2$
b) $x = \pm\sqrt{2}$, $x = \pm\sqrt{5}$
c) $x = 2$, $x = 5$
d) $x = \pm\sqrt{10}$
e) There are no solutions.

Answer: (e)
MC Moderate

0.4.22: Which of the following is/are <u>not</u> solution(s) to the equation $y = x^3 + 3x^2 - 9x - 27$?

(I) $x = 0$
(II) $x = -3$
(III) $x = 2$
(IV) $x = -\sqrt{3}$
(V) $x = 3$

Answer: (I), (III) and (IV)
ES Moderate

0.4.23: The equation $x^3 - \frac{16}{x} = -6x$ has as solutions:

a) $x = 0,\ x = \pm\sqrt{2}$
b) $x = -8,\ x = 2$
c) $x = \frac{\pm\sqrt{112}}{14}$
d) There are no solutions.
e) None of the above.

Answer: (e)
MC Moderate

0.4.24: Solve the equation $t^2 + 10t + 12 = 0$.

Answer: $t = -5 \pm \sqrt{13}$.
ES Moderate

0.4.25: Find the zeroes of the function $g(x) = 3x^2 + 6x - 8$.

Answer: $x = -1 \pm \frac{2}{3}\sqrt{33}$
ES Moderate

0.4.26: Determine the points of intersection of the graphs of the functions $y = \frac{1}{x^2}$ and $y = \frac{x^2}{3x^2 - 2}$.

Answer: $(1, 1)$, $(-1, 1)$, $\left(\sqrt{2}, 1/2\right)$ and $\left(-\sqrt{2}, 1/2\right)$.
ES Moderate

0.4.27: Find the zeroes of the expression $\frac{2x^2 + 7x - 4}{x + 4}$.

Answer: $x = \frac{1}{2}$.
ES Moderate

0.5.1: Calculate the following:

(a) $\left|2^{-1}\right|$

(b) $\left|-\dfrac{2}{3}\right|$

(c) $\left|\dfrac{2}{-3}\right|$

Answer:
(a) 2^{-1}

(b) $\dfrac{2}{3}$

(c) $\dfrac{2}{3}$

ES Moderate

0.5.2: Simplify: $8^{4/3}$

Answer: 16
ES Moderate

0.5.3: Simplify: 3285963^{0}

Answer: 1
ES Moderate

0.5.4: Simplify: $216^{2/3}$

Answer: 36
ES Moderate

0.5.5: Simplify: $\dfrac{27^{1/2}}{3^{8/3}} \cdot \dfrac{2^{7/3}}{4^{3/2}} \cdot \dfrac{9^{8/6}}{8^{2/3}}$

Answer: $\dfrac{3^{3/2}}{2^{8/3}}$
ES Moderate

0.5.6: Simplify: $\left(\left(8^{\ 1/3}\right)^{2}\right)^{3/4}$

Answer: $2\sqrt{2}$
ES Moderate

0.5.7: Simplify: i^{999}

Answer: $-i$
ES Moderate

0.5.8: Let $f(x) = \left(\frac{x + 3}{2}\right)^{-1/2} \sqrt{x + 1^3}$. Determine: $f(1)$, $f(-1)$, $f(2)$.

Answer: $f(1) = 2$; $f(-1) = 0$; $f(2) = \frac{3\sqrt{6}}{5}$
ES Moderate

0.5.9: Simplify: $\frac{x^7}{y^9 x^{-3}}$

Answer: $\frac{x^{10}}{y^9}$
ES Moderate

0.5.10: Simplify: $(9x)^{-3/2}$

Answer: $\frac{1}{27x\sqrt{x}}$
ES Moderate

0.5.11: Simplify: $\sqrt{x}\left(\frac{1}{8x^3}\right)^{1/3}$

Answer: $\frac{1}{2\sqrt{x}}$
ES Moderate

0.5.12: Simplify: $\sqrt[3]{x^2} \cdot \sqrt[5]{x^4}$

Answer: $x^{22/15}$
ES Moderate

0.5.13: Simplify: $\frac{1}{x^{-5}y^{-2}}$

Answer: x^5y^2
ES Moderate

0.5.14: The expression $\left(\left(\sqrt{x+5}\right)^3\right)^{1/4}$ is equivalent to:
a) $(x + 5)^{3/2}$ b) $(x + 5)^{3/4}$ c) $(x + 5)^{3/8}$ d) $(x + 5)^{1/2}$

Answer: (c)
MC Moderate

0.5.15: The expression $(t - 1)^{7/4} - (t - 1)^{3/4}$ is equivalent to:
a) $(t - 1)^{3/4}(t - 2)$
b) $(t - 1)^{5/2}$
c) $(t - 1)$
d) None of the above.

Answer: (a)
MC Moderate

0.5.16: The expression $\left((x + 3)^{4/3}(x - 2)^{-1/3}\right)^2$ is equivalent to:
a) $(x^2 + x - 6)^2$
b) $\frac{1}{(x^2 + x - 6)^{8/3}}$
c) $\frac{(x + 3)^{8/3}}{(x - 2)^{2/3}}$
d) $\frac{(x + 3)^{8/3}}{(x - 2)^3}$

Answer: (c)
MC Moderate

0.5.17: Let $f(x) = (x - 5)^{3/4}(x + 1)^{-1/5}$. The domain of this function is:
a) $x \neq -1$, $x \neq 5$
b) $x \geq 5$, $x \neq -1$
c) $x \leq 5$
d) All real numbers

Answer: (b)
MC Moderate

0.5.18: Let $P(x) = (3x - 5)^{1/3}$ and let $Q(x) = (3x - 5)^{1/4}$. Then $P(x) + Q(x)$ is equal to:
a) $(3x - 5)^{7/12}$
b) $(3x - 5)^{1/12}$
c) $(9x^2 - 30x + 25)^{1/12}$
d) none of the above.

Answer: (d)
MC Moderate

0.5.19: Let $F(t) = \left(\sqrt[3]{t} - 4\right)^{-1}$ and let $G(t) = t^{2/3}$. Calculate $F(G(t))$ and determine its domain.

Answer: $F(G(t)) = \frac{1}{t^2 - 4}$, $t \neq \pm 2$.
ES Easy

0.5.20: Let $f(x) = \sqrt[3]{\left(\frac{-x^5}{(x - 24)^{10/3}}\right)^{1/5}}$. Compute $f(32)$.

Answer: $f(32) = -2$.
ES Moderate

0.6.1: A baseball thrown straight up into the air has height $s(t) = 6 + 44t - 16t^2$ feet after t seconds. Its velocity at a time t is given by $v(t) = 44 - 32t$.
(a) At what time will the ball be 16ft above the ground?
(b) What is its velocity at that time?

Answer: (a) $t = 2.5$ sec.
(b) -36 ft/sec.
ES Moderate

0.6.2: A toy rocket is fired straight up in the air. After t seconds, its height above ground is $s(t) = 36t - 16t^2$ meters and its velocity is $v(t) = 36 - 32t$ meters per second. What is the rocket's velocity when it hits the ground? (Hint?: What time d it hit the ground?)
a) -36 m/sec.
b) 0 m/sec.
c) 9/8 m/sec.
d) 2.25 m/sec.

Answer: (a)
MC Moderate

0.6.3: Suppose a toy rocket is fired straight up in the air and that after t seconds its height (in inches) is given by the function $s(t) = -16t^2 + 64t + 80$.
(a) At what time(s), is the rocket on the ground?
(b) How far off the ground was the rocket when it was fired?

Answer: (a) $t = 5$ sec.
(b) 80 inches
ES Moderate

0.6.4: A warehouse is being designed. The brick walls will cost \$3 per square foot to build. The roof is to be flat and square and will cost \$24 per square foot to build. The volume of the building is to be 32,000 cubic feet. Which of the following expresses the cost of building the warehouse?

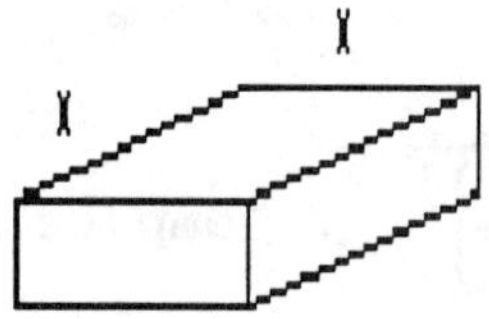

a) $x^2h = 32,000$

b) $12xh + 24x^2$

c) $4xh + x^2$

d) $4xh + 24x^2 = 32,000$

Answer: (b)
MC Moderate

0.6.5: A rectangular garden is to be fenced in and divided into three sections as shown below. The fencing for the boundary costs \$20 per foot while the fencing for the dividing fences costs \$15 per foot. The gardener has \$1300 to spend and wants the garden to enclose 150 square feet. Which of the following is/are true?

(I) $150 = 4w + 2l$

(II) $1300 = 70w + 40l$

(III) $1300 = w^4 \cdot l^2$

(IV) The gardener cannot afford the project if the garden is 15 feet long (ie, $l = 15$).

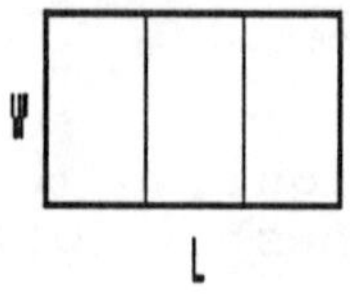

Answer: (II) only.
ES Moderate

0.6.6: A carpenter wishes to build a bookcase that is twice as long as it is high and has 2 adjustable shelves inside. The plywood to be used for the back costs \$1.50 per square foot. The rest of the bookcase will be made of pine boards costing \$5 per square foot. Write an equation expressing the fact that the total cost of materials is

Answer: $3h^2 + 10\ h \cdot w$; h = height, w = width of bookcase
ES Moderate

0.6.7: A closed rectangular box is to be constructed with square ends, a volume of 1000 cubic inches and surface area equal to 850 square inches. Which of the following is/are true?

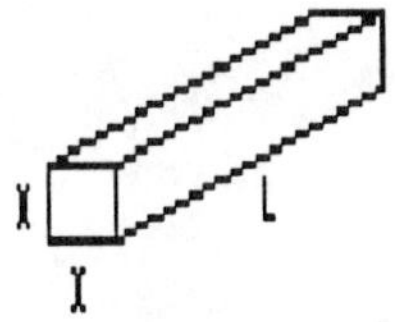

(I) $850 = 2x^2 + 4lx$

(II) $l = \frac{1000}{x^2}$

(III) $2x^2 + \frac{4000}{x} = 850$

(IV) $1000 = lx^2$

Answer: All are true.
ES Moderate

0.6.8: A chemical company wishes to build a cylindrical storage tank with a holding capacity of 500,000 cubic feet and a base of radius 50 feet. How tall will the tank be?

Answer: $\frac{200}{\pi}$ feet (about 63.7 feet)
ES Moderate

0.6.9: A 25 ft x 100 ft rectangular swimming pool is to be constructed so that its maximum depth is 4 times its minimum depth. The bottom of the pool is a steady slope. (See figure.) Which of the following is an expression for the volume of the pool? (Hint: What is the area of the shaded side?)

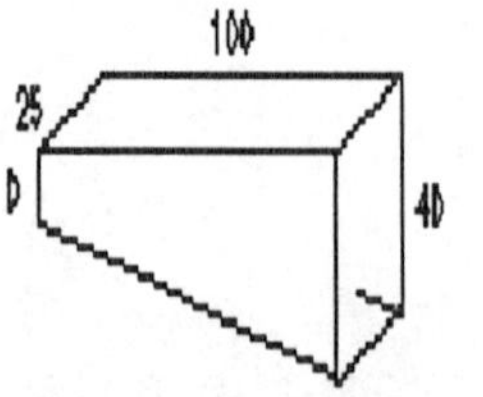

a) $25 \cdot 100 \cdot 4d - 25 \cdot 100 \cdot d$

b) $25 \cdot 100 \cdot d + \left(\frac{1}{2} \cdot 100 \cdot 3d\right) \cdot 25$

c) $\left(\frac{1}{2} \cdot 100 \cdot 4d\right) \cdot 25$

d) $25 \cdot 100 \cdot d$

Answer: (b)
MC Moderate

0.6.10: A manufacturer estimates that the hourly cost of producing x units of a product on an assembly line is $.2x^3 - x^2 + 30x + 105$ dollars.
(a) What is the cost of producing 3 units of the product?
(b) What is the additional cost if production is raised to 4 units?

Answer: (a) \$191.4
(b) \$30.4
ES Moderate

0.6.11: An entrepreneur is considering opening a shop to sell kites. She estimates the cost of making x kites to be $C(x) = 100 + 15x$ and plans to sell the kites for \$35 a piece. Which of the following statements is/are true?
(I) the sales revenue is given by $R(x) = 35x$.
(II) $C(10) = 350$.
(III) for \$400, she can produce 20 kites.
(IV) if she sells 4 kites, her profit will be \$140.

Answer: (I) and (III)
ES Moderate

0.6.12: A restauranteur determines that the average daily revenue per table is given by $R(x) = \frac{-1}{4}x + 15$ when there are x tables in the dining room. The average daily cost per table is \$5.00.

(a) Determine an expression for the average daily profit per table.

(b) How many tables are required to make the profit per table equal \$2.00?

Answer:
(a) $P(x) = \frac{-1}{4}x + 10$
(b) 32 tables
ES Moderate

0.6.13: Suppose the revenue received from the sale of x units of a product is given by $R(x) = 11x - 2x^2$ while the cost of producing those x units is $C(x) = \frac{1}{4}x^3 - x + \frac{1000}{x}$. What is the profit function P(x)?

a) $P(x) = \frac{1}{4}x^3 + 2x^2 - 12x + \frac{1000}{x}$

b) $P(x) = \frac{-x^4 - 8x^3 + 48x^2 - 4000}{4x}$

c) $P(x) =- \frac{1}{4}x^3 - 2x^2 + 12x - 1000$

d) $P(x) = \frac{-x^2 - 8x - 3952}{4x}$

Answer: (b)
MC Moderate

0.6.14: Consider the graph of the function V(x) below where V(x) gives the volume of a cube with sides x cm long.

(a) What is the volume of an 8-cm-sided cube?

(b) How long are the sides of an 8-cubic-cm cube?

(c) How much volume is gained or lost by changing the sides of a 64-cubic-cm cube to 8 cm?

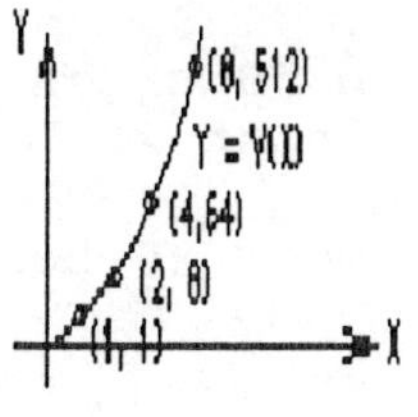

Answer: (a) 512 cm^3.
(b) 2 cm^3.
(c) 448 cm^3 gained.
ES Moderate

0.6.15: Consider the cost and revenue functions shown below, where the cost of producing x items is C(x) dollars and the revenue from the sale of x items is R(x) dollars. Which of the following statements is/are true?

(I) more than 300 items must be sold in order make a profit.

(II) No profit is made if less than 300 items are produced.

(III) It is better to produce even a very few items than to produce none at all.

(IV) The cost to produce 100 items is $400.

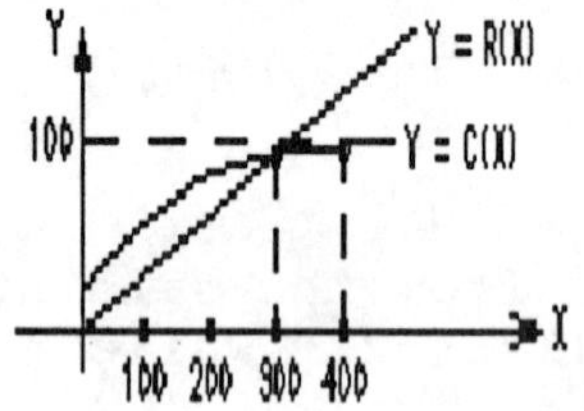

Answer: (I) and (II)
ES Moderate

0.6.16: An orange grower has plotted a graph of a function relating the number of trees in a grove to the productivity of the trees: if there are x trees in a grove, each tree will yield P(x) bushels of fruit. What is the significance of the point (50, 20)?

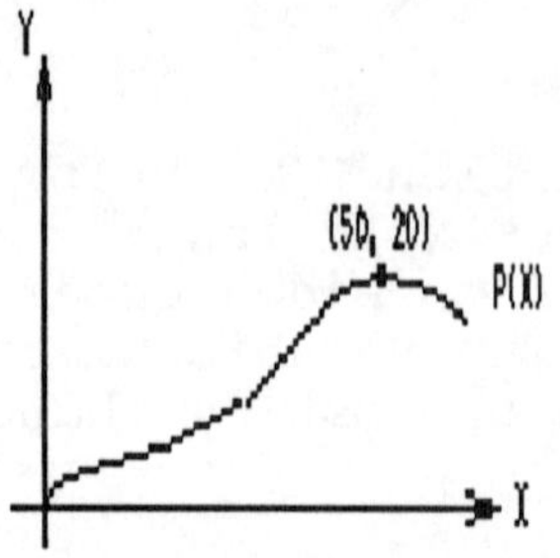

a) There can be no more than 50 trees in a grove.
b) The trees will yield at least 20 bushels apiece when the grove holds 50 trees.
c) The trees will yield at most 20 bushels of fruit apiece.
d) All of the above.

Answer: (c)
MC Moderate

0.6.17: The height of a toy rocket fired straight up in the air is s(t) meters after t seconds. The graph of s(t) is shown below.

(a) At what time does the rocket reach its highest point?

(b) How far does the rocket travel in its last second of flight?

(c) What was the height of the rocket's launch pad?

(d) Approximately how long does it take the rocket to fall from 256 ft to 192 ft?

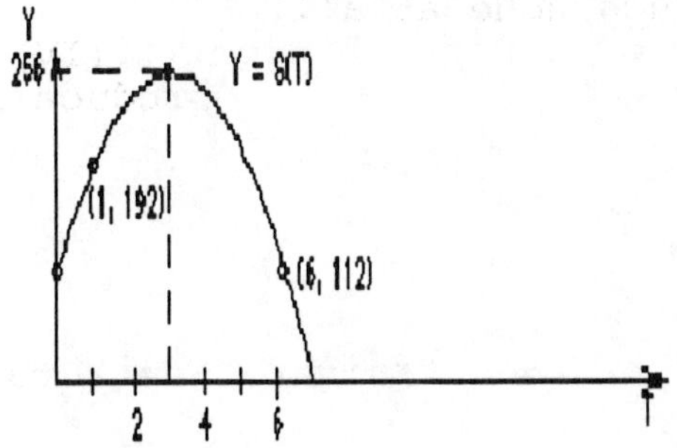

Answer: (a) t = 3 seconds

(b) 112 meters

(c) 112 meters

(d) 1 sec.

ES Moderate

1. The Derivative

1.1.1: Determine the equation of a line whose slope is 3 and y-intercept is 5.

Answer: $y = 3x + 5$
ES Moderate

1.1.2: Determine the equation of a line through (2, 3) and (4, 6).

Answer: $y = \frac{3}{2}x$
ES Moderate

1.1.3: Determine the equation of the straight line through the points $(\frac{1}{2}, 1)$ and (2, 0).

Answer: $y = -\frac{2}{3}(x - 2)$
ES Moderate

1.1.4: Determine the equation of the straight line through (-1, 3) and parallel to the line $5x - 3y = 7$.

Answer: $y - 3 = \frac{5}{3}(x + 1)$
ES Moderate

1.1.5: What is the slope of the line passing through the points (6, 1) and (3, -5)?

a) -2
b) $-\frac{4}{3}$
c) 2
d) $\frac{4}{3}$
e) none of the above

Answer: (c)
MC Moderate

1.1.6: Which of the following is (are) true of the lines $2x - 5y = -15$ and $5x + 2y = -6$?

I. They are parallel.

II. They are perpendicular.

III. They cross the x-axis at the same point.

IV. They cross the y-axis at the same point.

a) I
b) II
c) II and III
d) II and IV
e) none of the above

Answer: (b)
MC Moderate

1.1.7: Determine the slope of the line with equation $y = 4 - 5x$.

Answer: -5
ES Moderate

1.1.8: Determine the slope of the line with equation $2x + 4y = 5$.

Answer: -1/2
ES Moderate

1.1.9: Determine the slope of the line with equation $2x + 4 = 2(2y + 3)$.

Answer: 1/2
ES Moderate

1.1.10: Determine the slope of the line that passes through (2, 4) and (3, 7).

Answer: 3
ES Moderate

1.1.11: Determine the slope of the line that passes through (3, -7) and (-2, 1).

Answer: -8/5
ES Moderate

1.1.12: Write an equation of the line perpendicular to the line $5x = 7 - 8y$ with y-intercept -2.

Answer:
$$y = \frac{8}{5}x - 2$$
ES Moderate

1.1.13: The slope of the straight line passing through the points (2, -1) and (5, 0) is:

a) -3 b) 1/3 c) -1/3 d) 2/5

Answer: (b)
MC Moderate

1.1.14: What is the slope of the line with equation $3y + 2 = 5x - 2y$?

a) 5/3 b) 5 c) 1 d) -1/2

Answer: (c)
MC Moderate

1.1.15: Which of the following pairs of points lie on the line $x + 3y = 12$?

a) (4, 0) and (5, -3) b) (1, 3) and (0, 2)
c) (0, 12) and (2, 6) d) (-2, 10) and (0, 4)

Answer: (d)
MC Moderate

1.1.16: Consider 2 lines each having slope 4/3, one passing through the point (-1, 2) and the other through (3, 5). Which of the following is/are true?

(I) They are parallel.
(II) They are the same line.
(III) They cross the x-axis at the same point.
(IV) They intersect at the point (0, 3).

Answer: (I) only.
ES Moderate

1.1.17: Determine an equation for a line perpendicular to the line $3y - \frac{5}{2}x = 1$.

Answer: $y = \frac{-6}{5}x + \frac{1}{3}$.
ES Moderate

1.1.18: Write an equation of the line with slope $\frac{-5}{7}$ and passing through point $\left(-\frac{1}{2}, 0\right)$.

Answer: $y = \frac{-5}{7}\left(x + \frac{1}{2}\right)$.
ES Moderate

1.1.19: Write an equation of the straight line through the points $\left(\frac{3}{2}, -4\right)$ and $\left(\frac{-3}{2}, -4\right)$.

Answer: $y = -4$
ES Moderate

1.2.1: Determine:

(a) $\frac{dy}{dx}$ if $y = 4 - 6x$

(b) $g'(1)$ if $g(x) = 2x^2 - x + 3$

(c) $f'(x)$ if $f(x) = \frac{5}{x^2}$

Answer: (a) -6

(b) 3

(c) $-10x^{-3}$
ES Moderate

1.2.2: Find:

(a) $\frac{dy}{dx}$ if $y = (x^3 + 4x)^5$

(b) $\frac{d^2y}{dx^2}$ if $y = x^{3/2} + x^2$

Answer:

(a) $5(x^3 + 4x)^4(3x^2 + 4)$

(b) $\frac{3}{2}x^{1/2} + 2x$
ES Moderate

1.2.3: What is the equation of the tangent line to the graph of $y = x^2$ when $x = \frac{1}{3}$?

Answer: $y - \frac{1}{9} = \frac{2}{3}\left(x - \frac{1}{3}\right)$
ES Moderate

1.2.4: Find all points on the graph of $y = x^3$ where the curve has slope 12.

Answer: (2, 8) and (-2, -8)
ES Moderate

1.3.1: Find all points at which the tangent line to the graph of $y = x^3$ is parallel to the line $y = 27x + 5$.

Answer: (3, 27), (-3, -27)
ES Moderate

1.3.2: Consider the curve in the accompanying sketch.

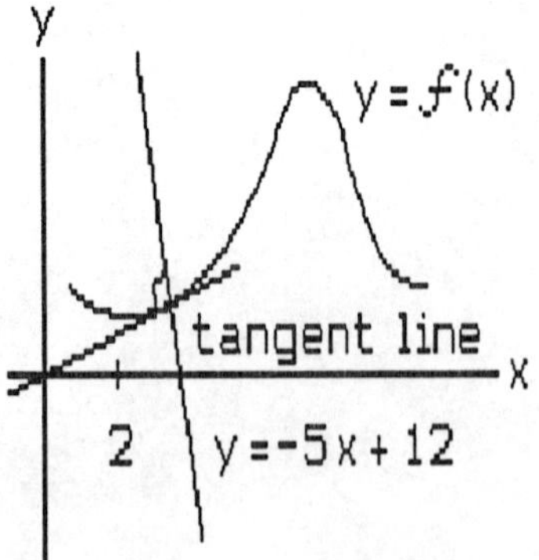

(a) Find: $f(2)$

(b) Find: $f'(2)$

Answer: (a) 2

(b) $\frac{1}{5}$
ES Moderate

1.3.3: Determine the derivative: $f(x) = 4x^{5/4}$

Answer: $f'(x) = 5x^{1/4}$
ES Moderate

1.3.4: Determine the derivative: $f(x) = \frac{1}{x^2}$

Answer: $f'(x) = -\frac{2}{x^3}$
ES Moderate

1.3.5: Determine the derivative: $f(x) = \frac{1}{\sqrt{x}}$

Answer: $f'(x) = -\frac{1}{2}x^{-3/2}$
ES Moderate

1.3.6: Determine the derivative: $f(x) = x^{6/5}$

Answer: $f'(x) = \frac{6}{5}x^{1/5}$

ES Moderate

1.3.7: Determine the equation of the line tangent to the curve

$y = \frac{1}{x + 2}$ at (2, 1/4).

Answer: $y = -\frac{1}{16}x + \frac{3}{8}$

ES Moderate

1.3.8: What is the derivative of $f(x) = x^{4/5}$ at the point (32, 16)?

Answer: $f'(x) = \frac{2}{5}$.

ES Moderate

1.3.9: The slope of the tangent line to the graph of $y = x^{5/2}$ is:

a) $5x^{3/2}$ b) $\frac{5}{2}x^{-1/2}$ c) $\frac{2}{5}x^{-1/2}$ d) $\frac{5}{2}x^{3/2}$

Answer: (d)

MC Moderate

1.3.10: Which of the following is/are parallel to the tangent line of graph of $y = x^3$ when x = 1?

(I) $y = -3x + 1$

(II) $y = 3x + 1$

(III) $y = -3x - 4$

(IV) $y = 3x + 1$

Answer: (I) and (III).

ES Moderate

1.3.11: If $y = \left(3\sqrt{x}\right)$, then f'(x) is equal to:

a) $\frac{5}{3x^2}$ b) $\frac{2}{3}x^{-4/3}$

c) $\frac{-1}{3x^{4/3}}$ d) none of the above.

Answer: (c)

MC Moderate

1.3.12: Which of the following is the equation of the line tangent to the graph of $y = \sqrt{x}$ at $x = 9$?

(I) $y = -x - \frac{1}{6}$

(II) $y = \frac{1}{2}x + 3$

(III) $y = \frac{1}{6}x + 3$

(IV) $y = \frac{1}{2}x^{-1/2} - \frac{1}{6}$

Answer: (III)
ES Moderate

1.4.1: Calculate the following limits if they exist.

(a) $\lim_{x \to -1} (x^3 - 2x + 5)$

(b) $\lim_{h \to 0} \frac{(h + 1)^{1/2} - 1}{h}$

(c) $\lim_{x \to 1} \frac{x^6 - 1}{x^3 - 1}$

Answer: (a) 6

(b) $\frac{1}{2}$

(c) 2
ES Moderate

1.4.2: Evaluate: $\lim_{x \to 5} \frac{x^2 - 8x + 15}{x^2 - 7x + 10}$

Answer: $\frac{2}{3}$
ES Moderate

1.4.3: Calculate: $\lim_{x \to \infty} \frac{3}{x^2 + 1}$

Answer: 0
ES Moderate

1.4.4: The limit $\lim_{x \to -1} \frac{x^2 - 1}{x + 1}$

a) equals -2
b) equals 0
c) does not exist
d) equals 1
e) none of the above

Answer: (a)
MC Moderate

1.4.5: The limit $\lim_{h \to 0} \frac{\sqrt{3 + h} - \sqrt{3}}{h}$ equals

a) $\left.\frac{dy}{dx}\right|_{x = 3}$ where $y = x$
b) $\left.\frac{dy}{dx}\right|_{x = 0}$ where $y = \sqrt{x}$
c) $\left.\frac{dy}{dx}\right|_{x = 3}$ where $y = \sqrt{x}$
d) 6
e) none of the above

Answer: (c)
MC Moderate

1.4.6: Determine: $\lim_{x \to 3} 4x^3$

Answer: 108
ES Moderate

1.4.7: Determine: $\lim_{x \to 2} (2x + 2)^2$

Answer: 36
ES Moderate

1.4.8: Determine: $\lim_{x \to -1} \left[(x + 1)^3 (x^2 + 1) \sqrt{2x + 1}\right]$

Answer: 0
ES Moderate

1.4.9: Determine: $\lim_{x \to -1} \frac{x^2 - 4x - 5}{x + 1}$

Answer: -6
ES Moderate

1.4.10: Determine: $\lim_{x \to -1} \frac{x^2 + x - 2}{x^3 + 1}$

Answer: undefined
ES Moderate

1.4.11: Let $f(x) = \frac{1}{2x}$. Calculate f'(3) using limits.

Answer: $-\frac{1}{18}$
ES Moderate

1.4.12: $\lim_{x \to \infty} 2 - \frac{1}{x^2 + 1}$ is equal to:

a) 0
b) 2
c) 1
d) ∞
e) none of the above

Answer: (b)
MC Moderate

1.4.13: If $f(t) = \frac{1}{x^{2/3}}$, then $\lim_{h \to 3} \frac{f(-8 + h) - f(-8)}{h}$ equals

(I) f'(8)
(II) f'(0)
(III) $\frac{1}{48}$
(IV) none of the above.

Answer: (III)
ES Moderate

1.4.14: Calculate the following limits if they exist.

(a) $\lim_{x \to \infty} \frac{1}{x^3 - 1} + 1$

(b) $\lim_{x \to 0} \frac{1}{x^3 - 1} + 1$

(c) $\lim_{x \to 1} \frac{1}{x^3 - 1} + 1$

Answer: (a) 1
(b) 0
(c) does not exist.
ES Moderate

1.4.15: Compute the derivative of $f(x) = (2x + 1)^2$.

Answer: $f'(x) = 8x + 4$
ES Moderate

1.4.16: Calculate $\lim_{x \to -2} \dfrac{x}{(x^3 + 8)^{-1}}$.

Answer: 0
ES Moderate

1.4.17: The limit $\lim_{h \to 3} \dfrac{\sqrt{x} - 4}{x^3 + 27}$.

a) Does not exist.
b) equals $\sqrt{3} - 4$
c) equals 0
d) none of the above.

Answer: (d)
MC Moderate

1.5.1: The graph of y = f(x) is shown below.

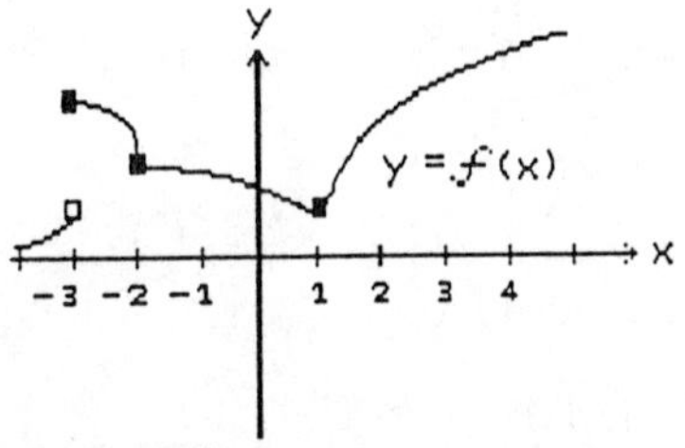

(a) For which values of x is f(x) not continuous?

(b) For which values of x is f(x) not differentiable?

Answer: (a) x = -3

(b) x = -3, -2, 1
SA Moderate

1.5.2: Which of the following is the best description of f'(t)?
a) f'(a) measures the rate of change of f(t) per unit change in t at the point t = a.
b) The derivative as a function is the best approximation of the tangent line to f(x).
c) It is approximately equal to $\dfrac{f(t + h) - f(t)}{h}$, as t gets very small.
d) It is a function which gives the slope of the secant line through any two points.
e) $f'(t) = \dfrac{f(t)}{t}$.

Answer: (a)
MC Moderate

1.5.3: The function whose graph is drawn below

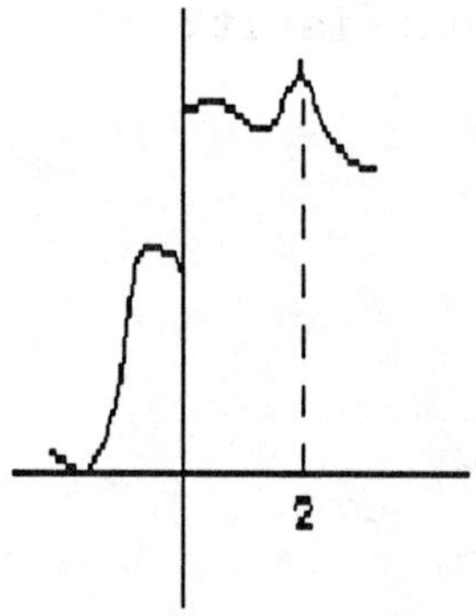

a) is continuous at x = 0
b) is differentiable at x = 2
c) is undefined at x = 0
d) none of the above

Answer: (d)
MC Moderate

1.5.4: Does $\lim_{x \to -a} \frac{x - a}{x^2 - a^2}$ (a ≠ 0) exist? If so, then what is it?

Answer: No
ES Moderate

1.5.5: The graph of y = f(x) is shown below.

Which of the following statements are true?
(I) f(x) is not defined at x = 4.
(II) f(x) is not differentiable at x = -4.
(III) f(x) is continuous at x = 1.
(IV) f(x) is continuous and differentiable at x = -2.

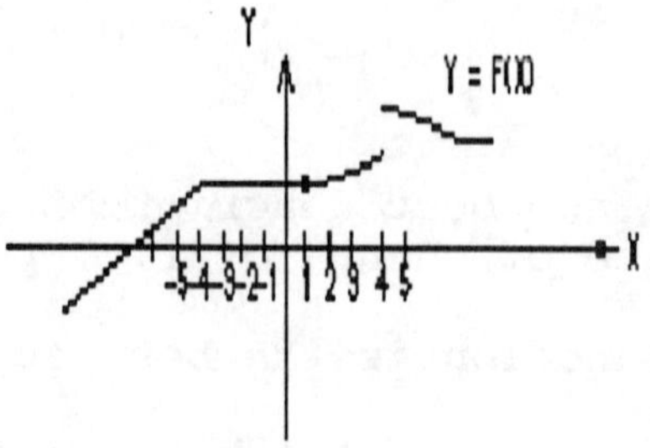

Answer: II, III, IV
ES Moderate

1.5.6: Let $f(x) = \frac{x^3 - 9x}{2x + 6}$.

(a) Does $\lim_{x \to -3} f(x)$ exist? If so, what is it?

(b) Is f(x) continuous at x = -3?

Answer: (a) yes; 9

(b) no

ES Moderate

1.6.1: Find:

(a) g'(3) if $g(x) = 2x^3 - 5x^2 + 1$

(b) $\frac{dy}{dx}$ if $y = \frac{5}{x^3}$

(c) f'(x) if $f(x) = \frac{1}{x^2 + 5}$

(d) F'(x) if $F(x) = \sqrt{3x + 1}$

Answer: (a) 24

(b) $-\frac{15}{x^4}$

(c) $-\frac{-2x}{(x^2 + 5)^2}$

(d) $\frac{3}{2}(3x + 1)^{-1/2}$

ES Moderate

1.6.2: Find the slope of the curve $y = (x^2 - 7)^3$ at x = 3.

Answer: 72

ES Moderate

1.6.3: Let $f(x) = 7(\frac{1}{2}x - 3)^2 + 5$. Which of the following statements is true?

a) f(a) = f'(a)

b) f'(a) = 2f(a)

c) f'(a) = f(a) - 5

d) f(a) = 2f(a) - 5

e) none of the above

Answer: (e)

MC Moderate

1.6.4: Determine the slope of the line tangent to the curve $f(x) = \frac{3}{(x + 2)^2}$ at $x = 3$.

Answer: $-\frac{6}{125}$

ES Moderate

1.6.5: Determine the slope of the line tangent to the curve $f(x) = 2(x^3 + 4)^2$ at $x = -1$.

Answer: 36

ES Moderate

1.6.6:

Determine the derivative: $y = \sqrt{x^4 + 1}$

Answer:

$$y' = \frac{2x^3}{\sqrt{x^4 + 1}}$$

ES Moderate

1.6.7:

Determine the derivative: $y = \sqrt{3x^2 + 4x}$

Answer:

$$y' = \frac{3x + 2}{\sqrt{3x^2 + 4x}}$$

ES Moderate

1.6.8:

Determine the derivative: $y = \frac{2}{\sqrt{2x + 1}}$

Answer:

$$y' = -2(2x + 1)^{-3/2}$$

ES Moderate

1.6.9:

Determine the derivative: $y = \frac{4}{3x^3 + x^2 + 4x}$

Answer:

$$y' = \frac{-36x^2 - 8x - 16}{(3x^3 + x^2 + 4x)^2}$$

ES Moderate

1.6.10: Determine the equation of the line tangent to the curve

$y = \sqrt{x^2 + 1}$ at $(2, \sqrt{5})$.

Answer:

$$y = \frac{2}{\sqrt{5}}x + \frac{\sqrt{5}}{5}$$

ES Moderate

1.6.11: Determine the derivative: $y = \frac{1}{\sqrt{x + 1}}$

Answer:

$$y' = -\frac{1}{2}(x + 1)^{-3/2}$$

ES Moderate

1.6.12: Determine the derivative: $y = \frac{1}{x^2 + 1}$

Answer:

$$y' = \frac{-2x}{(x^2 + 1)^2}$$

ES Moderate

1.6.13: Determine the derivative: $y = \frac{3}{4x^2 + 1}$

Answer:

$$y' = \frac{-24x}{(4x^2 + 1)^2}$$

ES Moderate

1.6.14: Determine the derivative: $y = \frac{4}{4x + 1}$

Answer:

$$y' = -\frac{16}{(4x + 1)^2}$$

ES Moderate

1.6.15: Determine the derivative: $y = \frac{x^2 + 4x + 3}{x + 1}$

Answer: $y' = 1,\ x \neq -1$

ES Moderate

1.6.16: Determine the derivative: $y = 3x^2(4x^4 + 3x^3)$

Answer: $y' = 72x^5 + 45x^4$

ES Moderate

1.6.17: Determine the derivative: $y = 3\sqrt{5x^2 + 2}$

Answer: $y' = \dfrac{15x}{\sqrt{5x^2 + 2}}$

ES Moderate

1.6.18: Determine the derivative: $y = -\dfrac{1}{2x^2}$

Answer: $y' = x^{-3}$

ES Moderate

1.6.19: Determine the derivative: $y = x^{15/19}$

Answer: $y' = \dfrac{15}{19}x^{-4/19}$

ES Moderate

1.6.20: Let $f(x) = 1 + 3x - x^2$.

(a) What is the slope of the graph of $f(x)$ when $x = 5$?

(b) What is the equation of the tangent line to the graph of $f(x)$ at $x = 5$?

Answer: (a) -7

(b) $y + 9 = -7(x - 5)$

ES Moderate

1.6.21: What is the slope of the curve $y = \sqrt{2x^2 + 1}$ at the point (2, 3)?

a) $\frac{4}{3}$

b) $\frac{1}{4}$

c) $\frac{3}{2}$

d) 12

e) none of the above

Answer: (a)

MC Moderate

1.6.22: Determine the derivative: $f(x) = \sqrt{x}$

Answer: $f'(x) = \dfrac{1}{2\sqrt{x}}$

ES Moderate

1.6.23: Determine the slope of the line tangent to the curve $f(x) = 3x^4 + 2x^3$ at $x = 1$.

Answer: 18

ES Moderate

1.6.24: Determine the equation of the line tangent to the curve $y = x^3 + 4x^2 + 4$ at $(1, 9)$.

Answer: $y = 11x - 2$

ES Moderate

1.6.25: Find the equation of the line tangent to the graph of $y = x^3 + 3x - 8$ at $x = 2$.

Answer: $y - 6 = 15(x - 2)$

ES Moderate

1.6.26: Find $f'(-2)$ if $f(x) = \dfrac{3x}{4} + (x^3 - 2x)^3$.

Answer: $\dfrac{1923}{4}$

ES Moderate

1.6.27: Find the slope of the graph of $y = x^9 - 2x + \left(\sqrt{5-x}\right)^3$ at the point $(1, 7)$.

Answer: 4

ES Moderate

1.6.28: The equation of the tangent line to the graph of $y = \dfrac{-3}{5x + 2}$ at $x = 3$ is:

a) $y = \dfrac{15}{(5x - 2)^2}$

b) $y = \dfrac{15}{169}(x - 3)$

c) $y = \dfrac{15}{289}(x - 3) - \dfrac{3}{17}$

d) $y - 3 = 15(5x - 2)^{-2}$

Answer: (c)

MC Moderate

1.6.29: For what value(s) of x is the slope of the tangent line to the graph of $y = \frac{2}{3}x^3 + x^2 - 4x$ equal to 5?

a) $x = 0,\ x = \frac{3 \pm 4\sqrt{6}}{4}$

b) $x = \frac{-1 \pm \sqrt{19}}{2}$

c) $x = 1,\ x = -2$

Answer: (b)
MC Moderate

1.6.30: The derivative of $y = -\frac{2}{3x^4}$ is given by:

a) $\frac{dy}{dx} = -\frac{24}{x^3}$ b) $y' = \frac{8}{3x^3}$ c) $y' = \frac{8}{(3x)^3}$ d) $\frac{dy}{dx} = -\frac{8}{3}x^3$

Answer: (b)
MC Moderate

1.6.31: $\frac{d}{dx}\left(\frac{2x - (x^2 + 1)^7}{3}\right)$ is equal to

a) $\frac{2}{3} - \frac{14}{3}x(x^2 + 1)^6$

b) $\frac{2 - x}{3(x^2 + 1)^6}$

c) $2 - 7(x^2 + 1)^6$

d) none of the above

Answer: (a)
MC Moderate

1.6.32: Let $g(x) = 3\left(\frac{2}{3}x - 1\right)^3 - 1$. Which of the following statements is/are true?

(I) $g'(a) = 2g(a)$

(II) $g'\left(\frac{3}{2}\right) = -1$

(III) $g'(a) = 2g(a)^2 + 1$

(IV) $g'(3) = 6$

Answer: (IV) only
ES Moderate

1.6.33: The limit $\lim_{h \to 0} \frac{\sqrt{5 + h} - \sqrt{5}}{2h}$ is equal to

a) $F'(5)$ where $F(x) = \sqrt{x}$

b) $F'(3)$ where $F(x) = \frac{\sqrt{x + 2}}{2}$

c) $\frac{1}{2}F(5)$ where $F(x) = \sqrt{x}$

d) none of the above.

Answer: (b)
MC Moderate

1.6.34: Let $G(x) = \frac{(x - 4)^{3/2}}{2}$. Then, which of the following is true?

a) $G'(4) = G(4)$

b) $G'(4)$ is undefined.

c) $G'(4) = \frac{1}{2}$

d) none of the above.

Answer: (a)
MC Moderate

1.6.35: The derivative of $h(x) = \frac{5}{x^3 - 4x^2 + 2}$ is equal to

a) $\frac{15x^2 - 40x}{x^3 - 4x^2 + 2}$

b) $\frac{3x^2 - 8x}{(x^3 - 4x^2 + 2)^2}$

c) $\frac{-15x^2 + 40x}{(x^3 - 4x^2 + 2)^2}$

d) $\frac{5}{(3x^2 - 8x)^2}$

Answer: (c)
MC Moderate

1.7.1: Let $f(t) = t^3 - \frac{9}{t}$. Find: $\left.\frac{d^2f}{dt^2}\right|_{t = 3}$

Answer: $\frac{52}{3}$
ES Moderate

1.7.2: Let $V = \frac{4}{3}\pi r^3$. What is $\left.\frac{dV}{dr}\right|_{r = 2}$?

a) $\frac{16\pi}{9}$

b) 8π

c) $\frac{32}{3}$

d) 16π

e) none of the above

Answer: (d)
MC Moderate

1.7.3: What is $\left.\frac{dy}{dx}\right|_{x = 4}$, where $y = (-4 + 3\sqrt{x})^4$?

a) 24

b) 6

c) $\frac{27}{16}$

d) -6

e) none of the above

Answer: (a)
MC Moderate

1.7.4: What is the second derivative of $y = \frac{1}{x^2} + 3$?

a) $\frac{1}{6x^4}$

b) $-6x^{-4}$

c) $2x^0$

d) $-\frac{1}{6x^4}$

e) none of the above

Answer: (e)
MC Moderate

1.7.5: Let $f(t) = 2\sqrt{t}$. What is $f''(4)$?

a) $\frac{1}{8}$

b) $-\frac{1}{8}$

c) $-\frac{1}{2}$

d) $\frac{1}{16}$

e) none of the above

Answer: (e)
MC Moderate

1.7.6: Let $y = (5s^3 - 4st + t)^5$. Compute $\frac{dy}{ds}$ and $\frac{dy}{dt}$.

Answer: $\frac{dy}{dt} = 5(5s^3 - 4st + t)^4(15s^2 - 4t)$

$\frac{dy}{dt} = 5(5s^3 - 4st + t)^4(-4s + 1)$

ES Moderate

1.7.7: Let $y = \sqrt{5t^2 - 6}$. Calculate $\left.\frac{dy}{dt}\right|_{t = 3}$.

Answer:

$\frac{15\sqrt{39}}{39}$

ES Moderate

1.7.8: Let $y = \frac{4\sqrt{r + 2t}}{3s}$. Compute $\frac{d^2y}{ds^2}$.

Answer:

$\frac{8\sqrt{r + 2t}}{3s^3}$

ES Moderate

1.7.9: Find $\frac{dz}{dt}$ where $z = 4t + \left(3 - \sqrt{2t + 1}\right)^3$.

Answer:
$4 - \frac{3\left(3 - \sqrt{2t + 1}\right)^3}{\sqrt{2t + 1}}$.

ES Moderate

1.7.10: Find $\left.\frac{d^2y}{dt^2}\right|_{t = 8}$ where $y = \frac{9t^{2/3}}{2}$.

Answer: $\frac{-1}{16}$

ES Moderate

1.8.1: A rock is thrown off a cliff. Its distance from the ground below at t seconds is $s(t) = -16t^2 - 16t + 96$ feet.

(a) What is the velocity after 1 second?

(b) When will it hit the ground?

(c) What is the velocity of the rock when it slams into the ground?

(d) How high was the cliff?

Answer: (a) -48 ft/sec

(b) after 2 seconds

(c) -80 ft/sec

(d) 96 ft

ES Moderate

1.8.2: During the month of February, a flu epidemic hit the University. The number of people sick at time t (measured in days) is given by the function P(t). The rate at which the epidemic is spreading on February 3 is 110 people per day. How is the information best represented mathematically?

a) $P'(3) = 110$

b) $\left.\frac{dP}{dt}\right|_{t = 3} = P'(3)$

c) $\left.\frac{dP}{dt}\right|_{t = 110}$

d) $P(3) = 110$

e) none of the above

Answer: (a)

MC Moderate

1.8.3: At time $t = 0$, a seed is planted. After t weeks, the height of the plant is $.3t^2 + .6t + .5$ inches. At what rate is the plant growing after 8 weeks?

a) 24.5 inches/week
b) $.6t + .6$ inches/week
c) 10.1 inches/week
d) 5.4 inches/week
e) none of the above

Answer: (d)
MC Moderate

1.8.4: A ball is thrown straight up. Its height at time t is represented by the equation $h = 30t - 50t^2$. Determine the instantaneous velocity of the ball at $t = 2$.

Answer: -70
ES Moderate

1.8.5: A ball is thrown straight up. Its height at time t is represented by the equation $h = 30t - 50t^2$. Determine the maximum height of the ball. (Hint: Consider the velocity of the ball at the moment the ball reaches its maximum height.)

Answer: 4.5
ES Moderate

1.8.6: A point P is moving along the x-axis. At any time t, the location of P on the x-axis is described by $x = t^3 - 4t^2 + 3t$. Determine the point's instantaneous velocity when $t = 5$.

Answer: 38
ES Moderate

1.8.7: A point P is moving along the x-axis. At any time t, the location of P on the x-axis is described by $x = t^3 - 4t^2 + 3t$. Determine the instantaneous acceleration at time $t = 5$ of the point P.

Answer: 22
ES Moderate

1.8.8: A manufacturer's profit from producing x units of a product is given by $P(x) = .002x^3 - .01x^2 + 0.5x$.

(a) What is the marginal profit when the production level is at 50 units?
(b) At what production level(s) will the marginal profit be \$9.30 per unit?

Answer: (a) \$14.50 per unit
(b) 40 units
ES Moderate

1.8.9: A winter storm front moves through campus. At t hours after the onset of the storm, the temperature is at $35 - 2t^2 + t$.

(a) What is the temperature 3 hours after the storm begins?

(b) At what rate is the temperature changing at this time?

Answer: (a) 20°

(b) -11° per hour

ES Moderate

1.8.10: Water is pouring into a tub such that after t minutes, there are $t^3 - t^2 + .3t$ gallons in the tub.

(a) What is the average rate at which water pours into the tub over the first 4 minutes.

(b) At what instantaneous rate is the water flowing when $t = 4$?

Answer: (a) 11.7 gal/min.

(b) 40.3 gal/min.

ES Moderate

1.8.11: An automobile's brakes are applied at time $t = 0$ when the vehicle is travelling at 48 ft/sec. The brakes cause the automobile to decelerate so that after t sec. the velocity is given by $v(t) = 48 - 16t$.

(a) At what rate is the vehicle decelerating after 1 sec?

(b) How long will it take for the vehicle to come to a complete stop?

Answer: (a) 16 ft/sec^2

(b) 3 sec.

ES Moderate

1.8.12: Suppose that t hours after being placed in a freezer, the temperature of a piece of meat is given by

$$f(t) = 70 - 12t + \frac{4}{t + 1}.$$

(a) What is the temperature of the meat after 3 hours?

(b) How fast is the temperature of the meat falling at this time?

Answer: (a) 35°

(b) 12.25° per hour.

ES Moderate

2. Applications of the Derivative

2.1.1: Sketch the graph of a function having the following properties:
(I) concave down for all x
(II) asymptotic to the line x = 0

Answer:

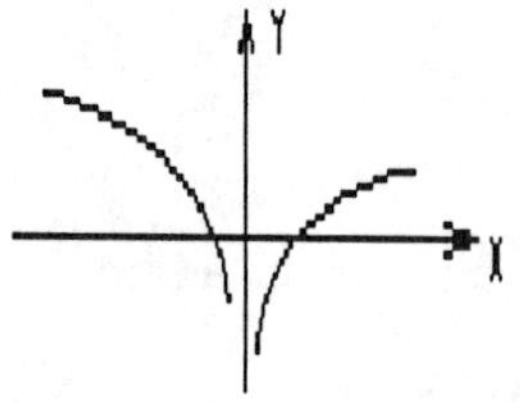

ES Moderate

2.1.2: Sketch the graph of a function having the given properties:
(I) defined for x ≥ -1
(II) horizontal asymptote at y = 3
(III) increasing for all x ≥ -1

Answer:

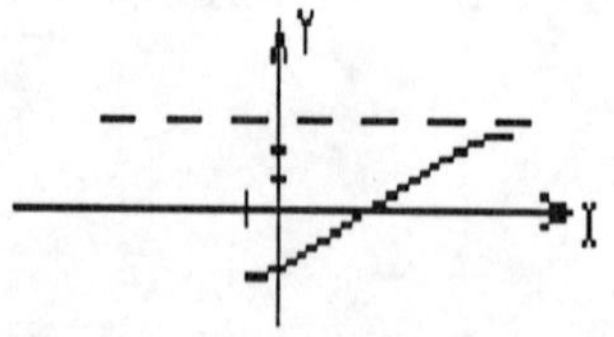

ES Moderate

2.1.3: Sketch the graph of a function having the given properties:
(I) asymptotic to the line y = (1/2)x - 2
(II) relative maximum at x = 0

Answer:

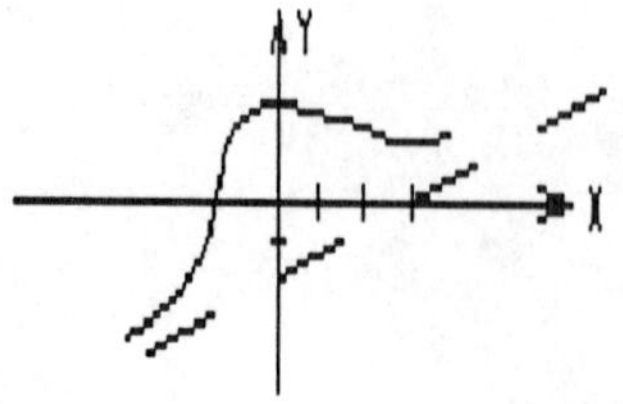

ES Moderate

2.1.4: Sketch the graph of the function having the given properties:
(I) x-intercept at $x = -2$
(II) absolute maximum at $x = -1$
(III) relative maximum at $x = 1$
(IV) concave up for $x \geq 2$

Answer:

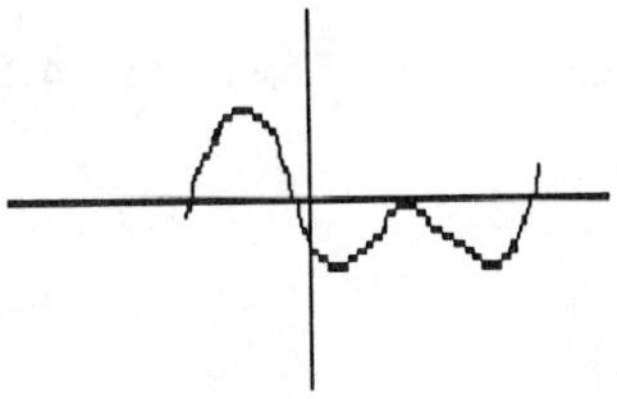

ES Moderate

2.1.5: Sketch the graph of the function having the given properties:
(I) inflection point at $x = 3$
(II) no relative maximum point
(III) asymptote at $x = 5$

Answer:

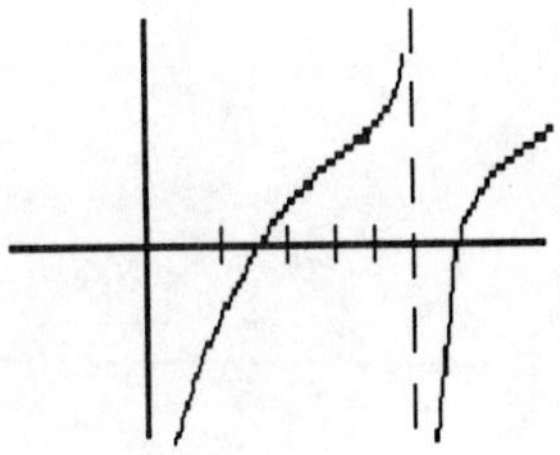

ES Moderate

2.2.1: Determine all values of x where relative maximum and minimum points of the function

$$f(x) = \frac{1}{3}x^3 - \frac{3}{2}x^2 - 10x$$

occur. Distinguish the maxima from the minima using the second derivative rule.

Answer: relative minimum at $x = 5$; relative maximum at $x = -2$
ES Moderate

2.2.2: (a) Find the maximum value of the function

$$f(x) = -x^3 + 6x^2 + 10$$

for $x \geq 0$.

(b) Prove that the value obtained in (a) is a maximum by using the second derivative rule.

Answer: (a) The relative maximum value of f(x) is 42 and it occurs at x = 4.

(b) $f''(4) < 0$

ES Moderate

2.2.3: Determine the values of x for which $f(x) = x^3 - 6x$ is concave down.

Answer: concave down for $x < 0$

ES Moderate

2.2.4: Points A, B, and C lie on the graph of a function f(x), as shown on the diagram.

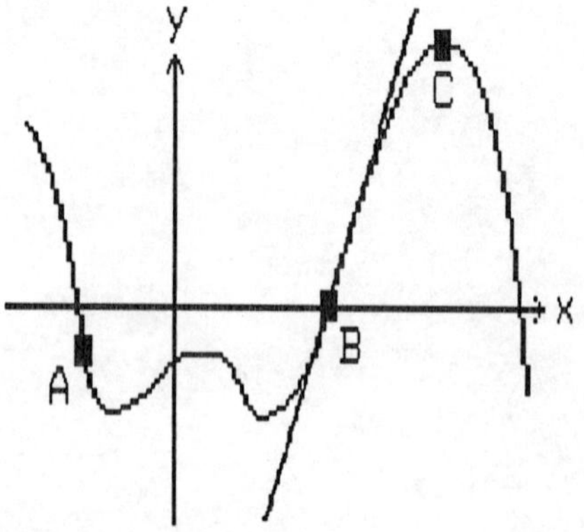

Fill in each blank of the table below with either "positive," "negative," or "0" to indicate the sign of f(x), f'(x), and f"(x) at A, B, and C.

	$f(x)$	$f'(x)$	$f''(x)$
A			
B			
C			

Answer: neg. neg. pos.
0 pos. zero
pos. 0 neg.
ES Moderate

2.2.5: Which functions graphed below have the property that the slope always increases as x increases?

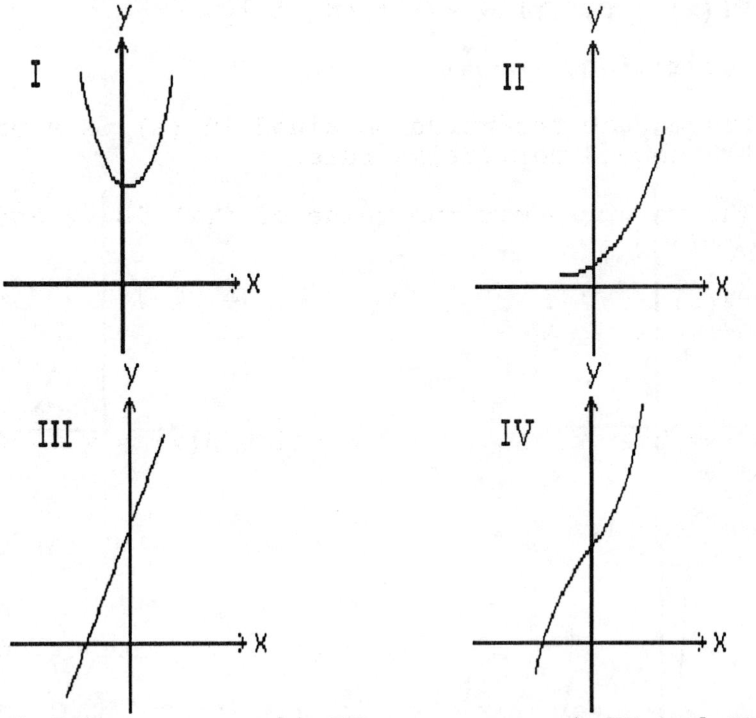

a) III b) IV c) II and IV d) I and II

Answer: (d)
MC Moderate

2.2.6: Which of the following graphs could represent a function with the following properties?

I. $f(x) > 0$, for $x < 0$

II. $f'(x) \le 0$, for all x, and

III. $f'(0) = 0$.

a)

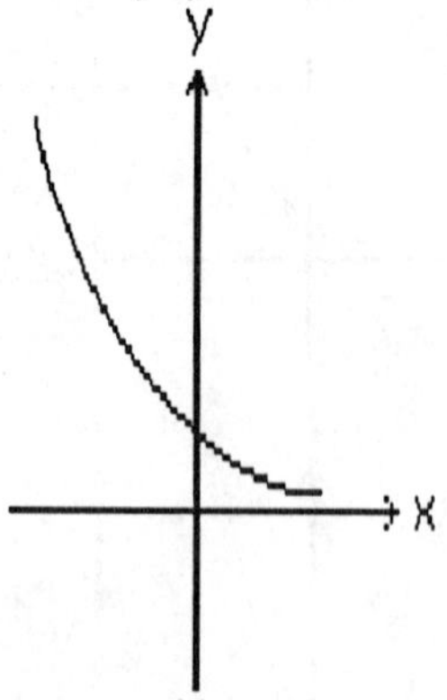

b)

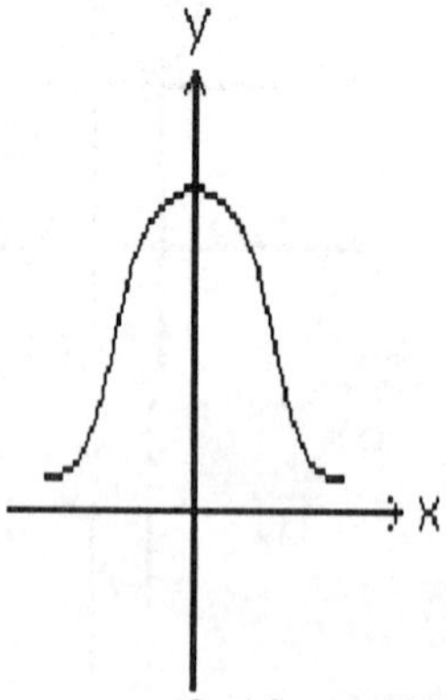

c)

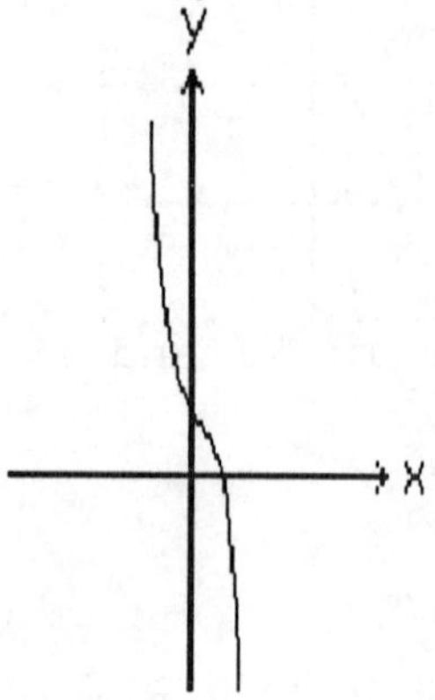

d) none of the above

Answer: (d)
MC Moderate

2.2.7: Where is the function $f(x) = x^4 - 6x^2 - 5$ concave up?

a) $x \ge -1$
b) $|x| \ge 1$
c) $x \ge 0$
d) none of the above

Answer: (b)
MC Moderate

2.2.8: Which of the following best pictures the graph of

$y = 2x^3 - 3x^2 - 12x + 17$ near the point (2, -3)?

a)

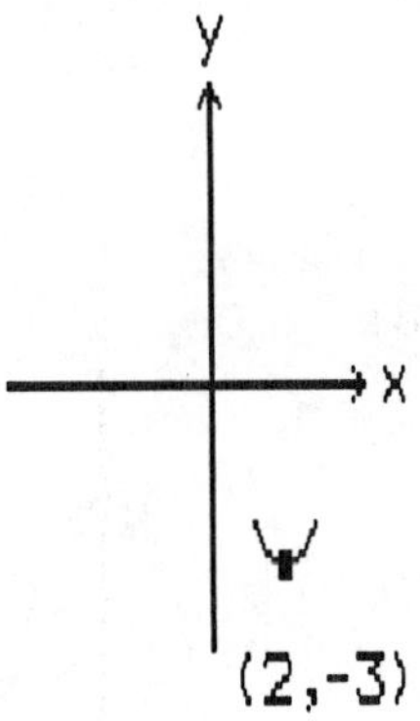

b)

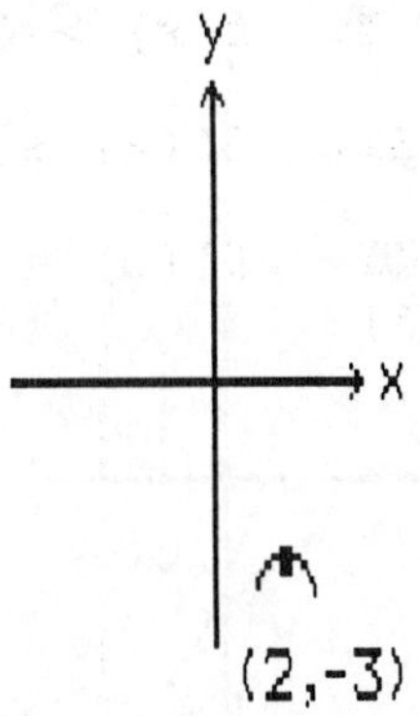

c)

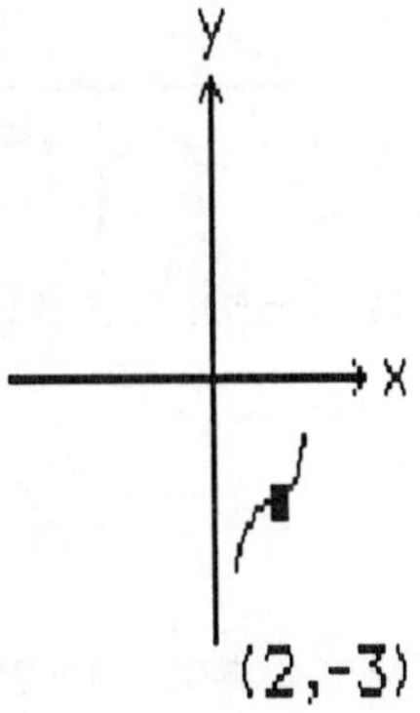

d)

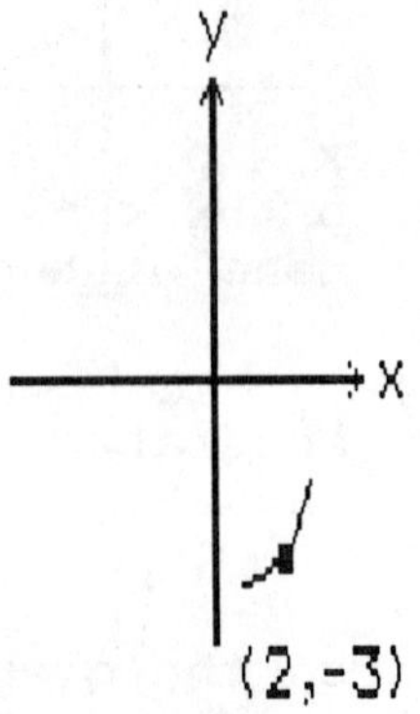

Answer: (a)
MC Moderate

2.2.9: Let a, b, and c be fixed numbers with $a > 0$ and let

$f(x) = ax^2 + bx + c$. Which of the following properties is true of the graph of f(x)?

a) f(x) has one inflection point
b) f(x) is always concave up
c) f(x) has one relative maximum
d) f(x) has either a relative maximum or inflection point
e) none of the above

Answer: (b)
MC Moderate

2.2.10: Let F(x) be the function graphed below. For what values is F'(x) > 0 ?

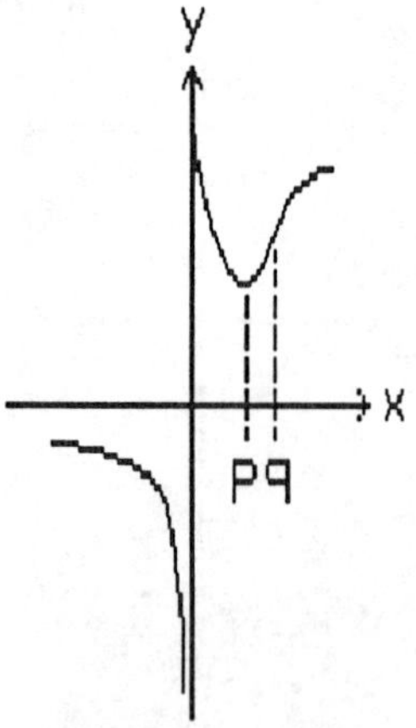

a) x > p
b) 0 < x < q
c) p < x < q
d) x < 0 and x > p
e) none of the above

Answer: (a)
MC Moderate

2.2.11:
Sketch the graph of a function having the given properties:
(I) $f(0) = -5$
(II) $f'(0) = 0$
(III) $f''(0) = -12$
(IV) $f(\sqrt{7}) = f(-\sqrt{7}) = 2$

Answer:

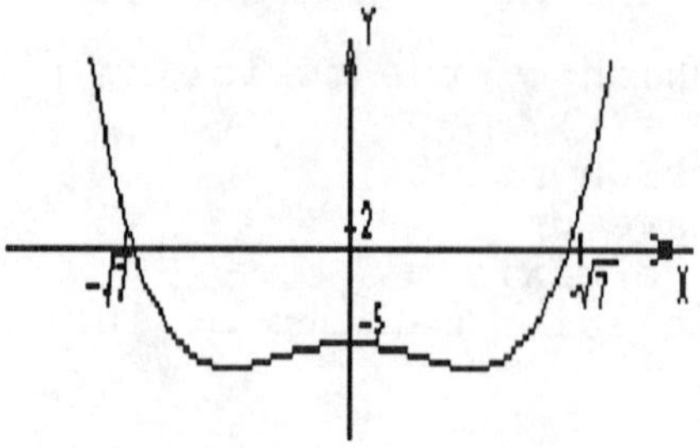

ES Moderate

2.2.12: Sketch the graph of a function having the given properties:

(I) $f'(x) > 0$ for all x

(II) $f''(x) > 0$ for $x < 0$, $f''(x) < 0$ for $x > 0$

(III) asymptotes at $y = \frac{\pi}{2}$, $y = \frac{-\pi}{2}$

Answer:

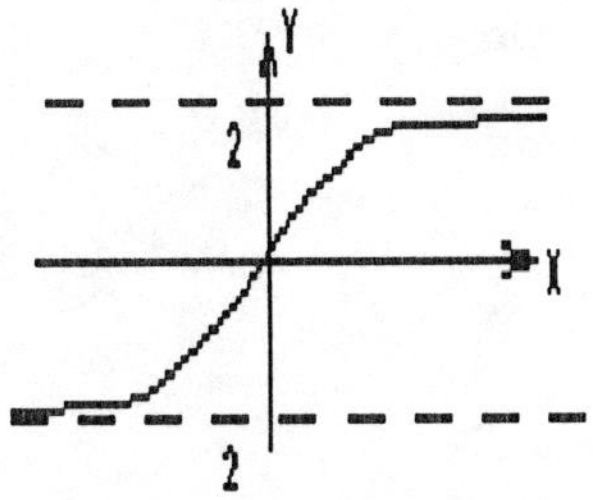

ES Moderate

2.2.13: Sketch the graph of a function having the given properties:

(I) $f'(x) < 0$ for all x

(II) $f''(x) > 0$ for all x

(III) (0, 1) is a point on the graph

Answer:

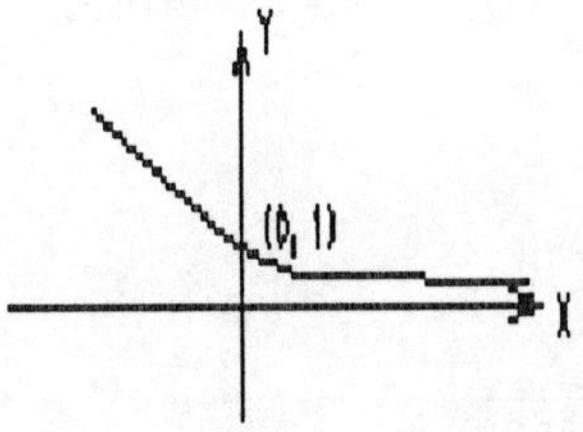

ES Moderate

2.2.14: Which of the following graphs could represent a function having the following properties?

(I) asymptotes x = 2 and x = -2

(II) f"(x) > 0 for all x

(III) f(x) > 0 for x > 4

a)

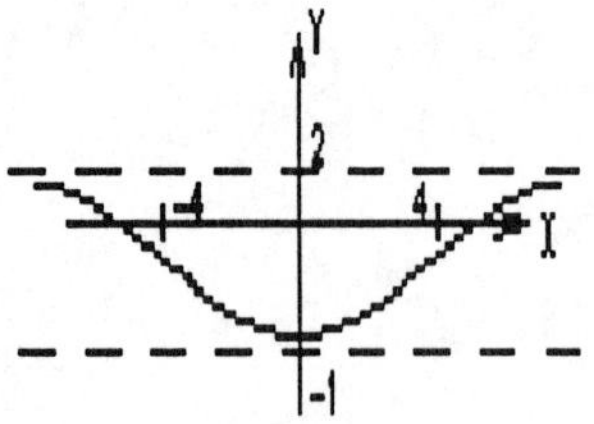

b)

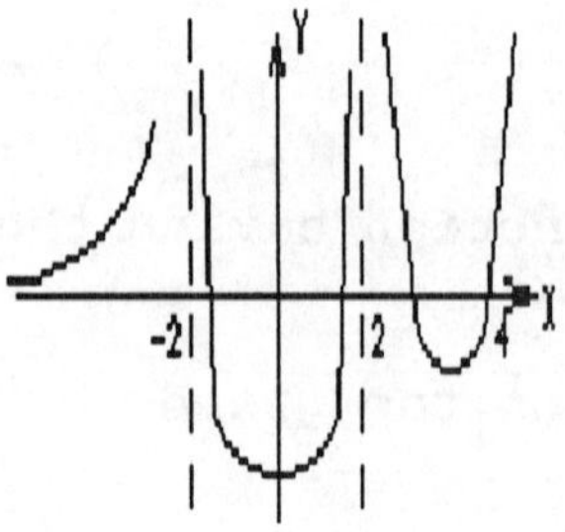

c)

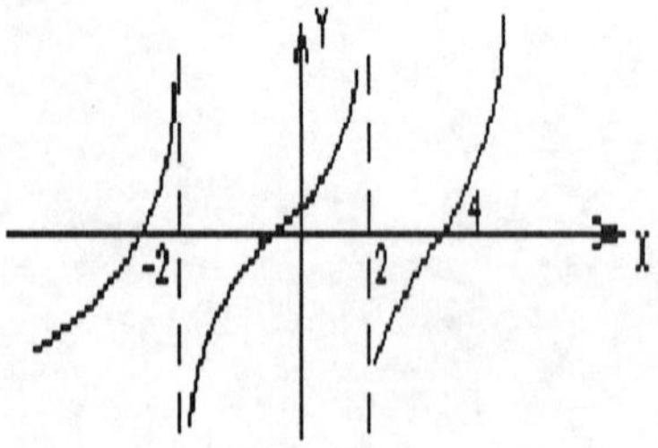

d)

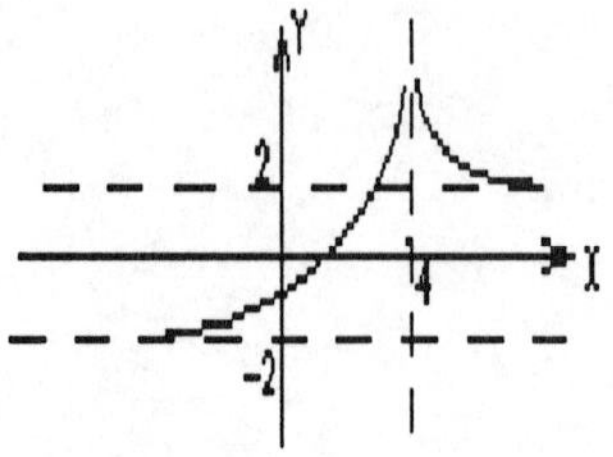

Answer: (b)

MC Moderate

2.2.15:

Where is the function $f(x) = \dfrac{5}{(2x - 4)^3}$ increasing?

a) for all x
b) for $x > 2$
c) at $x = 30$
d) none of the above

Answer: (d)
MC Moderate

2.2.16: Which of the following most resembles the graph of $y = 2x^9 - 5x^2$ near $x = 1$?

a)

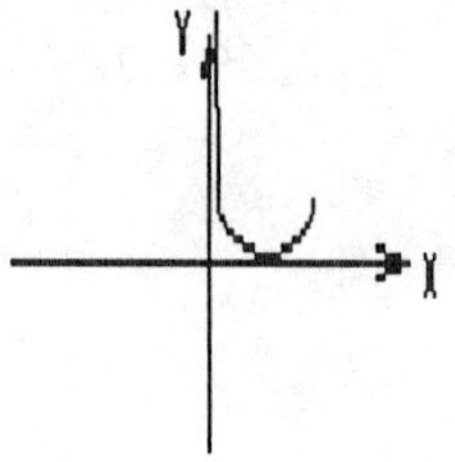

b)

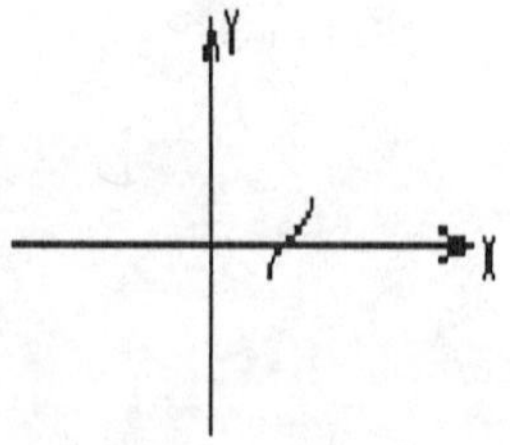

c)

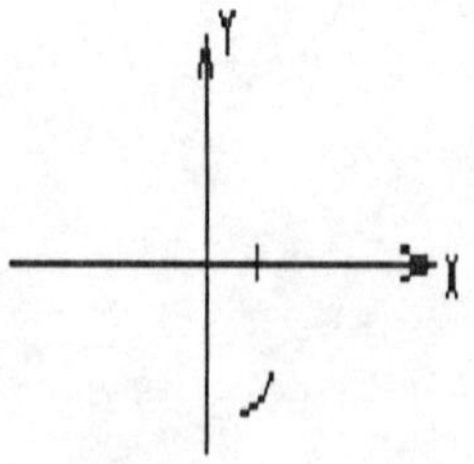

d)

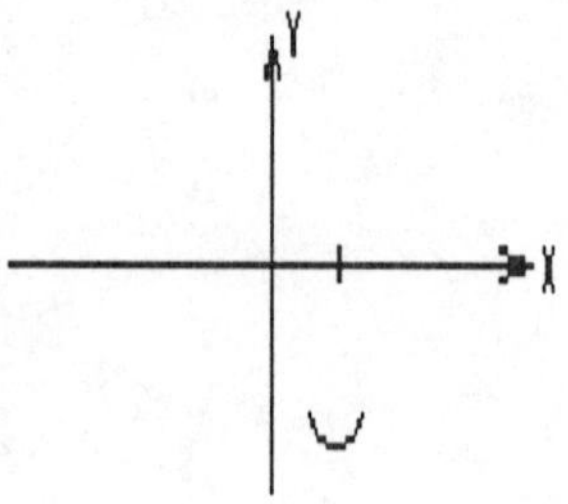

Answer: (c)
MC Moderate

2.2.17: Which of the following most resembles the graph of

$f(x) = \dfrac{1}{(2 - 3x)^3}$ near $x = \dfrac{2}{3}$?

a)

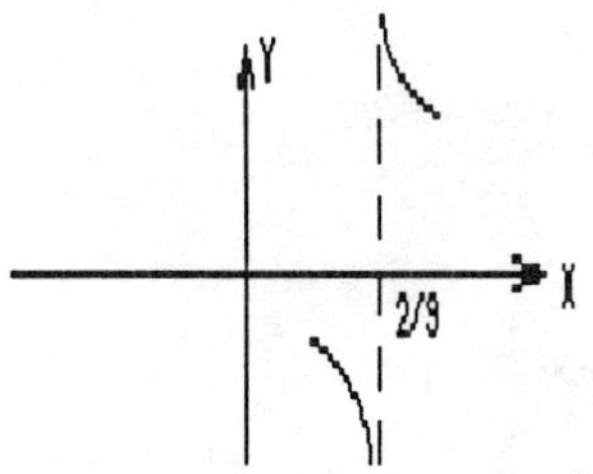

b)

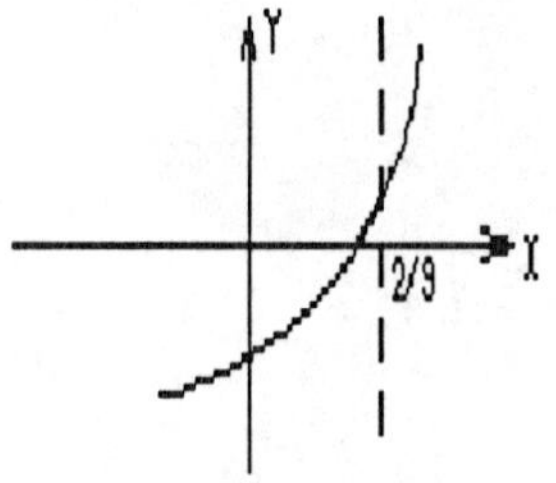

c)

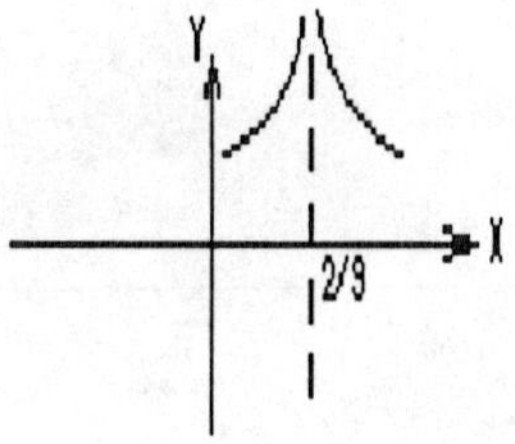

d)

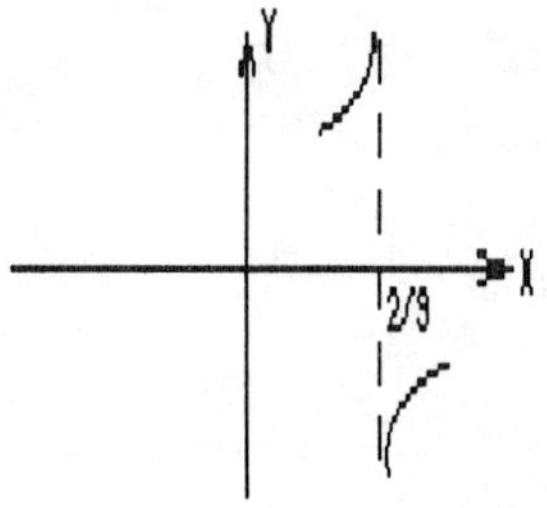

Answer: (d)
MC Moderate

2.2.18: Which of the following could represent a function having the following properties?

(I) increasing slope for $x < 4$

(II) $f'(x) > 0$ for all x, (x=4)

(III) asymptote at $x = 4$

a)

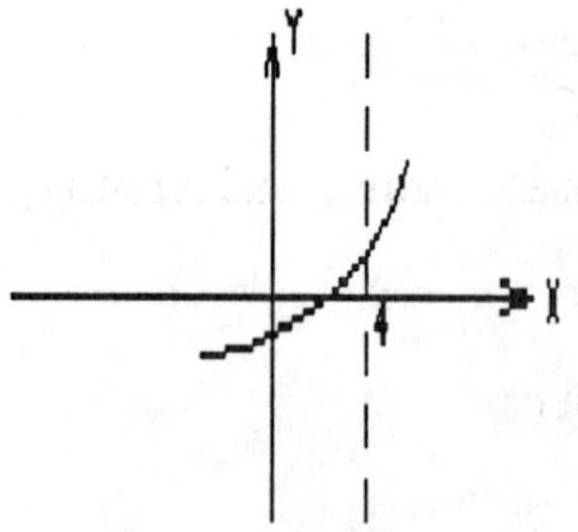

b)

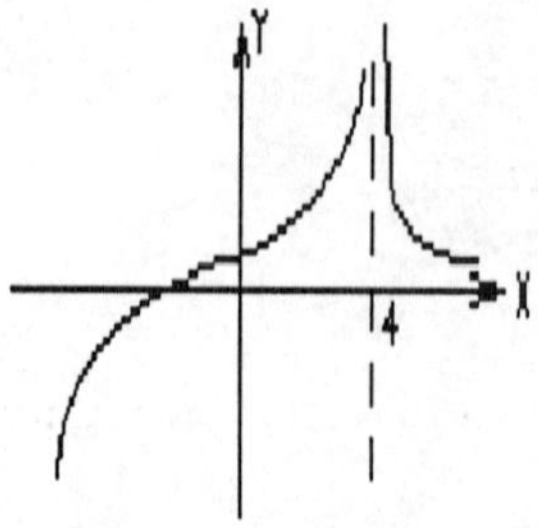

c)

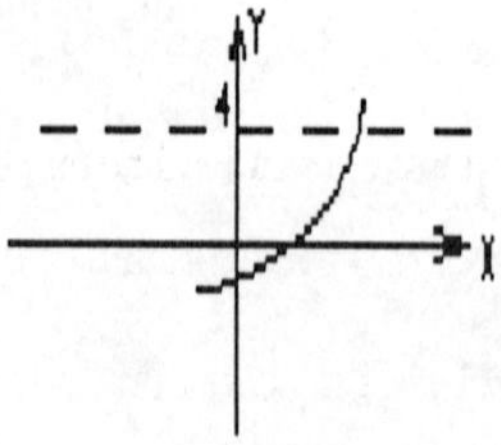

d)

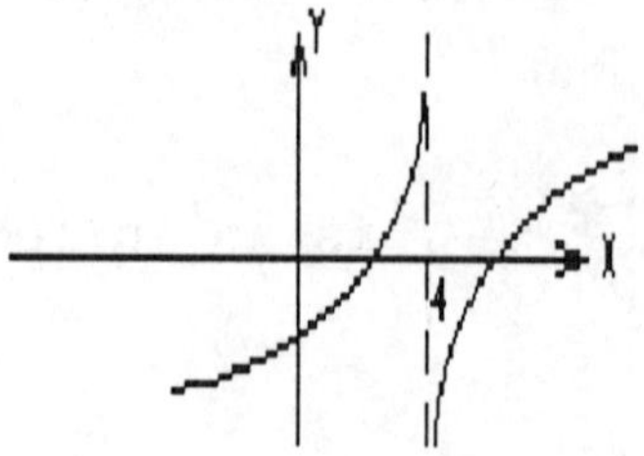

e) none of the above

Answer: (d)
MC Moderate

2.3.1: Determine all relative maximum, minimum, and inflection points of $f(x) = \frac{5000}{x} + \frac{1}{2}x$.

Answer: (100, 100) is a rel. min. point; (-100, -100) is a rel. max. point
ES Moderate

2.3.2: Determine all relative maximum, minimum, and inflection points of $f(x) = x^2 + 4x + 5$.

Answer: (-2, 1) is a minimum point
ES Moderate

2.3.3: Determine all relative maximum, minimum, and inflection points of $f(x) = -2x^2 + 4x + 1$.

Answer: (1, 3) is a maximum point
ES Moderate

2.3.4: Determine all relative maximum, minimum, and inflection points of $f(x) = x^3 - 3x^2 + 1$.

Answer: (2, -3) is a rel. min. point; (0, 1) is a relative maximum point; (1, -1) is an inflection point.
ES Moderate

2.3.5: Determine all relative maximum, minimum, and inflection points of $f(x) = x^3 - 6x^2 + 9x - 3$.

Answer: (1, 1) is a maximum; (3, -3) is a minimum; (2, -1) is an inflection point.
ES Moderate

2.3.6: Determine all relative maximum, minimum, and inflection points of $f(x) = xe^{-x}$.

Answer: $\left(1, \frac{1}{e}\right)$ is a maximum; $\left(2, \frac{2}{e^2}\right)$ is an inflection point.
ES Moderate

2.3.7: Determine all relative maximum, minimum, and inflection points of $f(x) = 2x^3 - 9x^2 + 12x - 1$.

Answer: (1, 4) is a maximum; (2, 3) is a minimum; $\left(\frac{3}{2}, \frac{7}{2}\right)$ is an inflection point.
ES Moderate

2.3.8: Graph: $f(x) = \frac{x^3}{3} - \frac{x^2}{2} - 6x$

Answer:

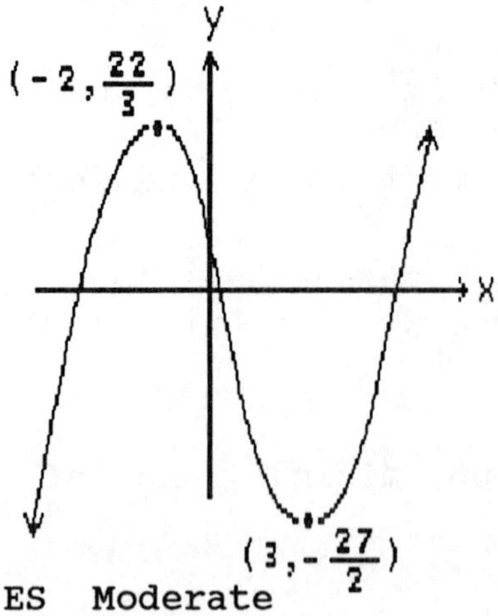

ES Moderate

2.3.9: Graph: $f(x) = (x + 2)^4$

Answer:

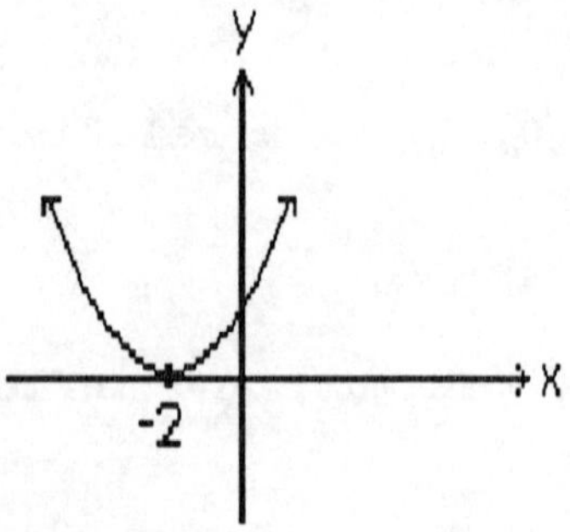

ES Moderate

2.3.10: Graph: $f(x) = x^3 - 3x^2 + 2$

Answer:

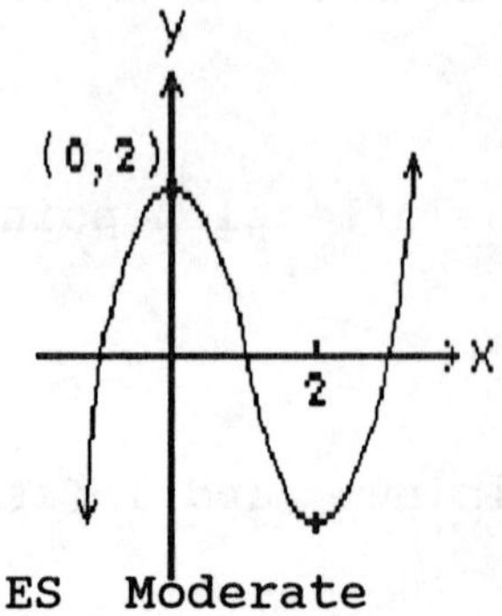

ES Moderate

2.3.11: Sketch the graph of the function $f(x) = x^3 + 3x^2 - 45x + 10$. Your calculations should include

(i) relative maximum and minimum points (<u>label your answers</u>),

(ii) values of x where f(x) is concave up and values of x where f(x) is concave down,

(iii) inflection points.

Answer: relative maximum at (-5, 185); relative minimum at (3, -71); concave up for x > -1; concave down for x < -1; inflection point (-1, 57)
ES Moderate

2.3.12: Sketch the graph of the function $f(x) = 2x^3 - 6x^2 + 6x - 6$.

Answer:

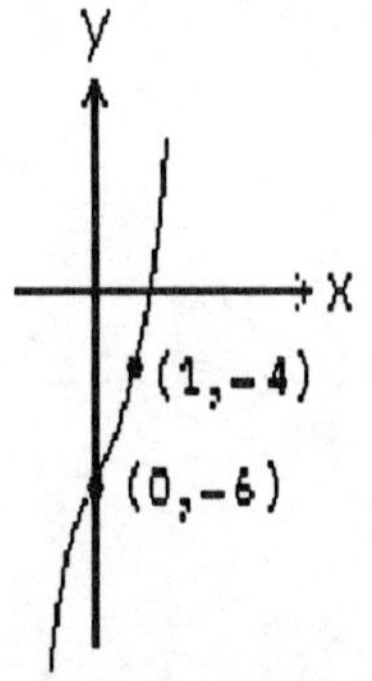

ES Moderate

2.3.13: Sketch the graph of the function $f(x) = -x^3 + 3x^2 + 9x - 15$.

Answer:

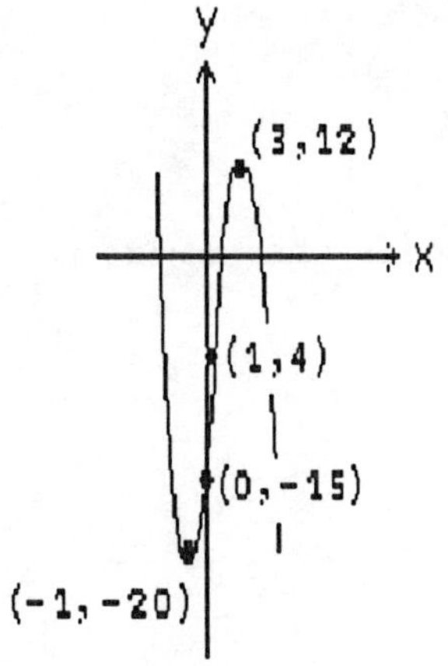

ES Moderate

2.3.14: What is (are) the inflection point(s) of
$y = 2x^3 - 3x^2 - 12x + 17$?

a) $x = 2$

b) $x = \frac{7}{2}, x = \frac{1}{2}$

c) $x = \frac{1}{2}, x = -\frac{1}{2}$

d) $x = \frac{1}{2}$

e) none of the above

Answer: (d)
MC Moderate

2.3.15: Which of the following is(are) true of $f(x) = \frac{1}{10}x^5 - \frac{7}{12}x^4 - \frac{1}{2}x^3 + 9x^2$?

(I) inflection point at x = 2
(II) inflection point at (0, 18)
(III) relative minimum point (0, 0)

Answer: (I) and (III)
ES Moderate

2.3.16: Which of the following is/are true of $h(x) = 32x^2 - \frac{8}{x} + 4$?

(I) relative minimum point $\left(-\frac{1}{2}, 28\right)$
(II) inflection point (0, 4)
(III) absolute minimum $\left(-\frac{1}{2}, 0\right)$
(IV) the graph is always concave up

Answer: (I) only
ES Moderate

2.3.17: Which of the following is/are true of $g(t) = (2t - 1)^{3/2} + 4$?

(I) concave down for $t < \frac{1}{2}$
(II) $g'(t)$ is positive for all t in the domain
(III) no inflection points
(IV) defined only for $t \geq \frac{1}{2}$

Answer: (II), (III) and (IV)
ES Moderate

2.3.18: Which of the following is/are true of $H(x) = 6x - 5x^{1/2} - 1$?

(I) no relative extrema

(II) $H''(x)$ is positive for all x in the domain

(III) inflection point at $x = \frac{5}{4}$

(IV) inflection point at $x = \frac{-3}{2}$

Answer: (II) only

ES Moderate

2.3.19: Which of the following is/are true of $F(t) = \frac{1}{t} - t^{1/2}$?

(I) $F(t) = 0$ for all $t > 0$

(II) $F'(t) < 0$ for $t > 0$

(III) the graph is concave up for $t > 0$

(IV) relative maximum point (0, 0)

Answer: (II) and (III)

ES Moderate

2.3.20: Which of the following is/are true of $y = 6t^4 - 8t^2 + 2$?

(I) relative maximum point (0, 2)

(II) relative minimum point $\left(-\sqrt{2/3}, -2/3\right)$

(III) inflection point at $t = \frac{\sqrt{2}}{3}$

(IV) the point (1, 0) lies on the graph

Answer: all of the above

ES Moderate

2.3.21: Which of the following could be represented by the graph below?

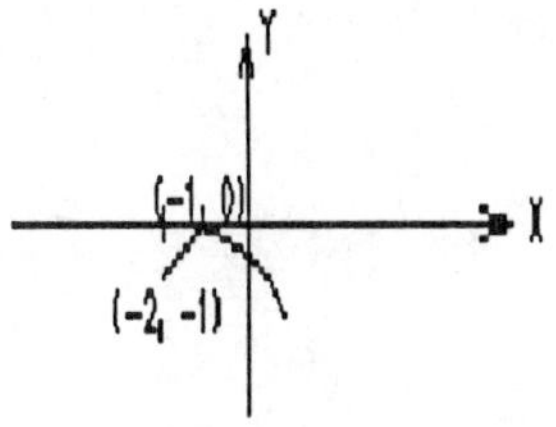

a) $y = -x^2 + 2x - 3$

b) $y = 2\sqrt{x + 2} - x - 3$

c) $y = -x^2 + \frac{2}{x + 2} - 3$

d) $y = 3x - 2\sqrt{x + 2} + 1$

Answer: (b)

MC Moderate

2.3.22: Which of the following could be represented by the graph below?

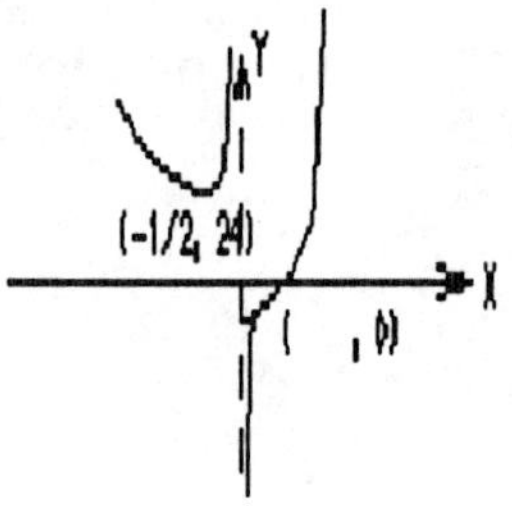

a) $y = \frac{8}{x} - 32x^2$

b) $y = 32x^2 - 8x$

c) $y = 64x^3 + 32x$

d) $y = 32x^2 - \frac{8}{x}$

Answer: (d)
MC Moderate

2.4.1: Graph: $f(x) = \frac{1}{x + 2}$

Answer:

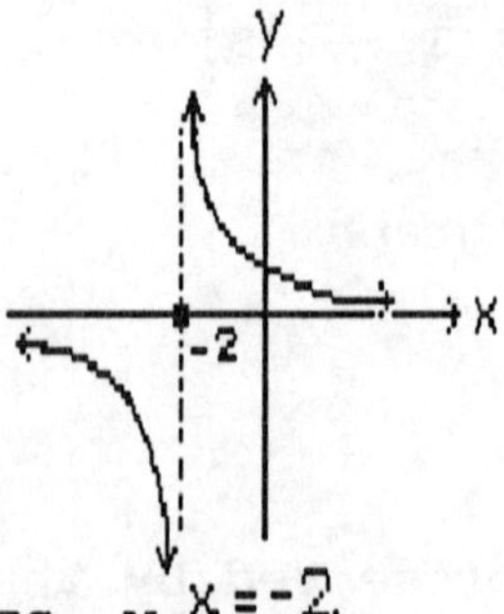

ES Moderate

2.4.2: Graph: $f(x) = (x - 2)^3 + 1$

Answer:

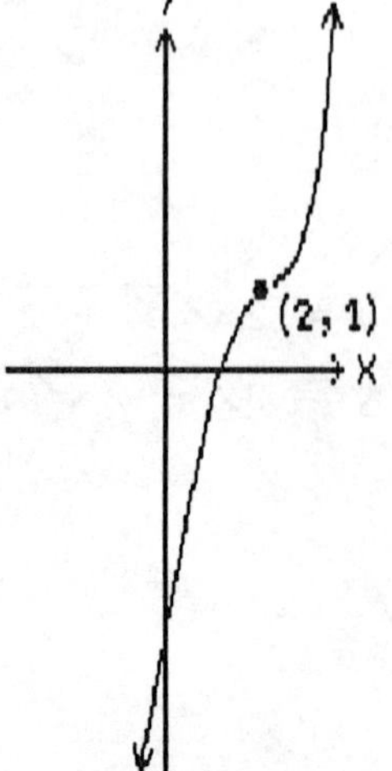

ES Moderate

2.4.3: Sketch the graph of the function $f(x) = 9x + 1 + \frac{1}{x}$, $x > 0$.

Answer:

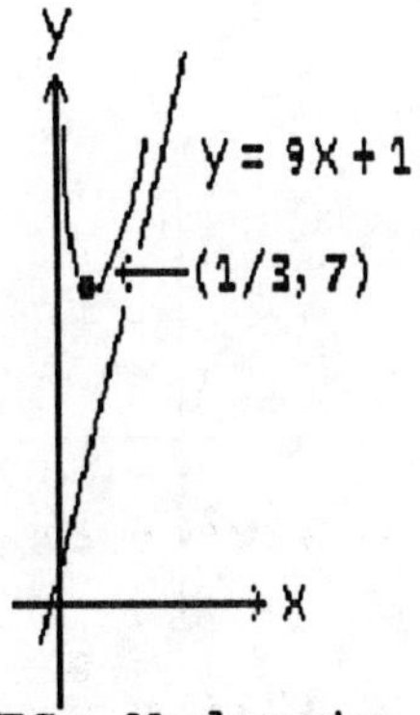

ES Moderate

2.4.4: Sketch the graph of the function $f(x) = \frac{12}{x^{1/2}} + x^{3/2}$ $(x > 0)$.

Answer:

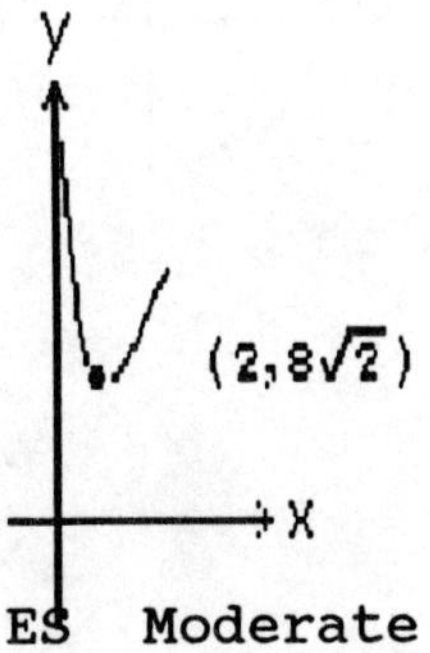

ES Moderate

2.4.5: Which of the following is the graph of $y = \frac{1}{x} + 2x + 5$?

(a)

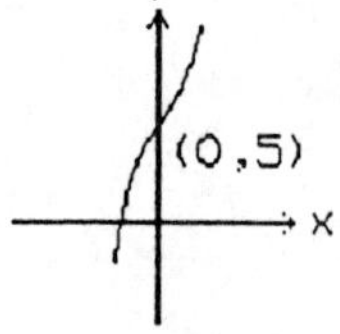

(b)

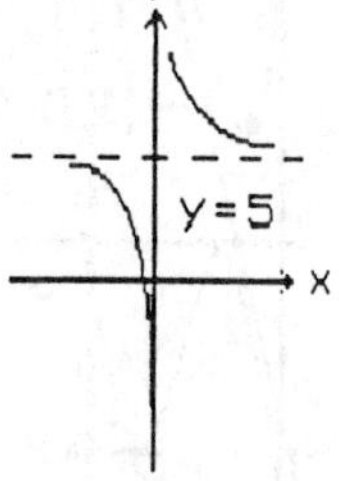

(c)

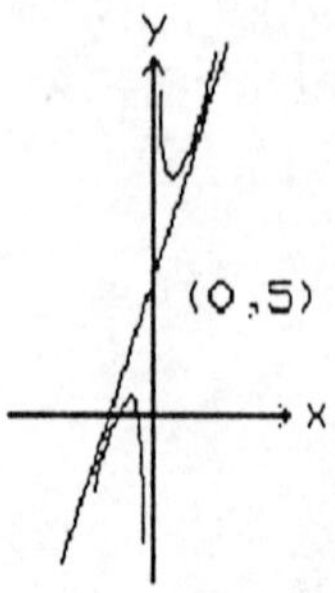

(d)

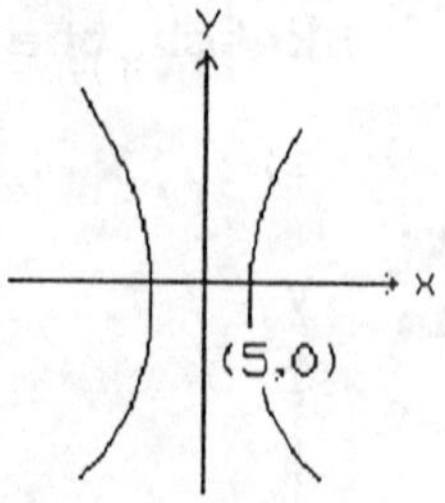

Answer: (c)
MC Moderate

2.4.6: What are the x-intercepts of the graph of $y = 3x^3 + 24$?

a) (2, 0)
b) (0, 24)
c) (24, 0)
d) (2, 24)
e) none of the above

Answer: (e)
MC Moderate

2.4.7: Suppose f(x) is the function graphed below. Which of the following is f(x)?

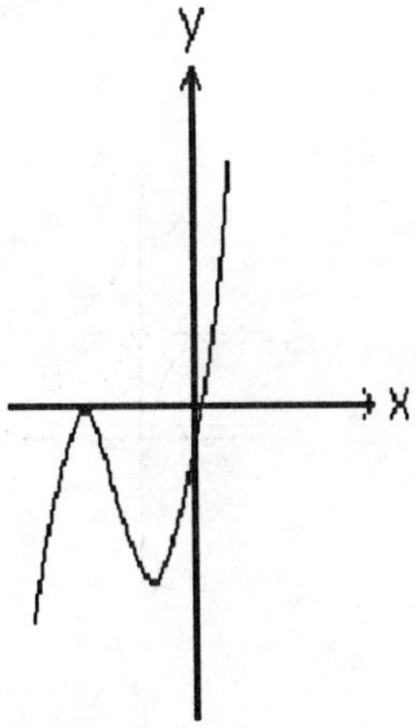

a) $f(x) = x^2 + 3x$

b) $f(x) = x^3 + 5$

c) $f(x) = \frac{1}{x} + x^2 + 3x$

d) $f(x) = x^3 + 6x^2 + 9x$

e) $f(x) = -x^2 + 2x + 5$

Answer: (d)
MC Moderate

2.4.8: Sketch the graph of $f(x) = \frac{1}{4}x^4 - x^3 + 2$, including any asymptotes, relative extrema and inflection points.

Answer:

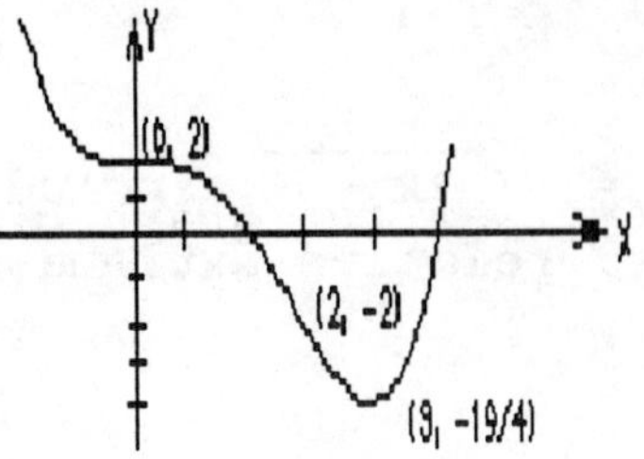

ES Moderate

2.4.9: Sketch the graph of $g(x) = 2x^3 + 3x^2 - 12x - 5$, including any asymptotes, relative extrema and inflection points.

Answer:

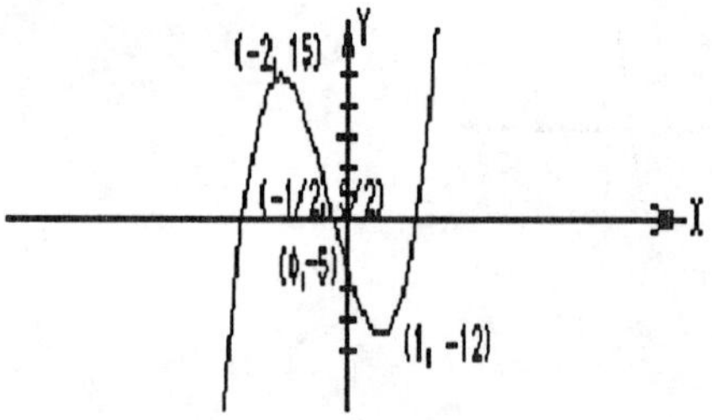

ES Moderate

2.4.10: Sketch the graph of $h(x) = x + 10 + \frac{9}{x}$, including any intercepts, asymptotes, relative extrema and inflection points.

Answer:

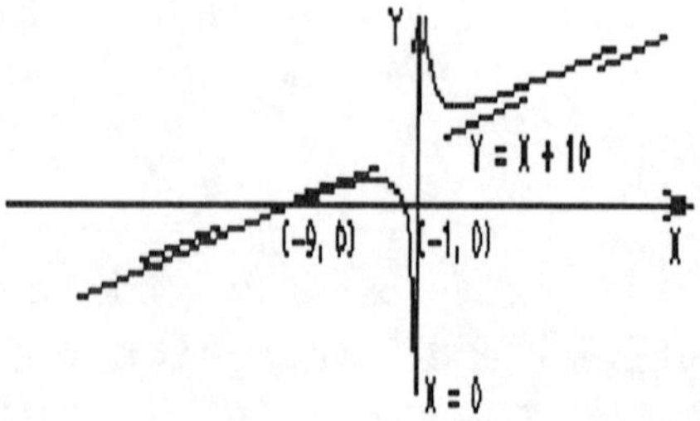

ES Moderate

2.4.11: Sketch the graph of $y = \sqrt{3x + 4}$, including any intercepts, asymptotes, relative extrema and inflection points.

Answer:

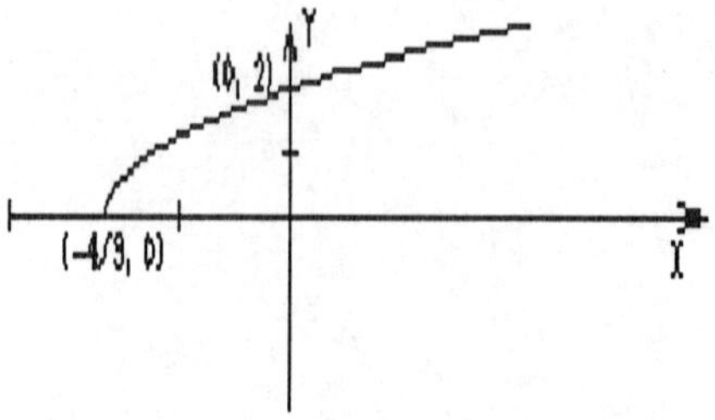

ES Moderate

2.4.12: Which of the following is the graph of $f(x) = x^4 - 9$?

a)

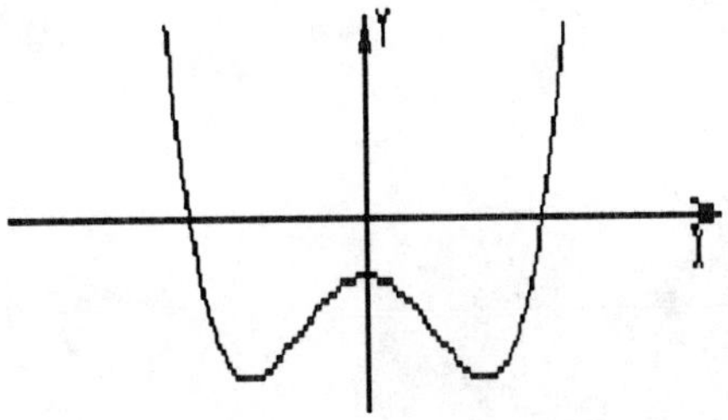

b)

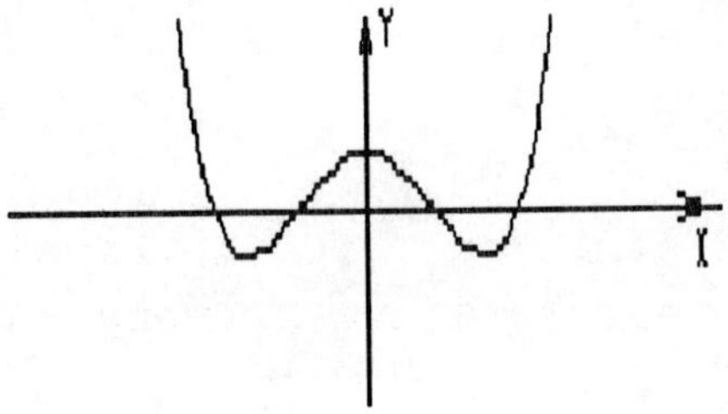

c)

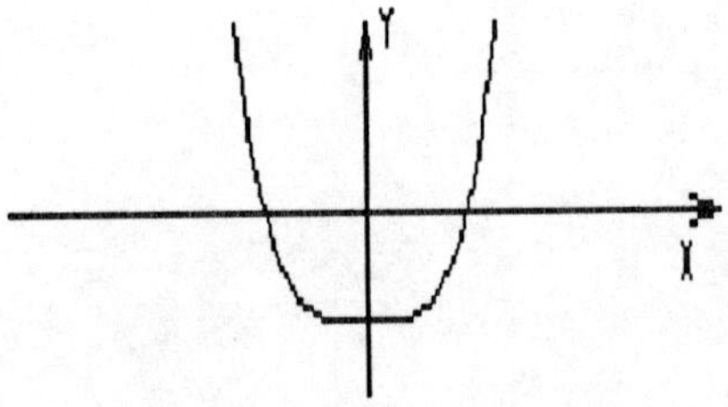

d)

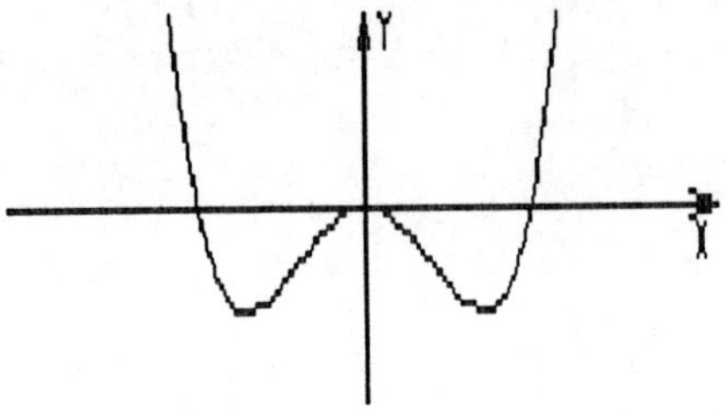

Answer: (c)
MC Moderate

2.4.13: The graph of a function f(x) is sketched below. What is f(x)?

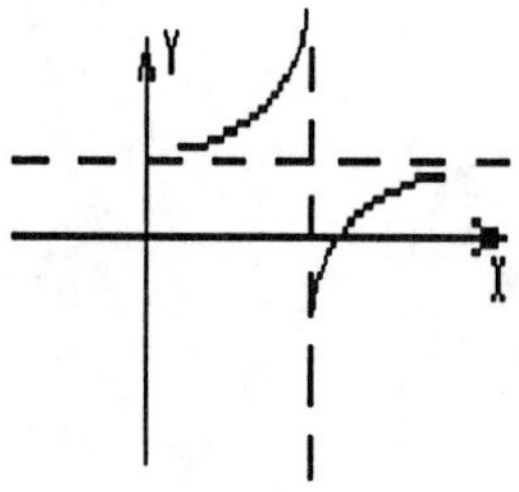

a) $f(x) = \frac{1}{5 - x} + 2$

b) $f(x) = \frac{1}{x - 5}$

c) $f(x) = \sqrt{x - 5} - 2$

d) $f(x) = x^2 - 5x - 2$

Answer: (a)
MC Moderate

2.4.14: The graph of a function f(x) is sketched below. What is f(x)?

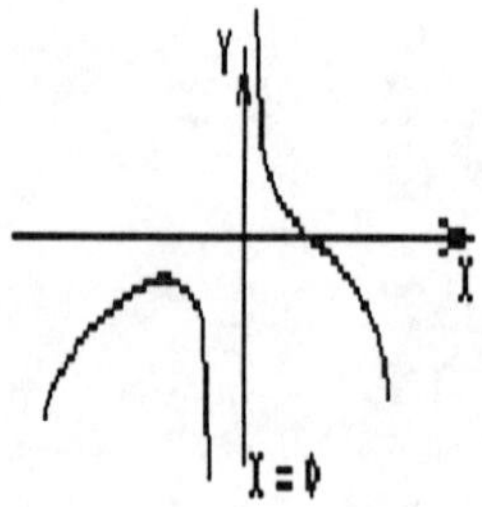

a) $f(x) = -x^2 - x^3$

b) $f(x) = \frac{1}{x} - x^2$

c) $f(x) = \frac{1}{x^2}$

d) $f(x) = \frac{1}{x + 2} - x$

Answer: (b)
MC Moderate

2.4.15: Which of the following is the graph of

$$F(x) = 2x - \frac{1}{x + 2} + 5?$$

a)

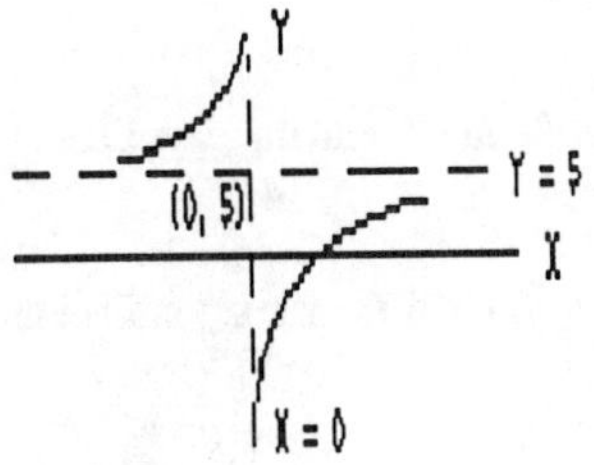

b)

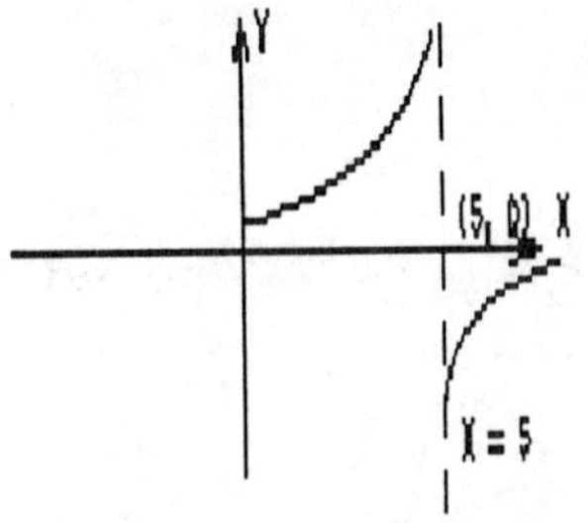

c)

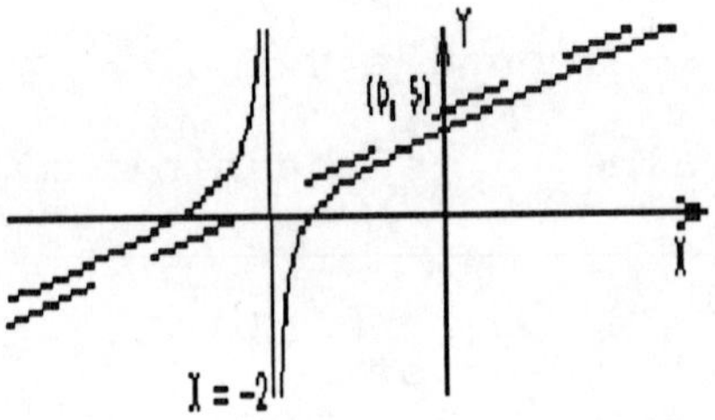

d)

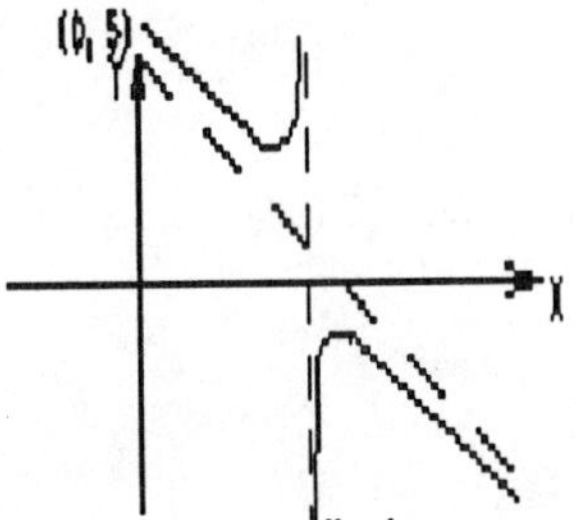

Answer: (c)
MC Moderate

2.5.1: Determine the minimum value of $x + y$ when $xy = 100$ ($x > 0$, $y > 0$).

Answer: 20
ES Moderate

2.5.2: Determine all maximum and minimum values of $f(x) = -x^3 + 3x^2 + 9x - 1$ on $-2 \le x \le 2$.

Answer: (-1, 6) is a minimum; (2, 21) is a maximum.
ES Moderate

2.5.3: Determine the maximum and minimum values of $f(x) = x^3 - 3x^2 - 9x$on $-4 \le x \le 4$.

Answer: 5 is the maximum value; -76 is the minimum value.
ES Moderate

2.5.4: Determine the minimum value of $f(x) = x^3 - 3x^2 + 2$ on $1 \le x \le 3$.

Answer: -2 is the minimum value.
ES Moderate

2.5.5: Determine the maximum value of $f(x) = -2x^3 + 6x + 3$ on $0 \le x \le 2$.

Answer: 7 is the maximum value.
ES Moderate

2.5.6: Determine the maximum and minimum values of $f(x) = 1 - x^3 + 3x + 21$ on $0 \le x \le 3$.

Answer: 24 is the maximum value; 4 is the minimum value.
ES Moderate

2.5.7: Determine the minimum value of $f(x) = (x^2 - 2x + 2)^2 + 2(x^2 - 2x + 2)$.

Answer: The minimum value is 3.
ES Moderate

2.5.8: Determine the minimum value of $f(x) = x + \dfrac{1}{x - 1}$ on $x > 1$.

Answer: 3 is the minimum value.
ES Moderate

2.5.9: A homebuilder's advertisement promises a house with a finished recreation room of 300 square feet. Two perpendicular walls of the room are to be panelled at a cost of $5 per running foot. A third side will be built out of windows at a cost of $10 per running foot. The fourth side will use the existing cinder block. What dimension should the room have to minimize the homebuilder's cost?

Answer: 30 ft x 10 ft (10 ft = window side)
ES Moderate

2.5.10: An airline flies 120,000 passengers per week to Florida when charging \$100 per flight. It estimates that for each \$1 increase in price it will lose 400 passengers. By how much should the fare be increased (or decreased) to maximize total revenue?

Answer: increase fares by \$100
ES Moderate

2.5.11: A hospital wishes to build an emergency water storage tower. The tower is to be cylindrical and have a capacity of 2,000,000 cubic feet of water. What should be the dimensions of the tower that require the minimum amount of materials? (For the sake of simplicity, you may assume that the amount of material is minimized when the surface area of the tower is minimized.)

Answer: radius $= \dfrac{100}{\pi^{1/3}}$ ft, height $= \dfrac{200}{\pi^{1/3}}$ ft
ES Moderate

2.5.12: A health food store stocks bottles of multivitamins. It orders equal quantities of stock from its wholesaler at equally spaced points throughout the year. The cost of placing each order is \$250. Moreover, the cost of keeping a jar of vitamins in inventory is \$1 per year. The store predicts that it will sell 12,500 bottles of vitamins in the next year. How many orders of how many bottles each will result in a minimum cost to the health food store?

Answer: 5 orders of 2500 bottles each
ES Moderate

2.5.13: A function f(x) has derivative $f'(x) = \dfrac{1}{1 + x^2}$. For what values of x does the graph of f(x) have a relative maximum or a relative minimum?

Answer: none
ES Moderate

2.5.14: The demand equation for a monopolist is $p = 1050 - .03x$ and the cost function is $C(x) = 150x + 750{,}000$, where x is the number of units produced.

(a) Find the value of x that maximizes the profit and determine the corresponding price and total profit for this level of production.

(b) Suppose that, in order to spur economic activity, the government reduces taxes so that the monopolist's costs are reduced by $1,000,000. What is the resulting change in the monopolist's profit?

Answer: Originally x = 15,000 units, price = 600/unit, profit = 5,000,000; after the tax cut, profits increase by 1,000,000 to 6,000,000.
ES Moderate

2.5.15: A manufacturer estimates that the profit from producing x units of a commodity is $-x^2 + 40x - 100$ dollars per week. What is the maximum profit he can realize in one week?

a) $400
b) $275
c) $300
d) $500
e) none of the above

Answer: (c)
MC Moderate

2.5.16: A toll road averages 36,000 cars per day when charging $1 (100 cents) per car. A survey concludes that changing the toll will result in 300 fewer cars for each cent of increase in price. Which of the following represents the revenue which will result from an increase of x cents in the price of the toll?

a) $36{,}000(100 + x) - 300x$
b) $(36{,}000 - 300x)(100 + x)$
c) $36{,}000 \cdot 100 - x(300 - x)$
d) $36{,}000 + 300x + (100 + x)$
e) none of the above

Answer: (b)
MC Moderate

2.5.17: Suppose a ball is thrown into the air and after t seconds has the height $h(t) = -16t^2 + 80t$ feet. When will it reach its maximum height?

a) 5 sec
b) 2.5 sec
c) 0.5 sec
d) 300 sec
e) none of the above

Answer: (b)
MC Moderate

2.5.18: A rectangular garden of area 50 square feet is to be surrounded on three sides by a fence costing \$2 per running foot and on one side by a brick wall costing \$6 per running foot. Let x be the length of the brick wall side. Which of the following represents the total cost of the material (expressed as a function of x)?

a) $\frac{3x}{50} + \frac{1}{50x}$

b) $3x + (50 - x)$

c) $6x + \frac{300}{x}$

d) $6x + \frac{50}{x}$

e) none of the above

Answer: (e)
MC Moderate

2.5.19: A rectangular garden is to be fenced in and divided into three parallel sections. The fencing for the boundary costs \$20 per foot while the fencing for the dividing fences costs \$5 per foot. Consider the problem of finding the dimensions of the largest garden possible if the gardener can spend \$2000 for the fencing.

(a) Draw a picture of the garden and label the variables.
(b) Determine the objective equation.
(c) Determine the constraint equation.
(d) Find the optimal dimensions of the garden.

Answer: (a)

(b)

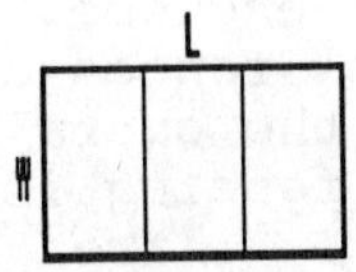

$A = w \cdot l$

(c) $2000 = 40l + 50w$

(d) $w = 20$ ft, $l = 25$ ft

ES Moderate

2.5.20: A rectangular corral with a total area of 60 square meters is to be fenced off and then divided into 2 rectangular sections by a fence down the middle. The fencing for the outside costs $9 per running meter, while that for the interior dividing fence costs $12 per running meter. Which of the following statements hold, if the cost (c) of the fencing is to be maximized?

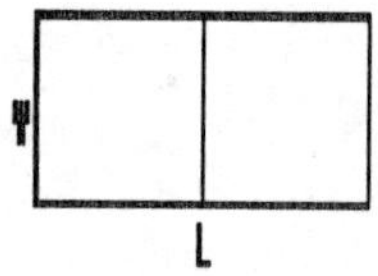

(I) The constraint equation is $3w + 2l = 60$.
(II) The objective equation is $2l \cdot w = 60$.
(III) The constraint equation is $w \cdot l = 60$.
(IV) The objective equation is $C = 30w + 18l$.
(V) The constraint equation is $C = 12w + 9wl$.
(VI) The objective equation is $C = 60 - lw$.

Answer: (III) and (IV)
ES Moderate

2.5.21: A large rectangular garden is to be enclosed by a fence and divided into 5 regions by 4 parallel fences across the interior of the garden as shown below. Find the overall dimensions of the garden, if a total of 1200 feet of fencing is to be used and the area maximized.

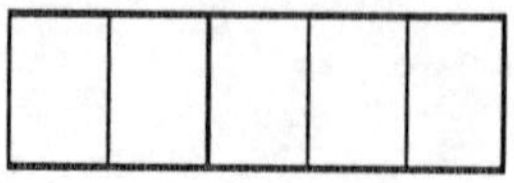

Answer: 300 ft x 100 ft
ES Moderate

2.5.22: A design for a bus stop has 2 square sides, a back and a top. The materials cost $7 per square foot for the top and $3 per square foot for the sides and back. If the volume is to be 1200 cubic feet, find the dimensions that minimize the cost.

Answer: 12 ft x 10 ft x 10 ft
ES Moderate

2.6.1: A fast food restaurant is establishing its inventory policy for ordering frozen french fries. In the coming year, they expect to sell 2500 lbs of french fries. It costs \$4 to place an order and the carrying costs for a year are \$2 per pound based on the average amount in storage. The restaurant wishes to determine how many pounds of fries to order at a time and the number of orders to place during the year in order to minimize its ordering and inventory costs.

(a) Define the variables in this problem.
(b) What is the objective equation?
(c) What is the constraint equation?
(d) Determine the optimal number of orders that should be placed and the optimal size of those orders.

Answer: (a) x = number of pounds of fries per order
r = number of orders per year
(b) $C = 4r + x$
(c) $2500 = x \cdot r$
(d) 25 orders @ 100 lbs/ order
ES Moderate

2.6.2: An open box with square ends is to contain 125 cubic inches. The bottom is to be made of a material which weighs twice as much per square inch as the material used for the sides. What should the dimensions of the box be in order to minimize its weight? (Note: the actual weights of the materials does not matter!)

Answer: 5 in x 5 in x 5 in.
ES Moderate

2.6.3: A sports retailer expects to sell 1200 sweat suits at a steady rate over the course of the coming year. The cost of placing an order with the wholesaler is \$60. The annual inventory cost per sweat suit is \$3 based on the average inventory level. Determine the economic order quantity for the suits (i.e., determine the order size that minimizes ordering and inventory expenses).

Answer: 40 suits per order.
ES Moderate

2.7.1: Suppose that 20,000 fans will go to a ball game when the price of a ticket is \$5.00, and that 500 fewer fans will go for each \$1.00 increase in ticket price. By how much should ticket prices be increased (or decreased) in order to maximize revenue?

a) increase price by \$45.00
b) decrease price by \$1.00
c) increase price by \$17.50
d) increase price by \$22.50

Answer: (c)
MC Moderate

2.7.2: A health club offers memberships at the rate of $300, provided that at least 50 people join. For each member in excess of 50, the membership fee will be reduced by $2. Due to space limitations, at most 125 memberships will be sold. How many memberships should the club sell in order to maximize its revenue?

a) 50 b) 75 c) 100 d) 125

Answer: (c)
MC Moderate

2.7.3: In planning a sidewalk cafe, it is estimated that, if there are 28 tables, the daily profit will be $8 per table and that, if the number of tables is increased by x, the profit per table will be reduced by $\frac{1}{4}x$ dollars (due to overcrowding). How many tables should be present in order to maximize the profit?

a) 30
b) 20
c) 10
d) can't do the problem without cost information.

Answer: (a)
MC Moderate

3. Techniques of Differentiation

3.1.1: Differentiate: $f(x) = \dfrac{x^2}{x^3 - 5x + 2}$

Answer: $\dfrac{2x(x^3 - 5x + 2) - x^2(3x^2 - 5)}{(x^3 - 5x + 2)^2}$

ES Moderate

3.1.2: Differentiate: $f(x) = (x^5 - 8x^2 + 5)(x^3 + \sqrt{x} - 1)$

Answer: $(5x^4 - 16x)(x^3 + \sqrt{x} - 1) + (x^5 - 8x^2 + 5)(3x^2 + \dfrac{1}{2\sqrt{x}})$

ES Moderate

3.1.3: Differentiate: $x^5 \cdot \dfrac{x - 1}{x + 1}$

Answer: $\dfrac{5x^6 + 2x^5 - 5x^4}{(x + 1)^2}$

ES Moderate

3.1.4: Let $f(x) = \dfrac{1}{(5x - 3)^2}$. What is $f'(1)$?

a) $\frac{5}{2}$

b) $\frac{5}{4}$

c) $-\frac{5}{2}$

d) $-\frac{5}{4}$

e) none of the above

Answer: (d)

MC Moderate

3.1.5: Let $f(x) = \sqrt{x}(x - 1)^3$. Then $f'(1)$ equals

a) $\frac{1}{2}$

b) -1

c) 1

d) 0

e) none of the above

Answer: (d)

MC Moderate

3.1.6: Let $F(x) = \frac{x}{x + 1}$. What is F"(x)?

a) $-\frac{2}{(x + 1)^3}$

b) $\frac{1}{(x + 1)^3}$

c) $-\frac{1}{(x + 1)^2}$

d) $\frac{x}{(x + 1)^3}$

e) none of the above

Answer: (a)
MC Moderate

3.1.7: At x = 1, the function $f(x) = \frac{1}{x^2 + 1}$

a) is concave up
b) is concave down
c) has an inflection point
d) has a relative extreme point
e) none of the above

Answer: (c)
MC Moderate

3.1.8: Let f(x) = (x + 1)(x + 2)(x + 3). Then f'(-2) is equal to

a) 1
b) 0
c) 6
d) -1
e) none of the above

Answer: (d)
MC Moderate

3.1.9: Let $f(x) = \frac{1}{x - 1}$. At x = 1 the graph of f(x)

a) is concave up
b) is concave down
c) has an inflection point
d) has a vertical asymptote
e) none of the above

Answer: (d)
MC Moderate

3.1.10: Let $f(x) = \frac{x^2 + 1}{x^2 - 1}$. At x = 0, the graph of f(x)

a) has a relative maximum point
b) has a relative minimum point
c) has an inflection point
d) has an asymptote
e) none of the above

Answer: (a)
MC Moderate

3.1.11: Differentiate: $f(x) = (x^2 + 1)^3(x^3 - 1)^2$

Answer: $f' = 6x^2(x^3 - 1)(x^2 + 1)^3 + 6x(x^3 - 1)^2(x^2 + 1)^2$
ES Moderate

3.1.12: Differentiate: $f(x) = (x^2 + x)(3x^2 + 4x - 1)$

Answer: $f' = (x^2 + x)(6x + 4) + (3x^2 + 4x - 1)(2x + 1)$
ES Moderate

3.1.13: Differentiate: $f(x) = (4x^2 + 4)(2x^2 + 2x)$

Answer: $f' = 8(x^2 + 1)(2x + 1) + 6x(x^2 + x)$
ES Moderate

3.1.14: Differentiate: $f(x) = 4(x^3 + 1)(2x^2 + 2x + 1)^4$

Answer: $f' = 16(x^3 + 1)(4x + 2)(2x^2 + 2x + 1)^3 + 12x^2(2x^2 + 2x + 1)^4$
ES Moderate

3.1.15: Differentiate: $f(x) = \frac{x}{x^2 + 1}$

Answer: $f' = \frac{1 - x^2}{(x^2 + 1)^2}$
ES Moderate

3.1.16: Differentiate: $f(x) = \frac{1}{x^2 + 1}$

Answer: $f' = \frac{-2x}{(x^2 + 1)^2}$
ES Moderate

3.1.17: Differentiate: $f(x) = \frac{x^2}{x + 1}$

Answer: $f' = \frac{x^2 + 2x}{(x + 1)^2}$
ES Moderate

3.1.18: Differentiate: $f(x) = \frac{6x - 7}{x^2 + 5}$

Answer:
$$f' = \frac{-6x^2 + 14x + 30}{(x^2 + 5)^2}$$
ES Moderate

3.1.19: Differentiate: $f(x) = 4(x^2 + 1)^3(x^2 + 1)^4$

Answer: $f' = 56x(x^2 + 1)^6$
ES Moderate

3.1.20: Differentiate: $f(x) = (x + 1)(x + 2) + (x + 3)(x + 4)$

Answer: $f' = 4x + 10$
ES Moderate

3.1.21: Differentiate: $f(x) = (x + 1)^2(x^2 + 1)^{-3}$

Answer:
$$\frac{-6x(x + 1)^2}{(x^2 + 1)^4} + \frac{2(x + 1)}{(x^2 + 1)^3}$$
ES Moderate

3.1.22: Differentiate: $f(x) = \frac{x^2 - 1}{x + 2}$

Answer:
$$f' = \frac{x^2 + 4x + 1}{x^2 + 4x + 4}$$
ES Moderate

3.1.23: If $F(r) = r^2(r - 1)(r + 1)^{-1}$, then $F'(r)$ equals...

a) $4r^3 - 2r$
b) $2r(r^2 + r - 1)$
c) $\frac{-2r}{(r + 1)^2}$
d) $(2r^3 + 2r^2 - 2r)(r + 1)^{-2}$

Answer: (d)
MC Moderate

3.1.24: If $y = \frac{x + 1}{x^2 - 1}$, then $\frac{dy}{dx}$ equals

a) $\frac{1}{2x}$

b) $-\frac{1}{(x - 1)^2}$

c) $\frac{x^3 + x^2 - x - 1}{(x^2 - 1)^2}$

d) none of the above.

Answer: (b)
MC Moderate

3.1.25: If $y = x^2 + 4x(x^{3/2})$, then $\frac{dy}{dx}$ equals...

a) $(x^2 + 4x)(3/2x^{1/2}) + (x^{3/2})(2x + 4)$

b) $6x(x^{1/2}) + 4x^{3/2}$

c) $2x + 10x^{3/2}$

d) $\frac{x^{3/2}(2x + 4) - (x^2 + 4x)(3/2x^{1/2})}{(x^{3/2})^2}$

Answer: (c)
MC Moderate

3.1.26: Let $T(x) = 3x^{-1/3}(x^3 + 1)$. Which of the following statements is/are true?

(I) $T(x) = \frac{3x^3 + 3}{x^{1/3}}$.

(II) $T'(x) = (8x^3 - 1)x^{-4/3}$.

(III) $T''(x) = \frac{-4}{3}(24x^2)x^{-5/3}$.

(IV) The point (0, 3) is a relative maximum.

Answer: (I) and (II)
ES Moderate

3.1.27: Let $G(t) = \frac{3t}{(3t + 3)}$. Which of the following statements is/are true?

(I) G(t) has an asymptote at $t = -1$.

(II) $\left(-\frac{1}{2}, -1\right)$ is a point on the graph.

(III) G(t) is increasing at $t = -\frac{1}{2}$.

(IV) G(t) is concave down for all $x > -1$.

Answer: all are true.

ES Moderate

3.1.28: Let $h(r) = r^3 \cdot \frac{1 - r}{1 + r}$. Which of the following is/are true?

(I) h(r) has 3 critical points.

(II) h(r) is increasing at $r = -1$.

(III) h(r) is decreasing at $r = \frac{1}{2}$.

(IV) h(r) is decreasing at $r = -\frac{1}{2}$.

Answer: (I).
ES Moderate

3.1.29: Let $F(p) = \frac{1}{(2p + 4)^3}$. Which of the following statements is/are true?

(I) $F(-2) = 0$.

(II) F(p) is concave up for all $x > 0$.

(III) F(p) has a relative minimum point.

(IV) F(p) has an inflection point at $p = -2$.

Answer: (II) only
ES Moderate

3.1.30: Consider the function $Q(x) = \frac{(x + 5)^2}{\sqrt{x}}$, $x > 0$. Which of the following is/are true?

(I) Q(x) is decreasing for all x.

(II) Q(x) has a relative minimum at $x = 1$.

(III) Q(x) is concave up for $x > 0$.

(IV) Q(x) has an asymptote at $x = 0$.

Answer: (III) and (IV)
ES Moderate

3.1.31: Which of the following is the graph of $y = \dfrac{t^3 - t^2 - 6t}{t}$?

a)

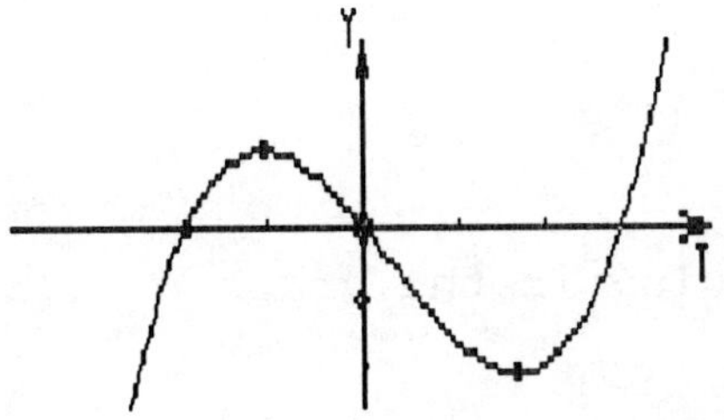

b)

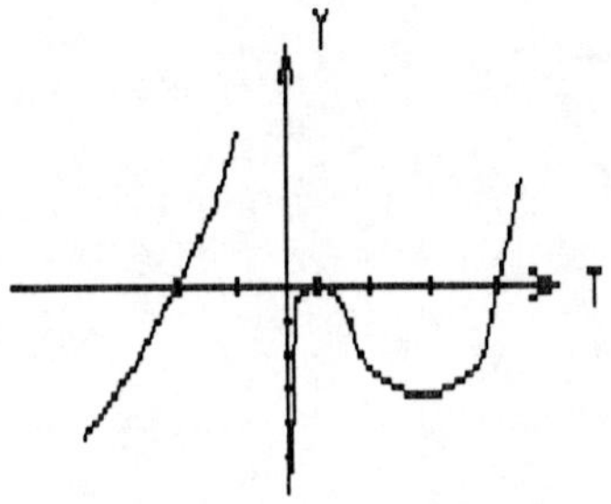

c)

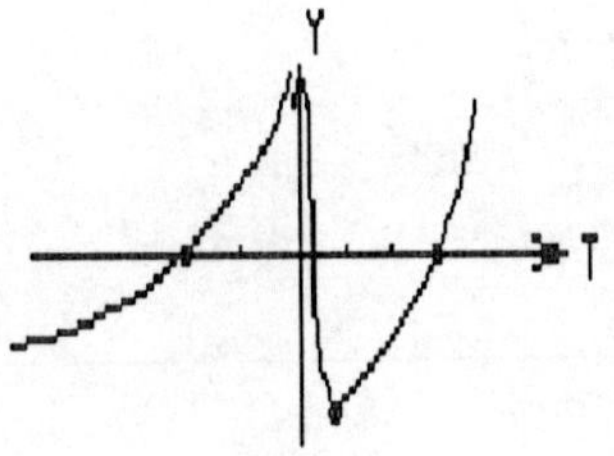

d)

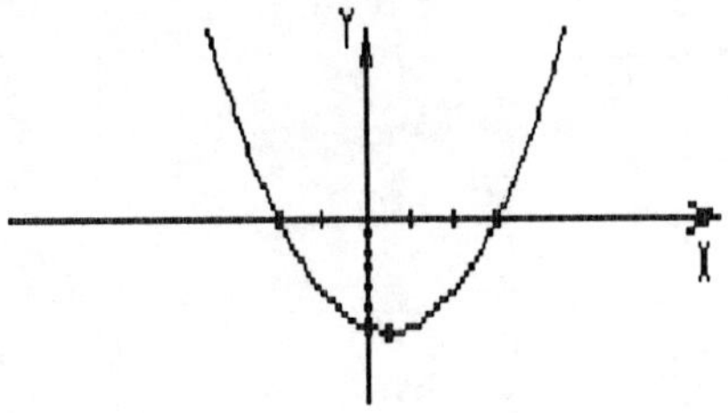

Answer: (d)
MC Moderate

3.1.32: Let $y = \dfrac{x^6 + 4x^3 + 1}{x^3 + 1}$. Determine $\dfrac{dy}{dx}$.

Answer: $\dfrac{3x^2(x^6 + 2x^3 + 3)}{(x^3 + 1)^2}$.

ES Moderate

3.1.33: Which of the following is the graph of $f(x) = \dfrac{(x - 2)^5}{(x - 4)^3}$?

a)

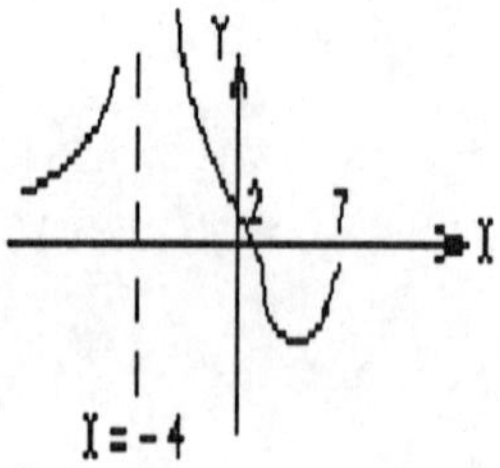

b)

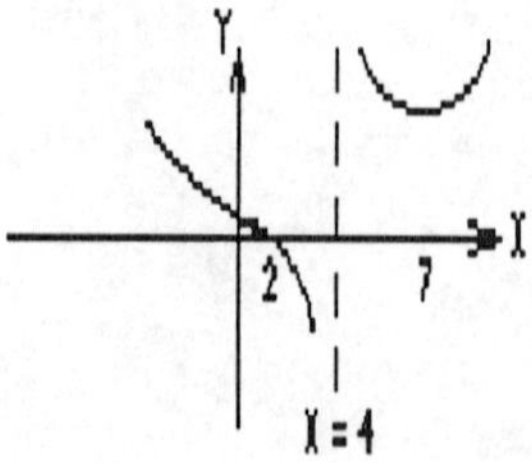

c)

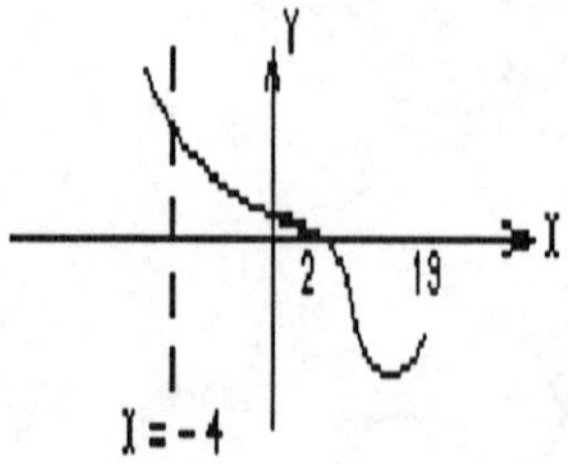

Answer: (b)

MC Moderate

3.1.34: Determine the x-coordinates of all possible relative extreme points of $P(x) = (2x - 3)^4(6x - 1)^{-2}$.

Answer: $x = \dfrac{3}{2}, -\dfrac{7}{6}$

ES Moderate

3.1.35: Determine the x-coordinates of all points on the graph of $Q(x) = \dfrac{x^3}{(x + 2)^5}$ where the tangent line is horizontal.

Answer: $x = 0$, $x = 3$.
ES Moderate

3.1.36: Let $f(x) = (2x^2 + 5)\left(2x - \sqrt{x}\right)$. Determine $f'(4)$.

Answer: $f'(4) = \dfrac{643}{4} = 160\ 3/4$
ES Moderate

3.1.37: Let $G(t) = \left(\dfrac{t + 4}{t^2}\right)(5t + 2)$. Find $G''(t)$.

Answer: $G''(t) = \dfrac{44t^5 - 48t^4}{t^8}$.
ES Moderate

3.1.38: Let $p(t) = \dfrac{6t^2 - t + 4}{t^{2/3}}$. Calculate $p'(8)$.

Answer: $p'(8) = \dfrac{95}{6} = 15\ 5/6$
ES Moderate

3.1.39: Determine the x-coordinates of the critical points of $H(x) = \dfrac{x^2 - 7x + 12}{2x + 1}$.

Answer: $x = \dfrac{-1 \pm 3\sqrt{7}}{2}$
ES Moderate

3.1.40: Let $g(x) = (4x^2 + 1)^3(2x - x^3)^5$. Determine $f'(1)$.

Answer: $f'(1) = -25$.
ES Moderate

3.2.1: Differentiate: $f(x) = ((x^5 + 2)^3 + 1))^4$

Answer: $4((x^5 + 2)^3 + 1)^3 \cdot 3(x^5 + 2)^2 \cdot 5x^4$
ES Moderate

3.2.2: Let $f(x) = \frac{x}{x+1}$. Determine: $f''(x)$

Answer:
$-2(x + 1)^{-3}$
ES Moderate

3.2.3: Let $f(x) = \frac{x^2 - x}{x}$, $g(x) = \frac{1}{\sqrt{x}}$. Calculate: $f(g(x))$

Answer: $\frac{1}{\sqrt{x}} - 1$
ES Moderate

3.2.4: Use the chain rule to compute the derivative of

$$\frac{(3x + 1)^3}{(3x + 1)^3 + 1}$$

Answer: Let $f(x) = \frac{x}{x + 1}$, $g(x) = (3x + 1)^3$. Then $f'(x) = \frac{1}{(x + 1)^2}$,

$g'(x) = 9(3x + 1)^2$ and $\frac{d}{dx} f(g(x)) =$

$$\frac{1}{((3x + 1)^3 + 1)^2} \cdot 9(3x + 1)^2 = \frac{9(3x + 1)^2}{((3x + 1)^3 + 1)^2}.$$

ES Moderate

3.2.5: Let $f(x) = \frac{1}{x^3 + x^2}$. Then $f'(x)$ equals

a) $\frac{1}{(x^3 + x^2)^2}$

b) $-\frac{3x^2 + 2x}{(x^3 + x^2)^2}$

c) $\frac{x^3 + x^2}{(x^3 + x^2)^2}$

d) $-\frac{2x + 1}{x^2(x + 1)^2}$

e) none of the above

Answer: (b)
MC Moderate

3.2.6: Let $f(x) = x^3$. Using the chain rule, determine an expression for $[f(g(x))]$.

a) $3[g(x)]^2 g'(x)$

b) $3x^2 g'(x^3)$

c) $3[g(x)]^2$

d) $g'(x^3)$

e) (e) none of the above

Answer: (a)
MC Moderate

3.2.7: Let $g(x) = \sqrt{x}$. Using the chain rule, find an expression for $[f(g(x))]'$.

a) $\sqrt{x}\, f'(x) + \frac{1}{2}x^{-1/2} f(x)$

b) $\frac{1}{2\sqrt{x}} f'(\sqrt{x})$

c) $\frac{1}{2\sqrt{x}} f(\sqrt{x})$

d) $\sqrt{x}\, f(\sqrt{x})$

e) none of the above

Answer: (b)
MC Moderate

3.2.8: Which of the following is the graph of $y = (3 - x)^3$?

a)

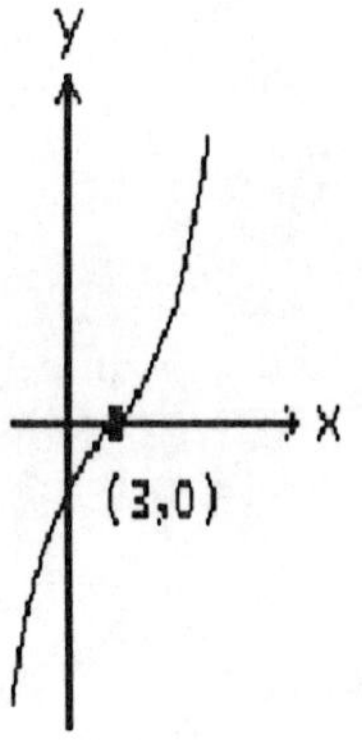

b)

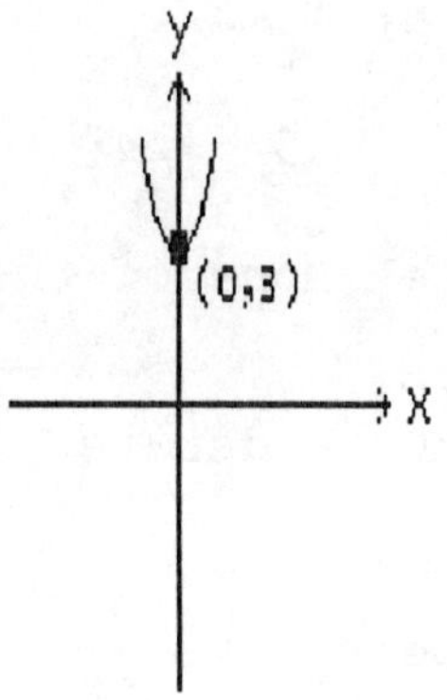

c)

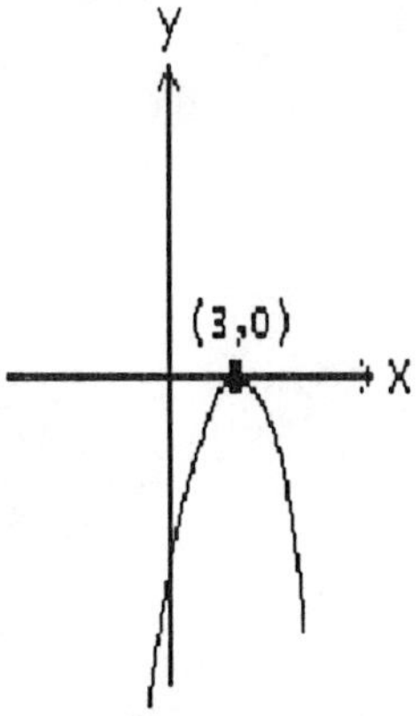

d)

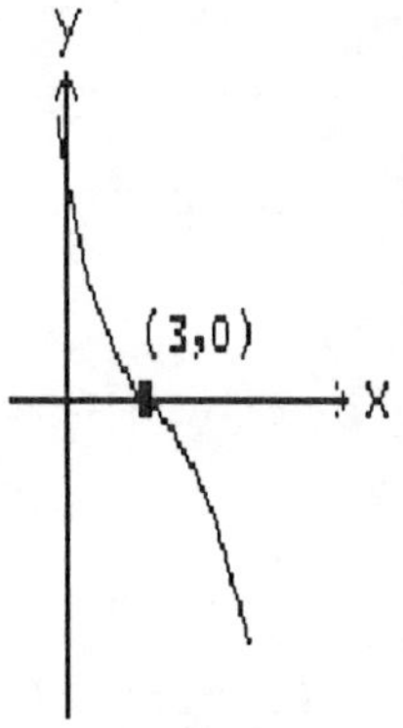

Answer: (d)
MC Moderate

3.2.9: Let $f(x) = (\sqrt[3]{x} + 1)^5$. Then $\frac{df}{dx}$ equals

a) $5(\sqrt[3]{x} + 1)^4$

b) $\frac{1}{3}(\sqrt[3]{x} + 1)^4 x^{-2/3}$

c) $\frac{5}{3}(\sqrt[3]{x} + 1)^4 x^{-2/3}$

d) $5(\sqrt[3]{x} + 1)^4 \sqrt[3]{x}$

e) none of the above

Answer: (c)
MC Moderate

3.2.10: Let $f(x) = \dfrac{\sqrt{x} - x}{\sqrt{x} + x}$. Then at x = 1 the graph of f(x)

a) is increasing

b) is decreasing

c) has a relative maximum

d) has a relative minimum

e) none of the above

Answer: (b)
MC Moderate

3.2.11: Let $f(x) = \dfrac{1}{\sqrt[4]{x}}$. Then $\dfrac{d^2f}{dx^2}$ equals

a) $\frac{1}{4}x^{-3/4}$

b) $-3x^{-2}$

c) $\frac{5}{16}x^{-9/4}$

d) $-\frac{3}{4}x^{-7}$

e) none of the above

Answer: (c)
MC Moderate

3.2.12: Determine $\frac{dy}{dx}$ where $y = u^2$ and $u = 3x + 4$.

Answer: $\frac{dy}{dx} = 18x + 24$
ES Moderate

3.2.13: Determine $\frac{dy}{dx}$ where $y = u^3$ and $u^2 = x$.

Answer: $\frac{dy}{dx} = \frac{3}{2}\sqrt{x}$
ES Moderate

3.2.14: Determine $\frac{dy}{dx}$ where $y = 4u^2 + 8u + 4$ and $u = 3x + 1$.

Answer: $\frac{dy}{dx} = 72x + 48$

ES Moderate

3.2.15: Determine $\frac{dy}{dx}$ where $y = \frac{u^2}{u^2 + 1}$ and $u = \sqrt{2x + 4}$.

Answer: $\frac{dy}{dx} = \frac{2}{(2x + 5)^2}$

ES Moderate

3.2.16: Determine $(f(g(x)))'$ where $f(x) = 8x^2 + 4x + 3$ and $g(x) = 2x + 3$.

Answer: $64x + 104$

ES Moderate

3.2.17: Let $f(x) = \frac{1}{2x^2}$ and let $g(x) = 4x^5 + 2$. Then $[g(f(x))]'$ equals

a) $\frac{20x^4}{x^3}$

b) $-\frac{5}{4x^{11}} - \frac{2}{x^3}$

c) $-\frac{20x^4}{(4x^5 + 2)^3}$

d) $10x^2 - \frac{4x^5 + 2}{x^3}$

Answer: (b)

MC Moderate

3.2.18: Let $f(x) = \frac{\sqrt{x}}{3}$ and let $g(x) = 4\sqrt{2x}$. Then $[f(g(x))]'$ equals

a) $\frac{1}{12(2x)^{1/4}} \cdot \frac{4}{(2x)^{1/2}}$

b) $\frac{1}{24\sqrt{2x}} \cdot \frac{2}{\sqrt{2x}}$

c) $\frac{1}{6\sqrt{x}} \cdot \frac{4}{\sqrt{2x}}$

d) $-\frac{1}{3x^{1/2}} \cdot \frac{2}{(2x)^{1/2}}$

Answer: (a)

MC Moderate

3.2.19: Let $f(x) = x^2 - 9$ and let $g(x) = x^2 - 16$. Then $\frac{d}{dx} g(f(x))$ equals

a) $4x^3 - 36x$

b) $4x^3 - 50x^2$

c) $(x^2 - 9)^2 - 16$

d) $((x^2 - 9)^2 - 16)(2x)$

Answer: (a)

MC Moderate

3.2.20: Let $y = \sqrt{u}$ and let $u = x^3 - 5x^2 + 1$. Then $\frac{dy}{dx}$ equals

a) $\sqrt{x^3 - 5x^2 + 1}$

b) $\sqrt{3x^2 - 10x}$

c) $(3x^2 - 10x)/\ 2\sqrt{x^3 - 5x^2 + 1}$

d) $\sqrt{3x^2 - 10x}(x^3 - 5x^2 + 1)$

Answer: (c)
MC Moderate

3.2.21: Let $y = 3u + 2$ and let $u = \frac{t}{t + 1}$. Then $\frac{dy}{dt}$ equals

a) $\frac{1}{(3u + 2)^2}$ b) $\frac{1}{9}$ c) $\frac{1}{(t + 1)^2}$ d) $\frac{3}{(t + 1)^2}$

Answer: (d)
MC Moderate

3.2.22: Let $y = u + 1 - 8^{1/3}$ and let $u = \frac{t}{6} + 1$. Then $\frac{dy}{dt}$ equals

a) $\frac{1}{6}$

b) $-2 \cdot 8^{-2/3}$

c) $-\frac{t^2}{6}$

d) $-\frac{1}{6t^2} \cdot 8^{-2/3}$

Answer: (a)
MC Moderate

3.2.23: Use the chain rule to find the derivative of $\sqrt{4x^3 - 2x}$.

Answer: Let $f(x) = \sqrt{x}$ and let $g(x) = 4x^3 - 2x$.

Then $f(g(x)) = \sqrt{4x^3 - 2x}$,

$f'(x) = \frac{1}{2\sqrt{x}}$, $g'(x) = 12x^2 - 2$

and $[f(g(x))]' = \frac{1}{2\sqrt{4x^3 - 2x}} \cdot (12x^2 - 2) = \frac{6x^2 - 1}{\sqrt{4x^3 - 2x}}$.

ES Moderate

3.2.24: When a manufacturer produces and sells x units per week its weekly profit is P dollars, where $P = 100(2000 + 120x - x^2)$. Production level t weeks from the present will be $x = \frac{t}{2} + 16$.

(a) Find the rate of change in profit with respect to x.
(b) Find the rate of change in profit with respect to t.

Answer: (a) $\frac{dP}{dx} = -200(x - 60)$

(b) $\frac{dP}{dt} = -50(t - 88)$

ES Moderate

3.2.25: Let $g(x) = x^{1/2}$ and let $h(x) = \sqrt{x^2 + 1}$. Then $\frac{d}{dx}(g(x)h(x))$ equals

a) $\frac{1}{2}\left(\sqrt{x^2 + 1}\right)^{-1/2} \cdot \frac{2x}{2\sqrt{x^2 + 1}}$

b) $\frac{2x^{1/2}}{2\sqrt{x + 1}} \cdot \frac{1}{2} x^{-1/2}$

c) $x^{1/2} \cdot \frac{2x}{2\sqrt{x^2 + 1}} + \frac{1}{2}x^{-1/2} \cdot \sqrt{x^2 + 1}$

d) $\dfrac{x^{1/2} \cdot \frac{2x}{2\sqrt{x^2 + 1}} - \frac{}{}x^{-1/2} \cdot \sqrt{x^2 + 1}}{(x^{1/2})^2}$

Answer: (c)

MC Moderate

3.2.26: Suppose that the cost of manufacturing x units of a product is $C(x) = 6x - 2\sqrt{x} + 1$ dollars and that the production level t weeks from the present is $x = 4t^2$.

(a) Find the rate of change in cost with respect to the production level, x.
(b) Find the rate of change in cost with respect to time.

Answer: (a) $C'(x) = 6 - \frac{1}{\sqrt{x}}$.

(b) $\frac{d}{dt} C(p(t)) = 48t - 4$

ES Moderate

3.3.1: Determine $\frac{dy}{dx}$ where $x^2 + y^2 = 4$.

Answer: $\frac{dy}{dx} = -\frac{x}{y}$

ES Moderate

3.3.2: Determine $\frac{dy}{dx}$ where $4x^3 + 4xy + y = 8$.

Answer:
$$\frac{dy}{dx} = \frac{-12x^2 - 4y}{4x + 1}$$
ES Moderate

3.3.3: Determine $\frac{dy}{dx}$ where $x^3 + y^3 = 2xy$.

Answer:
$$\frac{dy}{dx} = \frac{2y - 3x^2}{3y^2 - 2x}$$
ES Moderate

3.3.4: Determine $\frac{dy}{dx}$ where $\frac{x^2}{a^2} + \frac{y^2}{b^2} = 1$.

Answer:
$$\frac{dy}{dx} = -\frac{xb^2}{ya^2}$$
ES Moderate

3.3.5: Determine the equation of a line tangent to the circle

$x^2 + y^2 = 1$ at $\left(\frac{1}{2}, \frac{\sqrt{3}}{2}\right)$.

Answer:
$$y = -\frac{1}{\sqrt{3}}x + \frac{2\sqrt{3}}{3}$$
ES Moderate

3.3.6: Determine the rate of change of $\sqrt{2x + 4}$ with respect to

$\frac{2x}{x - 2}$ at $x = 1$.

Answer: $\frac{1}{-4\sqrt{6}}$
ES Moderate

3.3.7: Determine $\frac{dy}{dx}$ where $g'(x) = x^3$ and $y = g(4x)$.

Answer: $\frac{dy}{dx} = 192x^3$
ES Moderate

3.3.8: Determine the rate of change of y^2 with respect to x^2 where $2y = x - 4x^2$.

Answer: $\frac{1}{4} - 3x + 8x^2$
ES Moderate

3.3.9: The radius of a baseball increases at a rate of 1 mm/sec. How fast is the surface area increasing when the radius is 10 mm? (Note: The surface area S of a sphere of radius r is $S = 4\pi r^2$.)

Answer: 80π mm/sec
ES Moderate

3.3.10: Water runs into a conical tank at a constant rate of 2 ft^3/min. How fast is the water level rising when the water is 4 ft deep if, at that moment, the radius is 2 ft?

Answer: $\frac{1}{2\pi}$ ft/min
ES Moderate

3.3.11: A balloon is 100 ft off the ground and rises vertically at a constant rate of 10 ft/sec. Directly beneath the balloon is a bicycle which travels at a constant rate of 20 ft/sec. How fast does the distance between them change one second later?

Answer:

$6\sqrt{5}$ ft/sec
ES Moderate

3.3.12: Mr. Smith is 6 ft tall and walks at a constant rate of 2 ft/sec toward a street light that is 10 ft above the ground. At what rate is the length of his shadow changing when he is 6 ft from the base of the pole which supports the light?

Answer: 3 ft/sec
ES Moderate

3.3.13: Assume $x^3 + (2y + 1)^2 = y^2$. Then $\frac{dy}{dx}$ equals

a) $\frac{3x^2 + 4(2y + 1)}{2y}$

b) $3x^2 + 2(y + 1)$

c) $\frac{-3x^2}{6y + 1}$

d) $\frac{-3x^2}{2(3y + 2)}$

Answer: (d)
MC Moderate

3.3.14: Assume $\frac{x^2}{4} \quad \frac{y^3}{x} = 4$. Then $\frac{dy}{dx}$ equals

a) $-\frac{y}{3x}$, $y \neq 0$.

b) $\frac{8x^2}{3}$, $y \neq 0$.

c) $-\frac{2y}{3x}$

d) $\frac{16 - 2y}{3x}$

Answer: (a)
MC Moderate

3.3.15: Assume $4x^3 + 2xy^2 - y^3 = \frac{5}{2}$. The slope of the graph at the point (1/2, -1) is:

a) $-\frac{1}{2}$

b) 8

c) 1

d) $\frac{8}{5}$

Answer: (c)
MC Moderate

3.3.16: Assume $\frac{4}{x} + \sqrt{y} = x$. The slope of the graph at the point (-1, 9) is:

a) 30

b) 18

c) $3\sqrt{3}$

d) $4 + \sqrt{3}$

Answer: (a)
MC Moderate

3.3.17: Suppose that $15x^{1/3}y^{-2/3} = 50$, where x and y are both differentiable functions of t. Then $\frac{dy}{dt}$ equals:

a) $\frac{27}{2000y} \frac{dx}{dt}$

b) $\frac{10}{x^{2/3}y^{5/3}} \frac{dx}{dt}$

c) $\frac{25y^{1/3}}{2x^{2/3}}$

d) $\frac{-y}{x} \frac{dx}{dt}$

Answer: (a)
MC Moderate

3.3.18: Suppose $2x^3 - 3p^4 = 6$, where x and p are differentiable functions of t. Then $\frac{dp}{dt}$ equals:

a) $\frac{6 - 6x^2}{12p^3} \frac{dx}{dt}$

b) $\frac{x^2}{2p^3} \frac{dx}{dt}$

c) $6x^2 - 12p^3\frac{dp}{dt}$

d) $6x^2\frac{dx}{dt} - 12p^3$

Answer: (b)
MC Moderate

4. The Exponential and Natural Logarithm Functions

4.1.1: $8^{4/3} + 9(9^{-1}) =$

a) 17
b) 5
c) $\frac{9}{8}$
d) 25
e) none of the above

Answer: (a)
MC Moderate

4.1.2: Simplify: $4^{x} \cdot 2^{x/2}$

Answer: $2^{(5/2)x}$
ES Moderate

4.1.3: Simplify: $2^{x} \cdot 8^{x}$

Answer: 2^{4x}
ES Moderate

4.1.4: Simplify: $\left(\sqrt{3}\right)^{x} \cdot x^{x}$

Answer: $\left(x\sqrt{3}\right)^{x}$
ES Moderate

4.1.5: Simplify: $\frac{3^{x}}{6^{x}} \cdot 8^{x} \cdot \left(\frac{32}{4}\right)^{x}$

Answer: 2^{5x}
ES Moderate

4.1.6: Simplify: $16^{x} \cdot \frac{1}{8}x \cdot 4^{x}$

Answer: 2^{3x}
ES Moderate

4.1.7: Simplify: $7^{-x} \cdot 14^{x} \cdot 49^{8x}$

Answer: $7^{16x} \cdot 2^{x}$
ES Moderate

4.1.8: Solve for x: $5^{x} \cdot 5^{2x} \cdot 5^{3x} = 25$

Answer: $x = \frac{1}{3}$
ES Moderate

4.1.9: Solve for x: $7^{-5} \cdot 49 \cdot 7^{x^{2}} \cdot 49^{x} = 1$

Answer: $x = 1, -3$
ES Moderate

4.1.10: Solve for x: $\left(5^{x} \cdot 25\right)^{2} = 125 \cdot \left(\frac{1}{25}\right)^{x}$

Answer: $x = -\frac{1}{4}$
ES Moderate

4.1.11: Simplify: $9^{x} \cdot 81^{x} \cdot 243^{x}$

Answer: 3^{11x}
ES Moderate

4.1.12: Simplify: $2^{x}3^{x}5^{x}7^{x}$

Answer: 210^{x}
ES Moderate

4.1.13: Simplify: $\left(\frac{1}{4}\right)^{2x} \cdot \left(\frac{1}{27}\right)^{3x} \cdot \left(\frac{1}{64}\right)^{8x}$

Answer: $2^{-52x} \cdot 3^{-9x}$
ES Moderate

4.1.14: Simplify: $3^{10x} \cdot 3^{11x} \cdot 3^{12x} \cdot 3^{-10x}$

Answer: 3^{23x}
ES Moderate

4.1.15: Simplify: $y^{4x} \cdot y^{6x} \cdot y^{x} \cdot y^{4}$

Answer: y^{11x+4}
ES Moderate

4.1.16: Simplify: $\left(\frac{1}{9}\right)^{27x} \cdot \left(\frac{1}{3}\right)^{48x} \cdot \left(\frac{1}{81}\right)^{9x}$

Answer: 3^{-138x}
ES Moderate

4.1.17: Solve for x: $2^{4-x} \cdot 2^{8+2x} = 64$

Answer: x = -6
ES Moderate

4.1.18: Solve for x: $3^{5x} \cdot 3^{x^2} \cdot 3^{3} = 3^{-3}$

Answer: x = -2, -3
ES Moderate

4.1.19: The expression $(t^2)^x \cdot (t^4)^x \cdot (t^{1/3})^x$ is equal to:
a) $t^{8/3\ x}$ b) $t^{19/3\ x}$ c) $(t^{19/3})^{3x}$ d) $(t^8)^{x/3}$

Answer: (b)
MC Moderate

4.1.20: The expression $\left(\frac{1}{2}\right)^{t} \cdot \left(\frac{1}{8}\right)^{5t} \cdot (4)^{t - 2}$ is equal to:
a) $\left(\frac{1}{4}\right)^{7t - 2}$ b) $(2)^{-14t - 4}$
c) $\left(\frac{1}{2}\right)^{8t - 4}$ d) $\left(\frac{1}{2}\right)^{15t^2}_{3}\ (2)^{2t - 4}$

Answer: (b)
MC Moderate

4.1.21: The expression $(5)^{2p} \cdot (4)^{p} \cdot (9)^{p/2}$...
a) equals $(180)^{7/2\ p}$ b) equals $(300)^{p}$
c) equals $(30)^{5p}$ d) can not be simplified.

Answer: (b)
MC Moderate

4.1.22: The expression $(3)^{2p} \cdot (9)^{p} \cdot (4)^{p}$...
a) equals $(18)^{2p}$ b) equals $(108)^{4p}$
c) equals $(18)^{6p}$ d) cannot be simplified.

Answer: (a)
MC Moderate

4.1.23: The expression $(16)^{p/2} \cdot \left(\frac{1}{2}\right)^p \cdot \left(\frac{1}{2p}\right)$...

a) equals 1

b) equals 4^p

c) equals $\left(\frac{4}{p}\right)^{3/2\ p\ +\ 1}$

d) cannot be simplified.

Answer: (a)
MC Moderate

4.1.24: The expression $\left(\frac{1}{27}\right)^p \cdot \left(\frac{3}{27}\right)^p \cdot \left(\frac{1}{3}\right)$ equals:

a) $\frac{1}{3}\left(\frac{4}{27}\right)^p$

b) $\left(\frac{1}{3}\right)^{6p}$

c) $(3)^{-5p\ -\ 1}$

d) none of the above.

Answer: (c)
MC Moderate

4.1.25: If $\left(\frac{1}{4}\right)^{3x\ +\ 1} = 2^{6\ -\ 2x}$, then...

a) $x = \frac{1}{2}$

b) $x = \frac{1}{3}$

c) $x = 3$

d) $x = -2$

Answer: (d)
MC Moderate

4.1.26: If $(9)^t \cdot (3)^{4t} \cdot (9)^{-2t} = \sqrt{3}$, then...

a) $t = \frac{1}{2}$

b) $t = \frac{1}{4}$

c) $t = -1$

d) $t = -2$

Answer: (b)
MC Moderate

4.1.27: If $5^t + 5^t + 5^t = 75$, then...

a) $t = 25$

b) $t = 5$

c) $t = 2$

d) t can not be determined.

Answer: (c)
MC Moderate

4.1.28: Assume $\left(\frac{1}{9}\right)^{3x\ +\ 4} - 3 = 81$. Then x equals:

a) $-\frac{1}{2}$

b) $-\frac{3}{2}$

c) -3

d) none of the above

Answer: (d)
MC Moderate

4.1.29: Assume $5^{t} \cdot 5^{4t} \cdot 5^{-2t} = (25)^{-1/2\ t\ +\ 1}$. Then t equals:

a) $-\frac{1}{2}$ b) $\frac{1}{2}$ c) $\frac{4}{13}$ d) $\frac{2}{7}$

Answer: (b)
MC Moderate

4.2.1: Simplify: $(e^{3x})^2 \cdot e^{-x}$

Answer: e^{5x}
ES Moderate

4.2.2: Simplify: $(e^{x} + e^{-x})^2$

Answer: $e^{2x} + e^{-2x} + 2$
ES Moderate

4.2.3: Simplify: $(e^{x} + e^{-x})(e^{x} - e^{-x})$

Answer: $e^{2x} - e^{-2x}$
ES Moderate

4.2.4: Simplify: $(x + e^{-x})^2$

Answer: $x^2 + 2xe^{-x} + e^{-2x}$
ES Moderate

4.2.5: Simplify: $\dfrac{(e^{1/2x})^{-3/4}}{e^{2x-5}}$

Answer: $e^{(-19/8)x+5}$
ES Moderate

4.2.6: Simplify: $\left(2^{-x^2}\right)^{(x+1)/x}$

Answer: 2^{-x^2-x}
ES Moderate

4.2.7: Simplify: $\left(\frac{1}{3^x}\right)^{3x}$

Answer: 3^{-3x^2}
ES Moderate

4.2.8: What is $\sqrt{\frac{e^{(1/2)x} \cdot e^{(3/2)x}}{e^{-2x}}}$ simplified?

a) e^{-10x}
b) e^{-6x}
c) $\frac{1}{e^{2x}}$
d) $e^{(1/2)x}$
e) none of the above

Answer: (e)
MC Moderate

4.2.9: The expression $e^{1/2} \cdot 2e^2 \cdot 5e^3$ is equal to:

a) $10e^3$
b) $11e^{11/2}$
c) $10e^{11/2}$
d) can not be simplified

Answer: (c)
MC Moderate

4.2.10: The expression $\frac{1}{e^5} \cdot \frac{e^3}{2} \cdot e^{4x - 1}$ is equal to:

a) $\frac{1}{2}e^{4x - 3}$
b) $\frac{e^{12x - 3}}{2e^5}$
c) $\frac{1}{2}e^{12x - 8}$
d) cannot be simplified

Answer: (a)
MC Moderate

4.2.11: The expression $(e^{2x})^3 \frac{5}{e^{1/2\ x}}$ is equal to:

a) $\frac{5e^{2x + 3}}{e^{1/2\ x}}$
b) $5e^{3/2\ x + 3}$
c) $5e^{3x}$
d) $5e^{5/2\ x}$

Answer: (d)
MC Moderate

4.2.12: If $(e^x)^2 \cdot e^{2x} \cdot e = \frac{1}{e^2}$, then x equals:

a) -2/5
b) -3/4
c) -1 or -2
d) -2

Answer: (b)
MC Moderate

4.2.13: Assume $9(e^x)(e^x) + (3e^x)^2 = \left(\frac{1}{3e^{2x} - 4}\right)^{-2}$. Then x equals:

a) -8

b) $\frac{1 \pm \sqrt{17}}{2}$

c) -2 or 4

d) none of the above

Answer: (d)
MC Moderate

4.2.14: The derivative of $(1 - 4e^x)x^2$ is:

a) $-8e^x \cdot x$

b) $-8xe^x - 1$

c) $2x - 8xe^x - 4x^3e^x - 1$

d) $2x - 8xe^x - 4x^2e^x$

Answer: (d)
MC Moderate

4.2.15: If $y = \frac{e^x - 1}{e^x + 1}$, then $\frac{dy}{dx}$ equals:

a) $\frac{e^x - e^x(e^x - 1)}{(e^x + 1)}$

b) $\frac{2e^x}{(e^x + 1)^2}$

c) 0

d) 1

Answer: (b)
MC Moderate

4.3.1: Differentiate: $e^{2x} - x^2$

Answer: $2e^{2x} - 2x$
ES Moderate

4.3.2: Differentiate: $(4e^{3x} - 1)^5$

Answer: $5(4e^{3x} - 1)^4 \cdot 12e^{3x}$
ES Moderate

4.3.3: Differentiate: $e^{-x^2/4}$

Answer: $-\frac{x}{2}e^{-x^2/4}$
ES Moderate

4.3.4: Differentiate: $(x^3 + 1)e^{-4x}$

Answer: $-4(x^3 + 1)e^{-4x} + 3x^2e^{-4x}$
ES Moderate

4.3.5: Differentiate: $\left(e^{5x^2} + x\right)^3$

Answer: $3\left(e^{5x^2} + x\right)^2\left(10xe^{5x^2} + 1\right)$
ES Moderate

4.3.6: Differentiate: $e^{3e^{2x}}$

Answer: $6e^{3e^{2x}+2x}$
ES Moderate

4.3.7: Differentiate: $\dfrac{x}{1 + e^{-x}}$

Answer: $\dfrac{1 + e^{-x} + xe^{-x}}{(1 + e^{-x})^2}$
ES Moderate

4.3.8: Determine all functions y = f(x) which satisfy the differential equation y' = .05y, f(0) = 10.

Answer: $f(x) = 10e^{.05x}$
ES Moderate

4.3.9: Determine all functions y = f(x) which satisfy the differential equation y' = 20y.

Answer: $f(x) = Ce^{20x}$, C arbitrary constant
ES Moderate

4.3.10: Let $f(x) = \dfrac{e^t - e^{t/2}}{2}$. What is f"(0) ?

a) $\frac{1}{4}$

b) $\frac{3}{16}$

c) $\frac{3}{8}$

d) $\frac{3}{4}$

e) none of the above

Answer: (c)
MC Moderate

4.3.11: Which of the following properties are true of the graph of $y = 10e^{2x}$?

I. It is concave up.

II. The y-intercept is (0, 2).

III. It has a minimum at x = 0.

IV. y is positive for $x \geq 0$ and negative for $x < 0$.

a) I
b) I and II
c) I and III
d) III and IV
e) none of the above

Answer: (a)
MC Moderate

4.3.12: What is $\frac{d}{dx}\frac{1}{3}e^{3-2x}$?

a) $-\frac{2}{3}e^{3-2x}$ b) $-\frac{1}{6}e^{3-2x}$ c) $\frac{1}{12}e^{4-2x}$ d) e^{-2x}

Answer: (a)
MC Moderate

4.3.13: Let $y = e^{e^{2x+1}}$. What is $\frac{dy}{dx}$?

a) e^{e^2}
b) $2e^{2x+1}$
c) e^{2x+1}
d) e^2
e) none of the above

Answer: (e)
MC Moderate

4.3.14: What is the solution y = f(x) of the differential equation $y' = \frac{1}{10}y$, f(0) = -3 ?

a) $y = e^{(1/10)y} - 3$
b) $y = \frac{1}{20}y^2 - 3$
c) $y = \frac{1}{10}e^{-3x}$
d) $y = -3e^{(1/10)x}$
e) none of the above

Answer: (d)
MC Moderate

4.3.15: Which of the following gives the best graphical interpretation of the fact $\frac{d}{dx}(e^x) = e^x$?

a) The graph is approximately a straight line.

b) The slope of the secant through the points (h, e^h) and $(0, 1)$ is e^x.

c) The derivative as a function is the best approximation of the tangent line to f(x).

d) The slope of the curve $y = e^x$ at an arbitrary value of x is exactly equal to the function value at that point.

e) The function e^x is always positive.

Answer: (d)
MC Moderate

4.3.16: What is $\frac{d}{dx}(e^3)$?

a) e^3

b) $3e^2$

c) 0

d) $\frac{e^4}{4}$

e) none of the above

Answer: (c)
MC Moderate

4.3.17: If $f'(x) = e^{-x}$ what can you say about f(x) ?

a) It is always increasing.

b) It is always concave up.

c) It has an inflection point.

d) All of the above.

Answer: (a)
MC Moderate

4.3.18: What are the relative maximum and/or relative minimum point(s) of the function $f(x) = e^{-2x} + 2x$?

a) $\left(\frac{.69}{2}, f(\frac{.69}{2})\right)$, relative maximum

b) there are no relative maximum/minimum points

c) (0, f(0)), relative minimum point

d) $\left(\frac{e}{2}, f(\frac{e}{2})\right)$, relative maximum point

e) none of the above

Answer: (c)
MC Moderate

4.3.19: Which of the following is the graph of $y = e^{-x} + x$?

a)

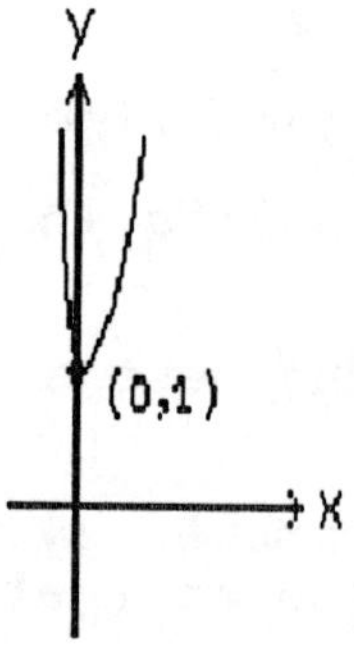

b)

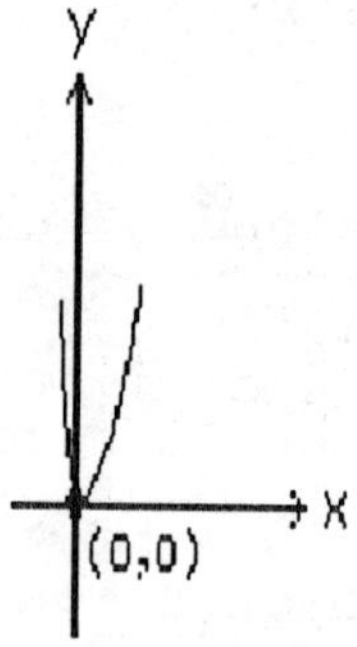

c)

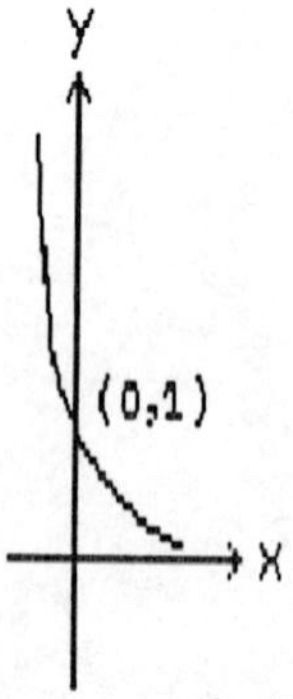

d)

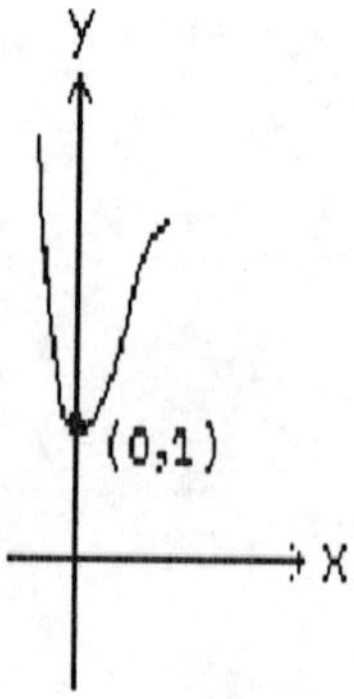

e) none of the above

Answer: (a)
MC Moderate

4.3.20: Find and classify the extreme point(s) of $f(x) = -2e^{x} + 6x + 5$.

a) (0, 3) relative minimum
b) (ln 3, 6 ln 3 - 1) relative maximum
c) $(\frac{1}{2}\ln 6, -1 + 3\ln 6)$ relative maximum
d) (ln 6, -7 + 6 ln 6) relative maximum
e) none of the above

Answer: (b)
MC Moderate

4.3.21: Differentiate: $f(x) = e^{x}$

Answer: $f' = e^{x}$
ES Moderate

4.3.22: Differentiate: $f(x) = 4e^{3x}$

Answer: $f' = 12e^{3x}$
ES Moderate

4.3.23: Differentiate: $f(x) = \frac{e^x + 1}{e^x - 1}$

Answer:
$f' = \frac{2e^x}{(e^x - 1)^2}$
ES Moderate

4.3.24: Differentiate: $f(x) = e^x + e^{2x} + \frac{1}{e^{-4x}}$

Answer: $f' = e^x + 2e^{2x} + 4e^{4x}$
ES Moderate

4.3.25: Differentiate: $f(x) = x^3 e^{-x^3}$

Answer:
$f' = -3x^5 e^{-x^3} + 3x^2 e^{-x^3}$
ES Moderate

4.3.26: Differentiate: $f(x) = e^{1/x}$

Answer: $-\frac{e^{1/x}}{x^2}$
ES Moderate

4.3.27: Differentiate: $f(x) = \frac{x}{e^x}$

Answer: $f' = \frac{1 - x}{e^x}$
ES Moderate

4.3.28: Determine all solutions of the differential equation
$y' = \frac{7}{6}y$.

Answer: $y = Ce^{7/6\ x}$, C any constant
ES Moderate

4.3.29: Determine all functions $y = f(x)$ that satisfy $y' = e \cdot y$.

Answer: $y = Ce^{ex}$, C any constant
ES Moderate

4.3.30: Determine all functions $y = f(x)$ that satisfy $y - 4y' = 0$ and $f(0) = \frac{2}{3}$.

Answer: $y = \frac{2}{3}e^{1/4\ x}$

ES Moderate

4.3.31: Determine all functions $y = f(x)$ satisfying $y' - y = 0$ and $f(1) = e$.

Answer: $y = e^x$

ES Moderate

4.3.32: Determine all functions $f(x)$ such that $f'(x) = \frac{-5}{2}f(x)$.

Answer: $f(x) = Ce^{-5/2\ x}$, C any constant

ES Moderate

4.3.33: Determine all functions $g(x)$ such that $g'(x) = .25g(x)$ and $g'(0) = 1$.

Answer: $g(x) = 4e^{.25x}$

ES Moderate

4.3.34: Let $P(t) = 2e^{-.5t}$. Determine $P''(0)$.

Answer: $P''(0) = \frac{1}{2}$

ES Moderate

4.3.35: Let $g(t) = 3e^{t^2 - 4}$. Determine $g''(-2)$.

Answer: $g''(-2) = 54$.

ES Moderate

4.3.36: Let $h(x) = e^{(x + 1)^{-1}}$. Determine $h''(x)$.

Answer: $\dfrac{e^{(x + 1)^{-1}}(2x + 3)}{(x + 1)^4}$

ES Moderate

4.3.37: Which of the following functions satisfy the differential equation $y' = -8y$?

(I) $y = -e^{-8x}$
(II) $y = e^{-4x} + 3$
(III) $y = 5e^{-4x}$
(IV) $y = 6e^{-8x}$

Answer: (I) and (IV)
ES Moderate

4.3.38: Which of the following functions $y = f(x)$ satisfy

$y' = 32y,\ f(0) = \frac{1}{2}$?

(I) $y = 32e^{1/2\ x}$
(II) $y = e^{16x}$
(III) $y = 1/2\ e^{32x}$
(IV) $y = 1/2x^{32}$

Answer: (III) only
ES Moderate

4.3.39: Which of the following functions $y = f(x)$ satisfy

$y' = \frac{-2}{3}y,\ f(1) = 1$?

(I) $y = e^{x}$
(II) $y = e^{-2/3\ x}$
(III) $y = e^{(-2/3)(x - 1)}$
(IV) $y = -\frac{2}{3}e^{x}$

Answer: (III) only
ES Moderate

4.3.40: What is $\frac{d}{dx}(6e^{2x} - x)^3$?

a) $3^{12e_x} - 1)_2$
b) $3(6e_{2x} - x)_2(12e_{2x} - 1)$
c) $3(6e^{2x} - x)^2(12e^{x})$
d) $3(12xe^{2x - 1} - 1)^2$

Answer: (b)
MC Moderate

4.3.41: If $f'(x) = e^{-3x} - 1$, what is true of $f(x)$?

(I) It is always decreasing.
(II) It has a critical point.
(III) It has an inflection point.
(IV) It has an x-intercepts.

Answer: (II) and (IV)
ES Moderate

4.3.42: Sketch the graph of $y = \frac{1}{2}e^{2x} - x$.

Answer:

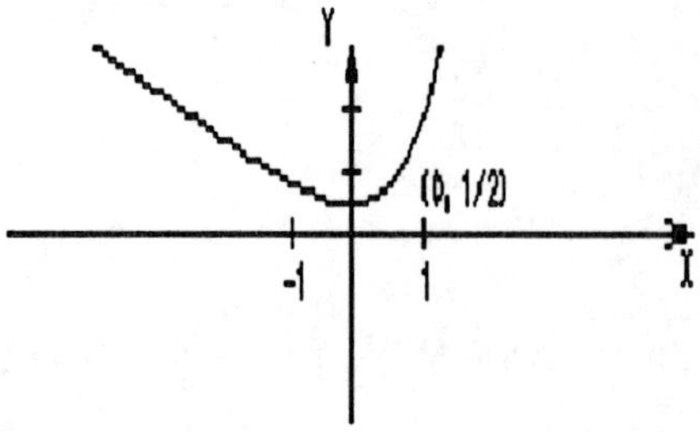

ES Moderate

4.3.43: Sketch the graph of $y = -3e^{2x} + 1 - 3e$.

Answer:

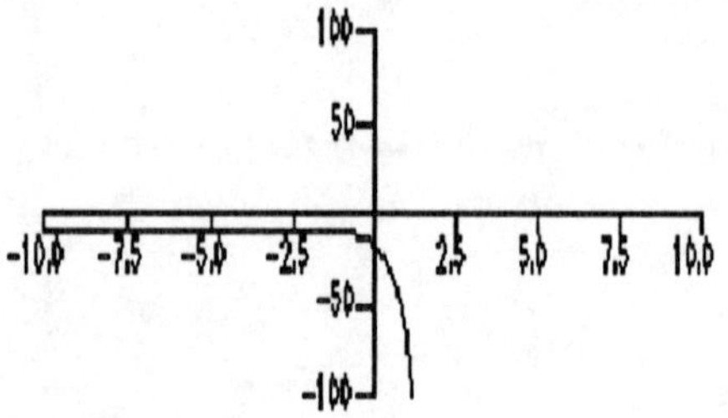

ES Moderate

4.4.1: Simplify: $\ln(x + 2) + \ln(x - 2)$

Answer: $\ln(x^2 - 4)$

ES Moderate

4.4.2: Simplify: $\ln xyz - \ln y^2/x$

Answer: $\ln \frac{x^2 z}{y}$

ES Moderate

4.4.3: Simplify: $e^{\ln 3 + \ln(2x)}$

Answer: 6x

ES Moderate

4.4.4: Simplify: $\ln e^{2x} - \ln e^{-x/2}$

Answer: $\frac{5x}{2}$

ES Moderate

4.4.5: Solve for x: $2e^{3x+1} = e^2$

Answer: $x = \frac{1}{3}(1 - \ln 2)$

ES Moderate

4.4.6: Solve for x: $\ln x^2 + (\ln x)^2 = 0$

Answer: $x = e^{-2}, 1$

ES Moderate

4.4.7: Solve for x: $\ln \sqrt{x} - \ln \sqrt[3]{x} = 2$

Answer: $x = e^{12}$

ES Moderate

4.4.8: Solve the following equation for x: $3 + \ln x = 0$

a) $x = \frac{1}{e^3}$

b) $x = -e^3$

c) $x = \ln(\frac{1}{3})$

d) $x = \ln(-3)$

e) none of the above

Answer: (a)

MC Moderate

4.4.9: Which of the following properties hold for the function $y = \ln x$?

I. $Y = \ln x$ is an increasing function <u>only</u> for $x \geq 1$.

II. $y = \ln x$ is not defined for $x \leq 0$.

III. The y-intercept of the graph of $y = \ln x$ is (0, 1).

IV. The graph of $y = \ln x$ is always concave down.

a) I and II

b) III and IV

c) II and IV

d) II

e) none of the above

Answer: (c)

MC Moderate

4.4.10: Let $y = (\ln x)^3 + e^{2+\ln x}$. What is $\frac{dy}{dx}$?

a) $3(\ln x)^2 + e^{1/x}$

b) $\frac{3(\ln x)^2 + e^{2+\ln x}}{x}$

c) $3x(\ln x)^2 + \frac{e^{2+\ln x}}{x}$

d) $3x^2(\ln x^3) + \frac{1}{x}e^{2+\ln x}$

e) none of the above

Answer: (b)
MC Moderate

4.4.11: If $y = \sqrt{\ln 3x}$, what is $\frac{dy}{dx}$?

a) $\frac{1}{3x\sqrt{\ln 3x}}$

b) $\frac{3}{2x\sqrt{\ln 3x}}$

c) $\frac{1}{6x\sqrt{\ln 3x}}$

d) $\frac{1}{6x}$

e) none of the above

Answer: (b)
MC Moderate

4.4.12: Let $f(x) = x^3 \ln x$. What is $\frac{df}{dx}$?

a) $3x^2 \ln x + x^2$

b) $(3x^1 + 1) \ln x$

c) $x^2 \ln x + x^2$

d) $3x^2 \ln x$

e) none of the above

Answer: (a)
MC Moderate

4.4.13: Let $f(x) = x \ln x$. Then

a) $f(x)$ has a relative maximum for $x = -e$

b) $f(x)$ has a relative maximum for $x = \frac{1}{e}$

c) $f(x)$ has a relative minimum for $x = -e$

d) $f(x)$ has a relative minimum for $x = \frac{1}{e}$

e) none of the above

Answer: (d)
MC Moderate

4.4.14: Which of the following is the largest number?

a) $\ln 6 - \ln 1$

b) $\frac{1}{3}\ln 27$

c) $\frac{1}{2}\ln 16$

d) $2\ln 2 + \ln 3$

e) $\frac{3}{2}\ln 16 - \ln 8$

Answer: (d)
MC Moderate

4.4.15: Simplify: $e^{\ln 2x}$

Answer: $2x$
ES Moderate

4.4.16: Simplify: $e^{x+2\ln x}$

Answer: $x^2 e^x$
ES Moderate

4.4.17: Simplify: $e^{\ln x - 2\ln y}$

Answer: $\frac{x}{y^2}$
ES Moderate

4.4.18: Simplify: $\ln\left(\frac{1}{e^x}\right)$

Answer: $-x$
ES Moderate

4.4.19: Simplify: $e^{\ln e^{\ln e^{\ln e^x}}}$

Answer: e^x
ES Moderate

4.4.20: Simplify: $e^{2\ln x}$

Answer: x^2
ES Moderate

4.4.21: Simplify: $\ln e^{1/x}$

Answer: $\frac{1}{x}$
ES Moderate

4.4.22: Simplify: $\frac{\ln e^{2x}}{\ln e^{4x^2}}$

Answer: $\frac{1}{2x}$
ES Moderate

4.4.23: Simplify: $e^{-\ln(x^4)}$

Answer: $\frac{1}{x^4}$
ES Moderate

4.4.24: Assume $\ln(5 - x^2) + \ln e = 1$. Then x equals:
a) ± 2 b) $\pm\sqrt{5}$ c) $\pm\sqrt{4 + e}$ d) $\pm\sqrt{5 - e}$

Answer: (a)
MC Moderate

4.4.25: If $5e^{\ln 2x} + (e^x)^3 = 10x + 8$, then x equals:
a) 0 b) ln2 c) $5 \cdot \ln 8$ d) $\sqrt[3]{\ln 8}$

Answer: (b)
MC Moderate

4.4.26: If $2 - \ln(x + 3) = \ln 4$, then x equals:
a) -3 b) 2e c) $\frac{1}{4}e^2 - 3$ d) $\ln 4 - 1$

Answer: (c)
MC Moderate

4.4.27: Suppose $\ln x^3 + 3\ln x = 0$. Then x equals:
a) $0, \pm\sqrt{3}$ b) $\sqrt{3}$ c) e^3 d) 1

Answer: (d)
MC Moderate

4.4.28: Assume $e^{(x^2 + 9)} \cdot e^{(6x)} = 1$. Then x equals:

a) 0 b) -3 c) $\pm\sqrt{3}$ d) $\ln \frac{1}{2}$

Answer: (b)
MC Moderate

4.4.29: Assume $e^{4x^2} + 3e^{(2x)^2} = 6$. Then x equals:

a) $\pm\frac{1}{2}\sqrt{\ln 6}$ b) $\pm\frac{1}{4}\sqrt{6}$ c) $\pm\frac{1}{4}\sqrt{\ln 6}$ d) $\frac{1}{2}\ln\frac{1}{2}$

Answer: (a)
MC Moderate

4.4.30: If $f'(x) = \frac{2\ln x^2}{x}$, which of the following are true of f(x)?

(I) f(x) is decreasing for $0 < x < 1$.
(II) f(x) has a critical point at $x = 1$.
(III) f(x) is always increasing.
(IV) f'(x) is undefined for $x \leq 0$.

Answer: (I) and (II)
ES Moderate

4.4.31: The expression $\ln(x^2 - 2) + e \cdot \ln(x^2 - 2)$ is equal to:

a) $\ln(x^2 - 2) + (x^2 - 2)$ b) $\pm\sqrt{3}$
c) $(1 + e)\ln(x^2 - 2)$ d) $(\ln + 1)(x^2 - 2)$

Answer: (c)
MC Moderate

4.4.32: The expression $e^{2\ln 5} + \ln(e^x \cdot e^4)$ is equal to:

a) $10 + 4x$ b) $25 + 4x$
c) $10 + x \cdot e^4$ d) $29 + x$

Answer: (d)
MC Moderate

4.4.33: Solve for x: $e^{\ln(3x) - \ln(4)} = 1$.

Answer: $x = \frac{4}{3}$
ES Moderate

4.4.34: Solve for x: $4e^{3x + 2} = 20$.

Answer: $x = \frac{\ln 5 - 2}{3}$
ES Moderate

4.4.35: Solve for x: $e^{e^{e^x}} = e^3$.

Answer: $x = \ln(\ln(3))$
ES Moderate

4.5.1: Differentiate: $\ln(2x^2 + 1)$

Answer: $\dfrac{4x}{2x^2 + 1}$
ES Moderate

4.5.2: Differentiate: $\dfrac{\ln x}{x^3}$

Answer: $\dfrac{1 - 3 \ln x}{x^4}$
ES Moderate

4.5.3: Differentiate: $x^4 \ln(x^2 + 1)$

Answer: $\dfrac{2x^5}{x^2 + 1} + 4x^3 \ln(x^2 + 1)$
ES Moderate

4.5.4: Differentiate: $e^x \ln 2x$

Answer: $e^x \ln 2x + \dfrac{2e^x}{x}$
ES Moderate

4.5.5: Differentiate: $e^{(\ln x)^2}$

Answer: $\dfrac{2 \ln x}{x} e^{(\ln x)^2}$
ES Moderate

4.5.6: Differentiate: $(\ln x)^5$

Answer: $\dfrac{5(\ln x)^4}{x}$
ES Moderate

4.5.7: Differentiate: $(x + 3 \ln x)^4$

Answer: $4(x + 3 \ln x)^3 (1 + \frac{3}{x})$

ES Moderate

4.5.8: Differentiate: $\dfrac{\ln 3x}{\ln x}$

Answer: $\dfrac{-\ln 3}{x(\ln x)^2}$

ES Moderate

4.5.9: Suppose that some garbage is dumped into a lake at time t = 0. The concentration of the dissolved oxygen in the water decreases at first and then begins to rise back to normal levels. Suppose that the oxygen concentration t weeks after garbage is dumped is given by

$$f(t) = 100 + 10e^{-t} - 10e^{-.2t}$$

At what value of t will the concentration of the dissolved oxygen be lowest? (Do not bother to check the second derivative.)

Answer: $t = \frac{5}{4} \ln 5$

ES Moderate

4.5.10: The weight w and the age t of an animal in some cases appear to satisfy an equation of the form

$$\ln w - \ln(M - w) = kt - a,$$

where M, k and a are positive constants. Solve this equation for w in terms of t.

Answer: $w = \dfrac{Me^{kt - a}}{1 + e^{kt - a}}$

ES Moderate

4.5.11: Sketch the curve $y = 2x - \ln x^2$, $x > 0$.

Answer:

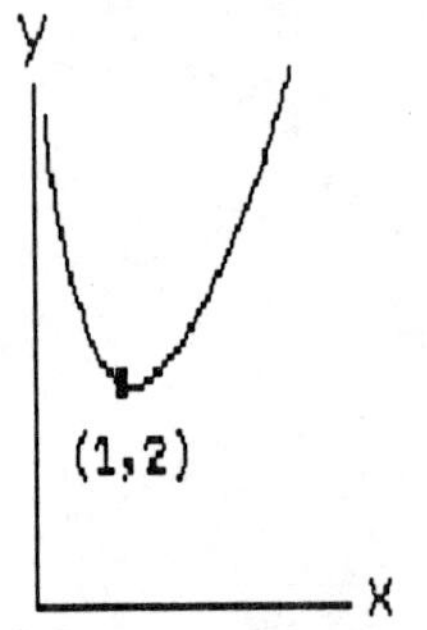

ES Moderate

4.5.12: Calculate: $\frac{d}{dx} \ln\left[\frac{\sqrt{x}e^x}{x^2 + 1}\right]$

Answer: $\frac{1}{2x} + 1 - \frac{2x}{x^2 + 1}$

ES Moderate

4.5.13: What is the slope of the curve $y = e^x$ at $x = 0$?

a) 0
b) e^x
c) e
d) 1
e) none of the above

Answer: (d)

MC Moderate

4.5.14: Suppose that $\ln(x + 3) - 2 \ln x = 1$. Then

a) $x = \frac{1 \pm \sqrt{1 + 12e}}{2e}$
b) $x = 0$
c) $x = \frac{1 \pm \sqrt{1 + 12e}}{e}$
d) $x = \frac{1 \pm \sqrt{e}}{3e}$
e) none of the above

Answer: (a)

MC Moderate

4.5.15: Let $y = \ln\left(\frac{x + 2}{x - 1}\right)$. then y' equals

a) $\frac{x - 1}{x + 2}$

b) $\frac{x - 1}{x + 2} \frac{d}{dx} \left(\frac{x + 2}{x - 1}\right)$

c) $\frac{(x - 1)}{(x + 2)(x + 3)}$

d) $\frac{3x}{(x + 2)(x - 1)}$

e) none of the above

Answer: (b)
MC Moderate

4.5.16: Let $f(x) = \ln[\ln x]$ then $f'(e)$ equals

a) e

b) 1

c) $\frac{1}{e}$

d) 0

e) none of the above

Answer: (c)
MC Moderate

4.5.17: At what value of x could the function $f(x) = \frac{\ln x + x}{x}$ have a possible relative maximum or minimum?

a) $\frac{1}{e}$

b) 1

c) e^2

d) e

e) none of the above

Answer: (d)
MC Moderate

4.5.18: Let $y = \ln\left[\frac{(x - 2)(3x + 2)(x + 1)}{x^2}\right]$. What is $\left.\frac{dy}{dx}\right|_{x - 1}$?

a) $-\frac{7}{5}$

b) $\frac{2}{3}$

c) $-\frac{6}{5}$

d) $-\frac{11}{3}$

e) none of the above

Answer: (e)
MC Moderate

4.5.19: Differentiate: $f(x) = \ln\left(3x^2\sqrt{x + 2}\right)$

Answer: $f' = \frac{2}{x} + \frac{1}{2(x + 2)}$
ES Moderate

4.5.20: Differentiate: $f(x) = x^3 \ln(2x)$

Answer: $f' = x^2 + 3x^2(\ln(2x))$
ES Moderate

4.5.21: Differentiate: $f(x) = (\ln x)^4$

Answer: $f' = \frac{r(\ln x)^3}{x}$
ES Moderate

4.5.22: Differentiate: $f(x) = e^{\ln x} + \ln x^e + e^{e^{\ln x}}$

Answer: $f' = 1 + \frac{e}{x} + e^x$
ES Moderate

4.5.23: Differentiate: $f(x) = \frac{\ln x}{e^x}$

Answer: $f' = \frac{1}{xe^x} - \frac{\ln x}{e^x}$
ES Moderate

4.5.24: Differentiate: $f(x) = x \ln (2x - x^2)$

Answer: $f' = x \cdot \frac{2 - 2x}{2x - x^2} + (\ln 2x - x^2)$
ES Moderate

4.5.25: An equation of the tangent line to the graph of $y = 2x + \ln\left(\frac{1}{x}\right)$ at $x = 1$ is:

a) $y - 1 = \left(2 - \frac{1}{x}\right)(x - 1)$

b) $y = 2(x - 1) + 1$

c) $y = x + 1$

d) $y - 2 = \left(2 - \frac{1}{x}\right)(x - 1)$

Answer: (c)
MC Moderate

4.5.26: The slope of the graph of $y = \ln(2x + 3)^{1/2}$ at the point (3, ln3) is:

a) $\frac{1}{9}$ b) $\frac{1}{\ln 3}$ c) $\frac{\ln 3}{2}$ d) $\frac{1}{2(\ln 3) + 3}$

Answer: (a)
MC Moderate

4.5.27: Let $y = (\ln(x^2 + 2))^3$. Then $\frac{dy}{dx}$ equals

a) $3(\ln(2x))^2$
b) $\frac{6x}{x^2 + 2}(\ln(x^2 + 2))^2$
c) $3\left(\frac{1}{x^2 + 2}\right)^2 \cdot 2x$
d) $\frac{1}{(\ln(x^2 + 2))^3} \cdot 2x$

Answer: (b)
MC Moderate

4.5.28: Let $f(x) = e^{x^2} + 2\ln(x^e)$. Then f'(x) equals:

a) $2xe^{x^2} + 2e\frac{1}{x}$
b) $x^2e^{x^2 - 1} + 2e\frac{1}{x^e} \cdot x^{e - 1}$
c) $2e^x + 2\frac{1}{\ln(x^e)} \cdot x^e$
d) none of the above

Answer: (a)
MC Moderate

4.5.29: What are the x-coordinates of the relative minimum point of the function $f(x) = x^2\ln x$?

a) $x = 0$
b) $x = 1$
c) $x = e^{-1/2}$
d) There are no critical points.

Answer: (c)
MC Moderate

4.5.30: At what point(s) is the slope of the graph of $y = \ln(e^{2x} - 4x)$ equal to 0?

a) $(1, \ln(e^2 - 4))$
b) $(0, 0)$ and $(\ln 2, 4 - 4\ln 2)$
c) $\left(\frac{1}{2}\ln 2, \ln(2 - 2\ln 2)\right)$
d) $(1, \ln(e^2 - 4))$ and $\left(\frac{1}{2}, \ln(e - 2)\right)$

Answer: (c)
MC Moderate

4.5.31: For what value(s) of x is the graph of $y = \ln\left(\frac{x}{x - 1}\right)$ concave down?

a) $x \le \frac{1}{2}$

b) $x \neq 0, 1$

c) $x < 0, x > 1$

d) $x < 0$

Answer: (d)
MC Moderate

4.5.32: Let $f(x) = \ln(x + 1)^2 - (\ln(x + 4))^3$. Then f'(x) equals:

a) $\frac{2\ln(x + 1)}{x + 1} - \frac{3(\ln(x + 4))^2}{x + 4}$

b) $\frac{1}{2\ln(x + 1)} - \frac{1}{3(\ln(x + 4))^2}$

c) $\frac{1}{(x + 1)^2} - \frac{3(x + 4)^2}{(x + 4)^3}$

d) $\frac{2}{x + 1} - \frac{3(\ln(x + 4))^2}{x + 4}$

Answer: (d)
MC Moderate

4.5.33: An equation of the tangent line to the graph of $y = x^3\ln(-2x)$ at the point $(-1, -\ln 2)$ is:

a) $y = (x + \ln 2) - 1$

b) $y - 1 = 4(x - \ln 2)$

c) $y + \ln 2 = 4(x - 1)$

d) $y = (1 + 3\ln 2)(x + 1) - \ln 2$

Answer: (d)
MC Moderate

4.5.34: Determine the point(s) at which the tangent line to the graph of $y = \ln\left(\frac{1}{3}x^3 - e^2x\right)$ is horizontal.

Answer: $\left(e, \ln\left(\frac{-2}{3}e\right)\right)$ and $\left(-e, \ln\left(\frac{2}{3}e\right)\right)$
ES Moderate

4.5.35: For what value(s) of x is the graph of $f(x) = \ln\left(\sqrt{x}\right)$ concave down?

Answer: $x > 0$
ES Moderate

4.5.36: Let $g(t) = \ln\left(\frac{t + e}{t - e}\right)$. Determine $g'(t)$.

Answer: $g'(t) = \frac{-2e}{t^2 - e^2}$.
ES Moderate

4.5.37: Let $H(t) = e^{2t^2} \cdot \ln(2t^2)$. Determine $H'(t)$.

Answer: $e^{2t^2}\left(4t\ln(2t^2) + \frac{2}{t}\right)$
ES Moderate

4.6.1: Calculate: $\frac{d}{dx}(3x + 1)^5(2x - 1)^{-2}(x + 3)^4$

Answer: $\left[\frac{15}{3x + 1} - \frac{4}{2x - 1} + \frac{4}{x + 3}\right](3x + 1)^5(3x - 1)^{-2}(x + 3)^4$
ES Moderate

4.6.2: Solve for t: $2e^{3t} = 8$

Answer: $\frac{1}{3}\ln 4$
ES Moderate

4.6.3: If $(2^x \cdot 2^{-3}) = 16$, what is x ?

a) $\frac{15}{4}$
b) $-\frac{4}{3}$
c) 4
d) 16
e) none of the above

Answer: (e)
MC Moderate

4.6.4: Solve the following equation for x: $e^{10} = 3^x$

a) $x = \frac{e^{10}}{3}$
b) $x = \frac{10}{\ln 3}$
c) $x = \sqrt[3]{e^{10}}$
d) $x = 10 + \ln 3$
e) none of the above

Answer: (b)
MC Moderate

4.6.5: Which of the following could be the graph of $y = \ln(x^2 + 3)$?

a)

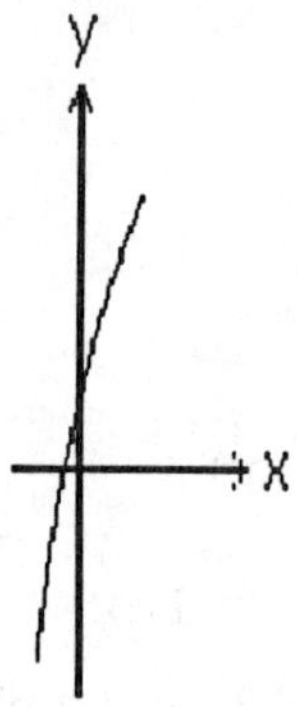

b)

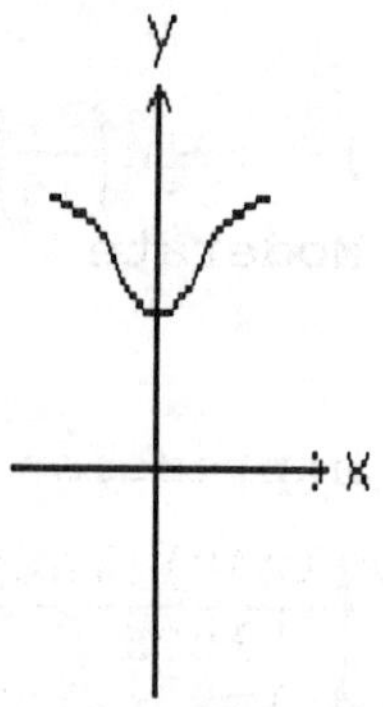

c)

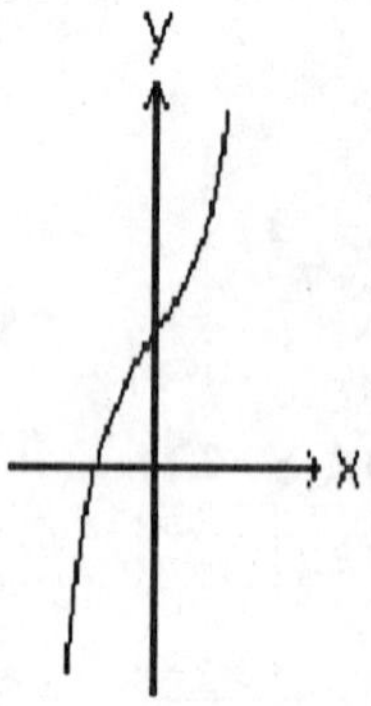

d)

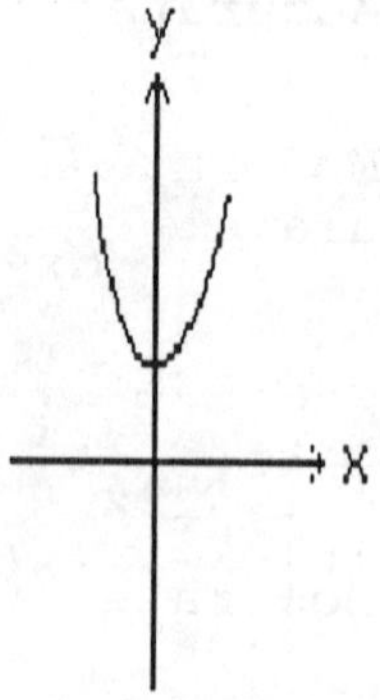

Answer: (b)
MC Moderate

4.6.6: Determine $\frac{dy}{dx}$ where $y = 3^x$.

Answer: $\frac{dy}{dx} = 3^x \ln 3$
ES Moderate

4.6.7: Determine $\frac{dy}{dx}$ where $y = 4^x \cdot 5^x \cdot 6x^3$.

Answer: $\frac{dy}{dx} = 4^x \cdot 5^x \cdot 6x^3\left(\ln 20 + \frac{3}{x}\right)$
ES Moderate

4.6.8: Determine $\frac{dy}{dx}$ where $y = e^x + \ln x^2 + 3^x$.

Answer: $\frac{dy}{dx} = e^x + \frac{2}{x} + 3^x \ln 3$
ES Moderate

4.6.9: Simplify: $\frac{1}{3}\ln 27 - 2\ln 4 + \ln 3 + (\ln 2)^2 - e^{\ln 6 + 1/4 \ln 81}$

Answer: $(\ln 2)^2 + \ln\left(\frac{9}{16}\right) - 18.$

ES Moderate

4.6.10: The expression $\ln\left(\frac{48x}{25}\right)$ is equal to:

a) $\frac{(\ln 48)(\ln x)}{\ln 25}$

b) $\frac{48}{25} \cdot \ln x$

c) $4\ln 2 + \ln 3 - 2\ln 5 + \ln x$

d) $\ln 48 + \ln 25 + \ln x$

Answer: (c)

MC Moderate

4.6.11: The expression $3\ln 5 - \frac{1}{2}\ln 4x^2 + (\ln x)(2\ln 3) - 5$ is equal to:

a) $\ln\left(\frac{125}{4x}\right) + (\ln x)(\ln 9) - 5$

b) $\ln\left(\frac{125}{20x}\right) + (\ln x)(\ln 6)$

c) $\ln\left(\frac{125}{20x}\right) + \ln(9x)$

d) $\ln\left(\frac{225}{4}\right)$

Answer: (a)

MC Moderate

4.6.12: The expression $\left(\frac{2}{3}\ln 8x - \frac{1}{4}\ln y^2 + \frac{3}{2}\ln(x - 4)\right) \cdot e^{2\ln x}$ is equal to:

a) $(\ln(8x^{2/3}(x - 4)^{3/2}) - \ln y^{1/2})e^{2x}$

b) $x^2 \cdot \ln\left(\frac{4x^{2/3}(x - 4)^{3/2}}{y^{1/2}}\right)$

c) $\left(\ln\left(\frac{4x^{2/3}(x - 4)^{2/3}}{y^{1/2}}\right)\right)e^{x^2}$

d) $e^{x^2} \cdot \ln\left(\frac{4x^2 - 16x}{y^{1/2}}\right)$

Answer: (b)

MC Moderate

4.6.13: If $y = e^{x - \ln x^2}$, what is $\left.\frac{dy}{dx}\right|_{x = -1}$?

a) $\frac{3}{e^2}$

b) $-3e^2$

c) $\frac{3}{e}$

d) 3

Answer: (c)

MC Moderate

5. Applications of Logarithm Functions

5.1.1: A colony of bacteria is growing at a rate proportional to the number of bacteria present. At the beginning of an experiment there were about 10^3 bacteria present. In two hours, the count rose to 3×10^3 bacteria.

(a) Find the specific growth law for this colony of bacteria.

(b) At what time will there be 6×10^3 bacteria present?

Answer:
(a) $P(t) = 10^3 e^{.55t}$
(b) 3.25 hours
ES Moderate

5.1.2: A parchment is offered for sale at a Paris flea market. The owner claims it to be at least 2000 years old. However, a carbon-dating test shows that $C^{14} - C^{12}$ ratio for the manuscript is 95% of the corresponding ratio for currently manufactured parchment. How old is the manuscript? (The decay constant of C^{14} is .00012.)

Answer: $\dfrac{\ln .95}{-.00012} \approx 427$ years
ES Moderate

5.1.3: A function P(t) satisfies $P'(t) = -\frac{1}{3}P(t)$, $P(0) = 20$. Solve for P(t).

Answer: $P(t) = 20e^{-(1/3)t}$
ES Moderate

5.1.4: A function Q(t) satisfies $Q'(t) = .01Q(t)$, $Q(5) = 10$. Solve for Q(t).

Answer: $Q(t) = 10e^{-.05 + .01t}$
ES Moderate

5.1.5: Potassium 42 has a half-life of 12 hours. How long will it take for a quantity of Potassium 42 to decay to 1/10 its original size?

Answer: 39.9 hours
ES Moderate

5.1.6: Barium 140 has a half-life of 13 days. After 25 days a given sample comprises 5 grams. How large was the original sample?

Answer: 18.96 grams
ES Moderate

5.1.7: Plutonium 239 has a half-life of 24,000 years. What is its decay constant?

Answer: .00002888
ES Moderate

5.1.8: Krypton 85 gas leaks into the reactor room of an electric power plant. Its half-life is 10 years. How long is it before 99.9% of the Krypton decays?

Answer: 99.66 years
ES Moderate

5.1.9: It is observed that the sales of a certain recording fall to 75% of their original level one month after advertising stops. What will be the sales after 4 months?

Answer: 31.64%
ES Moderate

5.1.10: Solve the following equation for t: $9e^{3t} = 27$

a) t = 1
b) t = .37
c) t = 1.1
d) t = .52
e) none of the above

Answer: (b)
MC Moderate

5.1.11: Let f(x) be the solution of the following differential equation:

$$f'(x) = .1f(x), \quad f(0) = .5$$

What is f(20), (approximately)?

a) 3.7
b) 2
c) 1
d) 2.5
e) none of the above

Answer: (a)
MC Moderate

5.1.12: A bacterial culture grows exponentially; that is, $P(t) = 100e^{kt}$, where P(t) is the size of the culture at time t hours. Suppose that after 2 hours, the size of the culture is 400. What is k?

a) 3
b) 06
c) $\frac{1.39}{2}$
d) .48
e) none of the above

Answer: (c)
MC Moderate

5.1.13: Suppose $e^a = b$, and $e^x = \frac{1}{b}$. What is x?

a) $\frac{1}{a}$

b) $\frac{b}{a}$

c) $\frac{1}{ab}$

d) $\frac{1}{e^a}$

e) none of the above

Answer: (e)
MC Moderate

5.1.14: In a certain country, the rate of increase of the population is proportional to the population P(t). In fact, $P'(t) = .23P(t)$. Suppose that initially the country's population is 50,000, and that 10 years later there are 500,000 people. Which of the following equations expresses this information mathematically?

a) $500{,}000 = e^{.23(10)}$

b) $500{,}000 = 50{,}000e^{2.3}$

c) $10 = e^{.23t}$

d) $500 = e^{.23(50)}$

e) none of the above

Answer: (b)
MC Moderate

5.1.15: The size of an insect colony t days after its formation is $P(t) = 1000e^{.2t}$. Approximately how many insects are there after 10 days?

a) 690

b) 54,598

c) 6,900

d) 7,389

e) none of the above

Answer: (d)
MC Moderate

5.1.16: Suppose that a school of fish in a pond grows according to the exponential law $P(t) = P_0e^{kt}$ and suppose that the size of the colony triples in 24 days. Determine k.

Answer: $k = \frac{\ln 3}{24}$
ES Moderate

5.1.17: Suppose that a school of fish in a pond grows according to the exponential law $P(t) = P_0e^{kt}$ and suppose that the size of the colony triples in 24 days. If the initial size of the school was 50, at what time will the school contain 200 fish?

Answer: $\frac{24 \ln 4}{\ln 3}$ days later
ES Moderate

5.1.18: The decay constant for strontium 90 is $\lambda = .0244$, where the time is measured in years. How long will it take for a quantity P_0 of strontium 90 to decay to 1/3 its original size?

Answer: $\frac{\ln(1/3)}{-.0244}$ years
ES Moderate

5.1.19: Carbon 14 has a half-life of 5730 years. Determine its decay constant.

Answer: $\lambda = \frac{\ln 1/2}{-5730}$
ES Moderate

5.1.20: A fossil was discovered that had about 70% of the C^{14} level found today in living matter. Given that the decay constant for C^{14} is -.00012, determine the age of the fossil.

Answer: $\frac{\ln .7}{-.00012}$
ES Moderate

5.1.21: Solve the differential equation $P'(t) = 48P(t)$, $P(0) = 32$.

Answer: $P(t) = 32e^{48t}$
ES Moderate

5.1.22: Suppose that at time t, a colony of fruit flies is growing at a rate of 50% per hour; that is, the colony is at a rate equal to one half the current size of the colony. Determine the differential equation which corresponds to the above information, then find a formula which gives the size of the colony at time t if there were originally 500 fruit flies present.

Answer: $y' = \frac{1}{2}y$; $y = 500e^{1/2\ t}$.
ES Moderate

5.1.23: A radioactive substance is observed to disintegrate at a rate such that 9/10 of the original amount remains after one year. What is the half life of the substance?

a) 5/9 yr.
b) $\frac{-\ln 2}{\ln 9 - \ln 10}$ yrs.
c) $\frac{\ln 2}{\ln 10}$ yr.
d) $\ln\frac{5}{9}$ yr.

Answer: (c)
MC Moderate

5.1.24: A certain radioactive substance is decaying at a rate proportional to the amount present. If 100 grams decays to 13.5 grams in 4 years, how long will it take for 90 grams to decay to 30 grams?

a) $\frac{\ln(.135)}{4}$ yrs.

b) $\frac{4\ln(.3)}{\ln(.135)}$ yrs.

c) $\frac{-4\ln 3}{\ln(.135)}$ yrs.

d) Problem cannot be solved as stated.

Answer: (c)
MC Moderate

5.1.25: Radioactive Carbon 11 has a half life of 20 minutes. If there are 200 grams present at the start of our experiment, how many grams will remain after 10 minutes?

a) 100 g.

b) 50 g.

c) $10 \ln\frac{1}{2}$ g.

d) $200e^{\frac{-\ln 2}{2}}$ g.

Answer: (d)
MC Moderate

5.1.26: The population of a colony of bacteria triples in 3 days. Assuming that the rate of growth is proportional to the size of the population, how long did it take for the colony to double in size?

a) 2 yrs.

b) $\frac{3 \ln 6}{\ln 3}$ days

c) 6 days

d) $\frac{3 \ln 2}{\ln 3}$ yrs.

e) none of the above

Answer: (e)
MC Moderate

5.1.27: A certain radioactive element has a half-life of 12 minutes. At what time is the substance decaying at a rate of -5 ln2 grams per minute if there are 120 grams present initially?

a) t = 12 min.

b) t = ln1/2 min.

c) $t = \frac{-\ln 2}{12}$ min.

d) t = 0

e) none of the above

Answer: (a)
MC Moderate

5.1.28: Sixteen pounds of a radioactive substance loses one fourth of its original mass in 2 days. The mass m(t) remaining at time t is given by:

a) $m(t) = 16e^{\frac{-\ln 4}{2} \cdot t}$

b) $m(t) = 16e^{\frac{-\ln 1/4}{2} \cdot t}$

c) $m(t) = 16e^{\frac{\ln 3/4}{2} \cdot t}$

d) $m(t) = 4e^{\frac{-\ln 4}{2} \cdot t}$

Answer: (c)
MC Moderate

5.1.29: Assume that a culture of bacteria grows at a rate proportional to its size such that if 10^6 bacteria are present initially, then there are 2×10^6 bacteria present after 3 hours. Determine a formula for the number of bacteria present after t hours in terms of powers of 2. (Hint: Recall that $b^x = e^{(\ln b)x}$.)

Answer: $P(t) = 10^6 \cdot 2^{1/3\ t}$
ES Moderate

5.1.30: The population of a certain region was 10 million in 1950. By 1970, it had increased to 13.5 million. Assuming exponential growth, estimate the population in the year 2000.

Answer: $10e^{\frac{5}{2}\ln 1.35} \approx 21.2$ (million)
ES Moderate

5.1.31: A colony of fruit flies grows at a rate proportional to its size. During an experiment, an initial population of 200 flies increased to 664 flies in 3 days.
(a) How long did it take for the colony to triple in size?
(b) Sketch the graph of the growth function y = P(t) for this problem and label the y-intercept.

Answer: (a) $t = \frac{3 \ln 3}{\ln 3.32} \approx 2.75$ days
(b)

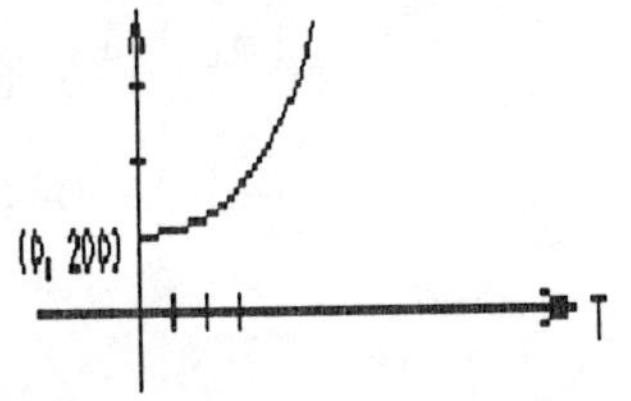

ES Moderate

5.2.1: What rate of interest will make an investment triple in 8 years if the interest is compounded continuously?

Answer: 13.73%
ES Moderate

5.2.2: Suppose that $500 is deposited in a bank certificate paying 10% interest compounded continuously. How much will the certificate be worth after 5 years?

Answer: $824.36
ES Moderate

5.2.3: What is the present value of an investment of $1000 payable at the end of 10 years at a 9% rate of interest compounded continuously?

Answer: $406.57
ES Moderate

5.2.4: Mr. Jones has two investments. The first is currently worth $50,000 and has an annual yield of 10% compounded continuously. The second is currently worth $70,000 and has an annual yield of 8% compounded continuously. Assuming that all earnings are reinvested at the same, respective rates, in how many years will the two investments be worth the same amount?

Answer: 16.82 years
ES Moderate

5.2.5: $1000 is invested at 6% interest compounded continuously. What is the value of the investment after 5 years?

a) $1000e^{.3}$
b) $1000(1.06)^{5}$
c) $1000e^{.06}$
d) $5000(1.06)^{5}$
e) none of the above

Answer: (a)
MC Moderate

5.2.6: Let P(t) be the quantity of strontium-90 remaining after t years. Suppose the half-life of strontium-90 is 28 years. Which of the following equations expresses the half-life information?

a) $28 = \frac{1}{2}P_0$
b) $P(28) = \frac{1}{2}P_0$
c) $28 = \frac{1}{2}P_0 e^{-kt}$
d) $P(\frac{1}{2}) = 28P_0$
e) none of the above

Answer: (b)
MC Moderate

5.2.7: \$1000 is invested at 6% interest (per annum) compounded semi-annually. What is the value of the investment after 5 years?

a) $1000e^{.03(5)}$
b) $1000(1.06)^{10}$
c) $1000(1.03)^{5}$
d) $5000(1.03)$
e) none of the above

Answer: (e)
MC Moderate

5.2.8: How long will it take for an investment to triple if interest is paid at 10% per annum, compounded continuously?

a) 8.6 years
b) 3 years
c) 30 years
d) 11 years
e) none of the above

Answer: (d)
MC Moderate

5.2.9: How much money has to be invested now at 8% continuous interest in order to have \$1000 after 5 years?

a) $1000e^{-.08(5)}$
b) $1000e^{.08(5)}$
c) $\frac{e^{.4}}{1000}$
d) $-1000e^{.4}$
e) none of the above

Answer: (a)
MC Moderate

5.2.10: Suppose that \$1000 is invested at 10% per year, with interest compounded annually. What is the compounded amount after 2 years?

Answer: \$1210.00
ES Moderate

5.2.11: Suppose that \$1000 is invested at 12% per year, with interest compounded monthly. What is the compounded amount after 2 months?

Answer: \$1020.10
ES Moderate

5.2.12: How long is required for an investment of \$2000 to double if the interest is 10% compounded continuously? An investment of \$1000?

Answer: 10 ln 2 years
ES Moderate

5.2.13: Suppose that the value in billions of dollars of a company is determined to be $f(t) = .5t + .2e^{-t}$ where t is measured in years. What is the percentage rate of growth of the company at time t = 0?

Answer: 150%
ES Moderate

5.2.14: If you had $2000 to invest, and bank A offered you an account with 8.2% interest compounded annually, and bank B offered you 8% interest per year, but compounded monthly, which bank would you put your money in?

Answer: B
ES Moderate

5.2.15: What is the annual yield of an account with 8% interest compounded
monthly?

Answer: ≈ 8.3%
ES Moderate

5.2.16: What is the annual yield of an account with 8% interest compounded quarterly?

Answer: ≈ 8.24%
ES Moderate

5.2.17: A high-yield saving pays 20% interest, compounded continuously. How long will it take an initial investment of $2500 to grow to $20,000?

a) $\frac{1}{2}\ln 8$ years
b) 15 ln2 yrs.
c) $5 \ln\frac{1}{8}$ yrs.
d) 50 ln8 yrs.
e) none of the above.

Answer: (b)
MC Moderate

5.2.18: A high school student deposits a $500 graduation gift in a bank account which pays 4.8% interest compounded continuously. How much will the account be worth after 18 months?

a) $500(1.004)^{216}$
b) $500(1.048)^{18}$
c) $500e^{.072}$
d) $500e^{.81}$
e) none of the above

Answer: (c)
MC Moderate

5.2.19: A savings account pays 7% interest, compounded continuously. How much should be deposited now in order to have $5000 in the account at the end of five years?

a) $5000e^{-.35}$
b) $5000(1.07)^5$
c) $5000e^{-35}$
d) $5000(1.07)^{35}$
e) none of the above

Answer: (a)
MC Moderate

5.2.20: A bank pays 2.5% interest on deposits. What is the return on a $1000 deposit after two years if interest is compounded continuously?

a) $1000e^{.5}$
b) $1000e^{-.25}$
c) $1000e^{.05}$
d) $1000e^{0.25}$
e) none of the above

Answer: (c)
MC Moderate

5.2.21: What rate of interest is required in order for a $100 investment to double in 3 years if the interest is compounded continuously?

a) $\frac{\ln 3}{2}$
b) $\frac{\ln 2}{36}$
c) $-\frac{\ln 3}{2}$
d) $\frac{\ln 2}{3}$
e) none of the above

Answer: (d)
MC Moderate

5.2.22: Eight years ago, $2000 was deposited in a savings account paying 3% interest compounded continuously. Three years ago, $500 was withdrawn from the account. What is the current value of the account?

a) $2000e^{.24} - 500$
b) $2000e^{.24} - 500e^{.09}$
c) $2000e^{.15} + 1500e^{.09}$
d) $2000e^{.15} + (2000e^{.15} - 500)e^{.09}$
e) none of the above

Answer: (d)
MC Moderate

5.2.23: A bank pays 5% interest on accounts worth less than $1000 and pays 8% interest on accounts with a balance of $1000 or more. Assume all interest is compounded continuously and an account is opened with an initial deposit of $500.
(a) How long will it take the account balance to reach $1000?
(b) What will the account be worth after 20 years?

Answer: (a) $t = 20\ln 2 \approx 13.9$ yrs.
(b) $1000 + 1000e^{1.6(1 - \ln 2)} \approx \$2633.89$
ES Moderate

5.2.24: What is the return on an investment of \$10,000 after 5 years if the interest rate is:
(a) 5.6%, compounded quarterly?
(b) 5.6%, compounded continuously?

Answer: (a) \$13,205.63
(b) $10{,}000e^{.28} \approx \$13{,}231.30$
ES Moderate

5.2.25: Twenty years ago, a couple opened a savings account for their grandchild with an initial deposit of \$700. That account payed 6% interest compounded monthly. Five years later, they transferred all of the money in the old account to a new account paying 6% interest compounded continuously. What is the grandchild's account worth now?

Answer: $944.20 + 944.20e^{.9} \approx \$3266.56$
ES Moderate

5.2.26: How long will it take for an investment of \$500 to triple if the interest is 7.3% compounded continuously? What will the investment be worth after 15 yrs.?

Answer: 15 yrs; \$1500
ES Moderate

5.3.1: The Dutch chemist Van't Hoff discovered that the "rate constant" K of certain chemical reactions depends upon the temperature T (in degrees Kelvin) and satisfies the differential equation

$$\frac{d}{dT}(\ln K) = \frac{a}{T^2}$$

where a is some constant which depends on the reacting chemicals and on the heat absorbed or released by the reaction. Solve this equation for ln K and then solve the resulting equation for K.

Answer: $\ln K = -\frac{-a}{T} + C$, C a constant; $K = e^{C-(a/T)}$
ES Moderate

5.3.2: Let $g(t) = 100 - 100e^{-.01t}$ be the number of cases of measles in a certain school t days after the first case is reported. Which of the following best describes the spread of the disease?
a) After the first day there are 100 cases, after which the number of cases decreases daily to 0.
b) Initially, the disease spreads quickly, but then the rate of increase slows so that the number of cases never exceeds 100.
c) The disease spreads quickly until the number of cases reaches 100; then gradually the number of cases decreases.
d) The number of cases decreases according to exponential decay.
e) none of the above

Answer: (b)
MC Moderate

5.3.3: Let $y = 6(1 - e^{-3x})$. What is $3(6 - y)$?

a) $18e^{-3x}$
b) $12 - 6e^{-3x}$
c) $6 - 6e^{-3x}$
d) $3e^{-3x}$
e) none of the above

Answer: (a)
MC Moderate

5.3.4: In a town of 10,000 people, the number of people who during each day first hear the news of a local tax increase is one-tenth the number of people who have not yet heard the news. If f(t) stands for the number of informed people in the town, what is the differential equation which f(t) satisfies?

a) $f(t) = \frac{10f'(t)}{10,000}$
b) $f'(t) = \frac{1}{10}f(t)$
c) $f'(t) = \frac{1}{10}(10,000 - f(t))$
d) $f'(t) + \frac{1}{10}f(t) = 10,000$
e) none of the above

Answer: (c)
MC Moderate

5.3.5: Let $A'(t) = .3(25,000 - A(t))$, $A(o) = 0$. Which of the following is the formula for A(t) ?

a) $A(t) = 25,000e^{.3t}$
b) $A(t) = 25,000e^{1-.3t}$
c) $A(t) = 25,000(1 - e^{-.3t})$
d) $A(t) = 25,000(1 - .3t)$
e) none of the above

Answer: (c)
MC Moderate

5.3.6: If the relative rate of change of a function f is always 5, what kind of equation will f be represented by? (There should be one constant in your answer.)

Answer: $f(t) = Ce^{5t}$
ES Moderate

5.3.7: Suppose that the value of a certain investment after t years can be approximated by the function $f(t) = 100,000e^{.12t^{2/3}}$.
(a) Use a logarithmic derivative to determine the percentage rate of increase in the value of the investment when t = 8 years.
(b) What is the dollar value of the investment after 8 years?

Answer: (a) 4% per yr
(b) $100,000e^{.48} = \$161,607.44$
ES Moderate

5.3.8: Suppose a manufacturer can sell $q = \frac{1000}{(p + 2)^2} - 6$ units of a product when the price is p dollars per unit.
(a) Determine the elasticity of demand, $E(p)$, when the price is $p = 8$ dollars.
(b) If the price is raised slightly, will revenue increase or decrease?

Answer: (a) $E(8) = 4$
(b) decrease
ES Moderate

5.4.1: Show that the function $f(t) = \frac{1}{1 + e^{-t}}$ satisfies the differential equation $y' = y(1 - y)$.

Answer: differentiate f(t)
ES Moderate

5.4.2: Which of the following functions satisfy the differential equation $y' = 3(12 - y)$?
(I) $y = 4(3 - e^{-3x})$
(II) $y = 12(1 - e^{-3x})$
(III) $y = 12 - e^{-3x}$
(IV) $y = 3(1 - e^{-12x})$

Answer: (I), (II) and (III)
ES Moderate

5.4.3: The function $f(t) = \frac{1}{2}(7 - e^{-3x})$ satisfies:

a) $y' = \frac{1}{2}(7 - y)$, $f(0) = 3$
b) $y' = -3\left(\frac{1}{2} - y\right)$, $f(0) = \frac{7}{2}$
c) $y' = 3\left(\frac{1}{2} - y\right)$, $f(0) = 3$
d) $y' = 7(3 - y)$, $f(0) = \frac{1}{2}$

Answer: (c)
MC Moderate

6. The Definite Integral

6.1.1: Find: $\int e^{-x/2}\,dx$

Answer: $-2e^{-x/2} + C$
ES Moderate

6.1.2: Find: $\int \left(4x^{3/2} - \dfrac{1}{2x^{3/2}}\right) dx$

Answer: $\dfrac{8}{5}e^{5/2} + \dfrac{1}{x^{1/2}} + C$
ES Moderate

6.1.3: What function y = f(x) satisfies the following conditions:

$$y' = -\frac{1}{4 - x}, \quad f(2) = 0\ ?$$

a) $y = -\ln|4 - x| + 2$
b) $y = \ln|4 - x| - \ln 2$
c) $-(4 - x)^{-2} + \dfrac{1}{\sqrt{2}}$
d) $\ln\left|\dfrac{1}{4 - x}\right| + \dfrac{1}{2}$
e) none of the above

Answer: (b)
MC Moderate

6.1.4: Suppose F(x) is an antiderivative of $\dfrac{2}{\sqrt{x}}$ and F(0) = 1. What is F(9) ?

a) 1
b) $-\dfrac{4}{3}$
c) 13
d) -4
e) none of the above

Answer: (c)
MC Moderate

6.1.5: Let $g''(x) = 3x^2 + 5$ and $g'(1) = 5$. What is a formula for all possible functions g(x) ?

a) $g(x) = x^3 + 5x - 1$
b) $g(x) = x^3 + 5x - 1 + C$
c) $g(x) = \dfrac{x^4}{4} + \dfrac{5}{2}x^2 - x + C$
d) $g(x) = \dfrac{x^4}{4} - \dfrac{5}{2}x^2 - 5x + C$
e) none of the above

Answer: (d)
MC Moderate

6.1.6:

What is $\int \frac{1}{(x+2)^2}\,dx$?

a) $-2(x+2)^{-3} + C$

b) $\frac{-1}{x+2} + C$

c) $\frac{1}{2(x+2)} + C$

d) $-\frac{1}{2(x+2)} + C$

e) none of the above

Answer: (b)
MC Moderate

6.1.7:

What is $\int \left(\frac{x^2}{4} - 4\right) dx$?

a) $\frac{x}{2} + C$

b) $\frac{3x^3}{4} - 4 + C$

c) $\frac{x^3}{12} - 4x + C$

d) $\frac{x^2}{4} - 4 + C$

e) none of the above

Answer: (c)
MC Moderate

6.1.8: Find: $\int x^3\,dx$

Answer: $\frac{x^4}{4} + C$
ES Moderate

6.1.9: Find: $\int y^4\,dy$

Answer: $\frac{y^5}{5} + C$
ES Moderate

6.1.10: Find: $\int (2x+1)^2\,dx$

Answer: $\frac{4x^3}{3} + 2x^2 + x + C$
ES Moderate

6.1.11: Find: $\int (3x + 2)^5 \, dx$

Answer: $\frac{(3x + 2)^6}{18} + C$

ES Moderate

6.1.12: Find: $\int x^5 \, dx$

Answer: $\frac{x^6}{6} + C$

ES Moderate

6.1.13: Find: $\int x^{15} \, dx$

Answer: $\frac{x^{16}}{16}$

ES Moderate

6.1.14: Calculate: $\int (x^5 + 2x^3 - 3x^2 + 6) \, dx$

Answer: $\frac{x^6}{6} + \frac{x^4}{2} - x^3 + 6x - C$

ES Moderate

6.1.15: Find: $\int (3 - 2x)^4 dx$

Answer: $\frac{(3 - 2x)^5}{-10} + C$

ES Moderate

6.1.16: Determine f(x) where $f'(x) = 3x^2 + 2x + 1$.

Answer: $x^3 + x^2 + x + C$

ES Moderate

6.1.17: Find: $\int (3x + 2)^2\, dx$

Answer: $\frac{(3x + 2)^3}{9}$

ES Moderate

6.1.18: Find: $\int (x^3 + 1)\, dx$

Answer: $\frac{x^4}{4} + x + C$

ES Moderate

6.1.19: Find: $\int \sqrt[4]{2x - 5}dx$

Answer: $\frac{2}{5}(2x - 5)^{5/4} + C$

ES Moderate

6.1.20: Determine f(x) where $\int f(x) = \ln|2x + 6| - e^2x + c$.

Answer: $f(x) = \frac{2}{2x + 6} - e^2$

ES Moderate

6.1.21: Determine g(x) where $\int g(x) = \frac{d}{dx}(e^{3x} - 2x^3)$.

Answer: $g(x) = 9e^{3x} - 12x$

ES Moderate

6.1.22: Let $f''(t) = \frac{1}{3}e^{3t} - t^2 + 3$ and assume $f'(0) = 1$. Which of the following is a formula for all possible functions f(t)?

a) $\frac{1}{9}e^{3t} - \frac{t^3}{3} + 3t + 1$

b) $\frac{1}{27}e^{3t} - \frac{t^4}{12} + \frac{3t^2}{2} + \frac{8t}{9} + C$

c) $3e^{3t} - 2$

d) $\frac{1}{9}e^{3t} - \frac{t^3}{3} + 3t + C$

Answer: (b)

MC Moderate

6.1.23: Assume $e^{7x} - \ln(x + 2)$ is an antiderivative of $f(x)$. What is $f(0)$?

a) $\frac{13}{2}$

b) $\frac{1}{7} - \frac{1}{4}(\ln 2)^2$

c) $\frac{-5}{14}$

d) $-\frac{1}{4}(\ln 2)^2$

Answer: (a)
MC Moderate

6.1.24: If $\frac{dy}{dx} = x^6 + \frac{x^4}{3} - x^3 + \frac{5}{3}x^2$ and $f(0) = e$, then the function $y = f(x)$ is given by:

a) $y = 6x^7 + \frac{4}{3}x^5 - 3x^4 + \frac{10}{3}x^3$

b) $y = \frac{x^7}{7} + \frac{x^5}{15} - \frac{x^4}{4} + \frac{5}{9}x^3 + e$

c) $y = 7e^7 + \frac{5}{3}e^5 - 4e^4 + 5e^3$

d) $y = \frac{x^7}{6} + \frac{x^5}{12} - \frac{x^4}{3} + \frac{5}{6}x^3 + ex$

Answer: (b)
MC Moderate

6.1.25: If $\frac{dy}{dx} = 2e^{2x} - \frac{4}{x} + \frac{3}{x^2}$ and $f(1) = -2$, then the function $y = f(x)$ is given by:

a) $y = e^{2x} - 4\ln|x| - \frac{6x}{x^3} + 3 + 4e$

b) $y = \frac{2}{3}e^{3x} - 4\ln|x| - \frac{1}{x} - \frac{5}{3}$

c) $y = \frac{2}{3}e^{3x} - 4x - \frac{1}{x} + \frac{5}{3}$

d) $y = e^{2x} - 4\ln|x| - \frac{3}{x} - (e^2-1)$

Answer: (d)
MC Moderate

6.1.26: $\int\left(4\sqrt{x} - \frac{1}{2\sqrt{x}}\right)dx$ is equal to:

a) $\frac{8}{3}x^{3/2} - x^{1/2} + C$

b) $\frac{2}{\sqrt{x}} + \frac{1}{4x^{3/2}} + C$

c) $2x^{3/2} + \frac{1}{4}x^{1/2} + C$

d) $6x^{3/2} - \frac{1}{2}\ln\left|\sqrt{x}\right| + C$

Answer: (a)
MC Moderate

6.1.27: $\int\left(\frac{6}{5}x^5 + 4e^{-2x}\right)dx$ is equal to:

a) $6x^4 - 8e^{-2x} + C$

b) $6x^6 - 2e^{-2x} + C$

c) $6x^6 - 2e^{-3x} + C$

d) $\frac{1}{5}x^6 - 2e^{-2x} + C$

Answer: (d)
MC Moderate

6.1.28: $\int\frac{x^2 + 1}{x + 1}dx$ is equal to:

a) $\dfrac{\frac{x^3}{3} + x}{x^2/2 + x}$

b) $\frac{x^3}{3} - \frac{x^2}{2} + C$

c) $\frac{x^2}{2} - x + 2\ln|x + 1| + C$

d) $\left(\frac{x^3}{3} + x\right)(\ln|x + 1|) + C$

Answer: (c)
MC Moderate

6.1.29: $\int 4x\sqrt{2x^2 + 1}dx$ is equal to:

a) $\frac{4}{3}x^2(2x^2 + 1)^{3/2} + C$

b) $\frac{2}{3}(2x^2 + 1)^{3/2} + C$

c) $\frac{4x^2}{\sqrt{2x^2 + 1}} + 4\sqrt{2x^2 + 1} + C$

d) $2x^2\sqrt{\frac{2}{3}x^3 + x} + C$

Answer: (b)
MC Moderate

6.1.30: $\int 4(x + 2)(x^2 + 4x + 5)^{1/3}dx$ is equal to:

a) $3(x + 2)(x^2 + 4x + 5)^{4/3} + 4\left(\frac{x^2}{2} + 2x\right)(x^2 + 4x + 5)^{1/3} + C$

b) $3\left(\frac{x^2}{2} + 2x\right)\left(\frac{x^3}{3} + 2x^2 + 5x\right)^{4/3} + C$

c) $\frac{3}{2}(x^2 + 4x + 5)^{4/3} + C$

d) $\frac{4}{3}(x + 2)(x^2 + 4x + 5)^{-2/3} + 4(x^2 + 4x + 5)^{1/3} + C$

Answer: (c)
MC Moderate

6.1.31: $\int \left(\frac{4}{t} - 3e^{-1/2\ t} + \frac{2 \ln t}{t} \right) dx$ is equal to:

a) $4 \ln t + 6e^{-1/2\ t} + (\ln t)^2 + C$

b) $\frac{2}{t^2} - 6e^{1/2\ t} + \frac{1}{t^3} + C$

c) $4t + \frac{3}{2}e^{-1/2\ t} + 2 + C$

d) $4 \ln t + \frac{3}{2}e^{-1/2\ t} + 2(\ln t)^2 + C$

Answer: (a)
MC Moderate

6.2.1: Use Riemann sums to approximate the area under the graph of y = 2x + 1 from x = 1 to x = 5. Use a partition with 4 subintervals and right endpoints as representative points.

Answer: 32

ES Moderate

6.2.2: Set up a Riemann sum to approximate the area under the graph of y = $e^x + x$ from x = 0 to x = 2, using 6 subintervals and left endpoints as representative points. Do not compute the sum.

Answer: $\frac{1}{3}\left[1 + \left(e^{1/3} + \frac{1}{3}\right) + \left(e^{2/3} + \frac{2}{3}\right) + (e + 1) + \left(e^{4/3} + \frac{4}{3}\right) + \left(e^{5/3} + \frac{5}{3}\right)\right]$

ES Moderate

6.2.3: Use Riemann sums to approximate the area under the graph of y = x^3 from x = 0 to x = 2, using 4 subintervals and right endpoints as representative points.

Answer: 6.25
ES Moderate

6.2.4: Set up a Riemann sum to approximate the area under the graph of y = ln(x + 1) from x = 0 to x = 1 using 3 subintervals and midpoints as representative points. Do not compute the sum.

Answer: $\frac{1}{3}\left(\ln\left(\frac{7}{6}\right) + \ln\left(\frac{3}{2}\right) + \ln\left(\frac{11}{6}\right)\right)$.
ES Moderate

6.2.5: The Riemann Sum $\frac{1}{50}\left(\frac{50}{100} + \frac{51}{100} + \frac{52}{100} + \ldots + \frac{99}{100}\right)$ is approximately equal to:

a) $\int_0^{49}\left(50 + \frac{x}{100}\right)dx$

b) $\int_{1/2}^{3/2} x\,dx$

c) $\int_1^2\left(\frac{1}{2}x\right)dx$

d) $\int_{50}^{100}\left(\frac{x}{100}\right)dx$

Answer: (c)
MC Moderate

6.2.6: The definite integral $\int_1^3\left(\sqrt{x} - \frac{1}{x}\right)dx$ is approximately equal which Riemann Sum?

a) $\frac{1}{100}\left(\frac{\sqrt{1}}{10} - \frac{100}{1}\right) + \frac{1}{100}\left(\frac{\sqrt{2}}{10} - \frac{100}{2}\right) + \ldots + \frac{1}{100}\left(\frac{\sqrt{300}}{10} - \frac{100}{300}\right)$

b) $\frac{1}{50}\left(\frac{\sqrt{102}}{10} - \frac{50}{51}\right) + \frac{1}{50}\left(\frac{\sqrt{104}}{10} - \frac{50}{52}\right) + \ldots + \frac{1}{50}\left(\frac{\sqrt{300}}{10} - \frac{50}{150}\right)$

c) $\frac{1}{50}\left(\sqrt{\frac{51}{50}} - \frac{50}{51}\right) + \frac{1}{50}\left(\sqrt{\frac{52}{50}} - \frac{50}{52}\right) + \ldots + \frac{1}{50}\left(\sqrt{\frac{100}{50}} - \frac{50}{100}\right)$

d) $\frac{2}{15}\left(\sqrt{\frac{2}{15}} - \frac{15}{2}\right) + \frac{2}{15}\left(\sqrt{\frac{4}{15}} - \frac{15}{4}\right) + \ldots + \frac{2}{15}\left(\sqrt{\frac{45}{15}} - \frac{15}{45}\right)$

Answer: (b)
MC Moderate

6.3.1: $\int_0^{1/2} (e^{2x} + 2x)\,dx =$

a) $2e - \frac{7}{4}$

b) $\frac{e}{2} - \frac{1}{4}$

c) $e + \frac{1}{4}$

d) $e^2 - e + 1$

e) none of the above

Answer: (b)
MC Moderate

6.3.2: Calculate: $\int_1^4 \sqrt{x}\,dx$

Answer: $\frac{14}{3}$
ES Moderate

6.3.3: Calculate: $\int_0^1 e^{3x-1}\,dx$

Answer: $\frac{1}{3}e^{2} - \frac{1}{3}e^{-1}$
ES Moderate

6.3.4: Calculate: $\int_1^2 \left(\frac{1}{x^2} - 3\right) dx$

Answer: $-\frac{5}{2}$
ES Moderate

6.3.5: Calculate: $\int_0^2 3e^{4-2x}\,dx$

Answer: $-\frac{3}{2} + \frac{3}{2}e^{4}$
ES Moderate

6.3.6: Suppose that at time t ($0 \le t \le 2$, t in months), the sales of a certain commodity are decreasing at a rate of $1000e^{-.05t}$ units per month. Calculate the total change in sales from t = 0 to t = 2.

Answer: $20{,}000(1 - e^{-.1})$ decrease
ES Moderate

6.3.7: Suppose that at time t, a bacteria culture is increasing at the rate of $500e^{.1t}$ bacteria per hour. Calculate the total increase in the number of bacteria from t = 0 to t = 1.

Answer: $5000(e^{.1} - 1)$
ES Moderate

6.3.8: Approximate the quantity

$$\frac{1^3 + 2^3 + 3^3 + \; . \; . \; . + 1000^3}{1000^3}$$

[Hint: This quantity closely resembles a Riemann sum.]

Answer: 250
ES Moderate

6.3.9: Approximate the quantity

$$\sqrt{\frac{1}{100}} \cdot \frac{1}{100} + \sqrt{\frac{2}{100}} \cdot \frac{1}{100} + \sqrt{\frac{3}{100}} \cdot \frac{1}{100} + \cdots +$$

$$\sqrt{\frac{99}{100}} \cdot \frac{1}{100} + \sqrt{\frac{100}{100}} \cdot \frac{1}{100}$$

by the appropriate integral.

Answer: $\int_0^1 \sqrt{x}\, dx = \frac{2}{3}$

ES Moderate

6.3.10: Find the area under the curve $y = x^3 + 2e^{-x}$ from $x = 0$ to $x = 3$.

Answer: $\frac{89}{4} - 2e^{-3}$

ES Moderate

6.3.11: What is the area under the curve $y = x^3 + x$ from $x = 1$ to $x = 2$?

062

a) $\frac{21}{4}$ b) $\frac{21}{2}$

c) $6\frac{1}{4}$ d) $\frac{6}{4}$

e) none of the above

Answer: (a)

MC Moderate

6.3.12: Suppose that during a controlled experiment, the temperature in a test tube at time t is rising at a rate of $6t^2 + 2$ degrees centigrade per minute. If the initial temperature is 0° C, what is the temperature in the test tube after 10 minutes?

a) 602 b) 120

c) 524 d) 2020

e) none of the above

Answer: (d)

MC Moderate

6.3.13: If A(t) denotes the annual rate of world consumption of oil at time t (with t = 0 corresponding to 1977), which of the following expressions represents the amount of oil consumed between 1977 and 1987?

a) $A'(10)$

b) $\int_0^{10} A(t)\,dt$

c) $\int_0^{10} A'(t)\,dt$

d) $\int_{1977}^{1987} A(t)\,dt$

e) none of the above

Answer: (b)
MC Moderate

6.3.14: Suppose the interval $0 \le x \le 2$ is subdivided into 20 subintervals of width $x = \frac{1}{10}$. Which of the following is the best approximation of the Riemann sum
$(2(.1) - 3)x + (2(.2) - 3)x + (2(.3) - 3)x + \ldots$
$+ (2(1.9) - 3)x + (2(2) - 3)x$?

a) $\int_0^2 (2(.1x) - 3)\,dx$

b) -2

c) 1

d) 4

e) 2

Answer: (b)
MC Moderate

6.3.15: $\int_0^1 \frac{2}{3 - 2x}\,dx =$

a) $\ln 3$

b) $-\ln 3$

c) $-e + \ln 3$

d) $\frac{4}{3}$

e) none of the above

Answer: (a)
MC Moderate

6.3.16: What is $\int_{-1}^{1} e^{-2x}\,dx$?

a) $e^2 - e^{-2}$

b) $\frac{1}{2}(e^{-2} - e^2)$

c) $\frac{1}{2}(e^2 - e^{-2})$

d) $-\frac{1}{2} e^{-2x} + C$

e) none of the above

Answer: (c)
MC Moderate

6.3.17: Calculate: $\int_1^2 5x\, dx$

Answer: $\frac{15}{2}$

ES Moderate

6.3.18: Calculate: $\int_0^2 (x^3 + 3x^2 + 3x + 1)\, dx$

Answer: 20

ES Moderate

6.3.19: Calculate: $\int_2^{10} \frac{dx}{x - 1}$

Answer: ln 9

ES Moderate

6.3.20: Calculate: $\int_{-100,000}^{100,000} x^3\, dx$

Answer: 0

ES Moderate

6.3.21: Calculate: $\int_1^3 (x^3 + 3)x^2\, dx$

Answer: $\frac{442}{3}$

ES Moderate

6.3.22: Determine the area under the curve $f(x) = 4x + 4$ from $x = 2$ to $x = 3$.

Answer: 14

ES Moderate

6.3.23: Determine the area under the curve $f(x) = e^{4x}$ from $x = 0$ to $x = 1$.

Answer: $\frac{e^4}{4} - 1$

ES Moderate

6.3.24: Determine the area under the curve $f(x) = \frac{1}{x}$ from $x = 1$ to $x = e$.

Answer: 1
ES Moderate

6.3.25: Determine the area between the curve $x = 4 - 2y^2$ and the y axis.

Answer: $\frac{16\sqrt{2}}{3}$
ES Moderate

6.3.26: Calculate: $\int_0^1 (2x^4 + 5x + 1)\,dx$

Answer: $\frac{39}{10}$
ES Moderate

6.3.27: Calculate: $\int_1^4 3\sqrt{x}\,dx$

Answer: 14
ES Moderate

6.3.28: Calculate: $\int_3^5 e^{5x}\,dx$

Answer: $\frac{e^{25} - e^{15}}{5}$
ES Moderate

6.3.29: Calculate: $\int \frac{dx}{x + 1}$

Answer: $\ln |x + 1| + C$
ES Moderate

6.3.30: What is $\int_0^1 \left(e^{3x} - \frac{1}{(x+1)^2}\right)dx$?

a) $\frac{1}{3}e^3 + \frac{7}{6}$

b) $\frac{1}{3}e + \frac{7}{6}$

c) $\frac{1}{3}e^3 - e - \frac{1}{2}$

d) $\frac{1}{3}e^3 - \frac{5}{6}$

e) none of the above

Answer: (d)
MC Moderate

6.3.31: Compute $\int_2^5 \left(e^{4x} - \frac{1}{x}\right)dx$.

Answer: $\frac{1}{4}(e^{20} - e^8) + \ln\frac{2}{5}$
ES Moderate

6.3.32: What is $\int_{-2}^{-1} (x^2 - 2x^{-3} + 3)dx$?

a) $\frac{7}{3} + \frac{1}{4} - 10$

b) $\frac{7}{3} - \frac{1}{4} + 4$

c) $\frac{-7}{3} + \frac{1}{4} - 2$

d) $\frac{-7}{3} - \frac{1}{4} - 10$

e) none of the above

Answer: (b)
MC Moderate

6.3.33: What is the area under the curve $f(x) = 3x^2 + 2x + 1$ from $x = \frac{-3}{2}$ to $x = \frac{-1}{2}$?

a) 18/8

b) 30/8

c) -30/8

d) 102/8

e) none of the above

Answer: (a)
MC Moderate

6.3.34: Find the area under the curve $y = \frac{1}{x} - 2x$ from $x = -3$ to $x = -2$.

Answer: $5 - \ln(3/2)$
ES Moderate

6.3.35: What is the area under the curve $y = \frac{2}{x}$ between $x = 1$ and $x = 3$?

a) $\frac{1}{2}\ln 3$
b) $2\ln 3 - 2$
c) $2\ln 3 - e$
d) $\ln 9$
e) none of the above

Answer: (d)
MC Moderate

6.3.36: What is the area under the curve $y = \frac{1}{\sqrt{x}}$ between $x = 1$ and $x = 2$?

a) $\frac{\sqrt{2} - 1}{2}$
b) $\ln\sqrt{2}$
c) $2\sqrt{2} - 2$
d) $\frac{\ln\sqrt{2}}{\sqrt{2}}$
e) none of the above

Answer: (c)
MC Moderate

6.3.37: Find the area under the curve $y = \frac{1}{e^x} + \frac{1}{x - 2}$ between $x = 3$ and $x = 5$.

Answer: $\frac{1}{e^3} - \frac{1}{e^5} + \ln 3$
ES Moderate

6.3.38: What is the area under the curve $y = \frac{1}{3} + 3e^{3x}$ between $x = \frac{-1}{3}$ and $x = \frac{1}{3}$.

a) $\frac{1}{e} - e$
b) $\frac{9e^2 + 2e + 9}{2e}$
c) $e + \frac{1}{e}$
d) $\frac{2}{9}$
e) none of the above

Answer: (b)
MC Moderate

6.3.39: A helicopter rises straight up in the air so that its velocity t seconds after take-off is $v(t) = t^{3/2} + 1/2\ t^{1/2} + 1$ feet per second. If the landing pad is 100 feet above the ground, which of the following gives the height of the helicopter at time t?

a) $h(t) = \frac{2}{3}t^{5/2} + \frac{1}{4}t^{3/2} + t + C$

b) $h(t) = \frac{3}{2}t^{1/2} + \frac{1}{4}t^{-1/2} + 100$

c) $h(t) = \frac{2}{5}t^{5/2} + \frac{1}{3}t^{3/2} + t + 100$

d) $h(t) = \frac{5}{3}t^{5/2} + \frac{3}{4}t^{3/2} + t - 100$

e) none of the above

Answer: (c)
MC Moderate

6.3.40: Suppose that the marginal cost of a manufacturer is $\frac{6}{11}x^2 - x + 10$ dollars per unit at production level x. If 30 units are currently being produced, the total cost of producing 10 additional units is equal to:

a) $\frac{d}{dx}\left(\frac{6}{11}x^2 - x + 10\right)\Big|_{x = 40}$

b) $\frac{d}{dx}\left(\frac{6}{11}x^2 - x + 10\right)\Big|_{x = 10}$

c) $\int_0^{10} \left(\frac{6}{11}x^2 - x + 10\right)dx$

d) $\int_{10}^{30} \left(\frac{6}{11}x^2 - x + 10\right)dx$

e) none of the above

Answer: (e)
MC Moderate

6.3.41: Suppose that the marginal revenue for a retailer is $6x^2 - \sqrt{x} + x$ dollars at sales level x. If 4 units are currently being sold, what is the extra revenue received from the sale of 5 additional units?

a) $\frac{11}{6} \approx \$1.83$

b) $1304 + \frac{259}{6} \approx \$1347.17$

c) \$152

d) $568 + \frac{42}{17} \approx \$570.47$

e) none of the above

Answer: (b)
MC Moderate

6.3.42: Use an appropriate integral to approximate the sum

$$\frac{1}{10}\left(\frac{11^3}{1000} + 8 + \frac{12^3}{1000} + 8 + \dots + \frac{20^3}{1000} + 8\right)$$

Answer: $\int_1^2 (x^3 + 8)dx = 12 - \frac{1}{4}$

ES Moderate

6.4.1: Consider the region above the graph of $y = x^2 - 15$, below the x-axis and also below the line $y = -2x$. Write down but do not evaluate the integral(s) that give(s) the area of this region.

Answer: $\int_{-\sqrt{15}}^{0} -(x^2 - 15)\ dx + \int_0^3 (-x^2 - 2x + 15)\ dx$

ES Moderate

6.4.2: Make a sketch of the region between the two curves $y = x^2$ and $y = x^2 - 10x + 25$ from $x = 0$ to $x = 4$. Write down (but do not evaluate) the integral(s) that give(s) the area of the region.

Answer: $\int_0^{2.5} (-10x + 25)\ dx + \int_{2.5}^4 (10x - 25)\ dx$

ES Moderate

6.4.3: Express the area between the graphs $y = x^3 - 3x + 1$ and $y = x^2 - x + 1$ in terms of an integral(s). [You need not evaluate the integral(s).]

Answer: $\int_{-1}^0 (x^3 - x^2 - 2x)\ dx + \int_0^2 (-x^3 + x^2 + 2x)\ dx$

ES Moderate

6.4.4: Refer to the information in the graph below. What is the integral(s) that represents the shaded area between $f(x) = x^2 - 4x + 4$ and $g(x) = x^2 + 4x + 4$?

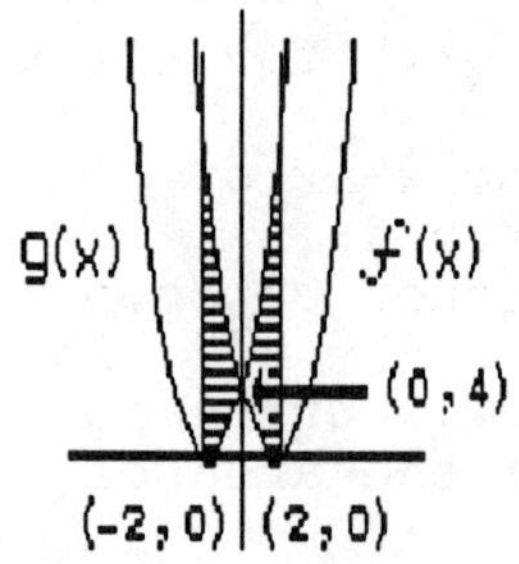

a. $\int_{-2}^{2} (x^2 - 4x + 4) - (x^2 + 4x + 4)\, dx$

b. $\int_{-2}^{4} (x^2 - 4x + 4) - (x^2 + 4x + 4)\, dx +$

c. $\int_{4}^{2} (x^2 + 4x + 4) - (x^2 - 4x + 4)\, dx$

d. $\int_{-2}^{0} -8x\, dx + \int_{0}^{2} 8x\, dx$

e. $\int_{-2}^{0} 8x\, dx + \int_{0}^{2} -8x\, dx$

f. none of the above

Answer: c
ES Moderate

6.4.5: $\int_{-1}^{1} \left(e^{x^2} + e^{x}\right) dx + \int_{-1}^{1} \left(e^{x} - e^{x^2}\right) dx =$

a) $2e - \frac{2}{e}$

b) $2e$

c) $\frac{e^2 + 1}{e}$

d) the answer cannot be found by the methods we know

e) none of the above

Answer: (a)
MC Moderate

6.4.6: $\int_a^b (f(x) - g(x))\,dx$ expresses the area between the curves $y = f(x)$ and $y = g(x)$ from $x = a$ to $x = b$ only if:

a) f(x) is greater than or equal to g(x) for all x between a and b.

b) f(x) is greater than or equal to g(x) for all x between a and b, and neither f(x) nor g(x) crosses the x-axis.

c) neither f(x) nor g(x) crosses the x-axis.

d) f(x) and g(x) do not cross each other between x = a and x = b.

e) none of the above

Answer: (a)
MC Moderate

6.4.7: Which of the following represents the area bounded by $y = x^2 + 3$ and $y = -5x - 3$?

a) $\int_2^3 (x^2 + 3) - (-5x - 3)\,dx$

b) $\int_{-3}^{-2} (-5x - 3) - (x^2 + 3)\,dx$

c) $\int_{-3}^{-2} (x^2 + 3) - (-5x - 3)\,dx$

d) $\int_2^3 (-5x - 3) - (x^2 + 3)\,dx$

e) none of the above

Answer: (b)
MC Moderate

6.4.8: Which of the following correctly expresses the shaded region below?

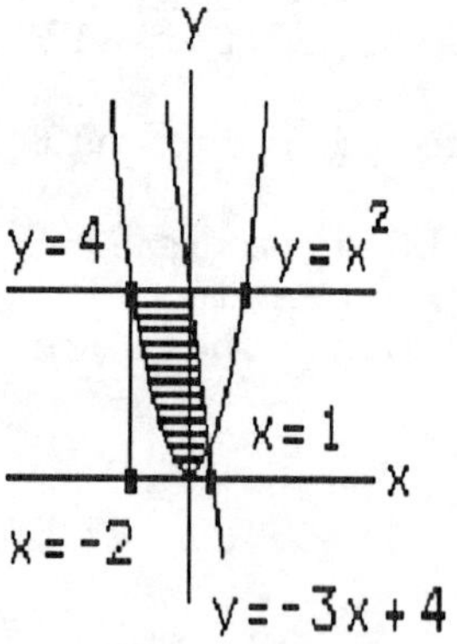

a) $\int_{-2}^{0} (4 - x^2)\, dx + \int_{0}^{1} 4 - (-3x + 4)\, dx$

b) $\int_{-2}^{-3x+4} (4 - x^2)\, dx$

c) $\int_{-2}^{1} 4 + (-3x + 4) - x^2\, dx$

d) $\int_{-2}^{0} (4 - x^2)\, dx + \int_{0}^{1} (-3x + 4 - x^2)\, dx$

e) none of the above

Answer: (d)
MC Moderate

6.4.9: Does $\int_{1}^{3} (e^x - 1)\, dx$ = area between the curve $y = e^x - 1$ and the x-axis between x = 1 and x = 3 ?

a) Yes, since $y = e^x - 1$ is nonnegative for all x between 1 and 3.

b) Yes, since $y = e^x - 1$ does not cross the x-axis between 1 and 3.

c) No, since $y = e^x - 1$ crosses the x-axis between 1 and 3.

d) No, since $\int_{1}^{3} (e^x - 1)\, dx$ is negative.

e) none of the above

Answer: (a)
MC Moderate

6.4.10: Determine the area between the curve $F(x) = 5x - 2x^2$ and the line y = 3.

Answer: $\frac{1}{24}$
ES Moderate

6.4.11: Determine the area between the curve $f(x) = x^2$ and the line $y = x$.

Answer: $\frac{1}{6}$

ES Moderate

6.4.12: Let $f(x) = \{x^2\ (0 \le x \le 1);\ 2x - x^2\ (1 \le x \le 2)\}$. Determine the area between $f(x)$ and the line $y = 0$.

Answer: 1

ES Moderate

6.4.13: Determine the area between the curve $y = -x^2 + 3$ and the line $y = 2x$.

Answer: $\frac{32}{3}$

ES Moderate

6.4.14: Determine the area of the shaded region.

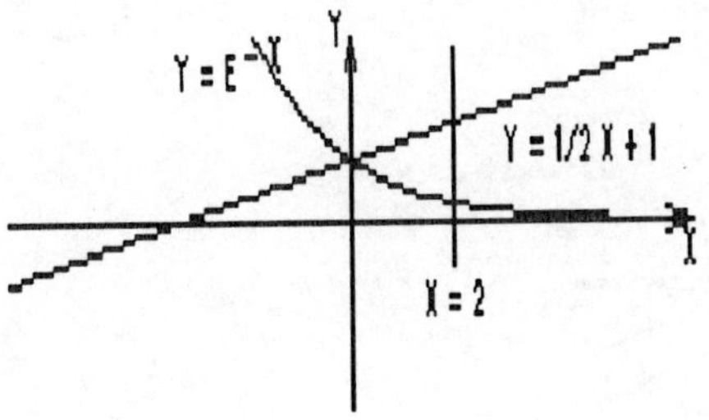

Answer: $2 + \frac{1}{e^2}$

ES Moderate

6.4.15: Determine the area of the region bounded by the curves $y = x^3 - \frac{8}{5}x + 1$ and $y = \frac{2}{5}x + 1$.

Answer: 2

ES Moderate

6.4.16: Write down, but do not evaluate, the integral(s) that give(s) the area of the region bounded by the curve $y = e^{2x} + \frac{4}{x^2}$, $x > 0$, and the lines $y = 0$, $x = 1$, and $x = 4$.

Answer: $\int_1^4 \left(e^{2x} + \frac{4}{x^2}\right) dx$
ES Moderate

6.4.17: Write down, but do not evaluate, the integral(s) that give(s) the area of the region enclosed by the curves $y = \sqrt{x}$, $y = 2\sqrt{x - 1}$ and the x-axis.

Answer: $\int_0^1 \sqrt{x}dx + \int_1^{4/3} \sqrt{x} - 2\sqrt{x - 1}dx$
ES Moderate

6.4.18: In the figure below, the region enclosed by the curves $y = \frac{-4}{x}$, $y = -x$ and $y = -x + 3$ has been shaded. Which of the following expressions represents the area of this region?

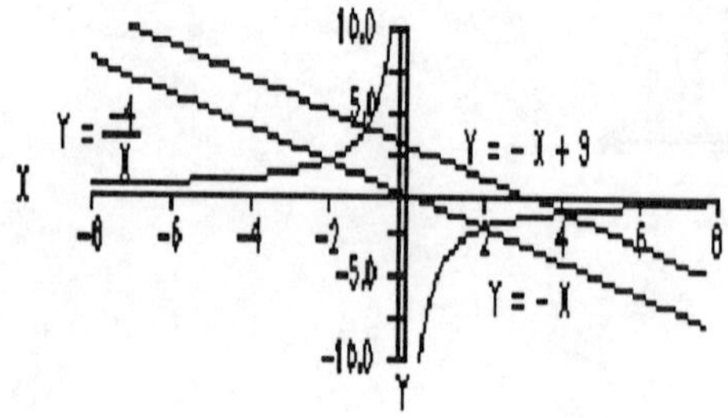

a) $\int_{-2}^{0} \left(-\frac{4}{x} + x\right) dx + \int_0^2 \left(-x + 3 + \frac{4}{x}\right) dx$
b) $\int_{-2}^{-1} \left(-\frac{4}{x} + x\right) dx + \int_{-1}^{2} 3dx + \int_2^4 \left(-x + 3 + \frac{4}{x}\right) dx$
c) $\int_{-2}^{4} 3dx$
d) $\int_{-2}^{2} 3dx + \int_2^4 \left(-x + 3 - \frac{4}{x}\right) dx$
e) none of the above

Answer: (b)
MC Moderate

6.4.19: What is the area of the region bounded by the curve $y = \left(\frac{1}{2}x + 3\right)^{-1}$ the y-axis and the line y = 1.

a) $4 - 2\ln 3$

b) $2\left(\ln\frac{7}{2} - \ln 3\right)$

c) $\ln 3$

d) $\ln\frac{7}{2} - \ln 3$

e) none of the above

Answer: (a)
MC Moderate

6.4.20: What is the area of the region between y = 3x - 1, the y-axis, and the lines y = 2 and y = 5?

a) $\frac{9}{2}$

b) $\frac{19}{2}$

c) $\frac{17}{2}$

d) $\frac{26}{3}$

e) none of the above

Answer: (a)
MC Easy

6.4.21: Which of the following expressions represents the area of the region bounded by $y = x^2 + 3$, $y = 4x$, and $y = 2x + 6$?

a) $\int_{-1}^{1} (x^2 - 2x - 3)\,dx + \int_{1}^{3} (-2x + 6)\,dx$?

b) $\int_{-1}^{3} (-x^2 + 2x + 3)\,dx$

c) $\int_{1}^{3} (-x^2 + 4x - 3)\,dx$

d) $\int_{-1}^{3} (-x^2 + 2x + 3)\,dx - \int_{-1}^{1} (-x^2 + 4x - 3)\,dx$

e) none of the above

Answer: (d)
MC Easy

6.5.1: Find the average value of the function $x - x^2$ as x ranges from 0 to 1.

Answer: $\frac{1}{6}$
ES Moderate

6.5.2: Suppose that each day a pension fund purchases a \$10,000 certificate of deposit paying 12% interest compounded continuously. Approximately how much money does the fund have at the end of 2 years?

Answer: $\int_0^2 10{,}000 \cdot 365e^{.12(2-t)}\, dt \approx 8{,}960{,}000$

ES Moderate

6.5.3: Suppose that the profit realized by a department store t days after its opening is given by the formula $4t^3 - 2t + 1$. What was the average profit per day of the store during the first five days?

a) \$121
b) \$218
c) \$225
d) \$97.80
e) none of the above

Answer: (a)

MC Moderate

6.5.4: Determine the average of $f(x) = \sqrt{3x + 2}$ where $4 \le x \le 10$.

Answer: $\dfrac{32^{3/2} - 14^{3/2}}{27}$

ES Moderate

6.5.5: Determine the average of $f(x) = e^x$ where $1 \le x \le 4$.

Answer: $\dfrac{e^4 - e}{3}$

ES Moderate

6.5.6: Determine the average of $f(x) = x^3 + 3x^2 + 3x + 1$ where $-1 \le x \le 3$.

Answer: 16

ES Moderate

6.5.7: Find the volume of the solid obtained by revolving the curve

$y = \dfrac{1}{\sqrt{x}}$ $(1 \le x \le 7)$ about the x-axis.

Answer: $\pi \ln 7$

ES Moderate

6.5.8: Suppose that a colony of fruit flies is growing exponentially with growth constant .04. If there are currently 30,000 flies present, what will be the average population over the next 6 months?

a) $30{,}000(e^{.02} - 1)$
b) $60{,}000(e^{.24} - 1)$
c) $150{,}000(e^{.02} - 1)$
d) $75{,}000(e^{.24} - 1)$
e) none of the above

Answer: (c)
MC Moderate

6.5.9: Suppose that $2500 are deposited in a savings account paying 5% interest, compounded continuously. What will be the average value of the account during the next 10 years?

a) $2500(e^{50} - 1)$
b) $50{,}000(e^{.05} - 1)$
c) $250(e^{5} - 1)$
d) $50{,}000(e^{.5} - 1)$
e) none of the above

Answer: (e)
MC Moderate

6.5.10: A certain commodity has demand curve $p = \dfrac{20}{x + 5} - 1$ at sales level x. What is the consumers' surplus if 5 units are currently being sold?

a) $\dfrac{1}{20}(\ln 3) - 20$
b) $20(\ln 10) - 10$
c) $20(\ln 15 - \ln 5) - 10$
d) $20(\ln 15 + \ln 5) - 10$
e) none of the above

Answer: (b)
MC Moderate

6.5.11: What is the consumers' surplus for the demand curve $p = 5 - \dfrac{x}{20}$ at sales level $x = 60$?

a) 90
b) 291
c) $200 - 20\ln 60$
d) $320 - \dfrac{1}{20}\ln 60$
e) none of the above

Answer: (a)
MC Moderate

6.5.12: Find the consumers' surplus for the demand curve $p = \sqrt{25 - .1x} - 3$ at sales level $x = 160$.

Answer: $173.33
ES Moderate

6.5.13: Suppose that money is deposited steadily into a savings account at the rate of \$3,000 per year. How long will it take for the balance to reach \$60,000 if the account pays 4% interest compounded continuously?

a) 20 yrs

b) $25 \ln \frac{9}{5}$ yrs

c) $25 \ln \frac{1}{5}$ yrs

d) 15.5 yrs

e) none of the above

Answer: (b)
MC Moderate

6.5.14: Suppose that money is deposited in a savings account at a steady rate of \$150 per month. If the account pays 2.5% interest compounded continuously, what will the account be worth at the end of 3 years?

Answer: $72{,}000(e^{.075} - 1) \approx \$5607.66$
ES Moderate

7. Functions of Several Variables

7.1.1: Suppose that the number of leather bags produced by a certain firm, utilizing x units of raw materials and y units of labor, is given by the formula $f(x, y) = 100x^{1/2}y^{1/2}$. What happens to production of bags if supplies of both raw materials and labor are tripled?

a) Production is increased by a factor of 6.
b) Production is increased by a factor of 3.
c) Production is increased by a factor of 9.
d) Production is increased by a factor of $\sqrt{3}$.
e) none of the above

Answer: (b)
MC Moderate

7.1.2: A rectangular garden is to be surrounded on three sides by a fence costing $5 per foot and on one side by a stone wall costing $15 per foot. Let x be the length of the side with the stone wall, and y the other dimension of the garden. Express the cost of enclosing the garden as a function of 2 variables.

a) $C(x, y) = (15x)(5y)$
b) $C(x, y) = (20x)(10y)$
c) $C(x, y) = 20x + 10y$
d) $C(x, y) = 15x + 5y$
e) none of the above

Answer: (c)
MC Moderate

7.2.1: Compute:

(a) $\frac{\delta z}{\delta x}$, when $z = x^2y - xy^2$

(b) $\frac{\delta f}{\delta x}$ and $\frac{\delta f}{\delta y}$, when $f(x, y) = (x^2 + x)e^{y-x}$

(c) $\frac{\delta f}{\delta x}(2, 3, 4)$, when $f(x, y, z) = \frac{x}{y} + \frac{z}{x}$

(d) $\frac{\delta^2 f}{\delta x^2}$, when $f(x, y) = (x + y)e^{xy}$

Answer: (a) $2xy - y^2$

(b) $\frac{\delta f}{\delta x} = -(x^2 + x)e^{y-x} + (2x + 1)e^{y-x}$; $\frac{\delta f}{\delta y} = (x^2 + x)e^{y-x}$

(c) $\frac{\delta f}{\delta x} = \frac{1}{y} - \frac{z}{x^2}$; $\frac{\delta f}{\delta x}(2, 3, 4) = -\frac{2}{3}$

(d) $\frac{\delta f}{\delta x} = (x + y)ye^{xy} + e^{xy}$; $\frac{\delta^2 f}{\delta x^2} = (x + y)y^2e^{xy} + 2ye^{xy}$

ES Moderate

7.2.2: (a) If $z = (x^2y^3 + x^3)^5$, find: $\frac{\delta z}{\delta x}$ and $\frac{\delta z}{\delta y}$

(b) If $f(x, y) = x^2e^{xy}$, what is $\frac{\delta f}{\delta x}$?

(c) If $f(x, y) = y(x + e^{y-x})$, find: $\frac{\delta f}{\delta x}$ (2, 3)

Answer: (a) $\frac{\delta z}{\delta x} = 5(x^2y^3 + x^3)^4(2xy^3 + 3x^2)$; $\frac{\delta z}{\delta y} = 5(x^2y^3 + x^3)^4(3x^2y^2)$

(b) $\frac{\delta f}{\delta x} = x^2ye^{xy} + 2xe^{xy}$

(c) $3(1 - e)$

ES Moderate

7.2.3: Let $f(x, y) = \ln(x + 2y)$. Calculate $\frac{\delta^2 f}{\delta x^2}$ and $\frac{\delta^2 f}{\delta x \delta y}$.

Answer: $\frac{\delta^2 f}{\delta x^2} = -\frac{1}{(x + 2y)^2}$; $\frac{\delta^2 f}{\delta x \delta y} = -\frac{2}{(x + 2y)^2}$

ES Moderate

7.2.4: It is determined that the demand for bicycles is a function $D(x, y)$ of their price x and the cost of gasoline y.

Explain why $\frac{\delta D}{\delta x} < 0$ and $\frac{\delta D}{\delta y} > 0$.

Answer: $\frac{\delta D}{\delta x} < 0$. If the cost of gasoline is held constant, then an increase in the price of bicycles will cause demand to fall.

$\frac{\delta D}{\delta y} > 0$. If the price of bicycles is held constant, then an increase in the cost of gasoline will cause demand to increase.

ES Moderate

7.2.5: Let $f(x, y) = x^3 y + e^{x+3y}$. What is $\frac{\delta^2 f}{\delta x \delta y}(1, 0)$?

a) $3 + 3e$
b) $6 + e$
c) 6
d) $3 + e$
e) none of the above

Answer: (a)

MC Moderate

7.2.6: The demand for a certain energy-inefficient home is given by $f(p_1, p_2)$, where p_1 is the price of the home and p_2 is the price of electricity. Which of the following best explains why $\frac{\delta f}{\delta p_2} < 0$?

a) The price of electricity keeps going up due to increased demand.
b) As the price of electricity goes up, demand for the home goes down.
c) As the price of electricity goes down, demand for the home goes down.
d) As the price of electricity increases, demand for the home depends more on the price of the home than on the price of electricity.
e) Homes are too expensive during the energy crisis.

Answer: (b)

MC Moderate

7.2.7: Let $f(x, y, z) = xyz(1 + e^{yz})$. What is $\frac{\delta f}{\delta z}$?

a) $xy(1 + e^{yz}) + xy^2ze^{yz}$
b) $xy(1 + ye^{yz})$
c) $xy(1 + e^{yz}) + xy(1 + ye^{yz})$
d) $xy(1 + e^{y})$
e) none of the above

Answer: (a)
MC Moderate

7.2.8: A certain manufacturer can produce $f(x, y) = 10(6x^3 + y^2)$ units of goods by utilizing x units of labor and y units of capital. Which of the following gives the marginal productivity of labor when x = 10 and y = 20 ?

a) 18,000 units
b) 22,000 units
c) 1800 units
d) 2200 units
e) none of the above

Answer: (a)
MC Moderate

7.2.9: Let f(x, y, z) be the amount of heat lost each day by a rectangular
building x feet wide, y feet long, and z feet high.
Suppose that $\frac{\delta f}{\delta z}(50, 80, 15) = 45$. Which one of the following conclusions can be drawn?

a) A building with dimensions 50 x 80 x 15 loses about 45 units of
heat each day.
b) A building with dimensions 50 x 80 x 15 loses about 45 units of
heat from the top of the building each day.
c) A building with dimensions 50 x 80 x 16 will lose about 45 more
units of heat each day than a building with dimensions 50 x 80 x 15.
d) A building with dimensions 51 x 81 x 16 will lose about 45 more
units of heat each day than a building with dimensions 50 x 80 x 15.
e) The marginal heat loss per day is 45 units of heat per square foot of surface area of the roof.

Answer: (c)
MC Moderate

7.2.10: Determine $\frac{\delta f}{\delta x}$ and $\frac{\delta f}{\delta y}$ where $f(x, y) = xy$.

Answer: $\frac{\delta f}{\delta x} = y$; $\frac{\delta f}{\delta y} = x$
ES Moderate

7.2.11: Determine $\frac{\delta f}{\delta x}$ and $\frac{\delta f}{\delta y}$ where $f(x, y) = 4y^2 - 2x^3 + 5xy^2$.

Answer: $\frac{\delta f}{\delta x} = -6x^2 + 5y^2$; $\frac{\delta f}{\delta y} = 8y + 10xy$

ES Moderate

7.2.12: Determine $\frac{\delta f}{\delta x}$ and $\frac{\delta f}{\delta y}$ where $f(x, y) = e^{x^2 y}$.

Answer: $\frac{\delta f}{\delta x} = e^{x^2 y} \cdot 2xy$; $\frac{\delta f}{\delta y} = e^{x^2 y} \cdot x^2$

ES Moderate

7.2.13: Determine $\frac{\delta f}{\delta x}$ and $\frac{\delta f}{\delta y}$ where $f(x, y) = \frac{\ln xy}{y}$.

Answer: $\frac{\delta f}{\delta x} = \frac{1}{xy}$; $\frac{\delta f}{\delta y} = \frac{1 - \ln xy}{y^2}$

ES Moderate

7.2.14: Determine $\frac{\delta f}{\delta x}$, $\frac{\delta f}{\delta y}$, and $\frac{\delta f}{\delta z}$ where $f(x, y, z) = \sqrt{xyz}$.

Answer: $\frac{\delta f}{\delta x} = \frac{yz}{2\sqrt{xyz}}$; $\frac{\delta f}{\delta y} = \frac{xz}{2\sqrt{xyz}}$; $\frac{\delta f}{\delta z} = \frac{xy}{2\sqrt{xyz}}$

ES Moderate

7.2.15: Determine $\frac{\delta f}{\delta x}$, $\frac{\delta f}{\delta y}$, and $\frac{\delta f}{\delta z}$ where $f(x, y, z) = e^{xy^2 z^3}$.

Answer: $\frac{\delta f}{\delta x} = y^2 z^3 e^{xy^2 z^3}$; $\frac{\delta f}{\delta y} = 2xyz^3 e^{xy^2 z^3}$; $\frac{\delta f}{\delta z} = 3xy^2 z^2 e^{xy^2 z^3}$

ES Moderate

7.2.16: Determine $\frac{\delta f}{\delta x}$, $\frac{\delta^2 f}{\delta x^2}$, and $\frac{\delta^2 f}{\delta x \delta y}$ where $f(x, y) = x^2 y + y^2 x + 2xy$.

Answer:

$\frac{\delta f}{\delta x} = 2xy + y^2 + 2y$; $\frac{\delta^2 f}{\delta x^2} = 2y$; $\frac{\delta^2 f}{\delta x \delta y} = 2x + 2y + 2$

ES Moderate

7.2.17: Determine $\frac{\delta f}{\delta x}$, $\frac{\delta^2 f}{\delta x^2}$, and $\frac{\delta^2 f}{\delta x \delta y}$ where $f(x, y) = \frac{x}{y + 1}$.

Answer: $\frac{\delta f}{\delta x} = \frac{1}{y + 1}$; $\frac{\delta^2 f}{\delta x^2} = 0$; $\frac{\delta^2 f}{\delta x \delta y} = -\frac{1}{(y + 1)^2}$

ES Moderate

7.2.18: Determine $\frac{\delta f}{\delta x}$, $\frac{\delta^2 f}{\delta x^2}$, and $\frac{\delta^2 f}{\delta x \delta y}$ where $f(x, y) = e^{x-y}$.

Answer: $\frac{\delta f}{\delta x} = e^{x-y}$; $\frac{\delta^2 f}{\delta x^2} = e^{x-y}$; $\frac{\delta^2 f}{\delta x \delta y} = -e^{x-y}$

ES Moderate

7.2.19: Determine $\frac{\delta f}{\delta x}$, $\frac{\delta^2 f}{\delta x^2}$, and $\frac{\delta^2 f}{\delta x \delta y}$ where $f(x,y) = (x+y)^2 - (x+y)^3$.

Answer: $\frac{\delta f}{\delta x} = 2(x + y) - 3(x + y)^2$; $\frac{\delta^2 f}{\delta x^2} = 2 - 6(x + y)$;

$\frac{\delta^2 f}{\delta xy} = 2 - 6(x + y)$

ES Moderate

7.2.20: Determine $\frac{\delta f}{\delta x}$ and $\frac{\delta f}{\delta y}$ where $f(x, y) = x^2 + y$.

Answer: $\frac{\delta f}{\delta x} = 2x$; $\frac{\delta f}{\delta y} = 1$

ES Moderate

7.2.21: Determine $\frac{\delta f}{\delta x}$ and $\frac{\delta f}{\delta y}$ where $f(x, y) = 3x^2 + 2xy$.

Answer: $\frac{\delta f}{\delta x} = 6x + 2y$; $\frac{\delta f}{\delta y} = 2x$

ES Moderate

7.2.22: Determine $\frac{\delta f}{\delta x}$ and $\frac{\delta f}{\delta y}$ where $f(x, y) = x^2 + 2xy + e^y$.

Answer: $\frac{\delta f}{\delta x} = 2x + 2y$; $\frac{\delta f}{\delta y} = 2x + e^y$

ES Moderate

7.2.23: Determine $\frac{\delta f}{\delta x}$ and $\frac{\delta f}{\delta y}$ where $f(x, y) = 4x^2 + 2y^{2/3} + e^{x^{4/3}}$.

Answer:

$\frac{\delta f}{\delta x} = 8x + \frac{4e^{x^{4/3}} \cdot x^{1/3}}{3}$; $\frac{\delta f}{\delta y} = \frac{4y^{-1/3}}{3}$

ES Moderate

7.2.24: Determine $\frac{\delta f}{\delta x}$ and $\frac{\delta f}{\delta y}$ where $f(x, y) = 4x^2 + 2y^2 + 3xy$.

Answer: $\frac{\delta f}{\delta x} = 8x + 3y$; $\frac{\delta f}{\delta y} = 4y + 3x$

ES Moderate

7.2.25: Determine $\frac{\delta f}{\delta x}$ and $\frac{\delta f}{\delta y}$ where $f(x, y) = 2y^3 + 4x^4 + 2xy$.

Answer: $\frac{\delta f}{\delta x} = 16x^3 + 2y$; $\frac{\delta f}{\delta y} = 6y^2 + 2x$

ES Moderate

7.2.26: Determine $\frac{\delta f}{\delta x}$, $\frac{\delta f}{\delta y}$, and $\frac{\delta f}{\delta z}$ where $f(x, y, z) = \ln(xy^2z^3)$.

Answer: $\frac{\delta f}{\delta x} = \frac{1}{x}$; $\frac{\delta f}{\delta y} = \frac{2}{y}$; $\frac{\delta f}{\delta z} = \frac{3}{z}$

ES Moderate

7.2.27: If $f(x, y) = \frac{1}{2}x^2 \cdot e^{x/y}$, then $\frac{\delta f}{\delta y}$ equals:

a) $\frac{1}{2}x^2e^{x/y}$

b) $\frac{x^2}{q}e^{x/y}$

c) $\frac{x^3}{y}e^{x/y} - 1$

d) $-\frac{x^3}{y^2} e^{\frac{x}{y}}$

e) none of the above

Answer: (d)

MC Moderate

7.2.28: If $G(x, r, t) = 2x^2t + \frac{1}{3}t^2 - r^3\sqrt{xt}$, then $\frac{\delta G}{\delta x}$ equals:

a) $4xt + \frac{1}{3}t^2 - r^3\frac{1}{2\sqrt{xt}}$

b) $4x - \frac{r^3}{2\sqrt{xt}}$

c) $4xt - \frac{r^3t}{2\sqrt{xt}}$

d) $4x - \frac{1}{3}t^2 - \frac{3r^2t}{2\sqrt{xt}}$

e) none of the above

Answer: (c)
MC Moderate

7.2.29: Let $P(x, y, z) = xy + 2x^3\sqrt{y^2 -1}$. What is $\frac{\delta P}{\delta z}(2, 1, 3)$?

a) 3

b) 0

c) 2

d) 1

e) none of the above

Answer: (b)
MC Moderate

7.2.30: Let $g(x, t, z) = \frac{1}{t}(x - z^2t)$. What is $\frac{\delta^2 g}{\delta z \delta t}$?

a) 0

b) $\frac{-2z}{t} + 1$

c) $-2zt + t^2$

d) $-2zt$

e) none of the above

Answer: (a)
MC Moderate

7.2.31: If $f(x, y, z) = y^3z - x^2y + \sqrt{xyz} - \frac{1}{x}$, what is $\frac{\delta f}{\delta y}$?

a) $3y^2 - 2x + 1$

b) $3y^2z - x^2 + \sqrt{xz}$

c) $3y^2z - x^2 + \frac{xz}{2\sqrt{xyz}}$

d) $3x^2z - x^2 + \frac{xz}{\sqrt{xyz}} + \frac{1}{x^2}$

e) none of the above

Answer: (c)
MC Moderate

7.2.32: If $H(x, y) = \frac{3xy}{x^2 - y}$, what is $\frac{\delta^2 H}{\delta y^2}$?

a) $\frac{6x^3}{(x^2 - y)^3}$

b) $\frac{3x^3}{(x^2 - y)^2}$

c) $\frac{6x^3y - 6x^5}{(x^2 - y)^4}$

d) $\frac{9x^4 - 6x^3 - 9x^2y}{(x^2 - y)^3}$

e) none of the above

Answer: (a)
MC Moderate

7.2.33: Let $F(x, y, z) = \frac{xz}{y^2z + x} - 5x^2y^3$. What is $\frac{\delta F}{\delta y}(-1, 0, 1)$?

a) 2

b) $-\frac{1}{2}$

c) $\frac{3}{5}$

d) 0

e) none of the above

Answer: (d)
MC Moderate

7.2.34: The productivity of a certain country is $P(x, y) = 1600x^{1/4}y^{3/4}$ units, where x and y are amounts of labor and capital utilized. What is the marginal productivity of capital when x = 16 and y = 625?

a) 5000

b) 2500

c) 6250

d) 1600

e) none of the above

Answer: (c)
MC Moderate

7.2.35: Suppose that a company can produce $P(x, y) = 50\sqrt{(x^3 + y^3)/10}$ items using x units of labor and y units of capital. What is the productivity of capital when x = 10 and y = 20?

a) 3000

b) 100

c) 2500

d) 500

e) none of the above

Answer: (b)
MC Moderate

7.2.36: Assume that a manufacturer has productivity function P(l, c) where l and c are the amounts of labor and capital utilized. Which of the following indicates that a slight increase in the amount of labor utilized will result in an increase in productivity of 3 units.

a) P(1, 75) = 3

b) P(101, 75) = 3

c) $\frac{\delta P}{\delta l}(100, 75) = 3$

d) $\frac{\delta P}{\delta c}(100, 75) = 3$

e) none of the above

Answer: (c)
MC Moderate

7.2.37: Suppose that a retailer sells f(p, a) units of an item, where p is the price per unit of the item and a is the amount of money spent on advertising that item. Which of the following indicates that at the amount spent on advertising is decreased, demand for the item also decreases?

a) $\frac{\delta f}{\delta p}(50, -1) = -5$

b) $\frac{\delta f}{\delta a}(50, 75) = -10$

c) $\frac{\delta f}{\delta p}(50, 25) = -1$

d) $\frac{\delta f}{\delta a}(50 ,25) = 10$

Answer: (b)
MC Moderate

7.3.1: Let $f(x, y) = x^2 + y^3 - 6y^2 + 6x - 15y$. Find all points (x, y) where f(x, y) may have a possible relative maximum or minimum. Use the second derivative test to determine, if possible, what happens to f(x, y) at these points.

Answer: Possible relative maxima or minima at (-3, -1) and (-3, 5). Test shows minimum at (-3, 5). No conclusion at (-3, -1).
ES Moderate

7.3.2: Let $f(x, y) = x^3 - y^2 - 3x + y + 5$.

(a) Find the point(s) where f(x, y) may have a possible relative maximum or minimum.

(b) Calculate the quantity $D(x, y) = \frac{\delta^2 f}{\delta^2 x} \cdot \frac{\delta^2 f}{\delta y^2} - \left(\frac{\delta^2 f}{\delta x \delta y}\right)^2$ and determine the nature of f(x, y) at the point(s) found in part (a).

Answer: (a) (1, 1/2), (-1, 1/2)

(b) D(1, 1/2) = -12, D(-1, 1/2) = 12. f(x, y) has a relative maximum at (-1, 1/2) and neither a relative maximum nor a relative minimum at (1, 1/2).

ES Moderate

7.3.3: Let $h(s, t) = s^2 + 2st + 5t^2 + 2s + 10t - 3$. At which point(s) does $h(s, t)$ have <u>possible</u> maximum/minimum values?

a) (-1, 17) and (0, 5)
b) (0, -1)
c) (-1, 0) and (0, 1)
d) (1, 0)
e) none of the above

Answer: (b)
MC Moderate

7.3.4: Let $f(x, y) = x^4 - y^2 - 2x^2 + 2y - 7$. Using the second derivative test for functions of 2 variables, classify (if possible) the points (0, 1) and (-1, 1) as relative maximum or minimum points.

a) (0, 1) relative maximum, (-1, 1) neither relative maximum nor minimum
b) (0, 1) no conclusion possible, (-1, 1) minimum
c) (0, 1) relative maximum, (-1, 1) relative minimum
d) (0, 1) neither relative maximum nor minimum, (-1, 1) maximum
e) none of the above

Answer: (a)
MC Moderate

7.3.5: Determine all possible relative extreme points of

$f(x, y) = 3xy - x^2 - y^2 - 2x - y + 3$.

Answer: $\left(\frac{7}{5}, \frac{8}{5}\right)$
ES Moderate

7.3.6: Determine all possible relative extreme points of

$f(x, y) = x^2 - 2y^2 + 4x - 6y + 8$.

Answer: $\left(-2, -\frac{3}{2}\right)$
ES Moderate

7.3.7: Determine all possible relative extreme points of

$f(x, y) = 2x^2 + 2y^3 - x - 6y + 14$.

Answer: $\left(\frac{1}{4}, 1\right)$
ES Moderate

7.3.8: Determine all possible relative extreme points of $f(x, y) = x^2 + xy + y^2 - x - y + 2$ as well as the nature of the point.

Answer: minimum at $\left(\frac{1}{3}, \frac{1}{3}\right)$
ES Moderate

7.3.9: Determine all possible relative extreme points of $f(x, y) = xy - 2x^2 + x - 4y + 1$ as well as the nature of the point.

Answer: saddle at (4, 15)
ES Moderate

7.3.10: Determine all possible relative extreme points of $f(x, y) = x^3 + y^3 - 6xy$ as well as the nature of the point.

Answer: saddle at (0, 0); minimum at (2, 2)
ES Moderate

7.3.11: Let $f(x, y) = x^2 - xy + y^2 + 2y - 4$. The point $\left(-\frac{2}{3}, \frac{2}{3}\right)$ is a:

a) relative maximum
b) relative minimum
c) saddle point
d) none of the above
e) can not be determined

Answer: (b)
MC Moderate

7.3.12: Let $Q(x, y) = x^2y + y^3x^4$. The point (1, 0) is:

a) a relative maximum
b) a relative minimum
c) a saddle point
d) none of the above
e) can not be determined

Answer: (e)
MC Moderate

7.3.13: Find the dimensions of the rectangular box of greatest volume that has girth equal to 60 inches.

Answer: 10" x 10" x 20"
ES Moderate

7.3.14: Let $f(x, y) = ye^x + xy^2$. $f(x, y)$ has a possible relative maximum at the point:

a) $(0, 0)$

b) $\left(\frac{1}{2}, \frac{-e^{1/2}}{2}\right)$

c) $\left(\frac{\sqrt{2}}{2}, \frac{-\sqrt{2}e^{\sqrt{2}/2}}{2}\right)$

d) $\left(\frac{1}{2}, -e^{1/2}\right)$

e) none of the above

Answer: (c)
MC Moderate

7.3.15: The function $H(x, y) = x^4 - 9y^2 - 2x^2y + 20y + 4$ has:

a) a relative maximum at the point $(-1, 1)$

b) a relative minimum at the point $(0, 0)$

c) neither a relative maximum nor minimum at $(1, 4)$

d) a relative maximum at the point $(0, 1)$

e) none of the above

Answer: (c)
MC Moderate

7.4.1: Find the points (x, y) where the function $8x - 2y$ may have a possible relative maximum or minimum, subject to the constraint

$$x^2 + \frac{1}{2}y = 18.$$

Use the method of Lagrange multipliers.

Answer: Possible relative maximum or minimum at $(-1, 34)$
ES Moderate

7.4.2: Let $f(x, y) = x^2 - 6xy + 10$. Find the point(s) where $f(x, y)$ may have a possible relative maximum or minimum, subject to the condition that $5x + 3y = 11$. Use the method of Lagrange multipliers. (No credit given for other methods.)

Answer: $(1, 2)$
ES Moderate

7.4.3: Find the points (x, y, z) where the function xyz may have a possible relative maximum or minimum, subject to the constraint $x + 6y + 3z = 36$ and where $x > 0$, $y > 0$, $z > 0$.

Answer: $(x, y, z) = (12, 2, 4)$
ES Moderate

7.4.4: Minimize $x + y$ subject to the constraint $xy = 100$, $x > 0$, $y > 0$. Use Lagrange multipliers.

Answer: $x = 10$, $y = 10$, minimum value = 20
ES Moderate

7.4.5: Maximize xy subject to the constraint $3x^2 + y = 1$, $x > 0$, $y > 0$. Use Lagrange multipliers.

Answer: $x = \frac{1}{3}$, $y = \frac{2}{3}$, maximum value $= \frac{2}{9}$
ES Moderate

7.4.6: A business produces two products A and B. Let x and y denote, respectively, the quantity of A and B to be produced. Limitations on the company resources require that $500x^2 + 100y$ be at most 100,000. Each unit of A yields a \$5000 profit and each unit of B yields a \$500 profit. What should x and y be to yield a maximum profit?

Answer: $x = 1$, $y = 995$
ES Moderate

7.4.7: Design a cylindrical can of volume 100 cu. in. which requires a minimum amount of aluminum. (That is, the can is to have a minimum surface area.)

Answer: radius $= \sqrt[3]{\frac{50}{\pi}}$, height $= 2\sqrt[3]{\frac{50}{\pi}}$
ES Moderate

7.4.8: An open rectangular box with square ends (as in the picture) is to be designed so that the surface area of the box is minimized. [Note: Surface area $= 2x^2 + 4xy$.] It is required that the volume be 32 cubic inches. Which of the following is the Lagrange function $F(x, y, \lambda)$ for this problem?

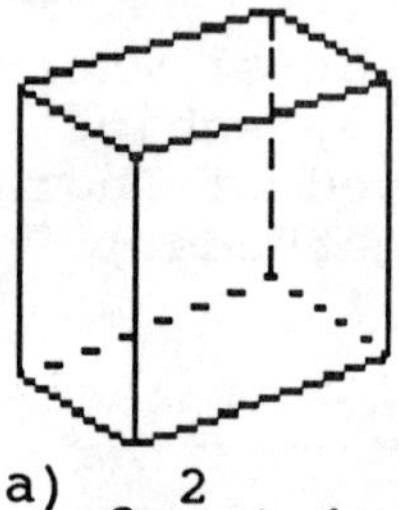

a) $2x^2 + 4xy + \lambda(x^2 y)$
b) $2x^2 + 4xy + \lambda(32)$
c) $(x^2 y - 32) + \lambda(2x^2 + 4xy)$
d) $x^2 y + \lambda(2x^2 + 4xy - 32)$
e) none of the above

Answer: (e)
MC Moderate

7.4.9: Suppose the partial derivatives of a Lagrange function $F(x, y, \lambda)$ are:

$$\frac{\delta F}{\delta x} = 2 - 8\lambda x, \qquad \frac{\delta F}{\delta y} = 1 - 2\lambda y, \qquad \frac{\delta F}{\delta \lambda} = 32 - 4x^2 - y^2.$$

What values of x and y minimize $F(x, y, \lambda)$? (Assume x and y are positive.)

a) $(2, 2\sqrt{2})$
b) (4, 2)
c) (2, 4)
d) $(2\sqrt{2}, \sqrt{4})$
e) none of the above

Answer: (c)
MC Moderate

7.4.10: Determine the minimum of $f(x, y) = x^2 + 2y^2$ where $x - 2y + 3 = 0$.

Answer: 3
ES Moderate

7.4.11: Determine the maximum value of $f(x, y) = 4 - x^2 - y^2$ where $y = 3x - 4$.

Answer: $\frac{12}{5}$
ES Moderate

7.4.12: Determine the minimum value of $f(x, y) = x^2 - xy + 2y^2 + 4$ where $x - y - 1 = 0$.

Answer: $\frac{39}{8}$
ES Moderate

7.4.13: Which of the following pairs of values (x, y) maximize(s) the function $g(x, y) = x + 3y$ subject to the constraint $x^2 + 9y^2 = 72$?

(I) (-6, -2)
(II) $\left(8, \sqrt{8/9}\right)$
(III) (2, -6)
(IV) (6, 2)
(V) $\left(-6, -2, \frac{1}{12}\right)$

Answer: (IV) only
ES Moderate

7.4.14: What is the minimum value of the function $f(x, y, z) = 3x^2 + 2y^2 + 4z^2$ subject to the constraint $3x + 4y - 4z = -45$?

a) 0 b) 135 c) 15 d) 225 e) -52

Answer: (b)
MC Moderate

7.4.15: Use the method of Lagrange multipliers to maximize $f(x, y) = 36 - 9x^2 - 4y^2$ subject to the constraint $2y + 3x - 3 = 0$.

Answer: $x = \frac{1}{2}$, $y = \frac{3}{4}$, min. value $= \frac{63}{2} = 31.5$
ES Moderate

7.4.16: Suppose the partial derivatives of a Lagrange function $F(X, y, \lambda)$ are $\frac{\delta F}{\delta x} = -18x + \lambda(2x)$, and $\frac{\delta F}{\delta y} = -8y + \lambda$ and $\frac{\delta F}{\delta \lambda} = y + x^2 - 6$. What positive values of x and y minimize $F(x, y, \lambda)$?

a) $\left(2, \frac{3}{2}\right)$
b) $\left(\sqrt{6}, 0\right)$
c) $\left(\frac{-81}{128} + \sqrt{1 + \frac{512}{27}}, \frac{-9}{6} + \sqrt{1 + \frac{512}{27}}\right)$
d) (1, 5)
e) none of the above

Answer: (a)
MC Moderate

7.4.17: The production function for a firm is $h(l, c) = l^2c$, where l and c are units of labor and capital utilized. Suppose that labor costs \$30 per unit and capital costs \$5 per unit, and that the firm decides to produce 3000 units of goods. Which of the following is the correct Lagrange function for minimizing cost subject to the stated constraint?

a) $F(l, c, \lambda) = l^2c + \lambda(3000 - 30l - 5c)$
b) $F(l, c, \lambda) = 3000 - l^2c + \lambda(30l + 5c)$
c) $F(l, c, \lambda) = 3000 - 30l - 5c + \lambda(l^2c)$
d) $F(l, c, \lambda) = 30l + 5c + \lambda(l^2c - 3000)$
e) none of the above

Answer: (d)
MC Moderate

7.4.18: An artist produces two items for sale. Each unit of item I costs \$50 to produce, while each unit of item II costs \$200. The revenue function is $R(x, y) = 40x + 7xy + 80y^2 + 10y$, where x is units of item I and y is units of item II. Suppose the artist has only \$1000 to spend on production. Which of the following is the Lagrange function the artist should use to determine what combination of production amounts (x, y) will yield maximum <u>profits</u> subject to the constraint that his costs must equal \$1000?

a) $40x + 7xy + 80y^2 + 10y + \lambda(50x + 200y - 1000)$
b) $-10x + 7xy + 80y^2 - 190y + \lambda(200x + 50y - 1000)$
c) $-10x + 7xy + 80y^2 - 190y + \lambda(1000 - 50x + 200y)$
d) $40x + 7xy + 80y^2 + 10y + \lambda(200x + 50y - 1000)$
e) none of the above

Answer: (c)
MC Moderate

7.4.19: Find the pair(s) (x, y) which give(s) the extreme values of $2x + 10y$, subject to the constraint $4x^2 + 5y^2 = 8400$, using the method of Lagrange multipliers.

Answer: (10, 40) and (-10, -40)
ES Moderate

7.5.1: Let $f(x, y) = \sqrt{x + y}$. Use the total differential to estimate $f(2.1, 1.95)$.

Answer: $f(2.1, 1.95) \approx 2\frac{1}{80}$
ES Moderate

7.5.2: Let $f(x, y) = x^5 + 5xy^2$. Using the fact that $f(1 + .01, 2 + .1) \approx$

$$f(1, 2) + \left[\frac{\delta f}{\delta x}(1, 2)\right] \cdot (.01) + \left[\frac{\delta f}{\delta y}(1, 2)\right] \cdot (.1),$$

estimate $f(1.01, 2.1) - f(1, 2)$.

a) 2.25
b) 2.1
c) 4.10
d) 4.21
e) none of the above

Answer: (a)
MC Moderate

7.5.3: Let $f(x, y) = 4x^2 + 2xy$. Approximate $f(1.1, 1.2)$.

Answer: 7.4
ES Moderate

7.5.4: Let $f(x, y) = e^{x} \cdot \ln y$. Approximate $f(.1, e + 1)$.

Answer: $\frac{1}{e} + \frac{11}{10}$

ES Moderate

7.6.1: Find the least squares line to fit the data (0, 3), (2, 5), (4, 5).
Use the method of partial derivatives.

Answer: $y = \frac{1}{2}t + \frac{10}{3}$

ES Moderate

7.6.2: Find the equation of the least squares line to fit the data (0, 6), (1, 3), (2, 3).

Answer: $y = -\frac{3}{2}t + \frac{11}{2}$

ES Moderate

7.6.3: Which of the following is the least squares error for the points (1, -3), (-2, 5), and (0, 10)?

a) $E^2 = (A + B + 3)^2 + (-2A + B - 5)^2 + (B - 10)^2$

b) $E^2 = (A + B - 3)^2 + (-2A + B - 5)^2 + (B - 10)^2$

c) $E^2 = (A - 3B)^2 + (-2A + 5B)^2 + (10B)^2$

d) $E^2 = (A - 3B + 3)^2 + (-2A + 5B - 5)^2 + (10B - 10)^2$

Answer: (a)

MC Moderate

7.6.4: Let $E = (2A + B - 3)^2 + (A + B + 2)^2 + (4A + B - 1)^2$. What is $\frac{\delta E}{\delta A}$?

a) $14A + 6B - 4$

b) $42A + 14B - 9$

c) $24A + 6B - 4$

d) $42A + 24B - 9$

e) none of the above

Answer: (e)

MC Moderate

7.6.5: Determine the equation of the line that best fits the points (0, 0), (1, 2), (2, 3).

Answer: $y = \frac{3}{2}x + \frac{1}{6}$

ES Moderate

7.6.6: The table below gives the height and weight of four randomly selected University undergraduate women. What is the least squares error for these data points?

height(cm)	150	155	165	170
weight(kg)	60	55	70	62

a) $E = 150(A + B - 60)^2 + 155(A + B - 55)^2 + 165(A + B - 70)^2 + 170(A + B - 62)^2$

b) $E = (150A + 60B)^2 + (155A + 55B)^2 + (165A + 70B)^2 + (170A + 62B)^2$

c) $E = (150A + B - 60)^2 + (155A + B - 55)^2 + (165A + B - 70)^2 + (170A + B - 62)^2$

d) $E = (150A - 60B)^2 + (155A - 55B)^2 + (165A - 70B)^2 + (170A - 62B)^2$

e) none of the above

Answer: (c)
MC Moderate

7.6.7: Which straight line best fits the data points (0, 1), (1, 3), (2, 7)?

a) $9x - 3y = -2$

b) $3x + y = \frac{2}{3}$

c) $y = \frac{2}{3}x + 3$

d) $y = 3Ax + B + \frac{2}{3}$

e) none of the above

Answer: (a)
MC Moderate

7.7.1: Calculate $\iint_R e^{x-y}\, dx\, dy$ where R is the rectangle

$0 \le x \le 1,\ 0 \le y \le 2.$

Answer: $(e - 1)(1 - e^{-2})$
ES Moderate

7.7.2: Calculate: $\int_1^2 \int_0^1 \frac{1}{xy + y}\, dx\, dy$

Answer: $(\ln 2)^2$
ES Moderate

7.7.3: Calculate: $\int_0^1 \int_0^1 (x^3 + y^2 + xy)\, dy\, dx$

Answer: $\frac{5}{6}$
ES Moderate

7.7.4: Calculate: $\int_1^2 \int_2^3 xy\, dy\, dx$

Answer: $\frac{15}{4}$
ES Moderate

7.7.5: What is $\iint_R (3x^2 + y)\,dxdy$, where R is the region bounded by $y = 2x + 3$, $y = x^2$ and the vertical lines $x = -1$ and $x = 1$?

a) $\int_{-1}^{1} \int_{2x+3}^{x^2} (3x^2 + y)\,dxdy = -\frac{4}{3}$

b) $\int_{-1}^{1} \int_{x^2}^{2x+3} (3x^2 + y)\,dxdy = \frac{4}{3}$

c) $\int_{-1}^{1} \int_{x^2}^{2x+3} (3x^2 + y)\,dydx = \frac{38}{5}$

d) $\int_{-1}^{1} \int_{2x+3}^{x^2} (3x^2 + y)\,dydx = \frac{-38}{5}$

Answer: (c)
MC Moderate

7.7.6: What is $\int_0^{\ln 2} \int_1^2 xy + ye^{xy}\,dxdy$?

a) $\frac{3}{2}\ln 2 - \frac{1}{2}e^2 - \frac{5}{2}$

b) $\frac{3}{4}(\ln 2)^2 - 1$

c) $\frac{3}{4}(\ln 2) + \frac{1}{2}$

d) $\frac{3}{2}\ln 2 - e^{2\ln 2} - \frac{3}{2}$

e) none of the above

Answer: (b)
MC Moderate

8. The Trigonometric Functions

8.1.1: Convert $\frac{9\pi}{2}$ radians to degrees.

Answer: 810°
ES Moderate

8.1.2: Convert the following angles to radians:

(a) 10°

(b) 6°

(c) -55°

(d) 760°

Answer: (a) $\frac{\pi}{18}$

(b) $\frac{\pi}{30}$

(c) $-\frac{11\pi}{36}$

(d) $\frac{38\pi}{9}$

ES Moderate

8.1.3: Construct an angle of $-\frac{3\pi}{4}$ radians.

Answer:

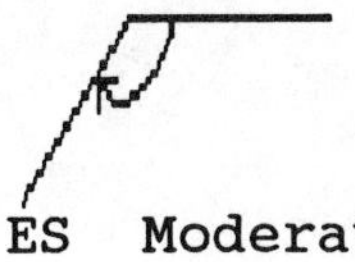

ES Moderate

8.1.4: Which of the following statements is false:

a) The terminal side of the angle $\theta = -3\pi/4$ lies in the fourth quadrant

b) The degree measure of $\theta = -3\pi/4$ is -135°.

c) The terminal side of the angle $\theta = 13\pi/6$ lies in the first quadrant

d) The degree measure of $\theta = 13\pi/6$ is 390°.

Answer: (a)
MC Moderate

8.1.5: Convert the following angles to radians
(i) $\theta = 15°$.
(ii) $\theta = -210°$.
(iii) $\theta = 780°$.

Answer: ($\pi/12$, $-7/6\ \pi$, $13/3\ \pi$)
ES Moderate

8.1.6: Convert the following angles to degrees
(i) $2/3\ \pi$
(ii) $-3/15\ \pi$
(iii) $\frac{29}{6}\pi$

Answer: (120°, -36°, 870°)
ES Moderate

8.1.7: Which of the following angles could represent the angle $-17\pi/12$?

a)

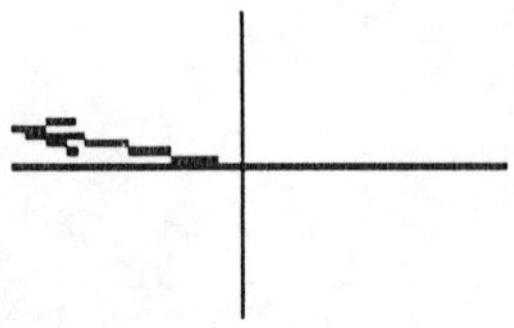

b)

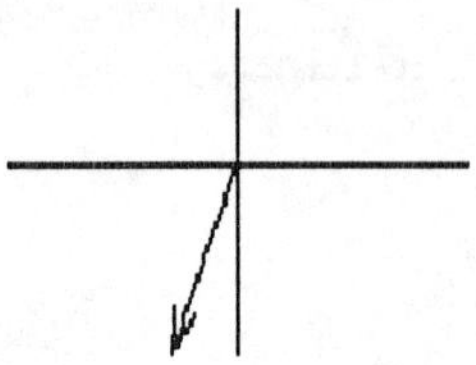

c)

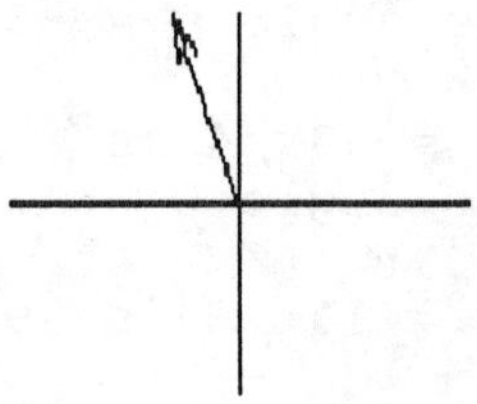

d) none of the above

Answer: (c)
MC Moderate

8.2.1: Find t such that $0 \le t \le \pi$ and $\cos t = \cos\left(-\frac{2\pi}{3}\right)$.

Answer: $\frac{2\pi}{3}$
ES Moderate

8.2.2: Determine: $\sin \frac{17\pi}{2}$, $\cos \frac{17\pi}{2}$, $\tan \frac{17\pi}{2}$

Answer: $\sin \frac{17\pi}{2} = 1$; $\cos \frac{17\pi}{2} = 0$; $\tan \frac{17\pi}{2}$ is undefined
ES Moderate

8.2.3: Find the value of t such that $0 \le t \le \pi$ and $\cos t = \cos(-\pi/3)$.
a) $\frac{\pi}{6}$
b) $-\frac{\pi}{3}$
c) $\frac{2\pi}{3}$
d) $\frac{\pi}{3}$
e) none of the above

Answer: (d)
MC Moderate

8.2.4: At what values of t do the maximum values of $y = \sin t$ occur?
a) $\ldots, -\frac{5\pi}{2}, -\frac{\pi}{2}, \frac{\pi}{2}, \frac{5\pi}{2}, \frac{9\pi}{2}, \ldots$
b) $\ldots, -\frac{5\pi}{2}, -\frac{3\pi}{2}, -\frac{\pi}{2}, \frac{\pi}{2}, \frac{3\pi}{2}, \frac{5\pi}{2}, \ldots$
c) $\ldots, -4\pi, -2\pi, 0, 2\pi, 4\pi, \ldots$
d) $\ldots, -\frac{7\pi}{2}, -\frac{3\pi}{2}, \frac{\pi}{2}, \frac{5\pi}{2}, \frac{9\pi}{2}, \ldots$
e) none of the above

Answer: (d)
MC Moderate

8.2.5: Find sin t, where t is the radian measure of the angle shown below.

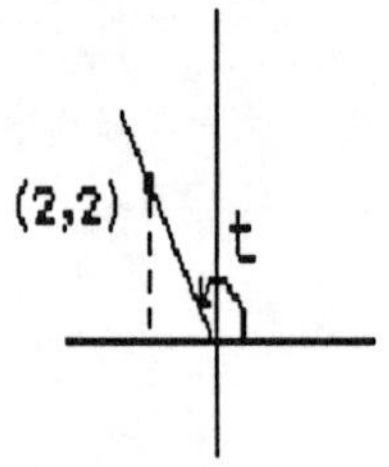

a) $\frac{2}{\sqrt{8}}$

b) $\frac{2}{-2}$

c) 2

d) -2

e) none of the above

Answer: (a)
MC Moderate

8.2.6: Suppose a surveyor wants to measure the distance across a river. He finds the distance along the bank (from points C to B) to be 100 ft, and, using a surveyor's transit, finds the angle ABC to be 35°. What is the approximate distance from points A to C ?
[Note: sin 35° = .574, cos 35° = .819, tan 35° = .700]

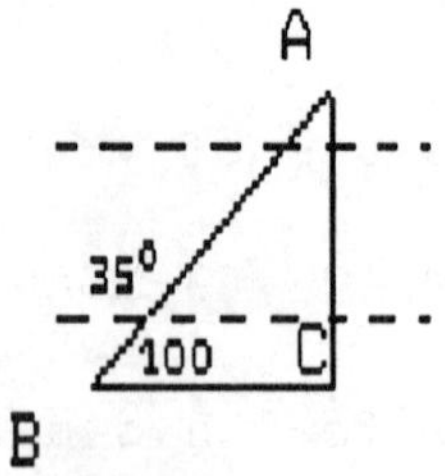

a) 70 ft

b) 81.9 ft

c) 57.4 ft

d) 143 ft

e) none of the above

Answer: (a)
MC Moderate

8.2.7: Suppose $\sin t = \frac{-5}{13}$ and cos t is negative. What is cos t ?

a) $-\frac{12}{13}$

b) $-\frac{8}{13}$

c) $-\frac{18}{13}$

d) $\frac{11}{13}$

e) none of the above

Answer: (a)
MC Moderate

8.2.8: Suppose cost = $\sqrt{3}/2$ and sint is negative. What is sint?

Answer: -1/2
ES Moderate

8.2.9: Suppose the terminal side of an angle θ is in the second quadrant. Which of the following statements is false?

a) $\tan \theta$ will be negative
b) $\sin \theta$ will be negative
c) $\cos \theta$ will be negative
d) $\cos(\theta + \pi/2)$ will be negative
e) $\sin(\theta + \pi/2)$ will be negative

Answer: (b)
MC Moderate

8.2.10: Find t so that $0 \le t \le \pi$ and $\cos t = \cos(-7\pi/6)$.

Answer: $t = 5\pi/6$
ES Moderate

8.2.11: Find t so that $-\pi/2 \le t \le \pi/2$ and $\sin t = \sin 10\pi/3$.

Answer: $t = -\pi/3$
ES Moderate

8.2.12: A child's slide has a 4 ft ladder and an 8 ft slide. What is the angle that the slide makes with the ground?

Answer: $\pi/6$
ES Moderate

8.2.13: Suppose $\sin \theta = -1/2$. Which of the following <u>must</u> be true?

a) $\cos \theta = \sqrt{3}/2$
b) The terminal angle of θ is in the fourth quadrant
c) $\sin \theta = \sin 11\pi/6$
d) $\theta = -\pi/6$ or $\theta = 7\pi/6$
e) All of the above must be true

Answer: (c)
MC Moderate

8.2.14: Suppose $\cos t = \sqrt{5}/3$ and $\sin t$ is negative. Find $\sin t$, $\tan t$, $\cos(-t)$, $\sin(-t)$.

Answer: $(-2/3, -2\sqrt{5}/5, \sqrt{5}/3, 2/3)$
ES Moderate

8.2.15: In the following two graphs indicate which graph could represent $y = \cos t$ and which represents $y = \sin t$. Also give the coordinates of the indicated points.

(a)

(b)

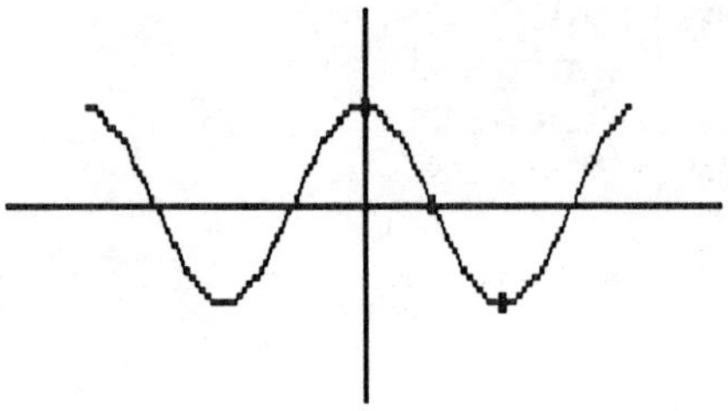

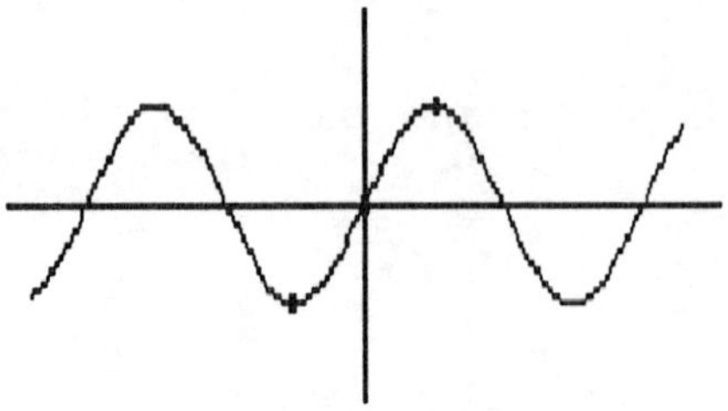

Answer: (a) $\cos t = y$

$(0, 1), (\pi/2, 0), (\pi, -1)$

(b) $y = \sin t$

$(-\pi/2, -1), (0, 0), (\pi/2, 1)$

ES Moderate

8.3.1: Find the equation of the line tangent to the graph of $y = 5t + \cos 2t$ at $t = \pi/4$.

Answer: $y = 3t + \frac{\pi}{2}$

ES Moderate

8.3.2: Find: $\int \sin(3t + 2)\, dt$

Answer: $-\frac{1}{3}\cos(3t + 2) + C$

ES Moderate

8.3.3: Find the area under the curve $y = \sin x + \cos x$ from $x = 0$ to $x = \frac{\pi}{4}$.

Answer: 1

ES Moderate

8.3.4: Differentiate: $\sin x^2$

Answer: $2x \cos x^2$
ES Moderate

8.3.5: Differentiate: $3 \cos x \sin 2x$

Answer: $6 \cos x \cos 2x - 3 \sin x \sin 2x$
ES Moderate

8.3.6: Differentiate: $\cos^2(x + 4)$

Answer: $-2 \cos(x + 4)\sin(x + 4)$
ES Moderate

8.3.7: Differentiate: $\sin^3 x^3$

Answer: $9x^2 \sin^2 x^3 \cos x^3$
ES Moderate

8.3.8: Evaluate: $\int_0^{\pi/2} \sin x \, dx$

Answer: 1
ES Moderate

8.3.9: Find all functions $y = f(x)$ for which $y' = 2 \cos x$ and $f(0) = -1$.

Answer: $2 \sin x - 1$
ES Moderate

8.3.10: The U.S. Geological Survey estimates that the normal stream flow of the Potomac River (in millions of gallons per day) is given by the function

$$f(t) = 4200 + 3000 \cos\left(\frac{\pi}{6}t + \frac{5\pi}{12}\right)$$

where t denotes the number of months after January 1. Determine the maximum and minimum normal daily stream flows of the Potomac River.

Answer: maximum 7200; minimum 1200
ES Moderate

8.3.11: The U.S. Geological Survey estimates that the normal stream flow of the Potomac River (in millions of gallons per day) is given by the function

$$f(t) = 4200 + 3000 \cos\left(\frac{\pi}{6}t + \frac{5\pi}{12}\right)$$

where t denotes the number of months after January 1. Determine when the maximum and minimum normal daily stream flows of the Potomac River occur.

Answer: maximum at $t = 9.5$; minimum at $t = 3.5$
ES Moderate

8.3.12: The size of an animal population at time t is given by

$$N(t) = 50{,}000 + 200 \sin \frac{2\pi t}{24} + 300 \cos \frac{2\pi t}{48}$$

where t is the number of months from June 1, 1980. At what rate is the animal population changing on June 1, 1981?

Answer: decrease of $\frac{175\pi}{6}$ (approximately 92) animals per month
ES Moderate

8.3.13: A person's blood pressure P at time t is given by $P = 90 + 30 \sin \frac{7\pi t}{3}$, where t is measured in seconds. What is the average blood pressure over a time interval of 60 seconds?

Answer: 90
ES Moderate

8.3.14: Let $y = (\sin 3t)^2$. What is $\frac{dy}{dt}$?

a) $2 \sin 3t$
b) $6 \sin 3t$
c) $2 \sin 3t \cos 3t$
d) $6 \sin 3t \cos 3t$
e) none of the above

Answer: (d)
MC Moderate

8.3.15: What is the equation of the line tangent to the graph of $y = \sin 2x$ at $x = \pi/2$?

a) $y = -(x - \frac{\pi}{2})$
b) $y - 1 = x - \frac{\pi}{2}$
c) $y = (\cos 2x)(x - \frac{\pi}{2})$
d) $y = -2x - \pi$
e) none of the above

Answer: (e)
MC Moderate

8.3.16: Let $f(x) = \sin^2 x$. What is $f''(x)$?

a) 1
b) 2
c) $-2 \sin^2 x + 2 \cos^2 x$
d) $(2 \cos x)(- \sin x)$
e) none of the above

Answer: (c)
MC Moderate

8.3.17: Let $f(x) = \ln(\sin t)$. What is $f'\left(\frac{\pi}{2}\right)$?

a) 1
b) 0
c) -1
d) ln 0
e) none of the above

Answer: (b)
MC Moderate

8.3.18: What is the area under the curve $y = \sin 2x$ from $x = 0$ to $x = \frac{\pi}{4}$?

a) -1
b) $\frac{1}{2}$
c) 0
d) $\frac{4\pi}{3}$
e) none of the above

Answer: (b)
MC Moderate

8.3.19: $\int 3 \sin(t - \pi)\, dt =$

a) $-3 \cos t + C$
b) $3\pi \cos(t - \pi) + C$
c) $3 \cos(t - \pi) + C$
d) $-3 \cos(t - \pi) + C$
e) none of the above

Answer: (d)
MC Moderate

8.3.20: Differentiate: $f(x) = \sin 3x$

Answer: $f' = 3 \cos 3x$
ES Moderate

8.3.21: Differentiate: $f(x) = \sin 3x^2$

Answer: $f' = 6x \cos 3x^2$
ES Moderate

8.3.22: Differentiate: $f(t) = 2\cos^2 4t$

Answer: $f' = -16\cos 4t \cdot \sin 4t$
ES Moderate

8.3.23: Differentiate: $f(t) = \ln(\sin 3t)$

Answer: $f' = 3\cot 3t$
ES Moderate

8.3.24: Differentiate: $f(t) = e^{\sin(\cos t)}$

Answer: $f' = -e^{\sin(\cos t)} \cdot \cos(\cos t) \cdot \sin(t)$
ES Moderate

8.3.25: Differentiate: $f(x) = 2\tan x^2$

Answer: $f' = 4x\sec^2 x^2$
ES Moderate

8.3.26: Determine the slope of $f(t) = \sin 2t$ at $t = \frac{\pi}{3}$.

Answer: -1
ES Moderate

8.3.27: Differentiate: $f(x) = 6x\cos^2(\sin x)$

Answer: $f' = -12x\cos(\sin x) \cdot \sin(\sin x) \cdot \cos x + 6\cos^2(\sin x)$
ES Moderate

8.3.28: Differentiate: $f(x) = \sin^3 3t$

Answer: $f' = 9\sin^2 3t \cdot \cos 3t$
ES Moderate

8.3.29: Differentiate cos(cosx).

Answer: sinx(sin(cosx))
ES Moderate

8.3.30: Differentiate $\frac{\cos x}{\sin x - 2}$

Answer: $\frac{2\sin x - 1}{(\sin x - 2)^2}$
ES Moderate

8.3.31: Differentiate $e^{x^3} \sin x^3$

Answer: $3x^2 e^{x^3} (\sin x^3 + \cos x^3)$
ES Moderate

8.3.32: Differentiate $\ln(1 + \sin x)$

Answer: $\frac{\cos x\ x}{(1 + \sin x)}$
ES Moderate

8.3.33: Find the equation of the tangent line to $y = \cos 3x + 2\sin x$ at $x = \pi/2$

Answer: $y - 2 = 3(x - \pi/2)$
ES Moderate

8.3.34: Find the slope of the tangent line to the curve $y = e^{3x}\cos x^3$ at the origin

Answer: $m = 3$
ES Moderate

8.3.35: Show that the function $y = \frac{e^{3x}}{2 + \sin x}$ has no relative extreme points.

Answer:
ES Moderate

8.3.36: Find the area under the curve $y = \sin(3t - \pi)$ for $\pi/3 \leq t \leq 2\pi/3$.

Answer: 2/3
ES Moderate

8.3.37: Find the area under the curve $y = \sin 2x + \cos x$ for $0 \leq x \leq \pi/4$.

Answer: $\frac{\sqrt{2} + 1}{2}$
ES Moderate

8.3.38: Find $\int 3\cos(2x)dx$

Answer: 3/2 sin(2x) + C
ES Moderate

8.3.39: Without integrating, show that $\int \frac{\cos x}{1 + \sin x} dx = \ln(1 + \sin x)$ + C. (Hint: Think of what the indefinite integral stands for.)

Answer:
ES Moderate

8.3.40: Suppose $y = 2e^{x^2} - \cos 3x$. Which of the following could be a graph of y near x = 0?

a)

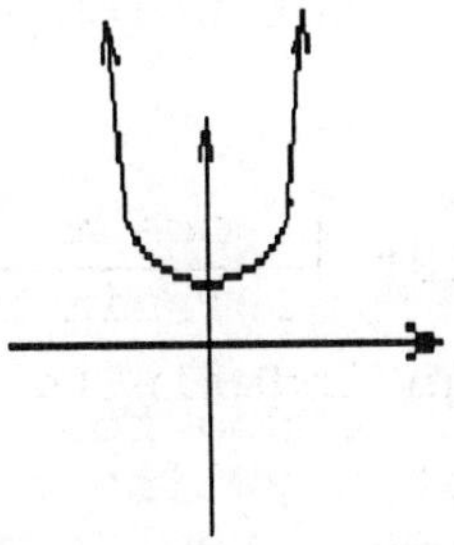

b)

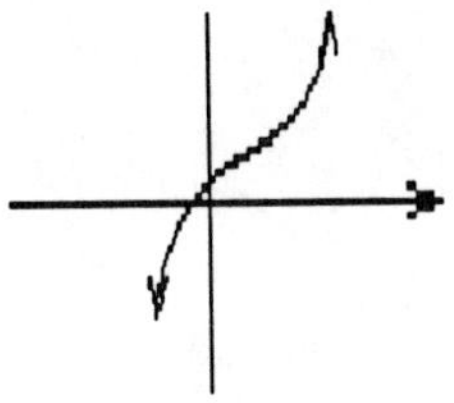

c)

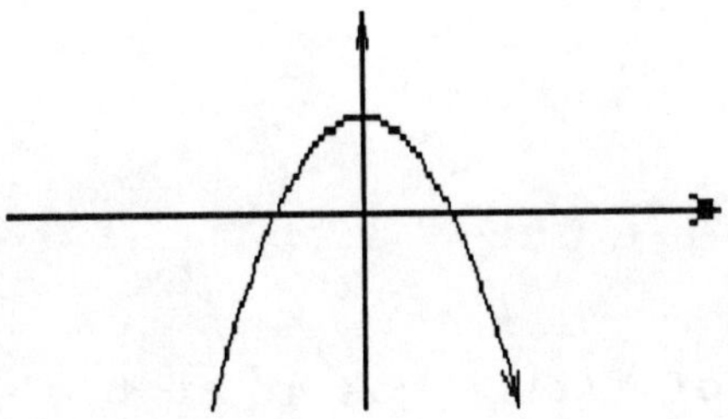

d)

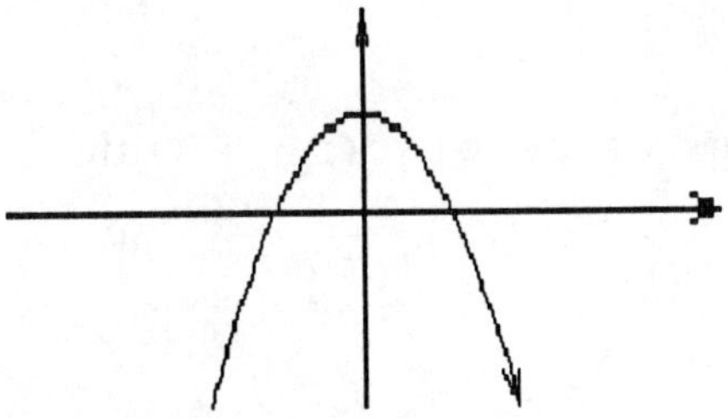

Answer: (a)
MC Moderate

8.3.41: Suppose a fire truck pumps water from its tank and the water pressure in the hose varies with time. If the pressure at maximum time t is given by $P(t) = 45 + 30\cos 3t$. Find the maximum value of P and give two values of t for which this maximum value occurs.

Answer: $\pi/18$, $5\pi/18$
ES Moderate

8.3.42: Suppose in a study of a prairie dog town it is discovered that the number of prairie dogs at any time t is given by $P(t) = 1000 + 500\cos(2\pi t/12)$ where t is measured in months from July 1, 1990. What is the average number of prairie dogs living in the town from July 1, 1990 to July 1, 1992?

Answer: $\cong 1040$
ES Moderate

8.4.1: Differentiate: $\ln(\sec t)$

Answer: $\tan t$
ES Moderate

8.4.2: Determine $\frac{dy}{dx}$ where $\csc y + \sec x = 2$.

Answer: $\frac{dy}{dx} = \frac{\sec x \cdot \tan x}{\csc y \cdot \cot y}$
ES Moderate

8.4.3: Determine the slope of $f(t) = \cos^2 4t$ at $t = \frac{\pi}{3}$.

Answer: $-2\sqrt{3}$
ES Moderate

8.4.4: Determine the minimum value of $f(x) = \tan x + \cot x$ where $0 \le x \le \frac{\pi}{2}$.

Answer: 2
ES Moderate

8.4.5: Determine all relative extreme points of $f(x) = \sin^2 x - \cos 4x$ and classify them as either maxima or minima, where $0 \le x \le 2\pi$.

Answer: $\left(0, -\frac{1}{4}\right)$, $\left(\pi, -\frac{1}{4}\right)$ are minima; $\left(\frac{\pi}{3}, \frac{7}{8}\right)$, $\left(\frac{2\pi}{3}, \frac{7}{8}\right)$ are maxima.
ES Moderate

8.4.6: Find: $\int \sin 4t \, dt$

Answer: $-\frac{1}{4}\cos 4t + C$
ES Moderate

8.4.7: Find: $\int 4 \sin 2t \, dt$

Answer: $-2 \cos 2t + C$
ES Moderate

8.4.8: Find: $\int \cos 2t \, dt$

Answer: $\frac{1}{2}\sin 2t + C$
ES Moderate

8.4.9: Differentiate: $f(x) = 2 \sec x$

Answer: $f' = 2 \sec x \tan x$
ES Moderate

8.4.10: Differentiate: $f(x) = 2 \cot^3 x$

Answer: $f' = -6 \cot^2 x \cdot \csc^2 x$
ES Moderate

8.4.11: Differentiate $f(x) = \tan(3x)\sin x$

Answer: $3\sec^2 3x \sin x + \cos x \tan 3x$
ES Moderate

8.4.12: Find the slope of the tangent line to the curve $y = e^{\tan x}$ at $x = 0$.

Answer: $m = 1$
ES Moderate

8.4.13: $\int e^x \sec^2(e^x)\,dx =$

a) $e^x \tan(e^x) + C$
b) $\tan(e^x) + C$
c) $\frac{\tan(e^x)}{e^x} + C$
d) $\tan^2(e^x) + C$
e) none of the above

Answer: (b)
MC Moderate

8.4.14: If $y = \sqrt{\tan x + \sin x}$. Find dy/dx.

Answer: $\frac{1}{2} \frac{(\sec^2 x + \cos x)}{\sqrt{\tan x + \sin x}}$

ES Moderate

8.4.15: Find f'(x) for $f(x) = e^{x^3} \tan 2x$

Answer: $3x^2 e^{x^3} \tan 2x + 2e^{x^3} \sec^2 2x$

ES Moderate

8.4.16: Find all relative extreme points x of $f(x) = \frac{2x - \tan 2x}{2}$ with $0 \le x \le \pi$.

Answer: $x = 0, \pi/2$

ES Moderate

8.4.17: For $y = e^{3x}\tan x^2$ then dy/dx is:

a) $3e^{3x}\sec^2 x^2$

b) $6xe^{3x}\sec^2 x^2$

c) $3e^{3x}\tan x^2 + e^{3x}\sec^2 x^2$

d) $3e^{3x}\tan x^2 + 2xe^{3x}\sec^2 x^2$

e) none of the above

Answer: (d)

MC Moderate

8.4.18: A river is 600 m. wide. A boat enters the water and is pushed by the current at an angle of 30° to its original position. How far downstream from its original position will the boat land?

Answer: $200\sqrt{3}$ m

ES Moderate

8.4.19: $\int_0^{\pi/4} \sec^2 x dx$

Answer: 1

ES Moderate

8.4.20: Let $f(x) = e^x \tan x$. If $f'(x) = e^x(\tan x + \sec^2 x)$ find
$\int_0^{\pi/4} e^x(\tan x + \sec^2 x)\,dx$.

Answer: $e^{\pi/4}$
ES Moderate

8.4.21: Find the equation of the tangent line to $y = \ln(\tan x)$ at $x = \pi/4$

Answer: $y = 2(x - \pi/4)$
ES Moderate

8.4.22: Give the value of tan t, sec t where t is the radian measure of the angle shown.

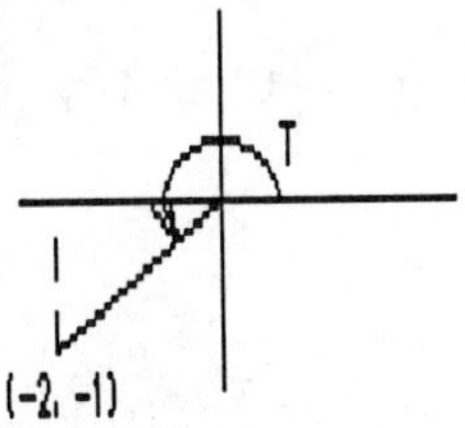

Answer: $\tan t = 1/2$

$\sec t = -\sqrt{5}/2$
ES Moderate

8.4.23: A ladder leaning against a building makes an angle of 60° with the ground. If the base of the ladder is 3 feet from the building, how far up the building will the ladder reach.

Answer:
$3\sqrt{3}$ ft.
ES Moderate

8.4.24: Differentiate $\ln(e^x \tan 2x)$

Answer: $1 + 2\dfrac{\sec^2 2x}{\tan 2x}$
ES Moderate

8.4.25: Find the area between the curves $y = \tan x$ and $y = \sin x$ for $0 \le x \le \pi/4$.

Answer:
$\sqrt{2}/2$
ES Moderate

8.4.26: Differentiate $\frac{\sin x^2}{\tan x}$

Answer: $\frac{2x\tan x\cos x^2 - \sec^2 x\sin x^2}{\tan^2 x}$

ES Moderate

8.4.27: Suppose $\pi/4 \le x \le 3\pi/4$. Which of the following statements can be made about tan x?

a) tan x is defined
b) tan x is positive
c) If tan x is defined then $|\tan x| \ge 1$
d) tan x is not defined
e) none of the above

Answer: (c)

MC Moderate

8.4.28: Find the area between the curves $y = 1$ and $y = \sec^2 x$ for $0 \le x \le \pi/4$

Answer: $-\pi/4$

ES Moderate

8.4.29: Find the average value of $\sec^2 x$ for $0 \le x \le \pi/4$.

Answer: $4/\pi$

ES Moderate

8.4.30: If $y = \ln(\tan x) + \tan x$ find dy/dx.

Answer: $\sec^2 x(\cot x + 1)$

ES Moderate

9. Techniques of Integration

9.1.1: Determine: $\int xe^{4-x^2}\,dx$

Answer: $-\frac{1}{2}e^{4-x^2} + C$

ES Moderate

9.1.2: Determine: $\int_4^5 x\sqrt{x^2 - 16}\,dx$

Answer: 9

ES Moderate

9.1.3: Determine: $\int (x - 1)e^{3x^2-6x}\,dx$

Answer: $\frac{1}{6}e^{3x^2-6} + C$

ES Moderate

9.1.4: Determine: $\int 5x \sin x^2\,dx$

Answer: $-\frac{5}{2}\cos x^2 + C$

ES Moderate

9.1.5: Determine: $\int \frac{e^x + xe^x}{4 + xe^x}\,dx$

Answer: $\ln|4 + xe^x| + C$

ES Moderate

9.1.6: Determine: $\int_0^2 x(x^2 - 1)^3\,dx$

Answer: 10

ES Moderate

9.1.7: Determine: $\int_0^{\pi/2} \sin^4 x \cos x \, dx$

Answer: $\frac{1}{5}$
ES Moderate

9.1.8: Determine: $\int \frac{\sin \sqrt{x}}{\sqrt{x}} \, dx$

Answer: $-2 \cos \sqrt{x} + C$
ES Moderate

9.1.9: Determine: $\int \frac{1}{x \ln x \ln(\ln x)} \, dx$
[Hint: Let $u = \ln(\ln x)$.]

Answer: $\ln(\ln(\ln x)) + C$
ES Moderate

9.1.10: $\int \tan^2 x \sec^2 x \, dx =$

a) $\frac{\tan^3 x}{3} + C$
b) $\sec x \tan x + C$
c) $\frac{\sec^2 x}{2} + C$
d) $\frac{\tan^2 x}{2} + C$
e) none of the above

Answer: (a)
MC Moderate

9.1.11: $\int (x + 1)(\sin x) \, dx =$

a) $x \sin x - \cos x + C$
b) $x \cos x - \sin x + C$
c) $-(x + 1) \cos x + \sin x + C$
d) none of the above

Answer: (c)
MC Moderate

9.1.12: $\int \frac{\ln 5x}{x}\, dx =$

a) $\frac{1}{2}(\ln x)(\ln 5x)^2 + C$

b) $\frac{1 - \ln 5x}{x^2} + C$

c) $\frac{1}{2}(\ln 5x)^2 + C$

d) $\frac{5}{2}(\ln 5x)^2 + C$

e) none of the above

Answer: (c)
MC Moderate

9.1.13: Find: $\int 2x(x^2 + 1)^3\, dx$

Answer: $\frac{(x^2 + 1)^4}{4} + C$
ES Moderate

9.1.14: Find: $\int 4x^5(x^6 + 100)^5\, dx$

Answer: $\frac{1}{9}\left(x^6 + 100\right)^6 + C$
ES Moderate

9.1.15: Find: $\int \frac{4x\, dx}{\sqrt{x^2 + 2}}$

Answer: $4\sqrt{x^2 + 2} + C$
ES Moderate

9.1.16: Find: $\int (3x^2 + 2)\sqrt{x^3 + 2x}\, dx$

Answer: $\frac{2}{3}\left(x^3 + 2x\right)^{3/2} + C$
ES Moderate

9.1.17: Find: $\int \frac{(\ln x)^5}{x}\, dx$

Answer: $\frac{(\ln x)^6}{6} + C$
ES Moderate

9.1.18: Find: $\int \frac{\sec^2(\ln x)\,dx}{x}$

Answer: $\tan(\ln x) + C$
ES Moderate

9.1.19: Find: $\int (x^3 - 3x^2)(6x^2 - 12x)\,dx$

Answer: $(x^3 - 3x^2)^2 + C$
ES Moderate

9.1.20: Find: $\int \frac{e^x - e^{-x}}{e^x + e^{-x}}\,dx$

Answer: $\ln|e^x + e^{-x}| + C$
ES Moderate

9.1.21: Find: $\int \cot x\,dx$

Answer: $\ln|\sin x| + C$
ES Moderate

9.1.22: Find: $\int xe^{x^2}\,dx$

Answer: $\frac{e^{x^2}}{2} + C$
ES Moderate

9.1.23: Find: $\int 2\cos^2 x \sin x\,dx$

Answer: $\frac{2}{3}\cos^3 x + C$
ES Moderate

9.1.24: Find: $\int \frac{\cos 2x}{\sin^3 2x}\,dx$

Answer: $-\frac{1}{4\sin^2 2x} + C$
ES Moderate

9.1.25: Find: $\int \tan^4 2x \sec^2 2x \, dx$

Answer: $\frac{\tan^5 2x}{10} + C$

ES Moderate

9.1.26: Find: $\int \sin x \cdot \sin(\cos x) \, dx$

Answer: $\cos(\cos x) + C$

ES Moderate

9.1.27: Find: $\int \frac{\tan x \, dx}{\ln \cos x}$

Answer: $-\ln(\ln(\cos x)) + C$

ES Moderate

9.1.28: Find $\int x\sqrt{x^2 - 4}dx$

Answer: $\frac{1}{3}(x^2 - 4)^{3/2} + C$

ES Moderate

9.1.29: Find $\int \tan xdx$

Answer: $-\ln |\cos x| + C$

ES Moderate

9.1.30: Simplify $\int \tan x \ln(\cos x)dx$

Answer: $\frac{-(\ln(\cos x))^2}{2} + C$

ES Moderate

9.1.31: Find $\int \frac{\cos x - \sin x}{\cos x + \sin x}dx$

Answer: $\ln|\cos x + \sin x| + C$

ES Moderate

9.1.32: Simplify $\int \cos x \sin^2 x dx$

Answer: $\frac{\sin^3 x}{3} + C$

ES Moderate

9.1.33: Find $\int \frac{\ln x^2}{x} dx$

Answer: $\frac{1}{4}(\ln x^2)^2 + C$

ES Moderate

9.1.34: Find $\int xe^{3x^2} dx$

Answer: $\frac{1}{6}(e^{3x} - 1)$

ES Moderate

9.1.35: Given that for $y = 2^x$, $dy/dx = (\ln 2)2^x$, find $\int (2x)^2 dx$

Answer: $\frac{(2^x)^2}{2\ln 2} = \frac{2^{2x}}{2\ln 2}$

ES Moderate

9.1.36: Simplify $\int \cos x e^{\sin x} dx$

Answer: $e^{\sin x} + C$

ES Easy

9.1.37: Simplify $\int e^x \sqrt{e^x + 2}\, dx$

Answer: $\frac{2}{3}(e^x + 2)^{3/2} + C$

ES Easy

9.1.38: $\int \frac{x^2 + 2}{\sqrt{x^3 + 6x}} dx$

Answer: $\frac{2}{3}\sqrt{x^3 + 6x} + C$

ES Moderate

9.1.39: Simplify $\int x(x^2 + 1)^{17} dx$

Answer: $\frac{1}{36}(x^2 + 1)^{18} + C$

ES Moderate

9.1.40: $\int \frac{\ln \sqrt{x}}{x} dx =$

a) $\frac{1}{2}\left(\ln\sqrt{x}\right)^2 + C$

b) $2\left(\ln\sqrt{x}\right)^2 + C$

c) $\left(\ln\sqrt{x}\right)^2 + C$

d) $(\ln x)^2 + C$

e) none of the above

Answer: (c)
MC Moderate

9.1.41: $\int e^{2x}\cos(e^{2x})\,dx$

Answer: $\frac{1}{2}\text{sine}^{2x} + C$
ES Moderate

9.1.42: Find $\int xe^{(x^2 - 2x)} - e^{(x^2 - 2x)}dx$

Answer: $\frac{1}{2}e^{(x^2 - 2x)} + C$
ES Moderate

9.1.43: Simplify $\int \frac{\sin xe^{(x + \cos x)}}{e^x}dx$

Answer: $-e^{\cos x} + C$
ES Moderate

9.1.44: $\int \frac{f'(x)}{f(x)}dx =$

a) $e^{f(x)}f'(x) + C$

b) $\ln |f(x)| + C$

c) $\ln |f'(x)/f(x)| + C$

d) $f(x) + C$

e) can't be determined from the information given

Answer: (b)
MC Moderate

9.1.45: $\int f'(x)\sqrt{3f(x) + 1}dx =$

a) $2/3\ (3f(x) + 1)^{3/2} + C$
b) $2/9\ (3f(x) + 1)^{3/2} + C$
c) $f(x)(2/3(3f(x) + 1)^{3/2}) + C$
d) $\dfrac{1}{3\sqrt{3f(x) + 1}} + C$
e) can't be determined

Answer: (b)
MC Moderate

9.1.46: If $\frac{d}{dx}\csc x = -\cot x\csc x$ find $\int \cot x \csc x\, e^{\csc x} dx$

Answer: $-e^{\csc x} + C$
ES Moderate

9.1.47: Find $\int \cos^3 x dx$. (Hint: Remember $\cos^2 x = 1 - \sin^2 x$)

Answer: $\sin x - \dfrac{\sin^3 x}{3} + C$
ES Moderate

9.2.1: Find: $\int x \cos 5x\, dx$

Answer: $\frac{1}{5}x \sin 5x + \frac{1}{25} \cos 5x + C$
ES Moderate

9.2.2: Find: $\int_0^5 (5 - t)e^{-t}\, dt$

Answer: $4 + e^{-5}$
ES Moderate

9.2.3: Find: $\int x^3 \ln x\, dx$

Answer: $\frac{1}{4}x^4 \ln x - \frac{1}{16}x^4 + C$
ES Moderate

9.2.4: Find: $\int xe^{-3x}\,dx$

Answer: $-\frac{1}{3}xe^{-3x} - \frac{1}{9}e^{-3x} + C$
ES Moderate

9.2.5: Find: $\int x \sec^2 x\,dx$

Answer: $x \tan x + \ln|\cos x| + C$
ES Moderate

9.2.6: Determine: $\int \frac{\cos x}{(3 \sin x + 1)^2}\,dx$

Answer: $-\frac{1}{3}(3 \sin x + 1)^{-1} + C$
ES Moderate

9.2.7: Decide whether integration by parts or substitution should be used to compute the indefinite integral. If substitution, indicate the value of u; if by parts, indicate f(x) and g(x) for the formula

$$\int f(x)g(x)\,dx = f(x)G(x) - \int G(x)f'(x)\,dx.$$

$\int \ln 4x\,dx$

a) $u = 4x$
b) $f(x) = \ln$, $g(x) = 4x$
c) $f(x) = \ln 4x$, $g(x) = 1$
d) $f(x) = 1$, $g(x) = \ln 4x$
e) none of the above

Answer: (c)
MC Moderate

9.2.8: Decide whether integration by parts or substitution should be used to compute the indefinite integral. If substitution, indicate the value of u; if by parts, indicate f(x) and g(x) for the formula

$$\int f(x)g(x)\,dx = f(x)G(x) - \int G(x)f'(x)\,dx.$$

$$\int \frac{\cos(\ln x)}{x}\,dx$$

a) $f(x) = \cos(\ln x),\ g(x) = \frac{1}{x}$

b) $u = \ln x$

c) $u = \frac{\ln x}{x}$

d) $f(x) = 1,\ g(x) = \frac{\cos(\ln x)}{x}$

e) none of the above

Answer: (b)
MC Moderate

9.2.9: Decide whether integration by parts or substitution should be used to compute the indefinite integral. If substitution, indicate the value of u; if by parts, indicate f(x) and g(x) for the formula

$$\int f(x)g(x)\,dx = f(x)G(x) - \int G(x)f'(x)\,dx.$$

$$\int \frac{-e^{1/x}}{x^2}\,dx$$

a) $f(x) = -\frac{1}{x^2},\ g(x) = e^{1/x}$

b) $f(x) = -e^{1/x},\ g(x) = \frac{1}{x^2}$

c) $u = \frac{1}{x}$

d) $u = x^{-2}$

e) none of the above

Answer: (c)
MC Moderate

9.2.10: Determine: $\int x^2 e^x\,dx$

Answer: $x^2 e^x - 2(xe^x - e^x) + C$
ES Moderate

9.2.11: Determine: $\int \ln x\,dx$

Answer: $x \ln x - x + C$
ES Moderate

9.2.12: Determine: $\int e^{x}\cos x\,dx$

Answer: $\dfrac{e^{x}\sin x + e^{x}\cos x}{2} + C$

ES Moderate

9.2.13: Determine: $\int x \ln x\,dx$

Answer: $(\ln x) \cdot \left(\dfrac{x^2}{2}\right) - \dfrac{x^2}{4} + C$

ES Moderate

9.2.14: Determine: $\int 3x \cos x\,dx$

Answer: $3x \sin x + 3 \cos x + C$

ES Moderate

9.2.15: Determine: $\int \sin(\ln x)\,dx$

Answer: $\dfrac{x \sin(\ln x) - x \cos(\ln x)}{2}$

ES Moderate

9.2.16: Simplify $\int \dfrac{xe^{-x}}{(x - 1)^2}dx$

Answer: $\dfrac{e^{-x}}{1 - x} + C$

ES Moderate

9.2.17: $\int xe^{3x}dx =$

a) $\dfrac{3xe^{3x} - e^{3x}}{9} + C$

b) $\dfrac{e^{3x}}{3} + C$

c) $\dfrac{xe^{3x} - e^{3x}}{3} + C$

d) $\dfrac{x^2e^{3x}}{6} + C$

e) none of the above

Answer: (a)

MC Moderate

9.2.18: If $\int x\sin x\,dx = \sin x - x\cos x + C$ find $\int x^2\cos x\,dx$

Answer: $x^2\sin x - 2\sin x + 2x\cos x + C$

ES Moderate

9.2.19: $\int x\sqrt{x + 3}dx$

Answer: $\frac{2}{3}x(x + 3)^{3/2} - \frac{4}{15}(x + 3)^{5/2} + C$
ES Moderate

9.2.20: Simplify $\int (x^2 + x)e^x dx$

Answer: $e^x(x^2 - x + 1) + C$
ES Moderate

9.2.21: Simplify $\int x(x + 2)^{2/3}dx$

Answer: $\frac{3}{5}x(x + 2)^{5/3} - \frac{15}{40}(x + 2)^{8/3} + C$
ES Moderate

9.2.22: Simplify $\int x\ln x dx$

Answer: $\frac{x^2}{2}\ln x - \frac{x^2}{4} + C$
ES Moderate

9.2.23: Which of the following accurately describes integration by parts?

a) $\int f(x)g(x)dx = f'(x)g'(x) - \int f'(x)g'(x)dx$
b) $\int f(x)g'(x)dx = f'(x)g(x) - \int f'(x)g(x)dx$
c) $\int f(x)g'(x)dx = f(x)g(x) - \int f'(x)g(x)dx$
d) $\int f(x)g(x)dx = f'(x)g(x) - \int f(x)g'(x)dx$
e) none of the above

Answer: (c)
MC Moderate

9.2.24: Simplify $\int x\cos x dx$

Answer: $x\sin x + \cos x + C$
ES Moderate

9.2.25: Simplify $\int \sin x\sec^2 x dx$

Answer: $\sin x\tan x + \cos x + C$
ES Moderate

9.2.26: If $\int \cot x dx = \ln|\sin x|$, find $\int x\csc^2 x dx$

Answer: $\ln|\sin x| - x\cot x + C$
ES Moderate

9.2.27: Simplify $\int (x + 2)\sin 2x\,dx$

Answer: $\frac{1}{4}\sin 2x - \frac{(x + 2)}{2}\cos 2x$

ES Moderate

9.2.28: $\int xf'(x)\,dx =$

a) $x^2/2\ f(x) + C$

b) $xf(x) + \int f'(x)\,dx$

c) $x^2 f'(x) - \int f(x)\,dx$

d) $xf(x) - \int f(x)\,dx$

e) $xf(x) - \int f'(x)\,dx$

Answer: (d)

MC Moderate

9.2.29: Simplify $\int \frac{\ln(\ln x)}{x}dx$

Answer: $\ln x(\ln(\ln x) - 1) + C$

ES Moderate

9.2.30: $\int \frac{\sqrt{x + 1}}{x}dx =$

a) $(\ln x)\sqrt{x + 1} - \frac{1}{2}\int \frac{\ln x}{\sqrt{x + 1}}dx$

b) $\frac{2\left(\sqrt{x + 1}\right)^3}{3x} + \int \frac{2\left(\sqrt{x + 1}\right)^3}{3x^2}dx$

c) neither (a) nor (b)

d) both (a) and (b)

Answer: (d)

MC Moderate

9.2.31: Simplify $\int \frac{x + 1}{e^{2x}}dx$

Answer: $\frac{-e^{-2x}}{4}(2x + 1) + C$

ES Moderate

9.3.1: $\int_0^1 xe^{x^2}\,dx =$

a) $e - 1$
b) e
c) $\frac{e}{2}$
d) 1
e) none of the above

Answer: (e)
MC Moderate

9.3.2: Calculate: $\int_0^1 \frac{2x^3\,dx}{(x^2 + 4)^2}$

Answer: $\ln 5 - \ln 4 - \frac{1}{5}$
ES Moderate

9.3.3: Calculate: $\int_0^{\frac{\pi}{2}} e^{\sec x + \tan x} \cdot \sec x\,(\tan x + \sec x)\,dx$

Answer: $e^{\sqrt{2}+1}$
ES Moderate

9.3.4: Which of the following is a correct substitution for the integral $\int_1^e \frac{\cos(\ln x)}{x}dx$

a) $u = \ln x,\ \int_1^e \cos u\,du$
b) $u = \frac{\ln x}{x},\ \int_1^e \cos u\,du$
c) $u = \ln x,\ \int_0^1 \cos u\,du$
d) $u = \ln x,\ \int_0^1 \sin u\,du$
e) $u = \ln x,\ \int_0^1 \frac{\cos u}{u}du$

Answer: (a)
MC Moderate

9.3.5: $\int_2^3 \frac{x^3}{(x^2 - 2)^2}dx$

Answer: $\frac{1}{4}\ln\left(\frac{77}{12}\right)$
ES Moderate

9.3.6: Find the area between the curves $y = e^x$ and $y = xe^x$ for $0 \le x \le 1$.

Answer: $\int_0^1 e^x - xe^x dx = e$

ES Moderate

9.3.7: Find the area under the curve $y = \sin x e^{\cos x}$ for $0 \le x \le \pi/2$

Answer: $e - 1$

ES Moderate

9.3.8: Evaluate $\int_0^1 \frac{2x + 1}{e^x} dx$

Answer: $\frac{3e - 5}{e}$

ES Moderate

9.3.9: Find the average value of the function $f(x) = \sin x \cos x$ on the interval $0 \le x \le \pi/2$.

Answer: $1/\pi$

ES Moderate

9.3.10: Evaluate $\int_1^{\sqrt{e}} \frac{\cos(\ln x^2)}{x} dx$

Answer: $\frac{1}{2}\sin 1$

ES Moderate

9.3.11: $\int_a^b f(x)f'(x)dx =$

a) $\frac{f^2(b)}{2} - \frac{f^2(a)}{2}$

b) $\int_a^b u du$ where $u = f(x)$

c) $\int_{f(a)}^{f(b)} u du$ where $u = f(x)$

d) a and c above

e) b and c above

Answer: (d)

MC Moderate

9.3.12: Find the area under the curve $y = \frac{x}{x^2 + 2}$ for $0 \le x \le 1$

Answer: $\frac{1}{2}\ln\frac{3}{2}$

ES Moderate

9.3.13: Find $\int_0^1 xe^{3x^2}\,dx$

Answer: $\frac{1}{6}(e^3 - 1)$

ES Moderate

9.4.1: (a) Use the trapezoidal rule with n = 4 to approximate the value of

$$\int_3^5 (2x - 5)^2\,dx.$$

(b) Evaluate the integral directly and calculate the error incurred using the trapezoidal rule.

Answer: (a) 21

(b) $20\frac{2}{3}$, error is $\frac{1}{3}$

ES Moderate

9.4.2: (a) Use the trapezoidal rule with n = 5 to approximate the value of

$$\int_1^3 \frac{1}{2x - 1}\,dx.$$

(b) Evaluate the integral directly and calculate the error incurred using the trapezoidal rule.

Answer: (a) $\left(1 + \frac{10}{9} + \frac{10}{13} + \frac{10}{17} + \frac{10}{21} + \frac{1}{5}\right)\frac{1}{5}x \approx .82895$

(b) $\frac{1}{2}\ln 5 \approx .80472$, error $\approx .02423$

ES Moderate

9.4.3: Approximate $\int_1^3 \frac{1}{x}\,dx$ using the trapezoidal rule, with n = 2.

a) $\frac{7}{3}$

b) $\frac{11}{6}$

c) $\frac{11}{12}$

d) $\frac{7}{6}$

e) none of the above

Answer: (d)

MC Moderate

9.4.4: Suppose the graph of the function f(x) passes through the points (2, 3), (3, 5), (4, 2), $\left(5, \frac{3}{2}\right)$, (6, 1). Estimate the area under the curve y = f(x) from x = 2 to x = 6 using the trapezoidal rule.

a) 3
b) 4.2
c) 5.25
d) 10.5
e) none of the above

Answer: (d)
MC Moderate

9.4.5: Calculate $\int_1^3 \frac{dx}{x}$ using the midpoint rule (n = 4).

Answer: 1.089755
ES Moderate

9.4.6: Calculate $\int_1^3 \frac{dx}{x}$ using the trapezoidal rule (n = 4).

Answer: 1.116667
ES Moderate

9.4.7: Calculate $\int_1^3 \frac{dx}{x}$ using Simpson's rule (n = 4).

Answer: 1.098726
ES Easy

9.4.8: Using the trapezoid rule with n = 2 estimate $\int_0^2 \frac{x}{3x + 1} dx$

Answer: $\frac{11}{28}$
ES Moderate

9.4.9: The shaded area in the diagram below represents an estimation of $\int_a^b f(x)dx$ using:

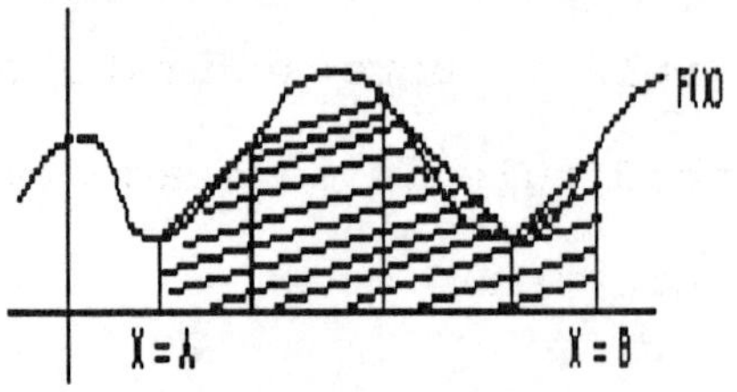

a) the midpoint rule with n = 4
b) the trapezoid rule with n = 4
c) Simpson's rule with n = 4
d) a Riemann Sum using left endpoints
e) a Riemann sum using right endpoints

Answer: (b)
MC Moderate

9.4.10: Use the midpoint rule with n = 4 to approximate $\int_0^8 x^2 + 2dx$.

Now obtain the exact value by integrating. How could the approximation be improved?

Answer: 182, 560/3
ES Moderate

9.4.11: A particle is set in motion and its velocity, v(t), is measured every 2 seconds to obtain the following table:

Time	(sec)	0	2	4	6	8	10	12	14	16	18	20
v(t)	(m/sec)	0	8	28	53	48	32	20	11	6	2	2

Find an approximation for S(20), the distance travelled in 20 seconds, keeping in mind $S(20) = \int_0^{20} v(t)dt$.

Answer: 418 m
ES Moderate

9.4.12: Use the midpoint rule, trapezoid rule and Simpson's rule with n = 2 to estimate $\int_0^4 \frac{1}{x^2 + 1}dx$

Answer: 6/5, 124/85, 282/255
ES Moderate

9.4.13: Which of the following statements is false:

a) In general, the error from the midpoint rule is less than the error from the trapezoid rule.

b) The error in the trapezoid rule decreases as the number of subintervals increases.

c) In general, Simpson's rule is more accurate than the midpoint rule.

d) In general, the trapezoid rule is more accurate than the midpoint rule.

e) All of the statements are true

Answer: (d)
MC Moderate

9.4.14: Estimate the area under the curve of $y = e^{x^2}$ between $-1 \le x \le 1$ by using the trapezoid rule with $n = 2$.

Answer: $e + 1$
ES Moderate

9.4.15: A home owner has fences on three sides of her property and a stream runs along the fourth side. She makes measurements of the distance to the stream every 10 feet as illustrated. What is the approximate area of her property?

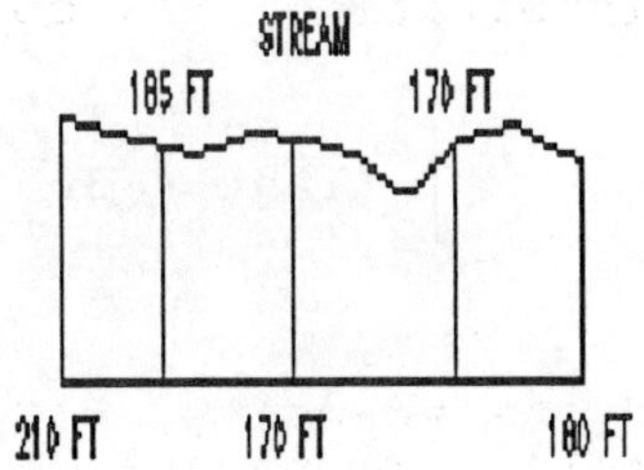

Answer: 7200 ft^2
ES Moderate

9.5.1: A company estimates that the rate of revenue produced by an investment will be K(t) thousand dollars per year at time t, where $K(t) = 9te^{-.2t}$. Find the present value of this stream of income over the next four years using a 10% interest rate.

Answer: approximately $64,000
ES Moderate

9.5.2: In 1940, the population density of Philadelphia was given by $60e^{-.4t}$ thousand people per square mile at a distance t miles from City Hall. How many people lived between 1 and 3 miles from City Hall?

Answer: approximately 650,000
ES Moderate

9.5.3: Suppose the annual rate of income from an investment at any time t is K(t) = -100 + 50t. What is the formula for the present value of the income over the next 5 years, at a 6% interest rate compounded continuously?

a) $\int_0^5 (50t - 100)e^{.06t}\,dt$

b) $\int_0^5 (50t - 100)e^{-.06t}\,dt$

c) $\int_1^5 50t^{-.06t}\,dt$

d) $\int_1^5 (50t - 100)e^{-.06t}\,dt$

e) none of the above

Answer: (b)
MC Moderate

9.5.4: A large conglomerate estimates that a new plant will yield an annual rate of return of 5,000 - 250t thousand dollars. If the interest rate is 9%, what is the present value of the income generated in the first three years of the plants operation?

Answer: $26,810 \times 10^3$ dollars
ES Moderate

9.5.5: If the annual rate of return from an investment is -3,000 + 125t find the present value of the income generated in the third year if interest rates are 8.5%.

a) $\int_0^3 (-3000 + 125t)e^{-.85t}dt$

b) $\int_2^3 (-3000 + 125t)e^{-.85t}dt$

c) $\int_0^3 (125t - 3000)e^{.85t}dt$

d) $\int_2^3 (125t - 3000)e^{.85t}dt$

e) none of the above

Answer: (b)
MC Moderate

9.6.1: Evaluate the improper integral $\int_{8/3}^{\infty} (3x + 1)^{-3/2}\,dx$.

Answer: $\frac{2}{9}$
ES Moderate

9.6.2: Find the area under the graph of $y = 4e^{-3x}$ for $x \geq 1$.

Answer: 0.06638
ES Moderate

9.6.3: Calculate: $\int_{3}^{\infty} e^{-x/2}\,dx$

a) e^{-1}

b) $e^{-3/2}$

c) $2e^{-3/2}$

d) divergent

e) none of the above

Answer: (c)
MC Moderate

9.6.4: Is $\int_{0}^{\infty} \frac{dx}{\sqrt{4x + 5}}$ convergent or divergent? If convergent, compute the integral.

Answer: divergent
ES Moderate

9.6.5: Is $\int_{0}^{\infty} \ln x^2\,dx$ convergent or divergent? If convergent, compute the integral.

Answer: divergent
ES Moderate

9.6.6: Is $\int_{0}^{\infty} \frac{(x + 1)\,dx}{(x + 1)^3}$ convergent or divergent? If convergent, compute the integral.

Answer: convergent; 1
ES Moderate

9.6.7: Consider $\lim_{b \to \infty} \frac{2b - 1}{b}$. Which of the following is true?

a) The limit exists and is equal to zero.
b) The limit exists and is equal to one.
c) The limit exists and is equal to two.
d) The limit diverges.

Answer: (c)
MC Moderate

9.6.8: Calculate $\int_1^{\infty} \frac{x}{(x^2 + 1)^2} dx$

Answer: $\frac{1}{4}$
ES Moderate

9.6.9: Is $\int_0^{\infty} e^{-2x} dx$ convergent or divergent? If it is divergent, explain why. If it is convergent, evaluate the integral.

Answer: convergent, 1/2
ES Moderate

9.6.10: Suppose $\int_1^{\infty} x^k dx$ is convergent and equals 1/3. Find k.

Answer: k = -4
ES Moderate

9.6.11: What is the area under the curve $y = xe^{-x^2}$ for $x \geq 0$?

Answer: 1/2
ES Moderate

9.6.12: Is $\int_0^{\infty} \frac{1}{\sqrt{x + 1}} dx$ convergent or divergent? If it is divergent, explain why. If it is convergent, evaluate the integral.

Answer: divergent, $\int_1^{\infty} \frac{1}{\sqrt{x + 1}} dx = \lim_{b \to \infty} \left(2\sqrt{b + 1} - 2\sqrt{2}\right) \to \infty$
ES Moderate

9.6.13: Evaluate $\int_0^{\infty} x(x + 1)^{-3} dx$

Answer: 1/2
ES Moderate

9.6.14: State whether the following expressions approach a limit as $b \to \infty$. If they do, give the value of the limit.
(i) $b(b + 3)^{-2}$

(ii) $\frac{2\sqrt{b} - 1}{}$

(iii) $e^{3b} + \frac{\sqrt{b}}{2}$

Answer: 0, 2, no limit
ES Moderate

9.6.15: Recall that the capital value of an asset is given by $\int_0^{\infty} K(t)e^{-rt}dt$. If the annual rate of earnings, K(t), of a certain machine is \$3000 and if the interest rate is 12%, find the capital interest of the machine.

Answer: \$25,000
ES Moderate

9.6.16: Evaluate $\int_{-\infty}^{0} e^{3t+1}dt$

Answer: e/3
ES Moderate

9.7.1: Determine: $\int \frac{dx}{\sqrt{4 + x^2}}$

Answer:

$$\ln \left| \frac{\sqrt{x^2 + 4}}{2} + \frac{x}{2} \right| + C$$

ES Moderate

9.7.2: Using the table of integrals, evaluate $\int \frac{x}{3x - 1}dx$

Answer: $\frac{x}{3} + \frac{1}{9}\ln|3x - 1| + C$
ES Easy

9.7.3: By using the table of integrals, determine which of the following statements is correct.

a) $\int x^2e^{-3x}dx = \frac{x^2e^{-3x}}{3} - \frac{2}{3}\int xe^{-3x}dx$

b) $\int x^2e^{-3x}dx = \frac{2}{3}\int xe^{-3x}dx - \frac{x^2e^{-3x}}{3}$

c) $\int x^2e^{-3x}dx = \frac{x^2e^{-3x}}{2} + \frac{3}{2}\int xe^{-3x}dx$

d) none of the above

Answer: (b)
MC Moderate

9.7.4: By first making the substitution u = cos x, use the table of integrals to evaluate $\int \tan x(2\cos x + 1)^{-1/2}dx$

Answer:

$$C - \ln\left|\frac{\sqrt{2\cos x + 1} - 1}{\sqrt{2\cos x + 1} + 1}\right|$$

ES Moderate

10. Differential Equations

10.1.1: Suppose $f(t)$ is a solution to the initial value problem $y' = e^{2t} - y$, $y(1) = 0$. What is $f(1)$, $f'(1)$?

Answer: 0, e^2
ES Moderate

10.1.2: Consider the differential equation $y' = y - y^2$. Which of the following statements are true?
a) The constant function $f(t) = 1$ is a solution to this differential equation.
b) This differential equation has infinitely many solutions.
c) The function $f(t) = \dfrac{1}{(1 + e^{-t})}$ is a solution to this differential with initial condition $y(0) = 1/2$.
d) If $f(t)$ is a solution to the differential equation satisfying the initial condition $y(0) = 0$ then $f'(0) = 0$.
e) All the above are true.

Answer: (e)
MC Moderate

10.1.3: Consider the differential equation $y' = t^3(y + 3)$
a) $f(t) = -3$ is a constant solution to this differential equation
b) $f(t) = 0$ is a constant solution to this differential equation
c) If $f(t)$ is a solution to the differential equation with initial conditions $y(1) = 0$ then $f'(1) = 3$.
d) (a) and (c) above
e) all of the above

Answer: (d)
MC Moderate

10.2.1: Solve the differential equation:

$$y' = \frac{3t^2 + 1}{2y}, \quad y(1) = -5$$

Answer:

$y = -\sqrt{t^3 + t + 23}$
ES Moderate

10.2.2: Solve the differential equation:

$$\frac{dy}{dx} = y^2 \ln x, \quad y(1) = \frac{1}{3}$$

Answer:
$y = \dfrac{1}{2 - x \ln x + x}$
ES Moderate

10.2.3: Solve the differential equation:

$$y' = y^2 - t^2y^2$$

Answer: $y = 0$ and $y = \dfrac{1}{\frac{1}{3}t^3 - t + C}$

ES Moderate

10.2.4: Solve the differential equation:

$$y' = 3t^2(y - 7)$$

Answer:

$y = 7 + Ae^{t^3}$, A any constant

ES Moderate

10.2.5: What are the constant solutions of the differential equation $y' = e^y - 1$?

a) $y = 0$

b) $y = 1$

c) $y = e^t - 1$

d) $y = e^{-1}$

e) none of the above

Answer: (a)

MC Moderate

10.2.6: Solve the following differential equation: $y'' = e^{3t}$, $y'(0) = \frac{1}{3}$, $y(0) = 0$

a) $y = e^{3t} + \frac{1}{3}t$

b) $y = \dfrac{e^{3t}}{3} + \dfrac{1}{3}$

c) $y = 9e^{3t} - \frac{8}{3}t - 9$

d) $y = \dfrac{e^{3t}}{9} - \dfrac{1}{9}$

e) none of the above

Answer: (d)

MC Moderate

10.2.7: Solve: $y' = y(t - 2)$, $y(0) = 0$

a) $y = 0$

b) $e^{(t^2/2)-2t} - 1$

c) $e^{(t^2/2)-2t} - 2t + 1$

d) $y = \dfrac{t^2}{2} - 2t$

e) none of the above

Answer: (b)

MC Moderate

10.2.8: Solve $y' = e^y \sin t$

a) $y = \ln(-\sin t) + C$
b) $y = -\ln(\cos t + C)$
c) $y = e^{-\cos t} + C$
d) $y = \cos(\ln t) + C$
e) none of the above

Answer: (b)
MC Moderate

10.2.9: Let t represent the number of hours that a packing machine is operated and p(t) represent the probability that the machine breaks down at least once during the t hours of operation. It has been observed that the rate of increase of the probability of a breakdown is proportional to the probability of not having a breakdown. A differential equation describing this situation is:

a) $y' = ky$, $y(0) = 0$
b) $y' = k(1 - y)$, $y(0) = 0$
c) $y' = k(1 - y)$, $y(0) = 0$
d) $y' = ky$. There is not enough information given to determine initial conditions.
e) $y' = k(1 - y)$. There is not enough information given to determine initial conditions

Answer: (c)
MC Moderate

10.2.10: Solve the differential equation $y' = e^{2t} - 1$

Answer: $y = e^{2t}/2 - t + C$, $y = 0$
ES Moderate

10.2.11: Solve the differential equation $(t^2 + 1)y' = yt$

Answer: $y = A\sqrt{t^2 + 1}$
ES Moderate

10.2.12: Solve the differential equation $y' = \frac{1}{ty}$

Answer: $y = \sqrt{2\ln|t| + C}$
ES Moderate

10.2.13: A cool object is to be heated to a maximum temperature M° C. At any time t, the rate at which the temperature rises is proportional to the difference between the actual temperature and the maximal temperature. If the object is originally 0° C, find and solve a differential equation describing this situation.

Answer: $y' = k(M - y)$, $y(0) = 0$, $y(t) = M - Me^{-kt}$
ES Moderate

10.2.14: Solve the differential equation $y' = \sqrt{\frac{t+1}{y}}$ subject to the initial conditions $y(0) = 4$

Answer: $y = ((t + 1)^{3/2} + 7)^{2/3}$
ES Moderate

10.2.15: Solve the differential equation $y' = t\cos t$ subject to the initial conditions $y(0) = 0$.

Answer: $y = t\sin t - \cos t + 1$
ES Moderate

10.2.16: Solve the differential equation $yy' = te^{t^2}$ with initial conditions $y(0) = 1$

Answer: $y = e^{t^2/2}$
ES Moderate

10.2.17: Suppose water is seeping from an underground storage facility at rate which is proportional to the square of the amount of water present. If $f(t) = y$ is the amount of water present at time t, which of the following is a differential equation describing the situation?

a) $y' = ky^2$, $k > 0$
b) $y' = ky^2$, $k < 0$
c) $y' = ky^2$, $k > 0$, $y(0) = 0$
d) $y' = ky^2$, $k < 0$, $y(0) = 0$
e) none of the above

Answer: (b)
MC Moderate

10.2.18: Solve the differential equation $ty' = \ln t$

Answer: $y = \frac{(\ln t)^2}{2}$
ES Moderate

10.2.19: What are the constant solutions to the differential equation $y' = y^2e^t - 2ye^t$

Answer: $y = 0$, $y = 2$
ES Moderate

10.2.20: Suppose $y = 2$ is a constant solution of the differential equation $yy' = 2yt - kt$. What is k?

Answer: $k = 4$
ES Moderate

10.2.21: Solve $y' = \tan t \sec^2 t$ if $y(0) = 1$

a) $y = \tan t + 1$

b) $y = \dfrac{\tan^2 t}{2} + 1$

c) $y = \dfrac{\sec^2 t}{3} + \dfrac{2}{3}$

d) $y = \ln|\tan t| + 1$

Answer: (b)
MC Moderate

10.2.22: Solve $y' = e^{-y}$, $y(0) = 0$

Answer: $y = \ln|t + 1|$
ES Moderate

10.2.23: Solve $y' = y - 2t$, $y(0) = 0$

a) $y = \dfrac{\sqrt{2}}{2}t$

b) $y = \sqrt{2t}$

c) $y = yt - t^2/2$

d) This differential equation can't be solved using the separation of variables technique

Answer: (d)
MC Moderate

10.2.24: Find all solutions to the differential equation $(t + 1)y' = yt^2 - y$

Answer: $y = Ae^{\frac{t^2}{2} - t}$
ES Moderate

10.2.25: Suppose the relationship between the price, p, of a product and the weekly sales, s, of the product is given by $\dfrac{dp}{ds} = -\dfrac{1}{10}\left(\dfrac{2s}{p + 1}\right)$ Then

a) As sales increase, the price increases.

b) The rate of decrease of the price is proportional to the sales.

c) s = 0 is a constant solution to this differential equation.

d) As the price increases the rate of change of the price also increases.

e) All of the above

Answer: (b)
MC Moderate

10.2.26: Solve the differential equation $\frac{dy}{dx} = \frac{x + e^x}{y}$

Answer: $y = \pm\sqrt{x^2 + 2e^x + C}$
ES Moderate

10.2.27: Solve the differential equation $\frac{dN}{dp} = 3p^2 + \sin p$ with

$N(0) = 2$

Answer: $N = p^3 - \cos p + 3$
ES Moderate

10.2.28: Solve the differential equation $y' = \frac{t\sin t^2}{y}$

Answer: $y = \pm\sqrt{C - \cos t^2}$
ES Moderate

10.3.1: Let f(t) be the solution of $y' = ty + 0.2$, $y(0) = 3$. Use Euler's method with $n = 2$ to estimate f(1) and f'(1).

Answer: $f(1) \approx 3.975$, $f'(1) \approx 4.175$
ES Moderate

10.3.2: Use Euler's method with $n = 4$ to approximate the solution f(t) to $y' = y - 4t$, $y(0) = 2$, for $0 \le t \le 2$. (Include a graph showing the coordinates of the appropriate points.)

Answer:

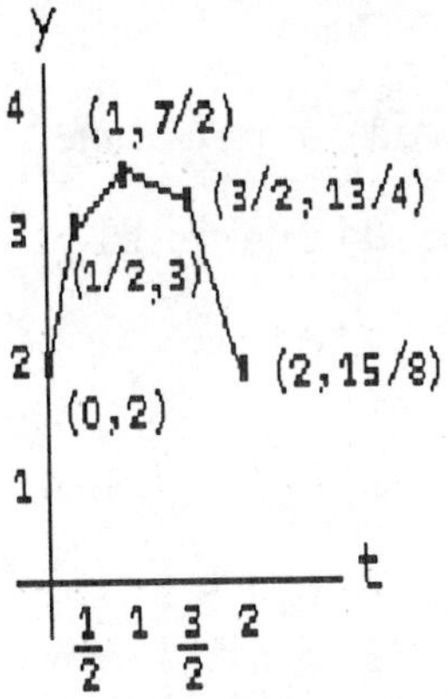

ES Moderate

10.3.3: Use Euler's method with $n = 5$ to approximate f(1) if $y = f(x)$ satisfies the differential equation $y' = y$, $y(0) = 1$. Compare this answer with the exact value of f(1).

Answer: Euler's method: $f(1) \approx 2.488$
Actual value: $f(x) = e \approx 2.718$
ES Moderate

10.3.4: Suppose that $y = f(x)$ satisfies the differential equation

$$y' = x^2 + 3y, \quad f(0) = 2.$$

Euler's method with $n = 10$ is used to construct an approximation $p(x)$ to $f(x)$ for $0 \le x \le 1$. Then $p(.1)$ equals

a) 2.2
b) 3.1
c) 6.0
d) 2.6
e) none of the above

Answer: (d)
MC Moderate

10.3.5: Use Euler's method with $n = 4$ to approximate the solution $f(2)$ to $y' = x + y - 1$ for $0 \le x \le 2$ where $y(0) = 2$.

Answer: $\frac{65}{8}$
ES Moderate

10.3.6: Use Euler's method with $n = 4$ to approximate the solution $f(3)$ to $y' = y^2 + xy - 3$ for $1 \le x \le 3$ where $y(1) = 2$.

Answer: 2587.1953
ES Moderate

10.3.7: Use Euler's method with $n = 5$ to approximate the solution $f(1)$ to $y' = 5 - y$ for $0 \le x \le 1$ where $y(0) = 1$.

Answer: 3.68928
ES Moderate

10.3.8: Suppose $f(t)$ is a solution to the initial value problem $y' = y^2t - y$, $y(0) = 1$.

a) $f(t)$ will be a constant solution of the differential equation
b) f is increasing at the origin
c) f is decreasing at the origin
d) $f'(1) = 0$

Answer: (c)
MC Moderate

10.3.9: Use Euler's method with $n = 2$ to find an approximation for $f(1)$ and $f'(1)$ where $f(t)$ is a solution to the initial value problem $y' = 2y - t$, $y(0) = 1$.

Answer: $f(1) \cong 15/4$, $f'(1) \cong 13/2$
ES Moderate

10.3.10: The following is a polygonal path obtained from Euler's method with $n = 4$ to approximate a solution $f(t)$ of a differential equation. Indicate whether the following statements are true or false:

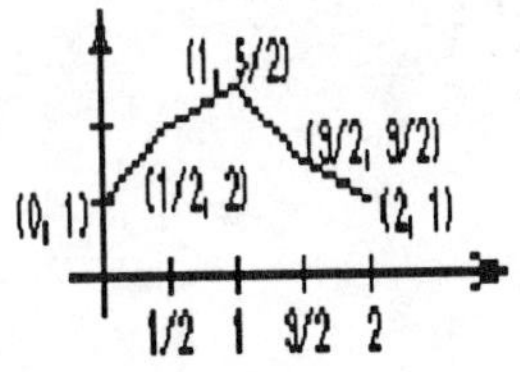

1) $f'(0) = 2$
2) $f(1) \cong 5/2$
3) $f'(3/2) \cong 3/2$
4) $f'(1/2) = 1$
5) $f'(1/2) \cong 1$
6) $f(0) = 1$

Answer: T, T, F, F, T, T
ES Moderate

10.3.11: Suppose $y = f(x)$ satisfies the differential equation
$$\frac{dy}{dx} = y^2x + y + e^x, \quad f(0) = 2$$
Euler's Method with $n = 4$ is used to approximate $f(x)$ for $0 \le x \le 2$. So $f(.5)$ is approximately

a) $6 + \sqrt{e}$
b) $2\left(4 + \sqrt{e}\right)$
c) $3 + e^2$
d) $y^2/2 + y + e^{1/2}$
e) none of the above

Answer: (a)
MC Moderate

10.3.12: Use Euler's method with $n = 2$ to estimate $f'(2)$ if $f(t)$ satisfies $yy' = y + t$, $f(0) = 1$

Answer: $f'(2) \cong 11/7$
ES Moderate

10.3.13: Suppose $f(t) = y$ is a solution of the initial value problem $y' = kt + y$, $f(1) = 1$. Suppose $f'(1) = 2$. What is k?

Answer: $k = 1$
ES Moderate

10.3.14: Use Euler's method with $n = 3$ to approximate $f(1)$ if $f(t) = y$ is a solution of the initial value problem $y' = 9t + y^2$, $y(0) = 0$.

Answer: 271/27
ES Moderate

10.4.1: Describe the qualitative behavior of solution of the differential equation by sketching the constant solution (if any) and the solution having the specified initial conditions. Indicate inflection points (if any). Include a y-z graph as well as a t-y graph.

$y' = 6 - 3y, \quad y(0) = -1, \quad y(0) = 3$

Answer:

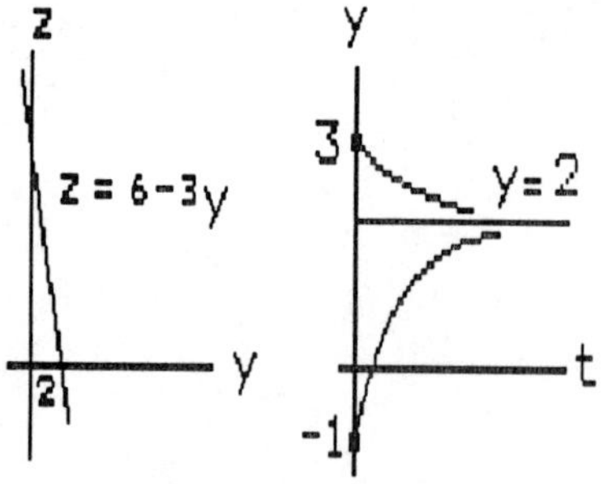

ES Moderate

10.4.2: Describe the qualitative behavior of solution of the differential equation by sketching the constant solution (if any) and the solution having the specified initial conditions. Indicate inflection points (if any). Include a y-z graph as well as a t-y graph.

$y' = 6 + 2y, \quad y(0) = -4, \quad y(0) = -2$

Answer:

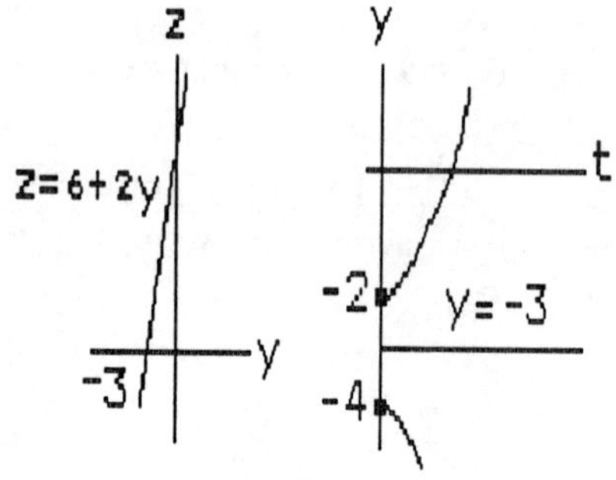

ES Moderate

10.4.3: Describe the qualitative behavior of solution of the differential equation by sketching the constant solution (if any) and the solution having the specified initial conditions. Indicate inflection points (if any). Include a y-z graph as well as a t-y graph.

$y' = y^2 - 9, \quad y(0) = -5, \quad y(0) = 2$

Answer:

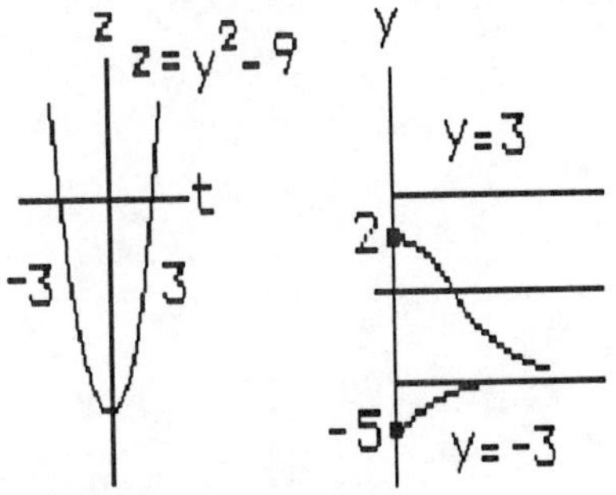

ES Moderate

10.4.4: Describe the qualitative behavior of solution of the differential equation by sketching the constant solution (if any) and the solution having the specified initial conditions. Indicate inflection points (if any). Include a y-z graph as well as a t-y graph.

$y' = y^2 - 2y - 8, \quad y(0) = -3, \quad y(0) = 3$

Answer:

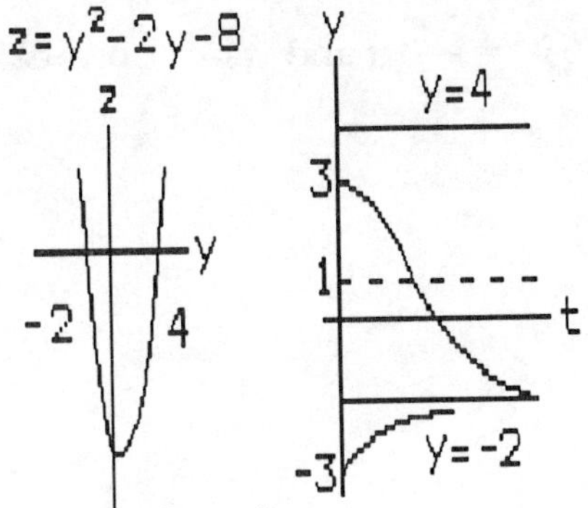

ES Moderate

10.4.5: Describe the qualitative behavior of solution of the differential equation by sketching the constant solution (if any) and the solution having the specified initial conditions. Indicate inflection points (if any). Include a y-z graph as well as a t-y graph.

$y' = y^3 - 3y^2, \quad y(0) = -1.5, \quad y(0) = 2.5$

Answer:

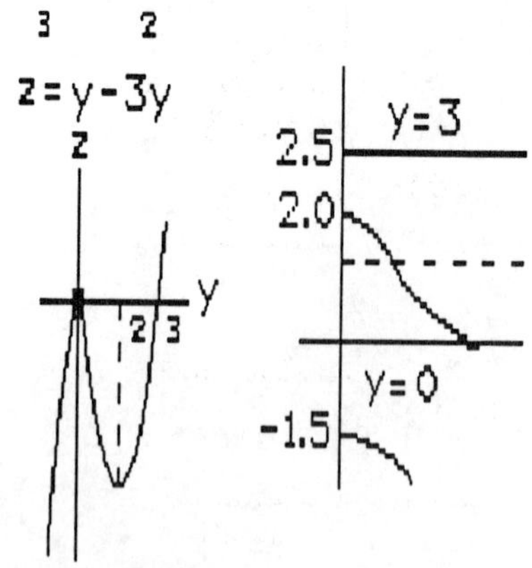

ES Moderate

10.4.6: Describe the qualitative behavior of solution of the differential equation by sketching the constant solution (if any) and the solution having the specified initial conditions. Indicate inflection points (if any). Include a y-z graph as well as a t-y graph.

$y' = \cos y, \quad y(0) = -\frac{\pi}{4}, \quad y(0) = \frac{5\pi}{4}$ (and two constant solutions)

Answer:

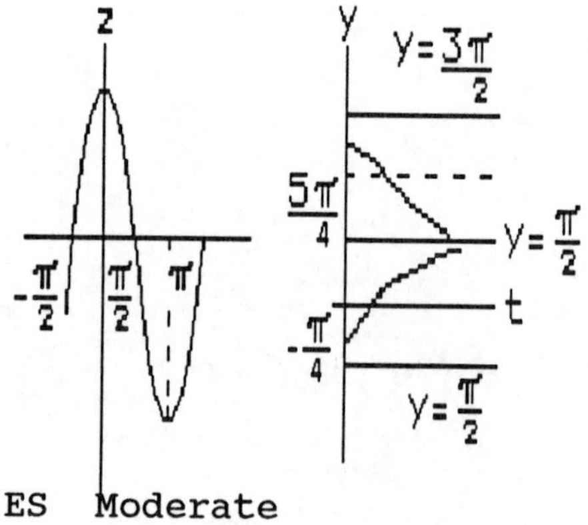

ES Moderate

10.4.7: Which of the following is a sketch of the solution to $y' = y^2 - 9$, $y(0) = 2$?

a)
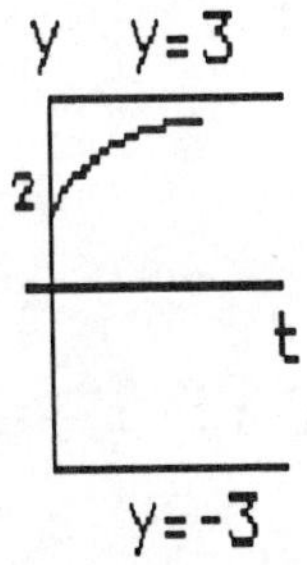

b)
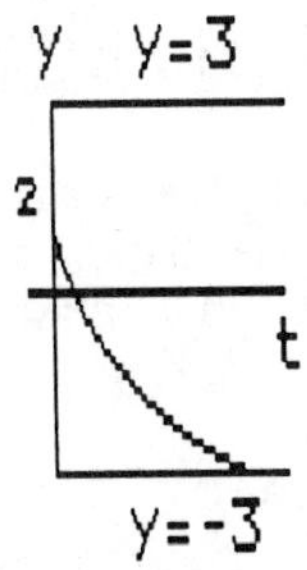

c)
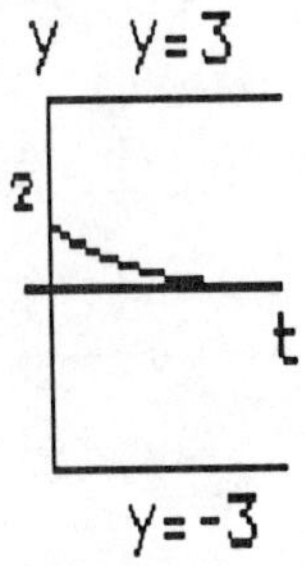

d)
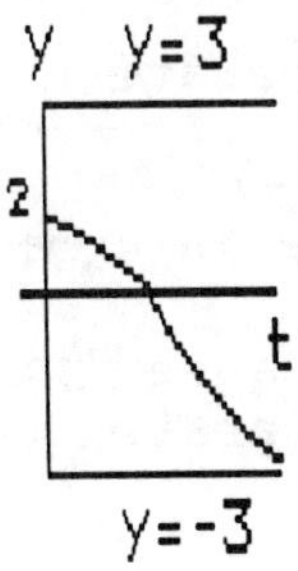

e) none of the above

Answer: (d)
MC Moderate

10.4.8: Let $y' = 2 - y$. Which of the following properties hold for the solution $y = f(t)$ determined by the initial condition $y(0) = 1$?

I. It is always concave down.

II. It is a constant solution.

III. It is always decreasing.

a) I
b) II
c) III
d) I and III
e) none of the above

Answer: (a)
MC Moderate

10.4.9: For what y value(s) does a solution of $y' = y^2 - 3y + 2$ have inflection points?

a) $y = 2$ and $y = 1$
b) $y = \frac{3}{2}$
c) $y = 2$
d) $y = 0$
e) none of the above

Answer: (b)
MC Moderate

10.4.10: Let $y' = y^3$. Which of the following properties hold for the solution $y = f(t)$ determined by the initial condition $y(0) = -2$?

I. It is always increasing.

II. It has an inflection point.

III. It is always concave down.

a) I
b) II
c) III
d) I & II
e) none of the above

Answer: (c)
MC Moderate

10.4.11: Is $y' = 3y + 2t$ autonomous? Explain.

Answer: No, because it is dependent upon time.
ES Moderate

10.4.12: Consider the differential equation $y' = g(y)$ where $g(y)$ is the function whose graph is shown below:

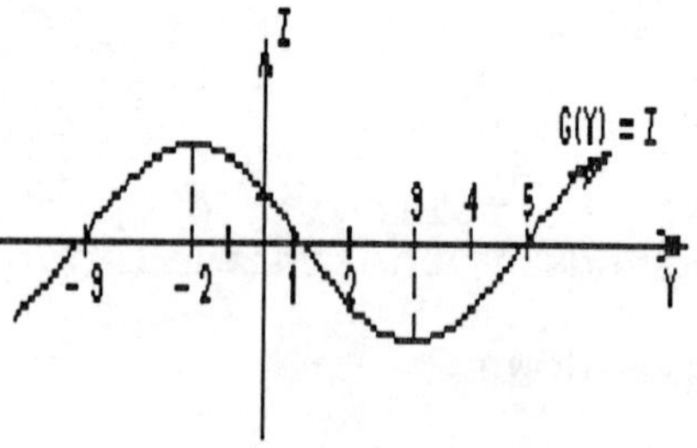

Indicate whether the following statements are true or false
1) $y = -3$, $y = 1$ and $y = 5$ are the constant solutions to $y' = g(y)$.
2) $y = 2$ is the only constant solution of $y' = g(y)$.
3) If the initial value $y(0)$ is greater than 5, then the corresponding solution will be an increasing function.
4) If the initial value of $y(0)$ is 4 then the corresponding solution has an inflection point.
5) If the initial value of $y(0)$ is 2 then the corresponding solution has an inflection point.

Answer: T, F, T, T, F
ES Moderate

10.4.13: Consider the differential equation $y' = y^2 - 4y + 3$. Which of the following could be a graph of solutions to this differential equation

a)

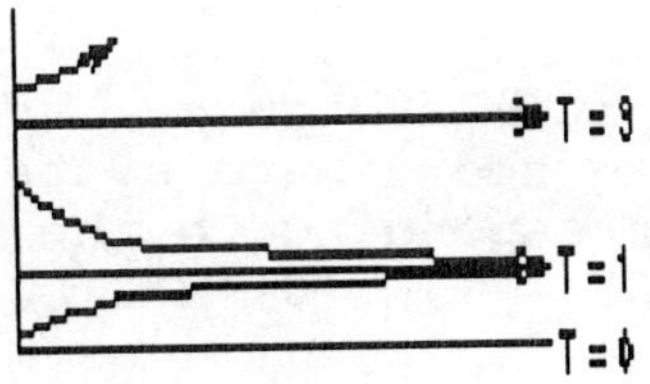

b)

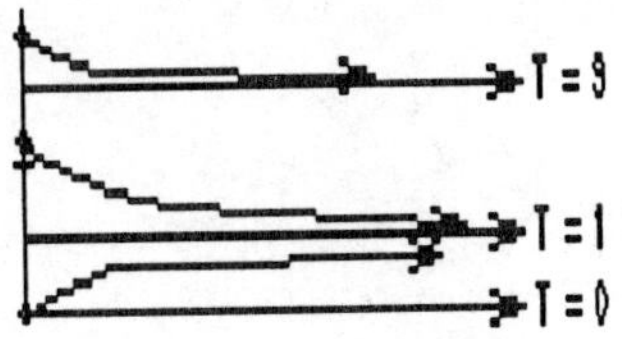

c)

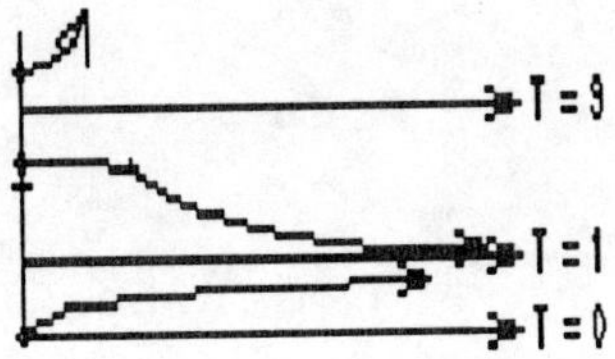

d)

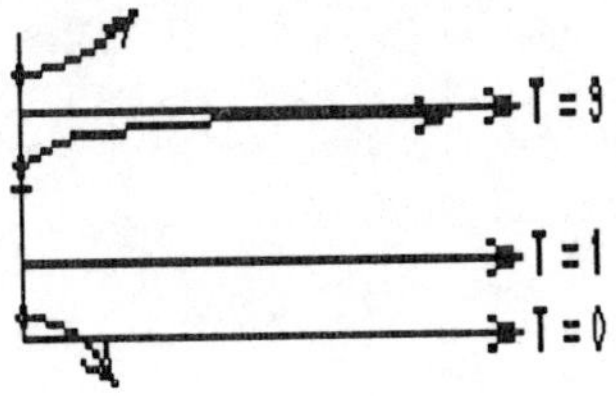

e)

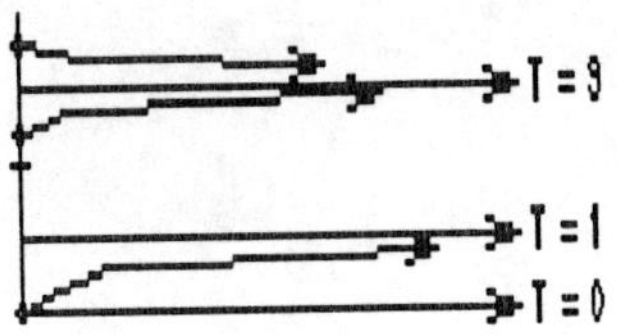

Answer: (a)
MC Moderate

10.4.14: On the yz - coordinate system sketch $z = y^2 - 4$. Suppose $y' = y^2 - 4$. Then sketch on a ty - coordinate system the solutions of the differential equation corresponding to the initial conditions $y(0) = -3$, $y(0) = -1$, $y(0) = 1$, $y(0) = 3$

Answer:

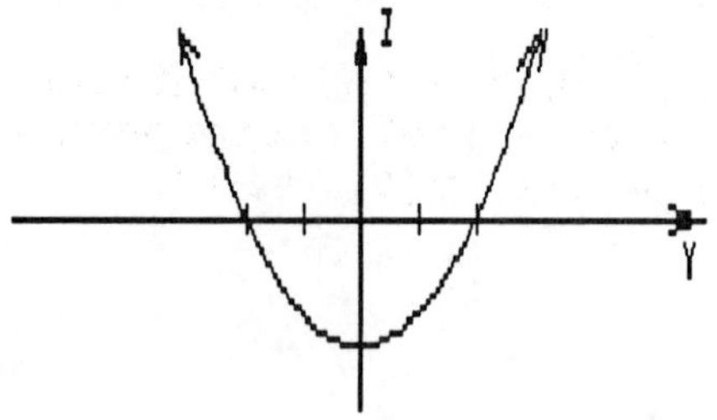

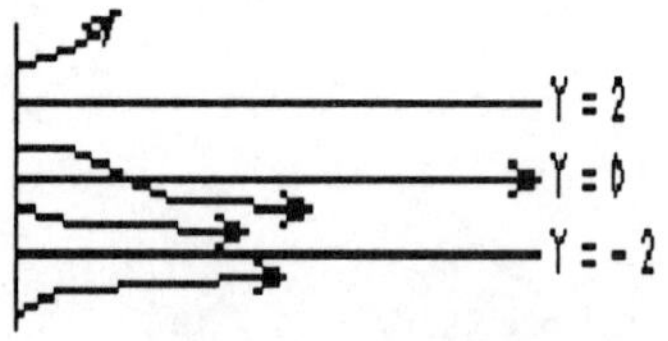

ES Moderate

10.4.15: If $y' = y - 3$ on a ty - coordinate system sketch the solutions corresponding to the initial conditions $y(0) = 0$, $y(0) = 4$

Answer:

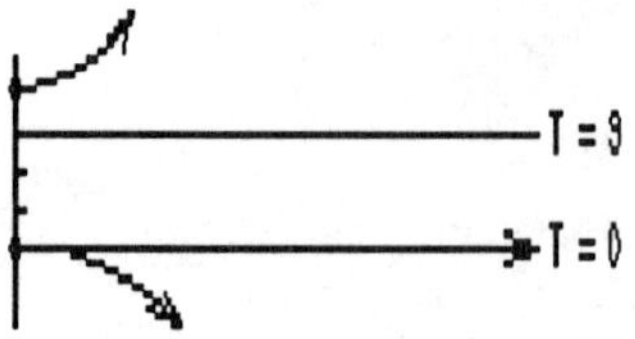

ES Moderate

10.4.16: Below is a sketch of $f(x) = (x - 1)e^x$.

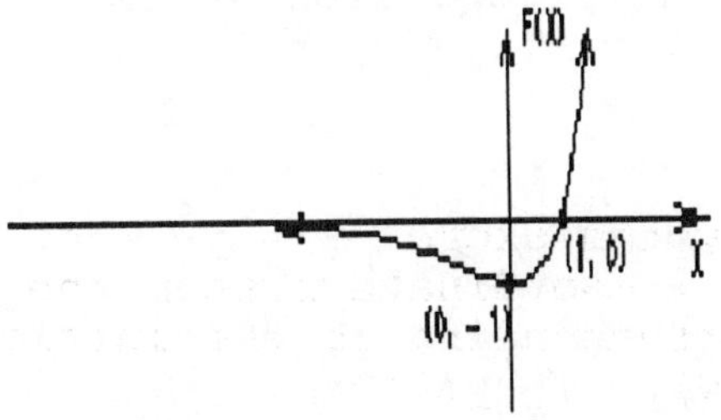

On a ty - coordinate system sketch the solutions to the differential equation $y' = (y - 1)e^y$ corresponding to the initial conditions $y(0) = 2$, $y(0) = 1/2$, $y(0) = -1/2$.

Answer:

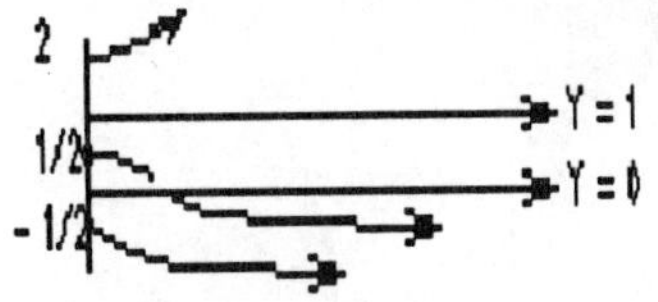

ES Moderate

10.4.17: Suppose the graph below gives a solution to the differential equation $\frac{dP}{ds} = g(P)$ where P is price of a product and s is the weekly sales.

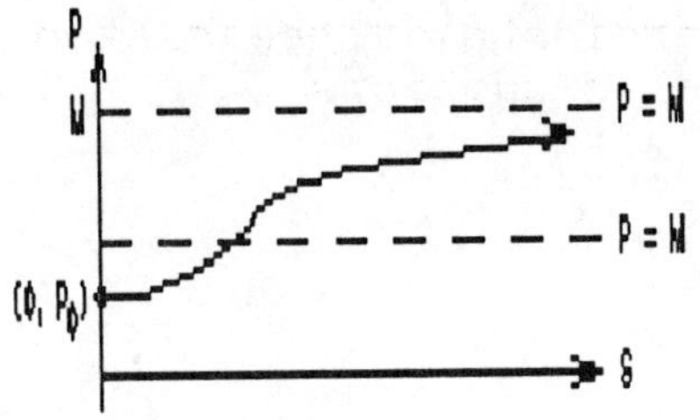

Which of the following statements is true?

I. $g(M) = 0$ III. $g'(m) = 0$
II. $g(m) = 0$ IV. $g(P_0) > 0$

a) I and IV
b) I, II, IV
c) I, III, IV
d) I
e) IV

Answer: c

ES Moderate

10.4.18: Suppose $y' = y(y + 3)$. On a ty - coordinate system graph the solutions of this differential equation satisfying the initial conditions $y(0) = -4$, $y(0) = -1$, $y(0) = 1$

Answer:

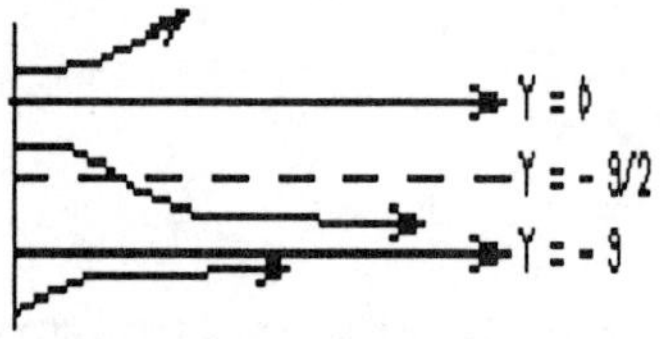

ES Moderate

10.4.19: Suppose the following is a graph of $z = g(y)$

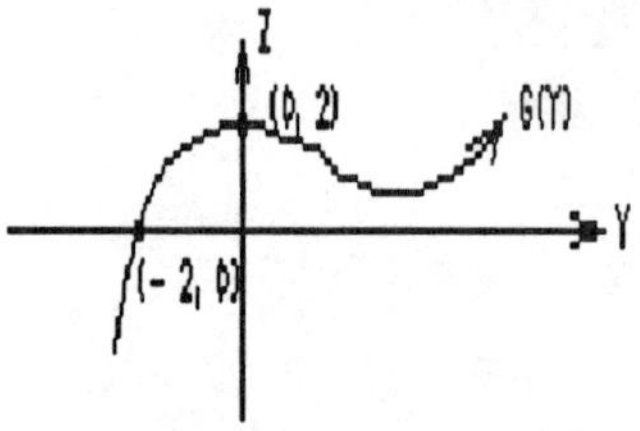

Which of the following can they be said about the solution $y = f(t)$ to the initial value problem $y' = g(y)$, $y(0) = -1$

I. $f(t)$ is an increasing function

II. $f(t)$ is always positive

III. $f(t)$ has an inflection point when $y = 2$

a) I
b) II
c) III
d) I, III
e) all of the above

Answer: (d)
MC Moderate

10.4.20: Suppose $y' = 1/y^2$. On a ty - coordinate system graph any constant solutions to the differential equation as well as the solutions corresponding to initial conditions
$y(0) = 1$, $y(0) = -1$

Answer:

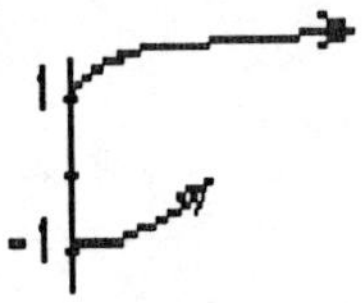

No constant solutions
ES Moderate

10.4.21: Suppose $y' = ky + b$ and a graph of several solutions of the differential equation are as below:

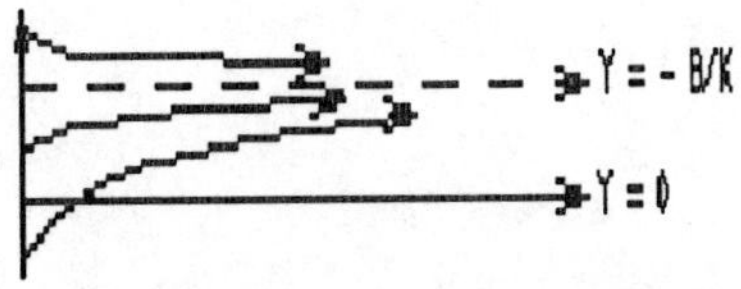

Then I. k is negative
II. k is positive
III. b is positive
IV. b is negative

a) I, III
b) I, IV
c) II, III
d) II, IV
e) not enough information given

Answer: (a)
MC Moderate

10.4.22: On a ty - coordinate system graph the solution to the initial value problems $y' = e^y - 1$, $y(0) = -1$, $y(0) = 1$

Answer:

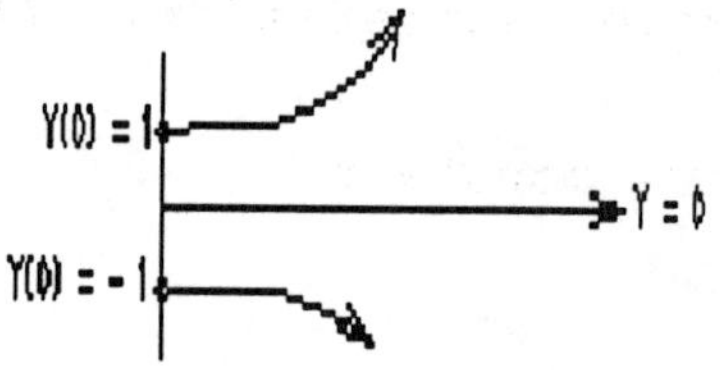

ES Moderate

10.4.23: The following could be graphs of solutions to which differential equation?

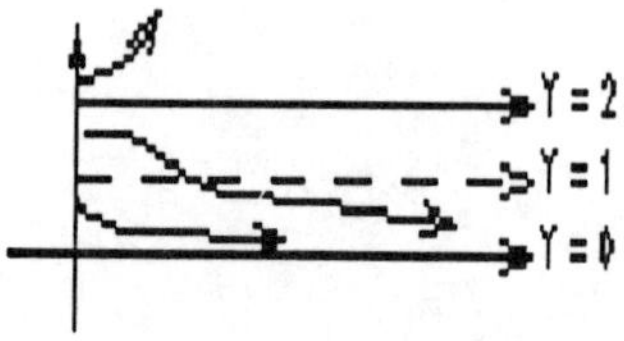

a) $y' = y(y + 2)$
b) $y' = 3y(y - 2)$
c) $y' = y^2 + 2$
d) $y' = (y - 2)e^y$
e) none of the above

Answer: (b)
MC Moderate

10.5.1: Suppose an infectious disease spreads through an elementary school at a rate proportional to the product of the percentage of pupils who have the disease and the percentage of pupils who have not yet contracted the disease. Suppose that at the beginning of the epidemic 5% of the pupils have the disease. Let f(t) be the percentage of pupils who have the disease at time t, and give the differential equation satisfied by f(t).

Answer: $y' = ky(100 - y)$, $y(0) = .05$, where k is a positive constant
ES Moderate

10.5.2: A certain drug is introduced into a person's bloodstream. Suppose that the rate of decrease of the concentration of the drug in the blood is directly proportional to the product of two quantities: (a) the amount of time elapsed since the drug was introduced, and (b) the square of the concentration. Let $y = f(t)$ denote the concentration of the drug in the blood at time t. Set up, but do <u>not</u> solve, a differential equation satisfied by $f(t)$.

Answer: $y' = kty^2$, where k is a negative constant
ES Moderate

10.5.3: A savings account earns 6% annual interest, compounded continuously. An initial deposit of $8500 is made, and thereafter money is withdrawn continuously at the rate of $480 per year. Make a qualitative analysis of an appropriate differential equation to determine what will happen to the savings account as time passes. (Include a y-z graph and a t-y graph.) After making this analysis, your conclusion need only be one sentence.

Answer: $y' = .06y - 480, \quad y(0) = 8500$

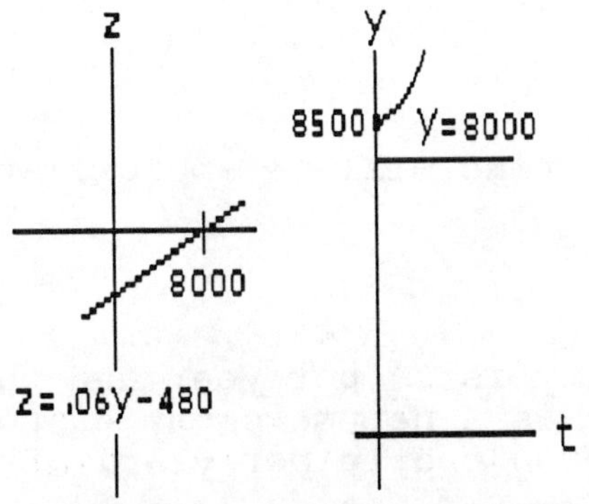

The savings account balance will grow (at an increasing rate) in spite of the steady withdrawals.
ES Moderate

10.5.4: A patient is receiving a steady infusion of glucose. Let y denote the concentration of glucose in the blood at time t, measured in milligrams of glucose per 100 cubic centimeters of blood, and suppose that y satisfies the differential equation $y' = 48 - .4y$.

(a) Make a qualitative analysis of this differential equation, sketching representative solutions. Include a y-z graph as well as a t-y graph.

(b) What will be the approximate concentration of glucose in the blood after a long period of time, provided the glucose infusion is continued at the same rate? (Include the appropriate units in your answer.)

Answer:

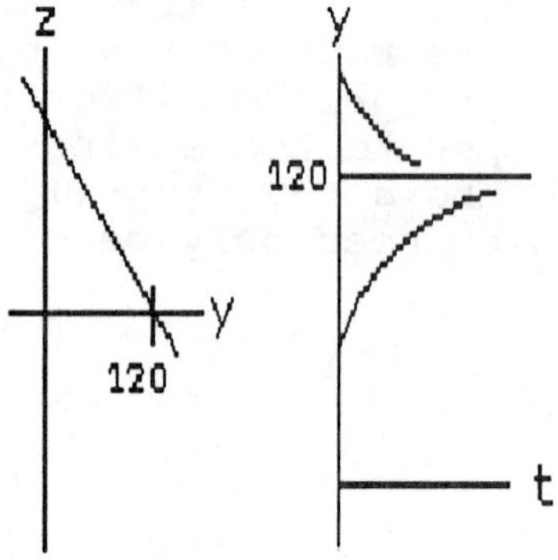

The concentration after a long time will be approximately 120 mg glucose per 100 cc blood.
ES Moderate

10.5.5: The birth rate in a certain city is 2% per year and the death rate is 2.5% per year. Also, there is a net movement of population into the city at the rate of 4000 people per year. Let $N = f(t)$ be the city's population at time t.

(a) Write the differential equation satisfied by f(t).

(b) Make a qualitative analysis of the equation in (a) to determine what will happen to the population over a long period of time (assuming the conditions described above do not change). Does the long-term situation depend on the initial size of the population?

Answer:

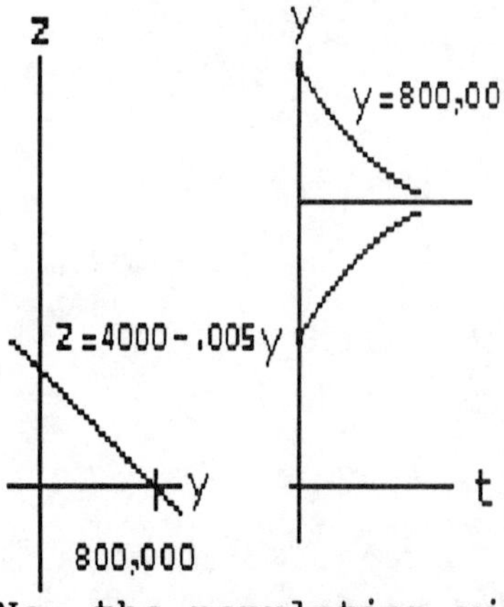

No, the population will approach 800,000 persons, regardless of the size of the initial population.
ES Moderate

10.5.6: A certain developing country has a population of 500,000. The yearly rate of increase of literacy among the people is proportional to the number of illiterate people in the population. Letting f(t) represent the number of literate people, determine the differential equation which f(t) satisfies: (Let k represent a positive constant.)

a) $f'(t) = 500{,}000k(1 - f(t))$
b) $f'(t) = k(500{,}000 - f(t))$
c) $f'(t) = 500{,}000 - kf(t)$
d) $f'(t) = \dfrac{kf(t)}{500{,}000}$
e) none of the above

Answer: (b)
MC Moderate

10.5.7: Suppose that an epidemic is spreading at a rate proportional to the square of the infected population. Let f(t) be the number of infected people at time t, and suppose y = f(t) satisfies a differential equation y' = g(y). Which of the following sets of curves could represent solutions of y' = g(y)? [Hint: First determine the differential equation y' = g(y).]

a)
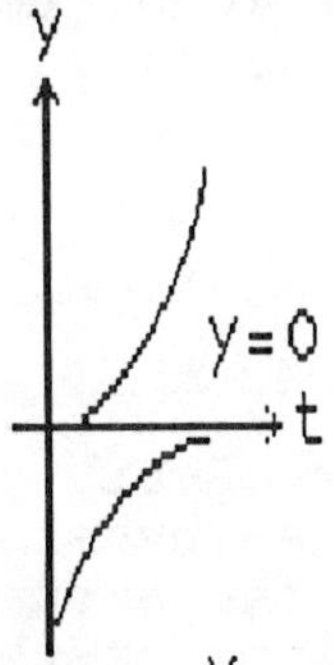

b)
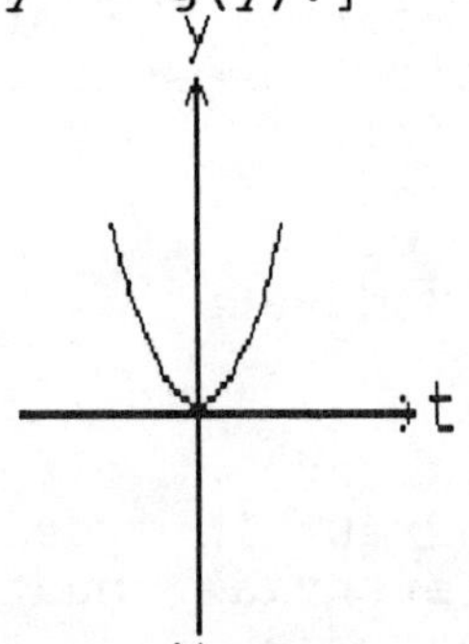

c)
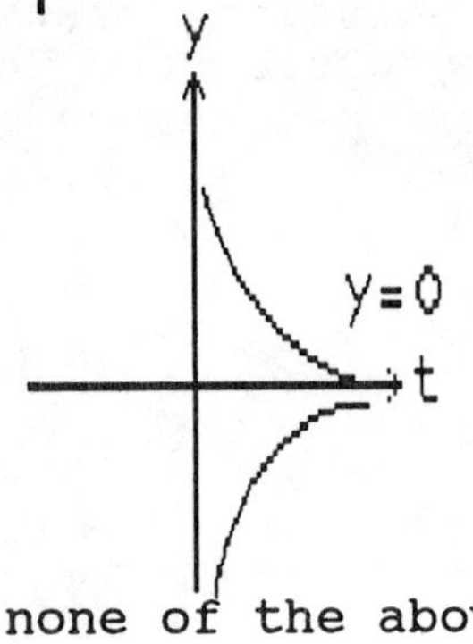

d)
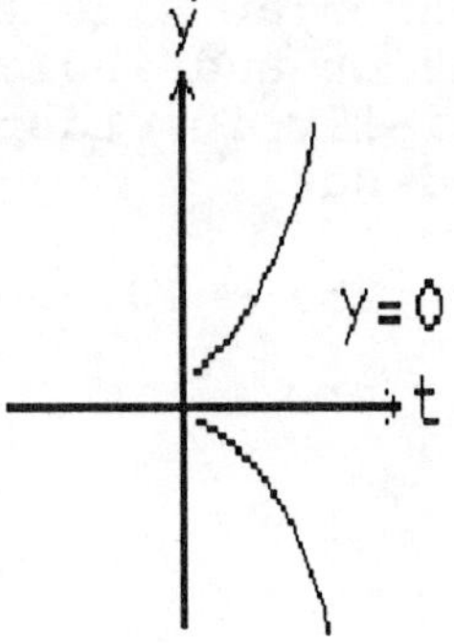

e) none of the above

Answer: (a)
MC Moderate

10.5.8: A man opens a savings account that earns interest at an annual rate of 6% compounded continuously. He plans to make continuous withdrawals at a rate of \$300 per year. What will happen if his initial deposit is \$5000? [Hint: Let f(t) be the savings account balance at time t, and determine the differential equation satisfied by f(t).]

a) The balance will increase indefinitely.
b) The balance will decrease until it runs out.
c) The balance will remain at \$5000 as long as the interest and withdrawals remain the same.
d) The balance will increase at an increasing rate until it reaches \$18,000, at which point it will increase at a decreasing rate.
e) none of the above

Answer: (c)
MC Moderate

10.5.9: A certain chemical vaporizes when exposed to the air. Suppose f(t) is the amount of chemical present. It is found that the rate of vaporization of the chemical is proportional to the amount of chemical present squared. Write a differential equation satisfied by f(t).

Answer: $y' = ky^2$, $k < 0$
ES Moderate

10.5.10: An investment earns 25% interest per year. Every year \$10,000 is withdrawn in order to pay dividends to the investors. Set up a differential equation satisfied by f(t), the amount of money invested at time t. Sketch typical solutions of the differential equation and indicate how much money would have to be invested initially in order for the investors to continue earning dividends.

Answer: $y' = .25y - 10,000$

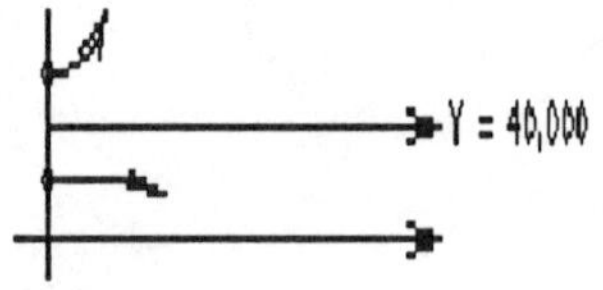

more than \$40,000 must be invested
ES Moderate

10.5.11: Depending on the type of soil there is a constant M which represents the maximum amount of water the soil can absorb per cubic ft. If the rate of absorption is proportional to the difference between the maximum amount of water that could be absorbed and the amount of water that has been absorbed, write a differential equation satisfied by $y = f(t)$ the amount of water in the soil at time t.

Answer: $y' = k(M - y)$, $k > 0$
ES Moderate

10.5.12: A sports enthusiast drinks 2 liters of water per hour. Water is eliminated from the body at a rate proportional to the amount of water in the body (due to perspiration). Write a differential equation satisfied by f(t), the amount of water in the body. Sketch several typical solutions

Answer: $y' = 2 - ky$

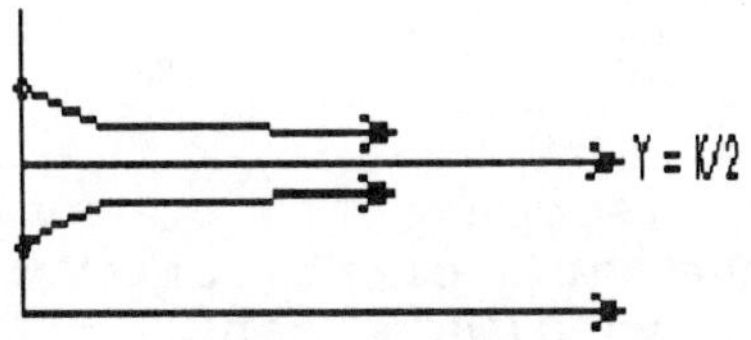

ES Moderate

10.5.13: After a baby whale is born, its weight gain at any time is proportional to the product of its weight and the difference between its weight and its weight at maturity. Give a differential equation satisfied by f(t), its weight at time t.

Answer: $y' = ky(M - y)$
M = weight at maturity
$k > 0$
ES Moderate

10.5.14: A millionaire wants to set up a trust for her grandchild. She wants to put a lump sum of money into an account earning 10% interest. She'd like her grandchild to be able to withdraw \$100 every month for the rest of the child's life. Write a differential equation satified by f(t), the amount of money in the account at time t. By graphing several typical solutions determine how much money the millionaire will have to put in the account.

Answer: $y' = .1y - 100$

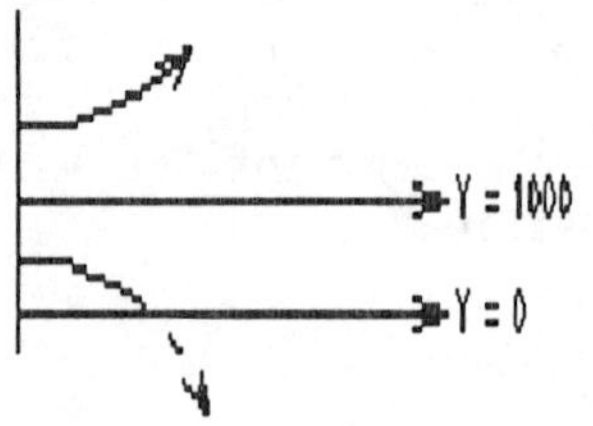

She must put over \$1000 in the bank.
ES Moderate

10.5.15: Suppose that a substance A is converted to substance B at a rate that is proportional to the cube of the amount of B present. The amount of A and B together is always constant, say M. Then if f(t) = y is the amount of A present at time t then the following differential equation describes the situation

a) $y' = ky^3$, $k > 0$
b) $y' = ky^3$, $k < 0$
c) $y' = k(M - y)^3$, $k < 0$
d) $y' = k(M - y^3)$, $k < 0$
e) None of the above

Answer: (c)
MC Moderate

11. Probability and Calculus

11.1.1: A car dealer records the number of Mercedes sold each week. During the past 50 weeks, there were 15 weeks with no sales, 20 weeks with one sale, 10 weeks with two sales, and 5 weeks with three sales. Let X be the number of Mercedes sold in a week selected at random from the past 50 weeks.

(a) Construct a relative frequency histogram for X.

(b) Compute E(X).

(c) Compute Var(X).

Answer: (a)

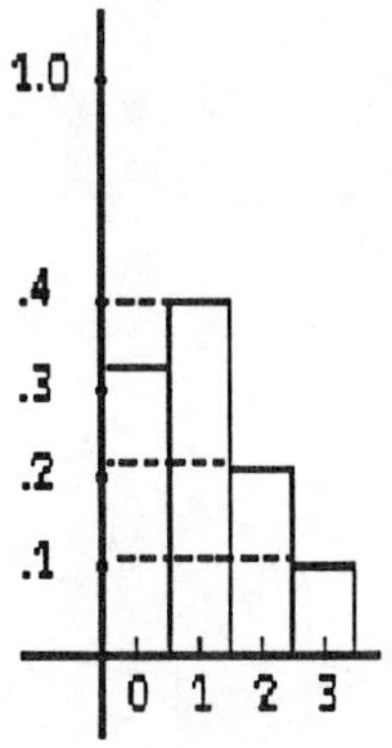

(b) 1.1

(c) .89
ES Moderate

11.1.2: A set of exam scores is 80, 75, 85, 90, 100, 70, 60. The standard deviation equals

a) $\sqrt{10}$ b) $\sqrt{50}$

c) 7 d) $\sqrt{20}$

e) none of the above

Answer: (e)
MC Moderate

11.1.3: Determine E(X) of the table below.

-1	0	1	2	Outcome
3/7	1/7	1/7	2/7	Probability

Answer: $\frac{2}{7}$
ES Moderate

11.1.4: Determine Var(X) of the table below.

-1	0	1	2	Outcome
3/7	1/7	1/7	2/7	Probability

Answer: $\frac{560}{343}$
ES Moderate

11.1.5: Determine the standard deviation of the table below.

-1	0	1	2	Outcome
3/7	1/7	1/7	2/7	Probability

Answer: 1.2778
ES Moderate

11.1.6: Determine E(X) of the table below.

-3	-2	-1	1	2	3	Outcome
.1	.1	.4	.3	.05	.05	Probability

Answer: -.35
ES Moderate

11.1.7: Determine Var(X) of the table below.

-3	-2	-1	1	2	3	Outcome
.1	.1	.4	.3	.05	.05	Probability

Answer: 2.5275
ES Moderate

11.1.8: Determine the standard deviation of the table below.

-3	-2	-1	1	2	3	Outcome
.1	.1	.4	.3	.05	.05	Probability

Answer: 1.5898
ES Moderate

11.1.9: A bag contains 4 white balls, 3 black balls, and 5 red balls. What is the probability that if 4 balls are picked, all 4 are red?

Answer: $\frac{1}{99}$
ES Moderate

11.1.10: A bag contains 4 white balls, 3 black balls, and 5 red balls. If 4 balls are picked, what is the probability that 2 are white and 2 are black?

Answer: $\frac{2}{55}$

ES Moderate

11.1.11: A bag contains 4 white balls, 3 black balls, and 5 red balls. If 4 balls are picked, what is the probability that 2 of the balls are white?

Answer: $\frac{56}{165}$

ES Moderate

11.1.12: A bag contains 4 white balls, 3 black balls, and 5 red balls. If 4 balls are picked, what is the probability that 2 balls are red, 1 is black, and 1 is white?

Answer: $\frac{8}{33}$

ES Moderate

11.1.13: John would like to place a two dollar bet on his favorite racehorse, Black Velvet. He can bet that Black Velvet will win or show (finish in the top three horses). If he bets correctly that Black Velvet wins, he wins $20. If he bets correctly that Black Velvet shows, he wins $7. John figures Black Velvet has a 20% chance of winning and a 70% chance of showing. If X is the amount of money John wins if he bets Black Velvet will win and Y is the amount of money he wins if Black Velvet will show, find E(X) and E(Y). What should John do?

Answer: E(X) = $2   E(Y) = $2.9

John should bet Black Velvet will show.

ES Moderate

11.1.14: Consider the table below:

-2	-1	0	1	2	Outcome
.2	.35	.15	.05	.25	Probability

Then E(X) and the standard deviation of X is

a) -.2, -.4
b) -.2, 2.32
c) -.2, 1.52
d) -.2, 1.49
e) none of the above

Answer: (b)

MC Moderate

11.1.15: A student taking five courses keeps a record of the number of assignments due each day in all her courses. Over the course of the 60-day semester she finds on 20 days no assignments are due, on 15 days an assignment is due in one course, on 15 days an assignment is due in two courses, on 9 days an assignment is due in three courses and once during the semester she has an assignment due in 4 courses. If X is the number of assignments due on a day selected at random from the semester, find E(X).

Answer: $E(X) = 0 \cdot 1/3 + 1 \cdot 1/4 + 2 \cdot 1/4 + 3 \cdot 3/20 + 4 \cdot 1/60 \cong 1.27$
ES Moderate

11.1.16: A carnival game costs $2 to play. A player draws a ball at random from a sack containing 1 white ball, 2 blue balls, 3 red balls, and 4 yellow balls. The payoff for drawing a particular color ball is as follows: white pays $5, blue pays $4, red pays $3 and yellow pays nothing. If X is the amount of money a player wins, construct a probability table for X and calculate E(X).

Answer:

-2	1	2	3	outcome
4/10	3/10	2/10	1/10	probability

E(X) = .20
ES Moderate

11.1.17: Find Var(X), where X has the following probability table:

40	50	60	70	80
.3	.15	.15	.2	.2

Answer: E(X) = 58.5
Var(X) = 232.75
ES Moderate

11.1.18: A Christmas tree grower anticipates a profit of $80,000 in a usual season. There is however a 10% chance of pine bark beetle infestation in which case 70% of the trees are destroyed and profit is reduced to $24,000. The grower can spray for beetles at the beginning of the season at a cost of $7,000. By computing E(X) , where X is the profit the grower will get if she does nothing to protect her trees, determine whether the grower should spray for beetles.

Answer: E(X) = 80,000(.9) + 24,000(.1) = 74,400
Since if grower sprays profit is $73,000, grower shouldn't spray.
ES Moderate

11.1.19: A relative frequency histogram is given for variable X below:

Find E(X).

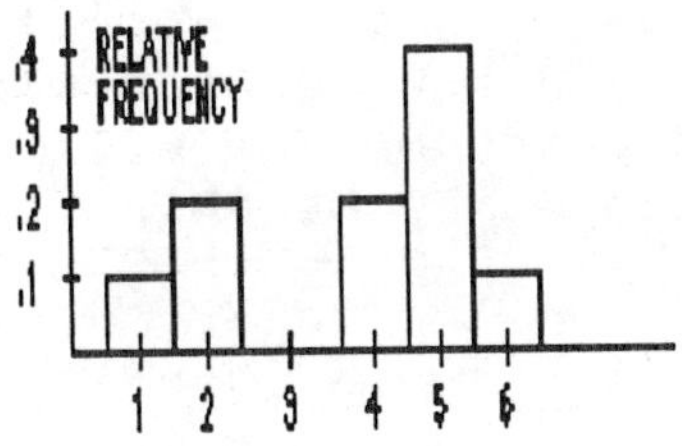

Answer: E(X) = 1(.1) + 2(.2) + 4(.2) + 5(.4) + 6(.1) = 3.9
ES Moderate

11.1.20: Compute E(X) and Var(X) where the probability table for random variable X is given below

outcome	0	2	4	6	8	10
probability	.1	.15	.3	.25	.1	.1

Answer: E(X) = 4.8, Var(X) ≅ 7.76
ES Moderate

11.1.21: The riders of the New Town Elementary school bus consists of 5 five year olds, 3 six year olds, 10 eight year olds, 1 nine year old, 4 eleven year olds and a twelve year old. A child is selected at random and her age is noted. Let X be the outcome. Construct a probability table of X and find E(X)

Answer:

5	6	8	9	11	12
5/24	3/24	10/24	1/24	4/24	1/24

E(X) = 188/24
ES Moderate

11.1.22: Determine the standard deviation of the random variable X whose probability table is given below.

outcome	1	2	3	4
probability	1/5	2/5	1/5	1/5

Answer: 7/5
ES Moderate

11.1.23: The probability table associated with random variable X is:

outcome	k	k + 1	k + 2	k + 3
probability	1/5	1/5	2/5	1/5

and E(X) = 18/5. What is k?

Answer: 2
ES Moderate

11.1.24: Joe has a lawn mowing job. If he completes the work he earns \$40. But there is a 30% chance it may rain, in which case he won't finish the job. He can pay Jane \$20 to help him and ensure that he finishes the job. If X is the amount Joe will get if he does not get Jane to help, calculate E(X) and thus decide whether Joe should hire Jane or not. (If it rains, assume Joe will make no money and if Joe hires Jane assume they will be able to finish the job befor it rains.)

Answer: E(X) = \$28. Joe should not hire Jane.
ES Moderate

11.2.1: Is $f(x) = \frac{1}{21}x^2$ a probability density function on the interval $1 \le x \le 4$? Why or why not?

Answer: Yes, because $f(x) \ge 0$ for $1 \le x \le 4$, and

$$\int_1^4 f(x)\,dx = \int_1^4 \frac{1}{21}x^2\,dx = \frac{1}{63}x^3\Big|_1^4 = 1$$

ES Moderate

11.2.2: Is $F(x) = x - x^2$ a cumulative distribution function on the interval $0 \le x \le 1$? Why or why not?

Answer: No, because $F(1) \ne 1$. (Another reason is that the derivative $F'(x)$ is not a density function because it is not nonnegative for $0 \le x \le 1$.)
ES Moderate

11.2.3: Suppose $f(x) = k(x^2 + 2x)$ is a probability density function for a continuous random variable on the interval $0 \le x \le 3$.

(a) Find the value of k.

(b) Find the corresponding cumulative distribution function.

(c) Compute E(X).

Answer: (a) $\frac{1}{18}$ (b) $F(x) = \frac{1}{54}x^3 + \frac{1}{18}x^2$ (c) $\frac{17}{8}$
ES Moderate

11.2.4: Suppose $f(x) = k \cos x$ is a density function for a random variable X on the interval $0 \le x \le \pi/2$.

(a) Find the value of k.

(b) Find the corresponding cumulative distribution function.

(c) Compute E(X).

Answer: (a) 1

(b) $F(x) = \sin x$

(c) $\frac{\pi}{2} - 1$

ES Moderate

11.2.5: Suppose $f(x) = kx^{-5}$ is a density function for a random variable X for $x \ge 2$.

(a) Find the value of k.

(b) Find the corresponding cumulative distribution function.

(c) Compute E(X).

Answer: (a) 64 (b) $F(x) = 1 - 16x^{-4}$ (c) $\frac{8}{3}$

ES Moderate

11.2.6: The density function for a random variable X is $f(x) = 3x^{-4}$, $x \ge 1$. Compute:

(a) $Pr(2 \le X)$ (b) E(X) (c) Var(X)

Answer: (a) $\frac{1}{8}$ (b) $\frac{3}{2}$ (c) $\frac{3}{4}$

ES Moderate

11.2.7: Let X be a continuous random variable on the interval $0 \le x \le 4$. Suppose that the probability that X is less than a specified value x is $\frac{1}{4}xe^{x-4}$, for $0 \le x \le 4$. Find the probability density function of X.

Answer: $f(x) = \frac{1}{4}(x + 1)e^{x-4}$

ES Moderate

11.2.8: Let X be a continuous random variable on the interval $1 \le x \le 2$, and suppose that $Pr(1 \le X \le a) = \frac{a^2 - 1}{3}$ for each a satisfying $1 \le a \le 2$. Compute E(X).

Answer: $\frac{14}{9}$
ES Moderate

11.2.9: Let F(x) be a cumulative distribution function of a continuous random variable X on the interval $A \le x \le B$. Use concepts of probability and calculus to show that F(X) is a non-decreasing function at each point on the interval $A \le x \le B$.

Answer: F'(x) must be a probability density function on $A \le x \le B$, and so $F'(x) \ge 0$ for $A \le x \le B$. By the first derivative rule, this implies that F(x) is non-decreasing at each x between A and B.
ES Moderate

11.2.10: A random variable X has the density function $f(x) = \frac{1}{\ln 16} \cdot \frac{1}{x}$, $1 \le x \le 16$. Determine the value of a for which $Pr(1 \le X \le a) = \frac{3}{4}$.

Answer: 8
ES Moderate

11.2.11: Missed work hours caused by one of a class of industrial accidents
has a probability density function

$$f(t) = \frac{1}{8}e^{-t} + \frac{3}{8}e^{-t/2} + \frac{1}{24}e^{-t/3}$$

where t is measured in hours.

(a) What proportion of these accidents result in 5 or fewer missed work hours?

(b) What proportion of these accidents result in more than 9 missed work hours?

Answer: (a) .91398 (b) .01457
ES Moderate

11.2.12: Which of the graphs below could not possibly be the graph of a probability density function f(x)?

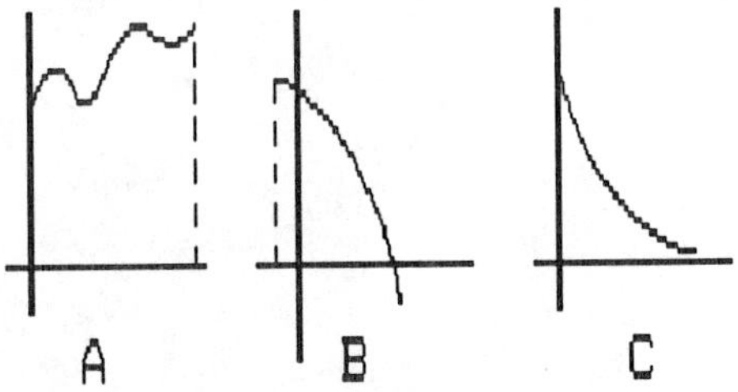

a) A and B
b) A and C
c) B and C
d) only B
e) none of the above

Answer: (d)
MC Moderate

11.2.13: Which of the graphs below could possibly be the graph of a cumulative distribution function F(x)?

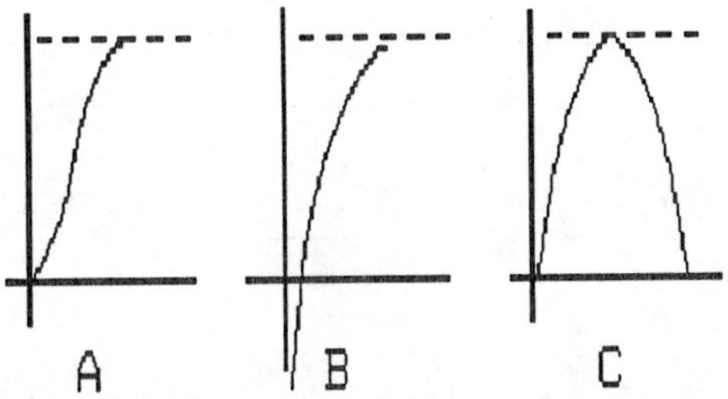

a) A
b) B
c) C
d) none of the above

Answer: (a)
MC Moderate

11.2.14: A random variable X has cumulative distribution function

$F(x) = 1 - e^{-x^2}$ $(x \geq 0)$. Then $\Pr(1 \leq X \leq 2)$ equals

a) $e^{-1} - e^{-2}$
b) $e^{-1} - e^{-2} - 2$
c) $1 - e^{-1} - e^{-2}$
d) $e^{-1} - e^{-4}$
e) none of the above

Answer: (d)
MC Moderate

11.2.15: A random variable has probability density function $f(x) = kx,\ 1 \le x \le 2$. Then k equals

a) $\frac{2}{5}$
b) $\frac{4}{5}$
c) $\frac{2}{3}$
d) $\frac{6}{5}$
e) none of the above

Answer: (c)
MC Moderate

11.2.16: A random variable has probability density function $f(x) = 30x^2(1 - x)^2 \quad (0 \le x \le 1)$. Then its cumulative distribution F(x) is given by

a) $10x^3 - 15x^4 + 6x^5$
b) $60x - 180x^2 + 120x^3$
c) $30x^2 - 60x^3 + 30x^4$
d) $30x(1 - x)$
e) none of the above

Answer: (a)
MC Moderate

11.2.17: The variance of the random variable whose cumulative distribution function is $f(x) = x \quad (0 \le x \le 1)$ is equal to

a) $\frac{1}{3}$
b) $\frac{1}{4}$
c) $\frac{5}{36}$
d) $\frac{1}{12}$
e) none of the above

Answer: (d)
MC Moderate

11.2.18: A random variable X has probability density function $f(x) = ke^{-kx} \quad (x \ge 1)$ for some constant k. Suppose that $\Pr(1 \le X \le 2) = \frac{1}{4}$. Then k equals

a) $\frac{3}{2} \ln 2$
b) $\frac{1}{2} \ln 2$
c) $\ln 2$
d) $\frac{1}{4}$
e) none of the above

Answer: (c)
MC Moderate

11.2.19: A random variable X has cumulative distribution function $F(x) = 1 - \frac{1}{x^2}$ $(x \geq 1)$. Then $Pr(a \leq X \leq 5)$ equals

a) $\frac{24}{25} - \frac{1}{a^2}$

b) $\frac{1}{a^2}$

c) $1 - \frac{1}{a^2}$

d) $\frac{1}{a^2} - \frac{1}{25}$

e) none of the above

Answer: (d)
MC Moderate

11.2.20: Determine the value of k that makes $f(x) = kx^3$ a probability density function where $0 \leq x \leq 1$.

Answer: $k = 4$
ES Moderate

11.2.21: Determine the value of k that makes $f(x) = k\sqrt{x}$ a probability density function where $4 \leq x \leq 9$.

Answer: $k = \frac{3}{38}$
ES Moderate

11.2.22: Determine the value of k that makes $f(x) = \frac{k}{\cos^2 x}$ a probability density function where $\frac{\pi}{6} \leq x \leq \frac{\pi}{3}$.

Answer: $k = \frac{\sqrt{3}}{2}$
ES Moderate

11.2.23: Determine the probability of an outcome of the probability density
function $f(x) = 4x^3$ where $0 \leq x \leq 1$ being between $\frac{1}{4}$ and $\frac{1}{2}$.

Answer: $\frac{15}{256}$
ES Moderate

11.2.24: Determine the probability of an outcome of the probability density function $f(x) = 12x^2 - 12x^3$ where $0 \leq x \leq 1$ being between $\frac{1}{2}$ and 1.

Answer: $\frac{11}{16}$
ES Moderate

11.2.25: Determine $Pr(1 \leq X \leq 2)$ of $f(x) = \frac{2 \ln x}{(\ln 4)^2 x}$, $1 \leq X \leq 4$, is a random variable.

Answer: $\frac{1}{4}$
ES Moderate

11.2.26: Dr. Smith's test score distribution is characterized by the probability density function $f(x) = \frac{x(10,000 - x^2)}{25,000,000}$, where $0 \leq x \leq 100$. What percentage of people are likely to get a 60 or above on the exam?

Answer: 40.96%
ES Moderate

11.2.27: Given the probability density function $f(x) = \frac{1}{3}$, determine the corresponding cumulative distribution function where $12 \leq x \leq 15$.

Answer: $\frac{1}{3}x - 4$
ES Moderate

11.2.28: Given the density function $f(x) = \frac{3}{4}(2x - x^2)$ where $0 \leq x \leq 2$, determine $Pr(0 \leq x \leq 1)$.

Answer: $\frac{1}{2}$
ES Moderate

11.2.29: Given the density function $f(x) = \frac{24}{x^3}$ where $3 \leq x \leq 6$, determine the value of b such that $Pr(x \leq b) = .4$.

Answer: $b = \sqrt{90/7}$
ES Moderate

11.2.30: Given the cumulative distribution function $f(x) = \frac{x}{5} - 2$ where $10 \le x < 15$, determine the value of a such that $Pr(a \le x \le 15) = \frac{2}{3}$.

Answer: $a = \frac{35}{3}$
ES Moderate

11.2.31: Find the value of k that makes $f(x) = kx^2$ a probability density function on $0 \le x \le 1$.

Answer: $k = 3$
ES Moderate

11.2.32: Show that $f(x) = 1/(x + 1)^2$ is a probability density function for $x \ge 0$. Find $P(X \ge 2)$

Answer: 1/3
ES Moderate

11.2.33: The probability density function of a continuous random variable X is $f(x) = 3/2\ x - 3/4\ x^2$, $0 \le x \le 2$. Sketch the graph of f(x) and shade the area corresponding to $Pr(1/2 \le x \le 3/2)$

Answer:

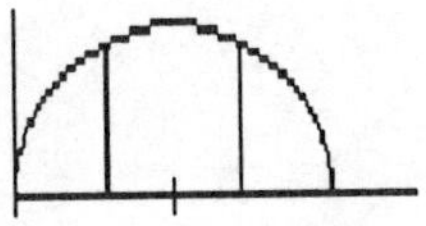

ES Moderate

11.2.34: If $f(x) = 1/8\ x$ is a probability density function for $0 \le x \le 4$ find F(x), the corresponding cumulative distribution function and $Pr(1 \le X \le 3)$

Answer: $F(x) = 1/16\ x^2$, $Pr(1 \le X \le 3) = 1/2$
ES Moderate

11.2.35: Find the value of k for which $F(x) = kx(x + 3)$ is a cumulative distribution function for $0 \le x \le 3$.

Answer: $k = 1/18$

ES Moderate

11.2.36: Consider a square with sides of length 2 as in the diagram below. An experiment consists of choosing a point at random from the square and noting its x - coordinate. If X is the x - coordinate of the point chosen, find the cumulative distribution function of X (Recall $F(x) = Pr(0 \le X \le x)$.)

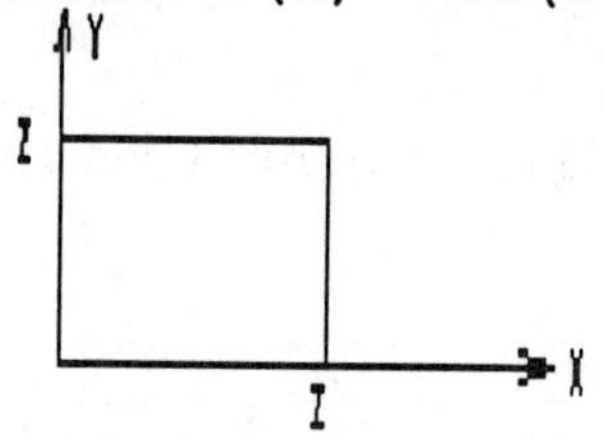

Answer: $F(x) = 2x$
ES Moderate

11.2.37: Let X be a continuous random variable $A \le X \le B$ and let $f(x)$ be its probability density function and $F(x)$ is cumulative distribution function. Indicate whether the following statements are true or false.

1. $Pr(a \le X \le b) = \int_a^b f(x)dx$
2. $\int_A^B F(x) = 1$
3. $Pr(A \le X \le b) = F(b)$
4. $f(A) = 0$, $f(B) = 1$
5. $F'(x) = f(x)$

Answer: T, F, T, F, T
ES Moderate

11.2.38: A random variable X has a probability density function $f(x) = 6x(1 - x)$, $0 \le x \le 1$. Find $F(x)$ and $Pr(1/2 \le X \le 1)$

Answer: $3x^2 - 2x^3 = F(x)$, 1/2
ES Moderate

11.2.39: If the probability density function of a continuous random variable is $f(x) = x/32$, $0 \le x \le 8$, find a so that the probability that X is greater than a is 1/4.

Answer:
$a = 4\sqrt{3}$
ES Moderate

11.2.40: Consider $F(x) = (x - 3)^2$. Show that F is a cumulative distribution function for a continuous random variable X, $3 \le X \le 4$. Find the corresponding probability density function.

Answer: $f(x) = 2x - 6$
ES Moderate

11.2.41: Suppose $f(x) = 3e^{-kx}$ is a probability density function for $x \geq 0$. What must k be?

Answer: $k = 3$
ES Moderate

11.2.42: Suppose $f(x) = 1/x^2$ is a probability density function for $x \geq 1$. Find $Pr(2 \leq X \leq 10)$

Answer: 2/5
ES Moderate

11.2.43: An experiment consists of selecting a point on a unit circle (radius 1) at random and noting the angle Θ, $0 \leq \Theta \leq 2\pi$, that it makes with the positive x-axis. Show that $f(\Theta) = (1/2\pi)\Theta$ is a probability density function for X and find the probability that a point picked at random has an angle of between $\pi/3$ and $2\pi/3$.

Answer: 1/6
ES Moderate

11.2.44: Which of the following functions could be a graph of a cumulative distribution function of a discrete random variable for $a \le X \le b$.

I.

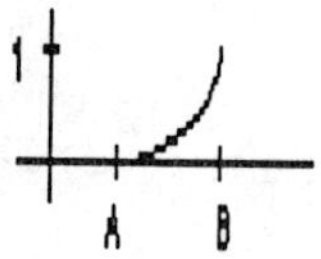

II.

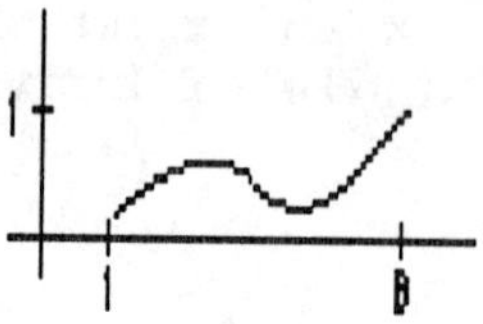

III.

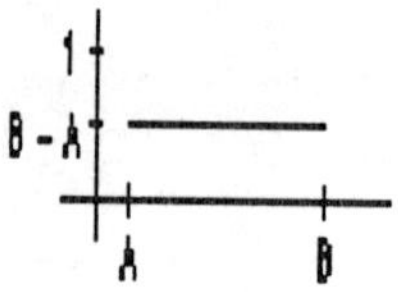

a) I
b) II
c) III
d) I and II
e) I, II, III

Answer: (a)
MC Moderate

11.2.45: A random variable X has a cumulative distribution function $F(x) = 1 - \dfrac{1}{(x + 1)^2}$ for $x \ge 0$. Find $Pr(1 \le X \le 4)$

Answer: 21/100
ES Moderate

11.2.46: Is $f(x) = (3/2)\ x - 1$ a probability density function for $0 \le x \le 2$? Why or why not?

Answer: f is not positive all x, $0 \le x \le 2$
ES Moderate

11.2.47: A random variable has a cumulative distribution function $f(x) = 1 - 1/x$, $x \geq 1$. Find the corresponding probability density function.

Answer: $\frac{1}{x^2}$
ES Moderate

11.3.1: The mean of the random variable whose density function is $f(x) = \frac{3}{8}x^2$ $(0 \leq x \leq 2)$ is equal to

a) $\frac{5}{8}$
b) $\frac{3}{8}$
c) $\frac{3}{2}$
d) $\frac{5}{2}$
e) none of the above

Answer: (c)
MC Moderate

11.3.2: A random variable X is normally distributed with density function

$$f(x) = \frac{1}{5\sqrt{2\pi}} e^{-(1/2)[(x - 50)/5]^2}$$

The variance of X is equal to

a) 25
b) 5
c) 50
d) $\frac{1}{5}$
e) none of the above

Answer: (a)
MC Moderate

11.3.3: Determine the expected value of $f(x) = \frac{1}{3}$ where $2 \leq x \leq 5$.

Answer: $\frac{7}{2}$
ES Moderate

11.3.4: Determine the variance of $f(x) = \frac{1}{3}$ where $2 \leq x \leq 5$.

Answer: $\frac{3}{4}$
ES Moderate

11.3.5: Determine the expected value of $f(x) = \frac{x^2}{3}$ where $-2 \le x \le 1$.

Answer: $-\frac{15}{12}$
ES Moderate

11.3.6: Determine the variance of $f(x) = \frac{x^2}{3}$ where $-2 \le x \le 1$.

Answer: .6375
ES Moderate

11.3.7: Given the density function $f(x) = \frac{3}{64}x^2$ $(0 \le x \le 4)$, determine the corresponding cumulative distribution function.

Answer: $\frac{x^3}{64}$
ES Moderate

11.3.8: Find the expected value of the random variable whose probability density function is $f(x) = 12x^2(1 - x)$, $0 \le x \le 1$.

Answer: 3/5
ES Moderate

11.3.9: A hardware store will cut lumber any length between 5 and 20 feet. Say X is the length of lumber requested by a customer. Then X is a uniform random variable with probability density function $f(x) = 1/15$. Find E(X) and Var(X). Interpret your answer for E(X).

Answer: $E(X) = 25/2$
$Var(X) = 75/4$
ES Moderate

11.3.10: Find the expected value of $f(x) = 2(x - 1)$, $1 \le x \le 2$.

Answer: 5/3
ES Moderate

11.3.11: Suppose random variable X has probability density function $f(x) = e^x$, $x \le 0$. Find E(X). (You may use the fact that $\lim_{b \to -\infty} be^b = 0$.

Answer: $E(X) = -1$
ES Moderate

11.3.12: Determine the variance of the random variable with probability density function $f(x) = 1/3,\ 1 \le x \le 4$.

Answer: $Var(X) = 3/4$
ES Moderate

11.3.13: The life of a battery is a random variable with probability density function $f(x)\ 3/56\ x^2,\ 2 \le x \le 4$ where x is time in months. Calculate E(X) and give an interpretation of your answer.

Answer: $E(X) = 45/14$
ES Moderate

11.3.14: If X is a random variable with probability density function $f(x)\ 4x - 1,\ 1/2 \le x \le 1$. Find the variance of X.

Answer: 1/64
ES Moderate

11.4.1: Find the expected value and the variance (by inspection) of the random variables with the following density functions:

(a) $f(x) = \dfrac{1}{4\sqrt{2\pi}} e^{-(1/2)[(x - 1)/4]^2},\ -\infty < x < \infty$

(b) $f(x) = .2e^{-.2x},\ x \ge 0$

Answer: (a) $E(X) = 1,\ Var(X) = 16$

(b) $E(X) = 5,\ Var(X) = 25$
ES Moderate

11.4.2: A survey shows that the time spent in a checkout line in a certain
supermarket is exponentially distributed with mean 5 minutes. What is the probability of spending 10 minutes or more in a checkout line?

Answer: .13534
ES Moderate

11.4.3: Let X be the time to failure of an electronic component, and suppose X is an exponential random variable with E(X) = 4 years. Find the median lifetime, i.e., find M such that $Pr(X \le M) = \frac{1}{2}$.

Answer: 2.77 years
ES Moderate

11.4.4: A lumber yard cuts 2" x 4" lumber into 8 foot studs. It is observed that the actual lengths of the studs are normally distributed with mean 8 feet and standard deviation .1 feet. What proportion of the studs are longer than 8.25 feet?

Answer: .00621
ES Moderate

11.4.5: Suppose that laboratory studies have shown that when rats are innoculated with a certain virus the recovery times of the rats are normally distributed with mean 5 days and standard deviation 1 day. A researcher innoculates a rat with the virus and then injects an experimental drug into the rat. If the rat recovers in
only 3 days, what is the likelihood that the drug had no special effect on the rat and the recovery was attributed to pure chance?

Answer: no answer
ES Moderate

11.4.6: Let $f(x) = 6x(1 - x)$ $(0 \le x \le 1)$ be the density function of a random variable X. Then $\Pr\left(\frac{1}{2} \le X \le 1\right)$ equals

a) $\frac{1}{4}$
b) $\frac{1}{2}$
c) $\frac{3}{4}$
d) $\frac{5}{8}$
e) none of the above

Answer: (b)
MC Moderate

11.4.7: A table saw cuts construction studding. Observation has shown that the lengths of the studs are normally distributed with mean 10 feet and standard deviation 6 inches. The probability that a randomly chosen stud exceeds 11 feet is equal to

a) $\frac{1}{\sqrt{2\pi}} \int_{9.5}^{10.5} e^{-x^2/2}\, dx$

b) $\frac{1}{\sqrt{2\pi}} \int_{11}^{\infty} e^{-x^2/2}\, dx$

c) $\frac{1}{\sqrt{\pi}} \int_{9.5}^{10.5} e^{-[(x - 10)/5]^2}\, dx$

d) $\frac{1}{.5\sqrt{2\pi}} \int_{11}^{\infty} e^{-(1/2)[(x - 10)/.5]^2}\, dx$

e) none of the above

Answer: (d)
MC Moderate

11.4.8: A random variable X is exponentially distributed with mean 2. Then $Pr(1 \le X \le 3)$ equals

a) $\frac{1}{2}e^{-1/2}$

b) $\frac{1}{2}e^{-3/2}$

c) $e^{-1/2} - e^{-3/2}$

d) $\frac{1}{2}\left(e^{-1} - e^{-3}\right)$

e) none of the above

Answer: (c)
MC Moderate

11.4.9: A random variable X is exponentially distributed with mean 10. Determine a so that $Pr(0 \le x \le a) = .75$.

a) $\ln \frac{3}{4}$

b) $\frac{1}{10} \ln 4$

c) $\frac{1}{10}$

d) $\ln .75$

e) none of the above

Answer: (e)
MC Moderate

11.4.10: A farmer has observed that the time to maturation of a certain crop is approximately normally distributed with a mean of 60 days and a standard deviation of 2 days. Find the percentage of plants that will mature in less than 55 days.

Answer: .0062
ES Moderate

11.4.11: Suppose X is a normal random variable with density function $f(x) = \frac{4}{\sqrt{2\pi}} e^{-1/2[4(x - 1)]^2}$

a) E(X) = 4, Var(X) = 1
b) E(X) = 1, Var(X) = 4
c) E(X) = 1, Var(X) = 16
d) E(X) = 1, Var(X) = 1/16
e) E(X) = 1/4, Var(X) = 1

Answer: (d)
MC Moderate

11.4.12: When mice are placed in a certain maze the amount of time it takes them to go through the maze is approximately normally distributed with a mean of 25 minutes and a standard deviation of 5 minutes. What is the normal density function for X, the amount of time it takes a mouse to complete the maze? What is the probability that a mouse will complete the maze in under 30 minutes?

Answer:

$$f(x) = \frac{1}{5\sqrt{2\pi}} e^{-\frac{1}{2}\left[\frac{(x - 25)}{5}\right]^2}$$

$\Pr(X \leq 30) = .5 + A(1) = .8413$
ES Moderate

11.4.13: Below is the graph of a density function of a normally distributed random variable. What is the expected value and standard deviation?

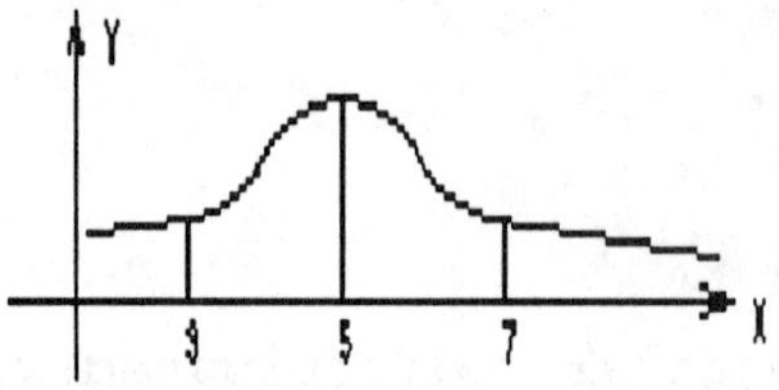

Answer: $E(X) = 5, \sqrt{\text{Var } x} = 2$
ES Moderate

11.4.14: By inspection give the mean and standard deviation of the random variables with the following density functions

(i) $f(x) = \frac{1}{2\sqrt{2\pi}} e^{-1/8\ x^2}$

(ii) $f(x) = e^{-.1x}/10$

(iii) $f(x) = \frac{7}{\sqrt{2\pi}} e^{-49/2(x - 3.9)^2}$

Answer:

(i) $E(X) = 0, \sqrt{Var(x)} = 1/2$

(ii) $E(X) = 10, \sqrt{Var(x)} = 10$

(iii) $E(X) = 3.9, \sqrt{Var(x)} = 1/7$

ES Moderate

11.4.15: It is estimated that the time between arrivals of visitors to a public library is an exponential random variable with expected value of 13 minutes. Find the probability that 30 minutes elapses without any arrivals.

Answer: $e^{-30/13} \cong .0995$

ES Moderate

11.4.16: An appliance comes with an unconditional money back guarantee for its first 6 months. It has been found that the time before the appliance experiences some sort of malfunction is an exponential random variable with mean 2 years. What percentage of appliances will malfunction during the warranty period?

Answer: $1 - e^{-.25} \cong .2212$ or 22.12%

ES Moderate

11.4.17: Let Z be the standard normal variable. Use the tables to compute

(i) $Pr(Z \le 1.3)$

(ii) $Pr(-3 \le Z \le 1)$

(iii) $Pr(Z \ge -1.1)$

Answer: i) $.5 + A(1.3) = .9032$

ii) $A(3) + A(1) = .84$

iii) $.5 + A(1.1) = .8643$

ES Moderate

11.4.18: When a road crew inspects a road that hasn't been worked on for several years then the distance between necessary repairs is an exponential random variable with a mean of .25 miles. What is the probability that the crew will find a mile long stretch of road that does not need repairs?

Answer: $e^{-4} \cong .0183$

ES Moderate

12. Taylor Polynomials and Infinite Series

12.1.1: Determine the third Taylor polynomial of $f(x) = \dfrac{1}{\sqrt{1 - x}}$ at $x = 0$.

Answer: $P_3(x) = 1 + \frac{1}{2}x + \frac{3}{8}x^2 + \frac{5}{16}x^3$
ES Moderate

12.1.2: Determine the first two Taylor polynomials of $\sin x^2$ at $x = 0$.

Answer: $P_1(x) = 0$; $P_2(x) = x^2$
ES Moderate

12.1.3: Determine the third Taylor polynomial of $f(x) = \ln(2 - x)$ at $x = 1$ and use it to estimate ln 1.3. [Hint: $\ln 1.3 = f(.7)$.]

Answer: $P_3(x) = -(x - 1) - \frac{1}{2}(x - 1)^2 - \frac{1}{3}(x - 1)^3$;
$\ln(1.3) \approx P_3(.7) = .264$
ES Moderate

12.1.4:

(a) Write down the fourth Taylor polynomial e^u at $u = 0$.

(b) Substitute $-x^2$ for u to obtain a polynomial that approximates e^{-x^2}.

(c) Use the polynomial in (b) to estimate the value of $\int_0^1 e^{-x^2}\,dx$.

Answer:

(a) $1 + u + \frac{1}{1 \cdot 2}u^2 + \frac{1}{1 \cdot 2 \cdot 3}u^3 + \frac{1}{1 \cdot 2 \cdot 3 \cdot 4}u^4$

(b) $1 - x^2 + \frac{1}{2}x^4 - \frac{1}{6}x^6 + \frac{1}{24}x^8$

(c) .74749
ES Moderate

12.1.5: Let $f(x) = \frac{1}{x + 1}$. Then the second Taylor polynomial $p_2(x)$ of $f(x)$ about $x = 0$ is given by

a) $1 - x + x^2$

b) $1 - x$

c) $1 + x - x^2$

d) $1 - 2x + 2x^2$

e) none of the above

Answer: (a)
MC Moderate

12.1.6: Let $f(x) = x^3 - 4x - 1$. Which of the following statements is true? (All Taylor polynomials are about $x = 0$.)

a) $p_3(1) = 7$

b) $p_2(-1) = 0$

c) $p_1(3) = -11$

d) $p_n = f(x)$ for all $n \geq 3$

e) none of the above

Answer: (d)
MC Moderate

12.1.7: The function $f(x) = \sin(x^2)$ is approximated by its second Taylor polynomial $p_2(x)$ about $x = 0$. Which of the following statements is <u>not</u> true?

a) $f''(0) = 2$

b) $p_2(x) = x^2$

c) $f'(0) = 0$

d) $p_2(x) = \frac{1}{3} + x^2$

e) none of the above

Answer: (d)
MC Moderate

12.1.8: A polynomial $f(x)$ for which $f(1) = -1$, $f'(1) = 2$, $f''(1) = -1$, $f'''(1) = -2$ is given by

a) $f(x) = x^3 - 2x^2 - 3x + 2$

b) $f(x) = -1 + 2(x - 1) - 1(x - 1)^2 - 2(x - 1)^3$

c) $f(x) = -1 + 2x - x^2 - 2x^3$

d) $f(x) = -1 + 2(x - 1) - \frac{1}{2}(x - 1)^2 - \frac{1}{3}(x - 1)^3$

e) none of the above

Answer: (d)
MC Moderate

12.1.9: Determine the first 4 terms of the Taylor series for $f(x) = \sin 2x$.

Answer:

$$f(x) = 2x - \frac{8x^3}{3!} + \frac{32x^5}{5!} - \frac{128x^7}{7!}$$

ES Moderate

12.1.10: Determine the first 4 terms of the Taylor series for $f(x) = \sqrt{4 - x}$.

Answer:

$$f(x) = 2 - \frac{x}{4} - \frac{x^2}{32 \cdot 2!} - \frac{3x^3}{256 \cdot 3!}$$

ES Moderate

12.1.11: Determine the first 4 terms of the Taylor series for $f(x) = e^x$.

Answer:

$$f(x) = 1 + x + \frac{x^2}{2!} + \frac{x^3}{3!}$$

ES Moderate

12.1.12: Determine the first 4 terms of the Taylor series for $f(x) = \ln(x + 1)$.

Answer:

$$f(x) = x - \frac{x^2}{2!} + \frac{2x^3}{3!} - \frac{6x^4}{4!}$$

ES Moderate

12.1.13: Determine the first 4 terms of the Taylor series for $f(x) = e^{-2x}$.

Answer:

$$f(x) = 1 - 2x + \frac{4x^2}{2!} - \frac{8x^3}{3!}$$

ES Moderate

12.1.14: Determine the first 3 terms of the Taylor series for $f(x) = e^x \sin x$.

Answer:

$$f(x) = x + x^2 + \frac{x^3}{3}$$

ES Moderate

12.1.15: If $f(x) = 2 + 3x - 2x^2 + 2x^3$, then what are $f''(0)$ and $f'''(0)$?

Answer: $f''(0) = -4$; $f'''(0) = 12$

ES Moderate

12.1.16: If $f(x) = 1 - 3(x - 2) + 4(x - 2)^2 + 6(x - 2)^3$, then what are $f''(2)$ and $f'''(2)$?

Answer: $f''(2) = 8$; $f'''(2) = 36$
ES Moderate

12.1.17: Below is a graph of function f(x). Which of the following could be the first Taylor polynomial of f(x) at x = 0?

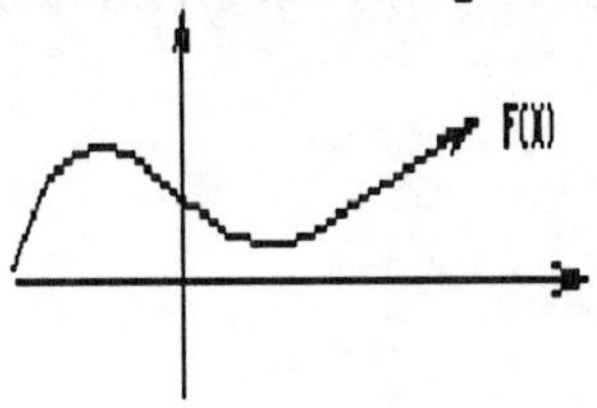

a) $p_1(x) = 3 + 4x$
b) $p_1(x) = 3 - 2x$
c) $p_1(x) = -2 - 3x$
d) $p_1(x) = -2 + 2x$
e) none of the above

Answer: (b)
MC Moderate

12.1.18: Below is a graph of function f(x). Which of the following could be the second Taylor polynomial of f(x) at x = a?

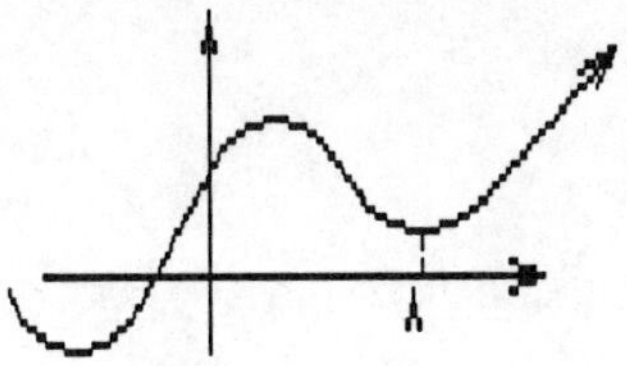

a) $p_2(x) = 13/3 - 5/3(x - a) + 2/3(x - a)^2$
b) $p_2(x) = 13/3 + 5/3(x - a)$
c) $p_2(x) = 13/3 + 2/3(x - a)^2$
d) $p_2(x) = 5/3(x - a) - 2/3(x - a)^2$
e) $p_2(x) = 13/3 - 2/3(x - a)^2$

Answer: (c)
MC Moderate

12.1.19: Suppose that the first Taylor polynomial of a function f(x) at x = 0 is $p_1(x) = 2 - 3x$. Which of the following could be a graph of f(x)?

a)

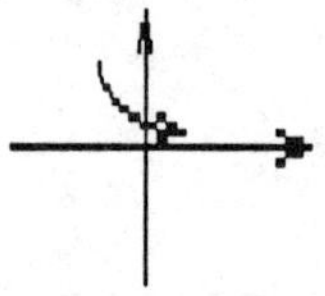

b)

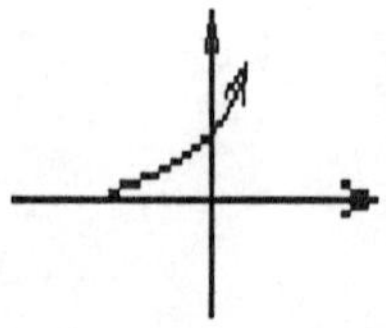

c)

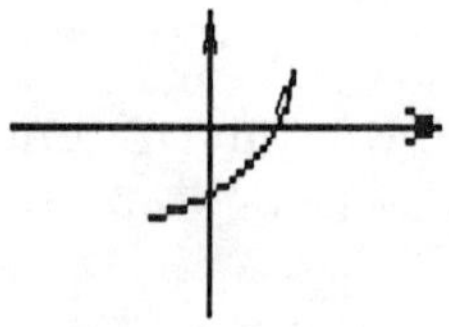

d)

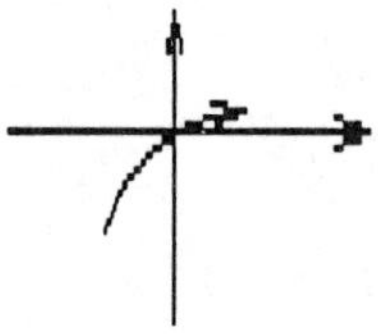

e)

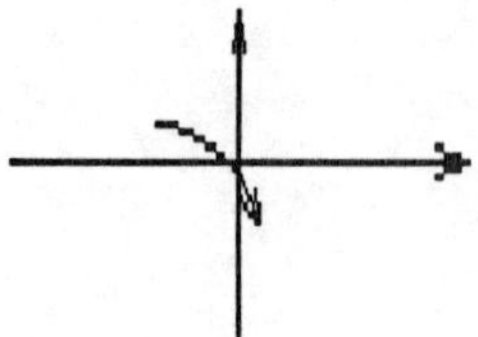

Answer: (a)
MC Moderate

12.1.20: Find the second Taylor polynomial for $f(x) = \dfrac{1}{\sqrt{x + 4}}$ at $x = 0$. Use it to approximate $1/\sqrt{4.1}$

Answer: $p_2(x) = 1/2 - 1/16\ x + 3/2!\ (128)x^2$

$$\frac{1}{\sqrt{4.1}} = f(.1) \cong \frac{1}{2} - \frac{1}{160} + \frac{3}{356000}$$

ES Moderate

12.1.21: Find the second Taylor polynomial for $f(x) = \sin x^2$ at $x = 0$. Use it to approximate the area under the curve $f(x)$ between 0 and $\pi/2$.

Answer:

$$p_2(x) = x^2 \qquad A \cong \int_0^{\pi/2} x^2 dx = \frac{1}{3}x^3 \Big|_0^{\pi/2} = \frac{\pi^3}{24}$$

ES Moderate

12.1.22: The area of a circle with radius 1 is π. If $f(x) = \sqrt{1 - x^2}$ gives the top half of this circle, as illustrated below, use the second Taylor polynomial of $f(x)$ at $y = 0$ to find an approximate value for π.

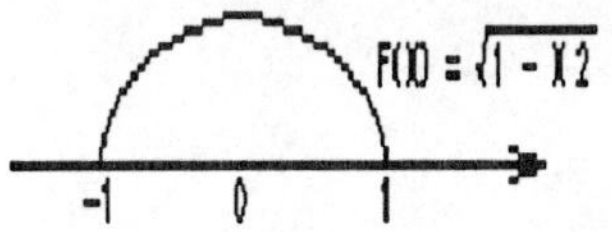

Answer: $p_2(x) = 1 - 1/2\ x^2$

$$\frac{\pi}{2} \cong \int_{-1}^{1} 1 - \frac{1}{2}x^2 dx = x - \frac{1}{6}x^3 \Big|_{-1}^{1} = \left(1 - \frac{1}{6}\right) - \left(-1 + \frac{1}{6}\right) = 2 - \frac{1}{3} =$$

$\pi \cong 3.\overline{3}$

ES Moderate

12.1.23: Find the third Taylor polynomial of $f(x) = \sin x$ at $x = 0$. Use it to approximate $\sin 1/2$.

Answer: $p_3(x) = x - x^3/3!$

$$\sin \frac{1}{2} \cong \frac{1}{2} - \frac{1}{54} \cong .4815$$

ES Moderate

12.1.24: Suppose $f(x) = x^4 - 7x^3 + 2$. Indicate which of the following statements are true:
1) The fifth Taylor polynomial of $f(x)$ at $x = 1$ is $p_5(x) = x^4 - 7x^3 + 2$
2) The fifth Taylor polynomial of $f(x)$ at $x = 0$ is $p_5(x) = x^4 - 7x^3 + 2$
3) The third Taylor polynomial of $f(x)$ at $x = 1$ is $p_3(x)$ $2- 7x^3$
4) The third Taylor polynomial of $f(x)$ at $x = 0$ is $p_3(x) = 2 - 7x^3$

Answer: (1), (2), (4) TRUE
ES Moderate

12.1.25: Determine all Taylor polynomials of $f(x) = x^3 - 2x + 4$ at $x = 1$.

Answer: $p_1(x) = 3 + (x - 1)$ $p_2(x) = 3 + (x - 1) + 3(x - 1)^2$
$f(x) = p_3(x) = 3 + (x - 1) + 3(x - 1)^2 + (x - 1)^3 = p_n(x)$, $n \geq 3$
ES Moderate

12.1.26: Find the third Taylor polynomial of e^x at $x = 0$ and use it to approximate e.

Answer: $p_3(x) = 1 + x + x^2/2 + x^3/6$

$e = e^1 \cong p_3(1) = 1 + 1 + 1/2 + 1/6 \cong 2.6\overline{7}$
ES Moderate

12.1.27: Sketch the graph of $y = e^x$ and of its first two Taylor polynomials on the same axis.

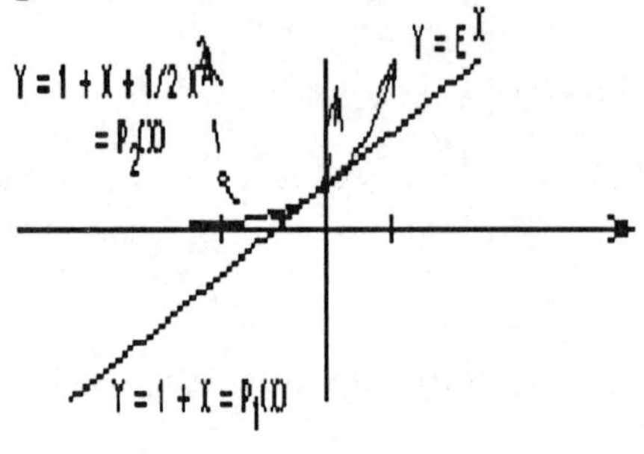

Answer:
ES Moderate

12.1.28: If the following graph is a graph of $f(x)$

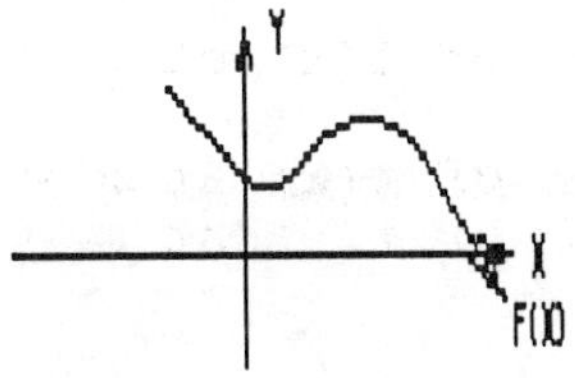

Which of the following could be the first Taylor polynomial of f?

a) $p_1(x) = 2 + 2x$
b) $p_1(x) = 2 - 2x$
c) $p_1(x) = 2x - 2$
d) $p_1(x) = 2x + 5$

Answer: (b)
MC Moderate

12.1.29: Find the third Taylor polynomials of $x^2 + \sin x$.

Answer: $x + x^2 - x^3/3!$
ES Moderate

12.1.30: Suppose $f(0) = 1$, $f'(0) = 1$, $f''(0) = -1$. Use a Taylor series to approximate $f(1/2)$.

Answer: $f(1/2) \cong 11/8$
ES Moderate

12.1.31: Determine all Taylor polynomials of $f(x) = x^3 - 3x$.

Answer: $p_1(x) = p_2(x) = -3x$, $p_3(x) = x^3 - 3x = p^n(x)$, $n > 3$
ES Moderate

12.1.32: Estimate $\int_0^1 e^{x^2} dx$ by using the second Taylor polynomial for $f(x) = e^{x^2}$.

Answer: $p_2(x) = 1 + x^2$, $\int_0^1 e^{x^2} dx \cong \frac{4}{3}$
ES Moderate

12.1.33: Find the second Taylor polynomial of $f(x) = \sqrt{x}$ at $x = 9$ and use it to approximate $\sqrt{9.1}$

Answer: $p_3(x) = 3 + \frac{1}{6}(x - 9) - \frac{1}{216}(x - 9)^2$, $p_3(9.1) \cong 3.02$
ES Moderate

12.1.34: Find the third Taylor polynomial of $\cos x$ at $x = \pi/2$.

Answer: $p_3(x) = 1/3!(x - \pi/2)^3 - (x - \pi/2)$
ES Moderate

12.1.35: Use the second Taylor polynomial about $x = 1$ to estimate $\int_1^2 \ln x^2 dx$

Answer: $p_2(x) = 2(x - 1) - (x - 1)^2$, $\int_1^2 \ln x^2 dx \cong \frac{2}{3}$
ES Moderate

12.1.36: Suppose the second Taylor polynomial for $f(x)$ at $x = 3$ is $p_3(x) = 2(x - 3) - 1/3(x - 3)^2$. Find $f(3)$, $f'(3)$, $f''(3)$.

Answer: $f(3) = 0$, $f'(3) = 2$, $f''(3) = -2/3$
ES Moderate

12.2.1:
(a) Show that $x^5 + x - 3$ has a zero between 1 and 2.

(b) Use two repetitions of the Newton-Raphson algorithm to approximate this zero.

Answer:
(a) If $f(x) = x^5 + x - 3$, then $f(1) = -1 < 0$ and $f(2) = 31 > 0$.

(b) If $x_0 = 1$, then $x_1 = \frac{7}{6}$, $x_2 = 1.1347$. If $x_0 = 2$, then $x_1 = 1.61728$, $x_2 = 1.34229$.

ES Moderate

12.2.2:
(a) By graphing $y = \sin x$ and $y = \frac{1}{2}x$ on a common coordinate system, show that the equation $\sin x = \frac{1}{2}x$ has exactly three solutions, with exactly one positive solution.

(b) Use the Newton-Raphson algorithm with two repetitions to estimate the positive solution of $\sin x = \frac{1}{2}x$. Set $x_0 = 2$.

Answer: (b) $x_0 = 2$; $x_1 = 1.90100$; $x_2 = 1.89551$
ES Moderate

12.2.3: Use three repetitions of the Newton-Raphson algorithm to estimate $5^{1/3}$.

Answer:
$x_0 = 1$; $x_1 = \frac{7}{4}$; $x_2 = 1.71088$; $x_3 = 1.709976$
ES Moderate

12.2.4: Suppose that when a drug is administered to a patient, the concentration of the drug in the blood at time t after the injection is approximated by the function $f(t) = 5te^{-.2t}$. At $t = 5$, the concentration reaches its maximum of 9.197. Estimate the time when the concentration will drop down to 6. Use the Newton-Raphson algorithm with two repetitions, starting with $t_0 = 10$.

Answer: $t_0 = 10$; $t_1 = 11.13313$; $t_2 = 11.142$
ES Moderate

12.2.5: The Newton-Raphson method is used to approximate the zero of $f(x) = x^3 + x - 5$ between $x = 1$ and $x = 2$. If x_0 is taken as 1, then x_1 equals

a) $\frac{3}{4}$
b) $\frac{1}{4}$
c) $\frac{7}{4}$
d) $\frac{7}{3}$
e) none of the above

Answer: (c)
MC Moderate

12.2.6: The Newton-Raphson method is applied to estimate $\sqrt{10}$. If $x_0 = 3$, then x_2 equals

a) $\frac{5}{3}$
b) $\frac{721}{228}$
c) $\frac{758}{521}$
d) $\frac{700}{237}$
e) none of the above

Answer: (b)
MC Moderate

12.2.7: The Newton-Raphson method is applied to estimate a zero of f(x) with $x_0 = 3$. Which of the following statements is true?

a) $x_1 = \frac{f'(3)}{f(3)}$
b) $x_1 = 3 - \frac{f(3)}{f'(3)}$
c) $x_1 = 3 - \frac{f'(3)}{f(3)}$
d) $x_1 = 3 + \frac{f'(3)}{f(3)}$
e) none of the above

Answer: (b)
MC Moderate

12.2.8: Use the Newton-Raphson algorithm with three repetitions to estimate $\sqrt{3}$.

Answer: 1.731698484
ES Moderate

12.2.9: Use the Newton-Raphson algorithm with three repetitions to locate the zero of $f(x) = e^{x} - 2$ near $x = 1$.

Answer: $x = .6931475811$
ES Moderate

12.2.10: Use the Newton-Raphson algorithm with three repetitions to locate the zero of $f(x) = \cos x + x - 2$ near $x = 2$.

Answer: $x = .233790961$
ES Moderate

12.2.11: Use the Newton-Raphson algorithm with three repetitions to approximate the solution to $e^{-x} = 2 - x$.

Answer: 1.955865069
ES Moderate

12.2.12: Below is a graph of a function $f(x)$. If x_0 is taken as the initial approximation of the zero of $f(x)$ then which of the following points could be given by the Newton-Raphson algorithm as the next approximation?

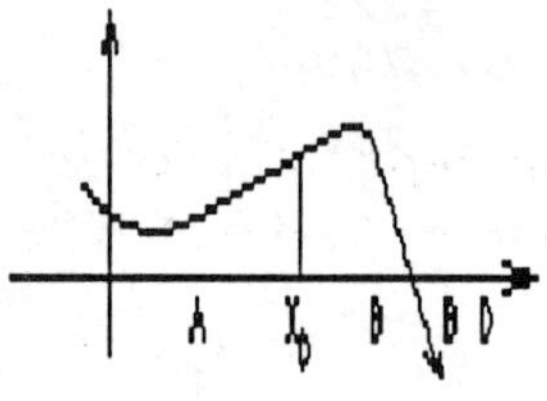

a) A b) B c) C d) D

Answer: (a)
MC Moderate

12.2.13: Below is a graph of functions h(x) and g(x). In using the Newton-Raphson algorithm to find where h(x) = g(x), which of the following statements is false?

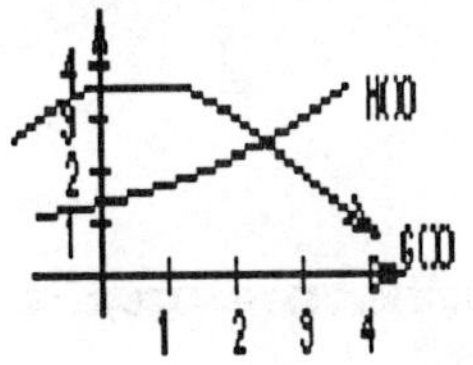

a) Use the Newton-Raphson algorithm to find the zeroes of f(x) = h(x) - g(x)
b) Use the Newton-Raphson algorithm to find the zeroes of f(x) = h(x) + g(x)
c) Use the Newton-Raphson algorithm to find the zeroes of f(x) = g(x) - h(x)
d) x_0 = 3 could be used as the initial approximation
e) x_0 = 4 could be used as the initial approximation

Answer: (b)
MC Moderate

12.2.14: Use two repetitions of the Newton-Raphson algorithm to approximate the value of x for which $e^x = 3x$. Use x = 0 as the first approximation.

Answer: $x_2 \cong .61$
ES Moderate

12.2.15: Use two repetitions of the Newton-Raphson algorithm to approximate the zero of f(x) = sin x - cos x near x = 0.

Answer: $x_2 = .782$

ES Moderate

12.2.16: Suppose x_0 is an initial approximation of a zero of the function f(x). Using the Newton-Raphson algorithm, a second approximation, x_1 is obtained. Which of the following must be true?
a) x_1 is closer to the zero of f(x) than x_0
b) $x_1 = x_0 - f'(x_0)/f(x_0)$
c) $f(x_1) = 0$
d) x_1 is the x-coordinate of the x-intercept of the tangent line to f(x) at $(x_0, f(x_0))$
e) All of the above are true

Answer: (d)
MC Moderate

12.2.17: Use two repetitions of the Newton-Raphson algorithm to estimate $\sqrt{15}$.

Answer: $x_0 = 4$, $x_1 = 4.033$, $x_2 = 3.876$
ES Moderate

12.2.18: Use two repetitions of the Newton-Raphson algorithm to find the value of x near zero for which cos x = x.

Answer: $x_0 = 0$, $x_1 = 1$, $x_2 = .75$
ES Moderate

12.2.19: Use two repetitions of the Newton-Raphson algorithm to find the value of x near zero for which $e^x = 2\cos x$

Answer: $x_0 = 0$, $x_1 = .6279$
ES Moderate

12.3.1: Determine the sum of the following series:

$$1 - \frac{1}{2^3} + \frac{1}{2^6} - \frac{1}{2^9} + \frac{1}{2^{12}} - \ldots.$$

Answer: $\frac{8}{9}$
ES Moderate

12.3.2: Determine the sum of the following series:

$$\frac{2^2}{5^3} + \frac{2^4}{5^5} + \frac{2^6}{5^7} + \frac{2^8}{5^9} + \frac{2^{10}}{5^{11}} + \ldots.$$

Answer: $\frac{4}{105}$
ES Moderate

12.3.3: The infinite series

$$\frac{2^2}{5^2} - \frac{2^3}{5^3} + \frac{2^4}{5^4} - \ldots$$

has the sum

a) $\frac{5}{7}$

b) $\frac{3}{5}$

c) $\frac{4}{35}$

d) $\frac{7}{5}$

e) none of the above

Answer: (c)
MC Moderate

12.3.4: Determine the sum of the geometric series $1 + \frac{1}{\sqrt{2}} + \frac{1}{2} + \ldots$.

Answer: $\frac{2}{2 - \sqrt{2}}$
ES Moderate

12.3.5: Determine the sum of the geometric series $2 + \frac{4}{3} + \frac{8}{9} + \ldots$.

Answer: 6
ES Moderate

12.3.6: Determine the sum of the geometric series $\sum_{k=0}^{\infty} (1 - \sqrt{2})^k$.

Answer: $\frac{\sqrt{2}}{2}$
ES Moderate

12.3.7: Determine the sum of the geometric series $\sum_{n=1}^{\infty} 2^n \left(\frac{1}{3}\right)^{n-1}$.

Answer: 6
ES Moderate

12.3.8: Determine the sum of the geometric series $\sum_{n=1}^{\infty} \frac{2^n + (-1)^n}{3^n}$.

Answer: $\frac{7}{4}$
ES Moderate

12.3.9: Determine the sum of the geometric series

$$\sum_{n=1}^{5} \lim_{m \to \infty} \sum_{x=n}^{n+m} \left(\frac{1}{2}\right)^x .$$

Answer: $\frac{31}{16}$
ES Moderate

12.3.10: Determine the sum of the geometric series $\sum_{n=1}^{\infty} \frac{2^n + (-1)^n}{3^n}$.

Answer: $\frac{7}{4}$
ES Moderate

12.3.11: Determine the fraction form of $.\overline{37}$.

Answer: $\frac{37}{99}$
ES Moderate

12.3.12: Determine the fraction form of $.4\overline{98}$.

Answer: $\frac{494}{990}$
ES Moderate

12.3.13: Consider the following geometric series $\frac{25}{2} - \frac{15}{2} + \frac{9}{2} - \frac{27}{10} + \ldots$. Which of the following statements is true?

a) This series diverges.
b) The ratio r of this series is 3/5
c) The sum of this series is 125/4
d) Another way writing this series is $\sum_{k=1}^{\infty} \left(-\frac{3}{5}\right)^k$
e) none of the above

Answer: (e)
MC Moderate

12.3.14: A student receives $100 at the start of each month from his parents. Every month the student spends 70% of all the money he has. If the only money the student receives is the money from his parents, estimate how much money the student will have at the beginning of each month after an extended period of time.

Answer: $\sum_{k=0}^{\infty} 100(.7)^k = \frac{1000}{3} \cong \$333.33$
ES Moderate

12.3.15: Find the rational number whose decimal expansion is $.185\underline{185}$

Answer: 5/27
ES Moderate

12.3.16: Find the rational number whose decimal expansion is $.196\underline{96}$

Answer: 13/66
ES Moderate

12.3.17: A patient receives M milligrams of a certain drug every hour. Each hour the body eliminates a fraction p of the amount of drug in the body. After an extended period of time which of the following series approximates the amount of drugs present in the patients body immediately before receiving an hourly dose?

a) $\sum_{k=0}^{\infty} Mp^k$

b) $\sum_{k=0}^{\infty} (Mp)^k$

c) $\sum_{k=1}^{\infty} Mp^k$

d) $\sum_{k=1}^{\infty} Mp^{-k}$

e) $M\sum_{k=0}^{\infty} p^k$

Answer: (c)
MC Moderate

12.3.18: Sum the following series 3 - 1.8 + 1.08 + .648 - ...

Answer: 1.875
ES Moderate

12.3.19: Sum the following series $\frac{27}{5} + \frac{18}{5} + \frac{12}{5} + \frac{8}{5} + \ldots$

Answer: $\frac{81}{5}$
ES Moderate

12.3.20: Sum the following series $\sum_{k=0}^{\infty} \left(\frac{1}{3}\right)^k (2)^{k+1}$

Answer: 6
ES Moderate

12.3.21: Sum the following series $\sum_{k=0}^{\infty} (-1)^k \frac{2}{7^k}$

Answer: $\frac{7}{4}$
ES Moderate

12.3.22: The series $\sum_{k=0}^{\infty} \left(\frac{1}{2}\right)^{2k}$ has sum

a) 2
b) 4/5
c) 4/3
d) 2/3
e) none of the above

Answer: (c)
MC Moderate

12.3.23: Give an example of a divergent geometric series.

Answer:
ES Moderate

12.3.24: Give an example of a convergent geometric series and give its sum.

Answer:
ES Moderate

12.3.25: Suppose $\sum_{k=0}^{\infty} \left(\frac{2}{m}\right)^{k}$ converges. What can you say about the value of m?

Answer: $-2 < m < 2$
ES Moderate

12.3.26: Find the sum of the following geometric series: $1 + (.25)^2 + (.25)^4 + (.25)^8 + \ldots$

Answer: $\frac{16}{15}$
ES Moderate

12.3.27: The geometric series $1 + (.2)^3 + (.2)^6 + (.2)^9 + \ldots$

I. converges

II. is equal to $\sum_{k=0}^{\infty} \left(\frac{1}{125}\right)^{k}$

III. is equal to $\sum_{k=0}^{\infty} (.2)^{3k}$

a) I, II
b) I, III
c) II, III
d) III
e) I, II, III

Answer: (e)
MC Moderate

12.3.28: Give the first four terms in the geometric series $\sum_{k=0}^{\infty} \ln 2(3^{-k})$. Does this series converge or diverge? If converges, give its sum.

Answer: $\ln 2 + \frac{\ln 2}{3} + \frac{\ln 2}{9} + \frac{\ln 2}{27} + \ldots$
sum = $(\ln 2)3/2$
ES Moderate

12.3.29: Write the following geometric series in sigma notation and give its sum $10 + 4 + \frac{8}{5} + \frac{16}{25} + \ldots$

Answer: $\sum_{k=1}^{\infty} 25(2/5)^k = 50/3$
ES Moderate

12.4.1: The <u>Morning Herald</u> surveys the number of typographical errors per page and finds that the probability of n errors is equal to

$$\frac{3^n}{n!}e^{-3}.$$

The probability that a given page has a smaller than average number is equal to

a) $1 + 12e^{-3}$

b) $\frac{17}{2}e^{-3}$

c) $1 + \frac{8}{3}e^{-3}$

d) $1 + \frac{4}{5}e^{-3}$

e) none of the above

Answer: (b)
MC Moderate

12.4.2: A quality control expert inspects carburators in an automobile assembly line. The probability that n nondefective carburators are observed before a defective one is spotted is equal to $\frac{1}{10}(.9)^n$. What is the average number of nondefective carburators observed prior to observing a defective one?

a) 9

b) 8

c) 7

d) 10

e) none of the above

Answer: (a)
MC Moderate

12.4.3: Use the integral test to determine whether $\sum_{k=1}^{\infty} \frac{1}{k\sqrt{k}}$ converges.

Answer: convergent
ES Moderate

12.4.4: It can be shown that $\int_0^{\infty} xe^{-x}dx = 1$. Use this fact and the integral test to construct an appropriate convergent infinite series.

Answer: $\sum_{k=0}^{\infty} ke^{-k}$
ES Moderate

12.4.5: Does $\sum_{k=1}^{\infty} \frac{1}{(k + 2)^2}$ converge or diverge. Explain.

Answer: convergent since $\int_1^{\infty} \frac{1}{(x + 2)^2}dx < \infty$
ES Moderate

12.4.6: Use the integral test to determine whether $\sum_{k=1}^{\infty} \frac{2}{2k + 1}$ converges or diverges. Then use the comparison test to determine whether $\sum_{k=1}^{\infty} \frac{4}{k + 1}$ converges or diverges.

Answer: both diverge
ES Moderate

12.4.7: Does $\sum_{k=1}^{\infty} \frac{k}{(k^2 + 2)^2}$ converge or diverge.

Answer: converges
ES Moderate

12.4.8: Find the limit of the following expressions as k ∞:
i) $\frac{2k + 1}{k}$ ii) $\frac{k!}{(k + 1)!}$ iii) $\frac{2^{2(k + 1)}}{2^{2k}}$

Answer: 2, 0, 4
ES Moderate

12.5.1: Does $\sum_{k=1}^{\infty} \frac{e^k}{2k}$ converge or diverge?

Answer: diverge
ES Moderate

12.5.2: Does $\sum_{k=1}^{\infty} \frac{2k!e^k}{4k^2}$ converge or diverge?

Answer: diverge
ES Moderate

12.5.3: Does $\sum_{k=1}^{\infty} \frac{k^2}{k!e^k}$ converge or diverge?

Answer: converge
ES Moderate

12.5.4: By applying the ratio test determine if the following series converges or not $\sum_{k=1}^{\infty} \frac{2^k}{k!}$

Answer: converges
ES Moderate

12.5.5: Consider the following infinite series where m is a positive number. $\sum_{k=1}^{\infty} m^k k$. Find $|a_{k+1}|/|a_k|$. For what values of m will the ratio test ensure that the series converges? For what values of m is the ratio test inconclusive?

Answer: $\frac{m(k+1)}{k}$, converges when $m < 1$
inconclusive for $m = 1$
ES Moderate

12.5.6: Use the ratio test to determine if the following series converges or not $\sum_{k=1}^{\infty} \frac{2k^2}{e^k}$

Answer: converges
ES Moderate

12.6.1: Determine the Taylor series for $f(x) = e^{3x}$ at $x = 0$.

Answer:

$$1 + 3x + \frac{3^2}{1 \cdot 2}x^2 + \frac{3^3}{1 \cdot 2 \cdot 3}x^3 + \frac{3^4}{1 \cdot 2 \cdot 3 \cdot 4}x^4 + \ldots$$

ES Moderate

12.6.2: (a) Write down the Taylor series expansion of $\frac{1}{1 + x}$ for $|x| < 1$.

(b) Multiply the series in (a) by x to obtain a series expansion of $\frac{x}{1 + x}$, and then use these two series to obtain a series expansion of $\frac{1 - x}{1 + x}$ for $|x| < 1$.

Answer: (a) $\frac{1}{1 + x} = 1 - x + x^2 - x^3 + x^4 - \ldots$

(b) $\frac{x}{1 + x} = x - x^2 + x^3 - x^4 + x^5 - \ldots;$

$\frac{1 - x}{1 + x} = 1 - 2x + 2x^2 - 2x^3 + 2x^4 - \ldots$

ES Moderate

12.6.3: The Taylor series of $f(x) = \frac{1}{1 - 3x}$ about $x = 0$ is equal to

a) $1 + \frac{3x}{1}x + \frac{9x}{1.2}x^2 + \ldots$

b) $1 + 3x + 9x^2 + 27x^3 + \ldots$

c) $1 - 3x + 9x^2 - 27x^3 + \ldots$

d) $1 + (3x)^2 + (3x)^4 + (3x)^6 + \ldots$

e) none of the above

Answer: (b)

MC Moderate

12.6.4: Let $f(x) = 1 + 2(x - 2) + 3(x - 2)^2 + 4(x - 2)^3 + \ldots$.

Then $f(\frac{5}{2}) =$

a) $\frac{1}{2}$

b) $\frac{1}{4}$

c) 2

d) 4

e) none of the above

Answer: (d)

MC Moderate

12.6.5: England decides to decrease taxes by 7 billion. It is estimated that of each pound received, a typical citizen will spend 90%. The level of economic activity generated by the tax cut is therefore estimated to be

$$7 \cdot (.90) + 7 \cdot (.90)^2 + 7 \cdot (.90)^3 + \ldots$$

billion pounds. This amount is equal to

a) 63 billion
b) 45 billion
c) 20 billion
d) 3.6 billion
e) none of the above

Answer: (a)
MC Moderate

12.6.6: Determine the first four terms of the Taylor series for $f(x) = \sin x^3$.

Answer:

$$f(x) = x^3 - \frac{x^9}{3!} + \frac{x^{15}}{5!} - \frac{x^{21}}{7!}$$

ES Moderate

12.6.7: Determine the first three terms of the Taylor series for $f(x) = x \cos x - \sin x$.

Answer:

$$-\frac{(3! - 2!)x^3}{2!3!} + \frac{(5! - 4!)x^5}{4!5!} - \frac{(7! - 6!)x^7}{6!7!}$$

ES Moderate

12.6.8: Determine the first four terms of the Taylor series for $f(x) = xe^{(1/2)x}$.

Answer:

$$f(x) = x + x^{3/2} + \frac{x^2}{2!} + \frac{x^{5/2}}{3!}$$

ES Moderate

12.6.9: Find the Taylor series expansion for $\frac{x}{1 - x}$. Which of the following is false?

a) $1/2 = 1/3 + (1/3)^2 + (1/3)^3 + \ldots$
b) $-3/5 = (-3/2) + (-3/2)^2 + (-3/2)^3 + \ldots$
c) $\frac{x}{1 - x} = x + x^2 + x^3 + x^4 + \ldots$
d) $-1/3 = (-1/2) + (-1/2)^2 + (-1/2)^3 + \ldots$
e) All the statements are true.

Answer: (b)
MC Moderate

12.6.10: Find the first four terms of the Taylor series for $f(x) = 1 + xe^x$.

Answer: $1 + x + x^2 + x^3/2!$
ES Moderate

12.6.11: Find the first three terms of the Taylor series for $f(x) = 3x + \log(1 - 2x)$

Answer: $x - 2x^2 - \dfrac{8x^3}{3}$
ES Moderate

12.6.12: Suppose the first three terms of the Taylor series for a function $f(x)$ are as follows: $f(x) = 1 - 2x^2 + 5x^4 + \ldots$ Find $f'(0)$ and $f''(0)$ without differentiating $f(x)$.

Answer: $f'(0) = 0$, $f''(0) = -4$
ES Moderate

12.6.13: Find Taylor series of $f(x) = x^2e^{2x}$. Use enough terms to calculate .25e to two decimal places of accuracy.

Answer: $f(x) = x^2 + 2x^3 + 2x^4 + 8x^5/3! + 16x^6/4! + \ldots$
$e/4 = f(1/2) \cong .67$
ES Moderate

12.6.14: Suppose $f(x) = x + 2x^2 + 4x^3 + 8x^4 + 16x^5 + \ldots$
Then $f(1/4) =$
a) 109/225
b) 1/2
c) 27/58
d) 5/8
e) none of the above

Answer: (b)
MC Moderate

12.6.15: The Taylor series of $f(x) = x\ln(1 + 2x)$ is
a) $2x - \dfrac{4x^2}{2!} + \dfrac{8x^3}{3!} - \dfrac{16x^4}{4!} + \ldots$
b) $2x^2 + \dfrac{4x^3}{2!} + \dfrac{8x^4}{3!} + \dfrac{16x^5}{4!} + \ldots$
c) $2x^2 - \dfrac{4x^3}{3!} + \dfrac{8x^4}{4!} - \dfrac{16x^5}{5!} + \ldots$
d) $2x^2 - \dfrac{4x^3}{2!} + \dfrac{8x^4}{3!} - \dfrac{16x^5}{5!} + \ldots$
e) none of the above

Answer: (d)
MC Moderate

12.6.16: Find the first four terms of the Taylor series of $f(x) = \cos 3x + \sin 2x$

Answer: $1 + 2x - 9x^2/2! - 8x^3/3! + \ldots$
ES Moderate

12.6.17: Use the ratio test to determine whether or not the following series converges for $x = 2$. $\sum_{k=1}^{\infty} \frac{e^k}{k!} x^k$

Answer: converges
ES Moderate

12.6.18: Consider the following series $\sum_{k=1}^{\infty} \frac{k}{2^k} x^k$. Use the ratio test to find a value of x for which the series diverges and one for which it converges.

Answer: converges $|x| < 2$, diverges $|x| > 2$
ES Moderate

12.7.1: A basketball player attempts successive free throws until he succeeds in making a basket. Suppose the probability of success on each attempt is .7; thus, the probability of exactly n failures before the first success is $(.3)^n(.7)$, $n \geq 0$. What is the probability that the number of failures before the first successful free throw is odd?

Answer: $\frac{21}{91} = .23077$
ES Moderate

12.7.2: A person throws a die until the side with two spots appears. The probability of throwing the die exactly n times before throwing a "2" is $\left(\frac{5}{6}\right)^n \left(\frac{1}{6}\right)$, $n \geq 0$. What is the probability that the number of throws before throwing a "2" is even?

Answer: $\frac{6}{11}$
ES Moderate

12.7.3: Suppose that during a certain part of the day, the number X of automobiles that arrive within any one minute at a tollgate is Poisson distributed, and

$$\Pr(X = k) = \frac{4^k e^{-4}}{1 \cdot 2 \cdot \ldots \cdot k}.$$

What is the average number of automobiles that arrive per minute?

Answer: 4
ES Moderate

12.7.4: A new car dealer observes that the number X of warranty claims for repairs on each new car sold is Poisson distributed, with an average of six claims per car. Compute the probability that a new car sold by the dealer will have no more than three warranty claims.

Answer: $61e^{-6} = .15120$
ES Moderate

12.7.5: In a certain office, the number of typewriters that break down during any given week is Poisson distributed with $\lambda = 2$. What is the probability that more than three typewriters break down during a week?

Answer: $1 - \frac{38}{6}e^{-2} = .14288$
ES Moderate

12.7.6: Suppose at a typesetting company the probability that n consecutive words will occur without a typographical error is $(.93)^n$. What is the average number of words between typographical errors?

Answer: approx 204
ES Moderate

12.7.7: Suppose a bag holds 3 blue balls and a red ball. We pull a ball from the bag at random, return it and then repeat the process. Suppose we continue pulling balls until the blue ball is drawn and then we observe the number of consecutive red balls we drew. What is the average number of red balls between occurrences of blue balls.

Answer:

$$E(X) = \sum_{n=1}^{\infty} n\frac{3^n}{4^{n+1}} = \frac{1}{4}\sum_{n=1}^{\infty} n\left(\frac{3}{4}\right)^n = \frac{1}{(1 - 3/4)^2} = 4$$

ES Moderate

12.7.8: Suppose the number of cars passing through a toll booth in a 10 minute interval is a Poisson random variable. If the average number of cars is 23 give an expression for the probability that n cars pass through the booth.

Answer: $p_n = \frac{(23)^n}{n!}e^{-23}$

ES Moderate

12.7.9: Suppose a small amount of blood is sampled and the number of white blood cells is counted. If the number of white blood cells is Poisson distributed with $\lambda = 6$, what is the probability that a person has more than 4 white blood cells? What is the average number of white blood cells?

Answer: .7149, 6

ES Moderate

12.7.10: Suppose X is a random variable whose probabilities are Poisson distributed with $p_n = \frac{(14)^n}{n!}e^{-14}$

a) The expected value of X is e^{-14}
b) the probability that x = 14 is approximately .1060
c) The standard deviation of X is 14
d) The probability that x = 0 is zero
e) All of the above

Answer: (b)

MC Moderate

Exercises 0.1

1. 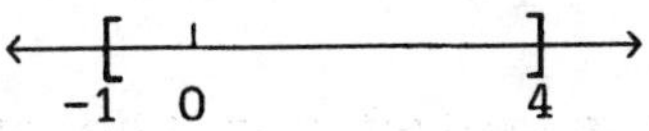

2. 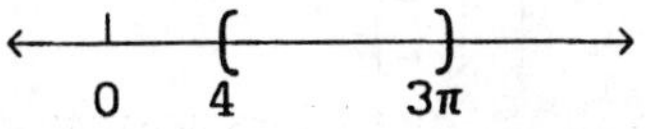

3. 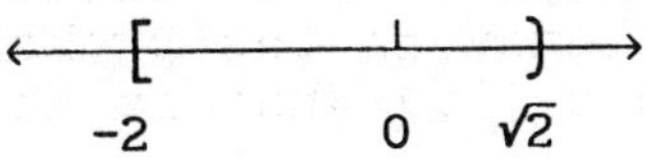

4. 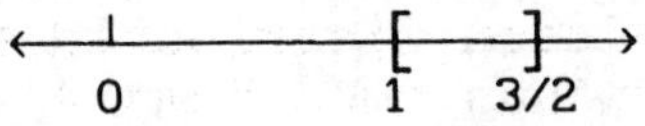

5. 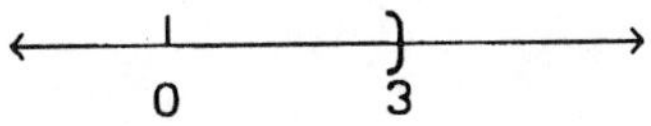

6. 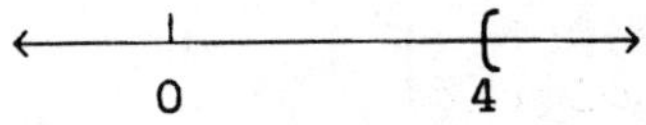

7. $[2, 3)$ 8. $\left[-1, \frac{3}{2}\right]$ 9. $[-1, 0)$ 10. $[-1, 8)$

11. $(-\infty, 3)$ 12. $[\sqrt{2}, \infty)$

13. $f(x) = x^2 - 3x;\ f(0) = 0^2 - 3(0) = 0$

$f(5) = 5^2 - 3(5) = 25 - 15 = 10;\ f(3) = 3^2 - 3(3) = 9 - 9 = 0$

$f(-7) = (-7)^2 - 3(-7) = 49 + 21 = 70$

14. $f(x) = 9 - 6x + x^2;\ f(0) = 9 - 6(0) + 0^2 = 9 - 0 + 0 = 9$

$f(2) = 9 - 6(2) + 2^2 = 9 - 12 + 4 = 1$

$f(3) = 9 - 6(3) + 3^2 = 9 - 18 + 9 = 0$

$f(-13) = 9 - 6(-13) + (-13)^2 = 9 + 78 + 169 = 256$

15. $f(x) = x^3 + x^2 - x - 1;\ f(1) = 1^3 + 1^2 - 1 - 1 = 0$

$f(-1) = (-1)^3 + (-1)^2 - (-1) - 1 = 0$

$f\left(\frac{1}{2}\right) = \left(\frac{1}{2}\right)^3 + \left(\frac{1}{2}\right)^2 - \left(\frac{1}{2}\right) - 1 = -\frac{9}{8};\ f(a) = a^3 + a^2 - a - 1$

16. $g(t) = t^3 - 3t^2 + t;\ g(2) = 2^3 - 3(2)^2 + 2 = 8 - 12 + 2 = -2$

$g\left(-\frac{1}{2}\right) = \left(-\frac{1}{2}\right)^3 - 3\left(-\frac{1}{2}\right)^2 + \left(-\frac{1}{2}\right) = -\frac{1}{8} - \frac{3}{4} - \frac{1}{2} = -\frac{11}{8}$

$g\left(\frac{2}{3}\right) = \left(\frac{2}{3}\right)^3 - 3\left(\frac{2}{3}\right)^2 + \left(\frac{2}{3}\right) = \frac{8}{27} - \frac{12}{9} + \frac{2}{3} = -\frac{10}{27}$

$g(a) = a^3 - 3a^2 + a$

17. $h(s) = \frac{s}{(1 + s)};\ h\left(\frac{1}{2}\right) = \frac{\frac{1}{2}}{\left(1 + \frac{1}{2}\right)} = \frac{\frac{1}{2}}{\frac{3}{2}} = \frac{1}{3}$

$$h\left(-\frac{3}{2}\right) = \frac{-\frac{3}{2}}{1 + \left(-\frac{3}{2}\right)} = \frac{\frac{3}{2}}{-\frac{1}{2}} = 3;\ h(a + 1) = \frac{a + 1}{1 + (a + 1)} = \frac{a + 1}{a + 2}$$

18. $f(x) = \dfrac{x^2}{x^2 - 1}$; $f\left(\frac{1}{2}\right) = \dfrac{\left(\frac{1}{2}\right)^2}{\left(\frac{1}{2}\right)^2 - 1} = \dfrac{\frac{1}{4}}{\frac{1}{4} - 1} = -\frac{1}{3}$;

$$f\left(-\frac{1}{2}\right) = \frac{\left(-\frac{1}{2}\right)^2}{\left(-\frac{1}{2}\right)^2 - 1} = \frac{\frac{1}{4}}{\frac{1}{4} - 1} = -\frac{1}{3}$$

$$f(a + 1) = \frac{(a + 1)^2}{(a + 1)^2 - 1} = \frac{a^2 + 2a + 1}{(a^2 + 2a + 1) - 1} = \frac{a^2 + 2a + 1}{a^2 + 2a}$$

19. $f(x) = x^2 - 2x$

$$f(a + 1) = (a + 1)^2 - 2(a + 1)$$
$$= (a^2 + 2a + 1) - 2a - 2 = a^2 - 1$$
$$f(a + 2) = (a + 2) - 2(a + 2) = (a + 4a + 4) - 2a - 4$$
$$= a^2 + 2a$$

20. $f(x) = x^2 + 4x + 3$

$$f(a - 1) = (a - 1)^2 + 4(a - 1) + 3$$
$$= (a^2 - 2a + 1) + (4a - 4) + 3 = a^2 + 2a$$
$$f(a - 2) = (a - 2)^2 + 4(a - 2) + 3$$
$$= (a^2 - 4a + 4) + (4a - 8) + 3 = a^2 - 1$$

21. (a) $f(0)$ represents the number of typewriters sold in 1990.

(b) $f(2) = 50 + 4(2) + \frac{1}{2}(2)^2 = 50 + 8 + 2 = 60$

22. $R(x) = \dfrac{100}{b + x}$, $x \geq 0$

(a) $b = 20$, $x = 60$; $R(60) = \dfrac{100(60)}{20 + 60} = 75\%$

(b) If $R(50) = 60$, then $60 = \dfrac{100(50)}{b + 50}$; $60b + 3000 = 5000$; $b = 33\frac{1}{3}$.

23. $f(x) = \dfrac{8x}{(x - 1)(x - 2)}$ All real numbers except 1 and 2.

24. $f(t) = \dfrac{1}{\sqrt{t}}$ All real numbers $t > 0$.

25. $g(x) = \dfrac{1}{\sqrt{3 - x}}$ All $x < 3$.

26. $g(x) = \dfrac{4}{x(x + 2)}$ All $x \neq 0, -2$.

27. function

28. not a function

29. not a function

30. not a function

31. not a function

32. function

33. 1 34. -1 35. 3 36. 0 37. positive

38. negative 39. positive 40. yes 41. -1, 5, 9

42. $-1 \le x \le 5,\ x \ge 9$ 43. .03 44. .03 45. .04 46. 3

47. $f(x) = \left(x - \frac{1}{2}\right)(x + 2);\ f(3) = \left(3 - \frac{1}{2}\right)(3 + 2) = 12\frac{1}{2}$

Thus, (3, 12) is not on the graph.

48. $f(x) = x(5 + x)(y - x);\ f(-2) = -2(5 + (-2))(4 - (-2)) = -36$

So (-2, 12) is not on the graph.

49. $g(x) = \dfrac{3x - 1}{x^2 + 1};\ g\left(\dfrac{2}{3}\right) = \dfrac{3\left(\frac{1}{2}\right) - 1}{\left(\frac{1}{2}\right)^2 + 1} = \dfrac{\frac{1}{2}}{\frac{5}{4}} = \dfrac{2}{5}$

So $\left(\frac{1}{2}, \frac{2}{5}\right)$ is on the graph.

50. $g(x) = \dfrac{(x^2 + 4)}{(x + 2)};\ g\left(\dfrac{2}{3}\right) = \dfrac{\left(\frac{2}{3}\right)^2 + 4}{\frac{2}{3} + 2} = \dfrac{\frac{40}{9}}{\frac{8}{3}} = \dfrac{5}{3}$

So $\left(\frac{2}{3}, \frac{5}{3}\right)$ is on the graph.

51. $f(x) = x^3$; $f(a + 1) = (a + 1)^3$

52. $f(x) = \left(\frac{5}{x}\right) - x$; $f(2 + h) = \dfrac{5}{(2 + h)} - (2 + h)$

$$= \frac{5 - (2 + h)^2}{(2 + h)} = \frac{1 - 4h - h^2}{2 + h}$$

53. $f(x) = \begin{cases} \sqrt{x} & \text{for } 0 \le x \le 2 \\ 1 + x & \text{for } 2 \le x \le 5 \end{cases}$

$f(1) = \sqrt{1} = 1$, $f(2) = 1 + 2 = 3$, $f(3) = 1 + 3 = 4$

54. $f(x) = \begin{cases} \frac{1}{x} & \text{for } 1 \le x \le 2 \\ x^2 & \text{for } 2 < x \end{cases}$

$f(1) = \frac{1}{1} = 1$, $f(2) = \frac{1}{2}$, $f(3) = 3^2 = 9$

55. $f(x) = \begin{cases} \pi x^2 & \text{for } x < 2 \\ 1 + x & \text{for } 2 \le x \le 2.5 \\ 4x & \text{for } 2.5 < x \end{cases}$

$f(1) = \pi(1)^2 = \pi$, $f(2) = 1 + 2 = 3$, $f(3) = 4(3) = 12$

56. $\begin{cases} \frac{3}{4 - x} & \text{for } x < 2 \\ 2x & \text{for } 2 \le x < 3 \\ \sqrt{x^2 - 5} & \text{for } 3 \le x \end{cases}$

$f(1) = \dfrac{3}{4 - 1} = 1$, $f(2) = 2(2) = 4$, $f(3) = \sqrt{3^2 - 5} = \sqrt{4} = 2$

57. $f(x) = \begin{cases} .06x & \text{for } 50 \le x \le 300 \\ .02x + 12 & \text{for } 300 < x \le 600 \\ .015x + 15 & \text{for } x > 600 \end{cases}$

58.

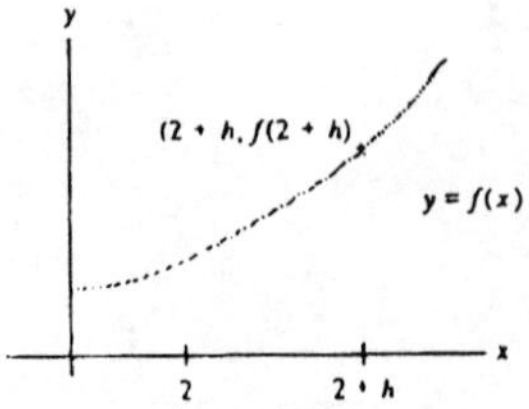

59.

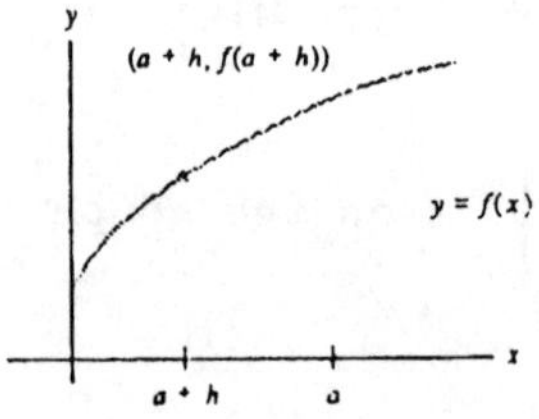

Exercises 0.2

1. f(x) = 2x - 1

x	y
1	1
0	-1
-1	-3

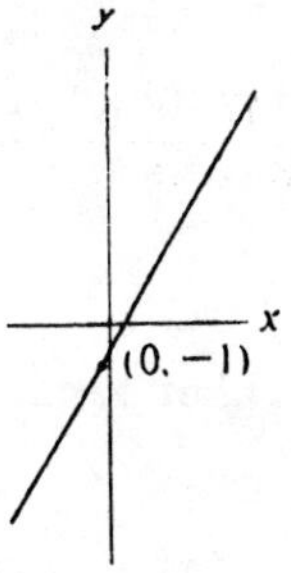

2. f(x) = 3

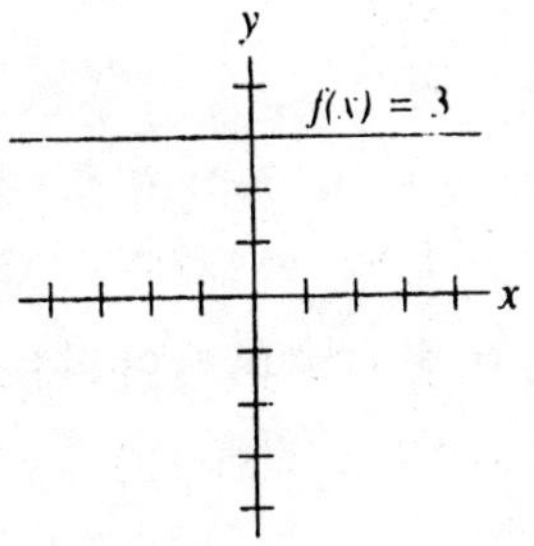

3. f(x) = 3x + 1

x	y
1	4
0	1
-1	-2

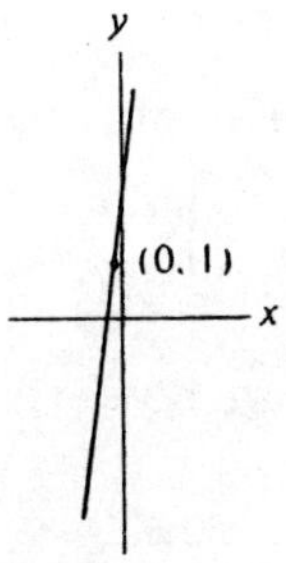

4. $f(x) = -\frac{1}{2}x - 4$

x	y
2	-5
0	-4
-2	-3

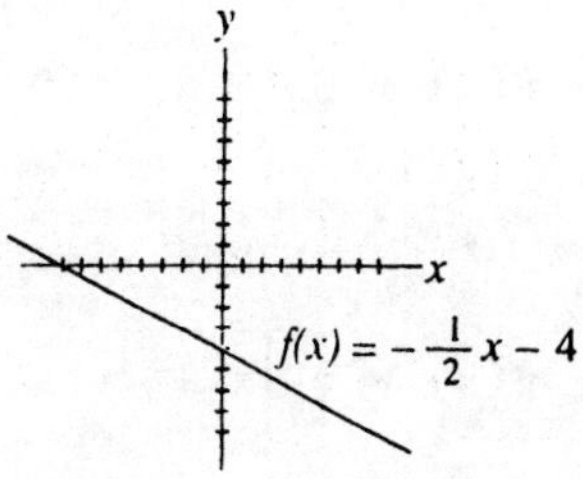

5. f(x) = -2x + 3

x	y
1	1
0	3
-1	5

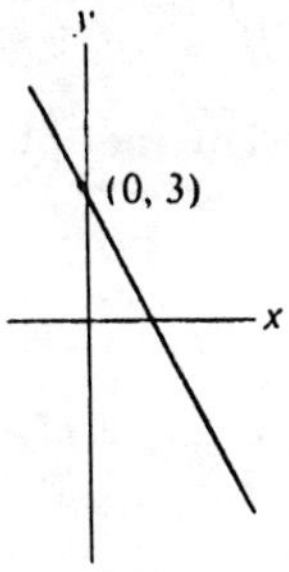

6. $f(x) = \frac{1}{4}$

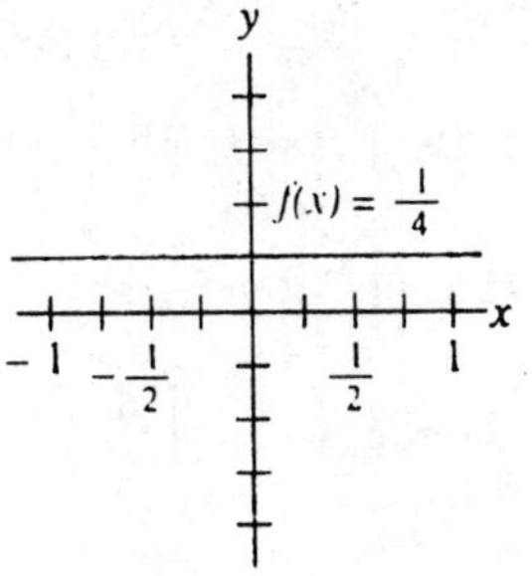

7. $f(x) = 9x + 3$, $f(0) = 9(0) + 3 = 3$ The y-intercept is (0, 3).

$9x + 3 = 0$, $9x = -3$, $x = -\frac{1}{3}$ The x-intercept is $\left(-\frac{1}{3}, 0\right)$.

8. $f(x) = -\frac{1}{2}x - 1$, $f(0) = -\frac{1}{2}(0) - 1 = -1$ y-intercept: (0, -1)

$-\frac{1}{2}x - 1 = 0$, $-\frac{1}{2}x = 1$, $x = -2$ x-intercept: (-2, 0)

9. $f(x) = 5$, y-intercept: (0, 5); There is no x-intercept.

10. $f(x) = 14$, y-intercept: (0, 14); There is no x-intercept.

11. $f(x) = -\frac{1}{4}x + 3$, $f(0) = -\frac{1}{4}(0) + 3 = 3$, y-intercept: (0, 3)

$-\frac{1}{4}x + 3 = 0$, $-\frac{1}{4}x = -3$, $x = 12$ x-intercept: (12, 0)

12. $f(x) = 6x - 4$, $f(0) = 6(0) - 4 = -4$, y-intercept: (0, -4)

$6x - 4 = 0$, $6x = 4$, $x = \frac{2}{3}$, x-intercept: $\left(\frac{2}{3}, 0\right)$

13. $f(x) = \left(\frac{K}{V}\right)x + \frac{1}{V}$

(a) $f(x) = -2x + 50$ We have $\frac{K}{V} = .2$ and $\frac{1}{V} = 50$.

If $\frac{1}{V} = 50$, then $V = \frac{1}{50}$. Now, $\frac{K}{V} = .2$ implies $\frac{K}{\frac{1}{50}} = .2$,

so $K = \frac{1}{5}\cdot\frac{1}{50} = \frac{1}{250}$.

(b) $y = \left(\frac{K}{V}\right)x + \frac{1}{V}$, $\left(\frac{K}{V}\right)\cdot 0 + \frac{1}{V} = \frac{1}{V}$, so the y-intercept is $\left(0, \frac{1}{V}\right)$.

Solving $\left(\frac{K}{V}\right)x + \frac{1}{V} = 0$, we get $\frac{K}{V}x = -\frac{1}{V}$, $x = -\frac{1}{K}$, so the

x-intercept is $\left(-\frac{1}{K}, 0\right)$.

14. From 13(b), $\left(-\frac{1}{K}, 0\right)$ is the x-intercept. From the experimental data, (-500, 0) is also the x-intercept. Thus $-\frac{1}{K} = -500$, $K = \frac{1}{500}$. Again from 13(b), $\left(0, \frac{1}{V}\right)$ is the y-intercept. From the experimental data, (0, 60) is also the y-intercept. Thus, $\frac{1}{V} = 60$, $V = \frac{1}{60}$.

15. (a) Cost is $\$(18 + 200(.2)) = \$58)$

(b) $f(x) = .2x + 18$

16. Let x be the volume of gas (in thousands of cubic feet) extracted. $f(x) = 5000 + (.10)x$

17. Let x be the number of days of hospital confinement. $f(x) = 150 + 135x$

18. $6x - 40 = 350$, $x = 65$ mph inches

19. $f(x) = \dfrac{50}{105 - x}$, $f(70) = 100$, $f(75) = 125$; the cost of removing an extra 5% is \$25 million. To remove the final cost $f(100) - f(95) = 1000 - 475 = \$525$ million. This costs 21 times as much.

20. (a) $f(85) = \dfrac{20(85)}{102 - 85} = \$100$ million;

(b) $f(100) - f(95) = 1000 - 271 = \$729$ million.

21. $y = 3x^2 - 4x$, $a = 3$, $b = -4$, $c = 0$

22. $y = \frac{x^2 - 6x + 2}{3} = \frac{1}{3}x^2 - 2x + \frac{2}{3}$, $a = \frac{1}{3}$, $b = -2$, $c = \frac{2}{3}$

23. $y = 3x - 2x^2 + 1$, $a = -2$, $b = 3$, $c = 1$

24. $y = 3 - 2x + 4x^2$, $a = 4$, $b = -2$, $c = 3$

25. $y = 1 - x^2$, $a = -1$, $b = 0$, $c = 1$

26. $y = \frac{1}{2}x^2 + \sqrt{3}x - \pi$, $a = \frac{1}{2}$, $b = \sqrt{3}$, $c = -\pi$

27. $f(x) = \begin{cases} 3x & \text{for } 0 \le x \le 1 \\ \frac{9}{2} - \frac{3}{2}x & \text{for } x > 1 \end{cases}$

$0 \le x \le 1$

x	y
0	0
1	3

$x > 1$

x	y
1	3
3	0

28. $f(x) = \begin{cases} 1 + x & \text{for } x \le 3 \\ 4 & \text{for } x > 3 \end{cases}$

$x \le 3$

x	y
3	4
0	1

$x > 3$

x	y
4	4
5	4

29. $f(x) = \begin{cases} 3 & \text{for } x < 2 \\ 2x + 1 & \text{for } x \ge 2 \end{cases}$

$x < 2$

x	y
1	3
0	3

$x \ge 2$

x	y
2	5
3	7

$y = 3$ for all $x < 2$.

30. $f(x) = \begin{cases} \frac{1}{2}x & \text{for } 0 \le x < 4 \\ 2x - 3 & \text{for } 4 \le x \le 5 \end{cases}$

$0 \le x < 4$

x	y
0	0
2	1
3	$\frac{3}{2}$

$4 \le x \le 5$

x	y
4	5
5	7

31. $f(x) = \begin{cases} 4 - x & \text{for } 0 \le x < 2 \\ 2x - 2 & \text{for } 2 \le x < 3 \\ x + 1 & \text{for } x \ge 3 \end{cases}$

$0 \le x < 2$

x	y
0	4
1	3

$2 \le x < 3$

x	y
2	2
$\frac{5}{2}$	3

$x \ge 3$

x	y
3	4
4	5

32. $f(x) = \begin{cases} 4x & \text{for } 0 \le x < 1 \\ 8 - 4x & \text{for } 1 \le x < 2 \\ 2x - 4 & \text{for } x \ge 2 \end{cases}$

$0 \le x < 1$

x	y
0	0
$\frac{1}{2}$	2

$1 \le x < 2$

x	y
1	4
$\frac{3}{2}$	2

$x \ge 2$

x	y
2	0
3	2

33. $f(x) = x^{100}$, $x = -1$, $f(-1) = 1$

34. $f(x) = x^5$, $x = \frac{1}{2}$, $f\left(\frac{1}{2}\right) = \left(\frac{1}{2}\right)^5 = \frac{1}{32}$

35. $f(x) = |x|$, $x = 10^{-2}$, $f(10^{-2}) = |10^{-2}| = 10^{-2}$

36. $f(x) = |x|$, $x = \pi$, $f(\pi) = |\pi| = \pi$

37. $f(x) = |x|$, $x = -2.5$, $f(-2.5) = |-2.5| = 2.5$

38. $f(x) = |x|$, $x = -\frac{2}{3}$, $f\left(-\frac{2}{3}\right) = \left|-\frac{2}{3}\right| = \frac{2}{3}$

Exercises 0.3

1. $f(x) + g(x) = (x^2 + 1) + 9x = x^2 + 9x + 1$

2. $f(x) - h(x) = (x^2 + 1) - (5 - 2x^2) = 3x^2 - 4$

3. $f(x)\cdot g(x) = (x^2 + 1)(9x) = 9x^3 + 9x$

4. $g(x)h(x) = (9x)(5 - 2x^2) = 45x - 18x^3$

5. $\dfrac{f(t)}{g(t)} = \dfrac{t^2 + 1}{9t} = \dfrac{t^2}{9t} + \dfrac{1}{9t} = \dfrac{t}{9} + \dfrac{1}{9t}$

6. $\dfrac{g(t)}{h(t)} = \dfrac{9t}{5 - 2t^2}$

7. $\dfrac{2}{x - 3} + \dfrac{1}{x + 2} = \dfrac{2(x + 2) + (x - 3)}{(x - 3)(x + 2)} = \dfrac{3x + 1}{x^2 - x - 6}$

8. $\dfrac{3}{x - 6} + \dfrac{-2}{x - 2} = \dfrac{3(x - 2) + (-2)(x - 6)}{(x - 6)(x - 2)} = \dfrac{x + 6}{x^2 - 8x + 12}$

9. $\dfrac{x}{x - 8} + \dfrac{-x}{x - 4} = \dfrac{x(x - 4) + (-x)(x - 8)}{(x - 8)(x - 4)} = \dfrac{4x}{x^2 - 12x + 32}$

10. $\dfrac{-x}{x + 3} + \dfrac{x}{x + 5} = \dfrac{(-x)(x + 5) + (x)(x + 3)}{(x + 3)(x + 5)} = \dfrac{-2x}{x^2 + 8x + 15}$

11. $\dfrac{x + 5}{x - 10} + \dfrac{x}{x + 10} = \dfrac{(x + 5)(x + 10) + (x)(x - 10)}{(x - 10)(x + 10)}$

$= \dfrac{2x^2 + 5x + 50}{x^2 - 100}$

12. $\dfrac{x + 6}{x - 6} + \dfrac{x - 6}{x + 6} = \dfrac{(x + 6)(x + 6) + (x - 6)(x - 6)}{(x - 6)(x + 6)} = \dfrac{2x^2 + 72}{x^2 - 36}$

13. $\dfrac{x}{x - 2} - \dfrac{5 - x}{5 + x} = \dfrac{x(5 + x) - (5 - x)(x - 2)}{(x - 2)(5 + x)} = \dfrac{2x^2 - 2x + 10}{x^2 + 3x - 10}$

14. $\dfrac{t}{t - 2} - \dfrac{t + 1}{3t - 1} = \dfrac{t(3t - 1) - (t - 2)(t + 1)}{(t - 2)(3t - 1)} = \dfrac{2t^2 + 2}{3t^2 - 7t + 2}$

15. $\dfrac{x}{x-2}\cdot\dfrac{5-x}{5+x} = \dfrac{5x - x^2}{x^2 + 3x - 10}$

16. $\dfrac{5-x}{5+x}\cdot\dfrac{x+1}{3x-1} = \dfrac{-x^2 + 4x + 5}{3x^2 + 14x - 5}$

17. $\dfrac{\dfrac{x}{x-2}}{\dfrac{5-x}{5+x}} = \dfrac{x}{x-2}\cdot\dfrac{5+x}{5-x} = \dfrac{5x + x^2}{-x^2 + 7x - 10}$

18. $\dfrac{\dfrac{s+1}{3s-1}}{\dfrac{s}{s-2}} = \dfrac{s+1}{3s-1}\cdot\dfrac{s-2}{s} = \dfrac{s^2 - s - 2}{3s^2 - s}$

19. $\dfrac{x+1}{(x+1)-2}\cdot\dfrac{5-(x+1)}{5+(x+1)} = \dfrac{x+1}{x-1}\cdot\dfrac{(-x+4)}{6+x} = \dfrac{-x^2 + 3x + 4}{x^2 + 5x - 6}$

20. $\dfrac{(x+2)}{(x+2)-2} + \dfrac{5-(x+2)}{5+(x+2)} = \dfrac{x+2}{x} + \dfrac{3-x}{x+7}$

$= \dfrac{(x+2)(x+7) + (3-x)(x)}{x(x+7)}$

$= \dfrac{12x + 4}{x^2 + 7x}$

21. $\dfrac{5-(x+5)}{5+(x+5)}\cdot\dfrac{(x+5)-2}{x+5} = \dfrac{-x}{10+x}\cdot\dfrac{x+3}{x+5} = \dfrac{-x^2 - 3x}{x^2 + 15x + 50}$

22. $\dfrac{\dfrac{1}{t}}{\dfrac{1}{t} - 2} = \dfrac{1}{t}\cdot\dfrac{t}{1-2t} = \dfrac{1}{1-2t}$

23. $\dfrac{5 - \dfrac{1}{u}}{5 + \dfrac{1}{u}} = \dfrac{5u-1}{u}\cdot\dfrac{u}{5u+1} = \dfrac{5u-1}{5u+1}$

24. $\dfrac{\dfrac{1}{x^2} + 1}{3\dfrac{1}{x^2} - 1} = \dfrac{1+x^2}{x^2}\cdot\dfrac{x^2}{3-x^2} = \dfrac{1+x^2}{3-x^2}$

25. $f\left(\dfrac{x}{1-x}\right) = \left(\dfrac{x}{1-x}\right)^6$

26. $h(t^6) = (t^6)^3 - 5(t^6)^2 + 1 = t^{18} - 5t^{12} + 1$

27. $h\left(\frac{x}{1 - x}\right) = \left(\frac{x}{1 - x}\right)^3 - 5\left(\frac{x}{1 - x}\right)^2 + 1$

28. $g(x^6) = \dfrac{x^6}{1 - x^6}$

29. $g(t^3 - 5t^2 + 1) = \dfrac{t^3 - 5t^2 + 1}{1 - (t^3 - 5t^2 + 1)} = \dfrac{t^3 - 5t^2 + 1}{-t^3 + 5t^2}$

30. $f(x^3 - 5x^2 + 1) = (x^3 - 5x^2 + 1)^6$

31. $(x + h)^2 - x^2 = x^2 + 2xh + h^2 - x^2 = 2xh + h^2$

32. $\dfrac{1}{x + h} - \dfrac{1}{x} = \dfrac{x - x - h}{x(x + h)} = \dfrac{-h}{x^2 + xh}$

33. $\dfrac{4(t + h) - (t + h)^2 - (4t - t^2)}{h}$

$= \dfrac{4t + 4h - (t^2 + 2th + h^2) - 4t + t^2}{h}$

$= \dfrac{4h - 2th - h^2}{h} = \dfrac{h\cdot(4 - 2t - h)}{h} = 4 - 2t - h$

34. $\dfrac{(t + h)^3 + 5 - (t^3 + 5)}{h} = \dfrac{t^3 + 3t^2h + 3th^2 + h^3 + 5 - t^3 - 5}{h}$

$= \dfrac{3t^2h + 3th^2 + h^3}{h} = \dfrac{h(3t^2 + 3th + h^2)}{h}$

$= 3t^2 + 3th + h^2$

35. (a) $C(A(t)) = 3000 + 80\left[20t - \frac{1}{2}t^2\right] = 3000 + 1600t - 40t^2$

(b) $C(2) = 3000 + 1600(2) - 40(2)^2 = 3000 + 3200 - 160 = 6040$

36. (a) $C(f(t)) = \frac{1}{10}(10t - 5)^2 + 25(10t - 5) + 200$

$= \frac{1}{10}(100t^2 - 100t + 25) + 250t - 125 + 200$

$= 10t^2 + 240t + 77.5$

(b) $c(4) = 10(4)^2 + 240(4) + 77.5 = 1197.50$

37. $h(x) = f(8x+1) = (1/8)(8x+1) = x + 1/8$; $h(x)$ converts from British to American sizes.

Exercises 0.4

1. $f(x) = 2x^2 - 7x + 6$, $2x^2 - 7x + 6 = 0$, $a = 2$, $b = -7$, $c = 0$

$\sqrt{b^2 - 4ac} = \sqrt{49 - 4(2)(6)} = \sqrt{1} = 1$

$$x = \frac{-b \pm \sqrt{b^2 - 4ac}}{2a} = \frac{7 \pm 1}{4} = 2,\ 1\tfrac{1}{2}$$

2. $f(x) = 3x^2 + 2x - 1$, $3x^2 + 2x - 1 = 0$

$$x = \frac{-b \pm \sqrt{b^2 - 4ac}}{2a} = \frac{-2 \pm \sqrt{2^2 - 4(3)(-1)}}{2(3)} = \frac{-2 \pm \sqrt{16}}{6}$$

$$= \frac{-2 \pm 4}{6} = \frac{1}{3},\ -1$$

3. $f(x) = 4x^2 - 12x + 9$, $4x^2 - 12x + 9 = 0$

$$x = \frac{-b \pm \sqrt{b^2 - 4ac}}{2a} = \frac{12 \pm \sqrt{(-12)^2 - 4(4)(9)}}{2(4)} = \frac{12 \pm \sqrt{0}}{8} = \frac{3}{2}$$

4. $f(x) = \frac{1}{4}x^2 + x + 1$, $\frac{1}{4}x^2 + x + 1 = 0$

$$x = \frac{-b \pm \sqrt{b^2 - 4ac}}{2a} = \frac{-1 \pm \sqrt{1^2 - 4(1/4)(1)}}{2\left(\frac{1}{4}\right)} = \frac{-1 \pm \sqrt{0}}{\frac{1}{2}} = -2$$

5. $f(x) = -2x^2 + 3x - 4,\ -2x^2 + 3x - 4 = 0$

$$x = \frac{-b \pm \sqrt{b^2 - 4ac}}{2a} = \frac{-3 \pm \sqrt{3^2 - 4(-2)(-4)}}{2(-2)} = \frac{-3 \pm \sqrt{-23}}{-4}$$

$\sqrt{-23}$ is undefined, so f(x) has no real zeros.

6. $f(x) = 11x^2 - 7x + 1,\ 11x^2 - 7x + 1 = 0$

$$x = \frac{-b \pm \sqrt{b^2 - 4ac}}{2a} = \frac{7 \pm \sqrt{(-7)^2 - 4(11)(1)}}{2(11)} = \frac{7 \pm \sqrt{5}}{22}$$

$$= \frac{7 + \sqrt{5}}{22},\ \frac{7 - \sqrt{5}}{22}$$

7. $5x^2 - 4x - 1 = 0$

$$x = \frac{-b \pm \sqrt{b^2 - 4ac}}{2a} = \frac{4 \pm \sqrt{(-4)^2 - 4(5)(-1)}}{2(5)} = \frac{4 \pm \sqrt{36}}{10}$$

$$= \frac{4 \pm 6}{10} = 1,\ -\frac{1}{5}$$

8. $x^2 - 4x + 5$

$$x = \frac{-b \pm \sqrt{b^2 - 4ac}}{2a} = \frac{4 \pm \sqrt{(-4)^2 - 4(1)(5)}}{2(1)} = \frac{4 \pm \sqrt{-4}}{2}$$

$\sqrt{-4}$ is undefined, so there is no real solution.

9. $15x^2 - 135x + 300 = 0$

$$x = \frac{-b \pm \sqrt{b^2 - 4ac}}{2a} = \frac{135 \pm \sqrt{(-135)^2 - 4(15)(300)}}{2(15)}$$

$$= \frac{135 \pm \sqrt{225}}{30} = \frac{135 \pm 15}{30} = 5,\ 4$$

10. $x^2 - \sqrt{2}x - \frac{5}{4} = 0$

$$x = \frac{-b \pm \sqrt{b^2 - 4ac}}{2a} = \frac{\sqrt{2} \pm \sqrt{(\sqrt{2})^2 - 4(1)(-5/4)}}{2(1)}$$

$$= \frac{\sqrt{2} \pm \sqrt{7}}{2} = \frac{\sqrt{2} + \sqrt{7}}{2},\ \frac{\sqrt{2} - \sqrt{7}}{2}$$

11. $\frac{3}{2}x^2 - 6x + 5 = 0$

$$x = \frac{-b \pm \sqrt{b^2 - 4ac}}{2a} = \frac{6 \pm \sqrt{(-6)^2 - 4(3/2)(5)}}{2\left(\frac{3}{2}\right)}$$

$$= \frac{6 \pm \sqrt{6}}{3} = 2 + \frac{\sqrt{6}}{3},\ 2 - \frac{\sqrt{6}}{3}$$

12. $9x^2 - 12x + 4 = 0$

$$x = \frac{-b \pm \sqrt{b^2 - 4ac}}{2a} = \frac{12 \pm \sqrt{(-12)^2 - 4(9)(4)}}{2(9)}$$

$$= \frac{12 \pm \sqrt{0}}{18} = \frac{2}{3}$$

13. $x^2 + 8x + 15 = (x + 5)(x + 3)$

14. $x^2 - 10x + 16 = (x - 2)(x - 8)$

15. $x^2 - 16 = (x - 4)(x + 4)$

16. $x^2 - 1 = (x + 1)(x - 1)$

17. $3x^2 + 12x + 12 = 3(x^2 + 4x + 4) = 3(x + 2)(x + 2) = 3(x + 2)^2$

18. $2x^2 - 12x + 18 = 2(x^2 - 6x + 9) = 2(x - 3)(x - 3) = 2(x - 3)^2$

19. $30 - 4x - 2x^2 = -2(-15 + 2x + x^2) = -2(x - 3)(x + 5)$

20. $15 + 12x - 3x^2 = -3(-5 - 4x + x^2) = -3(x - 5)(x + 1)$

21. $3x - x^2 = x(3 - x)$

22. $4x^2 - 1 = (2x + 1)(2x - 1)$

23. $6x - 2x^3 = -2x(x^2 - 3) = -2x(x + \sqrt{3})(x - \sqrt{3})$

24. $16x + 6x^2 - x^3 = -x(x^2 - 6x - 16) = -x(x - 8)(x + 2)$

25. $2x^2 - 5x - 6 = 3x + 4,\ 2x^2 - 8x - 10 = 0$

$$x = \frac{-b \pm \sqrt{b^2 - 4ac}}{2a} = \frac{8 \pm \sqrt{(-8)^2 - 4(2)(-10)}}{2(2)}$$

$$= \frac{8 \pm \sqrt{144}}{4} = \frac{8 \pm 12}{4} = 5,\ -1$$

$y = 3x + 4 = 15 + 4 = 19;\ y = -3 + 4 = 1$

points of intersection: (5, 19), (-1, 1)

26. $x^2 - 10x + 9 = x - 9,\ x^2 - 11x + 18 = 0$

$(x - 9)(x - 2) = 0,\ x = 9,\ 2;$

$y = x - 9 = 9 - 9 = 0,\ y = 2 - 9 = -7$

points of intersection: (9, 0), (2 - 7)

27. $y = x^2 - 4x + 4,\ y = 12 + 2x - x^2$

$x^2 - 4x + 4 = 12 + 2x - x^2,\ 2x^2 - 6x - 8 = 0,$

$2(x^2 - 3x - 4) = 0,\ 2(x - 4)(x + 1) = 0,\ x = 4,\ -1$

$y = x^2 - 4x + 4 = 4^2 - 4(4) + 4 = 4$

$y = (-1)^2 - 4(-1) + 4 = 9$

points of intersection: (4, 4), (-1, 9)

28. $y = 3x^2 + 9,\ y = 2x^2 - 5x + 3$

$3x^2 + 9 = 2x^2 - 5x + 3,\ x^2 + 5x + 6 = 0,\ (x + 3)(x + 2) = 0,$

$x = -3,\ -2$

$y = 3x^2 + 9 = 3(-3)^2 + 9 = 36;\ y = 3(-2)^2 + 9 = 21$

points of intersection: (-3, 36), (-2, 21)

29. $y = x^3 - 3x^2 + x,\ y = x^2 - 3x$

$x^3 - 3x^2 + x = x^2 - 3x,\ x^3 - 4x^2 + 4x = 0,\ x(x^2 - 4x + 4) = 0,$

$x(x - 2)(x - 2) = 0,\ x = 0,\ 2$

$y = x^2 - 3x = 0^2 - 3(0) = 0;\ y = 2^2 - 3(2) = 4 - 6 = -2$

points of intersection: (0, 0), (2, -2)

30. $y = \frac{1}{2}x^3 - 2x^2,\ y = 2x$

$\frac{1}{2}x^3 - 2x^2 = 2x,\ \frac{1}{2}x^3 - 2x^2 - 2x = 0,\ x\left(\frac{1}{2}x^2 - 2x - 2\right) = 0,$

$x = 0$ or $\frac{1}{2}x^2 - 2x - 2 = 0$

$$x = \frac{-b \pm \sqrt{b^2 - 4ac}}{2a} = \frac{2 \pm \sqrt{(-2)^2 - 4(1/2)(-2)}}{2\left(\frac{1}{2}\right)}$$

$$= \frac{2 \pm \sqrt{8}}{1} = 2 \pm 2\sqrt{2},\ 2 - 2\sqrt{2}$$

$y = 2x = 2(0) = 0;\ y = 2(2 + 2\sqrt{2}) = 4 + 4\sqrt{2};$

$y = 2(2 - 2\sqrt{2}) = 4 - 4\sqrt{2}$

points of intersection: $(0, 0)$, $(2 + 2\sqrt{2}, 4 + 4\sqrt{2})$, $(2 - 2\sqrt{2}, 4 - 4\sqrt{2})$

31. $y = \frac{1}{2}x^3 + x^2 + 5,\ y = 3x^2 - \frac{1}{2}x + 5$

$\frac{1}{2}x^3 + x^2 + 5 = 3x^2 - \frac{1}{2}x + 5;\ \frac{1}{2}x^3 - 2x^2 + \frac{1}{2}x = 0,$

$x\left(\frac{1}{2}x^2 - 2x + \frac{1}{2}\right) = 0,\ x = 0 \text{ or } \frac{1}{2}x^2 - 2x + \frac{1}{2} = 0$

$$x = \frac{-b \pm \sqrt{b^2 - 4ac}}{2a} = \frac{2 \pm \sqrt{(-2)^2 - 4(1/2)(1/2)}}{2\left(\frac{1}{2}\right)}$$

$$= \frac{2 \pm \sqrt{3}}{1} = 2 + \sqrt{3},\ 2 - \sqrt{3}$$

$y = 3x^2 - \frac{1}{2}x + 5 = 3(0)^2 - \frac{1}{2}(0) + 5 = 5$

$y = 3(2 + \sqrt{3})^2 - \frac{1}{2}(2 + \sqrt{3}) + 5) = 25 + \frac{23\sqrt{3}}{2}$

$y = 3(2 - \sqrt{3})^2 - \frac{1}{2}(2 - \sqrt{3}) + 5 = 25 - \frac{23\sqrt{3}}{2}$

points of intersection: $(0, 5)$, $\left(2 + \sqrt{3}, 25 + \frac{23\sqrt{3}}{2}\right)$, $\left(2 - \sqrt{3}, 25 - \frac{23\sqrt{3}}{2}\right)$

32. $y = 30x^2 - 3x^2,\ y = 16x^3 + 25x^2$

$30x^3 - 3x^2 = 16x^3 + 25x^2,\ 14x^3 - 28x^2 = 0,\ 14x^2(x - 2) = 0$

$x = 0$ or $x = 2$

$y = 30(0)^3 - 3(0)^2 = 0$

$y = 30(2)^3 - 3(2)^2 = 30(8) - 3(4) = 228$

points of intersection: $(0, 0)$, $(2, 228)$

33. $\frac{21}{x} - x = 4,\ 21 - x^2 = 4x,\ x^2 + 4x - 21 = 0,$

$(x + 7)(x - 3) = 0,\ x = -7,\ 3$

34. $x + \dfrac{2}{x - 6} = 3,\ x^2 - 6x + 2 = 3x - 18,\ x^2 - 9x + 20 = 0,$

$(x - 4)(x - 5) = 0,\ x = 4,\ 5$

35. $x + \dfrac{14}{x + 4} = 5,\ x^2 + 4x + 14 = 5x + 20,\ x^2 - x - 6 = 0,$

$(x - 3)(x + 2),\ x = 3,\ -2$

36. $1 = \dfrac{5}{x} + \dfrac{6}{x^2},\ 1 = \dfrac{5x + 6}{x^2},\ x^2 - 5x - 6 = 0,\ (x - 6)(x + 1) = 0,$

$x = 6,\ -1$

37. $\dfrac{x^2 + 14x + 49}{x^2 + 1} = 0,\ x^2 + 14x + 49 = 0,\ (x + 7)(x + 7) = 0,$

$x = -7$

38. $\dfrac{x^2 - 8x + 16}{1 + \sqrt{x}} = 0,\ x^2 - 8x + 16 = 0,\ (x - 4)(x - 4) = 0,$

$x = 4$

39. $C(x) = 275 + 12x;\ R(x) = 32x - .21x^2$

$C(x) = R(x) \Rightarrow 275 + 12x = 32x - .21x^2$

$\Rightarrow .21x^2 - 20x + 275 = 0$

Thus

$$x = \frac{20 \pm \sqrt{20^2 - 4(.21)275}}{.42}$$

$$= \$78571 \text{ or } \$16667$$

40. $x + (1/20)x^2 = 175,\ x^2 + 20x - 3500 = 0,$

$(x - 50)(x + 70) = 0,\ x = 50$ mph.

Exercises 0.5

1. $3^3 = 27$ 2. $(-2)^3 = -8$ 3. $1^{100} = 1$

4. $0^{25} = 0$ 5. $(.1)^4 = (.1)(.1)(.1)(.1) = .0001$

6. $(100)^4 = (100)(100)(100)(100) = 100,000,000$

7. $-4^2 = -16$ 8. $(.01)^3 = .000001$ 9. $(16)^{1/2} = \sqrt{16} = 4$

10. $(27)^{1/3} = \sqrt[3]{27} = 3$ 11. $(.000001)^{1/3} = \sqrt[3]{.000001} = .01$

12. $\left(\frac{1}{125}\right)^{1/3} = \sqrt[3]{\frac{1}{125}} = \frac{1}{5}$ 13. $6^{-1} = \frac{1}{6}$ 14. $\left(\frac{1}{2}\right)^{-1} = \frac{1}{\frac{1}{2}} = 2$

15. $(.01)^{-1} = \frac{1}{.01} = 100$ 16. $(-5)^{-1} = -\frac{1}{5}$

17. $8^{4/3} = (\sqrt[3]{8})^4 = 16$ 18. $16^{3/4} = (\sqrt[4]{16})^3 = 8$

19. $(25)^{3/2} = (\sqrt{25})^3 = 125$ 20. $(27)^{2/3} = (\sqrt[3]{27})^2 = 9$

21. $(1.8)^0 = 1$ 22. $9^{1.5} = 9^{3/2} = (\sqrt{9})^3 = 27$

23. $16^{.5} = 16^{1/2} = 4$ 24. $81^{.75} = 81^{3/4} = 27$

25. $4^{-1/2} = \frac{1}{\sqrt{4}} = \frac{1}{2}$ 26. $\left(\frac{1}{8}\right)^{-2/3} = \frac{1}{\left(\sqrt[3]{\frac{1}{8}}\right)^2} = 4$

27. $(.01)^{-1.5} = \frac{1}{(.01)^{3/2}} = \frac{1}{.001} = 1000$

28. $1^{-1.2} = \frac{1}{1^{1.2}} = 1$ 29. $5^{1/3} \cdot 200^{1/3} = 1000^{1/3} = 10$

30. $(3^{1/3} \cdot 3^{1/6})^6 = (3^{1/2})^6 = 27$ 31. $6^{1/3} \cdot 6^{2/3} = 6^1 = 6$

32. $(9^{4/5})^{5/8} = 9^{1/2} = 3$ 33. $\frac{10^4}{5^4} = 2^4 = 16$

34. $\dfrac{3^{5/2}}{3^{1/2}} = 3^{(5/2)-(1/2)} = 3^{4/2} = 9$

35. $(2^{1/3}\cdot 3^{2/3})^3 = (\sqrt[3]{2}\cdot\sqrt[3]{9})^3 = (\sqrt[3]{18})^3 = 18$

36. $20^{.5}\cdot 5^{.5} = (100)^{1/2} = 10$

37. $\left(\dfrac{8}{27}\right)^{2/3} = \dfrac{8^{2/3}}{27^{2/3}} = \dfrac{4}{9}$

38. $(125\cdot 27)^{1/3} = 125^{1/3}\cdot 27^{1/3} = 15$

39. $\dfrac{7^{4/3}}{7^{1/3}} = 7^{4/3}\cdot 7^{-1/3} = 7^{(4/3)-(1/3)} = 7$

40. $(6^{1/2})^0 = 6^{(1/2)(0)} = 6^0 = 1$

41. $(xy)^6 = x^6\cdot y^6$

42. $(x^{1/3})^6 = x^{(1/3)(6)} = x^2$

43. $\dfrac{x^4\cdot y^5}{xy^2} = x^4\cdot y^5\cdot x^{-1}\cdot y^{-2} = x^3y^3$

44. $\dfrac{1}{x^{-3}} = x^3$

45. $x^{-1/2} = \dfrac{1}{\sqrt{x}}$

46. $(x^3\cdot y^6)^{1/3} = x^{3(1/3)}\cdot y^{6(1/3)} = xy^2$

47. $\left(\dfrac{x^4}{y^2}\right)^3 = \dfrac{x^{4(3)}}{y^{2(3)}} = \dfrac{x^{12}}{y^6}$

48. $\left(\dfrac{x}{y}\right)^{-2} = \dfrac{1}{x^2}\cdot y^2 = \dfrac{y^2}{x^2}$

49. $(x^3y^5)^4 = x^{3(4)}\cdot y^{5(4)} = x^{12}y^{20}$

50. $\sqrt{1 + x}\,(1 + x)^{3/2} = (1 + x)^{1/2}(1 + x)^{3/2} = (1 + x)^{(1/2)+(3/2)}$

$= (1 + x)^2 = x^2 + 2x + 1$

51. $x^5\cdot\left(\dfrac{y^2}{x}\right)^3 = \dfrac{x^5\cdot y^{2(3)}}{x^3} = x^5\cdot y^6\cdot x^{-3} = x^2y^6$

52. $x^{-3}\cdot x^7 = x^{7-3} = x^4$

53. $(2x)^4 = 16x^4$

54. $\dfrac{-3x}{15x^4} = -\dfrac{1}{5x^3}$

55. $\dfrac{-x^3y}{-xy} = x^2$ 56. $\dfrac{x^3}{y^{-2}} = x^3y^2$ 57. $\dfrac{x^{-4}}{x^3} = \dfrac{1}{x^7}$

58. $(-3x)^3 = -27x^3$ 59. $\sqrt[3]{x}\cdot\sqrt[3]{x^2} = x^{1/3}\cdot x^{2/3} = x$

60. $(9x)^{-1/2} = \dfrac{1}{\sqrt{9x}} = \dfrac{1}{3\sqrt{x}}$ 61. $\left(\dfrac{3x^2}{2y}\right)^3 = \dfrac{3^3\cdot x^6}{2^3\cdot y^3} = \dfrac{27x^6}{8y^3}$

62. $\dfrac{x^2}{x^5y} = \dfrac{1}{x^3y}$ 63. $\dfrac{2x}{\sqrt{x}} = 2x\cdot x^{-1/2} = 2\sqrt{x}$

64. $\dfrac{1}{yx^{-5}} = \dfrac{x^5}{y}$ 65. $(16x^8)^{-3/4} = 16^{-3/4}\cdot x^{-6} = \dfrac{1}{8x^6}$

66. $(-8xy^9)^{2/3} = (-8)^{2/3}y^{9(2/3)} = 4y^6$

67. $\sqrt{x}\left(\dfrac{1}{4x}\right)^{5/2} = \dfrac{x^{1/2}}{4^{5/2}x^{5/2}} = \dfrac{x^{1/2}\cdot x^{-5/2}}{32} = \dfrac{1}{32x^2}$

68. $\dfrac{(25xy)^{3/2}}{x^2y} = \dfrac{(25)^{3/2}x^{3/2}y^{3/2}}{x^2y} = \dfrac{125\sqrt{y}}{\sqrt{x}}$

69. $\dfrac{(-27x^5)^{2/3}}{\sqrt[3]{x}} = \dfrac{(-27)^{2/3}x^{5(2/3)}}{x^{1/3}} = 9x^3$

70. $(-32y^{-5})^{3/5} = (-32)^{3/5}y^{-5(3/5)} = \dfrac{-8}{y^3}$

71. $\sqrt{x} - \dfrac{1}{\sqrt{x}} = \dfrac{1}{\sqrt{x}}(x - 1)$ 72. $2x^{2/3} - x^{-1/3} = x^{-1/3}(2x - 1)$

73. $x^{-1/4} + 6x^{1/4} = x^{-1/4}(1 + 6\sqrt{x})$

74. $\sqrt{\dfrac{x}{y}} - \sqrt{\dfrac{y}{x}} = \sqrt{xy}\left(\dfrac{1}{y} - \dfrac{1}{x}\right)$

75. $\sqrt{a}\cdot\sqrt{b} = \sqrt{ab}$, $a^{1/2}\cdot b^{1/2} = (ab)^{1/2}$ (law 5)

76. $\dfrac{\sqrt{a}}{\sqrt{b}} = \sqrt{\dfrac{a}{b}}$, $\dfrac{a^{1/2}}{b^{1/2}} = \left(\dfrac{a}{b}\right)^{1/2}$ (law 6)

77. $f(x) = x^2$, $f(4) = (4)^2 = 16$

78. $f(x) = x^3$, $f(4) = (4)^3 = 64$

79. $f(x) = x^{-1}$, $f(4) = (4)^{-1} = \frac{1}{4}$

80. $f(x) = x^{1/2}$, $f(4) = (4)^{1/2} = 2$

81. $f(x) = x^{3/2}$, $f(4) = (4)^{3/2} = 8$

82. $f(x) = x^{-1/2}$, $f(4) = (4)^{-1/2} = \frac{1}{2}$

83. $f(x) = x^{-5/2}$, $f(4) = (4)^{-5/2} = \frac{1}{32}$

84. $f(x) = x^0$, $f(4) = 4^0 = 1$

85. $709.26
86. $1295.65
87. $127,857.61
88. $28,515.22
89. $164.70
90. $522.97
91. $1592.75
92. $1795.80
93. $3268.00
94. $26,483.83
95. $\frac{125}{64}r^4 + \frac{125}{4}r^3 + \frac{375}{2}r^2 + 500r + 500$
96. $\frac{125}{2}r^4 + 500r^3 + 1500r^2 + 2000r + 1000$

Exercises 0.6

1.

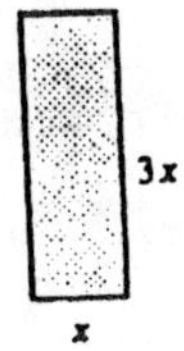

2.

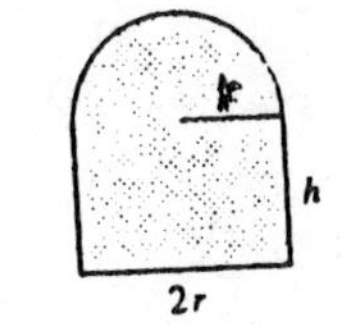

3.

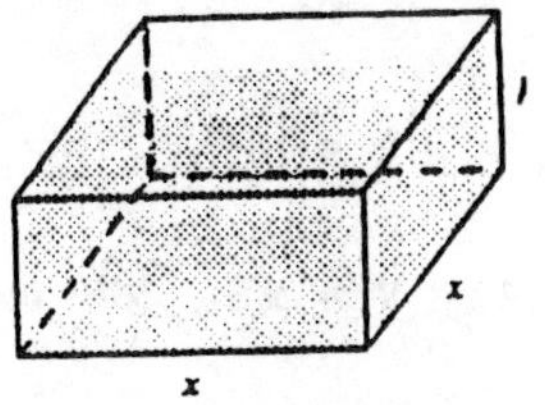

4.

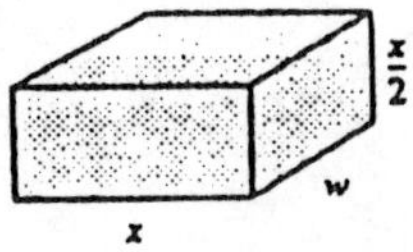

5.

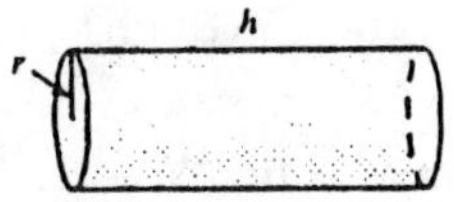

6.

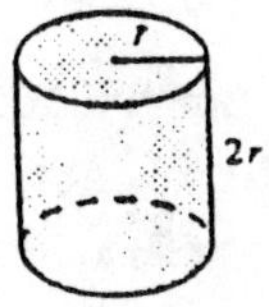

7. $P=8x$, $3x^2= 25$

8. $A = 3x^2$, $8x = 30$

9. $A = \pi r^2$, $2\pi r = 15$

10. $P = 2r + 2h + \pi r$, $2rh + (1/2)\pi r^2 = 2.5$

11. $V = x^2h$, $x^2 + 4xh = 65$

12. $S = 3xw + x^2$, $(1/2)wx^2 = 10$

13. $\pi r^2h = 100$, $C = 5\pi r^2 + 6\pi r^2 + 7(2\pi rh) = 11\pi r^2 + 14\pi rh$

14. $\pi r^2 + \pi r^2 + 2\pi r(2r) = 6\pi r^2 = 30\pi$, $V = \pi r^2(2r) = 2\pi r^3$

15. $2x + 3h = 5000, A = xh$

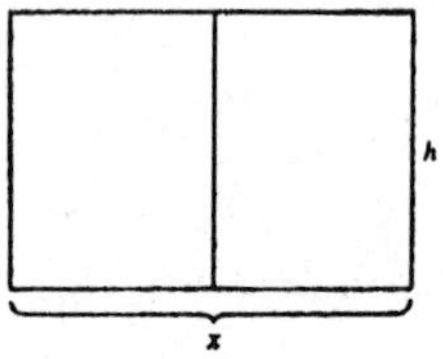

16. $xh = 2500$, $f = 4x + 2h$

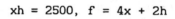

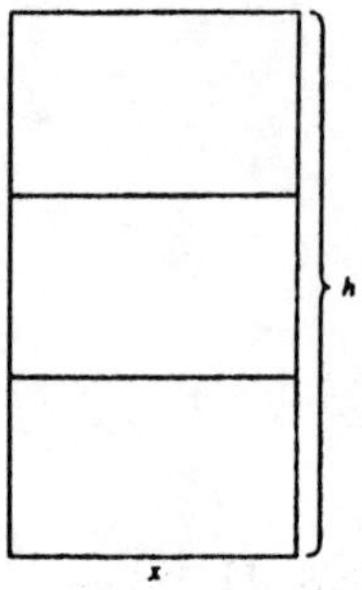

17. $C = 10(2x + 2h) + 8(2x) = 36x + 20h$

18. $5x^2 + 4(4xh) = 5x^2 + 16xh$

19. $8x = 40$, $x=5$; $A = 3x^2 = 3(25) = 75\ cm^2$

20. $V = 2\pi r^3 = 54\pi$, $r^3 = 27$, $r = 3$; By exercise 14 surface area is equal to $6\pi r^2$. thus in this example $= 6\pi 3^2 = 54\pi$ inches^2.

21. (a) $73 + 4x = 225$, $x = \$38$

(b) $C(40) = 233$, $C(50) =273$, cost rise $= 273 - 233 = \$40$

22. (a) $P(x) = 4x - C(x)$, $P(100) = 400 - (10 + 75) = \$315$
    (b) $P(101) = 404 - (10.1 + 75) = \$318.9$, increase is \$3.9

23. (a) $4x - 80$, $x = 80/.4 = 200$
    (b) $30 = .4x - 80$, $x = 275$
    (c) $40 = .4x - 80$, $x = 300$, so scoops is $300 - 275 = 25$

24. (a) $160 = 12x - 200$, $x = 30000$
    (b) $166 = 12x - 200$, $x = 30500$, number of new subscribers $= (30500 - 30000) = 500$

25. (a) $P(x) = R(x) - C(x) = 21x - 9x - 800 = 12x - 800$ dollars
    (b) $P(120) = 1440 - 800 = \$640$
    (c) $1000 = 12x - 800$, $x = 150$, so $R(150) = \$3150$

26. (a) $P(x) = 1200x - 550x - 6500 = 650x - 6500$ dollars, $P(12) = \$1300$
    (b) $C(x) = 14750 = 550x + 6500$, $x = 15$; $P(15) = 650(15) - 6500 = \$3250$

27. $f(6) = 270$ cents

28. From the graph $f(1) - f(6.87) = 330$, 1, 6.87

29. A 100 inch^3 cylinder of radius 3 inches costs \$1.62 to construct.

30. The least expensive cylinder has radius 3 inches and costs \$1.62 to construct.

31. $f(3) = \$1.62$, $f(6) = \$2.70$, so additional cost $= 2.7 - 1.62 = \$1.08$

32. $3.30 - 1.62 = \$1.68$

33. Revenue: \$1800; cost: \$1200   34. 20   35. 40

36. 1800 - 1200 =\$600 37. C(1000) = 4000

38. Find the x-coordinate of the point on the graph whose y-coordinate is 3500

39. Find the y-coordinate of the point whose x-coordinate is 400

40. 3136 - 2875 = 261

41. The greatest profit, \$52500, occurs when 2500 units of goods are produced.

42. P(1500) = 42500

43. Find the y-coordinate of the point on the graph whose y-coordinate is 30000.

44. Find the y-coordinate of the point on the graph whose x-coordinate is 2000.

45. Find h(3). Find the y-coordinate of the graph whose t-coordinate is 3.

46. Find t such that h(t) is as large as possible. Find the t-coordinate of the highest point of the graph.

47. Find the maximum value of h(t). Find the y-coordinate of the highest point of the graph.

48. Solve h(t) = 0. Find the t-intercept of the graph.

49. Solve h(t) = 100. Find the t-coordinate of the point whose y-coordinate is 100.

50. Find h(0). Find the y-intercept of the graph.

Chapter 0 Supplementary Exercises

1. $f(x) = x^3 + \frac{1}{x}$, $f(1) = 1^3 + \frac{1}{1} = 2$, $f(3) = 3^3 + \frac{1}{3} = 27\frac{1}{3}$

$f(-1) = (-1)^3 + \frac{1}{(-1)} = -2$, $f\left(-\frac{1}{2}\right) = \left(-\frac{1}{2}\right)^3 + -\frac{1}{8} - 2 = -2\frac{1}{8}$

$f(\sqrt{2}) = (\sqrt{2})^3 + \frac{1}{\sqrt{2}} = 2\sqrt{2} + \frac{1}{\sqrt{2}} = \frac{5\sqrt{2}}{2}$

2. $f(x) = 2x + 3x^2$, $f(0) = 2(0) + 3(0)^2 = 0$

$f\left(-\frac{1}{4}\right) = 2\left(-\frac{1}{4}\right) + 3\left(-\frac{1}{4}\right)^2 = -\frac{5}{16}$, $f\left(\frac{1}{\sqrt{2}}\right) = 2\left(\frac{1}{\sqrt{2}}\right) + 3\left(\frac{1}{\sqrt{2}}\right)^2 = \frac{3 + 2\sqrt{2}}{2}$

3. $f(x) = x^2 - 2$, $f(a - 2) = (a - 2)^2 - a = a^2 - 4a + 2$

4. $f(x) = \frac{1}{x + 1} - x^2$

$f(a + 1) = \frac{1}{(a + 1) + 1} - (a + 1)^2 = \frac{1}{a + 2} - (a^2 + 2a + 1)$

$= \frac{-(a^3 + 4a^2 + 5a + 1)}{a + 2}$

5. $f(x) = \frac{1}{x(x + 3)}$, $x \neq 0, -3$

6. $f(x) = \sqrt{x - 1}$, $x \geq 1$

7. $f(x) = \sqrt{x^2 + 1}$, all values of x

8. $f(x) = \frac{1}{\sqrt{3x}}$, $x > 0$

9. $h(x) = \frac{x^2 - 1}{x^2 + 1}$, $h\left(\frac{1}{2}\right) = \frac{\left(\frac{1}{2}\right)^2 - 1}{\left(\frac{1}{2}\right)^2 + 1} = -\frac{3}{5}$

So the point $\left(\frac{1}{2}, -\frac{3}{5}\right)$ is on the graph.

10. $k(x) = x^2 + \frac{2}{x}$, $k(1) = 1^2 + \frac{2}{1} = 2$

So $(1, -2)$ is not on the graph.

11. $5x^3 + 15x^2 - 20x = 5x(x^2 + 3x - 4) = 5x(x + 4)(x - 1)$

12. $3x^2 - 3x - 60 = 3(x^2 - x - 20) = 3(x - 5)(x + 4)$

13. $18 + 3x - x^2 = (-x - 3)(x - 6)$

14. $x^5 - x^4 - 2x^3 = x^3(x^2 - x - 2) = x^3(x - 2)(x + 1)$

15. $y = 5x^2 - 3x - 2$, set $5x^2 - 3x - 2 = 0$.

$$x = \frac{-b \pm \sqrt{b^2 - 4ac}}{2a} = \frac{3 \pm \sqrt{(-3)^2 - 4(5)(-2)}}{2(5)} = \frac{3 \pm 7}{10} = 1, -\frac{2}{5}$$

16. $y = -2x^2 - x + 2$, set $-2x^2 - x + 2 = 0$.

$$x = \frac{-b \pm \sqrt{b^2 - 4ac}}{2a} = \frac{1 \pm \sqrt{(-1)^2 - 4(-2)(2)}}{2(-2)}$$

$$= \frac{1 \pm \sqrt{17}}{-4} = \frac{-1 + \sqrt{17}}{4}, \frac{-1 - \sqrt{17}}{4}$$

17. Set $5x^2 - 3x - 2 = 2x - 1$, $5x^2 - 5x - 1 = 0$

$$x = \frac{-b \pm \sqrt{b^2 - 4ac}}{2a} = \frac{5 \pm \sqrt{(-5)^2 - 4(5)(-1)}}{2(5)} = \frac{5 \pm 3\sqrt{5}}{10}$$

$$y = 2x - 1 = 2\left(\frac{5 + 3\sqrt{5}}{10}\right) - 1 = \frac{3\sqrt{5}}{5}$$

$$y = 2x - 1 = 2\left(\frac{5 - 3\sqrt{5}}{10}\right) - 1 = \frac{-3\sqrt{5}}{5}$$

points of intersection: $\left(\frac{5 + 3\sqrt{5}}{10}, \frac{3\sqrt{5}}{5}\right)$, $\left(\frac{5 - 3\sqrt{5}}{10}, \frac{-3\sqrt{5}}{5}\right)$

18. $y = -x^2 + x + 1$, $y = x - 5$, set $-x^2 + x + 1 = x - 5$,
$x^2 - 6 = 0$

$$x = \frac{-b \pm \sqrt{b^2 - 4ac}}{2a} = \frac{0 \pm \sqrt{0^2 - 4(1)(-6)}}{2(1)} = \frac{\pm 2\sqrt{6}}{2} = \pm\sqrt{6}$$

$y = x - 5 = \sqrt{6} - 5$; $y = -\sqrt{6} - 5$

points of intersection: $(\sqrt{6}, \sqrt{6} - 5)$, $(-\sqrt{6}, -\sqrt{6} - 5)$

19. $f(x) + g(x) = (x^2 - 2x) + (3x - 1) = x^2 + x - 1$

20. $f(x) - g(x) = (x^2 - 2x) - (3x - 1) = x^2 - 5x + 1$

21. $f(x)h(x) = (x^2 - 2x)(\sqrt{x}) = x^2 \cdot x^{1/2} - 2x \cdot x^{1/2} = x^{5/2} - 2x^{3/2}$

22. $f(x)g(x) = (x^2 - 2x)(3x - 1) = 3x^3 - x^2 - 6x^2 + 2x$
$= 3x^3 - 7x^2 + 2x$

23. $\dfrac{f(x)}{h(x)} = \dfrac{x^2 - 2x}{\sqrt{x}} = x^{3/2} - 2x^{1/2}$

24. $g(x)h(x) = (3x - 1)\sqrt{x} = 3x \cdot x^{1/2} - x^{1/2} = 3x^{3/2} - x^{1/2}$

25. $f(x) - g(x) = \dfrac{x}{x^2 - 1} - \dfrac{1 - x}{1 + x} = \dfrac{x - (x - 1)(1 - x)}{x^2 - 1}$
$= \dfrac{x^2 - x + 1}{x^2 - 1}$

26. $f(x) - g(x + 1) = \dfrac{x}{x^2 - 1} - \dfrac{1 - (x + 1)}{1 + (x + 1)}$
$= \dfrac{x(x + 2) - (-x)(x^2 - 1)}{(x^2 - 1)(x + 2)} = \dfrac{x^3 + x^2 + x}{x^3 + 2x^2 - x - 2}$

27. $g(x) - h(x) = \dfrac{1 - x}{1 + x} - \dfrac{2}{3x + 1} = \dfrac{(1 - x)(3x + 1) - 2(1 + x)}{(1 + x)(3x + 1)}$
$= \dfrac{-3x^2 - 1}{3x^2 + 4x + 1}$

28. $f(x) + h(x) = \dfrac{x}{x^2 - 1} + \dfrac{2}{3x + 1} = \dfrac{x(3x + 1) + 2(x^2 - 1)}{(x^2 - 1)(3x + 1)}$
$= \dfrac{5x^2 + x - 2}{3x^3 + x^2 - 3x - 1}$

29. $g(x) - h(x - 3) = \dfrac{1 - x}{1 + x} - \dfrac{2}{3(x - 3) + 1}$

$$= \frac{(1 - x)(3x - 8) - 2(1 + x)}{(1 + x)(3x - 8)} = \frac{-3x^2 + 9x - 10}{3x^2 - 5x - 8}$$

30. $f(x) + g(x) = \frac{x}{x^2 - 1} + \frac{1 - x}{1 + x} = \frac{x + (1 - x)(x - 1)}{x^2 - 1}$

$$= \frac{-x^2 + 3x - 1}{x^2 - 1}$$

31. $f(g(x)),\ f(x) = x^2 - 2x + 4$

$$f\left(\frac{1}{x^2}\right) = \left(\frac{1}{x^2}\right)^2 - 2\left(\frac{1}{x^2}\right) + 4 = \frac{1}{x^4} - \frac{2}{x^2} + 4$$

32. $g(f(x)),\ g(x) = \frac{1}{x^2}$

$$g(x^2 - 2x + 4) = \frac{1}{(x^2 - 2x + 4)^2} = \frac{1}{x^4 - 4x^3 + 12x^2 - 16x + 16}$$

33. $g(h(x)),\ g(x) = \frac{1}{x^2}$

$$g\left(\frac{1}{\sqrt{x} - 1}\right) = \frac{1}{\left(\frac{1}{\sqrt{x} - 1}\right)^2} = \frac{1}{\frac{1}{x - 2\sqrt{x} + 1}}$$

$$= x - 2\sqrt{x} + 1 \text{ or } (\sqrt{x} - 1)^2$$

34. $h(g(x)),\ h(x) = \frac{1}{\sqrt{x} - 1}$

$$h\left(\frac{1}{x^2}\right) = \frac{1}{\sqrt{\frac{1}{x^2}} - 1} = \frac{1}{\frac{1}{x} - 1} = \frac{x}{1 - x}$$

35. $f(h(x)),\ f(x) = x^2 - 2x + 4$

$$f\left(\frac{1}{\sqrt{x} - 1}\right) = \left(\frac{1}{\sqrt{x} - 1}\right)^2 - 2\left(\frac{1}{\sqrt{x} - 1}\right) + 4$$

$$= \frac{1}{x - 2\sqrt{x} + 1} - \frac{2}{\sqrt{x} - 1} + 4 \text{ or } \frac{1}{(\sqrt{x} - 1)^2} - \frac{2}{\sqrt{x} - 1} + 4$$

36. $h(f(x)),\ h(x) = \dfrac{1}{\sqrt{x} - 1}$

$$h(x^2 - 2x + 4) = \frac{1}{(x^2 - 2x + 4)^{1/2} - 1}$$

37. $(81)^{3/4} = (\sqrt[4]{81})^3 = 27,\ 8^{5/3} = (\sqrt[3]{8})^5 = 2^5 = 32,$

$$(.25)^{-1} = \left(\frac{1}{4}\right)^{-1} = 4$$

38. $5^{-2} = \dfrac{1}{25},\ (100)^{3/2} = (\sqrt{100})^3 = 1000,\ (.001)^{3/2} = (\sqrt[3]{.001}) = .1$

39. $C(x)$ = carbon monoxide level corresponding to population x.

$P(t)$ = population of the city in t years

$C(x) = 1 + .4x;\ P(t) = 750 + 25t + .1t^2$

$$C(P(t)) = 1 + .4(750 + 25t + .1t^2) = 1 + 300 + 10t + .04t^2$$
$$= .04t^2 + 10t + 301$$

40. $R(x) = 5x - x^2,\ f(d) = 6\left(1 - \dfrac{200}{d + 200}\right) = \dfrac{6d}{d + 200}$

$$R(f(d)) = 5\left(\frac{6d}{d + 200}\right) - \left(\frac{6d}{d + 200}\right)^2$$
$$= \frac{30d(d + 200) - 36d^2}{(d + 200)^2} = \frac{-6d^2 + 6000d}{(d + 200)^2}$$

41. $(\sqrt{x + 1})^4 = (x + 1)^{4/2} = (x + 1)^2 = x^2 + 2x + 1$

42. $\dfrac{xy^3}{x^{-5}y^6} = \dfrac{x^6}{y^3}$

43. $\dfrac{x^{3/2}}{\sqrt{x}} = x^{3/2} \cdot x^{-1/2} = x$

44. $\sqrt[3]{x}\ (8x^{2/3}) = x^{1/3} \cdot 8x^{2/3} = 8x$

Exercises 1.1

1. $y = 2 - 5x$, slope $= -5$

2. $y = -5x$, slope $= -5$

3. $y = 2$, slope $= 0$

4. $y = \frac{1}{3}(x + 2) = \frac{1}{3}x + \frac{2}{3}$, slope $= \frac{1}{3}$

5. $y = \frac{2x - 1}{7} = \frac{2}{7}x - \frac{1}{7}$, slope $= \frac{2}{7}$

6. $y = \frac{1}{4}$, slope $= 0$

7. $2x + 3y = 6$, $3y = 6 - 2x$, $y = 2 - \frac{2}{3}x$, slope $= -\frac{2}{3}$

8. $x - y = 2$, $y = -2 + x$, slope $= 1$

9. slope $= 3$, y-intercept $(0, -1)$, $y = 3x - 1$

10. slope $= -\frac{1}{2}$, y-intercept $(0, 0)$, $y = -\frac{1}{2}x$

11. slope $= 1$, $(1, 2)$ on line. Let $(x, y) = (1, 2)$, $m = 1$.
$y - 2 = 1(x - 1)$, $y - 2 = x - 1$, $y = x + 1$

12. slope $= -\frac{1}{3}$, $(6, -2)$ on line. Let $(x, y) = (6, -2)$, $m = -\frac{1}{3}$.
$y - (-2) = -\frac{1}{3}(x - 6)$, $y + 2 = -\frac{1}{3}x + 2$, $y = -\frac{1}{3}x$

13. slope $= -7$, $(5, 0)$ on line. Let $(x, y) = (5, 0)$, $m = -7$.
$y - 0 = -7(x - 5)$, $y = -7x + 35$

14. slope $= -\frac{1}{2}$, $(2, -3)$ on line. Let $(x, y) = (2, -3)$, $m = \frac{1}{2}$.
$y - (-3) = \frac{1}{2}x - 1$, $y = \frac{1}{2}x - 4$

15. slope $= 0$, $(7, 4)$ on line. Let $(x, y) = (7, 4)$, $m = 0$.
$y - 4 = 0(x - 7)$, $y = 4$

16. slope $= -\frac{2}{5}$, $(0, 5)$ on line. Let $(x, y) = (0, 5)$, $m = -\frac{2}{5}$.
$y - 5 = -\frac{2}{5}(x - 0)$, $y = -\frac{2}{5}x + 5$

17. (2, 1) and (4, 2) on line.

$$\text{slope} = \frac{y_2 - y_1}{x_2 - x_1} = \frac{2 - 1}{4 - 2} = \frac{1}{2}$$

$y - 1 = \frac{1}{2}(x - 2)$, $y = \frac{1}{2}x$

18. (5, -3) and (-1, 3) on line.

$$\text{slope} = \frac{y_2 - y_1}{x_2 - x_1} = \frac{3 - (-3)}{-1 - 5} = \frac{6}{-6} = -1$$

$y - (-3) = -1(x - 5)$, $y + 3 = -x + 5$, $y = -x + 2$.

19. (0, 0) and (1, -2) on line.

$$\text{slope} = \frac{y_2 - y_1}{x_2 - x_1} = \frac{-2 - 0}{1 - 0} = -2$$

$y - 0 = -2(x - 0)$, $y = -2x$

20. (2, - 1) and (3, -1) on line.

$$\text{slope} = \frac{y_2 - y_1}{x_2 - x_1} = \frac{-1 - (-1)}{3 - 2} = \frac{0}{1} = 0$$

$y - (-1) = 0(x - 2)$, $y = -1$

21. parallel to $y = -2x + 1$, $\left(\frac{1}{2}, 5\right)$ on line. slope = m = -2

$y - y_1 = m(x - x_1)$, $y - 5 = -2\left(x - \frac{1}{2}\right)$, $y = -2x + 6$

22. parallel to $3x + y = 7$, (-1, 1) on line.

$y = 7 - 3x$, slope = m = -3

$y - (-1) = -3(x - (-1))$, $y + 1 = -3x - 3$, $y = -3x - 4$

23. parallel to $3x - 6y = 1$, (1, 0) on line.

$3x - 6y = 1$, $-6y = 1 - 3x$, $y = -\frac{1}{6} + \frac{1}{2}x$; slope = m = $\frac{1}{2}$

$y - 0 = \frac{1}{2}(x - 1)$, $y = \frac{1}{2}x - \frac{1}{2}$

24. parallel to $5x + 2y = -4$, (0, 17) on line.

$5x + 2y = -4$, $2y = -4 - 5x$, $y = -2 - \frac{5}{2}x$; slope = m = $-\frac{5}{2}$

$y - 17 = -\frac{5}{2}(x - 0)$, $y = -\frac{5}{2}x + 17$

25. (a) - (C) x and y intercepts are 1

(b) - (B) x-intercept is 1, y-intercept is -1

(c) - (D) x and y intercepts are -1

(d) - (A) x-intercept is -1, y-intercept is 1

26. Use (4.8, 3.6) and (4.9, 4.8)

$m = \dfrac{4.8 - 3.6}{4.9 - 4.8} = 12$; $y = mx + b$ so $6 = 12(5) + b$, $b = -54$

27. 1/2 =j/4, j units in y direction, j = 2

28. 2 = j/(1/4), j = 1/2

29. -3 = j/(1/4), j = -3/4

30. 1/5 = j/5, j = 1

31. slope = 2, (1, 3) on line.

$x_1 = 1$, $y_1 = 3$

$x = 2$

$y - 3 = 2(2 - 1)$, $y = 4 - 2 + 3 = 5$

$x = 3$

$y - 3 = 2(3 - 1)$, $y = 6 - 2 + 3 = 7$

$x = 0$

$y - 3 = 2(0 - 1)$, $y = 0 - 2 + 3 = 1$

32. slope = -3, (2, 2) on line.

$x_1 = 2$, $y_1 = 2$

$x = 3$

$y - 2 = -3(3 - 2)$, $y = -9 + 6 + 2 = -1$

$x = 4$

$y - 2 = -3(4 - 2)$, $y = -12 + 6 + 2 = -4$

$x = 1$

$y - 2 = -3(1 - 2),\ y = -3 + 6 + 2 = 5$

33. slope $= -\frac{1}{4}$, $(-1, -1)$ on line.

$x_1 = -1,\ y_1 = -1$

$x = 0$

$y - (-1) = -\frac{1}{4}(0 - (-1)),\ y = 0 - \frac{1}{4} - 1 = -\frac{5}{4}$

$x = 1$

$y - (-1) = -\frac{1}{4}(1 - (-1)),\ y = -\frac{1}{4} - \frac{1}{4} - 1, = -\frac{6}{4} = -\frac{3}{2}$

$x = -2$

$y - (-1) = -\frac{1}{4}(-2 - (-1)),\ y = \frac{2}{4} - \frac{1}{4} - 1 = -\frac{3}{4}$

34. slope $= \frac{1}{3}$, $(-5, 2)$ on line.

$x_1 = -5,\ y_1 = 2$

$x = -4$

$y - 2 = \frac{1}{3}(-4 - (-5)),\ y = -\frac{4}{3} + \frac{5}{3} + 2 = \frac{7}{3}$

$x = -3$

$y - 2 = \frac{1}{3}(-3 - (-5)),\ y = -\frac{3}{3} + \frac{5}{3} + 2 = \frac{8}{3}$

$x = -2$

$y - 2 = \frac{1}{3}(-2 - (-5)),\ y = -\frac{2}{3} + \frac{5}{3} + 2 = 3$

35. ℓ_1 36. ℓ_2

37. slope $= m = -2$

y-intercept: $(0, -1)$

Let $(x_1, y_1) = (0, -1)$.

$y - (-1) = -2(x - 0)$

$y = -2x - 1$

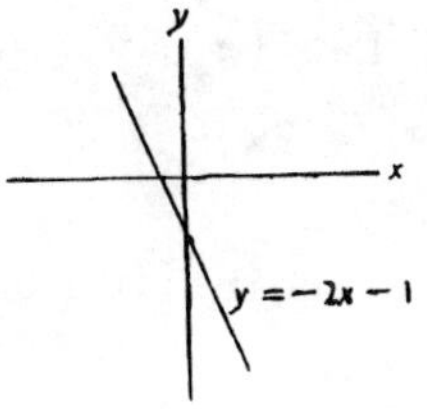

38. slope $= m = \frac{1}{3}$

y-intercept: $(0, 1)$

Let $(x_1, y_1) = (0, 1)$.

$y - 1 = \frac{1}{3}(x - 0)$

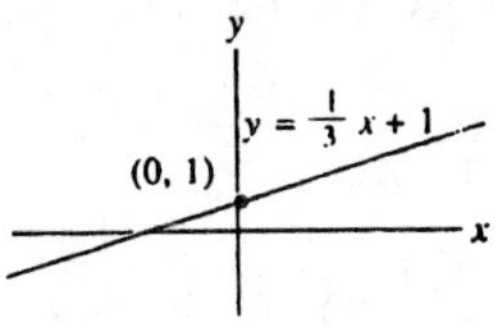

$y = \frac{1}{3}x + 1$

39. slope = m = $\frac{4}{5}$

Let $(x_1, y_1) = (2, 0)$.

$y - 0 = \frac{4}{5}(x - 2)$

$y = \frac{4}{5}x - \frac{8}{5}$

$y = \frac{4}{5}(x - 2)$

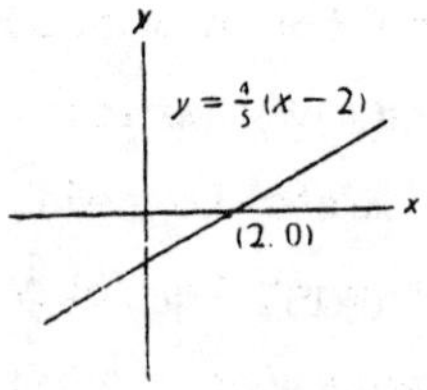

40. slope = m = 0

Let $(x_1, y_1) = (-1, 3)$.

$y - 3 = 0(x - (-1))$

$y = 3$

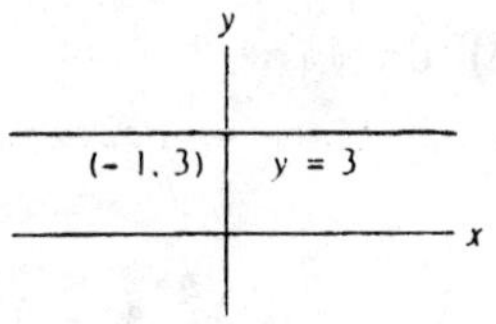

41. $y = mx + b$, $y = mx + 5$

42. $x = my + b$, $x = my + 9$

43. $y = -(4/5)x - 6/5$, $m = -4/5$, so the parallel lines

$y = -(4/5)x + C$

44. $y = C$

45. $y = -2x + C$

46. $x = C$

47. slope = m = 2x; slope of the tangent line at point $(1, 1) = 2(1) = 2$; Let $(x_1, y_1) = (1, 1)$.

$y - 1 = 2(x - 1)$, $y = 2x - 1$

48. slope = 2x; slope at point $(0, 0) = 2(0) = 0$

Let $(x_1, y_1) = (0, 0)$.

$y - 0 = 0(x - 0)$, $y = 0$

49. slope = 2x; slope at point $\left(-\frac{1}{2}, \frac{1}{4}\right) = 2\left(-\frac{1}{2}\right) = -1$

Let $(x_1, y_1) = \left(-\frac{1}{2}, \frac{1}{4}\right)$.

$y - \frac{1}{4} = -1\left(x - \left(-\frac{1}{2}\right)\right)$, $y = -x - \frac{1}{2} + \frac{1}{4}$, $y = -x - \frac{1}{4}$

50. weekly pay: $y = 5x + 60$, where x = units of goods sold. Each unit sold increases the pay by 5 dollars. The weekly pay is 60 dollars if no units are sold.

51. The demand equation for a monopolist is $y = -.02x + 7$, where x = number of units, y = price. If the monopolist wants to sell one more unit of goods, then the price per unit must be lowered by 2 units. No one will pay 7 dollars or more for a unit of goods.

52. x = degrees Fahrenheit, y = degrees Centigrade
$0 = 32m + b$ and $100 = 212m + b$ so $b = -32m$ and hence for m
$100 = 212m + -32m$ or $180m = 100$, $m = 5/9$ and $b = -160/9$

53.

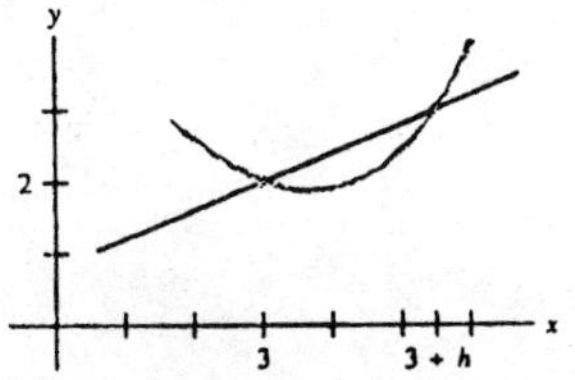

54. Let $y = mx + b$ and $y = m'x + b'$ be two distinct lines. We show that these lines are parallel if and only if $m = m'$. Since two lines are parallel if and only if they have no points in common, it suffices to show that $m = m'$ if and only if the equation $mx + b = m'x + b'$ has no solution in x. Suppose $m = m'$. Then $mx + b = m'x + b'$ implies $b = b'$; but since the lines are distinct, $b \neq b'$. Thus if $m = m'$, $mx + b = m' + b'$ has no solution. If $m \neq m'$, then $x = \dfrac{b' - b}{m - m'}$ is a solution to $mx + b = m'x + b'$. Thus, $mx + b = m'x + b'$ has no solution in x if and only if $m = m'$, and it follows that two distinct lines are parallel if and only if they have the same slope.

55. (See accompanying figure)

Let ℓ_1, ℓ_2, m_1, m_2, a and b be as in the diagram. Then m_1 is the slope of ℓ_1 and $-m_2$ is the slope of ℓ_2. From the pythagorean theorem, we have

$$a^2 + b^2 = (m_1 + m_2),\ 1^2 + m_1^2 = a^2 \text{ and } 1^2 + m_2^2 = b^2.$$

Combining these, we get

$$1 + m_1^2 + 1 + m_2^2 = (m_1 + m_2)^2 = m_1^2 + 2m_1m_2 + m_2^2.$$

Thus, $2 = 2m_1m_2$ and $m_1m_2 = 1$. Since the slope of ℓ_1 is m_1 and the slope of ℓ_2 is $-m_2$, the product of their slopes is therefore, -1.

Exercises 1.2

1.

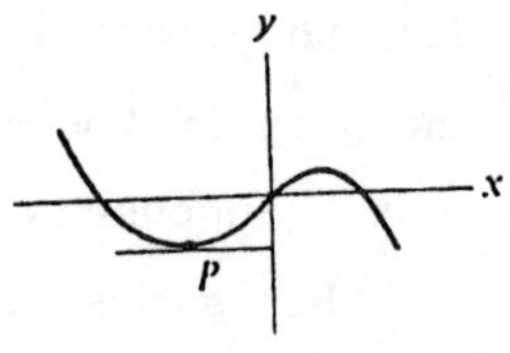

2.

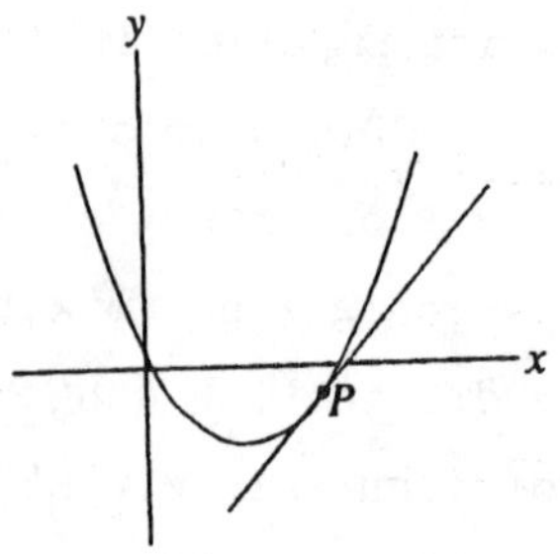

3.

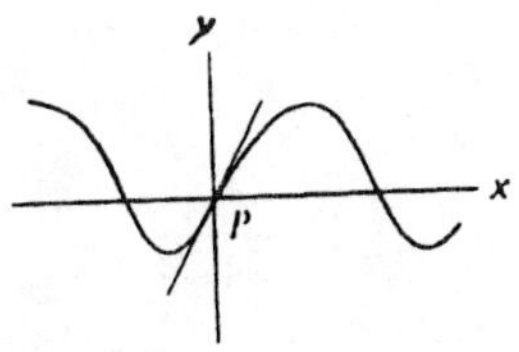

4.

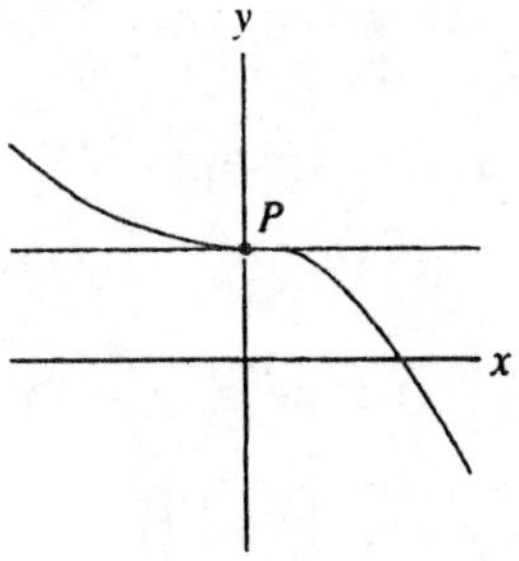

5.

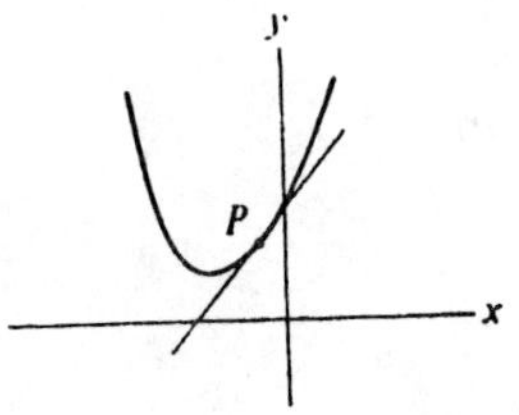

6.

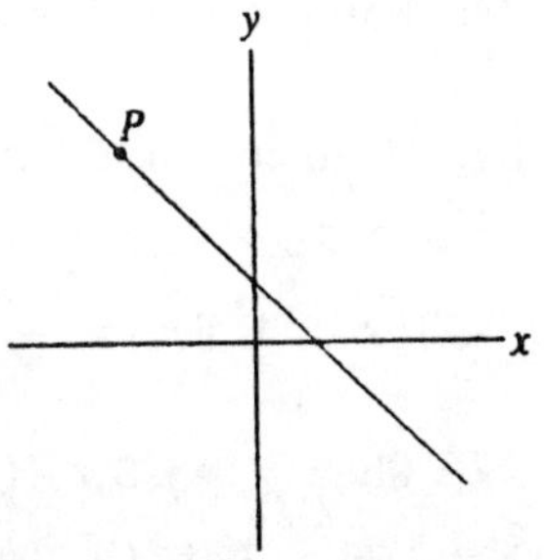

7. 1

8. $-\frac{1}{2}$

9. -3

10. 0

11. $\frac{2}{3}$

12. -1

13. small positive slope

14. large positive slope

15. zero slope

16. large positive slope

17. zero slope

18. small negative

19. The slope of the graph of $y = x^2$ at the point (x, y) is 2x.
So the slope at the point (-2, 4) is 2(-2) = -4.
Let $(x_1, y_1) = (-2, 4)$, m = -4.
$y - 4 = -4(x - (-2))$, $y = -4x - 4$.

20. The slope of the graph $y = x^2$ at point (x, y) is $2x$. At $(-.4, .16)$, $2x = 2(-.4) = -.8$.
Let $(x_1, y_1) = (-.4, .16)$, $m = -.8$.
$y - .16 = -.8(x - (-.4))$, $y = -.8x - .16$

21. At $\left(\frac{4}{3}, \frac{16}{9}\right)$, slope $= m = 2x = 2\left(\frac{4}{3}\right) = \frac{8}{3}$.
Let $(x_1, y_1) = \left(\frac{4}{3}, \frac{16}{9}\right)$.
$y - \frac{16}{9} = \frac{8}{3}\left(x - \frac{4}{3}\right)$, $y = \frac{8}{3}x - \frac{16}{9}$

22. slope $= 2x$; When $x = -\frac{1}{2}$, slope $= 2\left(-\frac{1}{2}\right) = -1$.

23. slope $= 2x$; When $x = 1.5$, slope $= 2(1.5) = 3$ and $y = (1.5)^2 = 2.25$. Let $(x_1, y_1) = (1.5, 2.25)$, $m = 3$.
$y - 2.25 = 3(x - 1.5)$, $y = 3x - 2.25$

24. slope $= 2x$; When $x = .6$, slope $= 2(.6) = 1.2$ and $y = (.6)^2 = .36$. Let $(x_1, y_1) = (.6, .36)$, $m = 1.2$.
$y - .36 = 1.2(x - .6)$, $y = 1.2x - .36$

25. slope $= 2x$, set $2x = \frac{5}{3}$, $x = \frac{5}{6}$
$y = x^2$, so when $x = \frac{5}{6}$, $y = \left(\frac{5}{6}\right)^2 = \frac{25}{36}$, so $\left(\frac{5}{6}, \frac{25}{36}\right)$ is the point.

26. slope $= 2x$, set $2x = -4$, $x = -2$
When $x = -2$, $y = (-2)^2 = 4$, so $(-2, 4)$ is the point.

27. $x + 2y = 4$, $y = -\frac{1}{2}x + 2$, so slope $= m = -\frac{1}{2}$.
Slope of the tangent line is $2x$.
Set $2x = -\frac{1}{2}$, $x = -\frac{1}{4}$.
$y = x^2$
When $x = -\frac{1}{4}$, $y = \frac{(-1)^2}{4} = \frac{1}{16}$, so $\left(-\frac{1}{4}, \frac{1}{16}\right)$ is the point.

28. $3x - y = 2$, $y = 3x - 2$, slope $= 3$.

The slope of the tangent line is 2x.

Set $2x = 3$, $x = \frac{3}{2}$.

$y = x^2$

When $x = \frac{3}{2}$, $y = \left(\frac{3}{2}\right)^2 = \frac{9}{4}$, so $\left(\frac{3}{2}, \frac{9}{4}\right)$ is the point.

29. slope $= 3x^2$; When $x = 2$, slope $= 3(2)^2 = 12$.

30. slope $= 3x^2$; When $x = \frac{3}{2}$, slope $= 3\left(\frac{3}{2}\right)^2 = \frac{27}{4}$.

31. slope $= 3x^2$; When $x = -\frac{1}{2}$, slope $= 3\left(-\frac{1}{2}\right)^2 = \frac{3}{4}$

32. The slope of the curve $y = x^3$ at the point (x, y) is $3x^2$.

When $x = \frac{1}{4}$, slope $= 3\left(\frac{1}{4}\right)^2 = \frac{3}{16}$.

33. $y = x^3$, slope $= 3x^2$; When $x = -1$, slope $= 3(-1)^2 = 3$,

$y = (-1)^3 = 1$. Let $(x_1, y_1) = (-1, -1)$.

$y - (-1) = 3(x - (-1))$, $y = 3x + 2$

34. $y = x^3$, slope $= 3x^2$; When $x = \frac{1}{2}$, slope $= 3\left(\frac{1}{2}\right)^2 = \frac{3}{4}$,

$y = \left(\frac{1}{2}\right)^3 = \frac{1}{8}$. Let $(x_1, y_1) = \left(\frac{1}{2}, \frac{1}{8}\right)$.

$y - \frac{1}{8} = \frac{3}{4}\left(x - \frac{1}{2}\right)$, $y = \frac{3}{4}x - \frac{1}{4}$

35. (a) $m = \frac{13 - 4}{5 - 2} = 3$, length is $13 - 4 = 9$

(b) Increase

36.

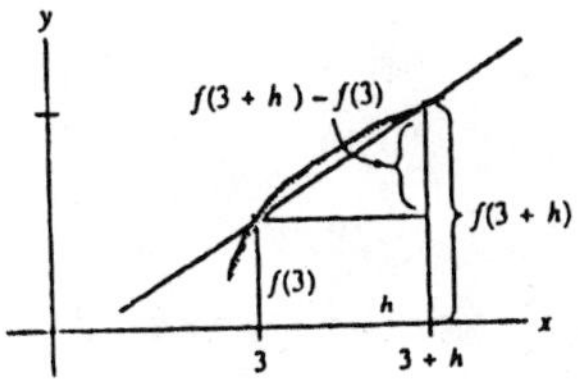

Exercises 1.3

1. $f(x) = 2x - 5$, $f'(x) = 2$

2. $f(x) = 3 - \frac{1}{2}x$, $f'(x) = -\frac{1}{2}$

3. $f(x) = x^8$, $f'(x) = 8x^7$

4. $f(x) = x^{75}$, $f'(x) = 75x^{74}$

5. $f(x) = x^{5/2}$, $f'(x) = \frac{5}{2}x^{3/2}$

6. $f(x) = x^{4/3}$, $f'(x) = \frac{4}{3}x^{1/3}$

7. $f(x) = \sqrt[3]{x} = x^{1/3}$, $f'(x) = \frac{1}{3}x^{-2/3}$

8. $f(x) = x^{3/4}$, $f'(x) = \frac{3}{4}x^{-1/4}$

9. $f(x) = x^{-2}$, $f'(x) = -2x^{-3}$

10. $f(x) = 5$, $f'(x) = 0$

11. $f(x) = x^{-1/4}$, $f'(x) = -\frac{1}{4}x^{-5/4}$

12. $f(x) = x^{-3}$, $f'(x) = -3x^{-4}$

13. $f(x) = \frac{3}{4}$, $f'(x) = 0$

14. $f(x) = \frac{1}{\sqrt[3]{x}} = x^{-1/3}$, $f'(x) = -\frac{1}{3}x^{-4/3}$

15. $f(x) = \frac{1}{x^3} = x^{-3}$, $f'(x) = -3x^{-4}$

16. $f(x) = \frac{1}{x^5} = x^{-5}$, $f'(x) = -5x^{-6}$

17. $f(x) = x$, at $x = -2$; $f'(x) = 6x$, $f'(-2) = 6(-2) \; = -192$

18. $f(x) = x^3$, at $x = \frac{1}{4}$; $f'(x) = 3x^2$, $f'\left(\frac{1}{4}\right) = 3\left(\frac{1}{4}\right)^2 = \frac{3}{16}$

19. $f(x) = \frac{1}{x}$, at $x = 3$, $f(x) = x^{-1}$; $f'(x) = -x^{-2}$,
$f'(3) = -\frac{1}{3^2} = -\frac{1}{9}$

20. $f(x) = 5x$, at $x = 2$; $f'(x) = 5$, $f'(2) = 5$

21. $f(x) = 4 - x$, at $x = 5$; $f'(x) = -1$, $f'(5) = -1$

22. $f(x) = x^{2/3}$, at $x = 1$; $f'(x) = \frac{2}{3}x^{-1/3} = \frac{2}{3\sqrt[3]{x}}$,

$f'(1) = \frac{2}{3\sqrt[3]{1}} = \frac{2}{3}$

23. $f(x) = x^{3/2}$, at $x = 9$; $f'(x) = \frac{3}{2}x^{1/2} = \frac{3}{2}\sqrt{x}$,

$f'(9) = \frac{3}{2}\sqrt{9} = \frac{3}{2}\cdot 3 = \frac{9}{2}$

24. $f(x) = \frac{1}{x^2}$, at $x = 2$, $f(x) = x^{-2}$; $f'(x) = -2x^{-3} = \frac{-2}{x^3}$,

$f'(2) = -\frac{2}{2^3} = -\frac{1}{4}$

25. $y = x^4$, slope $= y' = 4x^3$; at $x = 3$, $y' = 4(3)^3 = 108$

26. $y = x^5$, slope $= y' = 5x^4$; at $x = -2$, $y' = 5(-2)^4 = 80$

27. $y = \sqrt{x} = x^{1/2}$, slope $= y' = \frac{1}{2}x^{-1/2} = \frac{1}{2\sqrt{x}}$

at $x = 9$, $y' = \frac{1}{2\sqrt{9}} = \frac{1}{6}$

28. $y = x^{-3}$, slope $= y' = -3x^{-4} = -\frac{3}{x^4}$; at $x = 3$, $y' = -\frac{3}{3^4} = -\frac{1}{27}$

29. $f(x) = x^2$, $f(-5) = (-5)^2 = 25$

$f'(x) = 2x$, $f'(-5) = 2(-5) = -10$

30. $f(x) = x + 6$, $f(3) = 3 + 6 = 9$; $f'(x) = 1$, $f'(3) = 1$

31. $f(x) = \frac{1}{x^5} = x^{-5}$, $f(2) = \frac{1}{2^5} = \frac{1}{32}$

$f'(x) = -5x^{-6} = -\frac{5}{x^6}$, $f'(2) = -\frac{5}{2^6} = -\frac{5}{64}$

32. $f(x) = \frac{1}{x^2} = x^{-2}$, $f(5) = \frac{1}{5^2} = \frac{1}{25}$

$f'(x) = -2x^{-3} = -\frac{2}{x^3}$, $f'(5) = -\frac{2}{5^3} = -\frac{2}{125}$

33. $f(x) = x^{4/3}$, $f(8) = (8)^{4/3} = 16$

$f'(x) = \frac{4}{3}x^{1/3} = \frac{4}{3}x^{1/3} = \frac{4}{3}\sqrt[3]{x}$, $f'(8) = \frac{4}{3}\sqrt[3]{8} = \frac{8}{3}$

34. $f(x) = x^{3/2}$, $f(16) = (16)^{3/2} = 64$

$f'(x) = \frac{3}{2}x^{1/2} = \frac{3}{2}\sqrt{x}$, $f'(16) = \frac{3}{2}\sqrt{16} = \frac{12}{2} = 6$

35. $y = x^3$, slope $= y' = 3x^2$; at $x = 4$, $y' = 3(4)^2 = 48$

Let $(x_1, y_1) = (4, 64)$.

$y - 64 = 48(x - 4)$, $y = 48x - 192 + 64$, $y = 48x - 128$

36. $y = \sqrt{x} = x^{1/2}$, slope $= y' = \frac{1}{2}x^{-1/2} = \frac{1}{2\sqrt{x}}$;

at $x = 25$, slope $= \frac{1}{2\sqrt{25}} = \frac{1}{10}$

Let $(x_1, y_1) = (25, 5)$.

$y - 5 = \frac{1}{10}(x - 25)$, $y = \frac{1}{10}x - 2.5 + 5$, $y = \frac{1}{10}x + 2.5$

37. $\frac{d}{dx}(x^8) = 8x^7$

38. $\frac{d}{dx}(x^{-3}) = -3x^{-4}$

39. $\frac{d}{dx}(x^{3/4}) = \frac{3}{4}x^{-1/4}$

40. $\frac{d}{dx}(x^{-1/3}) = -\frac{1}{3}x^{-4/3}$

41. $y = 1$, $\frac{d}{dx}(1) = 0$

42. $y = x^{-4}$, $\frac{d}{dx}(x^{-4}) = -4x^{-5}$

43. $y = x^{1/5}$, $\frac{d}{dx}(x^{1/5}) = \frac{1}{5}x^{-4/5}$

44. $y = \frac{x - 1}{3} = \frac{1}{3}x - \frac{1}{3}$, $\frac{d}{dx}\left[\frac{x - 1}{3}\right] = \frac{1}{3}$

45. $f(x) = \frac{1}{3}x + 2$, $f(6) = \frac{1}{3}(6) + 2 = 4$

$f'(x) = \frac{1}{3}$, $f'(6) = \frac{1}{3}$

46. $f(x) = 4$, $f(1) = 4$; $f'(x) = 0$, $f'(1) = 0$

47. $y = f(x) = \sqrt{x} = x^{1/2}$, slope $= f'(x) = \frac{1}{2}x^{-1/2} = \frac{1}{2\sqrt{x}}$

The slope of the tangent line $y = \frac{1}{4}x + b$ is $\frac{1}{4}$.

First, find the value of a. Let $\frac{1}{4} = \frac{1}{2\sqrt{a}}$ and solve for a.

$2\sqrt{a} = 4$, $\sqrt{a} = 2$, $a = 4$

When $x = 4$, $f(4) = \sqrt{4} = 2$.

Let $(x_1, y_1) = (4, 2)$.

$y - 2 = \frac{1}{4}(x - 4)$, $y = \frac{1}{4}x - 1 + 2$, $y = \frac{1}{4}x + 1$; so $b = 1$.

48. $y = \frac{1}{x}$ When $x = 2$, $y = \frac{1}{2}$.

slope $= f'(x) = -x^{-2} = -\frac{1}{x^2}$, $f'(2) = -\frac{1}{4}$

To find the equation of the tangent line set

$$(x_1, y_1) = \left(2, \frac{1}{2}\right), \quad m = -\frac{1}{4}.$$

$y - \frac{1}{2} = -\frac{1}{4}(x - 2)$, $y = -\frac{1}{4}x + 1$

To find the value of a (which is the x-intercept),

let $-\frac{1}{4}x + 1 = 0$ and solve for x.

$-\frac{1}{4}x + 1 = 0$, $x = 4 = a$

49. At $x = a$, $y = 2.01a - .51$, or $y = 2.02a - .52$, so $.01a = .01$ $a = 1$, and $y = f(a) = 2.01 - .51 = 1.5$. $f'(a) = 2$ because the slope of the 'smallest' secant line is 2.01.

50. $$\frac{f(1 + .2) - f(1)}{.2} = \frac{1.1 - .8}{.2} = 1.5$$

51. $$\frac{(3 + h)^2 - 3^2}{h} = \frac{9 + 6h + h^2 - 9}{h} = 6 + h,$$ and since h approaches zero the answer is 6.

52. $\dfrac{\sqrt{4+h}-\sqrt{4}}{h}\cdot\dfrac{\sqrt{4+h}+\sqrt{4}}{\sqrt{4+h}+\sqrt{4}} = \dfrac{4+h-4}{h(\sqrt{4+h}+\sqrt{4})} = \dfrac{1}{\sqrt{4+h}+\sqrt{4}}$
as h approaches zero the answer is 1/4.

53. $\dfrac{f(x+h)-f(x)}{h} = 2x + 5 + h$, as $h \to 0$ the answer is $2x + 5$

54. $\dfrac{g(x+h)-g(x)}{h} = 8x + 1 + 4h$, as $h \to 0$ the answer is
$8x + 1$

55. $f(x) = x^3$, slope of the secant line

$$\frac{f(x+h)-f(x)}{h} = \frac{(x+h)^3 - x^3}{h} = \frac{x^3 + 3x^2h + 3xh^2 + h^3 - x^3}{h}$$
$$= \frac{3x^2h + 3xh^2 + h^3}{h} = \frac{h(3x^2 + 3xh + h^2)}{h}$$
$$= 3x^2 + 3xh + h^2$$

As h approaches zero (as the secant line approaches the tangent line), the quantity $3x^2 + 3xh + h^2$ approaches $3x^2$. Thus, we have $f'(x) = 3x^2$.

56. $f(x) = \sqrt{x}$, slope of the secant line

$$\frac{f(x+h)-f(x)}{h} = \frac{\sqrt{x+h}-\sqrt{x}}{h} = \frac{(\sqrt{x+h}-\sqrt{x})(\sqrt{x+h}+\sqrt{x})}{h(\sqrt{x+h}+\sqrt{x})}$$
$$= \frac{x+h-x}{h(\sqrt{x}+\sqrt{h}+\sqrt{x})} = \frac{h}{h(2\sqrt{x}+\sqrt{h})} = \frac{1}{2\sqrt{x}+\sqrt{h}}$$

As h approaches zero, $\dfrac{1}{2\sqrt{x}+\sqrt{h}}$ approaches $\dfrac{1}{2\sqrt{x}}$. So $f'(x) = \dfrac{1}{2\sqrt{x}}$.

57. The coordinates of A are $(4,5)$. Reading the graph of the derivative, we see that $f'(4) = \frac{1}{2}$, so the slope of the tangent line is $\frac{1}{2}$. By the point-slope formula, the equation of the tangent line is:

$$y - 5 = \frac{1}{2}(x - 4)$$

58. The coordinates of A are $(2, 1.75)$. Reading the graph of the derivative, we see that $f'(2) = \frac{1}{2}$, so the slope of the tangent

line is $\frac{1}{2}$. By the point-slope formula, the equation of the tangent line is:

$$y - 1.75 = \frac{1}{2}(x - 2)$$

Exercises 1.4

1. no limit
2. 2
3. 1
4. no limit
5. no limit
6. no limit

7. $\lim_{x\to 1} (1 - 6x) = 1 - 6(1) = -5$

8. $\lim_{x\to 2} \frac{x}{x - 2}$ is undefined.

9. $\lim_{x\to 3} \sqrt{x^2 + 16} = \sqrt{(3)^2 + 16} = \sqrt{25} = 5$

10. $\lim_{x\to 4} (x^3 - 7) = 4^3 - 7 = 57$

11. $\lim_{x\to 5} \frac{x^2 + 1}{5 - x}$ is undefined.

12. $\lim_{x\to 6} \left(\sqrt{6x} + 3x - \frac{1}{x}\right)(x^2 - 4)$

$$= \left(\lim_{x\to 6} \sqrt{6x} + \lim_{x\to 6} 3x - \lim_{x\to 6} \frac{1}{x}\right)\left(\lim_{x\to 6} x^2 - \lim_{x\to 6} 4\right)$$

$$= \left(6 + 18 - \frac{1}{6}\right)(36 - 4) = \frac{143}{6}\cdot 32 = \frac{2288}{3}$$

13. $\lim_{x\to 7} (x + \sqrt{x - 6})(x^2 - 2x + 1) = \lim_{x\to 7} (x + \sqrt{x - 6})(x - 1)^2$

$$= \left(\lim_{x\to 7} x + \lim_{x\to 7} \sqrt{x - 6}\right)\left(\lim_{x\to 7} x - \lim_{x\to 7} 1\right)^2$$

$$= (7 + 1)(7 - 1)^2 = 8\cdot 36 = 288$$

14. $$\lim_{x\to 8} \frac{\sqrt{5x - 4} - 1}{3x^2 + 2} = \frac{\lim_{x\to 8} \sqrt{5x - 4} - \lim_{x\to 8} 1}{\lim_{x\to 8} 3x^2 + \lim_{x\to 8} 2} = \frac{6 - 1}{192 + 2} = \frac{5}{194}$$

15. $$\lim_{x\to 9} \frac{\sqrt{x^2 - 5x - 36}}{8 - 3x} = \frac{\left(\lim_{x\to 9} x^2 - \lim_{x\to 9} 5x - \lim_{x\to 9} 36\right)^{1/2}}{\lim_{x\to 9} 8 - \lim_{x\to 9} 3x}$$

$$= \frac{(81 - 45 - 36)^{1/2}}{8 - 27} = \frac{\sqrt{0}}{-19} = 0$$

16. $\lim_{x \to 10} (2x^2 - 15x - 50)^{20} = \left[\lim_{x \to 10} 2x^2 - \lim_{x \to 10} 15x - \lim_{x \to 10} 50\right]^{20}$

$= (200 - 150 - 50)^{20} = 0^{20} = 0$

17. $\lim_{x \to 0} \frac{x^2 + 3x}{x} = \lim_{x \to 0} \frac{x(x + 3)}{x} = \lim_{x \to 0} x + 3 = 3$

18. $\lim_{x \to 1} \frac{x^2 - 1}{x - 1} = \lim_{x \to 1} \frac{(x - 1)(x + 1)}{(x - 1)} = \lim_{x \to 1} x + 1 = 2$

19. $\lim_{x \to 2} \frac{-2x^2 + 4x}{x - 2} = \lim_{x \to 2} \frac{-2x(x - 2)}{(x - 2)} = \lim_{x \to 2} -2x = -4$

20. $\lim_{x \to 3} \frac{x^2 - x - 6}{x - 3} = \lim_{x \to 3} \frac{(x - 3)(x + 2)}{(x - 3)} = \lim_{x \to 3} x + 2 = 5$

21. $\lim_{x \to 4} \frac{x^2 - 16}{4 - x} = \lim_{x \to 4} \frac{(x - 4)(x + 4)}{-(x - 4)} = \lim_{x \to 4} -x - 4$

$= \lim_{x \to 4} -x - \lim_{x \to 4} 4 = -4 - 4 = -8$

22. $\lim_{x \to 5} \frac{2x - 10}{x^2 - 25} = \lim_{x \to 5} \frac{2(x - 5)}{(x - 5)(x + 5)} = \frac{\lim_{x \to 5} 2}{\lim_{x \to 5} (x + 5)} = \frac{2}{5 + 5} = \frac{1}{5}$

23. $\lim_{x \to 6} \frac{x^2 - 6x}{x^2 - 5x - 6} = \lim_{x \to 6} \frac{x(x - 6)}{(x - 6)(x + 1)} = \frac{\lim_{x \to 6} x}{\lim_{x \to 6} (x + 1)}$

$= \frac{6}{6 + 1} = \frac{6}{7}$

24. $\lim_{x \to 7} \frac{x^3 - 2x^2 + 3x}{x^2} = \lim_{x \to 7} \frac{x(x^2 - 2x + 3)}{x^2} = \frac{\lim_{x \to 7} (x^2 - 2x + 3)}{\lim_{x \to 7} x}$

$= \frac{49 - 14 + 3}{7} = \frac{38}{7}$

25. $\lim_{x \to 8} \frac{x^2 + 64}{x - 8}$ is undefined.

26. $\lim_{x\to 9} \frac{1}{(x-9)^2} = \frac{\lim_{x\to 9} 1}{\lim_{x\to 9} (x-9)^2}$ is undefined.

27. $\lim_{x\to 0} \frac{-2}{\sqrt{x+16}+7} = \frac{\lim_{x\to 0} -2}{\lim_{x\to 0} \sqrt{x+16} + \lim_{x\to 0} 7} = \frac{-2}{\sqrt{16}+7} = -\frac{2}{11}$

28. $\lim_{x\to 0} \frac{4x}{x(x^2+3x+5)} = \frac{\lim_{x\to 0} 4}{\lim_{x\to 0} (x^2+3x+5)} = \frac{4}{0+0+5} = \frac{4}{5}$

29. $f(x) = x^2 + 1$

$$f'(3) = \lim_{h\to 0} \frac{f(3+h) - f(3)}{h} = \lim_{h\to 0} \frac{(3+h)^2 + 1 - (3^2+1)}{h}$$

$$= \lim_{h\to 0} \frac{9 + 6h + h^2 + 1 - 10}{h} = \lim_{h\to 0} \frac{h^2 + 6h}{h} = \lim_{h\to 0} h + 6 = 6$$

30. $f(x) = x^3$

$$f'(2) = \lim_{h\to 0} \frac{f(2+h) - f(2)}{h} = \lim_{h\to 0} \frac{(2+h)^3 - 2^3}{h}$$

$$= \lim_{h\to 0} \frac{8 + 12h + 6h^2 + h^3 - 8}{h} = \lim_{h\to 0} \frac{h(12 + 6h + h^2)}{h}$$

$$= \lim_{h\to 0} h^2 + 6h + 12 = 12$$

31. $f(x) = x^3 + 3x + 1$

$$f'(0) = \lim_{h\to 0} \frac{f(0+h) - f(0)}{h} = \lim_{h\to 0} \frac{h^3 + 3h + 1 - (-1)}{h}$$

$$= \lim_{h\to 0} \frac{h(h^2+3)}{h} = \lim_{h\to 0} h^2 + 3 = 3$$

32. $f(x) = x^2 + 2x + 2$

$$f'(0) = \lim_{h\to 0} \frac{(0+h) - f(0)}{h} = \lim_{h\to 0} \frac{h^2 + 2h + 2 - (-2)}{h}$$

$$= \lim_{h\to 0} \frac{h(h + 2)}{h} = \lim_{h\to 0} h + 2 = 2$$

33. $f(x) = \dfrac{1}{2x + 5}$

$$f'(3) = \lim_{h\to 0} \frac{f(3 + h) - f(3)}{h} = \lim_{h\to 0} \frac{\frac{1}{2(3 + h) + 5} - \frac{1}{2(3) + 5}}{h}$$

$$= \lim_{h\to 0} \frac{\frac{1}{6 + 2h + 5} - \frac{1}{11}}{h} = \lim_{h\to 0} \frac{\frac{11 - (2h + 11)}{121 + 22h}}{h}$$

$$= \lim_{h\to 0} \frac{-2h}{121 + 22h}\cdot\frac{1}{h} = \lim_{h\to 0} \frac{-2}{121 + 22h} = \frac{-2}{121}$$

34. $f(x) = \sqrt{2x - 1}$

$$f'(4) = \lim_{h\to 0} \frac{f(4 + h) - f(4)}{h} = \lim_{h\to 0} \frac{\sqrt{2(4 + h)} - \sqrt{2(4) - 1}}{h}$$

$$= \lim_{h\to 0} \frac{\sqrt{7 + 2h} - \sqrt{7}}{h} = \lim_{h\to 0} \frac{(\sqrt{7 + 2h} - \sqrt{7})(\sqrt{7 + 2h} + \sqrt{7})}{h(\sqrt{7 + 2h} + \sqrt{7})}$$

$$= \lim_{h\to 0} \frac{7 + 2h - 7}{h(\sqrt{7 + 2h} + \sqrt{7})} = \lim_{h\to 0} \frac{2}{\sqrt{7 + 2h} + \sqrt{7}} = \frac{2}{2\sqrt{7}} = \frac{1}{\sqrt{7}}$$

35. $f(x) = \sqrt{5 - x}$

$$f'(2) = \lim_{h\to 0} \frac{f(2 + h) - f(2)}{h} = \lim_{h\to 0} \frac{\sqrt{3 - h} - \sqrt{3}}{h}$$

$$= \lim_{h\to 0} \frac{(\sqrt{3 - h} - \sqrt{3})(\sqrt{3 - h} + \sqrt{3})}{h(\sqrt{3 - h} + \sqrt{3})} = \lim_{h\to 0} \frac{3 - h - 3}{h(\sqrt{3 - h} + \sqrt{3})}$$

$$= \lim_{h\to 0} -\frac{1}{\sqrt{3 - h} + \sqrt{3}} = -\frac{1}{2\sqrt{3}} = -\frac{\sqrt{3}}{6}$$

36. $f(x) = \dfrac{1}{7 - 2x}$

$$f'(3) = \lim_{h\to 0} \frac{f(3 + h) - f(3)}{h} = \lim_{h\to 0} \frac{\frac{1}{7 - 2(3 + h)} - \frac{1}{7 - 2(3)}}{h}$$

$$= \lim_{h\to 0} \frac{\frac{1}{1 - 2h} - 1}{h} = \lim_{h\to 0} \frac{1 - (1 - 2h)}{1 - 2h}\cdot\frac{1}{h} = \lim_{h\to 0} \frac{2}{1 - 2h} = 2$$

37. $f(x) = \sqrt{1 - x^2}$

$$f'(0) = \lim_{h\to 0} \frac{f(0 + h) - f(0)}{h} = \lim_{h\to 0} \frac{\sqrt{1 - h^2} - \sqrt{1 - 0}}{h}$$

$$= \lim_{h\to 0} \frac{\left[\sqrt{1 - h^2} - 1\right]\left[\sqrt{1 - h^2} + 1\right]}{h\left[\sqrt{1 - h^2} + 1\right]}$$

$$= \lim_{h\to 0} \frac{1 - h^2 - 1}{h\left[\sqrt{1 - h^2} + 1\right]} = \lim_{h\to 0} \frac{-h}{\sqrt{1 - h^2} + 1} = 0$$

38. $f(x) = (5x - 4)^2$

$$f'(2) = \lim_{h\to 0} \frac{f(2 + h) - f(2)}{h} = \lim_{h\to 0} \frac{[5(2 + h) - 4]^2 - [5(2) - 4]}{h}$$

$$= \lim_{h\to 0} \frac{(6 + 5h)^2 - 6^2}{h} = \lim_{h\to 0} \frac{36 + 60h + 25h^2 - 36}{h}$$

$$= \lim_{h\to 0} \frac{h(25h + 60)}{h} = \lim_{h\to 0} 25h + 60 = 60$$

39. $f(x) = (x + 1)^3$

$$f'(0) = \lim_{h\to 0} \frac{f(0 + h) - f(0)}{h} = \lim_{h\to 0} \frac{(h + 1)^3 - 1^3}{h}$$

$$= \lim_{h\to 0} \frac{h^3 + 3h^2 + 3h + 1 - 1}{h} = \lim_{h\to 0} \frac{h(h^2 + 3h + 3)}{h}$$

$$= \lim_{h\to 0} h^2 + 3h + 3 = 3$$

40. $f(x) = \sqrt{x^2 + x + 1}$

$$f'(0) = \lim_{h\to 0} \frac{f(0 + h) - f(0)}{h} = \lim_{h\to 0} \frac{\sqrt{h^2 + h + 1} - \sqrt{1}}{h}$$

$$= \lim_{h\to 0} \frac{\left[\sqrt{h^2 + h + 1} - 1\right]\left[\sqrt{h^2 + h + 1} + 1\right]}{h\left[\sqrt{h^2 + h + 1} + 1\right]}$$

$$= \lim_{h\to 0} \frac{h^2 + h + 1 - 1}{h\left[\sqrt{h^2 + h + 1} + 1\right]} = \lim_{h\to 0} \frac{h(h + 1)}{h\left[\sqrt{h^2 + h + 1} + 1\right]}$$

$$= \frac{1}{\sqrt{1} + 1} = \frac{1}{2}$$

41. $f'(a) = \lim_{h\to 0} \frac{f(a + h) - f(a)}{h}$, $a = 9$ and $f(x) = \sqrt{x}$

42. $a = 2$ and $f(x) = x^3$

43. $a = 10$ and $f(x) = x^{-1}$

44. $a = 64$ and $f(x) = x^{1/3}$

45. $a = 1$ and $f(x) = 3x^2 + 4$

46. $a = 1$ and $f(x) = x^{-1/2} = 1/\sqrt{x}$

47. $\lim_{x\to\infty} \frac{1}{x^2} = 0$

48. $\lim_{x\to-\infty} \frac{1}{x^2} = 0$

49. $\lim_{x\to\infty} \frac{1}{x - 8} = 0$

50. $\lim_{x\to\infty} \frac{1}{3x + 5} = 0$

51. $\lim_{x\to\infty} \frac{2x + 1}{x + 2} = \lim_{x\to\infty} \frac{2 + \frac{1}{x}}{1 + \frac{2}{x}} = 2$

52. $\lim_{x\to\infty} \frac{x^2 + x}{x^2 - 1} = \lim_{x\to\infty} \frac{1 + \frac{1}{x}}{1 - \frac{1}{x^2}} = 1$

Exercises 1.5

1. no 2. yes 3. yes 4. yes 5. no 6. no

7. no 8. no 9. yes 10. yes 11. no 12. no

13. $f(x) = x^2$; $\lim_{x\to 1} f(x) = \lim_{x\to 1} x^2 = 1$. Since $\lim_{x\to 1} f(x) = 1 = f(1)$, $f(x)$ is continuous at $x = 1$.

$$f'(1) = \lim_{h\to 0} \frac{f(1 + h) - f(1)}{h} = \lim_{h\to 0} \frac{(1 + h)^2 - (1)^2}{h}$$

$$= \lim_{h\to 0} \frac{1 + 2h + h^2 - 1}{h} = \lim_{h\to 0} \frac{h(2 + h)}{h} = 2$$

Therefore, $f(x)$ is continuous and differentiable at $x = 1$.

14. $f(x) = \frac{1}{x}$; $\lim_{x\to 1} f(x) = \lim_{x\to 1} \frac{1}{x} = 1$; $f(1) = \frac{1}{1} = 1$

Since $\lim_{x\to 1} f(x) = 1 = f(1)$, $f(x)$ is continuous at $x = 1$.

$$f'(1) = \lim_{h\to 0} \frac{f(1 + h) - f(1)}{h} = \lim_{h\to 0} \frac{\frac{1}{1 + h} - 1}{h}$$

$$= \lim_{h\to 0} \frac{1 - (1 + h)}{1 + h}\cdot\frac{1}{h} = \lim_{h\to 0} -\frac{1}{1 + h} = -1.$$

Therefore, $f(x)$ is continuous and differentiable at $x = 1$.

15. $$f(x) = \begin{cases} x + 2 & \text{for } -1 \le x \le 1 \\ 3x & \text{for } 1 < x < 5 \end{cases}$$

$\lim_{x\to 1} 3x = 3$; $\lim_{x\to 1} (x + 2) = 3$; $f(1) = 1 + 2 = 3$

Since $\lim_{x\to 1} f(x) = 3 = f(1)$, $f(x)$ is continuous at $x = 1$.

Since the graph of $f(x)$ at $x = 1$ does not have a tangent line, $f(x)$ is not differentiable at $x = 1$. Therefore, $f(x)$ is continuous but not differentiable at $x = 1$.

16. $$f(x) = \begin{cases} x & \text{for } 1 \le x \le 2 \\ x^3 & \text{for } 0 \le x < 1 \end{cases}$$

$\lim_{x\to 1} x^3 = 1$; $\lim_{x\to 1} x = 1$; $f(1) = 1$

Since $\lim_{x\to 1} f(x) = 1 = f(1)$, f(x) is continuous at x = 1.

Since the graph of f(x) at x = 1 does not have a tangent line, f(x) is not differentiable at x = 1. Therefore, f(x) is continuous but not differentiable at x = 1.

17. $f(x) = \begin{cases} 2x - 1 & \text{for } 0 \le x \le 1 \\ 1 & \text{for } 1 < x \end{cases}$

$\lim_{x\to 1} 1 = 1$; $\lim_{x\to 1} 2x - 1 = 1$; $f(1) = 2(1) - 1 = 1$

Since $\lim_{x\to 1} f(x) = 1 = f(1)$, f(x) is continuous at x = 1.

Since the graph of f(x) at x = 1 does not have a tangent line, f(x) is not differentiable at x = 1. Therefore, f(x) is continuous but not differentiable at x = 1.

18. $f(x) = \begin{cases} x & \text{for } x \ne 1 \\ 2 & \text{for } x = 1 \end{cases}$

$\lim_{x\to 1} x = 1$; $f(1) = 2$

Since $\lim_{x\to 1} f(x) = 1 \ne 2 = f(1)$, f(x) is not continuous at x = 1. By theorem 1, since f(x) is not continuous at x = 1, it is not differentiable.

19. $f(x) = \begin{cases} \dfrac{1}{x - 1} & \text{for } x \ne 1 \\ 0 & \text{for } x = 1 \end{cases}$

$\lim_{x\to 1} f(x) = \lim_{x\to 1} \dfrac{1}{x - 1}$ is undefined. Since $\lim_{x\to 1} f(x)$ does not exist, f(x) is not continuous at x = 1. By theorem 1, since f(x) is not continuous at x = 1, it is not differentiable.

20. $f(x) = \begin{cases} x - 1 & \text{for } 0 \le x < 1 \\ 1 & \text{for } x = 1 \\ 2x - 2 & \text{for } x > 1 \end{cases}$

$\lim_{x \to 1} (x - 1) = 0 = \lim_{x \to 1} (2x - 2)$; but $f(1) = 1$, so $f(x)$ is not continuous at $x = 1$. Therefore, $f(x)$ is not differentiable at $x = 1$.

21. $\dfrac{x^2 - 7x + 10}{x - 5} = \dfrac{(x - 5)(x - 2)}{x - 5} = x - 2$, so define $f(5) = 5 - 2 = 3$

22. $x^2 + x - 12 = (x + 4)(x - 3)$, so define $f(-4) = -7$

23. $\dfrac{x^3 - 5x^2 + 4}{x^2}$ it is not possible to define $f(x)$ at $x = 0$.

24. $\dfrac{x^2 + 25}{x - 5}$, it is not possible to define $f(x)$ at $x = 5$

25. $\dfrac{(6 + x)^2 - 36}{x} = 12 + 12x$, at $x = 0$ define $f(x) = 12$

26. $\dfrac{\sqrt{9 + x} - \sqrt{9}}{x} = \dfrac{\sqrt{9 + x} + \sqrt{9}}{\sqrt{9 + x} + \sqrt{9}} = \dfrac{1}{\sqrt{9 + x} + \sqrt{9}}$, at $x = 0$ define $f(x) = 1/6$.

Exercises 1.6

1. $y = x^3 + x^2$

$$\frac{dy}{dx} = \frac{d}{dx}(x^3 + x^2) = \frac{d}{dx}x^3 + \frac{d}{dx}x^2 = 3x^2 + 2x$$

2. $y = x^2 + \dfrac{1}{x} = x^2 + x^{-1}$

$$\frac{dy}{dx} = \frac{d}{dx}(x^2 + x^{-1}) = \frac{d}{dx}x^2 + \frac{d}{dx}x^{-1} = 2x - \frac{1}{x^2}$$

3. $y = x^2 + 3x - 1$

$$\frac{dy}{dx} = \frac{d}{dx}(x^2 + 3x - 1) = \frac{d}{dx}x^2 + \frac{d}{dx}3x - \frac{d}{dx}1 = 2x + 3 + 0$$
$$= 2x = 3$$

4. $y = x^3 + 2x + 5$

$$\frac{dy}{dx} = \frac{d}{dx}(x^3 + 2x + 5) = \frac{d}{dx}x^3 + \frac{d}{dx}2x + \frac{d}{dx}5 = 3x^2 + 2 + 0$$
$$= 3x^2 + 2$$

5. $f(x) = x^5 + \frac{1}{x} = x^5 + x^{-1}$

$$\frac{d}{dx}(x^5 + x^{-1}) = \frac{d}{dx}x^5 + \frac{d}{dx}x^{-1} = 5x^4 + (-1x^{-2}) = 5x^4 - \frac{1}{x^2}$$

6. $f(x) = x^8 - x$

$$\frac{d}{dx}(x^8 - x) = \frac{d}{dx}x^8 - \frac{d}{dx}x = 8x^7 - 1$$

7. $f(x) = x^4 + x^3 + x$

$$\frac{d}{dx}(x^4 + x^3 + x) = \frac{d}{dx}x^4 + \frac{d}{dx}x^3 + \frac{d}{dx}x = 4x^3 + 3x^2 + 1$$

8. $f(x) = x^5 + x^2 - x$

$$\frac{d}{dx}(x^5 + x^2 - x) = \frac{d}{dx}x^5 + \frac{d}{dx}x^2 - \frac{d}{dx}x = 5x^4 + 2x - 1$$

9. $y = 3x^2$

$$\frac{dy}{dx} = \frac{d}{dx}(3x^2) = 3\cdot\frac{d}{dx}x^2 = 3(2x) = 6x$$

10. $y = 2x^3$

$$\frac{dy}{dx} = 2\cdot\frac{d}{dx}x^3 = 2(3x^2) = 6x^2$$

11. $y = x^3 + 7x^2$

$$\frac{dy}{dx} = \frac{d}{dx}x^3 + 7\frac{d}{dx}x^2 = 3x^2 + 14x$$

12. $y = -2x$

$$\frac{dy}{dx} = -2\frac{d}{dx}x = -2(1) = -2$$

13. $y = \frac{4}{x^2} = 4x^{-2}$

$$\frac{dy}{dx} = 4\frac{d}{dx}x^{-2} = -\frac{8}{x^3}$$

14. $y = 2\sqrt{x} = 2x^{1/2}$

$$\frac{dy}{dx} = 2\frac{d}{dx}x^{1/2} = 2\cdot\frac{1}{2}x^{-1/2} = \frac{1}{\sqrt{x}}$$

15. $y = 3x - \frac{1}{x} = 3x - x^{-1}$

$$\frac{dy}{dx} = 3\frac{d}{dx}x - \frac{d}{dx}x^{-1} = 3 - (-1)\cdot x^{-2} = 3 + \frac{1}{x^2}$$

16. $y = -x^2 + 3x + 1$

$$\frac{dy}{dx} = -1\frac{d}{dx}x^2 + 3\frac{d}{dx}x + \frac{d}{dx}1 = -2x + 3 + 0 = -2x + 3$$

17. $f(x) = \frac{1}{3}x^3 - \frac{1}{2}x^2$

$$\frac{d}{dx}\left[\frac{1}{3}x^3 - \frac{1}{2}x^2\right] = \frac{1}{3}\frac{d}{dx}x^3 - \frac{1}{2}\frac{d}{dx}x^2 = \frac{1}{3}\cdot 3x^2 - \frac{1}{2}\cdot 2x = x^2 - x$$

18. $f(x) = 100x^{100}$

$$\frac{d}{dx}100x^{100} = 100(100x^{99}) = 10,000x^{99}$$

19. $f(x) = -\frac{1}{5x^5} = -\frac{1}{5}x^{-5}$

$$\frac{d}{dx}\left[-\frac{1}{5}x^{-5}\right] = -\frac{1}{5}\frac{d}{dx}x^{-5} = -\frac{1}{5}\cdot -5x^{-6} = \frac{1}{x^6}$$

20. $f(x) = x^2 - \frac{1}{x^2} = x^2 - x^{-2}$

$$\frac{d}{dx}(x^2 - x^{-2}) = \frac{d}{dx}x^2 - \frac{d}{dx}x^{-2} = 2x - (-2)x^{-3} = 2x + \frac{2}{x^3}$$

21. $f(x) = 1 - \sqrt{x} = 1 - x^{1/2}$

$$\frac{d}{dx}(1 - x^{1/2}) = \frac{d}{dx}1 - \frac{d}{dx}x^{1/2} = 0 - \frac{1}{2}x^{-1/2} = \frac{-1}{2\sqrt{x}}$$

22. $f(x) = -3x^2 + 7$

$$\frac{d}{dx}(-3x^2 + 7) = -3\frac{d}{dx}x^2 + \frac{d}{dx}7 = -3(2x) + 0 = -6x$$

23. $f(x) = (3x + 1)^{10}$

$$\frac{d}{dx}(3x + 1)^{10} = 10(3x + 1)^9 \cdot \frac{d}{dx}(3x + 1) = 10(3x + 1)^9 \cdot 3$$
$$= 30(3x + 1)^9$$

24. $f(x) = \dfrac{1}{x^2 + x + 1} = (x^2 + x + 1)^{-1}$

$$\frac{d}{dx}(x^2 + x + 1)^{-1} = -1(x^2 + x + 1)^{-2}\frac{d}{dx}(x^2 + x + 1)$$
$$= -(x^2 + x + 1)^{-2}(2x + 1)$$
$$= \frac{-(2x + 1)}{(x^2 + x + 1)^2}$$

25. $f(x) = 5\sqrt{3x^3 + x} = 5(3x^3 + x)^{1/2}$

$$\frac{d}{dx}\left[5(3x^3 + x)^{1/2}\right] = 5\frac{d}{dx}(3x^3 + x)^{1/2}$$
$$= 5 \cdot \frac{1}{2}(3x^3 + x)^{-1/2} \cdot \frac{d}{dx}(3x^3 + x)$$
$$= \frac{5(9x^2 + 1)}{2\sqrt{3x^3 + x}} = \frac{45x^2 + 5}{2\sqrt{3x^3 + x}}$$

26. $y = \dfrac{1}{(x^2 - 7)^5} = (x^2 - 7)^{-5}$

$$\frac{dy}{dx} = \frac{d}{dx}(x^2 - 7)^{-5} = -5(x^2 - 7)^{-6} \cdot \frac{d}{dx}(x^2 - 7) = -\frac{5(2x)}{(x^2 - 7)^6}$$
$$= -\frac{10}{(x^2 - 7)^6}$$

27. $y = (2x^2 - x + 4)^6$

$$\frac{dy}{dx} = \frac{d}{dx}(2x^2 - x + 4)^6 = 6(2x^2 - x + 4)^5 \cdot \frac{d}{dx}(2x^2 - x + 4)$$
$$= 6(2x^2 - x + 4)^5 \cdot (4x - 1)$$
$$= 6(4x - 1)(2x^2 - x + 4)^5$$

28. $y = \sqrt{-2x + 1} = (-2x + 1)^{1/2}$

$$\frac{dy}{dx} = \frac{d}{dx}\left[(-2x + 1)^{1/2}\right] = \frac{1}{2}(-2x + 1)^{-1/2}\frac{d}{dx}(-2x + 1)$$
$$= \frac{-2}{2\sqrt{-2x + 1}} = \frac{-1}{\sqrt{-2x + 1}}$$

29. $y = \frac{x}{3} + \frac{3}{x} = \frac{1}{3}x + 3x^{-1}$

$$\frac{dy}{dx} = \frac{d}{dx}\left[\frac{1}{3}x + 3x^{-1}\right] = \frac{1}{3}\cdot\frac{d}{dx}x + 3\cdot\frac{d}{dx}x^{-1} = \frac{1}{3} + 3(-1)x^{-2}$$
$$= \frac{1}{3} - 3x^{-2}$$

30. $y = \frac{2x - 1}{5} = \frac{1}{5}(2x - 1)$

$$\frac{dy}{dx} = \frac{d}{dx}\left[\frac{1}{5}(2x - 1)\right] = \frac{1}{5}\cdot\frac{d}{dx}(2x - 1) = \frac{1}{5}\cdot 2 = \frac{2}{5}$$

31. $y = \frac{2}{1 - 5x} = 2(1 - 5x)^{-1}$

$$\frac{dy}{dx} = \frac{d}{dx}\left[2(1 - 5x)^{-1}\right] = 2\cdot\frac{d}{dx}\left[(1 - 5x)^{-1}\right]$$
$$= 2\cdot(-1)(1 - 5x)^{-2}\cdot\frac{d}{dx}(1 - 5x)$$
$$= -2(1 - 5x)^{-2}\cdot(-5) = 10(1 - 5x)^{-2}$$

32. $y = \frac{4}{3\sqrt{x}} = \frac{4}{3}x^{-1/2}$

$$\frac{dy}{dx} = \frac{d}{dx}\left[\frac{4}{3}x^{-1/2}\right] = \frac{4}{3}\cdot\frac{d}{dx}x^{-1/2} = \frac{4}{3}\cdot\left(-\frac{1}{2}x^{-3/2}\right) = -\frac{2}{3}x^{-3/2}$$

33. $y = \dfrac{1}{1 - x^4} = (1 - x^4)^{-1}$

$$\frac{dy}{dx} = \frac{d}{dx}\left[(1 - x^4)^{-1}\right] = -1(1 - x^4)^{-2}\cdot\frac{d}{dx}(1 - x^4)$$
$$= -(1 - x^4)^{-2}\cdot(-4x^3) = 4x^3(1 - x^4)^{-2}$$

34. $y = \left(x^3 + \dfrac{x}{2} + 1\right)^5$

$$\frac{dy}{dx} = \frac{d}{dx}\left(x^3 + \frac{x}{2} + 1\right)^5 = 5\left(x^3 + \frac{x}{2} + 1\right)^4\cdot\frac{d}{dx}\left(x^3 + \frac{x}{2} + 1\right)$$
$$= 5\left(x^3 + \frac{x}{2} + 1\right)^4\cdot\left(3x^2 + \frac{1}{2}\right)$$
$$= 5\left(3x^2 + \frac{1}{2}\right)\left(x^3 + \frac{x}{2} + 1\right)^4$$

35. $f(x) = \dfrac{4}{\sqrt{x^2 + x}} = 4(x^2 + x)^{-1/2}$

$$\frac{d}{dx}4(x^2 + x)^{-1/2} = 4\cdot\frac{d}{dx}(x^2 + x)^{-1/2}$$
$$= 4\cdot\left(-\frac{1}{2}\right)(x^2 + x)^{-3/2}\cdot\frac{d}{dx}(x^2 + x)$$
$$= -2(x^2 + x)^{-3/2}\cdot(2x + 1)$$
$$= -2(2x + 1)(x^2 + x)^{-3/2}$$

36. $f(x) = \dfrac{6}{x^2 + 2x + 5} = 6(x^2 + 2x + 5)^{-1}$

$$\frac{d}{dx}6(x^2 + 2x + 5)^{-1} = 6\frac{d}{dx}(x^2 + 2x + 5)^{-1}$$
$$= 6(-1)(x^2 + 2x + 5)^{-2}\cdot\frac{d}{dx}(x^2 + 2x + 5)$$
$$= -6(x^2 + 2x + 5)^{-2}(2x + 2)$$
$$= -6(2x + 2)(x^2 + 2x + 5)^{-2}$$

37. $f(x) = \left(\dfrac{\sqrt{x}}{2} + 1\right)^{3/2}$

$$\frac{d}{dx}\left(\frac{\sqrt{x}}{2} + 1\right)^{3/2} = \frac{3}{2}\left(\frac{\sqrt{x}}{2} + 1\right)^{1/2}\cdot\frac{d}{dx}\left(\frac{\sqrt{x}}{2} + 1\right)$$

$$= \frac{3}{2}\left(\frac{\sqrt{x}}{2} + 1\right)^{1/2} \cdot \frac{1}{2} \cdot \frac{1}{2} \cdot x^{-1/2} = \frac{3}{2}\left(\frac{\sqrt{x}}{2} + 1\right)^{1/2}\left(\frac{1}{4}x^{-1/2}\right)$$

$$= \frac{3}{8\sqrt{x}}\left(\frac{\sqrt{x}}{2} + 1\right)^{1/2}$$

38. $f(x) = \left(4 - \frac{2}{x}\right)^3$

$$\frac{d}{dx}\left(4 - \frac{2}{x}\right)^3 = 3\left(4 - \frac{2}{x}\right)^2 \cdot \frac{d}{dx}\left(4 - \frac{2}{x}\right)$$

$$= 3\left(4 - \frac{2}{x}\right)^2\left(\frac{d}{dx}4 - 2\frac{d}{dx}x^{-1}\right)$$

$$= 3\left(4 - \frac{2}{x}\right)^2\left(2x^{-2}\right) = \frac{6}{x^2}\left(4 - \frac{2}{x}\right)^2$$

39. $f(x) = 3x^2 - 2x + 1,\ (1, 2)$

$\text{slope} = f'(x) = \frac{d}{dx}(3x^2 - 2x + 1) = 6x - 2$

$f'(x) = 6(1) - 2 = 4$

40. $f(x) = x^{10} + 1 + \sqrt{1 - x},\ (0, 2)$

$\text{slope} = f'(x) = \frac{d}{dx}(x^{10} + 1 + \sqrt{1 - x}) = 10x^9 + \frac{d}{dx}(1 - x)^{1/2}$

$$= 10x^9 + \left(\frac{1}{2}(1 - x)^{-1/2} \cdot (-1)\right) = 10x^9 - \frac{1}{2\sqrt{1 - x}}$$

$f'(0) = 10(0)^9 - \frac{1}{2\sqrt{1 - 0}} = -\frac{1}{2}$

41. $y = x^3 + 3x - 8$

$\text{slope} = y' = \frac{d}{dx}(x^3 + 3x - 8) = 3x^2 + 3$

$f'(2) = 3(2)^2 + 3 = 15$

42. $y = x^3 + 3x - 8$

$f'(x) = 3x^2 + 3,\ f'(2) = 15$

To find the equation of the tangent line, let

$(x_1, y_1) = (2, 6)$ and the slope = 15.

$y - 6 = 15(x - 2),\ y = 15x - 30 + 6,\ y = 15x - 24$

43. $y = (x^2 - 15)^6$

$$\text{slope} = \frac{dy}{dx} = \frac{d}{dx}(x^2 - 15)^6 = 6(x^2 - 15)^5 \cdot \frac{d}{dx}(x^2 - 15)$$

$$= 6(x^2 - 15)^5 \cdot 2x = 12x(x^2 - 15)^5$$

$\text{slope} = f'(x) = 12x(x^2 - 15)^5;\ f'(4) = 12(4)(16 - 15)^5 = 48$

$y = f(x) = (x^2 - 15)^6;\ f(4) = (4^2 - 15)^6 = 1$

Let $(x_1, y_1) = (4, 1)$, slope = 48.

$y - 1 = 48(x - 4),\ y = 48x - 192 + 1,\ y = 48x - 191$

44. $y = f(x) = \dfrac{8}{x^2 + x + 2},\ f(2) = \dfrac{8}{2^2 + 2 + 2} = 1$

$$\text{slope} = f'(x) = \frac{d}{dx} 8(x^2 + x + 2)^{-1}$$

$$= 8(-1)(x^2 + x + 2)^{-2} \cdot \frac{d}{dx}(x^2 + x + 2)$$

$$= -8(x^2 + x + 2)(2x + 1) = \frac{-8(2x + 1)}{(x^2 + x + 2)}$$

$$f'(2) = \frac{-8(4 + 1)}{(4 + 2 + 2)^2} = -\frac{40}{64} = -\frac{5}{8}$$

Let $(x_1, y_1) = (2, 1)$.

$y - 1 = -\frac{5}{8}(x - 2),\ y = -\frac{5x}{8} + \frac{10}{8} + 1,\ y = -\frac{5x}{8} + \frac{9}{4}$

45. $f(x) = (3x^2 + x - 2)^2$

(a) $\frac{d}{dx}(3x^2 + x - 2)^2 = 2(3x^2 + x - 2) \cdot \frac{d}{dx}(3x^2 + x - 2)$

$$= 2(3x^2 + x - 2) \cdot (6x + 1)$$

$$= 2(6x + 1)(3x^2 + x - 2)$$

(b) $(3x^2 + x - 2)(3x^2 + x - 2)$

$$= 9x^4 + 3x^3 - 6x^2 + 3x^3 + x^2 - 2x - 6x^2 - 2x + 4$$

$$= 9x^4 + 6x^3 - 11x^2 - 4x + 4$$

$\frac{d}{dx}(9x^4 + 6x^3 - 11x^2 - 4x + 4) = 36x^3 + 18x^2 - 22x - 4$

46. $\frac{d}{dx}\left[f(x) - g(x)\right] = \frac{d}{dx}f(x) - \frac{d}{dx}g(x)$

$= \frac{d}{dx}f(x) + \frac{d}{dx}(-1)g(x)$ (sum rule)

$= \frac{d}{dx}f(x) - \frac{d}{dx}g(x)$ (const. mult. rule)

47. $f(1) = .6(1) + 1 = 1.6$, so $g(1) = 3f(1) = 4.8$

$f'(1) = .6$ (slope of the line), $g'(1) = 3f'(1) = 1.8$

48. $h(1) = f(1) + g(1) = -.4(1) + 2.6 + .26(1) + 1.1 = 3.56$

$h'(1) = f'(1) + g'(1) = -.4 + .26 = -.14$

49. $h(5) = 3(2) + 2(4) = 14$, $h'(5) = 3(3) + 2(1) = 11$

50. $f(x) = 2[g(x)]^3$, $f'(x) = 6[g(x)]^2 g'(x)$

$f(3) = 2[g(3)]^3 = 2(2)^3 = 16$, $f'(3) = 6(2)^2 4 = 96$

51. $f(x) = 5\sqrt{g(x)}$, $f'(x) = \frac{5}{2\sqrt{g(x)}}g'(x)$

$f(1) = 5\sqrt{4} = 10$, $f'(1) = \frac{5}{2\sqrt{4}}3 = 15/4$

52. $\frac{dy}{dx} = 3x^2 - 12x - 34 = 2$, $3x^2 - 12x - 34 = 0$

$x^2 - 4x - 12 = 0$, $(x - 6)(x + 2) = 0$, $x = -2$ or $x = 6$

Fit -2 and 6 back into $y \Rightarrow$ points $(-2,27)$ and $(6.-213)$

53. $\frac{dy}{dx} = x^2 - 8x - 18 = 3$, since the slope of $6x - 2y = 1$ is 3

$x^2 - 8x - 15 = 0$, $(x - 5)(x - 33 = 0$, $x = 3$ or $x = 5$

Fit 3 and 5 back into $y \Rightarrow$ points $(3,49)$ and $(5.161/3)$

54. $y = \frac{1}{2}x^2 - 4x + 10$, $f(x) = \frac{1}{2}x^2 - 4x + 10$

$f(6) = \frac{1}{2}(6)^2 - 4(6) + 10 = 18 - 24 + 10 = 4$

slope $= f'(x) = \frac{d}{dx}\left(\frac{1}{2}x^2 - 4x + 10\right) = \frac{1}{2}\cdot 2x - 4 = x - 4$

$f'(6) = 6 - 4 = 2$

Let $(x_1, y_1) = (6, 4)$.

$y - 4 = 2(x - 6)$, $y = 2x - 12 + 4$, $y = 2x - 8$

To find the value of b, let x = 0 and solve for y,

$y = 2(0) - 8 = -8$.

55. $y = f(x)$

$\text{slope} = \dfrac{3 - 5}{0 - 4} = \dfrac{1}{2}$; Let $(x_1, y_1) = (4, 5)$.

$y - 5 = \frac{1}{2}(x - 4)$, $y = \frac{1}{2}x - 2 + 5$, $y = \frac{1}{2}x + 3$

$y = f(x) = \frac{1}{2}x + 3$, $f(4) = \frac{1}{2}(4) + 3 = 2 + 3 = 5$

$f'(x) = \dfrac{d}{dx}\left[\frac{1}{2}x + 3\right] = \frac{1}{2}$, $f'(4) = \frac{1}{2}$

Exercises 1.7

1. $f(t) = (t^2 + 1)^5$

$\dfrac{d}{dt}(t^2 + 1)^5 = 5(t^2 + 1)^4 \cdot \dfrac{d}{dt}(t^2 + 1) = 5(t^2 + 1)^4(2t)$

$= 10t(t^2 + 1)^4$

2. $f(P) = P^4 - P^3 + 4P^2 - P$

$\dfrac{d}{dP}(P^4 - P^3 + 4P^2 - P) = 4P^3 - 3P^2 + 8P - 1$

3. $u = \sqrt{2t - 1} = (2t - 1)^{1/2}$

$\dfrac{du}{dt} = \dfrac{d}{dt}(2t - 1)^{1/2} = \frac{1}{2}(2t - 1)^{-1/2} \cdot \dfrac{d}{dt}(2t - 1)$

$$= \frac{1}{2}(2t - 1)^{-1/2}\cdot 2 = (2t - 1)^{-1/2}$$

4. $g(z) = (z^3 - z + 1)^2$

$$\frac{d}{dz}(z^3 - z + 1)^2 = 2(z^3 - z + 1)\cdot\frac{d}{dz}(z^3 - z + 1)$$
$$= 2(z^3 - z + 1)(3z^2 - 1)$$
$$= 2(3z^2 - 1)(z^3 - z + 1)$$

5. $y = (T^3 + 5T)^{2/3}$

$$\frac{dy}{dT} = \frac{d}{dT}(T^3 + 5T)^{2/3} = \frac{2}{3}(T^3 + 5T)^{-1/3}\cdot(T^3 + 5T)$$
$$= \frac{2}{3}(T^3 + 5T)^{-1/3}\cdot(3T^2 + 5)$$
$$= \frac{2}{3}(3T^2 + 5)(T^3 + 5T)^{-1/3}$$

6. $s = \sqrt{t} + \dfrac{1}{\sqrt{t}} = t^{1/2} + t^{-1/2}$

$$\frac{ds}{dt} = \frac{d}{dt}(t^{1/2} + t^{-1/2}) = \frac{1}{2}t^{-1/2} + \left(-\frac{1}{2}t^{-3/2}\right) = \frac{t^{-1/2}}{2} - \frac{t^{-3/2}}{2}$$

7. $\dfrac{d}{dP}\left[3P^2 - \frac{1}{2}P + 1\right] = 6P - \dfrac{1}{2}$

8. $\dfrac{d}{dz}\left[\sqrt{z^2 - 1}\right] = \dfrac{d}{dz}(z^2 - 1)^{1/2} = \dfrac{1}{2}(z^2 - 1)^{-1/2}\cdot\dfrac{d}{dz}(z^2 - 1)$

$$= \frac{1}{2}(z^2 - 1)^{-1/2}\cdot(2z) = z(z^2 - 1)^{-1/2}$$

9. $\dfrac{d}{dt}(a^2t^2 + b^2t + c^2) = 2a^2t + b^2 + 0 = 2a^2t + b^2$

10. $\dfrac{d}{dx}(x^3 + t^3) = 3x^2$

11. $f(x) = \frac{1}{2}x^2 - 7x + 2;\ f'(x) = \frac{1}{2}\cdot 2x - 7 = x - 7;\ f''(x) = 1$

12. $y = \dfrac{1}{x^2} + 1 = x^{-2} + 1;\ y' = -2x^{-3} = \dfrac{-2}{x^3};\ y'' = 6x^{-4} = \dfrac{6}{x^4}$

13. $y = \sqrt{x} = x^{1/2}$; $y' = \frac{1}{2}x^{-1/2}$; $y'' = -\frac{1}{4}x^{-3/2}$

14. $f(t) = t^{100} + t + 1$; $f'(t) = 100t^{99} + 1$; $f''(t) = 9900t^{98}$

15. $f(r) = \pi hr^2 + 2\pi r$; $f'(r) = 2\pi hr + 2\pi = 2\pi(hr + 1)$; $f''(r) = 2\pi h$

16. $v = t^{3/2} + t$; $v' = \frac{3}{2}t^{1/2} + 1$; $v'' = \frac{3}{4}t^{-1/2}$

17. $g(x) = 2 - 5x$; $g'(x) = -5$; $g''(x) = 0$

18. $V(r) = \frac{4}{3}\pi r^3$; $V'(r) = 4\pi r^2$; $V''(r) = 8\pi r$

19. $f(P) = (3P + 1)^5$

$$f'(P) = 5(3P + 1)^4\cdot\frac{d}{dP}(3P + 1) = 5(3P + 1)^4\cdot 3 = 15(3P + 1)^4$$

$$f''(P) = 60(3P + 1)^3\cdot\frac{d}{dP}(3P + 1) = 60(3P + 1)^3\cdot 3 = 180(3P + 1)^3$$

20. $u = \frac{t^6}{30} - \frac{t^4}{12}$; $u' = \frac{1}{5}t^5 - \frac{1}{3}t^3$; $u'' = t^4 - t^2$

21. $\frac{d}{dx}(2x^2 - 3)\Big|_{x=5} = 4x\Big|_{x=5} = 4(5) = 20$

22. $\frac{d}{dt}(1 - 2t - 3t^2)\Big|_{t=-1} = (-2 - 6t)\Big|_{t=-1} = -2 - 6(-1) = 4$

23. $\frac{d}{dz}(z^2 - 4)^3\Big|_{z=1} = \left[3(z^2 - 4)^2\cdot\frac{d}{dz}(z^2 - 4)\right]\Big|_{z=1}$

$$= 6z(z^2 - 4)^2\Big|_{z=1} = 6(1)(1^2 - 4)^2 = 54$$

24. $\frac{d}{dT}\left[\frac{1}{3T + 1}\right]\Big|_{T=2} = \frac{d}{dT}(3T + 1)^{-1} = -(3T + 1)^{-2}\cdot\frac{d}{dT}(3T + 1)$

$$= -(3T + 1)^{-2}\cdot 3 = -3(3T + 1)^{-2}$$

$$\frac{d}{dT}\Big|_{T=2} = \frac{-3}{(3(2) + 1)^2} = -\frac{3}{49}$$

25. $\dfrac{d^2}{dx^2}(3x^3 - x^2 + 7x - 1)\bigg|_{x=2}$

$\dfrac{d}{dx}(3x^3 - x^2 + 7x - 1) = 9x^2 - 2x + 7$

$\dfrac{d}{dx}(9x^2 - 2x + 7) = 18x - 2$

$\dfrac{d^2}{dx^2}\bigg|_{x=2} = 18(2) - 2 = 34$

26. $\dfrac{d}{dt}\left(\dfrac{dv}{dt}\right)$, where $v = 2t^{-3}$

$\dfrac{dv}{dt} = -6t^{-4}$; $\dfrac{d}{dt}\left(\dfrac{dv}{dt}\right) = \dfrac{d}{dt}(-6t^{-4}) = 24t^{-5}$

27. $\dfrac{d}{dP}\left(\dfrac{dy}{dP}\right)$, where $y = \dfrac{k}{2P - 1} = k(2P - 1)^{-1}$

$\dfrac{dy}{dP} = -k(2P - 1)^{-2}\cdot\dfrac{d}{dP}(2P - 1) = -k(2P - 1)^{-2}\cdot 2 = -2k(2P - 1)^{-2}$

$\dfrac{d}{dP}\left(\dfrac{dy}{dP}\right) = \dfrac{d}{dP}\left[-2k(2P - 1)^{-2}\right] = 4k(2P - 1)^{-3}\cdot\dfrac{d}{dP}(2P - 1)$

$= 4k(2P - 1)^{-3}\cdot 2 = 8k(2P - 1)^{-3}$

28. $\dfrac{d^2V}{dr^2}\bigg|_{r=2}$, where $V = ar^3$

$\dfrac{dV}{dr} = 3ar^2$; $\dfrac{d^2V}{dr^2} = 6ar$; $\dfrac{d^2V}{dr^2}\bigg|_{r=2} = 6a(2) = 12a$

29. $f(x) = \sqrt{10 - 2x} = (10 - 2x)^{1/2}$

$f'(x) = \frac{1}{2}(10 - 2x)^{-1/2}\cdot\dfrac{d}{dx}(10 - 2x) = \frac{1}{2}(10 - 2x)^{-1/2}\cdot(-2)$

$= -(10 - 2x)^{-1/2}$

$f'(3) = -(10 - 2(3))^{-1/2} = -\frac{1}{2}$

$f''(x) = \frac{1}{2}(10 - 2x)^{-3/2}\cdot\dfrac{d}{dx}(10 - 2x) = \frac{1}{2}(10 - 2x)^{-3/2}\cdot(-2)$

$= -(10 - 2x)^{-3/2}$

$f''(3) = -\frac{1}{8}$

30. $g(T) = (3T - 5)^{10}$; $g'(T) = 10(3T - 5)^9 \cdot \frac{d}{dT}(3T - 5)$

$= 10(3T - 5)^9 \cdot 3 = 30(3T - 5)^9$

$g'(2) = 30(3(2) - 5)^9 = 30$

$g''(T) = 270(3T - 5)^8 \cdot \frac{d}{dT}(3T - 5) = 270(3T - 5)^8 \cdot 3$

$= 810(3T - 5)^8$

$g''(2) = 810(3(2) - 5)^8 = 810$

31. $R = 1000 + 80x - .02x^2$, for $0 \le x \le 2000$

$\frac{dR}{dx} = 80 - .04x$; $\left.\frac{dR}{dx}\right|_{x=1500} = 80 - .04(1500) = 20$

32. $V = 20\left[1 - \frac{100}{100 + t^2}\right]$, $0 \le t \le 24$

$V = 20 - 2000(100 + t^2)^{-1}$

$\frac{dV}{dt} = 2000(100 + t^2)^{-2} \cdot \frac{d}{dt}(100 + t^2) = 2000(100 + t^2)^{-2} \cdot 2t$

$= 4000t(100 + t^2)^{-2}$

$\left.\frac{dV}{dt}\right|_{t=10} = \frac{4000(10)}{(100 + 10^2)^2}$

33. (a) $f(x) = x^5 - x^4 + 3x$; $f'(x) = 5x^4 - 4x^3 + 3$;

$f''(x) = 20x^3 - 12x^2$; $f'''(x) = 60x^2 - 24x$

(b) $f(x) = 4x^{5/2}$; $f'(x) = 10x^{3/2}$; $f''(x) = 15x^{1/2}$

$f'''(x) = \frac{15}{2}x^{-1/2} = \frac{15}{2\sqrt{x}}$

34. (a) $f(t) = t^{10}$; $f'(t) = 10t^9$; $f''(t) = 90t^8$; $f'''(x) = 720t^7$

(b) $f(z) = \frac{1}{z + 5} = (z + 5)^{-1}$; $f'(z) = -(z + 5)^{-2}$;

$f''(z) = 2(z + 5)^{-3}$; $f'''(z) = -6(z + 5)^{-4} = -\frac{6}{(z + 5)^4}$

Exercises 1.8

1. (a) Average rate of change is: $\frac{\Delta y}{\Delta x} = \frac{16}{2} = 8$, (0 to 2).

$\frac{\Delta y}{\Delta x} = \frac{4}{1} = 4$, (0 to 1).

$\frac{\Delta y}{\Delta x} = \frac{4(.5)^2}{.5} = 2$, (0 to .5).

(b) $f'(0) = 0$

2. (a) $\frac{\Delta y}{\Delta x} = \frac{-6/2 + 6/1}{2 - 1} = 3$, (1 to 2).

$\frac{\Delta y}{\Delta x} = \frac{-6/1.5 + 6/1}{1.5 - 1} = 4$, (1 to 1.5).

$\frac{\Delta y}{\Delta x} = \frac{-6/1.2 + 6/1}{1.2 - 1} = 5$, (1 to 1.2).

(b) $f'(1) = 6/(1)^2 = 6$

3. (a) $\frac{\Delta y}{\Delta x} = \frac{f(6) + f(5)}{6 - 5} = 14$

(b) $f'(5) = 2(5) + 3 = 13$

4. (a) $\frac{\Delta y}{\Delta x} = \frac{f(3) + f(2)}{3 - 2} = 5$

(b) $f'(2) = 3 + 12/(2)^2 = 6$

5. $f(t) = 60t + t^2 - \frac{1}{12}t^3$ is the units produced after t hours of work, $0 \le t \le 8$. The rate of production is $f'(t)$.

$$f(t) = 60t + t^2 - \frac{1}{12}t^3;\ f'(t) = 60 + 2t - \frac{1}{4}t^2$$

When $t = 2$, $f'(2) = 60 + 2(2) - \frac{1}{4}(2)^2 = 60 + 4 - 1 = 63$.

Therefore, the rate of production after 2 hours of work is 63 units per hour.

6. $f(t) = 5t - t^{1/2}$ is the gallons of liquid in a large vat after t hours. The rate of the liquid flowing into the vat is $f'(t)$.

$f(t) = 5t - t^{1/2};\ f'(t) = 5 - \frac{1}{2}t^{-1/2} = 5 - \frac{1}{2\sqrt{t}}$

When $t = 4$, $f'(4) = 5 - \frac{1}{2\sqrt{4}} = \frac{19}{4}$.

Therefore, the liquid is flowing at $\frac{19}{4}$ gallons per hour.

7. The weight of the tumor at time t is $W(t) = .1t^2$.

(a) The rate of growth is $W'(t)$. $W(t) = .1t^2$; $W'(t) = .2t$

When $t = 5$, $W'(5) = .2(5) = 1.0$. Therefore, the tumor is growing at 1 gram per week after five weeks.

(b) $W'(t) = 5$, $.2t = 5$, $t = 25$

So after 25 weeks, the tumor is growing at 5 grams per week.

8. The daily sales after t days is $f(t) = -3t^2 + 32t + 100$.

(a) The rate of change after t days is $f'(t)$.

$f(t) = -3t^2 + 32t + 100$; $f'(t) = -6t + 32$

When $t = 2$, $f'(2) = -6(2) + 32 = 20$. Therefore, 2 days after the end of advertising, the sales are increasing 20 units per day.

(b) $f'(t) = 2$, $-6t + 32 = 2$, $-6t = -30$, $t = 5$

Therefore, 5 days after the end of advertising, the sales are increasing at the rate of 2 units per day.

9. $f(t)$ is the amount of oxygen in the lake t days after the sewage started flowing into the lake.

$f(t) = 1 - \frac{10}{t + 10} + \frac{100}{(t + 10)^2}$

(a) The rate of change of the oxygen content of the lake at t days is $f'(t)$.

$f(t) = 1 - 10(t + 10)^{-1} + 100(t + 10)^{-2}$

$f'(t) = 10(t + 10)^{-2} - 200(t + 10)^{-3}$

When $t = 5$, $f'(5) = 10(5 + 10)^{-2} - 200(5 + 10)^{-3}$

$= \frac{10}{225} - \frac{200}{3375} = \frac{-2}{135} = -.015$ units per day.

When $t = 15$, $f'(15) = 10(15 + 10)^{-2} - 200(15 + 10)^{-3}$

$= \frac{10}{25^2} - \frac{200}{25^3} = .0032$ units per day.

Therefore, the oxygen content of the lake is decreasing at .015 units per day after 5 days and increasing .0032 units per day after 15 days.

(b) The oxygen content is increasing after 15 days.

10. $P(t) = 20t - t^2$ is the number of sick students at time t days after the beginning of the epidemic.

(a) The rate of spread is $P'(t) = 20 - 2t$.

$P'(t) = 20 - 2(1) = 18$ students per day.

(b) Let $P'(t) = 8$ and solve for t. $20 - 2t = 8$, $t = 6$ days. To find the number of sick students, let $t = 6$ and solve for $P(t)$.

$P(6) = 20(6) - 6^2 = 120 - 36 = 84$ students.

11. $f(a + \Delta x) - f(a) \cong f'(a).\Delta x$, $f(a + \Delta x) \cong f(a) + f'(a).\Delta x$

This means the statement is true.

12. true

13. true

14. true

15. $f(101) \cong f(100) + f'(100).1 = 5010.$

$f(100.5) \cong f(100) + f'(100).1/2 = 5005.$

$f(99) \cong f(100) + f'(100).-1 = 4990.$

$f(98) \cong f(100) + f'(100).-1 = 4980.$

$f(99.75) \cong f(100) + f'(100).(.25) = 4997.5$

16. $f(27) \cong f(25) + f'(25).2 = 6.$

$f(26) \cong f(25) + f'(25).1 = 8.$

$f(25.25) \cong f(25) + f'(25).(.25) = 9.5$

$f(24) \cong f(25) + f'(25).(-1) = 12.$

$f(23.5) \cong f(25) + f'(25).(-1.5) = 13.$

17. The manufacture of 2000 radios costs $50000 and at that level the cost of manufacturing 1 additional radio is about $10.

18. $P(99) \cong P(100) + P'(100).1 = 90000 - 1200 = \$88800$

To find the production level where the revenue is $1800, set $R(x) = 1800$ and solve for x.

$.01x^2 - 3x = 1800, \ .01x^2 - 3x - 1800 = 0$

$$x = \frac{-b \pm \sqrt{b^2 - 4ac}}{2a} = \frac{3 \pm \sqrt{9 - 4(.01)(-1800)}}{.02} = \frac{3 \pm 9}{.02}$$

$$= 600, \ -300$$

Therefore, at 600 units production level, the revenue is $1800.

25. $C(x) = .1x^3 - 6x^2 + 136x + 200$ is the hourly cost of producing x units.

(a) $C(21) = .1(21)^3 - 6(21)^2 + 136(21) + 200$

$= 926.1 - 2646 + 2856 + 200 = 1336.1$

$C(20) = .1(20)^3 - 6(20)^2 + 136(20) + 200$

$= 800 - 2400 + 2720 + 200 = 1320$

$C(21) - C(20) = 1336.1 - 1320 = 16.1$

Therefore, the extra cost of raising the production from 20 to 21 is $16.10.

(b) The marginal cost is $C'(x)$.

$C(x) = .1x^3 - 6x^2 + 136x + 200; \ C'(x) = .3x^3 - 12x + 136$

When $x = 20$, $C'(20) = .3(20)^2 - 12(20) + 136$

$= 120 - 240 + 136 = 16.$

Therefore, the marginal cost is 16 dollars per unit.

26. $P(x) = .003x^3 + .01x$ is the profit from producing x units of a product.

(a) $P(101) = .003(101)^3 + .01(101) = 3090.903 + 1.01$

$= 3091.913$

$P(100) = .003(100)^3 + .01(100) = 3000 + 1 = 3001$

$P(101) - P(100) = 3091.913 - 3001 = 90.913$

Therefore, 90.913 dollars is the additional profit gained from increasing sales from 100 to 101 units.

(b) $P'(x)$ is the marginal profit.

$P(x) = .003x^3 + .01x$; $P'(x) = .009x^2 + .01$

When $x = 100$, $P'(100) = .009(100)^2 + .01 = 90 + .01 = 90.01$. Therefore, the marginal profit at a production level of 100 units is 90.01 dollars.

27. A - b, B - e, C - f, D - d, E - a, F - c, G - g.

28. A - d, B - b, C - a, D - c.

29. $$\frac{47.4 - 45}{1.05 - 1} = 48 \text{ mph}$$

$$\frac{45.4 - 45}{1.01 - 1} = 40 \text{ mph}$$

30. a - B, b - A, c - C

31. The height of the rocket after t seconds is $s(t) = 160t - 16t^2$ feet.

(a) The velocity of the rocket is $s'(t)$.

$s(t) = 160t - 16t^2$; $s'(t) = 160 - 32t$

The velocity of the rocket when $t = 0$ is

$s'(0) = 160 - 32(0) = 160$ feet per second.

(b) The velocity when $t = 2$ is

$s'(2) = 160 - 32(2) = 96$ feet per second.

(c) The acceleration of the rocket is $s'(t) = -32$ feet per second per second. The acceleration when $t = 3$ is $s''(3) = -32$ feet/sec^2.

(d) Set $s(t) = 0$ and solve for t.

$160t - 16t^2 = 0$, $t(160 - 16t) = 0$, $t = 0$ or $t = 10$

Therefore, after 10 seconds, the rocket will hit the ground.

(e) Let $t = 10$ and solve for $s'(t)$.

$s'(10) = 160 - 32(10) = -160$ feet per second.

The velocity of the rocket is 160 feet per second when it smashes into the ground.

32. The position of a car a time t is $s(t) = 50t - \frac{7}{(t + 1)}$.

The velocity of the car is $s'(t)$.

$$s(t) = 50t - \frac{7}{(t + 1)} = 50t - 7(t + 1)^{-1}$$

$$s'(t) = 50 + 7(t + 1)^{-2} = 50 + \frac{7}{(t + 1)^2}$$

At $t = 0$, $s'(0) = 50 + \frac{7}{(0 + 1)^2} = 57$ kilometers per hour.

The acceleration of the car is $s''(t) = -14(t + 1)^{-3}$

$$= \frac{-14}{(t + 1)^3}.$$

At $t = 0$, $s''(0) = \frac{-14}{(0 + 1)^3} = -14$ kilometers per hour per hour.

Therefore, at $t = 0$, the velocity of the car is 57 km/hr and the acceleration is -14 km/hr^2.

33. The distance an object travels in t hours is $s(t) = \frac{1}{2}t^2 + 4t$.

(a) The velocity of the object at t hours is $s'(t)$.

$s(t) = \frac{1}{2}t^2 + 4t$; $s'(t) = t + 4$

When $t = 6$, $s'(6) = 6 + 4 = 10$ kilometers per hour.

(b) $s(6) = \frac{1}{2}(6)^2 + 4(6) = 18 + 24 = 42$ kilometers

(c) $t + 4 = 6$, $t = 2$

Therefore, after 2 hours, the object is traveling at 6 kilometers per hour.

34. The distance of the helicopter from the ground t seconds after takeoff is $s(t) = t^2 + t$ feet.

(a) Let s(t) = 20 and solve for t.

$t^2 + t = 20$, $t^2 + t - 20 = 0$, $(t + 5)(t - 4) = 0$, $t = -5, 4$

It will take 4 seconds for the helicopter to rise 20 feet.

(b) The velocity of the helicopter after 4 seconds is $s'(4) = v(4) = 2(4) + 1 = 9$ feet per second. The acceleration of the helicopter after 4 seconds is $s''(4) = v'(4) = 2$ feet per second per second.

35. Use $f(a + \Delta x) \cong f(a) + f'(a).\Delta x$

(a) $f(237) \cong f(235) + f'(235).2 = 4600 + 2(-100) = 4400$

$f(234) \cong f(235) + f'(235).(-1) = 4600 - 1(-100) = 4700$

$f(240) \cong f(235) + f'(235).1 = 4600 - 500 = 4100$

$f(232) \cong f(235) + f'(235).1 = 4600 - 3(-100) = 4900$

36. (a) $f(16.5) \cong f(16) + f'(16).(-5) = 351.57 - (4.94)1/2 = 354.04$

(b) $f(15) \cong f(16) + f'(16).(-1) = 351.57 - 4.94 = 346.63$

37. (a) An increase in the interest rate from 6% to 7% would produce about $168.95 additional interest.

(b) $f(6.2) \cong f(6) + f'(6).(2) = 1790.85 - (168.95)1/2$
$= \$1824.64$

(c) $f(5.7) \cong f(6) + f'(6).(-.3) = 1790.85 - (168.95)(-.3)$
$= \$1740.17$

38. $f'(x) = 3x^2$; $(2.01)^3 \cong 2^3 + 3(2)^2(.01) = 8 + .12 = 8.12$

39. $f'(x) = (1/3)x^{-2/3} = \dfrac{1}{3x^{2/3}}$; $f(8.024) \cong \sqrt[3]{8} + \dfrac{1}{3(8)^{2/3}}.(.024)$

$= 2.002$

40. Area $= f(x) = x^2$, $f'(x) = 2x$

$f(5.001) \cong f(5) + f'(5).(.001) = 25 + 2(5)(.001) = 25.01$, so the increase in area is $.01\ m^2$.

41.a. Consult the graph of $s(t)$. Since $s(3.5) = 60$, after 3.5 seconds the vehicle has traveled 60 ft.

b. Consult the graph of $s'(t)$. Since $s'(2) = 20$ ft/sec, the velocity after 2 seconds is 20 ft/sec.

c. Consult the graph of $s''(t)$. Since $s''(1) = 10$ ft/sec^2, the acceleration after 1 second is 10 ft/sec^2

d. We seek to solve the equation $s(t) = 120$. Consulting the graph of $s(t)$, we see that the vehicle will have traveled 120 ft after 5.5 sec

e. Consult the right side of the graph of $s'(t)$. The height of the graph equals 20 when $t = 7$. So the vehicle will be traveling 20 ft/sec after 7 sec.

f. Locate the point at the peak of the graph of $s'(t)$. Its coordinates are $(4.5, 30)$So the greatest velocity is 30 ft/sec and is achieved at $t = 4.5$ sec. The distance the vehicle has traveled at this time is 90 ft.

42.a. Consult the graph of $f(t)$. \$150 billion

b. Determine $f'(20)$. \$25 billion/yr

c. Determine the point on the graph of $f(t)$ with y-coordinate 375. The corresponding x-coordinate is 25. Answer:1985

d. Determine the point on the graph of $f'(t)$ with y-coordinate 100. The corresponding x-coordinate is 33. Answer: 1993

Chapter 1 Supplementary Exercises

1. Let $(x_1, y_1) = (0, 3)$.

   $y - 3 = -2(x - 0)$

   $y = 3 - 2x$

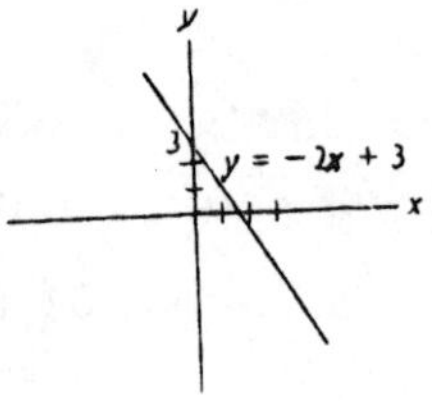

2. Let $(x_1, y_1) = (0, -1)$.

   $y - (-1) = \frac{3}{4}(x - 0)$

   $y = \frac{3}{4}x - 1$

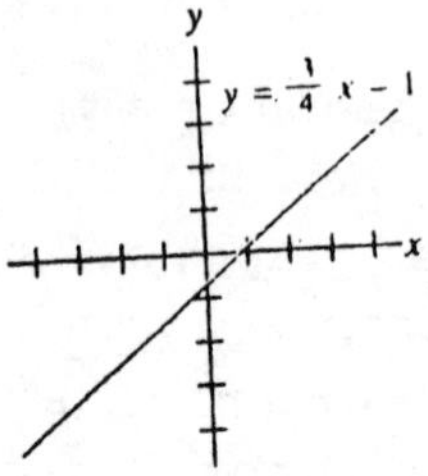

3. Let $(x_1, y_1) = (2, 0)$.

   $y - 0 = 5(x - 2)$

   $y = 5x - 10$

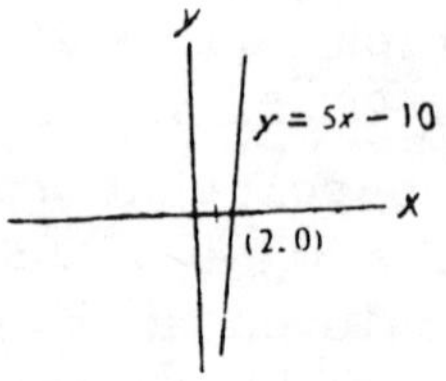

4. Let $(x_1, y_1) = (1, 4)$.

$y - 4 = -\frac{1}{3}(x - 1)$

$y = \frac{13 - x}{3}$

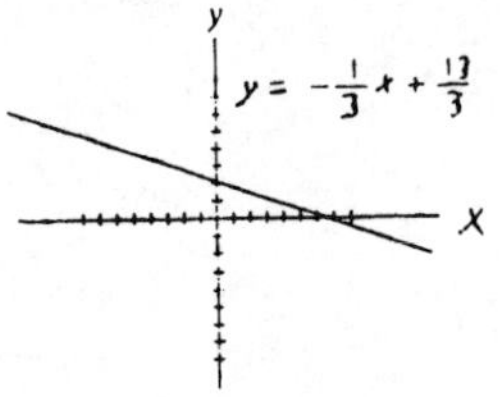

5. $y = -2x$, slope $= -2$

Let $(x_1, y_1) = (3, 5)$.

$y - 5 = -2(x - 3)$

$y = 11 - 2x$

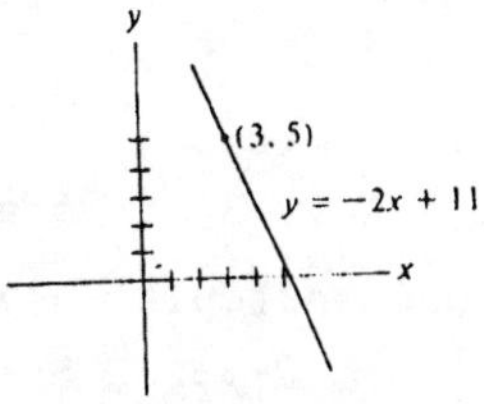

6. $-2x + 3y = 16$

$y = 2 + \frac{2}{3}x$, slope $= \frac{2}{3}$

Let $(x_1, y_1) = (0, 1)$.

$y - 1 = \frac{2}{3}(x - 0)$

$y = \frac{2}{3}x + 1$

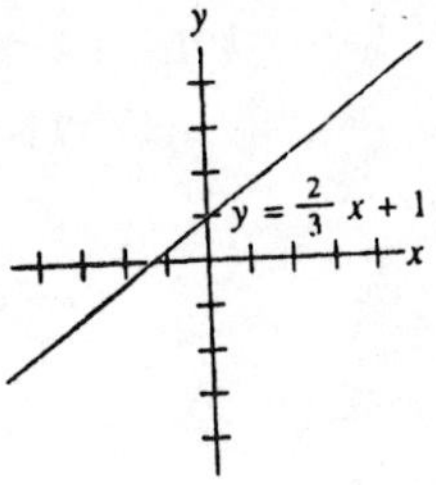

7. slope $= \frac{7 - 4}{3 - (-1)} = \frac{3}{4}$

Let $(x_1, y_1) = (3, 7)$.

$y - 7 = \frac{3}{4}(x - 3)$

$y = \frac{3}{4}x + \frac{19}{4}$

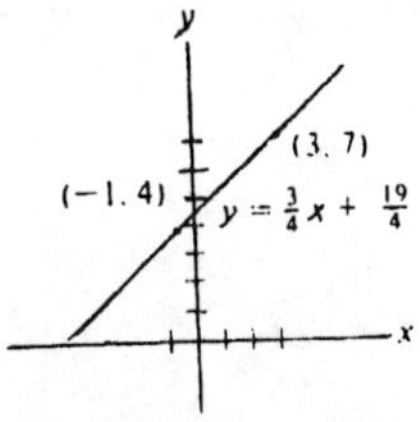

8. slope = $\frac{1 - 1}{5 - 2} = 0$

Let $(x_1, y_1) = (2, 1)$.

$y - 1 = 0(x - 2)$

$y = 1$

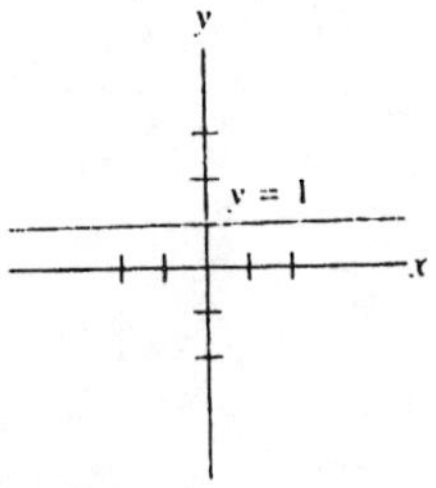

9. Slope of y = 3x + 4 is 3, thus a perpendicular line has slope of −1/3. The perpendicular line through (1,2) is y − 2 = (−1/3)(x − 1), y = −1/3x + 7/3.

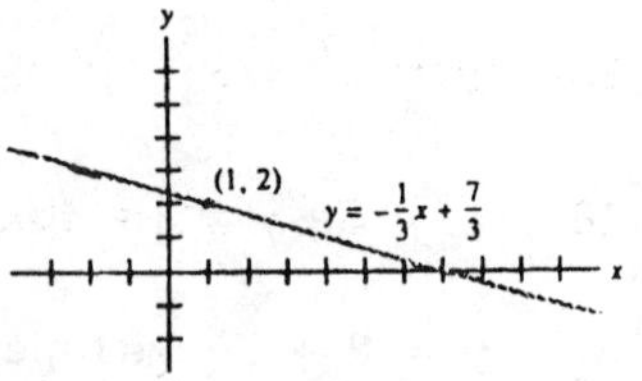

10. Slope of 3x + 4y = 4 is −3/4 since y = −3/4x + 1, thus a perpendicular line has slope of 4/3. The perpendicular line through (1,2) is y − 2 = (−1/3)(x − 1) or y = −1/3x + 7/3.

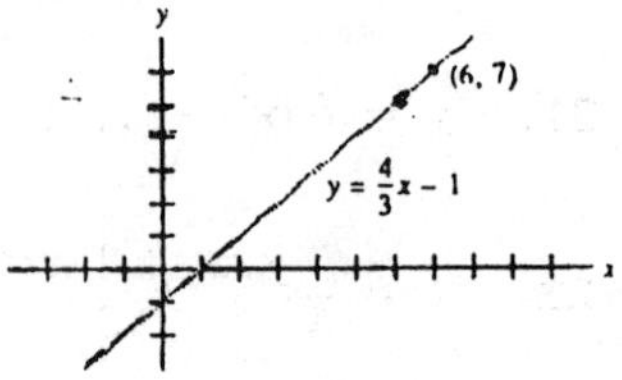

11. The equation of the x-axis is y = 0, so the equation of this line is y = 3.

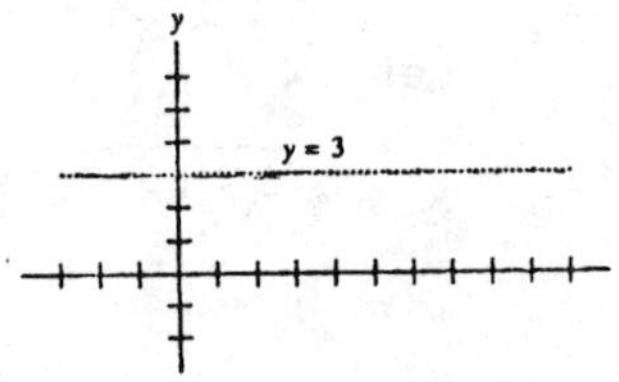

12. The equation of the y-axis is $x = 0$, so 4 units to the right is $x = 4$.

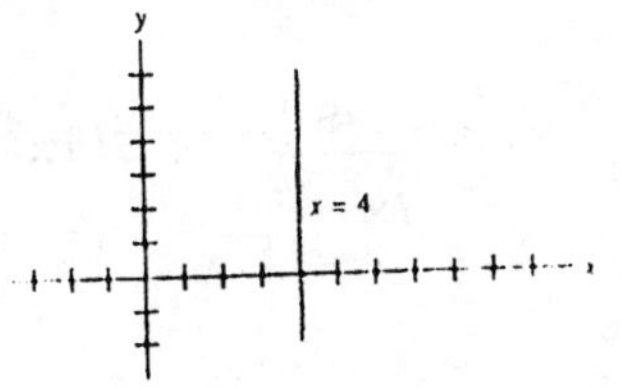

13.

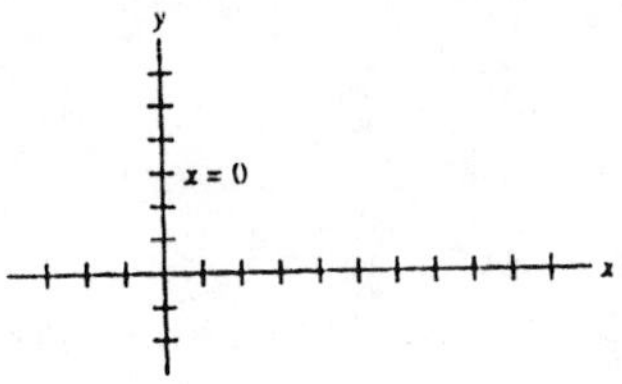

14.

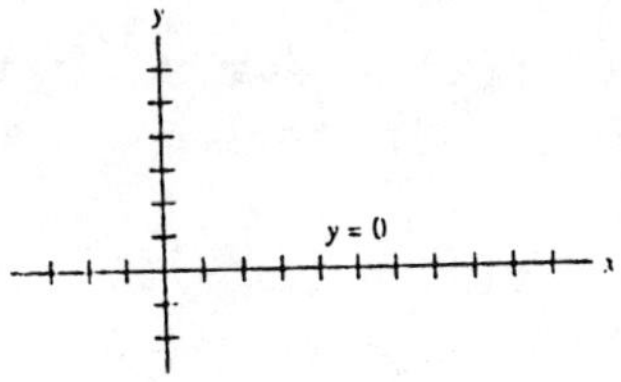

15. $y = x^7 + x^3$; $y' = 7x^6 + 3x^2$

16. $y = 5x^8$; $y' = 40x^7$

17. $y = 6\sqrt{x} = 6^{1/2}$; $y' = 3x^{-1/2} = \dfrac{3}{\sqrt{x}}$

18. $y = x^7 + 3x^5 + 1$; $y' = 7x^6 + 15x^4$

19. $y = \dfrac{3}{x} = 3x^{-1}$; $y' = -3x^{-2} = \dfrac{3}{x^2}$

20. $y = x^4 - \dfrac{4}{x} = x^4 - 4x^{-1}$; $y' = 4x^3 + 4x^{-2} = 4x^3 + \dfrac{4}{x^2}$

21. $y = (3x^2 - 1)^8$; $y' = 8(3x^2 - 1)^7(6x) = 48x(3x^2 - 1)^7$

22. $y = \dfrac{3}{4}x^{4/3} + \dfrac{4}{3}x^{3/4}$; $y' = x^{1/3} + x^{-1/4}$

23. $y = \dfrac{1}{5x - 1} = (5x - 1)^{-1}$; $\dfrac{dy}{dx} = -(5x - 1)^{-2}(5) = \dfrac{-5}{(5x - 1)^2}$

24. $y = (x^3 + x^2 + 1)^5$; $y' = 5(x^3 + x^2 + 1)^4(3x^2 + 2x)$

25. $y = \sqrt{x^2 + 1} = (x^2 + 1)^{1/2}$; $y' = \frac{1}{2}(x^2 + 1)^{-1/2}(2x)$

$= x(x^2 + 1)^{-1/2}$

26. $y = \dfrac{5}{7x^2 + 1} = 5(7x^2 + 1)^{-1}$; $\dfrac{dy}{dx} = -5(7x^2 + 1)^{-2}(14x)$

$= \dfrac{-70x}{(7x^2 + 1)^2}$

27. $f(x) = \dfrac{1}{\sqrt[4]{x}} = x^{-1/4}$; $f'(x) = -\frac{1}{4}x^{-5/4}$

28. $f(x) = (2x + 1)^3$; $f'(x) = 3(2x + 1)^2(2) = 6(2x + 1)^2$

29. $f(x) = 5$; $f'(x) = 0$

30. $f(x) = \dfrac{5x}{2} - \dfrac{2}{5x} = \frac{5}{2}x - \frac{2}{5}x^{-1}$; $f'(x) = \frac{5}{2} + \frac{2}{5}x^{-2}$

31. $f(x) = [x^5 - (x - 1)^5]^{10}$

$f'(x) = 10\left[x^5 - (x - 1)^5\right]^9\left[5x^4 - 5(x - 1)^4\right]$

32. $f(t) = t^{10} - 10t^9$; $f'(t) = 10t^9 - 90t^8$

33. $g(t) = 3\sqrt{t} - \dfrac{3}{\sqrt{t}} = 3t^{1/2} - 3t^{-1/2}$; $g'(t) = \frac{3}{2}t^{-1/2} + \frac{3}{2}t^{-3/2}$

34. $h(t) = 3\sqrt{2}$; $h'(t) = 0$

35. $f(t) = \dfrac{2}{t - 3t^3} = 2(t - 3t^3)^{-1}$; $f'(t) = -2(t - 3t^3)^{-2}(1 - 9t^2)$

$= \dfrac{-2(1 - 9t^2)}{(t - 3t^3)^2}$

36. $g(P) = 4P^{.7}$; $g'(P) = 2.8P^{-.3}$

37. $h(x) = \frac{3}{2}x^{3/2} - 6x^{2/3}$; $h'(x) = \frac{9}{4}x^{1/2} - 4x^{-1/3}$

38. $f(x) = \sqrt{x + \sqrt{x}} = (x + x^{1/2})^{1/2}$

$f'(x) = \frac{1}{2}(x + x^{1/2})^{1/2}\left(1 + \frac{1}{2}x^{-1/2}\right) = \frac{1}{2}(x + \sqrt{x})^{-1/2}\left(1 + \frac{1}{2\sqrt{x}}\right)$

39. $f(t) = 3t^3 - 2t^2$; $f'(t) = 9t^2 - 4t$; $f'(2) = 36 - 8 = 28$

40. $V(r) = 15\pi r^2$; $V'(r) = 30\pi r$; $V'\left(\frac{1}{3}\right) = 10\pi$

41. $g(u) = 3u - 1$; $g(5) = 15 - 1 = 14$; $g'(u) = 3$; $g'(5) = 3$

42. $h(x) = -\frac{1}{2}$; $h(-2) = -\frac{1}{2}$; $h'(x) = 0$; $h'(-2) = 0$

43. $f(x) = x^{5/2}$; $f'(x) = \frac{5}{2}x^{3/2}$; $f''(x) = \frac{15}{4}x^{1/2}$; $f''(4) = \frac{15}{2}$

44. $g(t) = \frac{1}{4}(2t - 7)^4$; $g'(t) = (2t - 7)^3(2) = 2(2t - 7)^3$

$g''(t) = 6(2t - 7)^2(2) = 12(2t - 7)^2$

$g''(3) = 12[2(3) - 7]^2 = 12$

45. $y = (3x - 1)^3 - 4(3x - 1)^2$

slope $= y' = 3(3x - 1)^2(3) - 8(3x - 1)(3)$

$= 9(3x - 1)^2 - 24(3x - 1)$

When $x = 0$, slope $= 9 + 24 = 33$.

46. $y = (4 - x)^5$

slope $= y' = 5(4 - x)^4(-1) = -5(4 - x)^4$

When $x = 5$, slope $= -5$.

47. $\frac{d}{dx}(x^4 - 2x^2) = 4x^3 - 4x$

48. $\frac{d}{dt}(t^{5/2} + 2t^{3/2} - t^{1/2}) = \frac{5}{2}t^{3/2} + 3t^{1/2} - \frac{1}{2}t^{-1/2}$

49. $\frac{d}{dP}(\sqrt{1 - 3P}) = \frac{d}{dP}(1 - 3P)^{1/2} = \frac{1}{2}(1 - 3P)^{-1/2}(-3)$

$= -\frac{3}{2}(1 - 3P)^{-1/2}$

50. $\frac{d}{dn}(n^{-5}) = -5n^{-6}$

51. $\frac{d}{dz}(z^3 - 4z^2 + z - 3)\Big|_{z=-2} = (3z^2 - 8z + 1)\Big|_{z=-2}$

$= 12 + 16 + 1 = 29$

52. $\frac{d}{dx}(4x - 10)^5\Big|_{x=3} = \left[5(4x - 10)^4(4)\right]\Big|_{x=3} = 320$

53. $\frac{d^2}{dx^2}(5x + 1)^4 = \frac{d}{dx}\left[4(5x - 1)^3(5)\right] = 60(5x - 1)^2(5)$

$= 300(5x - 1)^2$

54. $\frac{d^2}{dt^2}(2\sqrt{t}) = \frac{d^2}{dt^2}2t^{1/2} = \frac{d}{dt}t^{-1/2} = -\frac{1}{2}t^{-3/2}$

55. $\frac{d^2}{dt^2}(t^3 + 2t^2 - t)\Big|_{t=-1} = \frac{d}{dt}(3t^2 + 4t - 1)\Big|_{t=-1}$

$= (6t + 4)\Big|_{t=-1} = -2$

56. $\frac{d^2}{dP^2}(3P + 2)\Big|_{P=4} = \frac{d}{dP}3\Big|_{P=4} = 0$

57. $\frac{d^2y}{dx^2}(4x^{3/2}) = \frac{dy}{dx}(6x^{1/2}) = 3x^{-1/2}$

58. $\frac{d}{dt}\left(\frac{1}{3t}\right) = \frac{d}{dt}\left(\frac{1}{3}t^{-1}\right) = -\frac{1}{3}t^{-2}$ or $-\frac{1}{3t^2}$

$\frac{d}{dt}\left(-\frac{1}{3}t^{-2}\right) = \frac{2}{3}t^{-3}$ or $\frac{2}{3t^3}$

59. $f(x) = x^3 - 4x^2 + 6$; slope $= f'(x) = 3x^2 - 8x$

When $x = 2$, slope $= 3(2)^2 - 8(2) = -4$.

When $x = 2$, $y = 2^3 - 4(2)^2 + 6 = -2$.

Let $(x_1, y_1) = (2, -2)$.

$y - (-2) = -4(x - 2)$, $y = 6 - 4x$

60. $y = \dfrac{1}{(3x - 5)} = (3x - 5)^{-1}$, $y' = -(3x - 5)^{-2}(3) = \dfrac{-3}{(3x - 5)^2}$

When $x = 1$, slope $= -\dfrac{3}{(3(1) - 5)^2} = -\dfrac{3}{4}$.

When $x = 1$, $y = \dfrac{1}{(3(1) - 5)} = -\dfrac{1}{2}$.

Let $(x_1, y_1) = \left(1, -\dfrac{1}{2}\right)$.

$y - \left(-\dfrac{1}{2}\right) = -\dfrac{3}{4}(x - 1)$, $y = \dfrac{1}{4} - \dfrac{3}{4}x$

61. $y = x^2$; slope $= y' = 2x$

When $x = \dfrac{3}{2}$, slope $= 2\left(\dfrac{3}{2}\right) = 3$.

Let $(x_1, y_1) = \left(\dfrac{3}{2}, \dfrac{9}{4}\right)$.

$y - \dfrac{9}{4} = 3\left(x - \dfrac{3}{2}\right)$, $y = 3x - \dfrac{9}{4}$

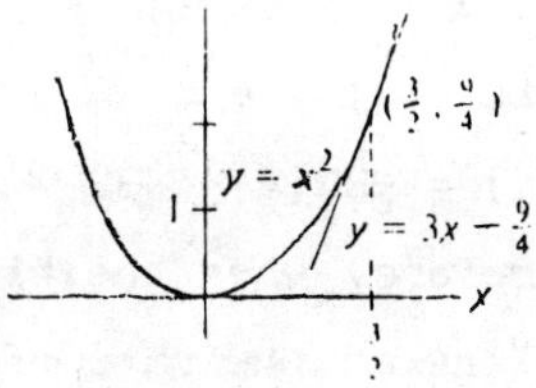

62. $y = x^2$; slope $= y' = 2x$

When $x = -2$, slope $= 2(-2) = -4$.

Let $(x_1, y_1) = (-2, 4)$.

$y - 4 = -4(x + 2)$

$y = -4x - 4$

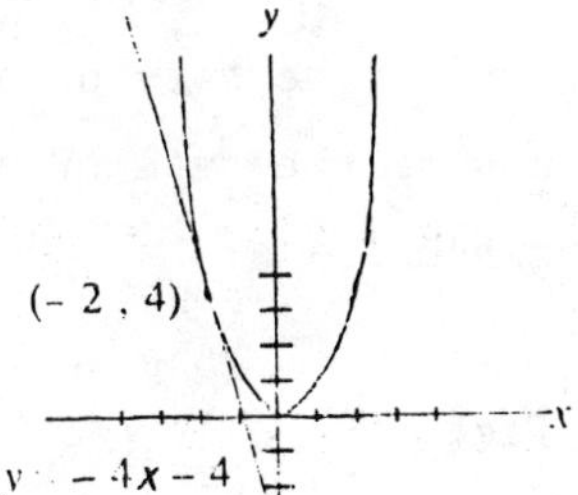

63. $y = 3x^2 - 5x^2 + x + 3$; slope $= y' = 9x^2 - 10x + 1$

When $x = 1$, slope $= 9(1)^2 - 10(1) + 1 = 0$.

When $x = 1$, $y = 3(1)^3 - 5(1)^2 + 1 + 3 = 2$.

Let $(x_1, y_1) = (1, 2)$. $y - 2 = 0(x - 1)$, $y = 2$.

64. $y = (2x^2 - 3x)^3$; slope $= y' = 3(2x^2 - 3x)^2(4x - 3)$

When $x = 2$, slope $= 3(2(2)^2 - 3(2))^2(4(2) - 3) = 60$.

When $x = 2$, $y = (2(2)^2 - 3(2))^3 = 8$.

Let $(x_1, y_1) = (2, 8)$. $y - 8 = 60(x - 2)$, $y = 60x - 112$

65. slope $= \dfrac{y_2 - y_1}{x_2 - x_1}$; $-1 = \dfrac{y_2 - 0}{2 - 5}$, $y_2 = 3$

Let $(x_1, y_1) = (2, 3)$. $y - 3 = -1(x - 2)$, $y = 5 - x$.

$y = f(x) = 5 - x$, $f(2) = 5 - 2 = 3$, $f'(x) = -1$, $f'(2) = -1$

66. The tangent line contains the points $(0, 2)$ and (a, a^3) and has slope $= 3a^2$. Thus, $\dfrac{a^3 - 2}{a} = 3a^2$, $a^3 - 2 = 3a^3$, $a = -1$.

67. $s'(t) = -32t + 32$

The binoculars will hit the ground when $s(t) = 0$, i.e.,

$s(t) = -16t^2 + 32t + 128 = 0$, $-16(t^2 - 2t - 8) = 0$,

$-16(t - 4)(t + 2) = 0$, $t = 4$ or $t = -2$.

$s'(4) = -32(4) + 32 = -96$ feet/sec.

Therefore, when the binoculars hit the ground, they will be falling at the rate of 96 feet/sec.

68. $40t + t^2 - \frac{1}{15}t^3$ tons is the total output of a coal mine after t hours. The rate of output is $40 + 2t - \frac{1}{5}t^2$ tons per hour. At $t = 5$, the rate of output is $40 + 2(5) - \frac{1}{5}(5)^2 = 45$ tons/hour.

69. 11 feet

70. $\dfrac{\Delta y}{\Delta x} = \dfrac{6 - 1}{4 - 1} = 5/3$ ft/sec

71. Slope of the tangent line is 5/3 so 5/3 ft/sec.

72. t=6

73. (a) positive

(b) limousine

(c) $f(25005) \cong f(25000) + f'(25000).5$,

so $f(25005) - f(25000) = (.04)(5) = .2$ gallons

74. (a) After 10 hours, lava was spewing at the rate of 300 tons per hour.

(b) After 10 hours the eruption had ceased.

(c) Impossible. Lava can't flow back into the volcano.

75. $f(12.5) - f(12) \cong f'(12)(.5) = (1.5)(.5) = .75$ inches.

76. $f\left(7 + \frac{1}{2}\right) - f(7) \sim f'(7)\frac{1}{2} = (25.06)\frac{1}{2} = 12.53$

$12.53 is the additional money earned if the bank paid $7\frac{1}{2}\%$ interest.

77. The limit does not exist.

78. The limit does not exist.

79. The limit does not exist.

80. $\lim\limits_{x\to 5} \frac{x - 5}{x^2 - 7x + 2} = \frac{5 - 5}{25 - 35 + 2} = 0.$

81. $f'(5) = \lim\limits_{h\to 0} \frac{f(5 + h) - f(5)}{h}$

If $f(x) = \frac{1}{2x}$, then $f(5 + h) - f(5) = \frac{1}{2(5 + h)} - \frac{1}{2(5)}$

$$= \frac{1}{2(5 + h)}\cdot\frac{5}{5} - \frac{1}{2(5)}\cdot\left(\frac{5 + h}{5 + h}\right)$$

$$= \frac{5 - (5 + h)}{10(5 + h)} = \frac{-h}{10(5 + h)}.$$

Thus, $f'(5) = \lim\limits_{h\to 0} [f(5 + h) - f(5)]\cdot\frac{1}{h}$

$$= \lim_{h\to 0} \frac{-h}{10(5 + h)}\cdot\frac{1}{h} = \lim_{h\to 0} \frac{-1}{10(5 + h)} = -\frac{1}{50}.$$

82. $f'(3) = \lim\limits_{h\to 0} \frac{f(3 + h) - f(3)}{h}$ If $f(x) = x^2 - 2x + 1$,

then $f(3 + h) - f(3) = (3 + h)^2 - 2(3 + h) + 1 - (9 - 6 + 1)$

$$= h^2 + 4h.$$

Thus, $f'(3) = \lim\limits_{h\to 0} \frac{f(3 + h) - f(3)}{h} = \lim\limits_{h\to 0} \frac{h^2 + 4h}{h}$

$$= \lim_{h\to 0} h + 4 = 4.$$

83. The slope of the tangent line at (3,9); x = 3, f(x) = 9

84. $$\frac{\frac{1}{2+h} - \frac{1}{2}}{h} = \frac{\frac{2-2-h}{2(2+h)}}{h} = \frac{-1}{2(2+h)}; \text{ as } h \to 0, \frac{-1}{2(2+h)} \to \frac{-1}{4}$$

Exercises 2.1

1. a, e, f   2. c, d   3. b, c, d   4. a, e

5. Decreasing for $x < -2$, relative minimum point at $x = -2$, minimum value $= -2$, increasing for $x > -2$, concave up, y-intercept (0, 0), x-intercepts (0, 0) and (-3.6, 0).

6. Increasing for $x < -1$, relative maximum point at $x = -1$, decreasing for $-1 < x < 5$, relative minimum point at $x = 5$, increasing for $x > 5$, concave down for $x < 3$, inflection point at (3, 0), concave up for $x > 3$, y-intercept (0, 2.8), x-intercepts (-5, 0), (3, 0), (7, 0).

7. Decreasing for $x < 0$, relative minimum point at $x = 0$, increasing for $0 < x < 2$, relative maximum point at $x = 2$, decreasing for $x > 2$, concave up for $x < 1$, concave down for $x > 1$, inflection point at (1, 3), y-intercept at (0, 2), x-intercept (3.4, 0).

8. Endpoint maximum value at $x = 0$. Decreasing for $x > 0$, concave down for $0 < x < 3$, inflection point (3, 3), concave up for $x > 3$, y-intercept (0, 6). The graph approaches the x-axis as a horizontal asymptote.

9. Decreasing for $x < 2$, relative minimum at $x = 2$, minimum value $= 3$, increasing for $x > 2$, concave up for all x, no inflection point, defined for $x > 0$, the line $y = x$ is an asymptote, the y-axis is an asymptote.

10. Increasing for all x, concave down for $x < 4$, inflection point (4, 4), concave up for $x > 4$, y-intercept (0, 1), x-intercept (.7, 0).

11. Decreasing for $1 \le x < 3$, relative minimum at $x = 3$,

increasing for $x > 3$, maximum value = 6 (at $x = 1$), minimum value = 1 (at $x = 3$), inflection point at $x = 4$, concave up for $1 \le x < 4$, concave down for $x > 4$, the line $y = 4$ is an asymptote.

12. Increasing for $2 < x < 4$, decreasing for $4 < x < 6$, concave down for $2 < x < 6$. Endpoint minimum value at $x = 2$, and $x = 6$. Maximum point and maximum value at $x = 4$.

13. Slope increases for all x.

14. Slope decreases for $x < 3$, increases for $x > 3$.

15. Slope decreases for $x < 3$, increases for $x > 3$. Minimum slope occurs at $x = 3$.

16. Slope decreases for $x < 4$, increases for $x > 4$.

17. (a) C, F (b) A, B, F (c) C

18. (a) A, E (b) D (c) E

19.

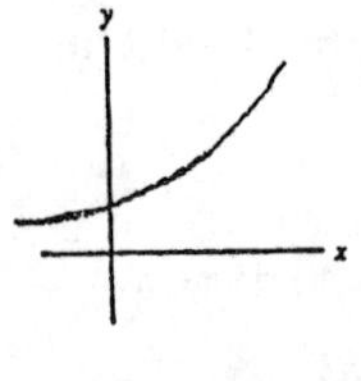

20.

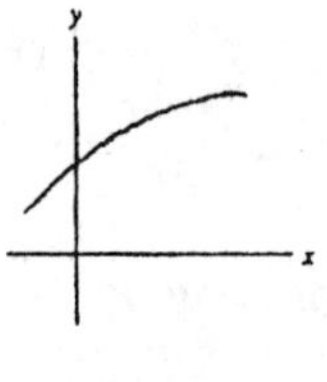

21.

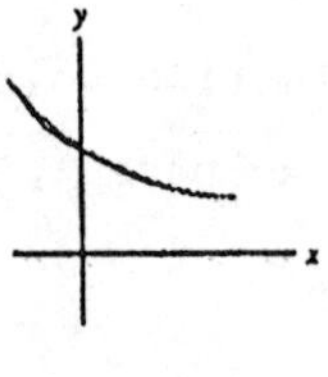

22.

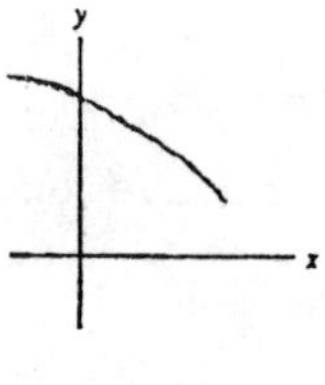

23.

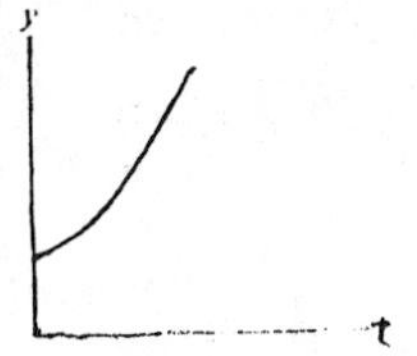

24.

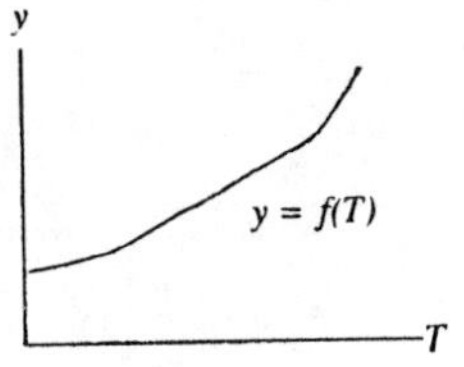

25.

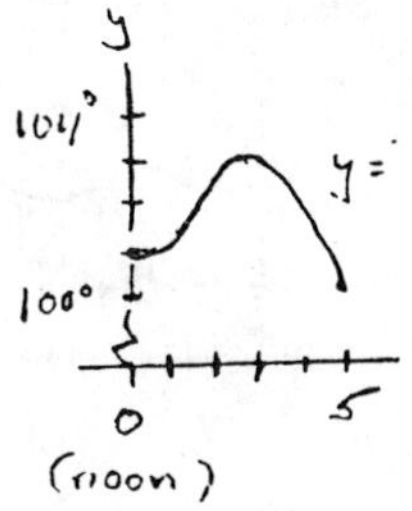

26.

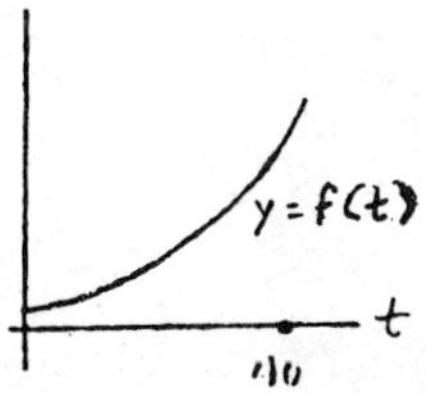

27.

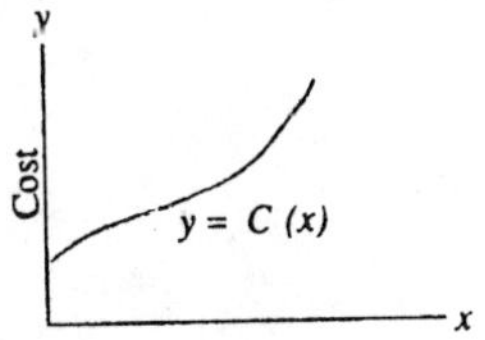

28.

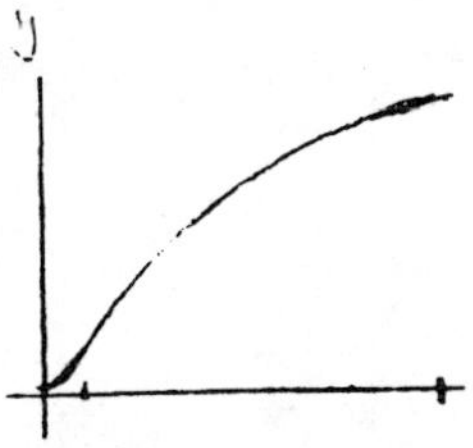

29. Oxygen content decreases until time a, at which time it reaches a minimum. After a, oxygen content steadily increases. The rate of increase increases until b, and then decreases. Time b is the time when oxygen content is increasing fastest.

30. 1975

31. 1960

32. 1980, 1986

33. The parachutist's speed levels off to 15 ft/sec.

34. Bacteria population stabilizes at 25,000,000.

35.

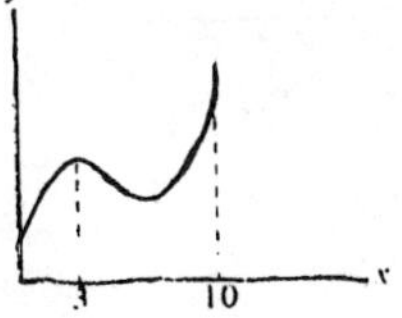

36.

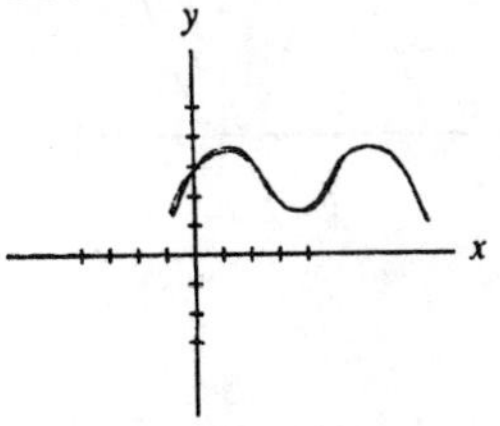

37.

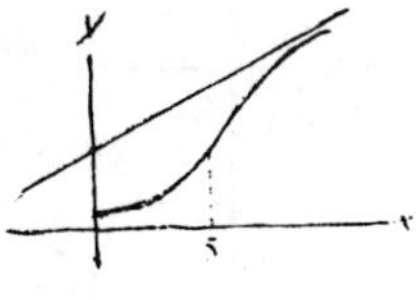

38.

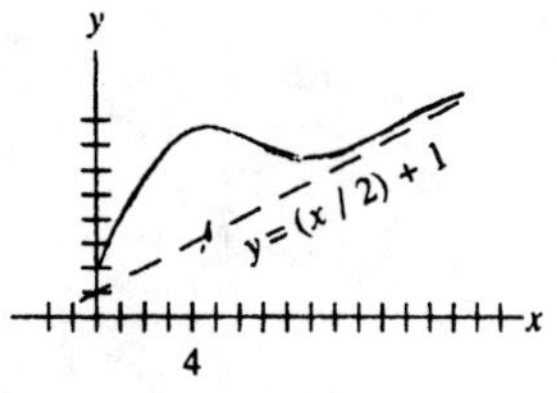

39. (a) yes (b) yes

40. No

41. Relatively low

42. I

Exercises 2.2

1. b, c, f
2. d
3. d, e, f
4. a, b
5. d
6. c

7.

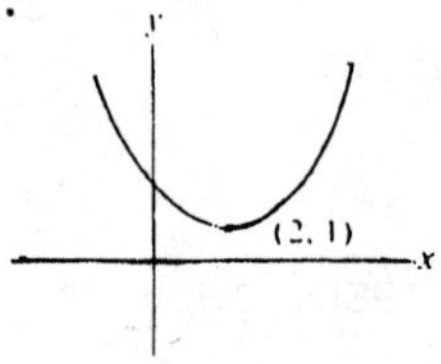

8.

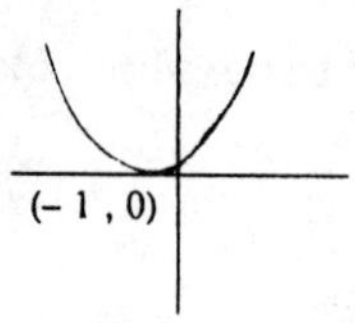

9.

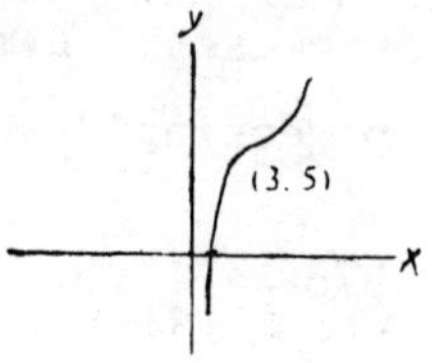

10.

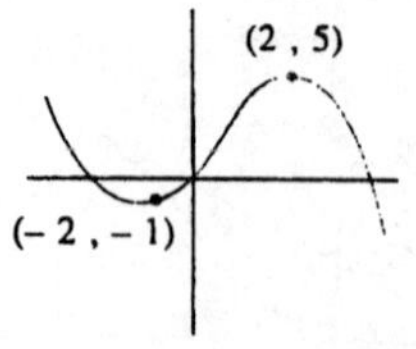

11.

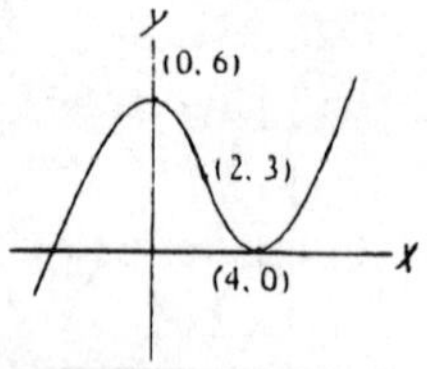

12.

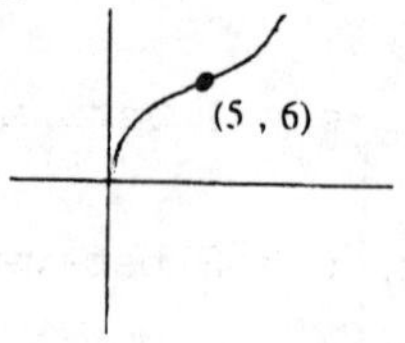

13.

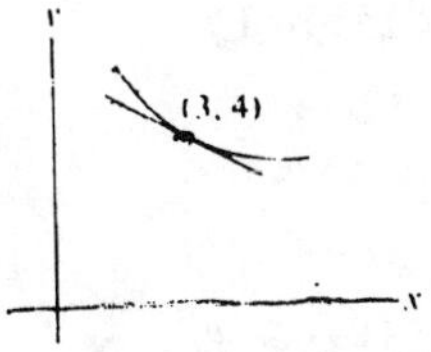

14.

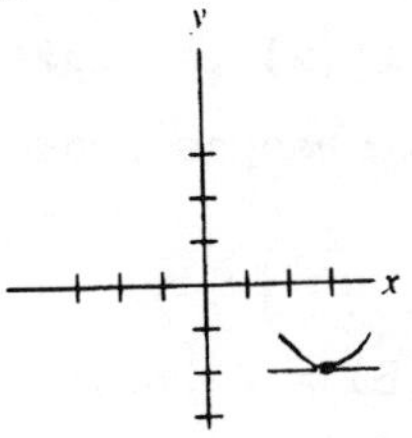

15.

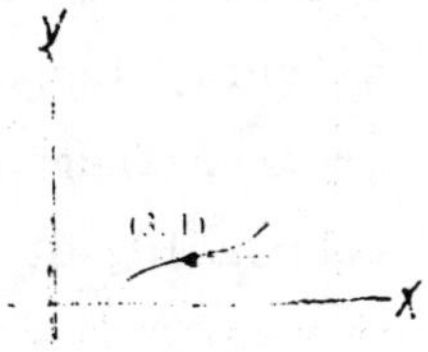

16.

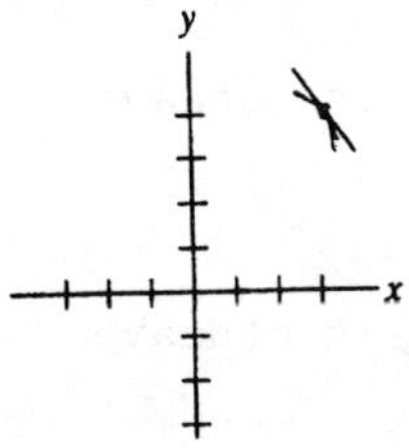

17.

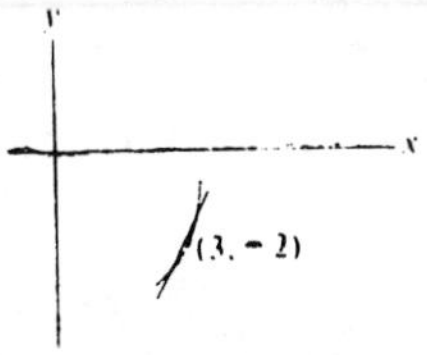

18.

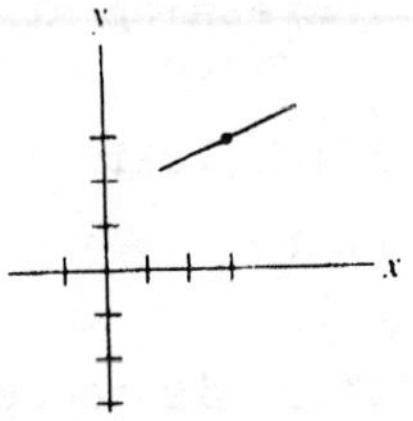

19.

	f	f'	f"
A	POS	POS	NEG
B	0	NEG	0
C	NEG	0	POS

20. (a) $f'(x) = 0$ at $x = 2$ or $x = 4$; but $f''(x) = 0$ at $x = 4$, so there is an extreme point at $x = 2$.

(b) $f''(x) = 0$ at $x = 3$ or $x = 4$, so there are inflection points at $x = 3$or $x = 4$.

21. $t = 1$ because the slope is more positive at $t = 1$.

22. $t = 2$ because the slope is more positive at $t = 2$.

23. The slope is positive because $f'(6) = 2$.

24. The slope is negative because $f'(4) = -1$.

25. The slope is 0 because $f'(3) = 0$. Also $f'(x)$ is positive for x slightly less than 3, and $f'(x)$ is negative for x slightly greater than 3. Hence $f(x)$ changes from increasing to decreasing at $x = 3$.

26. The slope is 0 because $f'(5) = 0$. Also $f'(x)$ is negative for x slightly less than 5, and $f'(x)$ is positive for x slightly greater than 5. Hence $f(x)$ changes from decreasing to increasing at $x = 5$.

27. $f'(x)$ is increasing at $x = 0$, so the graph of $f(x)$ is concave up.

28. $f'(x)$ is decreasing at $x = 2$, so the graph of $f(x)$ is concave down.

29. At $x = 1$, $f'(x)$ changes from increasing to decreasing, so the slope of graph of $f(x)$ changes from increasing to decreasing. Or $(f'(x))' = 0$ so $f''(x) = 0$, (slope of $f'(x) = 0$).

30. At $x = 4$, $f'(x)$ changes from decreasing to increasing, so the slope of graph of $f(x)$ changes from decreasing to increasing.

31. $f'(x) = 2$, so $m = 2 \Rightarrow y - 3 = 2(x - 6)$

32. $f(6.5) \cong f(6) + f'(6)(.5) = 8 + 1/2(2) = 9$

33. $f(.25) \cong f(0) + f'(0)(.25) = 3 + 1/4(1) = 3.25$

34. $f(0) = 3$, $f'(0) = 1$, so $y - 3 = 1(x - 0)$

35. (a) $h(100.5) \cong h(100) + h'(100)(.5)$

The change $= h(100.5) - h(100) \cong h'(100)(.5) = 1/3.1/2 = 1/6$ inches.

(b)(ii) Because the water level is falling.

36. (a) $T(10) - T(10.75) \cong T'(10)(.75) = 4.3/4 = 3$ degrees.

(b)(ii) Because the is falling.

37. $f'(x) = (3x^2 + 1)^3(x) = 24x(3x^2 + 1)^3$

II cannot be the graph of $f(x)$ since $f'(x)$ is always positive for $x > 0$.

38. $f'(x) = 3x^2 - 8x + 24 = 3(x^2 - 6x + 8)$

I cannot be the graph since it does not have horizontal tangents at $x = 2, 4$.

39. $f'(x) = \dfrac{1}{2\sqrt{x}} = \frac{1}{2}x^{-1/2}$; $f''(x) = -\frac{1}{4}x^{-3/2} = -\dfrac{1}{4\sqrt{x^3}}$

II could be the graph of $f(x)$.

40. $f'(x) = \frac{5}{2}x^{3/2}$; $f''(x) = \frac{15}{4}x^{1/2}$

I could be the graph of $f(x)$.

41. a. 3
   b. $t = 4, 6$
   c. $t = 1$
   d. $t = 5$
   e. $f'(7.5) = 1$
   f.

$$f'(t) = -1$$
$$t = 2.5, 3.5$$

   g. $f'(t)$ has a minimum. $t = 3$
   h. $f'(t)$ has a maximum. $t = 7$

42. a. $t = 4, 16$
   b. $(4, 10)$
   c. $f'(2) =$ 2 units/sec
   d. 8
   e.

$$f'(t) = 3$$
$$t = 1,\ t = 15$$
$$f(1) = 6,\ f(15) = 6$$

So the point is $(1,6)$.

f. $f'(t)$ has a maximum, $t = 16.5$. So the points are $(16.5, 12)$, $(15, 6)$.

43. a. $f'(9) < 0$, so the graph is decreasing.
b. $(2, 9)$
c. $(10, 1)$
d. $f''(2) < 0$, so the graph is concave down
e. $f''(x) = 0$, so that $x = 6$. Since $f(6) = 5$, the coordinates of the inflection point are $(6, 5)$.
f.

$$f'(x) = 6$$
$$x = 15$$
$$f(15) = 14$$

The point is $(15, 14)$.

44. a. Never
b. 1.8
c. $f'(x) = 2$ has no solutions. Never
d. *f* is not growing, it is decreasing
e. *f* is never increasing

45. a. 1947 corresponds to $t = 22$. $f'(t) = -.075$, so the number of farms was declining at the rate of 75,000 per year
b. $f(t) = 6$ has solution $t = 20$. So the year was 1945.
c. $f'(t) = -.08$ has solution $t = 25, 50$. So the years were 1950, 1975.
d. The graph of $f'(t)$ has a minimum at $t = 36$. So the number was declining the fastest in 1961.

46. a. $f(3) = 80$
b. $f'(7) = 200$ students/day
c. $f(t) = 400$ has the solution $t = 1$. After 19 days
d. In the later stages of the epidemic, the equation $f'(t) = 20$ has the solution $t = 15$. On the 15th day
e. The graph of $f'(t)$ has a maximum at $t = 11$. On day 11. Since $f(11) = 240$, 240 students will have contracted the flu at that time.

47. a. $f(60) = 900$ trillion kilowatt-hours
b. $f'(90) = 35$ trillion kilowatt-hours/yr
c. $f'(t) = 2300$ has solution $t = 80$. In 1980.
d. $f'(t) = 10$ has solution $t = 35$. In 1935.
e. $f'(t)$ has a maximum at $t = 70$ and $f(70) = 1600$. In 1970, 1600 trillion kilowatt-hours.

Exercises 2.3

1.

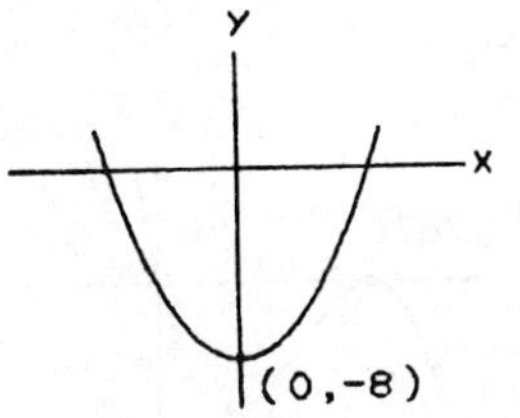

2.

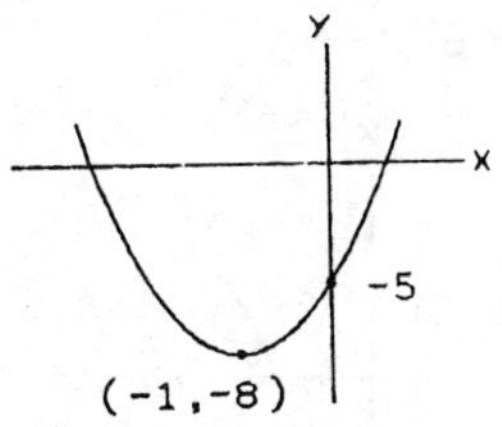

3.

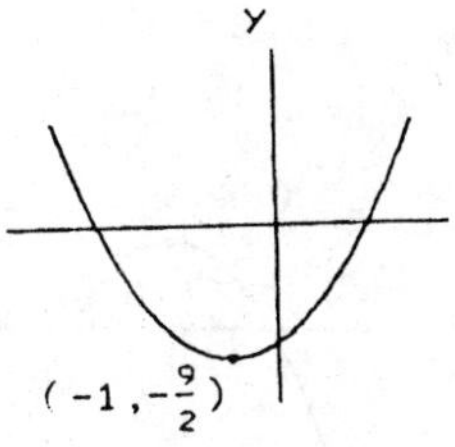

4.

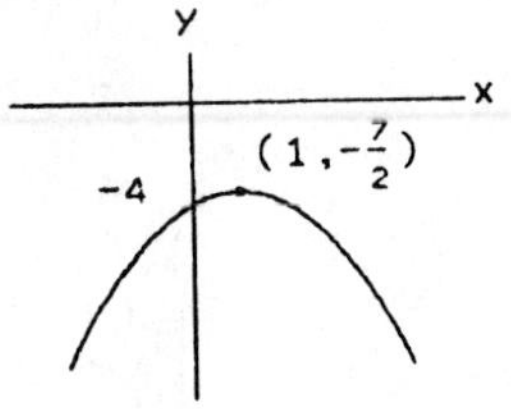

5.

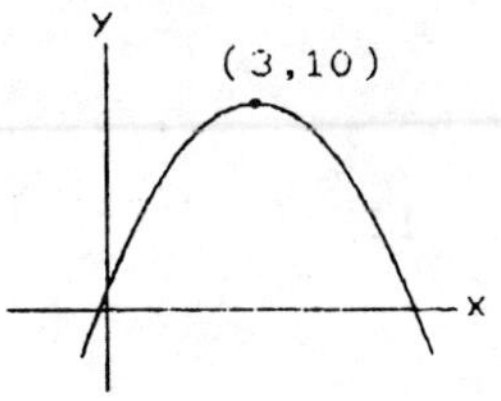

6.

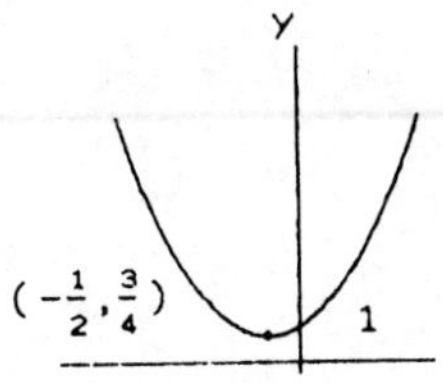

7.

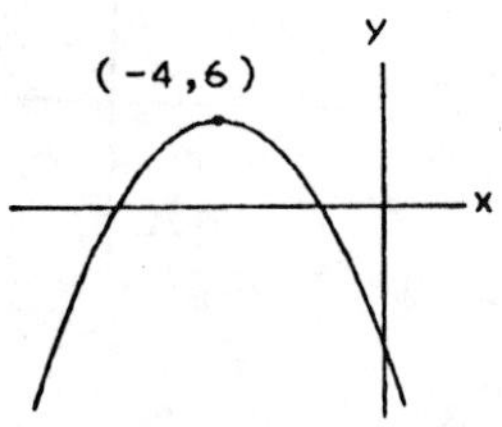

8.

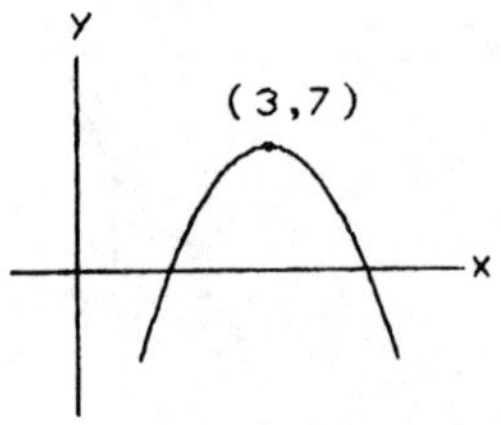

9.

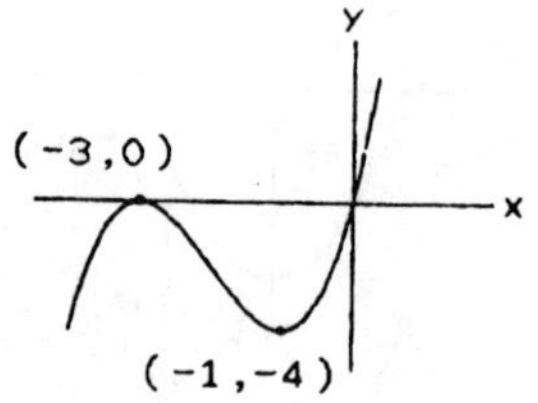

10.

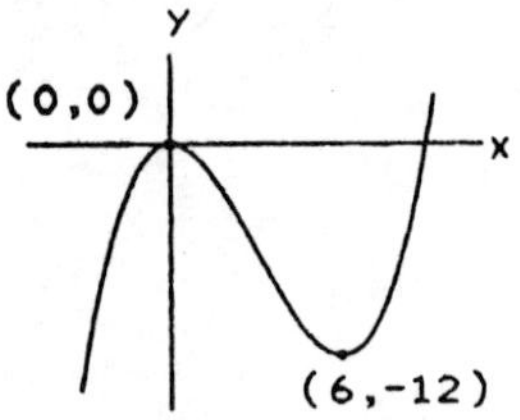

11.

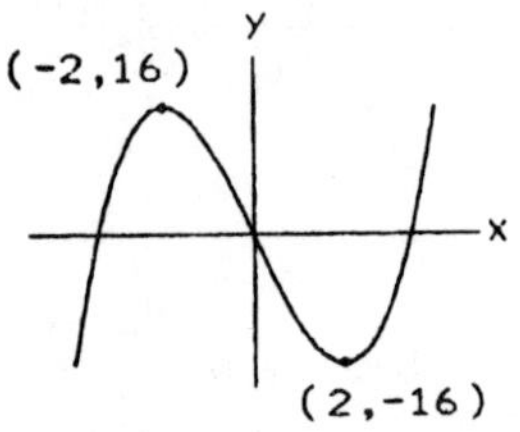

12.

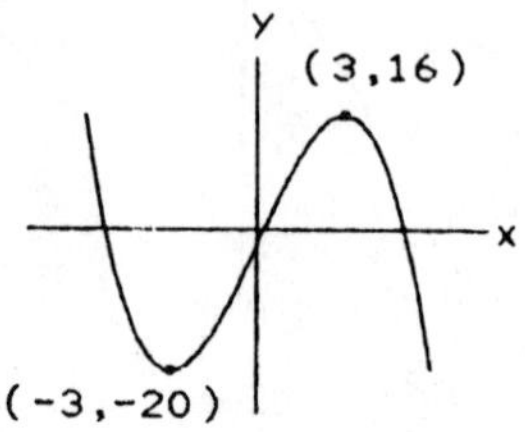

13.

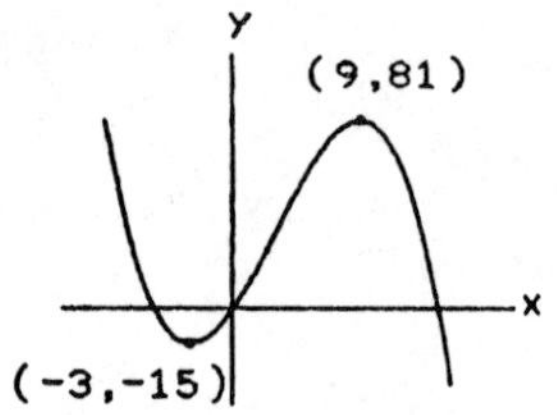

14.

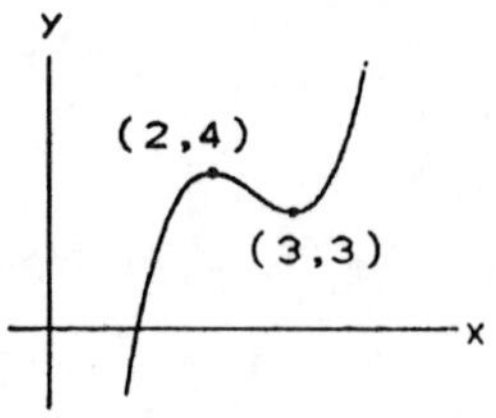

15.

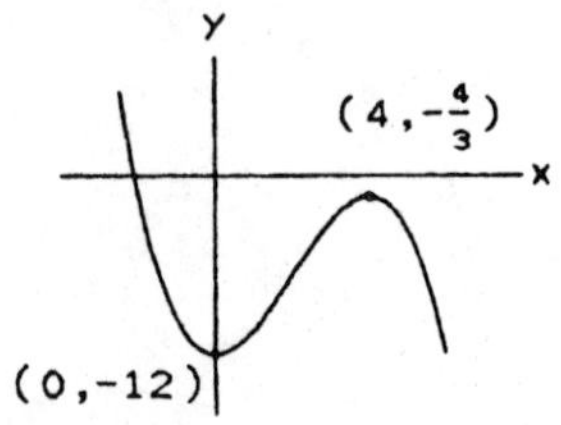

16.

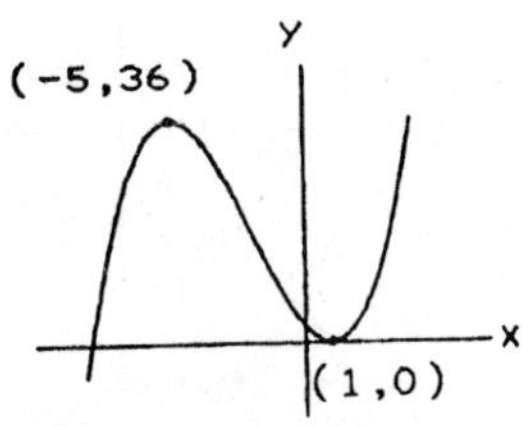

17.

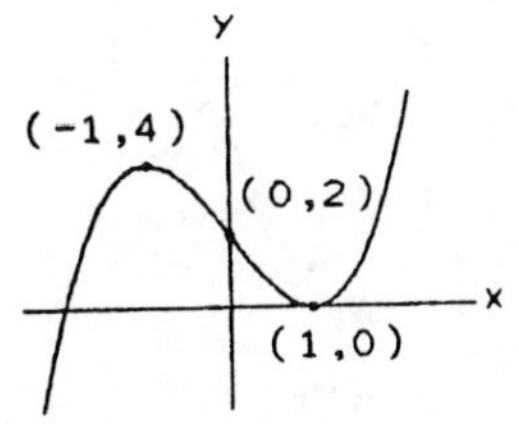

18.

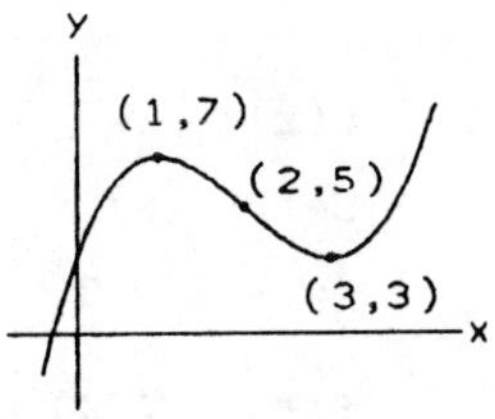

19.

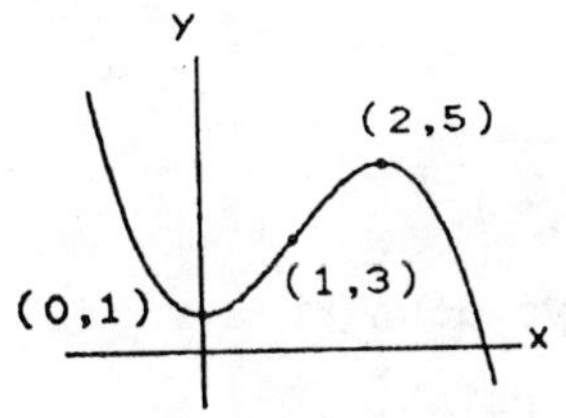

20.

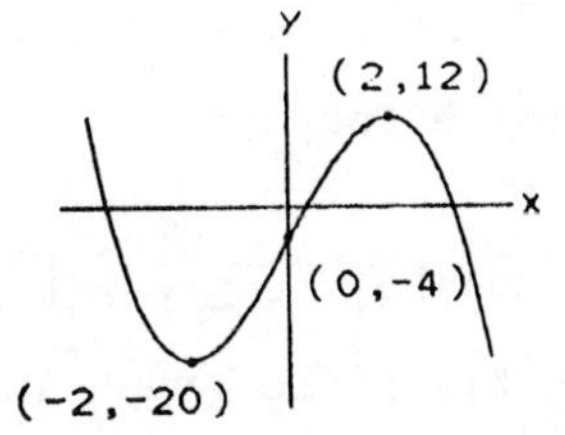

21.

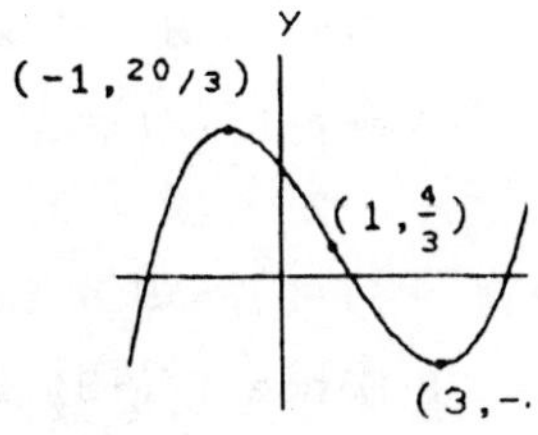

22.

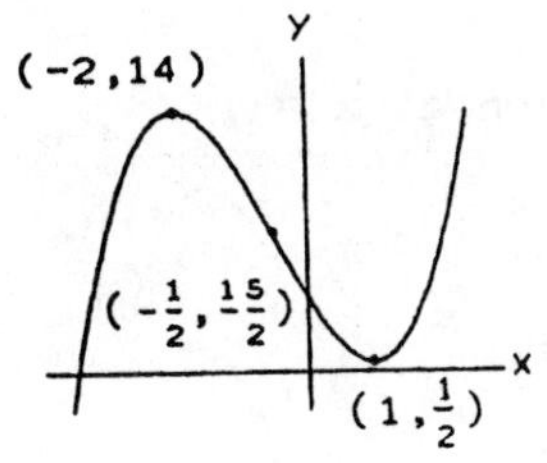

23.

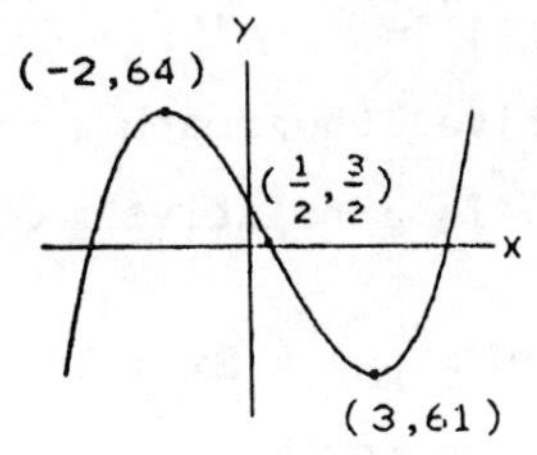

24.

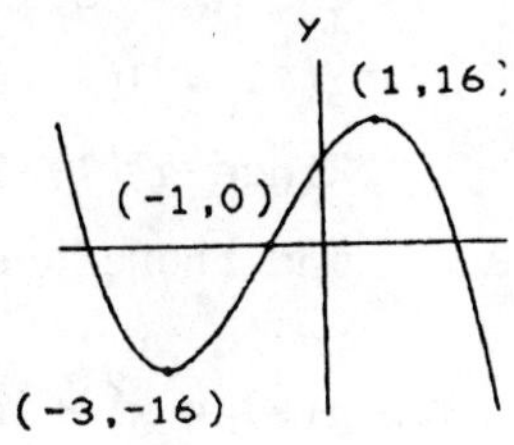

25. $f'(x) = 2ax + b$; $f''(x) = 2a$

It is not possible for the graph of f(x) to have an inflection point since $f''(x) = 2a \neq 0$.

26. $f'(x) = 3ax^2 + 2bx + c$; $f''(x) = 6ax + 2b$

No, $f''(x)$ is a linear function of x and hence can be zero for at most one value of x.

27. $f(x) = \frac{1}{4}x^2 - 2x + 7$; $f'(x) = \frac{1}{2}x - 2$; $f''(x) = \frac{1}{2}$.
Set $f'(x) = 0$ and solve for x,
$\frac{1}{2}x - 2 = 0$, $x = 4$.
$f(4) = \frac{1}{4}(4)^2 - 2(4) + 7 = 3$; $f''(4) = \frac{1}{2}$.
Since $f''(4)$ is positive, the graph is concave up at x = 3 and therefore (4, 3) is a relative minimum point.

28. $f(x) = 5 - 12x - 2x^2$; $f'(x) = -12 - 4x$; $f''(x) = -4$
Set $f'(x) = 0$ and solve for x.
$-12 - 4x = 0$, $x = -3$.
$f(-3) = 5 - 12(-3) - 2(-3)^2 = 23$; $f''(-3) = -4$.
Since $f''(-3)$ is negative, the graph is concave down at x = -3 and therefore (-3, 23) is a relative maximum point.

29. $g(x) = 3 + 4x - 2x^2$; $g'(x) = 4 - 4x$; $g''(x) = -4$
Set $g'(x) = 0$ and solve for x.
$4 - 4x = 0$, $x = 1$.
$g(1) = 3 + 4(1) - 2(1)^2 = 5$; $g''(1) = -4$.
Since $g''(1)$ is negative, the graph is concave down at x = 1 and therefore (1, 5) is a relative maximum point.

30. $g(x) = x^2 + 10x + 10$; $g'(x) = 2x + 10$; $g''(x) = 2$
Set $g'(x) = 0$ and solve for x.
$2x + 10 = 0$, $x = -5$.
$g(-5) = (-5)^2 + 10(-5) + 10 = -15$; $g''(-5) = 2$.
Since $g''(-5)$ is positive, the graph is concave up at x = -5 and therefore (-5, -15) is a relative minimum point.

31. $f(x) = 5x^2 + x - 3$; $f'(x) = 10x + 1$; $f''(x) = 10$
Set $f'(x) = 0$ and solve for x.
$10x + 1 = 0$, $x = -\frac{1}{10} = -.1$.

$f\left(-\frac{1}{10}\right) = 5\left(-\frac{1}{10}\right)^2 + \left(-\frac{1}{10}\right) - 3 = -3.05$; $f''(-.1) = 10$.

Since $f''(-.1)$ is positive, the graph is concave up at $x = -.1$ and therefore $(-.1, -3.05)$ is the relative minimum point.

32. $f(x) = 30x^2 - 1800x + 29{,}000$; $f'(x) = 60x - 1800$; $f''(x) = 60$
Set $f'(x) = 0$ and solve for x.
$60x - 1800 = 0$, $x = 30$.
$f(30) = 30(30)^2 - 1800(30) + 29{,}000 = 2000$; $f''(30) = 60$.
Since $f''(30)$ is positive, the graph is concave up at $x = 30$ and therefore $(30, 2000)$ is the relative minimum point.

33. $f(x) = g'(x)$. Since the zeros of $f(x)$ corresponds to the two extreme points of $g(x)$. $g(x) \neq f'(x)$, because $f(x)$ has only one extreme point and $g(x)$ has 3 zeros.

34. $g(x) = f'(x)$, because $f(x)$ has constant slope, $f'(x) = C = g(x)$. $f(x) \neq g'(x) = 0$.

35. a. f has a relative minimum
b. f has an inflection point

36. a. $f(40) = 20$ million
b. $f'(90) = 1.3$. Increasing at a rate of 1.3 million/yr
c. $f(t) = 90$ has solution $t = 110$. In 1890.
d. $f'(t) = 1.1$ has solution $t = 80$. In 1860.
e. $f'(t)$ has a maximum at $t = 140$ and $f'(140) = 2.1$. In 1920, rate of increase = 2.1 million per year, population = 150 million

37. a. $f(10) = .7 = 70\%$
b. $f(t) = .1$ has solution $t = 4$. In 1984.
c. $f'(t) = .05$. Increasing at 5% per year
d. $f'(t) = .05$ has the solution $t = 10$. In 1990.
e. $f'(t)$ has a maximum at $t = 7$, $f(7) = .4$. In 1987, 40%.

Exercises 2.4

1. $y = x^2 - 3x + 1$

$$x = \frac{-(-3) \pm \sqrt{(-3)^2 - 4(1)(1)}}{2(1)} = \frac{3 \pm \sqrt{5}}{2}$$

The x-intercepts are $\left(\frac{3 + \sqrt{5}}{2}, 0\right)$ and $\left(\frac{3 - \sqrt{5}}{2}, 0\right)$.

2. $y = x^2 + 5x + 5$

$$x = \frac{-5 \pm \sqrt{5^2 - 4(1)(5)}}{2(1)} = \frac{-5 \pm \sqrt{5}}{2}$$

The x-intercepts are $\left(\frac{-5 + \sqrt{5}}{2}, 0\right)$ and $\left(\frac{-5 - \sqrt{5}}{2}, 0\right)$.

3. $y = 2x^2 + 5x + 2$

$$x = \frac{-5 \pm \sqrt{5^2 - 4(2)(2)}}{2(2)} = \frac{-5 \pm 3}{4} = -\frac{1}{2}, -2$$

The x-intercepts are $\left(-\frac{1}{2}, 0\right)$ and $(-2, 0)$.

4. $y = 4 - 2x - x^2$

$$x = \frac{-(-2) \pm \sqrt{(-2)^2 - 4(-1)(4)}}{2(-1)} = \frac{2 \pm 2\sqrt{5}}{-2} = -1 \pm \sqrt{5}$$

The x-intercepts are $(-1 + \sqrt{5}, 0)$ and $(-1 - \sqrt{5}, 0)$.

5. $y = 4x - 4x^2 - 1$

$$x = \frac{-4 \pm \sqrt{4^2 - 4(-4)(-1)}}{2(-4)} = \frac{-4 \pm 0}{-8}$$

The x-intercept is $\left(\frac{1}{2}, 0\right)$.

6. $y = 3x^2 + 7x + 2$

$$x = \frac{-7 \pm \sqrt{7^2 - 4(3)(2)}}{2(3)} = \frac{-7 \pm 5}{6}$$

The x-intercepts are $\left(-\frac{1}{3}, 0\right)$ and $(-2, 0)$.

7. $f(x) = \frac{1}{3}x^3 - 2x^2 + 5x;\ f'(x) = x^2 - 4x + 5$

$$x = \frac{-(-4) \pm \sqrt{(-4)^2 - 4(1)(5)}}{2(1)} = \frac{4 \pm \sqrt{-4}}{2}$$

Since $f'(x)$ has no zeros, $f(x)$ has no relative extreme points.

8. $f(x) = 5 - 11x + 6x^2 - \frac{4}{3}x^3$

$f'(x) = -11 + 12x - 4x^2 < 0$ for all x.

9.

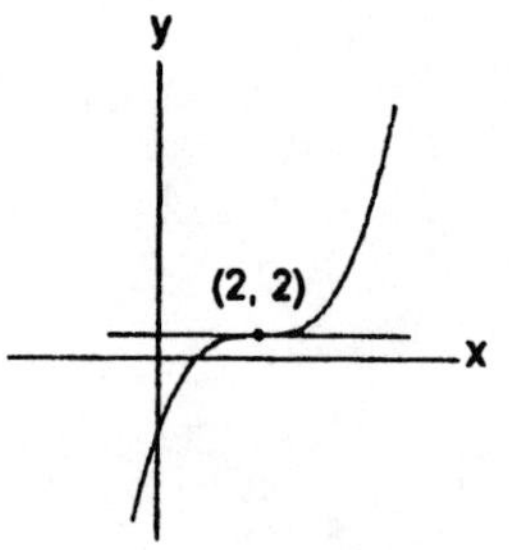

10.

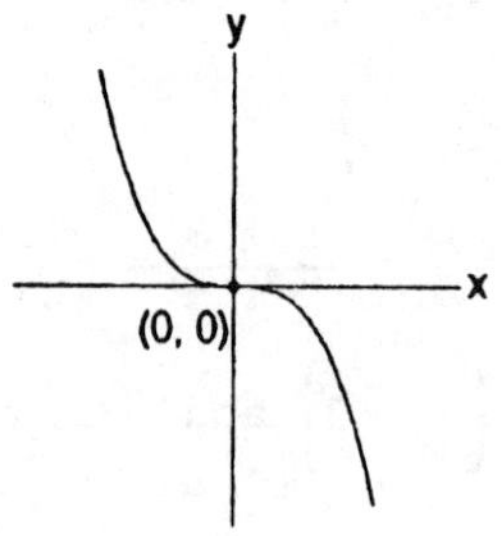

11.

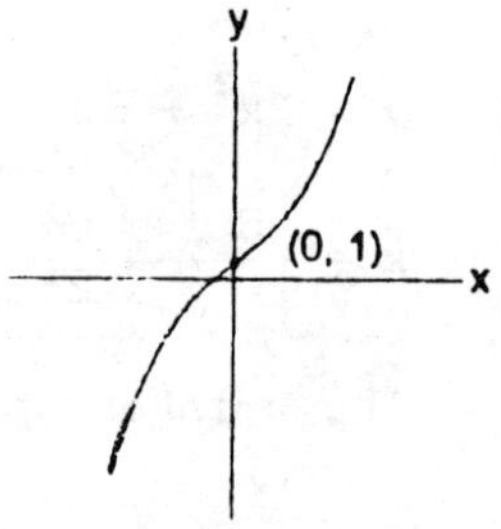

12.

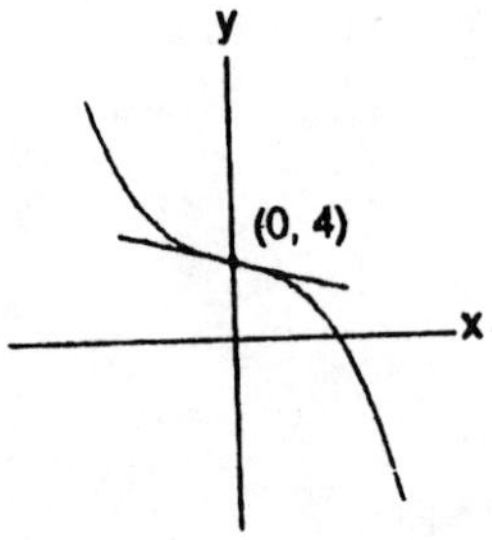

13.

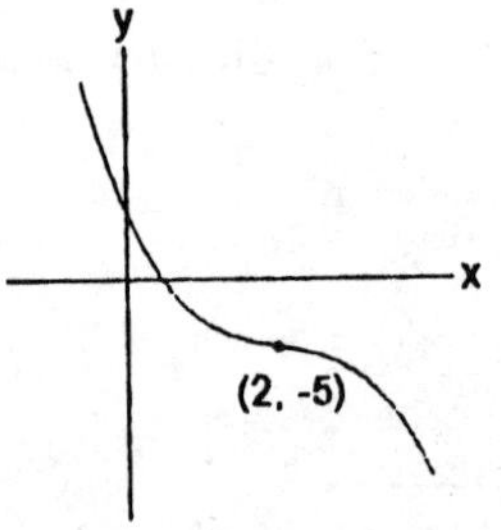

14.

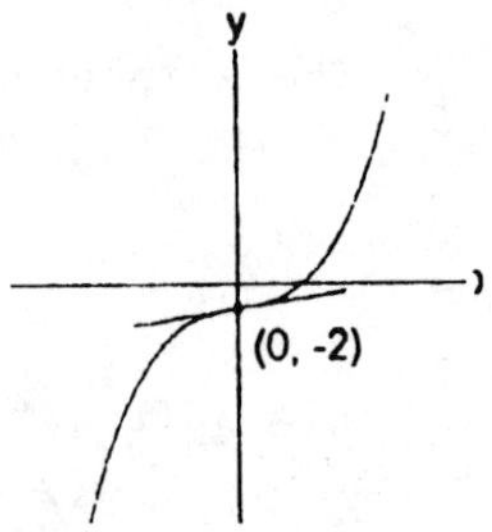

15.

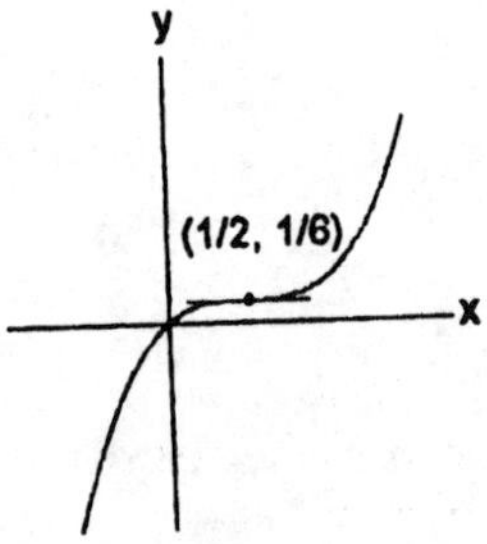

16.

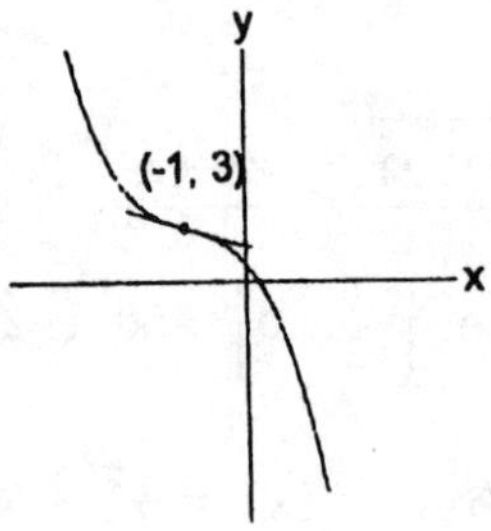

17.

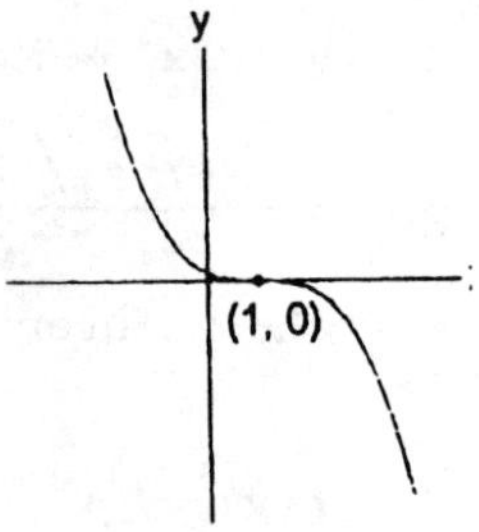

18.

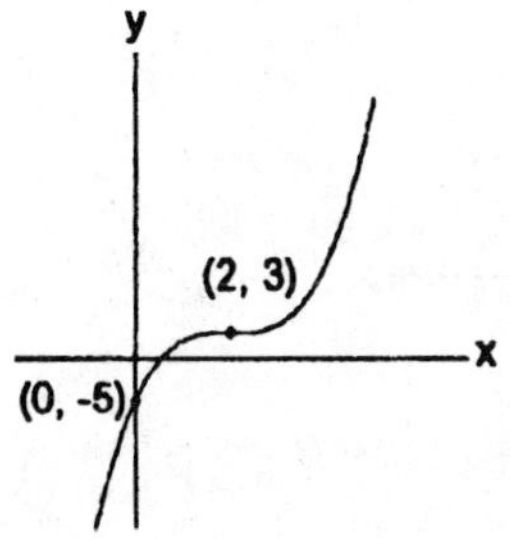

19.

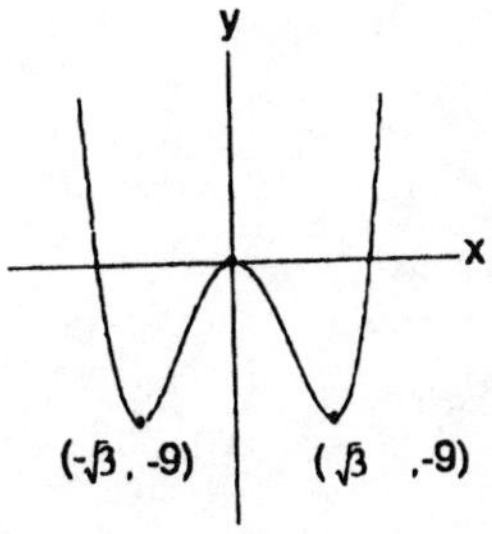

20.

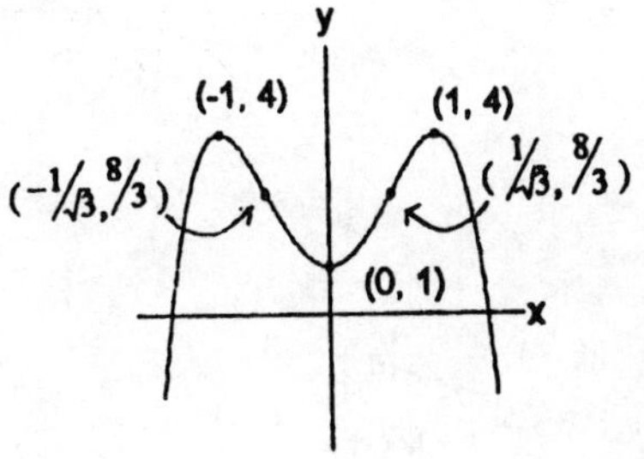

21.

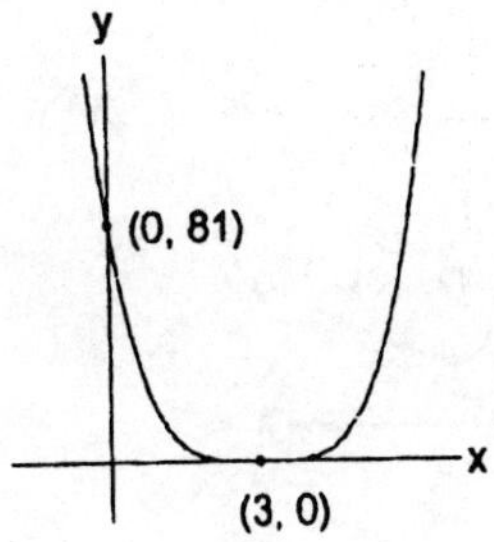

22.

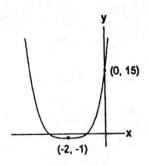

23.

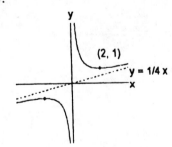

24.

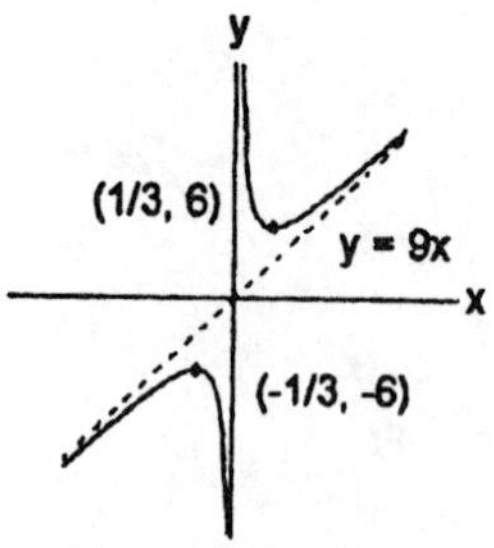

25.

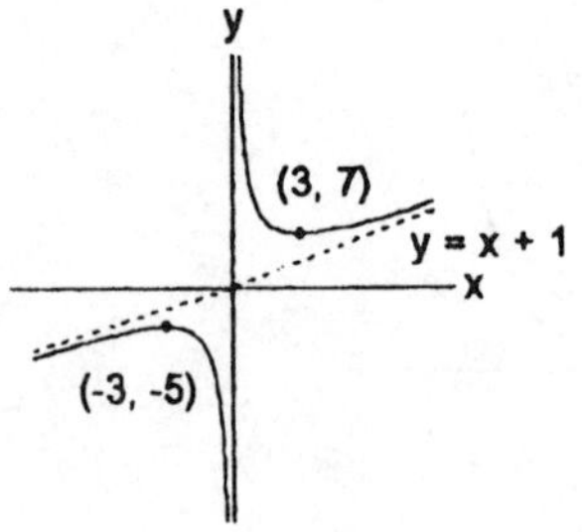

26.

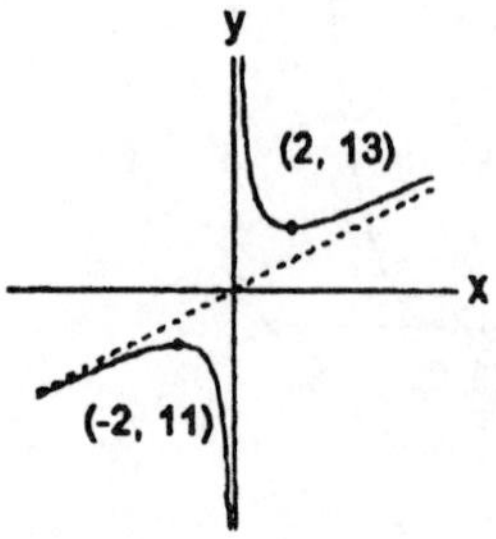

27.

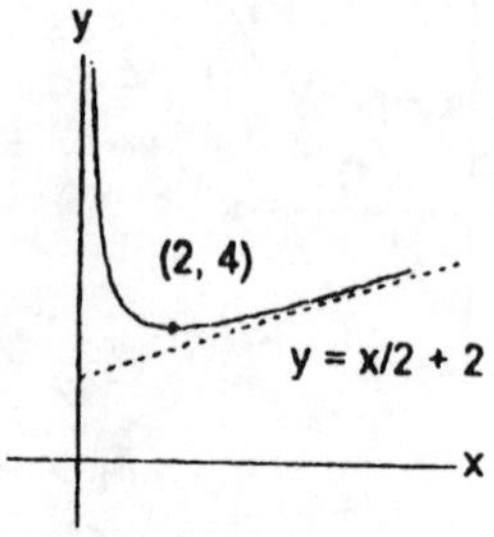

28.

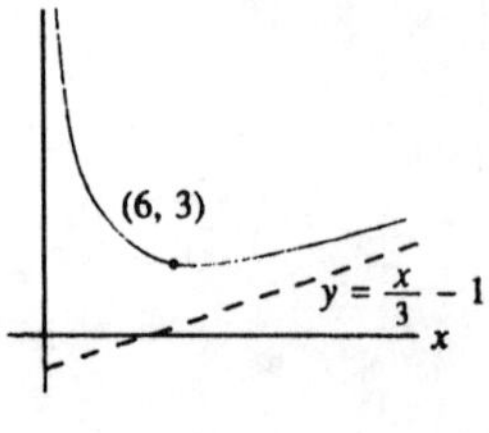

29.

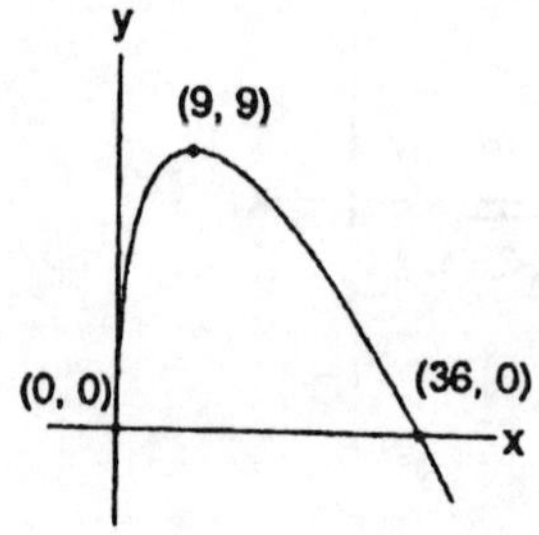

30.

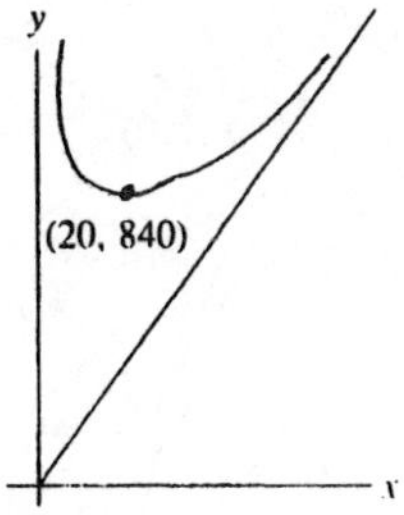

31. $g(x) = f'(x)$. The 3 zeros of $g(x)$ correspond to the 3 extreme points of $f(x)$. $f(x) \neq g'(x)$, the zeros of $f(x)$ don't correspond with the extreme points of $g(x)$.

32. $g(x) = f'(x)$. The zeros of $g(x)$ correspond to the extreme points of $f(x)$. But the zeros of $f(x)$ correspond to the extreme points of $g(x)$.

Start at the left hand side, $f'(x)$ is negative to the value of x at the first extreme point also $g(x)$ is negative. Then $f'(x)$ is positive to the next extreme point and so is $g(x)$. Continue this argument. Apply the same argument to show $f(x) \neq g'(x)$. In fact $f(x)$ could be the sinx and $g(x)$ could be the cosx.

Exercises 2.5

1. $g(x) = 10 + 40x - x^2$

   $g'(x) = 40 - 2x$

   $g''(x) = -2$

   The maximum value of g(x) occurs at $x = 20$; $g(20) = 410$.

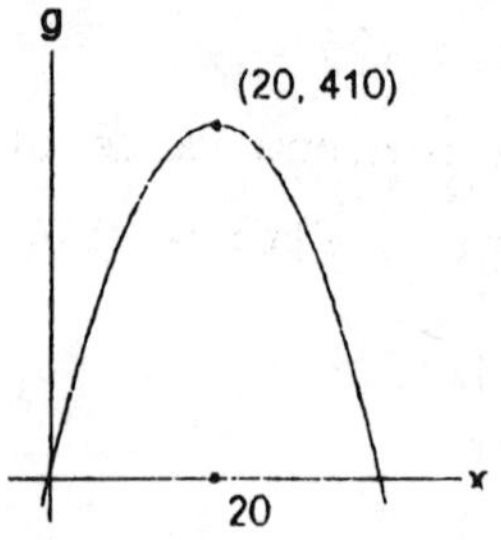

2. $f(x) = 12x - x^2$

   $f'(x) = 12 - 2x$

   $f''(x) = -2$

   The maximum value of f(x) occurs at $x = 6$; $f(6) = 36$.

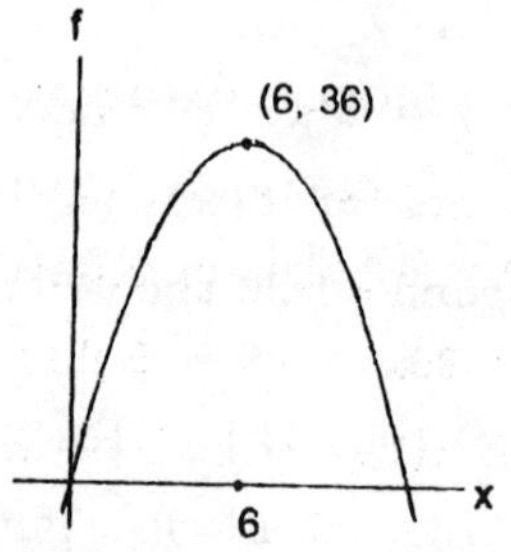

3. $f(t) = t^3 - 6t^2 + 40$

   $f'(t) = 3t^2 - 12t$

   $f''(t) = 6t - 12$

   The minimum value for $t \geq 0$ occurs at $t = 4$; $f(4) = 8$.

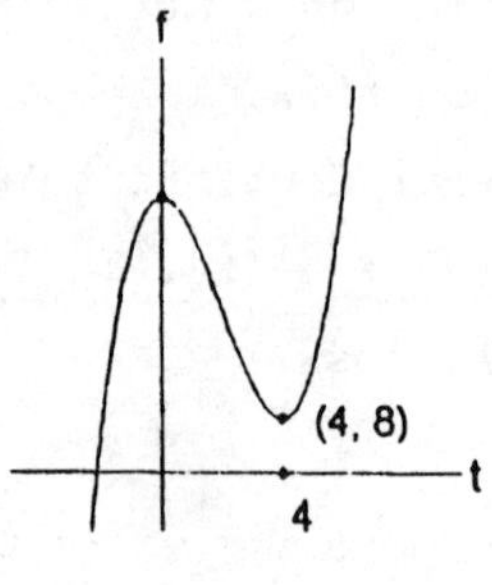

4. $f(t) = t^2 - 24t$

$f'(t) = 2t - 24$

$f''(t) = 2$

The minimum value of $f(t)$ occurs at

$t = 12$; $f(t) = -144$.

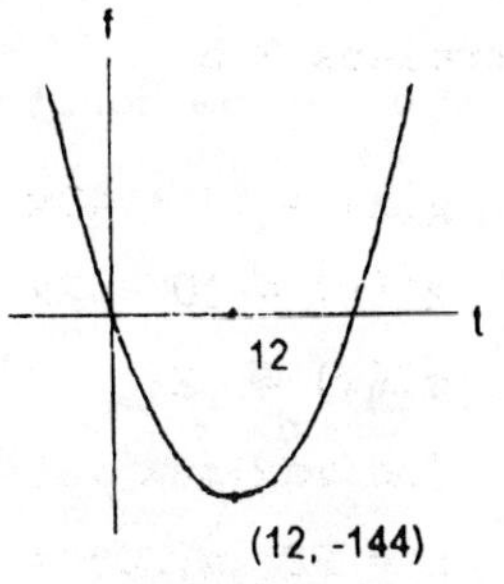

5. Let A = area.

(a) Objective equation: $A = xy$

Constraint equation: $8x + 4y = 320$

(b) Solving constraint equation for y in terms of x gives

$y = 80x - 2x$. Substituting into objective equation yields

$A = x(80x - 2x) = -2x^2 + 80x$

(c) $\dfrac{dA}{dx} = -4x + 80$ $\qquad \dfrac{d^2A}{dx^2} = -4$

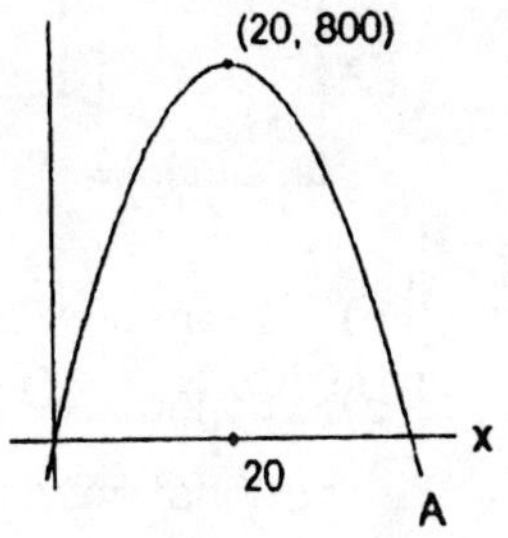

The maximum value of A occurs at $x = 20$. Substituting this value into the equation for y in part (b) gives $y = 80 - 40 = 40$. Answer: $x = 20$, $y = 40$.

6. Let S = surface area.

(a) Objective equation: $S = x^2 + 4xh$

Constraint: $x^2h = 32$

(b) From constraint equation, $h = \dfrac{32}{x^2}$. Thus $S = x^2 + 4x\left[\dfrac{32}{x^2}\right]$

$= x^2 + \dfrac{128}{x}$

(c) $\dfrac{ds}{dx} = 2x - \dfrac{128}{x^2}$ $\qquad \dfrac{d^2s}{dx^2} = 2 + \dfrac{256}{x^3}$

The minimum value of S for $x > 0$ occurs at $x = 4$. Solving for h gives $h = \dfrac{32}{4^2} = 2$.

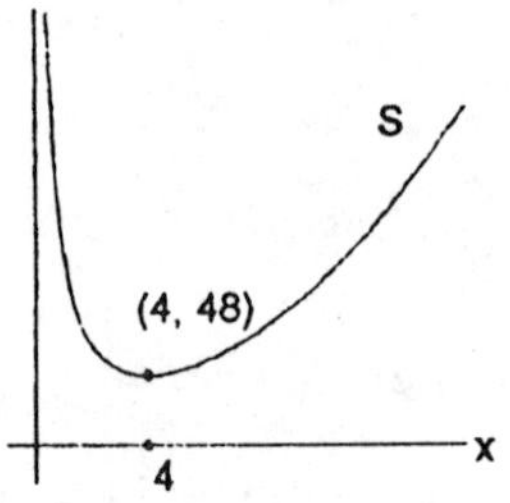

Answer: $x = 4$, $h = 2$.

7. (a)

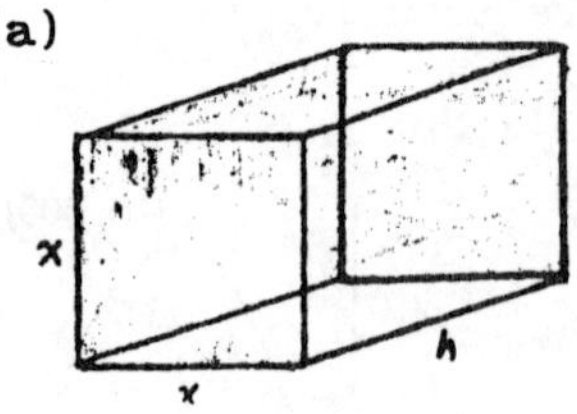

(b) length + girth = $h + 4x$

(c) Objective equation: $V = x^2h$

Constraint equation: $h + 4x = 84$ or $h = 84 - 4x$.

(d) Substituting $h = 84 - 4x$ into the objective equation, we have

$V = x^2(84 - 4x) = -4x^3 + 84x^2$.

(e) $V' = -12x^2 + 168x$

$V'' = -24x + 168$.

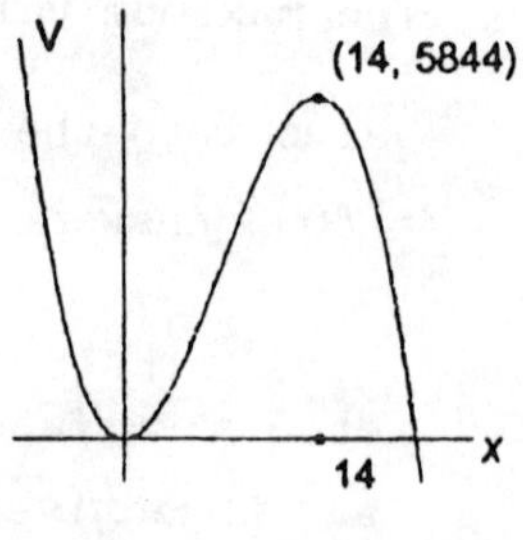

8. (a)

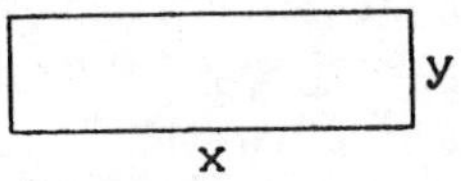

(b) Let P = perimeter.
Objective: $P = 2x + 2y$
Constraint: 100xy

(c) From the constraint, $y = \frac{100}{x}$. So $P = 2x + 2\left(\frac{100}{x}\right)$

$= 2x + \frac{200}{x}$; $\frac{dP}{dx} = 2 - \frac{200}{x^2}$; $\frac{d^2P}{dx^2} = 400x^3$

The minimum value of P for $x > 0$ occurs at $x = 10$. Solving for y gives $y = \frac{100}{10} = 10$.

Answer: $x = 10$, $y = 10$.

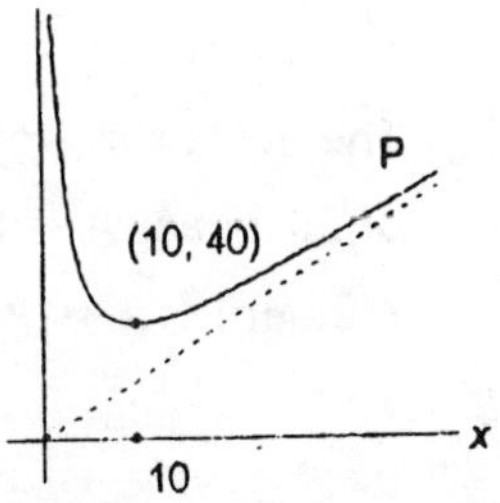

9.

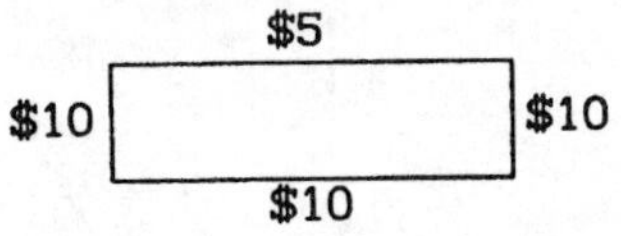

Let C = cost of materials.
Constraint: $xy = 75$
Objective: $C = 15x + 20y$

Solving constraint and substituting gives $C = 15x + 20\left(\frac{75}{x}\right)$

$= 15x + \frac{1500}{x}$; $\frac{dC}{dx} = 15 - \frac{1500}{x^2}$; $\frac{d^2C}{dx^2} = \frac{3000}{x^3}$

The minimum value for $x > 0$ occurs at $x = 10$.

Answer: $x = 10$, $y = 7.5$

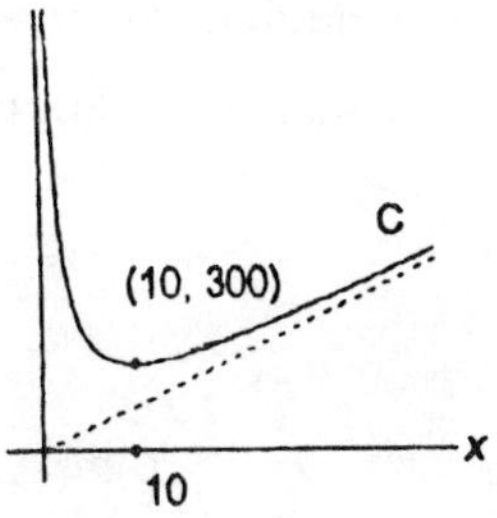

10.

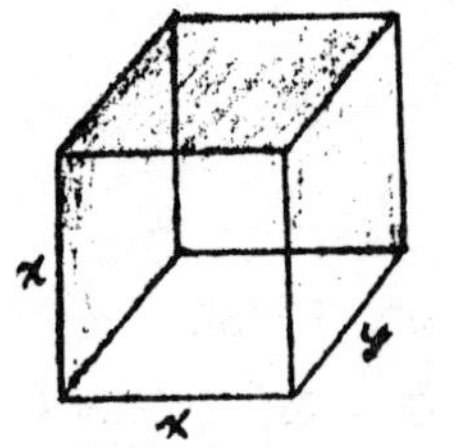

Let C = cost of materials

Constraint: $x^2y = 12$

Objective: $C = 2x^2 + 4xy + x^2$

$= 3x^2 + 4x\left(\frac{12}{x^2}\right)$

$= 3x^2 + \frac{48}{x}$

$\frac{dC}{dx} = 6x - \frac{48}{x^2}$; $\frac{d^2C}{dx^2} = 6 + \frac{96}{x^3}$

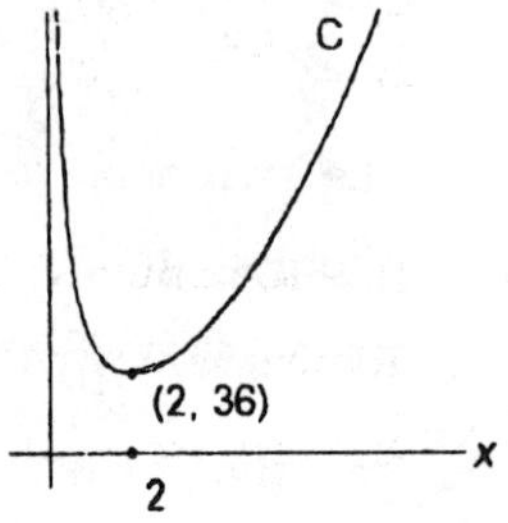

The minimum value of C for x > 0 occurs at x = 2.

Answer: x = 2ft., y = 3ft.

11. Let x = length of base, y = height, M = surface area

Constraint: $x^2y = 8000$

Objective: $M = 2x^2 + 4xy$

$= 2x^2 + 4x\left(\frac{8000}{x^2}\right)$

$= 2x^2 + \frac{32000}{x}$

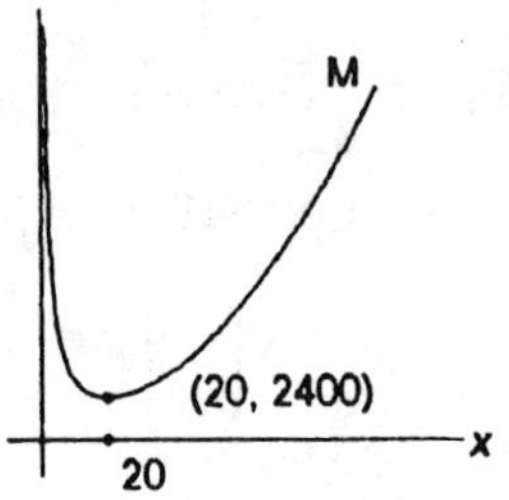

$\frac{dM}{dx} = 4x - \frac{32000}{x^2}$; $\frac{d^2M}{dx^2} = 4 + \frac{64000}{x^2}$

The minimum value of M for x > 0 occurs at x = 20.

Answer: 20cm. x 20cm. x 20cm.

12.

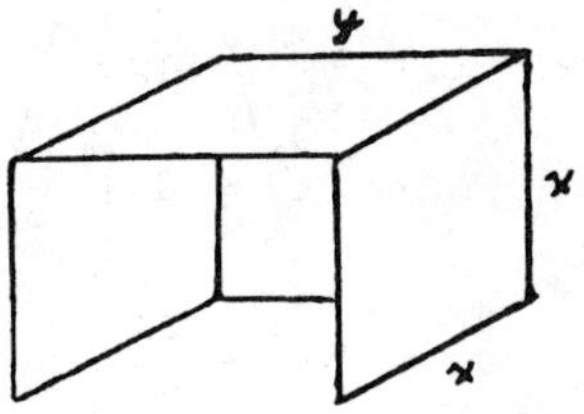

Constraint: $x^2y = 250$

Objective: $C = 2x^2 + 2xy$

$= 2x^2 + \frac{500}{x}$

$$\frac{dC}{dx} = 4x - \frac{500}{x^2}; \qquad \frac{d^2C}{dx^2} = 4 + \frac{1000}{x^3}$$

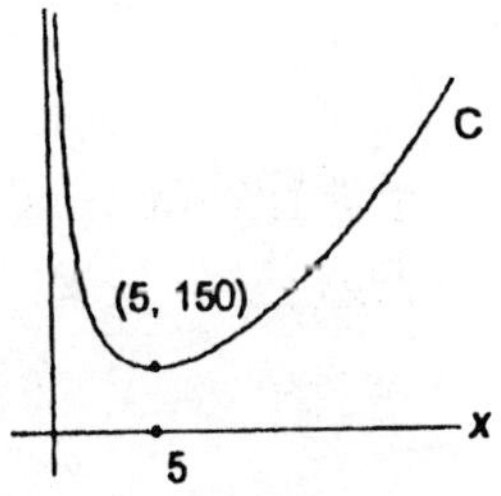

The minimum value of C for $x > 0$ occurs at $x = 5$.

Answer $x = 5$ft., $y = 10$ft.

13. Objective: $A = xy$

Constraint: $6x + 15y = 1500$

$$A = x\left[-\frac{2}{5}x + 100\right] = -\frac{2}{5}x^2 + 100x$$

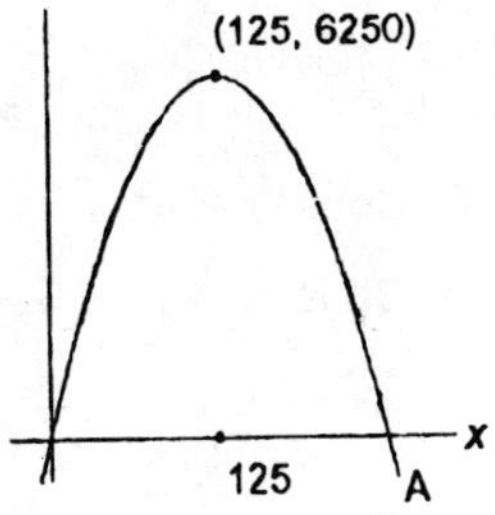

Answer: $x = 125$, $y = 50$

14. Let x = length, y = width of garden.

Constraint: $2x + 2y = 300$

Objective: $A = xy$

$= x(150 - x)$

$= -x^2 + 150x$

$\frac{dA}{dx} = -2x + 150; \quad \frac{d^2A}{dx^2} = -2$

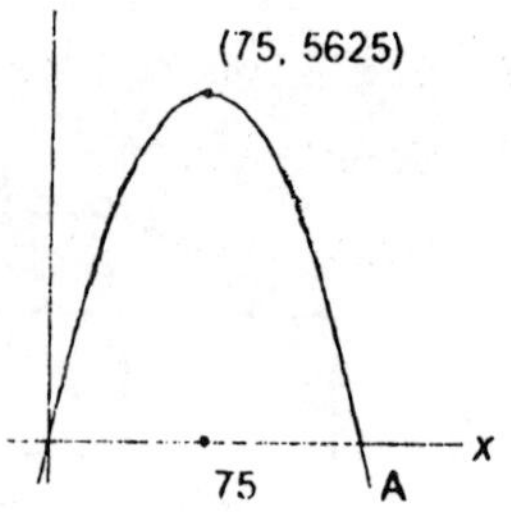

The maximum value of A occurs at $x = 75$.

Answer: 75ft. x 75 ft.

15. Constraint: $x + y = 100$

Objective: $P = xy$

$= x(100 - x)$

$= -x^2 + 100x$

$\frac{dP}{dx} = -2x + 100; \quad \frac{d^2P}{dx^2} = -2$

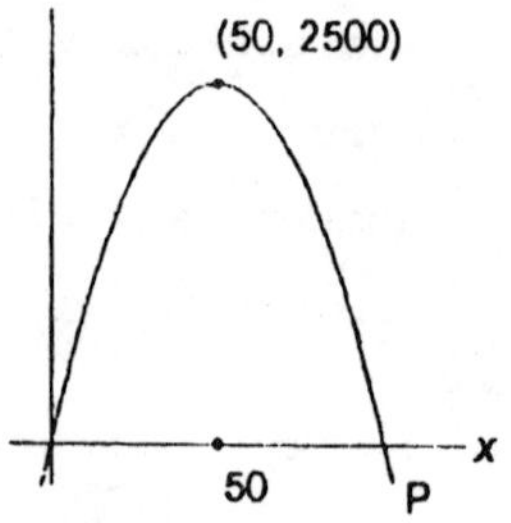

The maximum value of P occurs at $x = 50$.

Answer: $x = 50$, $y = 50$

16. Constraint: $xy = 100$

Objective: $S = x + y$

$= x + \frac{100}{x}$

$\frac{dS}{dx} = 1 - \frac{100}{x^2}; \quad \frac{d^2S}{dx^2} = \frac{200}{x^3}$

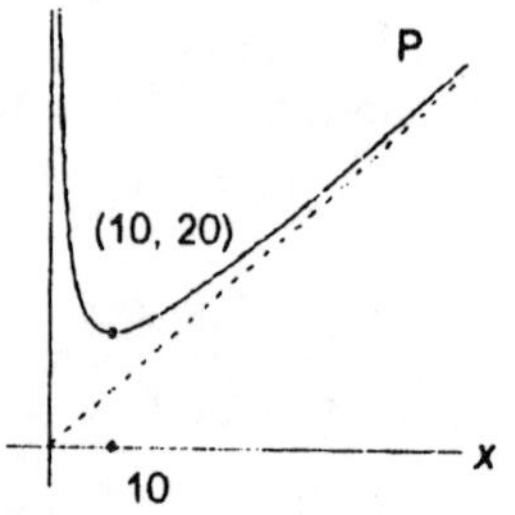

The minimum value of S for $x > 0$ occurs at $x = 10$.

Answer: $x = 10$, $y = 10$

17. See figure 12(a)

Constraint: $2x + 2h + \pi x = 14$ or $(2 + \pi)x + 2h = 14$

Objective: $A = 2xh + \frac{\pi}{2}x^2$

$$= 2x\left(7 - \frac{2 + \pi}{2}\right)x + \frac{\pi}{2}x^2$$

$$= \left(\frac{-\pi}{2} - 2\right)x^2 + 14x$$

$\frac{dA}{dx} = (-\pi - 4)x + 14;$ $\frac{d^2A}{dx^2} = -\pi - 4$

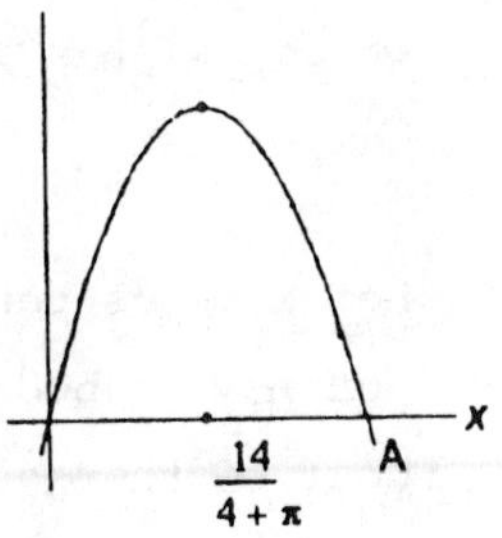

The maximum value of A occurs at

$x = \frac{14}{4 + \pi}$

Answer: $x = \frac{14}{4 + \pi}$ ft.

18. Let S = surface area.

Constraint: $\pi x h = 16\pi$ or $x h = 16$

Objective: $S = 2\pi x^2 + 2\pi xh$

$$= 2\pi x^2 + 2\pi x\left(\frac{16}{x^2}\right)$$

$$= 2\pi\left(x^2 + \frac{16}{x}\right)$$

$\frac{dS}{dx} = 2\pi\left(2x - \frac{16}{x^2}\right);$ $\frac{d^2S}{dx^2} = 2\pi\left(2 + \frac{32}{x^3}\right)$

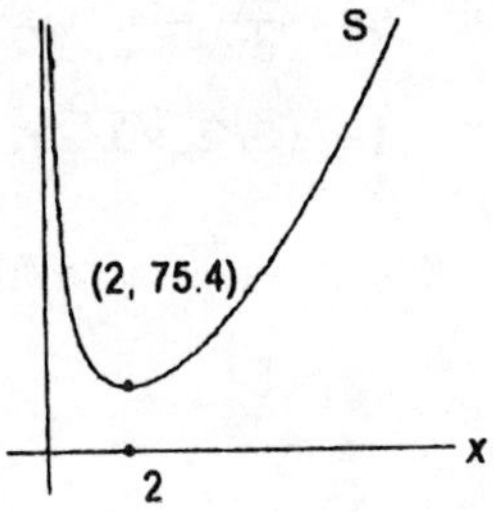

The minimum value of S for $x > 0$ occurs at $x = 2$.

Answer: $x = 2$ in., $h = 4$ in.

19. $A = 20w - \frac{1}{2}w^2$; $\frac{dA}{dw} = 20 - w;$ $\frac{d^2A}{dw^2} = -1$

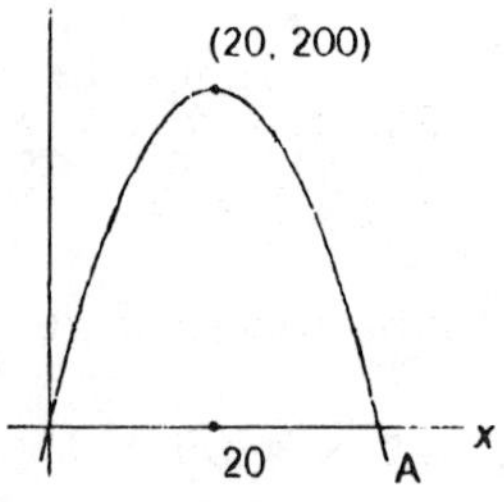

The maximum value of A occurs at $w = 20$.

$x = 20 - \frac{1}{2}w = 20 - \frac{1}{2}(20) = 10$

20. Let x miles per hour be the speed. d = s.t, so time of the journey is 500/x hours. Cost per hour is $5x^2 + 2500$ dollars. Cost of the journey is

$C = (5x^2 + 2500).(500/x) = 2500x + 1000000/x.$

$$\frac{dC}{dx} = 2500 - 1000000/x^2$$

set $\frac{dC}{dx} = 0$, and we obtain $x^2 = 400$, $x = 20$

21. Distance $= \sqrt{(x - 2)^2 + y^2}$; by hint we minimize $(x - 2)^2 + y^2 = (x - 2)^2 + x$, since $y = \sqrt{x}$.

$$\frac{dD}{dx} = 2(x - 2) + 1$$

set $\frac{dD}{dx} = 0$ to give: $2x = 2$, or $x = 3/2$, $y = \sqrt{3/2}$. So the point is $(3/2, \sqrt{3/2})$.

Exercises 2.6

1. Let x be the number of prints the artist sells. Then his revenue = [price]·[quantity]

$$\begin{cases} (400 - 5(x - 50))x & \text{if } x > 50 \\ 400x & \text{if } x \le 50 \end{cases}$$

For $x > 50$, $r(x) = -5x^2 + 650x$, $r'(x) = -10x + 650$, $r'(65) = 0$.

The maximum value of $r(x)$ occurs at $x = 65$. The artist should sell 65 prints.

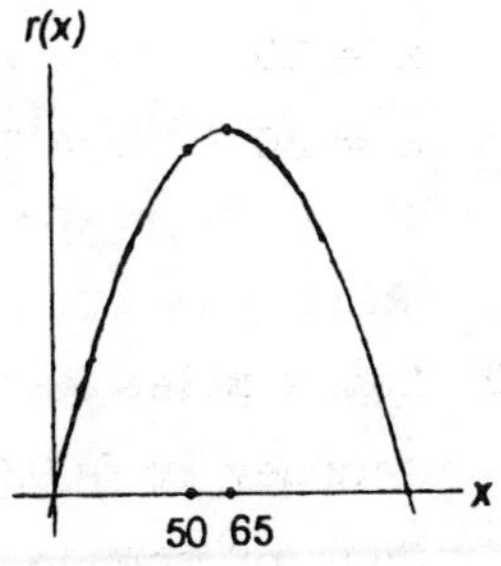

2. Let x be the number of memberships the club sells. Then their

$$\text{revenue} = r(x) = \begin{cases} 200x \\ (200 - (x - 100))x \end{cases}$$

$$= \begin{cases} 200x & \text{if } x \le 100 \\ -x^2 + 300x & \text{if } 100 < x \le 160 \end{cases}$$

For $100 < x \le 160$, $r'(x) = -2x + 300$, $r'(150) = 0$.

The maximum value of $r(x)$ occurs at $x = 150$. The club should try to sell 150 memberships.

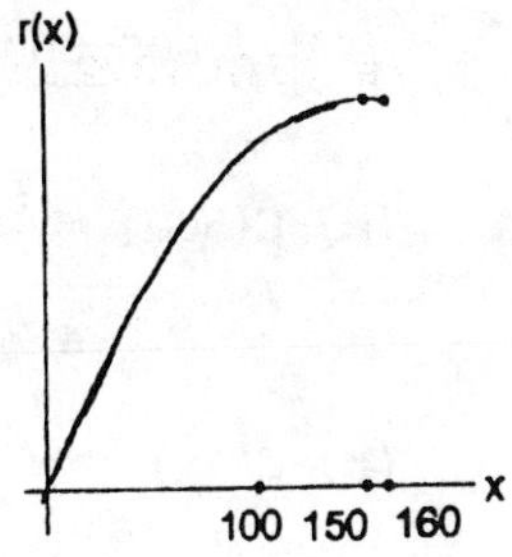

3. Let $P(x)$ be the profit from x tables. Then

$P(x) = (10 - (x - 12).5)x = -.5x^2 + 16x$ for $x \ge 12$

$P'(x) = 16 - x$; $P'(16) = 0$

The maximum value of $P(x)$ occurs at $x = 16$. The cafe should provide 16 tables.

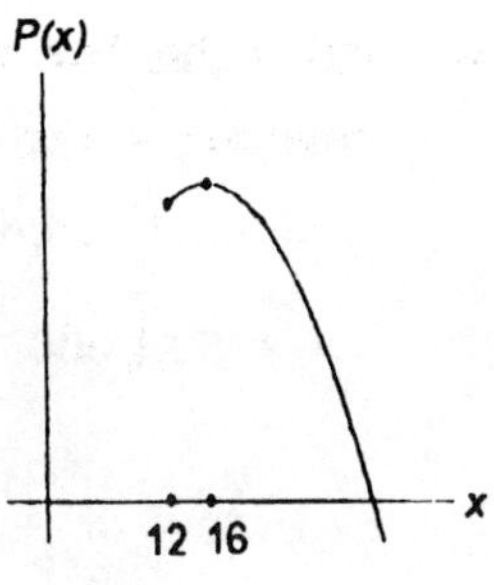

4. The revenue function is $R(x) = (36{,}000 - 300(x - 100))x$, where x is the price in cents and $x \geq 100$.

$R(x) = 66{,}000x - 300x^2$, $R'(x) = 66{,}000 - 600x$, $R'(110) = 0$

The maximum value occurs at $x = 100$. The toll should be $1.10.

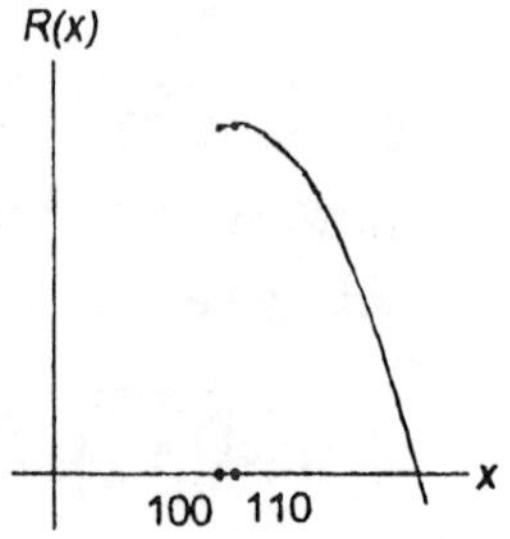

5. Let x be the order quantity and r the number of orders placed in the year. Then the inventory cost is $C = 80r + 5x$. The constraint is $rx = 10{,}000$, so $r = \dfrac{10{,}000}{x}$ and we can write

$C(x) = \dfrac{800{,}000}{x} + 5x$

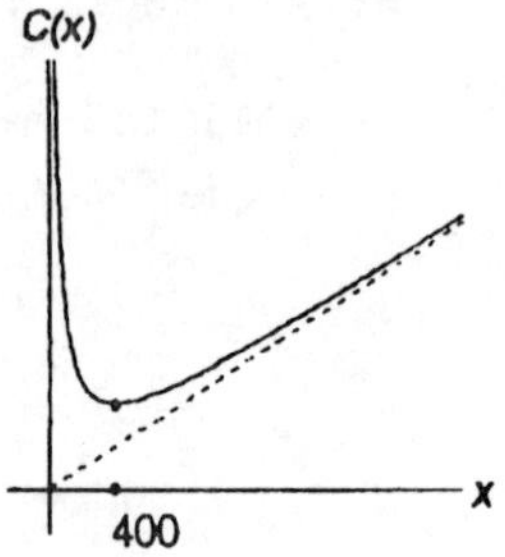

(a) $C(500) = \dfrac{800{,}000}{500} + 5(500)$

$= 4100$

(b) $C'(x) = -\dfrac{800{,}000}{x^2} + 5$

$C'(400) = 0$

The maximum value of $C(x)$ occurs at $x = 400$.

6. Let x be the number of tires produced in each production run, and let r be the number of runs in the year. Then the production cost is $C = 15{,}000r + 2.5x$. The constraint is $rx = 600{,}000$, so $x = \dfrac{600{,}000}{r}$ and we can write

$C(r) = 15{,}000r + \dfrac{1{,}500{,}000}{r}$

(a) $C(10) = 15{,}000(10) + \dfrac{1{,}500{,}000}{10}$

$= 300{,}000$

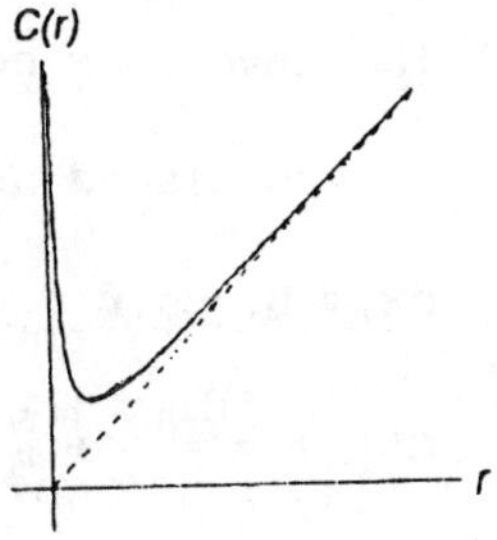

(b) $C(x) = \dfrac{15{,}000(600{,}000)}{x} + 2.5x$

$C'(x) = \dfrac{-9 \cdot 10^9}{x^2} + 2.5,$

$C'(60{,}000) = 0$ Each run should produce 60,000 tires.

7. Let x be the number of microscopes produced in each run and let r be the number of runs. The objective is

$C = 2500r + 15x + 20\left(\dfrac{x}{2}\right)$. The constraint is $xr = 1600$,

$x = \dfrac{1600}{r}$, so $C(r) = 2500r + \dfrac{40{,}000}{r}$. $C'(r) = 2500 - \dfrac{40{,}000}{r^2}$,

$C'(4) = 0$. C has a minimum at $r = 4$.

There should be 4 production runs.

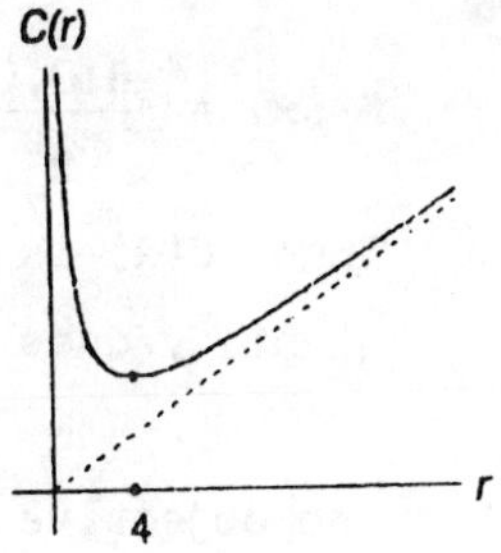

8. Let x be the size of each order and let r be the number of orders placed in the year. Then the inventory cost is

$C = 40r + 2x$ and $rx = 8000$, so $x = \dfrac{8000}{r}$, $C(r) = 40r + \dfrac{1600}{r}$;

$C'(r) = 40 - \dfrac{16{,}000}{r^2}$, $C'(20) = 0$.

The minimum value for C occurs at $r = 20$ (for $r > 0$).

9. The inventory cost is $C = hr + \frac{s}{r}x$ where r is the number of orders placed and x is the order size. The constraint is $rx = Q$, so $r = \frac{Q}{x}$ and we can write $C(x) = \frac{hQ}{x} + \frac{s}{2}$.

$C'(x) = \frac{-hQ}{x^2} + \frac{s}{2}$. Setting $C'(x) = 0$ gives $\frac{-hQ}{x^2} + \frac{s}{2}$,

$x^2 = \frac{2hQ}{s}$, $x = \pm\sqrt{\frac{2hQ}{s}}$. The positive value $\sqrt{\frac{2hQ}{s}}$ gives the minimum value for $C(x)$ for $x > 0$.

10. In this case, the inventory cost becomes

$$C = \begin{cases} 75r + 4x & \text{for } x < 600 \\ (75 - (x - 600))r + 4x & \text{for } x \geq 600 \end{cases}$$

Since $r = \frac{1200}{x}$, $C(x) = \begin{cases} \frac{90,000}{x} + 4x & \text{for } x < 600 \\ \frac{810,000}{x} + 4x - 1200 & \text{for } x \geq 600 \end{cases}$

Now the function $f(x) = \frac{810,000}{x} + 4x - 1200$ has

$f'(x) = -\frac{810,000}{x^2} + 4$, $f'(450) = 0$ and $f'(x) > 0$ for $x > 450$.

Thus, $C(x)$ is increasing for $x > 600$ and the optimal order quantity does not change.

11. The objective is $A = (x + 100)w$ and the constraint is $x + (x + 100) + 2w = 2x + 2w + 100 = 400$; or $x + w = 150$, $w = 150 - x$.

$A(x) = (x + 100)(150 - x) = -x^2 + 50x + 15,000$;

$A'(x) = -2x + 50$, $A'(25) = 0$.

The maximum value of A occurs at $x = 25$. Thus the optimal values are $x = 25$, $w = 150 - 25 = 125$.

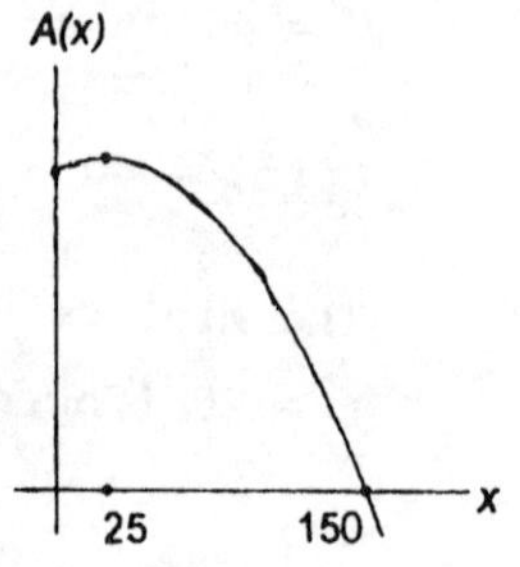

12. The objective remains $A = (x + 100)w$, but the constraint becomes $2x + 2w + 100 = 200$; or $x + w = 50$, so $A(x) = (x + 100)(50 - x) = -x^2 - 50x + 5000$, $A'(x) = -2x - 50$, $A'(-25) = 00$. In this case, the maximum value of A occurs at $x = -25$, and $A(x)$ is decreasing for $x > -25$. Thus, the best non-negative value for x is $x = 0$. The optimal dimensions are $x = 0$, $w = 50$.

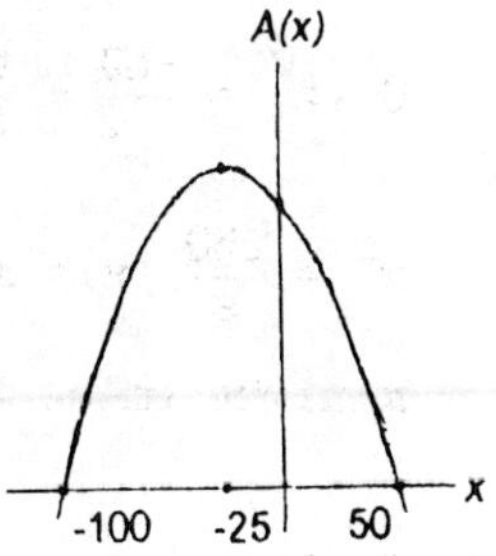

13. The objective is $F = 2x + 3w$, and the constraint is $xw = 54$, or $w = \frac{54}{x}$, so $F(x) = 2x + \frac{162}{x}$, $F'(x) = 2 - \frac{162}{x^2}$, $F'(9) = 0$.

The minimum value of F for $x > 0$ is $x = 9$. The optimal dimensions are thus $x = 9$, $w = 6$.

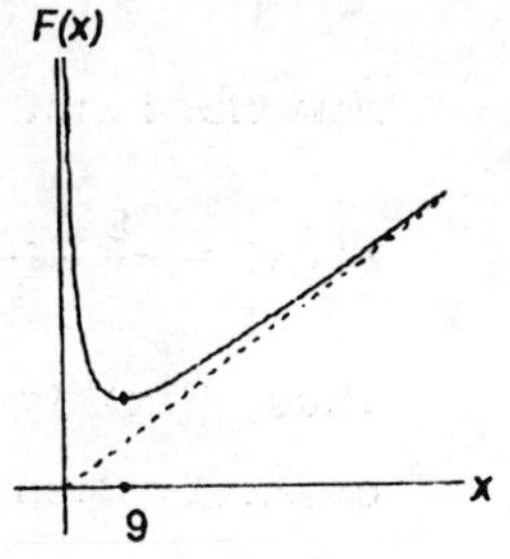

14. The objective is $C = 2 \cdot 5x + 2 \cdot 5w + 2w = 10x + 12w$. The constraint is $xw = 54$, so $w = \frac{54}{x}$ and $C(x) = 10x + \frac{648}{x}$.

$C'(x) = 10 - \frac{648}{x^2}$, $C'\left(\frac{18}{\sqrt{5}}\right) = 0$

The optimal dimensions are

$x = \frac{18}{\sqrt{5}}$, $w = \frac{3}{\sqrt{5}}$.

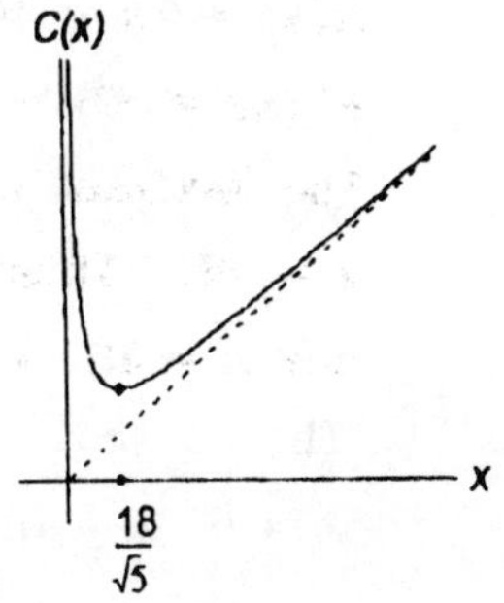

15. The revenue function is

$R(x) = (800 - 20(x - 12))x$, $12 \le x \le 25$, where x is the tour-group size. $R(x) = 1040x - 20x^2$, $R'(x) = 1040 - 40x$, $R'(26) = 0$. The best size between 12 and 25 is 25.

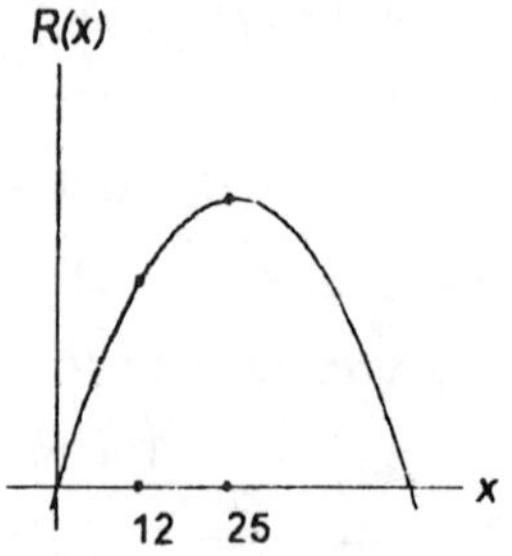

16. The objective is $S = 2x^2 + 3xy$ where x and y are the dimensions of the box. The constraint is $x^2y = 36$, so $y = \frac{36}{x^2}$

and $S(x) = 2x^2 + 3x\left(\frac{36}{x^2}\right) = 2x^2 + \frac{108}{x}$. $S'(x) = 4x - \frac{108}{x^2}$,

$S'(3) = 0$. The optimal dimensions are 3 x 3 x 4 (in.).

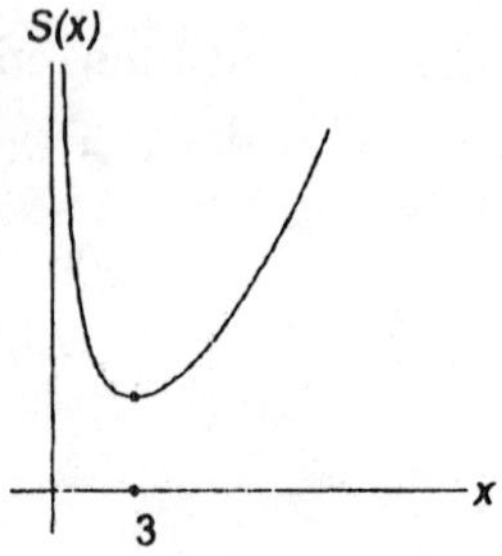

17. Let x be the lengths of the base and let y be the height of the shed. The objective is

$C = 4x^2 + 2x^2 + 4 \cdot 2.5xy = 6x^2 + 10xy$. The constraint is

$x^2y = 150$, $y = \frac{150}{x^2}$, $C(x) = 6x^2 + \frac{1500}{x}$, $C'(x) = 12x - \frac{1500}{x^2}$,

$C'(5) = 0$

The optimal dimensions are

$x = 5$, $y = 6$.

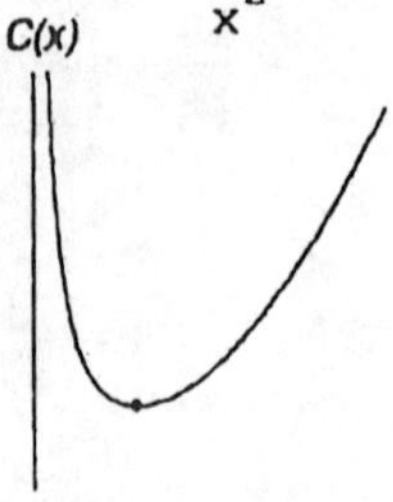

18. Let x be the length of the front of the building and let y be the other dimension. The objective is

$C = 70x + 2.50y + 50x = 120x + 100y$ and the constraint is

$xy = 12{,}000$, $y = \dfrac{12{,}000}{x}$. So $C(x) = 120x + \dfrac{1{,}200{,}000}{x}$,

$C'(x) = 120 - \dfrac{1{,}200{,}000}{x^2}$,

$C'(100) = 0$. The optimal dimensions are

$x = 100$, $y = 120$.

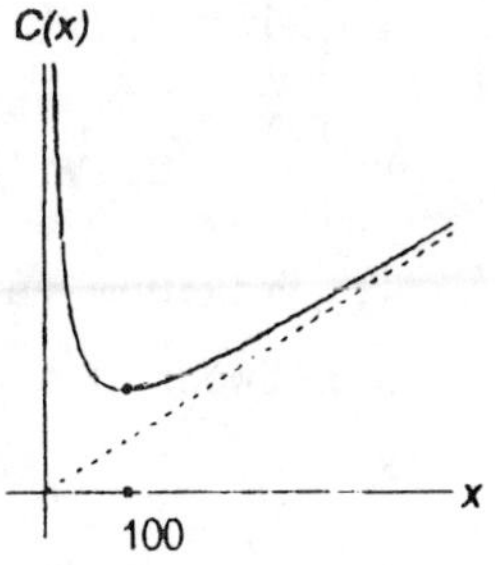

19. Let x be the length of the square end and let h be the other dimensions. The objective is $V = x^2h$ and the constraint is

$2x + h = 120$, $h = 120 - 2x$. $V(x) = 120x^2 - 2x^3$,

$V'(x) = 240x - 6x^2$, $V'(0) = 0$,

$V'(40) = 0$. The maximum value of V for $x > 0$ occurs at

$x = 40$, $h = 40$.

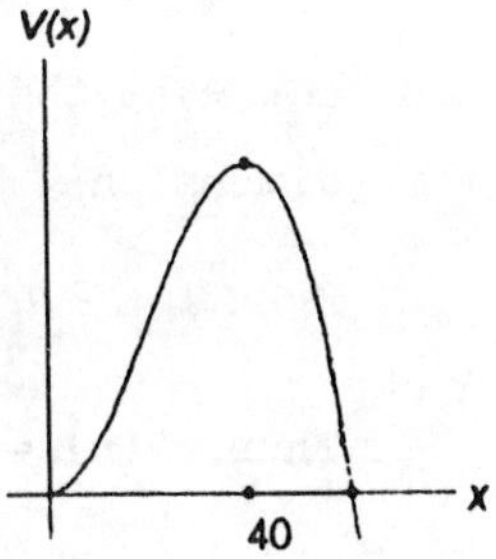

20. The objective is $A = xh$ and the constraint is $\pi x + 2h = 440$,

$h = 220 - \frac{\pi}{2}x$. $A(x) = x\left(220 - \frac{\pi}{2}x\right) = 220x - \frac{\pi}{2}x^2$,

$A'(x) = 220 - \pi x$, $A'\left(\frac{220}{\pi}\right) = 0$.

The optimal dimensions are

$x = \dfrac{220}{\pi}$ yds., $h = 110$ yds.

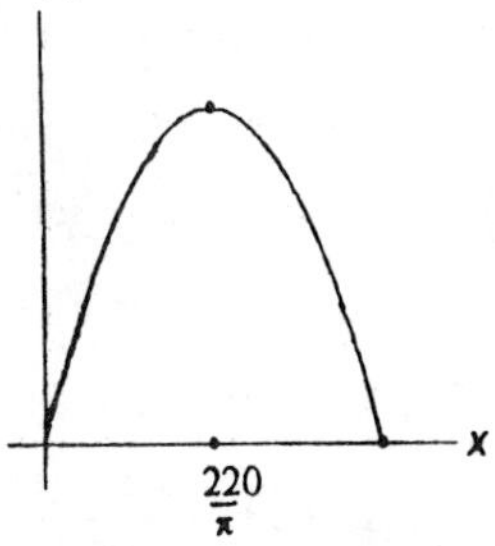

21. The objective equation is $V = w^2x$ and the constraint is $w + 2x = 16$, $w = 16 - 2x$.

$V(x) = (16 - 2x)^2x = 4x^3 - 64x^2 + 256x$,

$V'(x) = 12x^2 - 128x + 256$,

$V'\left(\frac{8}{3}\right) = 0$, $V'(8) = 0$,

$V''\left(\frac{8}{3}\right) < 0$, $V''(8) > 0$.

The maximum value of V for x between 0 and 8 occurs at $x = \frac{8}{3}$.

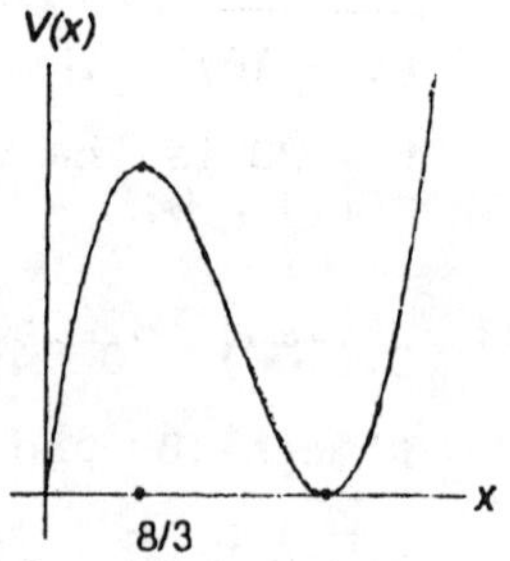

22. Let x be the width of the base and let h be the other dimension. The objective is $V = 2x^2$ and the constraint is $2(2x^2) + 2xh + 2(2xh) = 27$, or $4x^2 + 6xh = 27$, $h = \frac{27 - 4x^2}{6x}$.

Thus, $V(x) = \frac{2x^2(27 - 4x^2)}{6x} = 9x - \frac{4}{3}x^3$. $V'(x) = 9 - 4x^2$,

$V'\left(\frac{3}{2}\right) = 0$. The optimal values are $x = \frac{3}{2}$, $h = 2$. The dimensions should be $\frac{3}{2} \times 3 \times 2$.

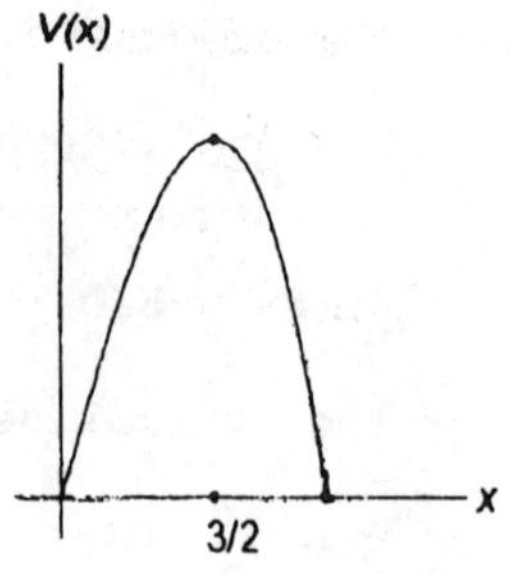

23. We want to find the maximum value of $f'(t)$.

$f'(t) = \frac{10}{(t + 10)^2} - \frac{200}{(t + 10)^3}$; $f''(t) = \frac{-20}{(t + 10)^3} + \frac{600}{(t + 10)^4}$;

$f'''(t) = \dfrac{60}{(t + 10)^4} - \dfrac{2400}{(t + 10)^5}$. Setting $f''(t) = 0$ gives

$\dfrac{20}{(t + 10)^3} = \dfrac{600}{(t + 10)^4}$, $20 = \dfrac{600}{(t + 10)}$, $t = 20$, $f'''(20) < 0$, so $t = 20$ is the maximum value of $f'(t)$.

24. We want to find the maximum value of $f'(t) = 40 + 2t - \frac{1}{5}t^2$.

$f''(t) = 2 - \frac{2}{5}t$, $f''(5) = 0$.

The maximum rate of output occurs at $t = 5$. The maximum output rate is $f'(5) = 45$ hrs/day.

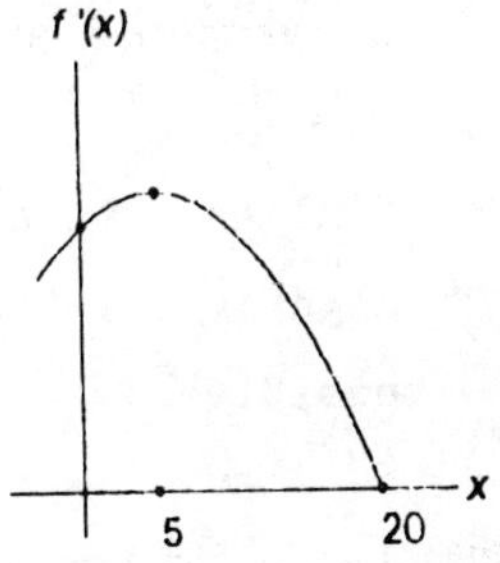

25. Let (x, y) be the top right-hand corner of the window. The objective is $A = 2xy$ and the constraint is $y = 9 - x^2$. Thus, $A(x) = 2x(9 - x^2) = 18x - 2x^3$, $A'(x) = 18 - 6x^2$, $A'(\sqrt{3}) = 0$.

The maximum value of A for $x > 0$ occurs at $x = \sqrt{3}$. Thus, the window should be 6 units high and $2\sqrt{3}$ units wide.

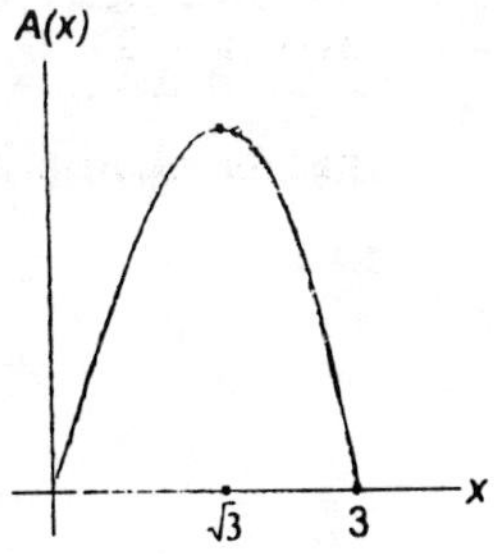

26. We want to find the minimum value of

$f'(t) = \dfrac{-1000}{(t + 8)^2} + \dfrac{8000}{(t + 8)^3}$; $f''(t) = \dfrac{2000}{(t + 8)^3} - \dfrac{24{,}000}{(t + 8)^4}$;

$f'''(t) = \dfrac{-6000}{(t + 8)^4} + \dfrac{96,000}{(t + 8)^5}$. Setting $f''(t) = 0$ gives

$\dfrac{2000}{(t + 8)^3} = \dfrac{24,000}{(t + 8)^4}$, $2000 = \dfrac{24,000}{(t + 8}$, $t = 4$.

$f'''(4) > 0$, so $t = 4$ gives the minimum value of $f'(t)$.

27. $A = x^2 + 5xh$ (Area: where x is the length of the square base and h is the height). $V = x^2h = 400 \Rightarrow h = 400/(x^2)$, so $A = x^2 +$ $2000/x$, and $\dfrac{dA}{dx} = 2x - 2000/x^2$.
Setting $\dfrac{dA}{dx} = 0$ gives $2x^3 = 2000$, or $x = 10$ which in turn yields $h = 4.1$.

28. f(x) has its greatest value at zero since f'(x) becomes more negative as there is no extreme point for f'(x) to be positive.

Exercises 2.7

1. The marginal cost function is $M(x) = C'(x) = 3x^2 - 12x + 13$.
$M'(x) = 6x - 12$, $M'(2) = 0$
The minimum value of M(x) occurs at x = 2. The minimum marginal cost is
M(2) = $1.

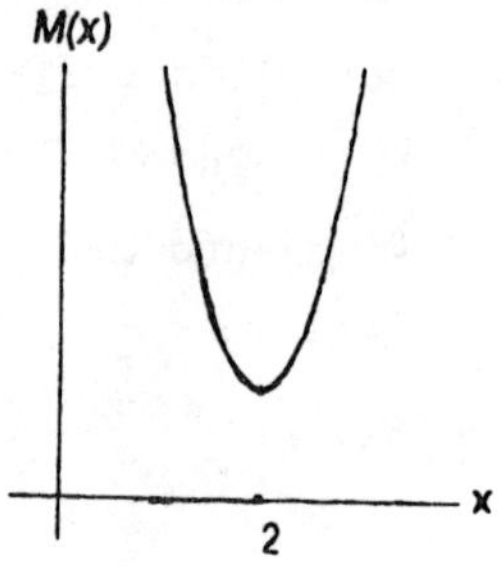

2. $M(x) = C'(x) = .0003x^2 - .12x + 12$. $M'(x) = .0006x - .12$, $M'(100) = -.06 < 0$, so the marginal cost is decreasing at $x = 100$. $M'(200) = 0$.
The minimum marginal cost is $M(200) = 0$.

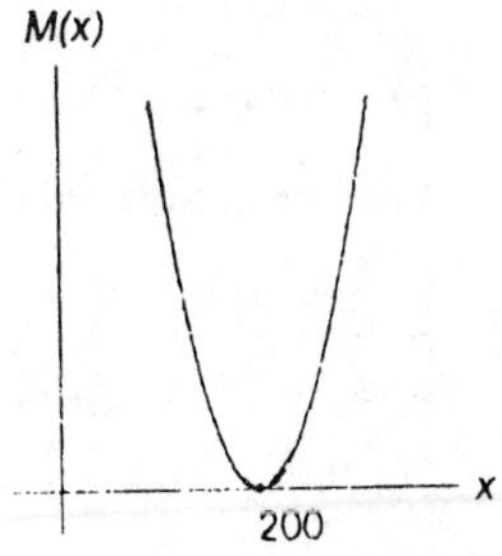

3. $R(x) = 200 - \dfrac{1600}{x + 8} - x$, $R'(x) = \dfrac{1600}{(x + 8)^2} - 1$,

$R''(x) = -\dfrac{3200}{(x + 8)^3}$, $R'(32) = 0$.
The maximum value of $R(x)$ occurs at $x = 32$.

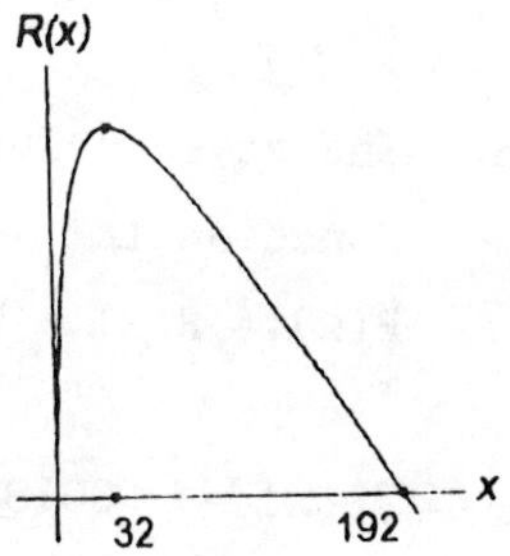

4. $R(x) = 4x - .0001x^2$, $R'(x) = 4 - .0002x$, $R'(20{,}000) = 0$.
The maximum value of $R(x)$ occurs at $x = 20{,}000$. The maximum possible revenue is $R(20{,}000) = 40{,}000$.

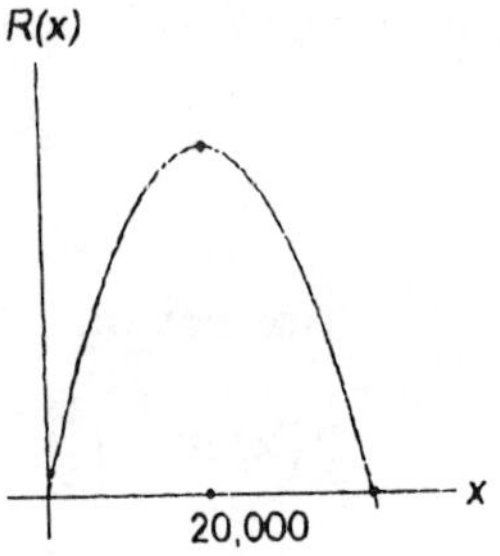

5. The profit function is

$P(x) = R(x) - C(x) = 28x - (x^3 - 6x^2 + 13x + 15)$

$= -x^3 + 6x^2 + 15x - 15$

$P'(x) = -3x^2 + 12x + 15$, $P''(x) = -6x + 12$, $P'(5) = 0$.

The maximum value of $P(x)$ for $x > 0$ occurs at $x = 5$.

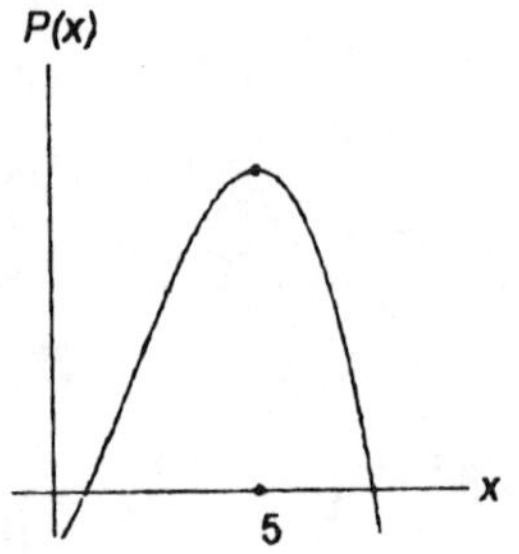

6. The revenue function is $R(x) = 3.5x$. Thus, the profit function is

$P(x) = R(x) - C(x) = 3.5x - (.0006x^3 - .03x^2 + 2x + 20)$

$= -.0006x^3 + .03x^2 + 1.5x - 20.$

$P'(x) = -.0018x^2 + .06x + 1.5$, $P''(x) = -.0036x + .06$, $P'(50) = 0$. Thus, the maximum value of $P(x)$ for $x > 0$ occurs at $x = 50$.

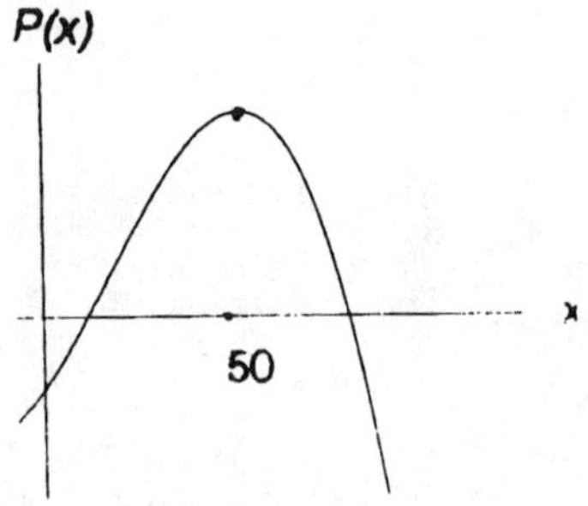

7. The revenue function is

$R(x) = x\left(\frac{1}{2}x^2 - 10x + 300\right) = \frac{1}{12}x^3 - 12x^2 + 300x.$

$R'(x) = \frac{1}{4}x^2 - 20x + 300$, $R''(x) = \frac{1}{2}x - 20$. $R'(20) = 0$, $R'(60) = 0$; $R''(20) < 0$, $R''(60) > 0$.

The maximum value of $R(x)$ occurs at $x = 20$. The corresponding price is $133\frac{1}{3}$.

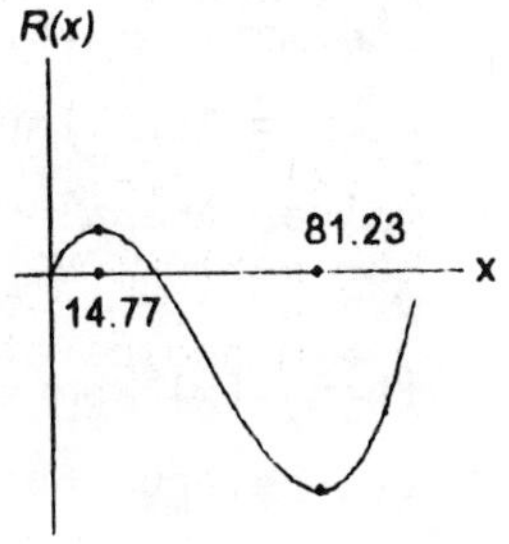

8. The revenue function is $R(x) = x(2 - .001x) = 2x - .001x^2$. $R'(x) = 2 - .002x$, $R'(1000) = 0$. The maximum value of $R(x)$ occurs at $x = 1000$. The corresponding price is $p = 2 - .001(1000) = 1$.

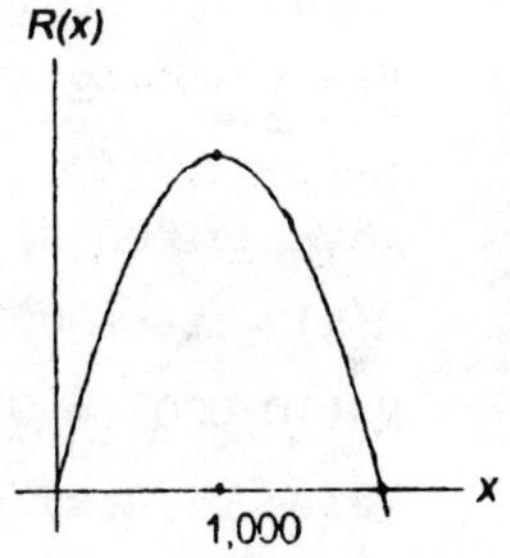

9. The revenue function is $R(x) = x(256 - 50x) = 256x - 50x^2$. Thus, the profit function is

$$P(x) = R(x) - C(x) = 256x - 50x^2 - 182 - 56x$$
$$= -50x^2 + 200x - 182$$

$P'(x) = -100x + 200$, $P'(2) = 0$. The maximum profit occurs at $x = 2$ (million tons). The corresponding price is $256 - 50(2) = 156$ (million dollars).

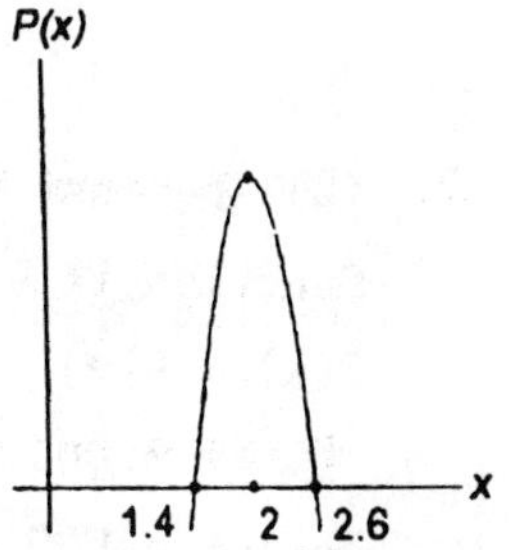

10. The objective is $A = xy$ and the constraint is $y = 30 - x$. $A(x) = x(30 - x) = 30x - x^2$, $A'(x) = 30 - 2x$, $A(15) = 0$. The maximum value of $A(x)$ occurs at $x = 15$. Thus, the optimal values are $a = 15$, $b = 15$. If $y = 30 - x$ is a demand curve, then $A(x)$ above corresponds to the revenue function $R(x)$ and the optimal value a, b correspond to the revenue-maximizing quantity and price, respectively.

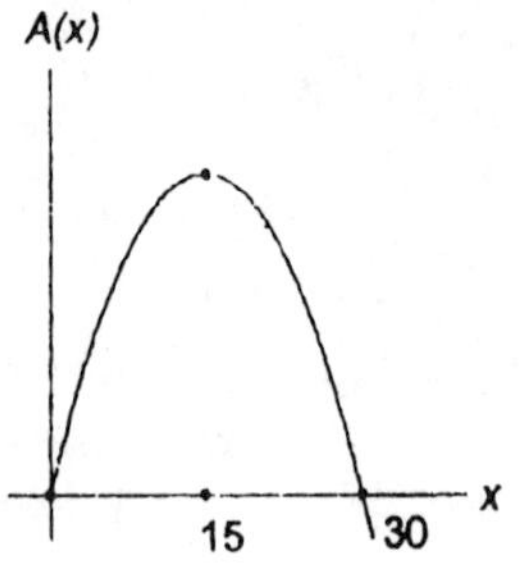

11. (a) Let p stand for the price of hamburgers and let x be the quantity. Using the point-slope equation,

$$p - 2 = \frac{2.4 - 2}{8000 - 10,000}(x - 10,000) \text{ or } p = -.0002x + 4.$$

The revenue function is thus

$R(x) = x(-.0002x + 4) = -.0002x^2 + 4x$, $R'(x) = -.0004x + 4$.

$R'(10,000) = 0$. The maximum value of $R(x)$ occurs at $x = 10,000$. The optimal price is thus, \$2.

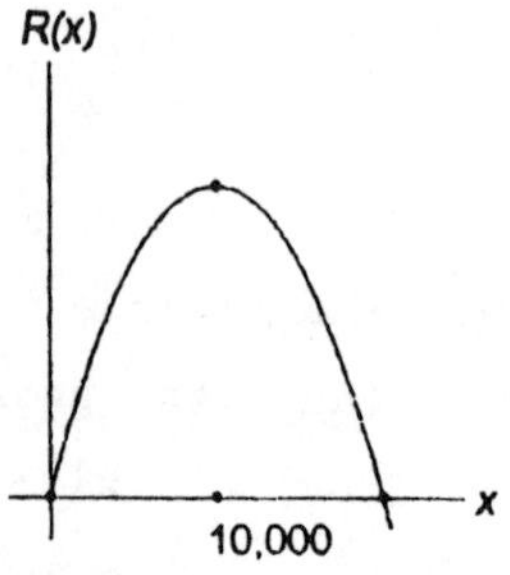

(b) The cost function is $C(x) = 1000 + .6x$, so the profit function is

$$\begin{aligned} P(x) &= R(x) - C(x) \\ &= -.0002x^2 + 4x - 1000 - .6x \\ &= -.0002x^2 + 3.4x - 1000. \end{aligned}$$

$P'(x) = -.0004x + 3.4$, $P'(8500) = 0$.

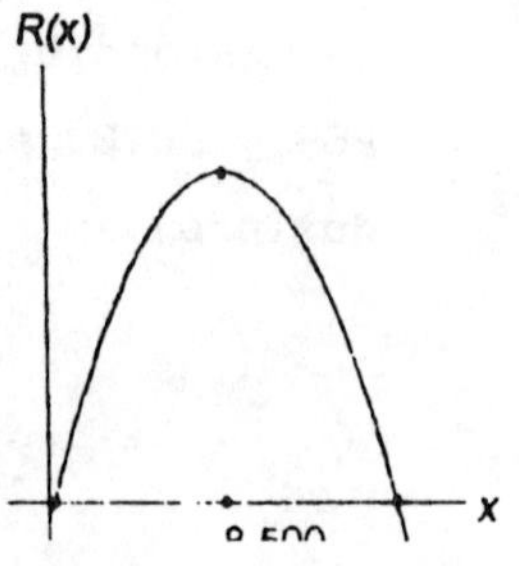

The maximum value of $P(x)$ occurs at $x = 8500$. The optimal price is $-.0002(8500) + 4 = \$2.30$.

12. Let $50 + x$ denote the ticket price and y the attendance. Since a \$2 increase in price lowers the attendance by 200, we have

$$y = 4000 - 100x$$

We now have

$$\begin{aligned}
\text{Revenue} = R &= [\text{price}] \times [\text{attendance}] \\
&= (50 + x)(4000 - 100x) \\
&= -100x^2 - 1000x + 20000 \\
R' &= -200x - 1000 = 0 \\
x &= -5 \\
R &= (50 - 5)(4000 - 100(-5)) \\
&= 202,500
\end{aligned}$$

Answer: Charge \$45 per ticket , Revenue = \$202,500

13. (a) $R(x) = x(60 - 10^{-5}x) = 60x - 10^{-5}x^2$; so the profit function is

$$\begin{aligned}
P(x) = R(x) - C(x) &= 60x - 10^{-5}x^2 - 7\cdot 10^6 - 30x \\
&= -10^{-5}x^2 + 30x - 7\cdot 10^6
\end{aligned}$$

$P'(x) = -2\cdot 10^{-5}x + 30$, $P'(1.5\cdot 10^6) = 0$.

The maximum value of $P(x)$ occurs at $x = 1.5\cdot 10^6$ (thousand kh).

The corresponding price is $p - 60 - 10^{-5}(1.5\cdot 10^6) = 45$.

This represents \$45,000/thousand kh.

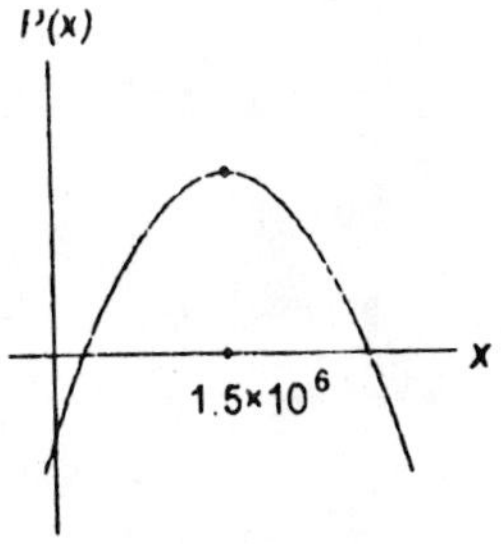

$G'(T) = 30 - \frac{1}{2}T$, $G'(60) = 0$.

The maximum value of G(T) occurs at T = 60. Thus a tax of $60/unit will maximize the government's tax revenue.

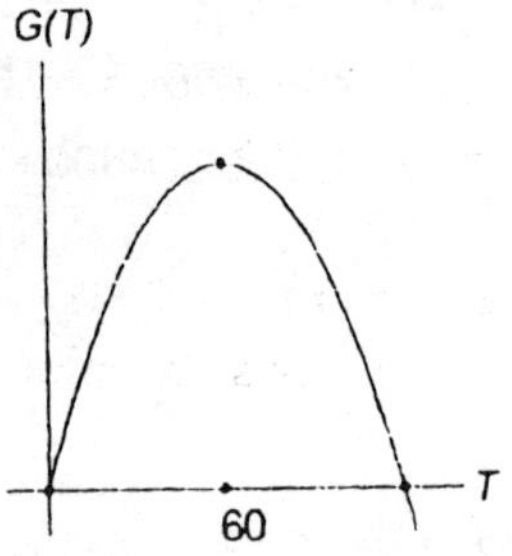

15. Total deposit is $1000000r. Total interest paid out is $1000000r^2$. Total interest received on the loans of 1000000r is 10000000r.

$$P = 10000000r - 1000000r^2$$

$$\frac{dP}{dr} = 10000000 - 2000000r$$

Set $\frac{dP}{dx} = 0$ and obtain r = 5%.

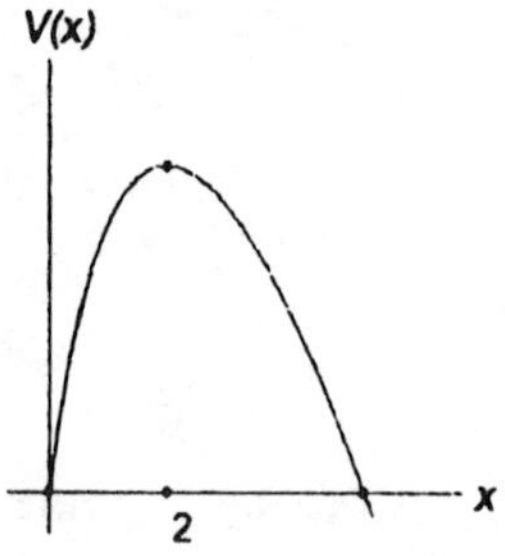

16. (a) P(0) is the profit with no advertising budget.

(b) As money is spent on advertising, the marginal profit initially increases. However, at some point the marginal profit begins to decrease.

(c) Additional money spent on advertising is most advantageous at the inflection point.

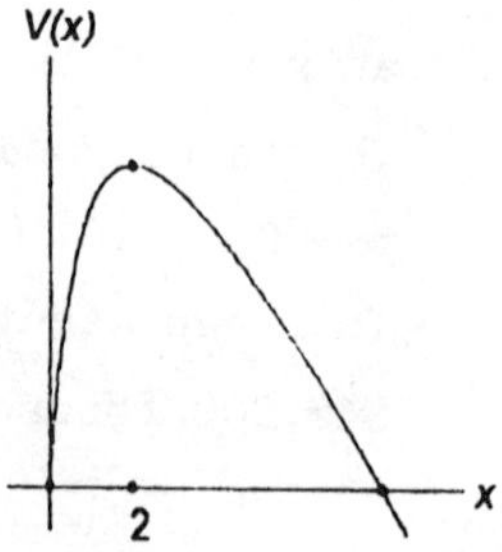

17.

a. \$79 thousand
b. \$2 thousand
c. 5
d. $f'(x) = -2$ has solution $x = 34$.
e. f has a maximum at $x = 28$.

18.

a. $f(14) = 1,000$
b. $f'(22) = 7$
c. $f(x) = 1300$ has solution $x = 50$.
d. $f'(x) = 17$ has solutions $x = 12$ and $x = 52$.
e. $f'(x)$ has a minimum at $x = 32$, marginal cost $= f'(32) = 5$, cost $= f(32) = 1150$

Chapter 2 Supplementary Exercises

1. (b) Because the population is increasing at an increasing rate.

2. (b) Because the velocity is increasing at an increasing rate.

3.

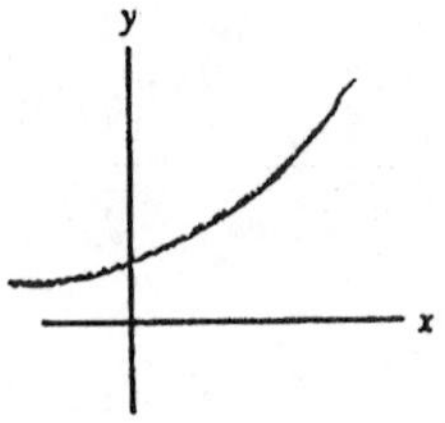

4.

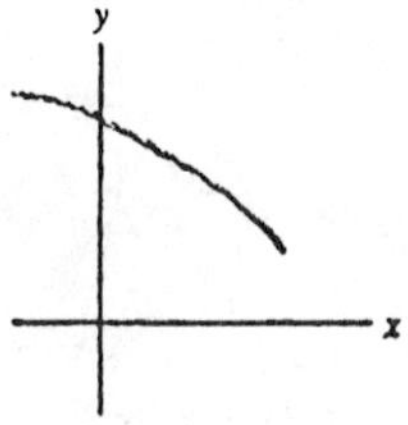

5.

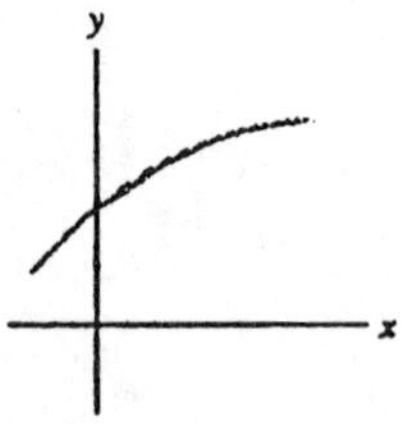

6

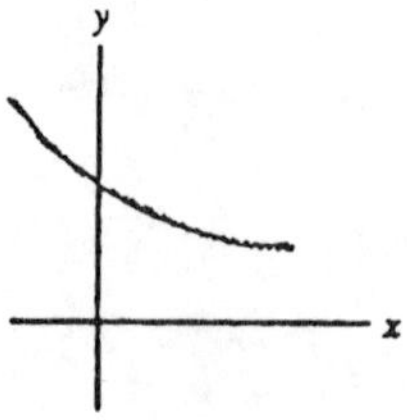

7. d, e 8. b 9. c, d 10. a 11. e 12. b

13. Graph goes through (1, 2), increasing at x = 1.

14. Graph goes through (1, 5), decreasing at x = 1.

15. Increasing and concave up at x = 3.

16. Decreasing and concave down at x = 2.

17. (10, 2) is a relative minimum.

18. Graph goes through (4, -2), increasing and concave down at x = 4.

19. Graph goes through (5, -1), decreasing at x = 5.

20. (0, 0) is a relative minimum.

21. (−2, 0) is a relative maximum.

22. Graph goes through (4, −2), increasing and concave down at $x = 4$.

23.

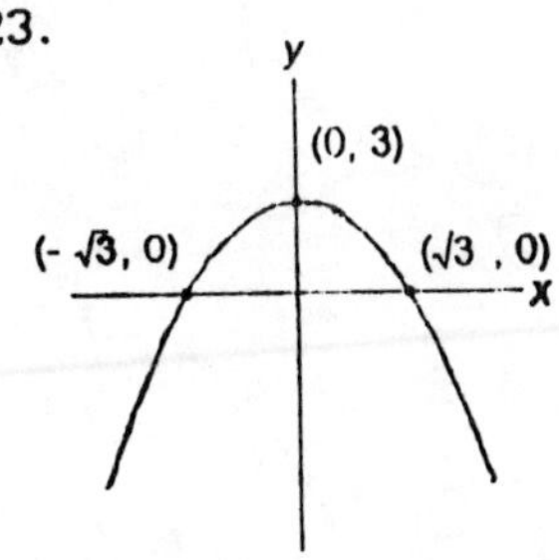

24.

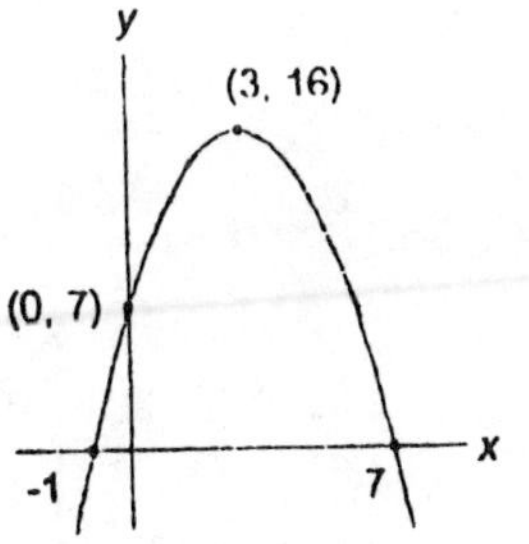

25.

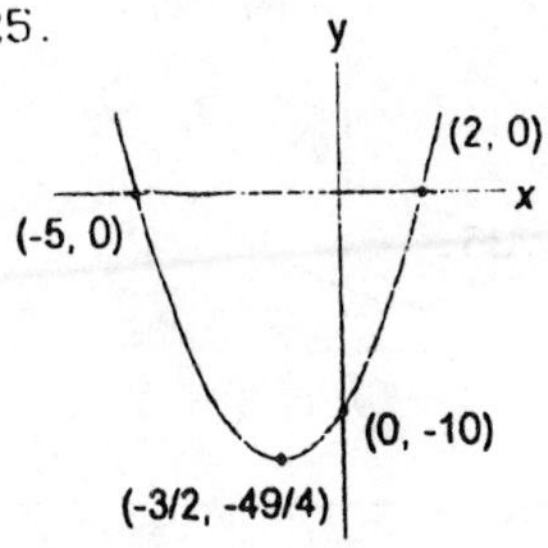

26.

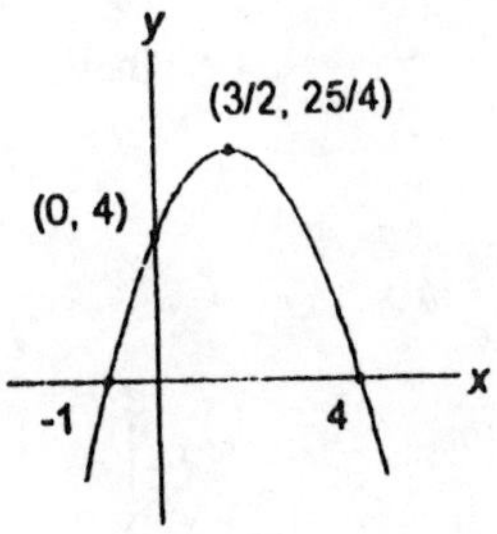

27.

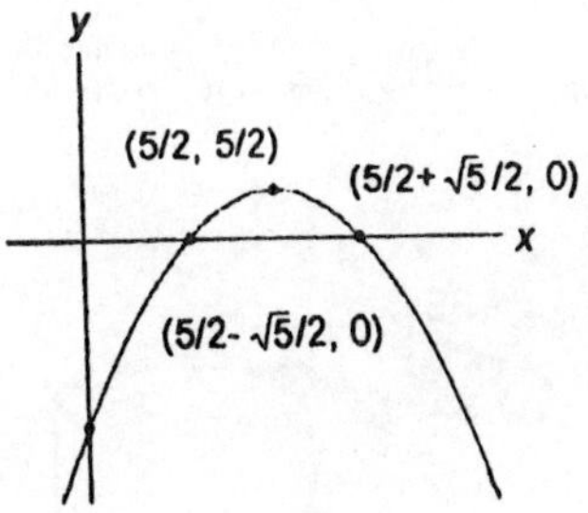

28.

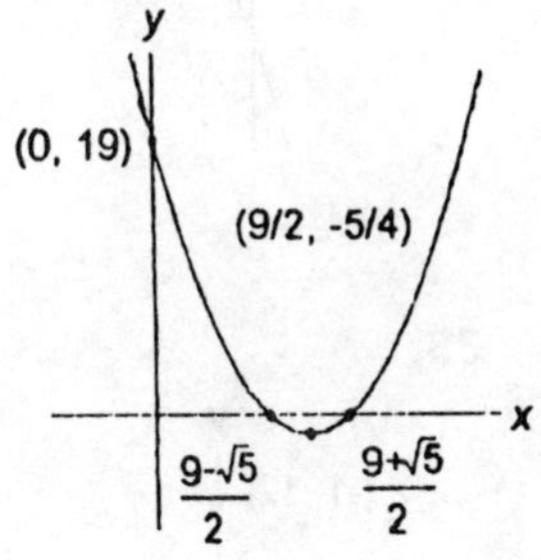

29.

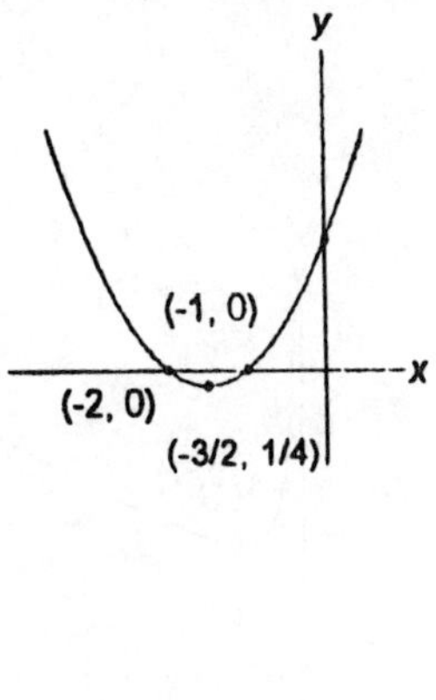

30.

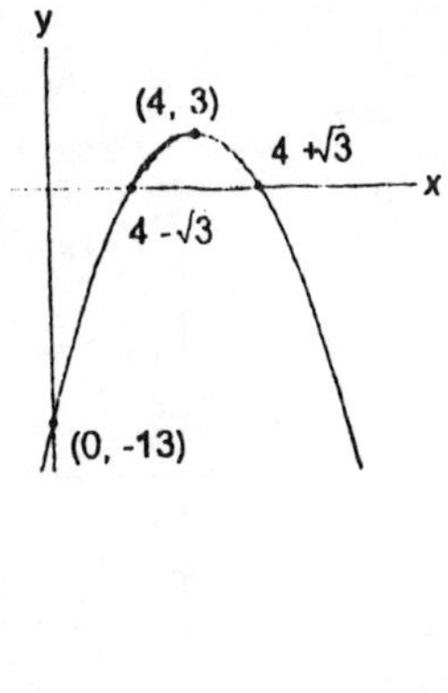

31.

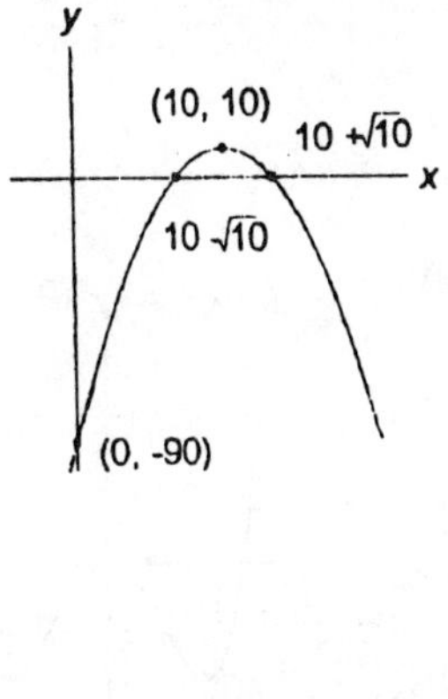

32.

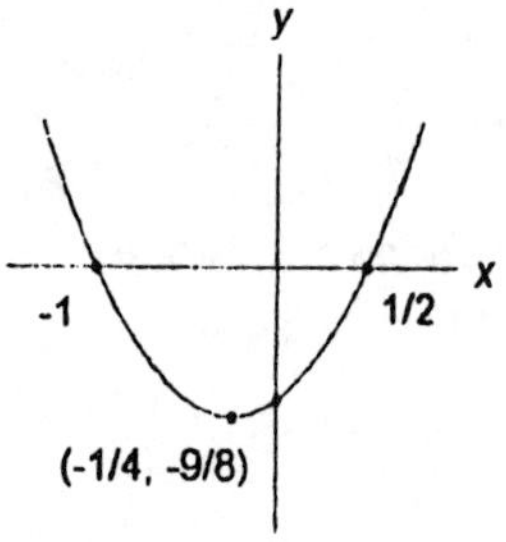

33.

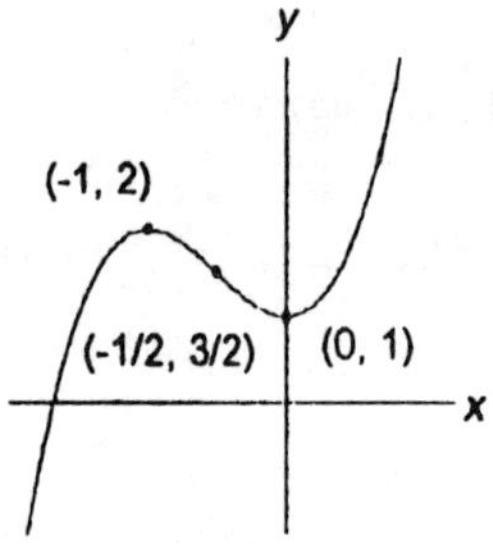

34.

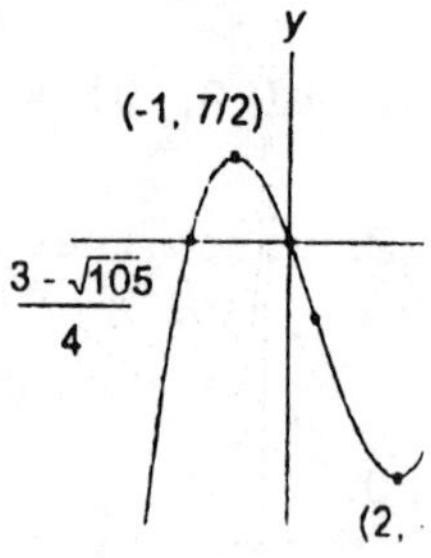

35.

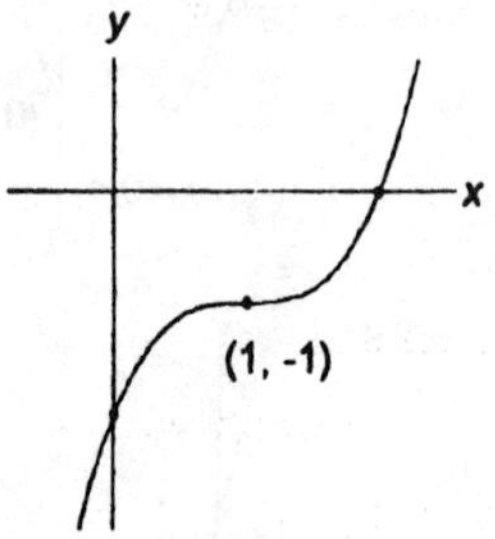

36.

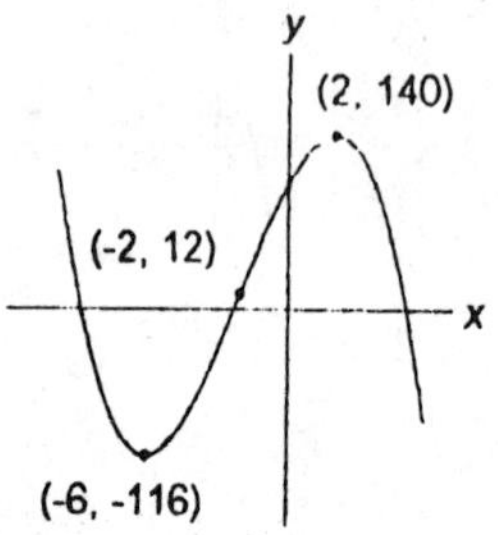

37.

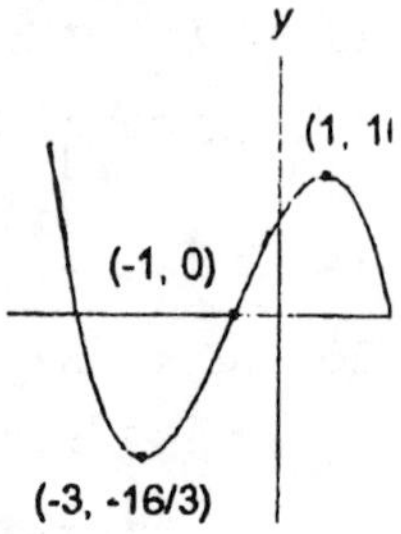

38.

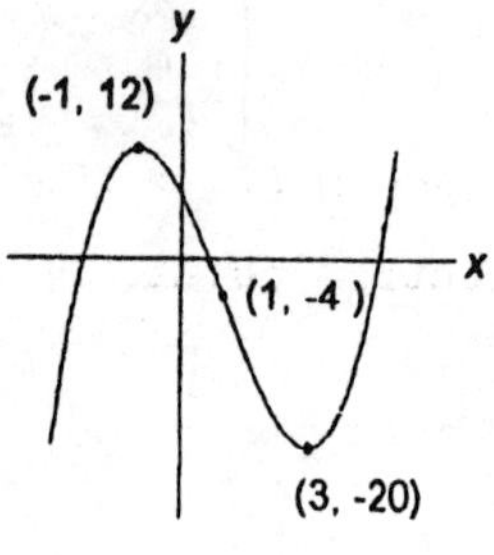

39.

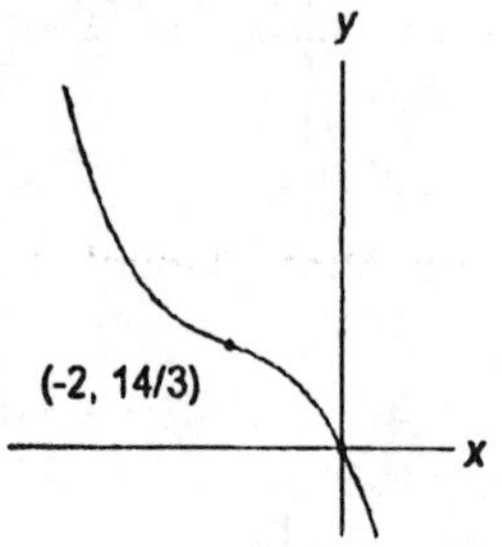

40.

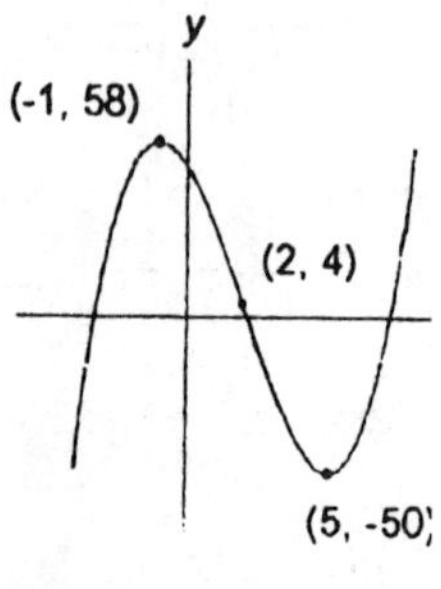

41.

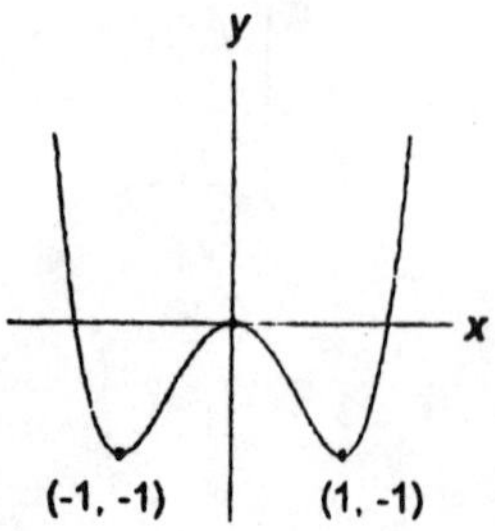

42.

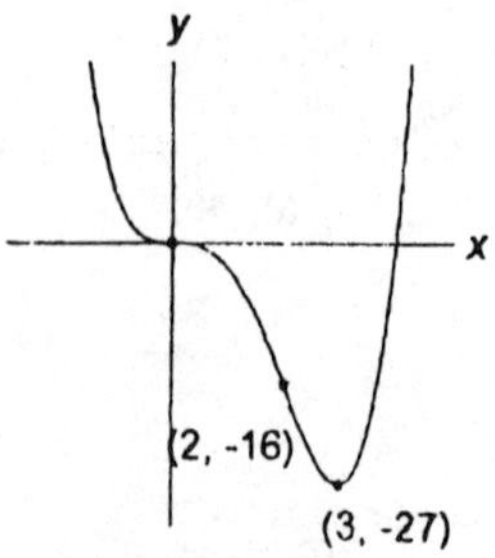

43.

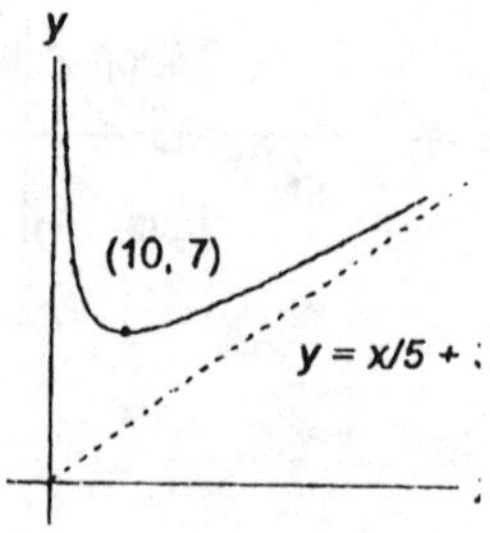

44.

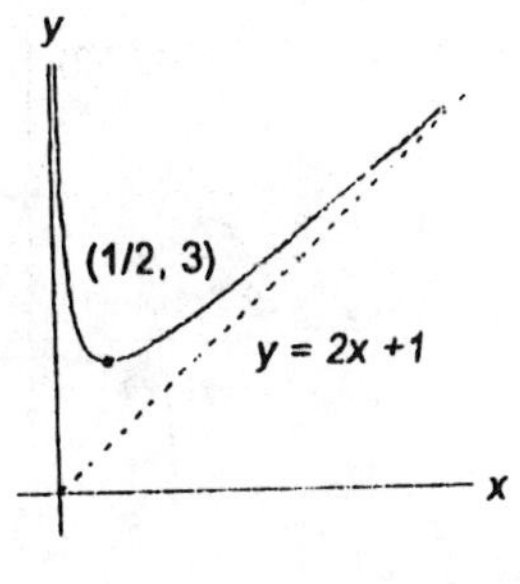

45. $f'(x) = \frac{3}{2}(x^2 + 2)^{1/2}(2x)$ Since $f'(0) = 0$, f has a possible extreme value at $x = 0$.

46. $f'(x) = \frac{3}{2}(2x^2 + 3)^{1/2}(4x)$; $2x^2 + 3 > 0$ for all x, so the sign of $f'(x)$ is determined by the sign of 4x. Therefore, $f'(x) > 0$, if $x > 0$, $f'(x) < 0$ if $x < 0$. This means that f(x) is decreasing for $x < 0$ and increasing for $x > 0$.

47. $f''(x) = \dfrac{-2x}{(1 + x^2)^2}$, so $f''(0) = 0$. Since $f'(x) > 0$ for all x, it follows that 0 must be an inflection point.

48. $f''(x) = \frac{1}{2}(5x^2 + 1)^{-1/2}(10x)$, so $f''(0) = 0$. Since $f'(x) > 0$ for all x, it follows that 0 must be an inflection point.

49. A - c, B - e, C - f, D - b, E - a, F - d

50. A - c, B - e, C - f, D - b, E - a, F - d

51. (a) The number of people living between 10 + 0 and 10 miles from the center of the city.

(b) If so, f(x) would be decreasing at x = 10.

52. $f(x) = \frac{1}{4}x^2 - x + 2 \quad (0 \le x \le 8)$

$f'(x) = \frac{1}{2}x - 1$, $f'(1) = 0$; $f''(x) = \frac{1}{2}$.

Since f′(1) is a relative minimum, the maximum value of f(x) must occur at one of the endpoints. f(0) = 2, f(8) = 10. 10 is the maximum value, attained x = 8.

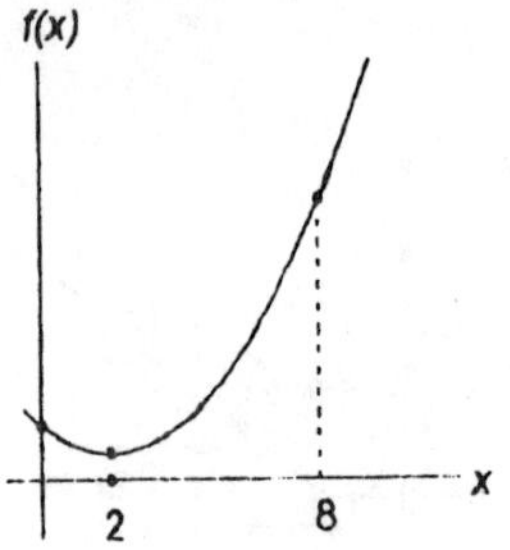

53. $f(x) = 2 - 6x - x^2 \quad (0 \le x \le 5)$

$f'(x) = -6 - 2x$; Since $f'(x) < 0$ for all $x > 0$, f(x) is decreasing on the interval [0, 5]. Thus, the maximum value occurs at x = 0. The maximum value is f(0) = 2.

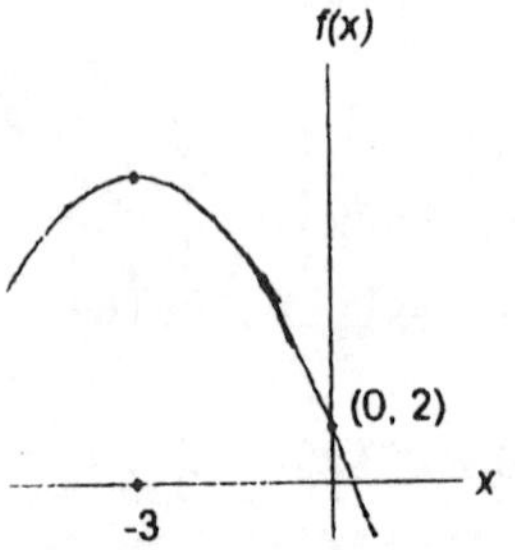

54. $g(t) = t^2 - 6t + 9 \quad (1 \le t \le 6)$

$g'(t) = 2t - 6$, $g'(3) = 0$

$g''(t) = 2$ The minimum value of g(t) is g(3) = 0.

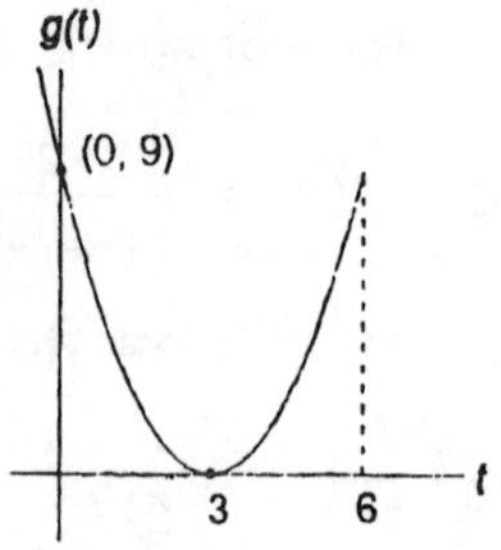

55. Let x and y be the other dimensions of the box. The objective is $S = 2xy + 4x + 8y$ and the constraint is $4xy = 200$, $y = \frac{50}{x}$.

Thus, $S(x) = 2x\left[\frac{50}{x}\right] + 4x + \frac{400}{x} = 100 + 4x + \frac{400}{x}$.

$S'(x) = 4 - \frac{400}{x^2}$; $S(10) = 0$.

The minimum value of S(x) for x > 0 occurs at x = 10. Thus, the dimensions of the box should be 10 x 5 x 4.

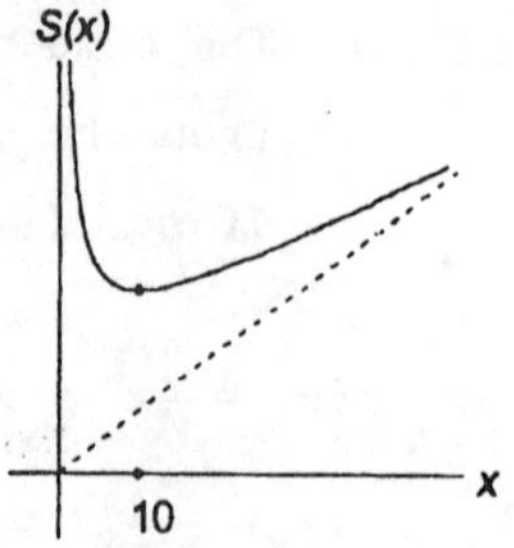

56. Let x be the length of the base of the box and let y be the other dimension. The objective is $V = x^2y$ and the constraint is $3x^2 + x^2 + 4xy = 48$, $y = \frac{48 - 4x^2}{4x} = \frac{12 - x^2}{x}$.

$V(x) = x^2 \cdot \frac{12 - x^2}{x} = 12x - x^3$; $V'(x) = 12 - 3x^2$, $V'(2) = 0$.

The maximum value of x for x > 0 occurs at x = 2. The optimal dimensions are thus 2 x 2 x 4 ft.

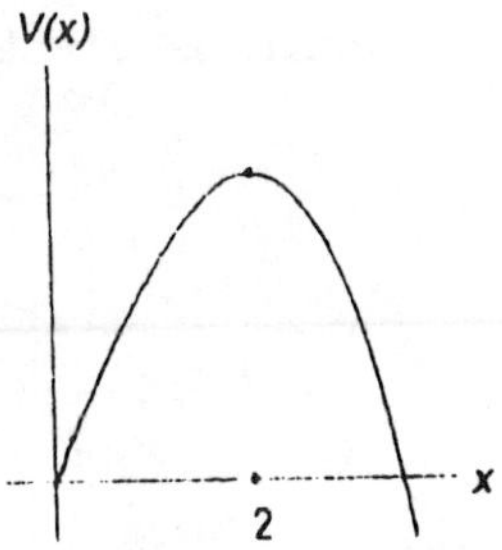

57. Let x be the number of inches turned up on each side of the gutter. The objective is $A(x) = (30 - 2x)x$ (A is the cross-sectional area of the gutter--maximizing this will maximize the volume). $A'(x) = 30 - 4x$, $A'\left(\frac{30}{4}\right) = 0$. $x = \frac{30}{4}$ inches gives the maximum value of A.

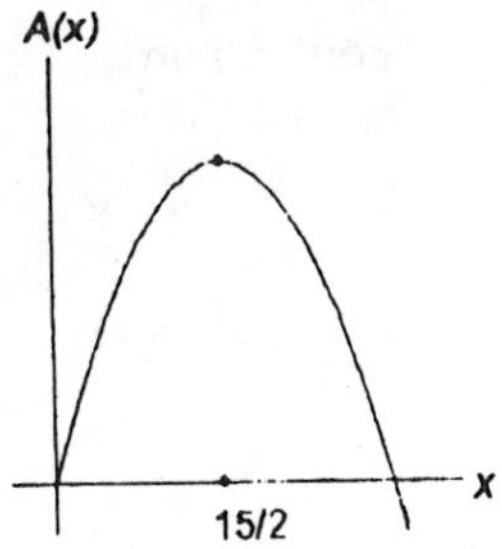

58. Let x be the number of trees planted. The objective is

$f(x) = \left[25 - \frac{1}{2}(x - 40)\right]x \quad (x \geq 40)$. $f(x) = 45x - \frac{1}{2}x^2$;

$f'(x) = 45 - x$, $f'(45) = 0$.

The maximum value of f(x) occurs at x = 40. Thus, 40 trees should be planted.

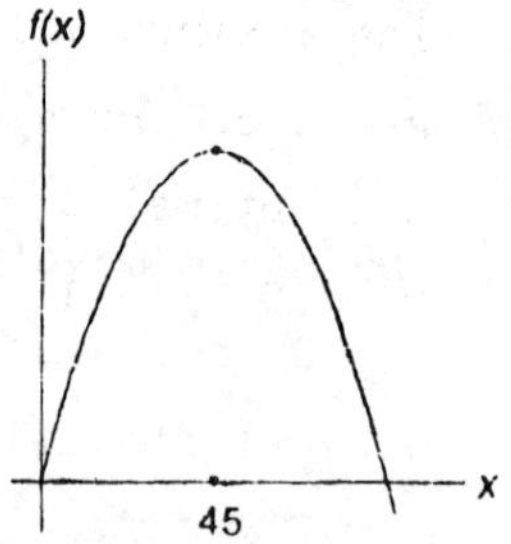

59. Let r be the number of production runs and let x be the lot size. Then the objective is $C = 1000r + .5\left(\frac{x}{2}\right)$ and the constraint is $rx = 400{,}000$, $r = \frac{400{,}000}{x}$, so

$C(x) = \frac{4 \cdot 10^8}{x} + \frac{x}{4}$; $C'(x) = \frac{-4 \cdot 10^8}{x^2} + \frac{1}{4}$, $C'(4 \cdot 10^4) = 0$.

The minimum value of C(x) for $x > 0$ occurs at $x = 4 \cdot 10^4$. Thus the economic lot size is 40,000 books/run.

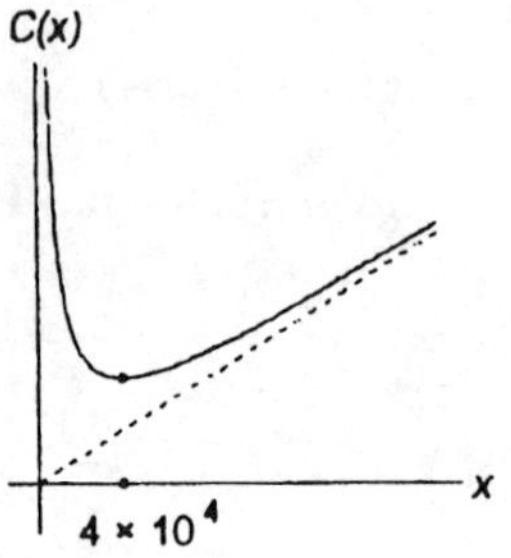

60. Let x be the width of the poster and let y be its height. Then the objective is $A = (x - 4)(y - 5)$ and the constraint is $xy = 125$, $y = \frac{125}{x}$.

$A(x) = (x - 4)\left(\frac{125}{x} - 5\right) = 125 - 5x - \frac{500}{x} + 20 = 145 - 5x - \frac{500}{x}$

$A'(x) = -5 + \frac{500}{x^2}$; $A'(10) = 0$

The maximum value of $A(x)$ for $x > 0$ occurs at $x = 10$. Thus the optimal dimensions are $x = 10$, $y = 12.5$ inches.

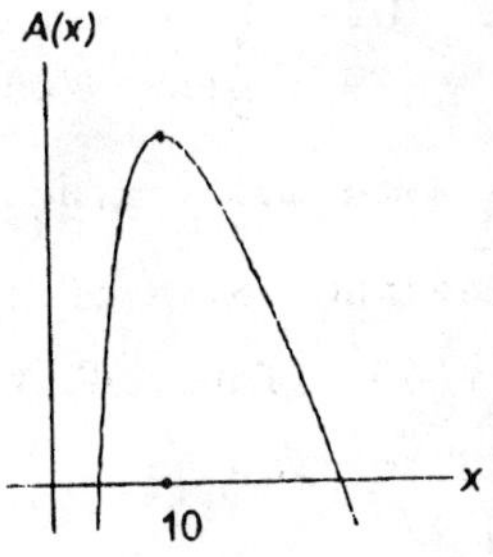

61. The revenue function is $R(x) = (150 - .02x)x = 150x - .02x^2$. Thus, the profit function is $P(x) = (150x - .02x^2) - (10x + 300) = -.02x^2 + 140x - 300$. $P'(x) = -.04x + 140$, $P'(3500) = 0$. The maximum value of $P(x)$ occurs at $x = 3500$.

Exercises 3.1

1. $\frac{d}{dx}\left[(x + 1)(x^3 + 5x + 2) \right] = (x + 1)(3x^2 + 5) + (x^3 + 5x + 2)$

$$= 4x^3 + 3x^2 + 10x + 7$$

2. $\frac{d}{dx}\left[(2x - 1)(x^2 - 3) \right] = (2x - 1)(2x) + (x^2 - 3)(2)$

$$= 6x^2 - 2x - 6$$

3. $\frac{d}{dx}\left[(3x^2 - x + 2)(2x^2 - 1) \right] = (3x^2 - x + 2)(4x) +$

$$(6x - 1)(2x^2 - 1)$$

$$= 24x^3 - 6x^2 + 2x + 1$$

4. $\frac{d}{dx}\left[(x^4 + 1)(3x + 5) \right] = (x^4 + 1)(3) + (3x + 5)(4x^3)$

$$= 15x^4 + 20x^3 + 3$$

5. $\frac{d}{dx}\left[(2x - 7)(x -1)^5 \right] = (2x - 7)(5)(x - 1)^4 + (x - 1)^5(2)$

$$= (x - 1)^4(12x - 37)$$

6. $\frac{d}{dx}\left[(x + 1)^2(x - 3)^3 \right] = (x + 1)^2(3)(x - 3)^2 +$

$$(x - 3)^3(2)(x + 1)$$

$$= (x + 1)(x - 3)^2(5x - 3)$$

7. $\frac{d}{dx}\left[(x^2 + 3)(x^2 - 3)^{10} \right] = (x^2 + 3)(10)(x^2 - 3)^9 +$

$$(x^2 - 3)^{10}(2x)$$

$$= 2x(x^2 - 3)^9(11x^2 + 27)$$

8. $\frac{d}{dx}\left[(x^2 - 4)(x^2 + 3)^6 \right] = (x^2 - 4)(6)(x^2 + 3)^5 + (x^2 +3)^6(2x)$

$$= 2x(x^2 + 3)^5(7x^2 - 21)$$

9. $$\frac{d}{dx}\left[\frac{1}{3}(4 - x)^3(4 + x)^3\right] = \frac{1}{3}\Big[(4 - x)^3(3)(4 + x)^2 + (4 + x)^3(3)(4 - x)^2(-1)\Big]$$
$$= -2x(4 - x)^2(4 + x)^2$$

10. $$\frac{d}{dx}\left[\frac{1}{6}(3x + 1)^4(4x - 1)^3\right] = \frac{1}{6}\Big[(3x + 1)^4(3)(4x - 1)^2(4) + (4x - 1)^3(4)(3x + 1)^3(3)\Big]$$
$$= 14x(3x + 1)^3(4x - 1)^2$$

11. $$\frac{d}{dx}\left[\frac{4 - x}{4 + x}\right] = \frac{(4 + x)(-1) - (4 - x)(1)}{(4 + x)^2} = \frac{-8}{(4 + x)^2}$$

12. $$\frac{d}{dx}\left[\frac{x - 2}{x + 2}\right] = \frac{(x + 2)(1) - (x - 2)(1)}{(x + 2)^2} = \frac{4}{(x + 2)^2}$$

13. $$\frac{d}{dx}\left[\frac{x^2 - 1}{x^2 + 1}\right] = \frac{(x^2 + 1)(2x) - (x^2 - 1)(2x)}{(x^2 + 1)^2} = \frac{4x}{(x^2 + 1)^2}$$

14. $$\frac{d}{dx}\left[\frac{1 + x^3}{1 - x^3}\right] = \frac{(1 - x^3)(3x^2) - (1 + x^3)(-3x^2)}{(1 - x^3)^2} = \frac{6x^2}{(1 - x^3)^2}$$

15. $$\frac{d}{dx}\left[\frac{1}{5x^2 + 2x + 5}\right] = \frac{(5x^2 + 2x + 5)(0) - (1)(10x + 2)}{(5x^2 + 2x + 5)^2}$$
$$= \frac{-10x - 2}{(5x^2 + 2x + 5)^2}$$

16. $$\frac{d}{dx}\left[\frac{2x - 1}{x}\right] = \frac{x(2) - (2x - 1)(1)}{x^2} = \frac{1}{x^2} \text{ or } x^{-2}$$

17. $$\frac{d}{dx}\left[\frac{x^2 + 2x}{x + 1}\right] = \frac{(x + 1)(2x + 2) - (x^2 + 2x)(1)}{(x + 1)^2} = \frac{x^2 + 2x + 2}{(x + 1)^2}$$

18. $$\frac{d}{dx}\left[\frac{3x^3 - 2x}{x + 3}\right] = \frac{(x + 3)(6x - 2) - (3x^2 - 2x)(1)}{(x + 3)^2}$$
$$= \frac{3x^2 + 18x - 6}{(x + 3)^2}$$

19. $$\frac{d}{dx}\left[\frac{3x^2 + 5x + 1}{3 - x^2}\right] = \frac{(3 - x^2)(6x + 5) - (3x^2 + 5x + 1)(-2x)}{(3 - x^2)^2}$$
$$= \frac{5(x + 1)(x + 3)}{(3 - x^2)^2}$$

20. $$\frac{d}{dx}\left[\frac{1}{x^2 + 1}\right] = \frac{d}{dx}\left[(x^2 + 1)^{-1}\right] = (-1)(x^2 + 1)^{-2}(2x)$$
$$= \frac{-2x}{(x^2 + 1)^2}$$

21. $$\frac{d}{dx}\left[\frac{x}{(x^2 + 1)^2}\right] = \frac{(x^2 + 1)^2(1) - x(2)(x^2 + 1)(2x)}{(x^2 + 1)^4}$$
$$= \frac{1 - 3x^2}{(x^2 + 1)^3}$$

22. $$\frac{d}{dx}\left[\frac{x - 1}{(2x + 1)^2}\right] = \frac{(2x + 1)^3(1) - (x - 1)(2)(2x + 1)(2)}{(2x + 1)^4}$$
$$= \frac{5 - 2x}{(2x + 1)^3}$$

23. $$\frac{d}{dx}\left[\frac{(x - 1)^4}{(x + 2)^2}\right] = \frac{(x + 2)^2(4)(x - 1)^3 - (x - 1)^4(2)(x + 2)}{(x + 2)^4}$$
$$= \frac{2(x - 1)^3(x + 5)}{(x + 2)^3}$$

24. $\frac{d}{dx}\left[\frac{(x+3)^3}{(x+4)^2}\right] = \frac{(x+4)^2(3)(x+3)^2 - (x+3)^3(2)(x+4)}{(x+4)^4}$

$= \frac{(x+3)^2(x+6)}{(x+4)^3}$

25. $\frac{d}{dx}\left[\frac{x^4 - 4x^2 + 3}{x}\right] = \frac{d}{dx}\left[x^3 - 4x + \frac{3}{x}\right] = 3x^2 - 4 - \frac{3}{x^2}$

26. $\frac{d}{dx}\left[(x^2+9)\left(x - \frac{3}{x}\right)\right] = (x^2+9)\left(1 + \frac{3}{x^2}\right) + \left(x - \frac{3}{x}\right)(2x)$

$= 3x^2 + 6 + \frac{27}{x^2}$

27. $\frac{d}{dx}\left[2\sqrt{x}\,(3x^2-1)^3\right] = 2\sqrt{x}\,(3)(3x^2-1)^2(6x) +$

$(3x^2-1)^3(2)\left(\frac{1}{2}\right)x^{-1/2}$

$= \frac{(3x^2-1)^2(39x^2-1)}{\sqrt{x}}$

28. $\frac{d}{dx}\left[4\sqrt{x}\,(5x-1)^4\right] = 4\sqrt{x}\,(4)(5x-1)^3(5) +$

$(5x-1)^4(4)\left(\frac{1}{2}\right)x^{-1/2}$

$= \frac{2(5x-1)^3(45x-1)}{\sqrt{x}}$

29. $\frac{d}{dx}\left[(x+3)\sqrt{2x-3}\right] = (x+3)\left(\frac{1}{2}\right)(2x-3)^{-1/2}(2) +$

$(2x-3)^{1/2}(1)$

$= \frac{3x}{\sqrt{2x-3}}$

30. $\frac{d}{dx}\left[(2x - 5)(4x - 1)^{1/2} \right] = (2x - 5)\left(\frac{1}{2}\right)(4x - 1)^{-1/2}(4) +$

$(4x - 1)^{1/2}(2)$

$= \frac{12(x - 1)}{\sqrt{4x - 1}}$

31. $y = (x - 2)^5(x + 1)^2$

$\frac{dy}{dx} = (x - 2)^5(2)(x + 1) + (x + 1)^2(5)(x - 2)^4$

$= (x - 2)^4(x + 1)(7x - 5)$

$\left.\frac{dy}{dx}\right|_{x=3} = 88.$ Equation of tangent line: $y - 16 = 88(x - 3)$

32. $y = (x + 1)/(x - 1)$

$\frac{dy}{dx} = \frac{(x - 1)(1) - (x + 1)(1)}{(x - 1)^2} = \frac{-2}{(x - 1)^2}$

$\left.\frac{dy}{dx}\right|_{x=2} = -2.$ Equation of tangent line: $y - 3 = -2(x - 2)$

33. $y = (x^2 - 4)^3(2x + 5)^5$

$\frac{dy}{dx} = (x^2 - 4)^3(5)(2x^2 + 5)^4(4x) + (2x^2 + 5)^5(3)(x^2 - 4)^2(2x)$

$= (2x)(x^2 - 4)^2(2x^2 + 5)(16x^2 - 25)$

$\frac{dy}{dx} = 0$ for $x = 0, \pm 2, \pm\frac{5}{4}.$

34. $y = (3x - 8)^4(2x - 3)^3$

$\frac{dy}{dx} = (3x - 8)^4(3)(2x - 3)^2(2) + (2x - 3)^3(4)(3x - 8)^3(3)$

$= 6(3x - 8)^3(2x - 3)^2(7x - 14)$

$\frac{dy}{dx} = 0$ for $x = \frac{8}{3}, \frac{3}{2}, 2.$

35. $y = (x - 2)^5/(x - 4)$

$$\frac{dy}{dx} = \frac{(x - 4)^3(5)(x - 2)^4 - (x - 2)^5(3)(x - 4)^2}{(x - 4)^6}$$

$$= \frac{(x - 2)^4(2x - 14)}{(x - 4)^4}$$

The tangent line is horizontal when $\frac{dy}{dx} = 0$. This happens for $x = 2$ or 7.

36. $y = x^4/(x - 1)$

$$\frac{dy}{dx} = \frac{(x - 1)(4x^3) - x^4(1)}{(x - 1)^2} = \frac{3x^4 - 4x^3}{(x - 1)^2}$$

$\frac{dy}{dx} = 0$ for $x = 0, \frac{4}{3}$.

37. $y = (x^2 + 3x - 1)/x$

$$\frac{dy}{dx} = \frac{x(2x + 3) - (x^2 + 3x + 1)(1)}{x^2} = \frac{x^2 - 1}{x^2}$$

$\frac{dy}{dx} = 5$ for $x = \pm\frac{1}{2}$. Points on the graph are $\left(\frac{1}{2}, \frac{3}{2}\right), \left(-\frac{1}{2}, \frac{9}{2}\right)$.

38. $y = (2x^4 + 1)(x - 5)$

$$\frac{dy}{dx} = (2x^4 + 1)(1) + (x - 5)(8x^3) = 10x^4 - 40x^3 + 1$$

$\frac{dy}{dx} = 1$ when $x^3(10x - 40) = 0$, i.e. for $x = 0, 4$.

The points on the graph are $(0, -5)$, $(4, -513)$.

39. Let x be the width of the box and let h be its height.

Constraint: $16 = \underbrace{2xh + 2(3h)}_{\text{sides}} + \underbrace{3x}_{\text{bottom}}$ $\quad \left(h = \frac{16 - 3x}{2x + 6}\right)$

Objective: $V = 3xh = 3x\left(\frac{16 - 3x}{2x + 6}\right) = \frac{48x - 9x^2}{2x + 6}$

$$\frac{dV}{dx} = \frac{(2x + 6)(48 - 18x) - (48x - 9x^2)(2)}{(2x + 6)^2}$$

$$= \frac{-18(x + 8)(x - 2)}{(2x + 6)^2} \text{ or } \frac{-18(x^2 + 6x - 16)}{(2x + 6)^2}$$

$$\frac{d^2V}{dx^2} = -18\left[\frac{(2x + 6)^2(2x + 6) - (x^2 + 6x - 16)(2)(2x + 6)(2)}{(2x + 6)^4}\right]$$

$$-18\left(\frac{16x + 100}{(2x + 6)^4}\right)$$

The maximum value of V occurs at $x = 2$. Answer: $x = 2$, $h = 1$ i.e. optimal dimensions are $3 \times 2 \times 1$.

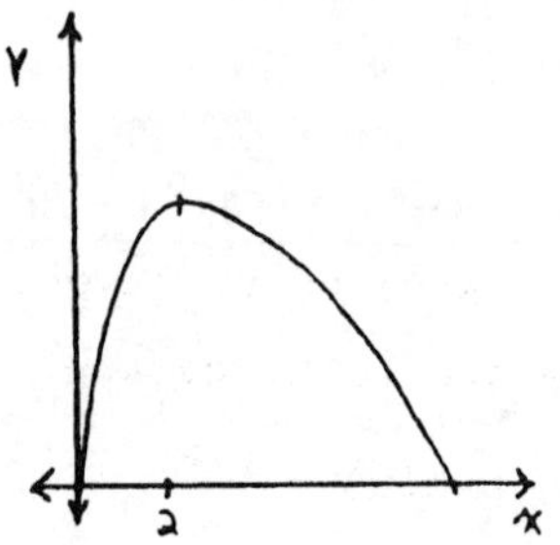

40. Let x be the length of the other side of the box and let h be its height.

Constraint: $240 = (1)x(20) + (1)x(10) + 2xh(10) + (2)(1)h(10)$

$$= 30x + 20xh + 20h \qquad \left(h = \frac{24 - 3x}{2x + 2}\right)$$

Objective: $V = xh = x\left(\frac{24 - 3x}{2x + 2}\right) = \frac{24x - 3x^2}{2x + 2}$

$$\frac{dV}{dx} = \frac{(2x + 2)(24 - 6x) - (24x - 3x^2(2)}{(2x + 2)^2}$$

$$= \frac{-6(x + 4)(x - 2)}{(2x + 2)^2} \text{ or } \frac{-6(x^2 + 2x - 8)}{(2x + 2)^2}$$

$$\frac{d^2V}{dx^2} = -6\left[\frac{(2x + 2)^2(2x + 2) - (x^2 + 2x - 8)(2)(2x + 2)(2)}{(2x + 2)^4}\right]$$

$$= -6\left(\frac{36}{(2x + 2)^3}\right)$$

The maximum value for V occurs at $x = 2$. Answer: $x = 2$, $h = 3$, i.e. optimal dimensions are $1 \times 2 \times 3$.

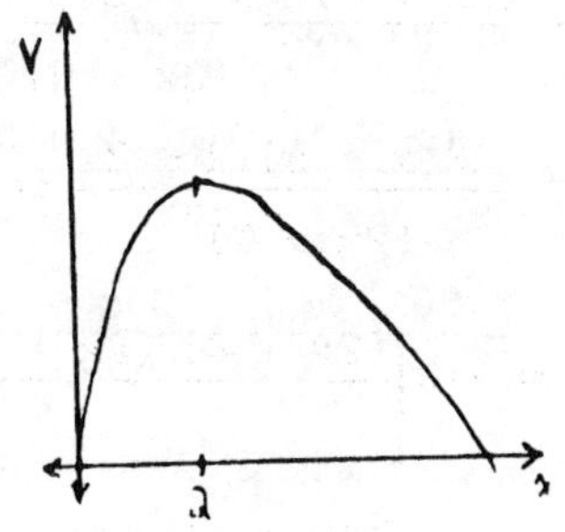

41. Let A(x) be the average cost of producing x units and let C(x) be the total cost.

Then $A(x) = \dfrac{C(x)}{x} = \dfrac{.1x^2 + 5x + 2250}{x}$.

$$A'(x) = \frac{x(.2x + 5) - (.1x^2 + 5x + 2250)(1)}{x^2} = \frac{.1x^2 - 2250}{x^2}$$

$$= \frac{(.1)(x + 150)(x - 150)}{x^2}$$

$$A''(x) = \frac{x^2(.2x) - (-.1x^2 - 2250)(2x)}{x^4} = \frac{.2x^3 - .2x^2 + 4500x}{x^4}$$

The minimum value of A(x) occurs at $x = 150$. The marginal cost function is $C'(x) = .2x + 5$. At the level of 150 units, we have $A(150) = 35 = .2(150) + 5 = C'(150)$.

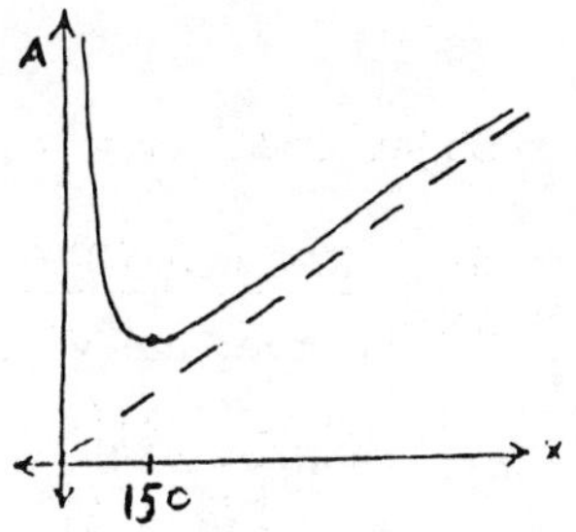

42. The cost function is $C(x) = 50x(x + 200)/(x + 100)$

$$= \frac{50x^2 + 10{,}000x}{(x + 100)}$$

The average cost is $AC = \frac{C(x)}{x} = \frac{50x^2 + 10,000x}{x(x + 100)} = \frac{50x + 10,000}{x + 100}$

$$C'(x) = \frac{(x + 100)(100x + 10,000) - (50x^2 + 10,000x)(1)}{(x + 100)^2}$$

$$= \frac{50x^2 + 10,000x + 1,000,000}{(x + 100)^2} > 0 \text{ for all } x > 0.$$

Thus cost increases as output increases.

$$AC' = \frac{(x + 100)(5) - (50x + 10,000)(1)}{(x + 100)^2}$$

$$= \frac{-5000}{(x + 100)^2} < 0 \text{ for all } x.$$

Thus the average cost decreases as output increases.

43. Recall that the marginal revenue, MR, is defined by $MR = R'(x)$. The average revenue is maximized when $\frac{d}{dx}(AR) = 0$.

$$\frac{d}{dx}(AR) = \frac{d}{dx}\left[\frac{R(x)}{x}\right] = \frac{xR'(x) - R(x)(1)}{x^2}$$

If $\frac{xR(x) - R(x)}{x^2} = 0$ then $xR'(x) - R(x) = 0$

$$xR'(x) = R(x)$$

$$R(x) = \frac{R(x)}{x}$$

i.e. $MR = AR$

44. If the average velocity, $\bar{v}(t) = s(t)/t$, is maximized at time t_0, then $\bar{v}'(t_0) = 0$. Using the quotient rule,

$$\bar{v}'(t) = \frac{d}{dt}\left[\frac{s(t)}{t}\right] = \frac{ts'(t) - s(t)(1)}{t^2}$$

Thus if $\bar{v}'(t_0) = 0$, we must have $t_0 s'(t) = s(t_0)$ or

$$s'(t_0) = \frac{s(t_0)}{t_0}.$$

45. $A = w(t)L(t)$: $\frac{dA}{dt} = w(t)L'(t) + w'(t)L(t) = 5.4 + 3.6 = 38$

46. $f(x) = \frac{3x}{105 - x}$, $f'(x) = \frac{3(105 - x) - 3x(-1)}{(105 - x)^2}$

$$= \frac{315}{(105 - x)^2}$$

at $x = 90$, $f'(x) = 1.4$

47. $\frac{d}{dx}\left[\frac{x^4}{x^2 + 1}\right] = \frac{(x^2 + 1)2x^3 - x^4(2x)}{(x^2 + 1)^2} = \frac{2x^5 - 2x^3 - 2x^5}{(x^2 + 1)^2}$

$$= \frac{-2x^3}{(x^2 + 1)^2} \neq \frac{4x^3}{2x} = 2x^2$$

48. $(5x^3)(2x^4) = 10x^7$: $\frac{d}{dx}\left[10x^7\right] = 70x^6 \neq (15x^2)(8x^3) = 90x^5$.

49. $\frac{d}{dx}\left[\frac{f(x)}{1 + x^2}\right] = \frac{(1 + x^2)f'(x) - f(x)\frac{d}{dx}(1 + x^2)}{(1 + x^2)^2}$

$$= \frac{(1 + x^2)f'(x) - f(x)2x}{(1 + x^2)^2}$$

Substituting the value $f'(x) = \frac{1}{1 + x^2}$ into the derivative we have

$$\frac{d}{dx}\left[\frac{f(x)}{1 + x^2}\right] = \frac{(1 + x^2)\left(\frac{1}{1 + x^2}\right) - f(x)2x}{(1 + x^2)^2} = \frac{1 - 2xf(x)}{(1 + x^2)^2}.$$

50. $\frac{d}{dx}\left[f(x)g(x)\right]\Big|_{x=1} = f(1)g'(1) + g(1)f'(1) = (2)(5) + (4)(3)$

$$= 22$$

51. $\frac{d}{dx}\left[\frac{f(x)}{g(x)}\right]\Bigg|_{x=1} = \frac{g(1)f'(1) - f(1)g'(1)}{g(1)^2} = \frac{(4)(3) - (2)(5)}{4^2} = \frac{1}{8}$

52. $\frac{d}{dx}\left[xf(x) - x\right] = xf'(x) + f(x) - 1 = x\left(\frac{1}{x}\right) + f(x) - 1$

$= 1 + f(x) - 1 = f(x)$

53. $f(x) = \frac{1}{x}$, $g(x) = x^3$

(a) Using the product rule, $\frac{d}{dx}\left[\left(\frac{1}{x}\right)x^3\right] = \frac{1}{x}(3x^2) + x^3(-1)x^{-2}$

$= 3x - x$

$= 2x$

$= \frac{d}{dx}\left[x^2\right]$

(b) $f'(x) = -\frac{1}{x^2}$ and $g'(x) = 3x^2$, so $f'(x)g'(x) = -3$.

Now $f(x)g(x) = \left(\frac{1}{x}\right)x^3 = x^2$ which have derivative $2x$.

$f'(x)g'(x) \neq (f(x)g(x))'$.

54. Using the quotient rule, $\frac{d}{dx}\left[\frac{x^3 - 4x}{x}\right] = \frac{x(3x^2 - 4) - (x^3 - 4x)}{x^2}$

$= \frac{2x^2}{x^2} = 2x.$

55. $\frac{d}{dx}\left[f(x)g(x)h(x)\right] = \frac{d}{dx}\left[f(x)\left(g(x)h(x)\right)\right]$

$= f(x)\frac{d}{dx}\left[g(x)h(x)\right] + \left(g(x)h(x)\right)f'(x)$

$= f(x)\left(g(x)h'(x) + h(x)g'(x)\right) + g(x)h(x)f'(x)$

$= f(x)g(x)h'(x) + f(x)g'(x)h(x) + f'(x)g(x)h(x)$

56. Let f and g be differentiable functions. From the identity in the statement of the problem,

$$\frac{d}{dx}\left[\, fg \,\right] = \frac{d}{dx}\left[\frac{1}{4}((f + g)^2 - (f - g)^2)\right]$$
$$= \frac{1}{4}\left[\frac{d}{dx}(f + g)^2 - \frac{d}{dx}(f - g)^2\right]$$

and by the special case

$$= \frac{1}{4}\left[2(f + g)(f' + g') - 2(f - g)(f' - g')\right]$$
$$= \frac{1}{2}\left[ff' + fg' + f'g + gg' - ff' + fg' + f'g - gg'\right]$$
$$= fg' + f'g$$

Exercises 3.2

1. $f(x) = \frac{x}{x + 1}$, $g(x) = x^3$; $f(g(x)) = \frac{x^3}{x^3 + 1}$

2. $f(x) = x\sqrt{x + 1}$, $g(x) = x^4$; $f(g(x)) = x^4\sqrt{x^4 + 1}$

3. $f(x) = x^5 + 3x$, $g(x) = x^2 + 4$; $f(g(x)) = (x^2 + 4)^5 + 3(x^2 + 4)$

4. $f(x) = \frac{x}{(x + 1)^4}$, $g(x) = 5 - 3x$

$$f(g(x)) = \frac{5 - 3x}{(5 - 3x + 1)^4} = \frac{5 - 3x}{(6 - 3x)^4}$$

5. $f(g(x)) = (x^3 + 8x - 2)^5$; $f(x) = x^5$, $g(x) = x^3 + 8x - 2$

6. $f(g(x)) = (9x^2 + 2x - 5)^7$; $f(x) = x^7$, $g(x) = 9x^2 + 2x - 5$

7. $f(g(x)) = \sqrt{4 - x^2}$; $f(x) = \sqrt{x}$, $g(x) = 4 - x^2$

8. $f(g(x)) = (5x^2 + 1)^{-1/2}$; $f(x) = x^{-1/2}$, $g(x) = 5x^2 + 1$

9. $f(g(x)) = \dfrac{1}{x^3 - 5x^2 + 1}$; $f(x) = \dfrac{1}{x}$, $g(x) = x^3 - 5x^2 + 1$

10. $f(g(x)) = (4x - 3)^3 + \dfrac{1}{4x - 3}$; $f(x) = x^3 + \dfrac{1}{x}$, $g(x) = 4x - 3$

11. $\dfrac{d}{dx}(x^2 + 5)^{15} = 15(x^2 + 5)^{14}(2x) = 30x(x^2 + 5)^{14}$

12. $\dfrac{d}{dx}(x^4 + x^2)^{10} = 10(x^4 + x^2)^9(4x^3 + 2x)$

13. $\dfrac{d}{dx}6x^2(x - 1)^3 = 12x(x - 1)^3 + 3(x - 1)^2(1)6x^2$

$= 12x(x - 1)^3 + 18x^2(x - 1)^2$

14. $\dfrac{d}{dx}5x^3(2 - x)^4 = 15x^2(2 - x)^4 + 4(2 - x)^3(-1)5x^3$

$= 15x^2(2 - x)^4 - 20x^2(2 - x)^3$

15. $\dfrac{d}{dx}\left[2(x^3 - 1)(3x^2 + 1)^4\right] = 2\left[3x^2(3x^2 + 1)^4 + 4(3x^2 + 1)^3(6x)\right]$

$= 6x^2(3x^2 + 1)^4 + 48x(3x^2 + 1)$

16. $\dfrac{d}{dx}\left[2(2x - 1)^{5/4}(2x + 1)^{3/4}\right]$

$= 2\left[\frac{5}{4}(2x - 1)^{1/4}(2)(2x + 1)^{3/4} + \frac{3}{4}(2x + 1)^{-1/4}(2)(2x - 1)^{5/4}\right]$

$= 6(2x - 1)^{1/4}(2x + 1)^{3/4} + 3(2x + 1)^{-1/4}(2x - 1)^{5/4}$

17. $\dfrac{d}{dx}\left(\dfrac{4}{1 - x}\right)^3 = 4^3 \cdot \dfrac{d}{dx}(1 - x)^{-3} = 4^3 \cdot (-3)(1 - x)^{-4}(-1)$

$= 192(1 - x)^{-4}$

18. $\frac{d}{dx}\left[\frac{4x^2 + x}{\sqrt{x}}\right] = \frac{(8x + 1)x^{1/2} - \frac{1}{2}x^{-1/2}(4x^2 + x)}{x}$

19. $\frac{d}{dx}\left(\frac{4x - 1}{3x + 1}\right)^3 = 3\left(\frac{4x - 1}{3x + 1}\right)^3 \cdot \frac{(3x + 1)(4) - (4x - 1)(3)}{(3x + 1)^2}$

$= \frac{21(4x - 1)^2}{(3x + 1)^4}$

20. $\frac{d}{dx}\left[\frac{x}{x^2 + 1}\right]^2 = 2\left[\frac{x}{x^2 + 1}\right] \cdot \frac{(1)(x^2 + 1) - 2x(x)}{(x^2 + 1)^2} = \frac{2x(1 - x^2)}{(x^2 + 1)^3}$

21. $\frac{d}{dx}\left[\frac{4 - x}{x^2}\right]^2 = 3\left[\frac{4 - x}{x^2}\right]^2 \cdot \frac{(-1)(x^2) - (4 - x)(2x)}{x^4}$

$= \frac{3(4 - x)^2(x - 8)}{x^7}$

22. $\frac{d}{dx}\left(\frac{1 - x^2}{x}\right)^3 = 3\left(\frac{1 - x^2}{x}\right)^2 \cdot \frac{(-2x)(x) - (1)(1 - x^2)}{x^2}$

$= \frac{-3(1 - x^2)^2(1 + x^2)}{x^4}$

23.

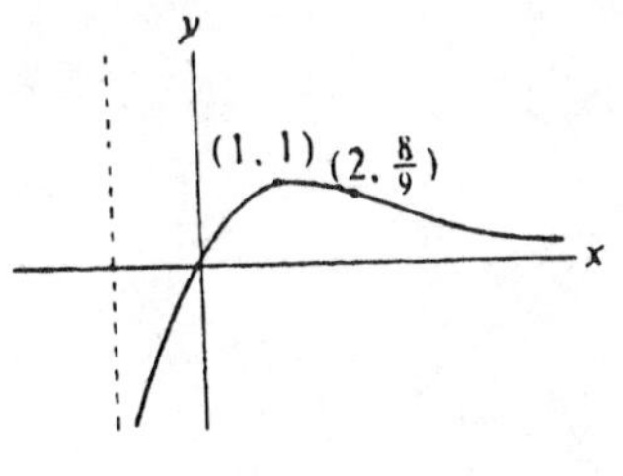

24.

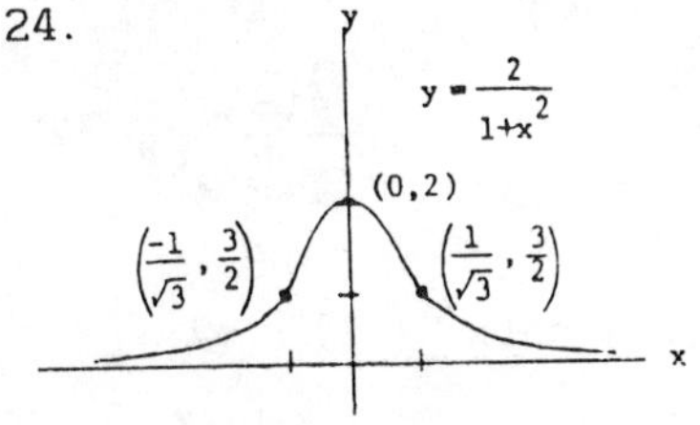

25. $f(x) = x^5$, $g(x) = 6x - 1$

$$\frac{d}{dx} f(g(x)) = \frac{d}{dx} (6x - 1)^5 = 5(6x - 1)^4(6) = 30(6x - 1)^4$$

26. $f(x) = x^2$, $g(x) = x^3 + 1$

$$\frac{d}{dx} f(g(x)) = \frac{d}{dx} (x^3 + 1)^2 = 2(x^3 + 1)(3x^2) = 6x^2(x^3 + 1)$$

27. $f(x) = \frac{1}{x}$, $g(x) = 1 - x^2$

$$\frac{d}{dx} f(g(x)) = \frac{d}{dx} \frac{1}{1 - x^2} = \frac{0(1 - x^2) - (-2x)(1)}{(1 - x^2)^2} = \frac{2x}{(1 - x^2)^2}$$

28. $f(x) = x^{10}$, $g(x) = x^2 + 3x$

$$\frac{d}{dx} f(g(x)) = \frac{d}{dx} (x^2 + 3x)^{10} = 10(x^2 + 3x)^9(2x + 3)$$

29. $f(x) = x^4 - x^2$, $g(x) = x^2 - 4$

$$\frac{d}{dx} f(g(x)) = \frac{d}{dx} \left[(x^2 - 4)^4 - (x^2 - 4)^2\right]$$

$$= 4(x^2 - 4)(2x) - 2(x^2 - 4)(2x)$$

$$= 8x(x^2 - 4)^3 - 4x(x^2 - 4)$$

30. $f(x) = \frac{4}{x} + x^2$, $g(x) = 1 - x^4$

$$\frac{d}{dx} f(g(x)) = \frac{d}{dx} \left[\frac{4}{1 - x^4} + (1 - x^4)^2\right]$$

$$= \frac{-4(-4x^3)}{(1 - x^4)^2} + 2(1 - x^4)(-4x^3)$$

31. $f(x) = (x^3 + 1)^2$, $g(x) = x^2 + 5$

$$\frac{d}{dx} f(g(x)) = \frac{d}{dx} \left[(x^2 + 5)^3 + 1)\right]^2$$

$$= 2\left[(x^2 + 5)^3 + 1\right]3(x^2 + 5)^2(2x)$$

$$= 12x\left[(x^2 + 5)^3 + 1\right](x^2 + 5)^2$$

32. $f(x) = x(x - 2)^4$, $g(x) = x^3$

$$\frac{d}{dx} f(g(x)) = \frac{d}{dx}\left[x^3(x^3 - 2)^4\right]$$

$$= 3x^2(x^3 - 2)^4 + 12x^5(x^3 - 2)^3$$

$$= (x^3 - 2)^3(3x^2(x^3 - 2) + 12x^5)$$

$$= (x^3 - 2)^3(15x^5 - 6x^2)$$

33. $u = (4x + 1)$, $y = u^{3/2}$

$$\frac{dy}{du} = \frac{3}{2}u^{1/2};\ \frac{du}{dx} = 4;\ \frac{dy}{dx} = 6u^{1/2} = 6(4x + 1)^{1/2}$$

34. $y = u^5$, $u = 2 + \sqrt{x}$

$$\frac{dy}{du} = 5u^4;\ \frac{du}{dx} = \frac{1}{2}x^{-1/2};\ \frac{dy}{dx} = 5u^4\left(\frac{1}{2}x^{-1/2}\right) = 5(2 + \sqrt{x})^4\left(\frac{1}{2}x^{-1/2}\right)$$

35. $y = \frac{4}{u^2} + \frac{u^2}{4}$, $u = x - 3x^2$

$$\frac{dy}{du} = \frac{-2u(4)}{u^4} + \frac{2u(4)}{16} = -\frac{8}{u^3} + \frac{u}{2};\ \frac{du}{dx} = 1 - 6x$$

$$\frac{dy}{dx} = \left[-8(x - 3x^2)^{-3} + \frac{1}{2}(x - 3x^2)\right](1 - 6x)$$

36. $y = \frac{1}{2}u^2 + 2u^{1/2}$, $u = 1 - 3x$

$$\frac{dy}{du} = u + u^{-1/2};\ \frac{du}{dx} = -3$$

$$\frac{dy}{dx} = -3\left[(1 - 3x) + (1 - 3x)^{-1/2}\right]$$

37. $y = u(u + 1)^5$, $u = x^2 + x$

$$\frac{dy}{du} = 1(u + 1)^5 + 5(u + 1)^4(u)(1) = (u + 1)^5 + 5u(u + 1)^4$$

$$\frac{du}{dx} = 2x + 1$$

$$\frac{dy}{dx} = (x^2 + x)^4(6x^2 + 6x + 1)(2x + 1)$$

38. $y = (u^2 + 1)^3,\ u = x - x^2$

$$\frac{dy}{du} = 3(u^2 + 1)^2(2u) = 6u(u^2 + 1)^2;\ \frac{du}{dx} = 1 - 2x$$

$$\frac{dy}{dx} = 6(x - x^2)\left[(x - x^2) + 1\right]^2(1 - 2x)$$

39. $y = \frac{u - 1}{u + 1},\ u = 2 + \sqrt{x}$

$$\frac{dy}{du} = \frac{(u - 1) - (u + 1)}{(u + 1)^2} = \frac{-2}{(u + 1)^2};\ \frac{du}{dx} = \frac{1}{2}x^{-1/2}$$

$$\frac{dy}{dx} = \frac{-2}{(2 + \sqrt{x} + 1)^2}\left(\frac{1}{2}x^{-1/2}\right) = \frac{1}{\sqrt{x}(3 + \sqrt{x})^2}$$

40. $y = \frac{u}{1 + u^2},\ u = 5x + 1$

$$\frac{dy}{du} = \frac{(1 + u^2) - 2u^2}{(1 + u^2)^2};\ \frac{du}{dx} = 5$$

$$\frac{dy}{dx} = 5\left[\frac{(1 + (5x + 1)^2) - 2(5x + 1)^2}{(1 + (5x + 1)^2)^2}\right]$$

41. $\frac{dy}{du} = 2(x - 4)^6 + 12x(x - 4)^3;\ \left.\frac{dy}{du}\right|_{x=5} = 62$

The tangent line is $y - 1 = 2(x - 5)$

42. $$\frac{dy}{du} = \frac{\sqrt{2 - x^2}\,1 - x(1/2[2 - x^2]^{-1/2}\cdot(-2x))}{2 - x^2}$$

$\left.\frac{dy}{du}\right|_{x=1} = \frac{1 + 1}{1} = 2$; the tangent line is $y - 1 = 2(x - 1)$

43. A horizontal tangent line has $\frac{dy}{du} = 0$

$$\frac{dy}{du} = 3(-x^2 + 4x - 3)^3(-2x + 4) = 0 \Rightarrow -2x + 4 = 0 \text{ and}$$

$-x^2 + 4x - 3 = 0 \Rightarrow x = 2$ or $(x - 3)(x - 1) = 0$; i.e., $x = 1,\ x = 3$

44. $f'(x) = 1/2(x^2 - 6x + 10)^{-1/2}(2x - 6)$

$f'(x) = 0$ when $x = 3$, so $y = \sqrt{9 - 18 + 10} = 1$

The relative minimum point is (3, 1).

45. (a) $\frac{dV}{dt} = \frac{dV}{dx}\cdot\frac{dx}{dt}$

(b) By the chain rule $\frac{dV}{dt} = 3x^2\frac{dx}{dt}$

We want $\frac{dV}{dt} = 12\frac{dx}{dt}$; so $3x^2 = 12$ or $x = 2$

46. $\frac{dw}{dt} = 3(446)x^2\frac{dx}{dt}$, $x = .4 \Rightarrow \frac{dx}{dt} = .2$, so $\frac{dw}{dt} = 42.816$ grams/yr.

47. (a) $\frac{dy}{dx}, \frac{dP}{dy}, \frac{dP}{dt}$ (b) $\frac{dP}{dt} = \frac{dP}{dy}\cdot\frac{dy}{dt}$

48. (a) $\frac{dx}{dy}, \frac{dQ}{dy}, \frac{dQ}{dx}$ (b) $\frac{dQ}{dy} = \frac{dQ}{dx}\cdot\frac{dx}{dy}$

49. $P = \frac{200x}{100 + x^2}$, $x = 4 + 2t$

(a) $\frac{dP}{dx} = \frac{200(100 + x^2) - (2x)(200x)}{(100 - x^2)^2}$

$$= \frac{20{,}000 + 200x^2 - 400x^2}{(100 + x^2)^2} = \frac{200(100 - x^2)}{(100 + x^2)^2}$$

(b) $\frac{dx}{dt} = 2$

$$\frac{dP}{dt} = \left[\frac{200(100 - (4 + 2t)^2)}{(100 + (4 + 2t)^2)^2}\right]2 = \frac{400(100 - (4 + 2t)^2)}{(100 + (4 + 2t)^2)^2}$$

(c) When $t = 8$,

$$\frac{dP}{dt} = \frac{400(100 - (4 + 16)^2)}{(100 + (4 + 16)^2)^2} = -48.$$

50. $C = 3x + 4\sqrt{x} + 2, \quad x = 6200 + 100t$

(a) $\frac{dC}{dx} = 3 + 2x^{-1/2}$

(b) $\frac{dx}{dt} = 100$

$$\frac{dC}{dt} = \left[3 + 2(6200 + 100t)^{-1/2}\right]100$$

(c) When $t = 2$,

$$\frac{dC}{dt} = \left[3 + 2(6200 + 200)^{-1/2}\right]100 = 302.5 \text{ dollars per week.}$$

51. $L = 10 + .4x + .0001x^2, \quad x = 752 + 23t + .5t^2$

(a) $\frac{dL}{dx} = .4 + .0002x$ (b) $\frac{dx}{dt} = 23 + t$

52. When $t = 5$, $x = .05(5^2) + 2(5) + 5 = 16.25$. Thus,

$$\left.\frac{dP}{dt}\right|_{t=5} = \left[\left.\frac{dP}{dx}\right|_{x=16.25}\right]\left[\left.\frac{dx}{dt}\right|_{t=5}\right]$$

$\frac{dP}{dx} = .002x + .1$, so $\left.\frac{dP}{dx}\right|_{x=16.25} = (.002)16.25 + 1 = .1325$

$\frac{dx}{dt} = .1t + 2$, so $\left.\frac{dx}{dt}\right|_{t=5} = 2.5$

Therefore, $\left.\frac{dP}{dt}\right|_{t=5} = 2.5(.1325) = .3314.$

53. $\frac{d}{dx} f(g(x)) = f'(g(x))\cdot g'(x) = 3x^2\cdot f'(x^3 + 1)$

So $g(x) = x^3 + 1$.

54. $\frac{d}{dx} f(g(x)) = f'(g(x))\cdot g'(x) = \frac{g'(x)}{g(x)} = \frac{2x + 5}{x^2 + 5x - 4}$

So $g(x) = x^2 + 5x - 4$.

55. $\left.\frac{d}{dx} f(g(x))\right|_{x=1} = f'(g(1))\cdot g'(1) = f'(5)\cdot 6 = 4\cdot 6 = 24.$

56. $\left.\frac{d}{dx} g(f(x))\right|_{x=1} = g'(f(1))\cdot f'(1) = g'(2)\cdot 3 = 7\cdot 3 = 21.$

Exercises 3.3

1. $x^2 - y^2 = 1,\ 2x - 2y\frac{dy}{dx} = 0,\ \frac{dy}{dx} = \frac{x}{y}$

2. $x^2 + y^3 - 6 = 0,\ 3x^2 + 3y^2\frac{dy}{dx} - 0 = 0,\ \frac{dy}{dx} = -\frac{x^2}{y^2}$

3. $y^5 - 3x^2 = x,\ 5y^4\frac{dy}{dx} - 6x = 1,\ \frac{dy}{dx} = \frac{1 + 6x}{5y^4}$

4. $x^4 + (y + 3)^4 = x^2,\ 4x^3 + 4(y + 3)^3\frac{dy}{dx} = 2x,$

   $\frac{dy}{dx} = \frac{2x - 4x^3}{4(y + 3)^3} = \frac{x - 2x^3}{2(y + 3)^3}$

5. $y^4 - x^4 = y^2 - x^2,\ 4y^3\frac{dy}{dx} - 4x^3 = 2y\frac{dy}{dx} - 2x,$

   $\frac{dy}{dx}(4y^3 - 2y) = 4x^3 - 2x,\ \frac{dy}{dx} = \frac{2x^3 - x}{2y^3 - y}$

6. $x^3 + y^3 = x^2 + y^2,\ 3x^2 + 3y^2\frac{dy}{dx} = 2x + 2y\frac{dy}{dx},\ \frac{dy}{dx} = \frac{2x - 3x^2}{3y^2 - 2y}$

7. $2x^3 + y = 2y^3 + x,\ 6x^2 + \frac{dy}{dx} = 6y^2\frac{dy}{dx} + 1,$

   $\frac{dy}{dx}(1 - 6y^2) = 1 - 6x^2,\ \frac{dy}{dx} = \frac{1 - 6x^2}{1 - 6y^2}$

8. $x^4 + 4y = x - 4y^3$, $4x^3 + 4\frac{dy}{dx} = 1 - 12y^2\frac{dy}{dx}$,

$\frac{dy}{dx}(4 + 12y^2) = 1 - 4x^3$, $\frac{dy}{dx} = \frac{1 - 4x^3}{4 + 12y^2}$

9. $xy = 5$, $1(y) + x\frac{dy}{dx} = 0$, $\frac{dy}{dx} = -\frac{y}{x}$

10. $xy^3 = 2$, $(1)y^3 + 3y^2\frac{dy}{dx}(x) = 0$, $\frac{dy}{dx} = -\frac{y^3}{3xy^2} = -\frac{y}{3x}$

11. $x(y + 2)^5 = 8$, $(1)(y + 2)^5 + 5(y + 2)^4(x)\frac{dy}{dx} = 0$,

$\frac{dy}{dx} = -\frac{(y + 2)^5}{5x(y + 2)^4} = -\frac{y + 2}{5x}$

12. $x^2y^3 = 6$, $2xy^3 + 3y^2(x^2)\frac{dy}{dx} = 0$, $\frac{dy}{dx} = -\frac{2xy^3}{3y^2x^2} = -\frac{2y}{3x}$

13. $x^3y^2 - 4x^2 = 1$, $3x^2y^2 + 2yx^3\frac{dy}{dx} - 8x = 0$, $\frac{dy}{dx} = \frac{8 - 3xy^2}{2yx^2}$

14. $(x + 1)^2(y - 1)^2 = 1$,

$2(x + 1)(y - 1)^2 + 2(y - 1)(x + 1)^2\frac{dy}{dx} = 0$,

$\frac{dy}{dx} = -\frac{2(x + 1)(y - 1)^2}{2(x + 1)^2(y - 1)} = -\frac{y - 1}{x + 1}$

15. $x^3 + y^3 = x^3y^3$, $3x^2 + 3y^2\frac{dy}{dx} = 3x^2y^3 + 3y^2x^3\frac{dy}{dx}$,

$\frac{dy}{dx}(3y^2 - 3y^2x^3) = 3x^2y^3 - 3x^2$, $\frac{dy}{dx} = \frac{3x^2y^3 - 3x^2}{3y^2 - 3y^2x^3}$,

$\frac{dy}{dx} = \frac{x^2(y^3 - 1)}{y^2(1 - x^3)}$

16. $x^2 + 4xy + 4y = 1$, $2x + 4\left(y + x\frac{dy}{dx}\right) + 4\frac{dy}{dx} = 0$,

$4\frac{dy}{dx}(x + 1) = -2x - 4y$, $\frac{dy}{dx} = \frac{-2(x + 2y)}{4(x + 1)}$, $\frac{dy}{dx} = \frac{-x - 2y}{2(x + 1)}$

17. $x^2y + y^2x = 3$, $2xy + 2yx\frac{dy}{dx} + x^2\frac{dy}{dx} + y^2 = 0$,

$x\frac{dy}{dx}(2y + x) = -2xy - y^2$, $\frac{dy}{dx} = \frac{-2xy - y^2}{x(2y + x)}$, $\frac{dy}{dx} = \frac{-2xy - y^2}{2xy + x^2}$

18. $x^3y + xy^3 = 4$, $3x^2y + x^3\frac{dy}{dx} + y^3 + 3y^2x\frac{dy}{dx} = 0$,

$x\frac{dy}{dx}(x^2 + 3y^2) = -3x^2 - y^3$, $\frac{dy}{dx} = \frac{-3x^2y - y^3}{x(x^2 + 3y^2)}$,

$\frac{dy}{dx} = \frac{-3x^2y - y^3}{x^3 + 3xy^2}$

19. $4y^3 - x^2 = -5$, $12y^2\frac{dy}{dx} - 3x = 0$, slope $= \frac{dy}{dx} = \frac{2x}{12y^2} = \frac{x}{6y^2}$

When $x = 3$, $y = 1$,

slope $= \frac{dy}{dx} = \frac{3}{6(1)^2} = \frac{1}{2}$.

20. $y^2 = x^3 + 1$, $2y\frac{dy}{dx} = 3x^2 + 0$, slope $= \frac{dy}{dx} = \frac{3x^2}{2y}$

When $x = 3$, $y = -3$,

slope $= \frac{dy}{dx} = \frac{3(2)^2}{2(-3)} = -2$.

21. $xy^3 = 2$, $(1)y^3 + 3y^2x\frac{dy}{dx} = 0$, slope $= \frac{dy}{dx} = \frac{-y^3}{3y^2x} = \frac{-y}{3x}$

When $x = -\frac{1}{4}$, $y = -21$,

slope $= \frac{dy}{dx} = \frac{-(-2)}{3\left(-\frac{1}{4}\right)} = -\frac{8}{3}$.

22. $\sqrt{x} + \sqrt{y} = 7$, $\frac{1}{2}x^{-1/2} + \frac{1}{2}y^{-1/2}\frac{dy}{dx} = 0$, $\frac{dy}{dx} = \frac{-\frac{1}{2}x^{-1/2}}{\frac{1}{2}y^{-1/2}}$,

$$\text{slope} = \frac{dy}{dx} = -\frac{\sqrt{y}}{\sqrt{x}}$$

When $x = 9$, $y = 16$,

$$\text{slope} = \frac{dy}{dx} = -\frac{\sqrt{16}}{\sqrt{9}} = -\frac{4}{3}.$$

23. $xy + y^3 = 14$, $(1)y + x\frac{dy}{dx} + 3y^2\frac{dy}{dx} = 0$, slope $= \frac{dy}{dx} = \frac{-y}{x + 3y^2}$

When $x = 3$, $y = 2$,

$$\text{slope} = \frac{dy}{dx} = \frac{-2}{3 + 3(2)^2} = -\frac{2}{15}.$$

24. $y^2 = 3xy - 5$, $2y\frac{dy}{dx} = 3(1)y + 3x\frac{dy}{dx} - 0$, slope $= \frac{dy}{dx} = \frac{3y}{2y - 3x}$

When $x = 2$, $y = 1$,

$$\text{slope} = \frac{dy}{dx} = \frac{3(1)}{2(1) - 3(2)} = -\frac{3}{4}.$$

25. $x^2y^4 = 1$, $2xy^4 + 4y^3x^2\frac{dy}{dx} = 0$, slope $= \frac{dy}{dx} = \frac{-2xy^4}{4y^3x^2} = -\frac{y}{2x}$

When $x = 4$, $y = \frac{1}{2}$, slope $= -\frac{\frac{1}{2}}{2(4)} = -\frac{1}{16}$.

Let $(x_1, y_1) = \left(4, \frac{1}{2}\right)$. $y - \frac{1}{2} = -\frac{1}{16}(x - 4)$

When $x = 4$, $y = -\frac{1}{2}$, slope $= -\frac{-\frac{1}{2}}{2(4)} = \frac{1}{16}$.

Let $(x_1, y_1) = \left(4, -\frac{1}{2}\right)$. $y + \frac{1}{2} = \frac{1}{16}(x - 4)$

26. $x^4y^2 = 144$, $4x^3y^2 + 2yx^4\dfrac{dy}{dx} = 0$, $\dfrac{dy}{dx} = -\dfrac{4x^3y^2}{2yx^4}$

$$\text{slope} = \frac{dy}{dx} = -\frac{2y}{x}$$

When $x = 2$, $y = 3$, slope $= \dfrac{-2(3)}{2} = -3$.

Let $(x_1, y_1) = (2, 3)$. $y - 3 = -3(x - 2)$

When $x = 2$, $y = -3$, slope $= \dfrac{-2(-3)}{2} = 3$.

Let $(x_1, y_1) = (2, -3)$. $y + 3 = 3(x - 2)$

27. (a) $x^4 + 2x^2y^2 + y^4 = 4x^2 - 4y^2$

$$4x^3 + 4xy^2 + 4x^2y\frac{dy}{dx} + 4y^3\frac{dy}{dx} = 8x - 8y\frac{dy}{dx}$$

$$\frac{dy}{dx}(4x^2y + 4y^3 + 8y) = 8x - 4x^3 - 4xy^2$$

$$\frac{dy}{dx} = \frac{4(2x - x^3 - xy^2)}{4(x^2y + y^3 + 2y)} = \frac{2x - x^3 - xy^2}{x^2y + y^3 + 2y}$$

(b) $\text{slope} = \dfrac{dy}{dx} = \dfrac{2x - x^3 - xy^2}{x^2y + y^3 + 2y}$

When $x = \dfrac{\sqrt{6}}{2}$, $y = \dfrac{\sqrt{2}}{2}$, slope $= \dfrac{\sqrt{6} - \dfrac{\sqrt{6^3}}{8} - \dfrac{\sqrt{6}}{2}\cdot\dfrac{2}{4}}{\dfrac{6}{4}\cdot\dfrac{\sqrt{2}}{2} + \dfrac{\sqrt{2^3}}{8} + \sqrt{2}} = 0.$

28. (a) $x^4 + 2x^2y^2 + y^4 = 9x^2 - 9y^2$

$$4x^3 + 4xy^2 + 4x^2y\frac{dy}{dx} + 4y^3\frac{dy}{dx} = 18x - 18y\frac{dy}{dx}$$

$$\frac{dy}{dx}(4x^2y + 4y^3 + 18y) = 18x - 4x^3 - 4xy^2$$

$$\frac{dy}{dx} = \frac{18x - 4x^3 - 4xy^2}{2x^2y + 2y^3 + 9y} = \frac{9x - 2x^3 - 2xy^2}{2x^2y + 2y^3 + 9y}$$

(b) slope $= \frac{dy}{dx} = \frac{9x - 2x^3 - 2xy^2}{2x^2y + 2y^3 + 9y}$

When $x = \sqrt{5}$, $y = -1$, slope $= \frac{9\sqrt{5} - 10\sqrt{5} - 2\sqrt{5}}{-10 - 2 - 9} = \frac{3\sqrt{5}}{21} = \frac{\sqrt{5}}{7}$.

29. $30x^{1/3}y^{2/3} = 1080$, $10x^{-2/3}y^{2/3} + 20x^{1/3}y^{-1/3}\frac{dy}{dx} = 0$,

$\frac{dy}{dx} = \frac{-10y^{2/3}x^{-2/3}}{20x^{1/3}y^{-1/3}} = -\frac{y}{2x}$

When $x = 16$, $y = 54$, $\frac{dy}{dx} = \frac{-54}{2(16)} = -\frac{54}{32} = -\frac{27}{16}$.

30. $10x^{1/2}y^{1/2} = 600$, $5x^{-1/2}y^{1/2} + 5y^{-1/2}x^{1/2}\frac{dy}{dx} = 0$,

$\frac{dy}{dx} = \frac{-x^{-1/2}y^{1/2}}{y^{-1/2}x^{1/2}} = -\frac{y}{x}$

When $x = 50$, $y = 72$, $\frac{dy}{dx} = -\frac{72}{50} = -\frac{36}{25}$.

31. $x^4 + y^4 = 1$, $4x^3\frac{dx}{dt} + 4y^3\frac{dy}{dt} = 0$, $\frac{dy}{dt} = -\frac{4x^3}{4y^3}\cdot\frac{dx}{dt}$, $\frac{dy}{dt} = -\frac{x^3}{y^3}\frac{dx}{dt}$

32. $y^4 - x^2 = 1$, $4y^3\frac{dy}{dt} - 2x\frac{dx}{dt} = 0$, $\frac{dy}{dt} = \frac{x}{2y^3}\frac{dx}{dt}$

33. $3xy - 3x^2 = 4$, $3\frac{dx}{dt}y + 3x\frac{dy}{dt} - 6x\frac{dx}{dt} = 0$, $\frac{dy}{dt} = \frac{6x - 3y}{3x}\frac{dx}{dt}$,

$\frac{dy}{dt} = \frac{2x - y}{x}\frac{dx}{dt}$

34. $y^2 = 8 + xy$, $2y\frac{dy}{dt} = 0 + y\frac{dx}{dt} + x\frac{dy}{dt}$, $\frac{dy}{dt} = \frac{y}{2y - x}\frac{dx}{dt}$

35. $x^2 + 2xy = y^3$, $2x\frac{dx}{dt} + 2y\frac{dx}{dt} + 2x\frac{dy}{dt} = 3y^2\frac{dy}{dt}$,

$$\frac{dy}{dt} = \frac{-2x - 2y}{2x - 3y^2}\frac{dx}{dt}$$

36. $x^2y^2 = 2y^3 + 1$, $2x\frac{dx}{dt}y^2 + 2y\frac{dy}{dt}x^2 = 6y^2\frac{dy}{dt} + 0$,

$$\frac{dy}{dt} = \frac{-2xy^2}{2yx^2 - 6y^2}\frac{dx}{dt} = \frac{xy}{3y - x^2}\frac{dx}{dt}$$

37. $x^2 - 4y^2 = 9$, $2x\frac{dx}{dt} - 8y\frac{dy}{dt} = 0$, $\frac{dy}{dt} = \frac{2x}{8y}\frac{dx}{dt}$, $\frac{dy}{dt} = \frac{x}{4y}\frac{dx}{dt}$

When $x = 5$, $y = -2$, $\frac{dx}{dt} = 3$ per sec., $\frac{dy}{dt} = \frac{5}{-8}(3) = \frac{-15}{8}$ units per sec.

38. $x^3y^2 = 200$, $3x^2\frac{dx}{dt}y^2 + 2y\frac{dy}{dt}x^3 = 0$, $\frac{dy}{dt} = -\frac{3x^2y^2}{2yx^3}\frac{dx}{dt} = \frac{-3y}{2x}\frac{dx}{dt}$

When $x = 2$, $y = 5$, $\frac{dx}{dt} = -4$ units per min.,

$\frac{dy}{dt} = \frac{-3(5)}{2(2)}(-4) = 15$ units per min.

39. $2p^3 + x^2 = 4500$, $6p^2\frac{dp}{dt} + 2x\frac{dx}{dt} = 0$, $\frac{dx}{dt} = -\frac{6p^2}{2x}\frac{dp}{dt}$

When $p = 10$, $x = 50$, $\frac{dp}{dt} = -.5$ per week.

$\frac{dx}{dt} = -\frac{6(100)}{100}(-.5) = 3.0$

The sales are rising at 3 thousand units per week.

40. $6p + x + xp = 94$, $6\frac{dp}{dt} + \frac{dx}{dt} + p\frac{dx}{dt} + x\frac{dp}{dt} = 0$, $\frac{dx}{dt} = \frac{-x - 6}{p + 1}\frac{dp}{dt}$

When $x = 4$, $p = 9$, $\frac{dp}{dt} = 2$, $\frac{dx}{dt} = \frac{-4 - 6}{9 + 1}(2) = -2$.

The demand is falling at the rate of 2 thousand units per week.

41. $A = 6\sqrt{x^2 - 400}$, $x \ge 20$, $A = 6(x^2 - 100)^{1/2}$

$$\frac{dA}{dx} = 3(x^2 - 100)^{-1/2}(2x) = \frac{6x}{\sqrt{x^2 - 400}}$$

$$\frac{dA}{dt} = \frac{dA}{dx}\cdot\frac{dx}{dt} = \frac{6x}{\sqrt{x^2 - 400}}\cdot\frac{dx}{dt}$$

When $x = 25$, $\frac{dx}{dt} = 2$, $\frac{dA}{dt} = \frac{6(25)}{\sqrt{25^2 - 400}}(2) = 20.$

The revenue is increasing at the rate of 20 thousand dollars per month.

42. $px + 7x + 8p = 328$

$$x\frac{dp}{dt} + p\frac{dx}{dt} + 7\frac{dx}{dt} + 8\frac{dp}{dt} = 0, \quad \frac{dp}{dt} = \frac{-p - 7}{x + 8}\frac{dx}{dt}$$

When $p = 25$, $x = 4$, $\frac{dx}{dt} = -.3$, $\frac{dp}{dt} = \frac{-25 - 7}{4 + 8}(-.3) = .8.$

The price is increasing at the rate of .8 dollars per crate.

43. $P^5V^7 = k$

$$5P^4\frac{dP}{dt}V^7 + 7V^6\frac{dV}{dt}P^5 = 0, \quad \frac{dV}{dt} = \frac{-5V}{7P}\frac{dP}{dt}$$

When $V = 4$, $P = 200$, $\frac{dP}{dt} = 5$, $\frac{dV}{dt} = \frac{-5(4)}{7(200)}(5) = -\frac{1}{14}.$

The volume is decreasing at $\frac{1}{14}$ liters per sec.

44. $V = \frac{\pi x^3}{6}$, $\frac{dV}{dx} = \frac{3\pi x^2}{6} = \frac{\pi x^2}{2}$, $\frac{dV}{dt} = \frac{dV}{dx}\frac{dx}{dt} = \frac{\pi x^2}{2}\frac{dx}{dt}$

When $x = 10$, $\frac{dx}{dt} = .4$, $\frac{dV}{dt} = \frac{\pi(10)^2}{2}(.4) = 62.8.$

The tumor is growing at the rate of 62.8 mm^3 per day.

45. (a) $x^2 + y^2 = (10)^2$, $x^2 + y^2 = 100$

(b) $2x\frac{dx}{dt} + 2y\frac{dy}{dt} = 0$, $\frac{dy}{dt} = \frac{-x}{y}\frac{dx}{dt}$

When $x = 8$, $y = 6$, $\frac{dx}{dt} = 3$, $\frac{dy}{dt} = -\frac{8}{6}(3) = -4$.

The top end is sliding down at 4 feet per second.

46. (a) $x^2 + (5000)^2 = y^2$, $x^2 + 25{,}000{,}000 = y^2$,

$y^2 - x^2 = 25{,}000{,}000$

(b) When $y = 13{,}000$, $x^2 + 25{,}000{,}000 = (13{,}000)^2$, $x = 12{,}000$.

(c) $2y\frac{dy}{dx} = 2x + 0$, $\frac{dy}{dx} = \frac{x}{y}$, $\frac{dy}{dt} = \frac{dy}{dx}\frac{dx}{dt} = \frac{x}{y}\frac{dx}{dt}$

When $x = 12{,}000$, $y = 13{,}000$, $\frac{dx}{dt} = 390$.

$\frac{dy}{dt} = \frac{12{,}000}{13{,}000}(390) = 360$ ft/sec

47. $x^2 + 90^2 = y^2$; differentiate with respect to t

$2x\frac{dx}{dt} = 2y\frac{dy}{dt}$. If $x = 45$ and $\frac{dx}{dt} = 22$ and by the equation

above $y = \sqrt{45^2 + 90^2} = 45\sqrt{5}$, so

$$\frac{dy}{dt} = \frac{45.22}{45\sqrt{5}} = \frac{22}{\sqrt{5}} \text{ ft/sec}$$

Chapter 3 Supplementary Exercises

1. $$\frac{d}{dx}\left[(4x - 1)(3x + 1)^4\right] = 4(3x + 1)^4 + 4(3x + 1)^3(3)(4x - 1)$$

$$= 4(3x + 1)^3((3x + 1) + 3(4x - 1))$$

$$= 4(3x + 1)^3(15x - 2)$$

2. $$\frac{d}{dx}\left[2(5-x)^3(6x-1)\right] = 2\left[3(5-x)^2(-1)(6x-1) + 6(5-x)^3\right]$$

$$= 2\left[-3(5-x)^2(6x-1) + 6(5-x)^3\right]$$

$$= 6\left[2(5-x)^3 - (5-x)^2(6x-1)\right]$$

$$= 6(5-x)^2\left[10 - 2x - 6x + 1\right]$$

$$= 6(5-x)^2(11-8x)$$

3. $$\frac{d}{dx}\left[x(x^5-1)^3\right] = (1)(x^5-1)^3 + 3(x^5-1)^2(x)(5x^4)$$

$$= (x^5-1)^2(16x^5-1)$$

4. $$\frac{d}{dx}\left[(2x+1)^{5/2}(4x-1)^{3/2}\right]$$

$$= \frac{5}{2}(2x+1)^{3/2}(2)(4x-1)^{3/2} + \frac{3}{2}(4x-1)^{1/2}(4)(2x+1)^{5/2}$$

$$= 5(2x+1)^{3/2}(4x-1)^{3/2} + 6(4x-1)^{1/2}(2x+1)^{5/2}$$

$$= (2x+1)^{3/2}(4x-1)^{1/2}(5(4x-1) + 6(2x+1))$$

$$= (2x+1)^{3/2}(4x-1)^{1/2}(32x+1)$$

5. $$\frac{d}{dx}\left[5(\sqrt{x}-1)^4(\sqrt{x}-2)^2\right]$$

$$= 5\left[4(\sqrt{x}-1)^3\left(\frac{1}{2}x^{-1/2}\right)(\sqrt{x}-2)^2 + 2(\sqrt{x}-2)\left(\frac{1}{2}x^{-1/2}\right)(\sqrt{x}-1)^4\right]$$

$$= 5\left[2x^{-1/2}(\sqrt{x}-1)^3(\sqrt{x}-2)^2 + x^{-1/2}(\sqrt{x}-2)(\sqrt{x}-1)^4\right]$$

$$= 5x^{-1/2}(\sqrt{x}-1)^3(\sqrt{x}-2)(2(\sqrt{x}-2) + (\sqrt{x}-1))$$

$$= 5x^{-1/2}(\sqrt{x}-1)^3(\sqrt{x}-2)(3\sqrt{x}-5)$$

6. $$\frac{d}{dx}\left[\frac{\sqrt{x}}{\sqrt{x}+4}\right] = \frac{\frac{1}{2}x^{-1/2}(x^{1/2}+4) - \frac{1}{2}x^{-1/2}(x^{1/2})}{(x^{1/2}+4)^2}$$

$$= \frac{\frac{1}{2}x^{-1/2}(x^{1/2} + 4 - x^{1/2})}{(x^{1/2} + 4)^2} = \frac{2}{x^{1/2}(x^{1/2} + 4)^2}$$

7. $\frac{d}{dx}\left[3(x^2 - 1)^3(x^2 + 1)^5\right]$

$$= 3\left[3(x^2 - 1)^2(2x)(x^2 + 1)^5 + 5(x^2 + 1)^4(2x)(x^2 - 1)^3\right]$$

$$= 3(x^2 - 1)^2(2x)(x^2 + 1)^4(3(x^2 + 1) + 5(x^2 - 1))$$

$$= 12x(x^2 - 1)^2(x^2 + 1)^4(4x^2 - 1)$$

8. $\frac{d}{dx}\left[\frac{1}{(x^2 + 5x + 1)^6}\right] = \frac{0 - 6(x^2 + 5x + 1)^5(2x + 5)}{(x^2 + 5x + 1)^2}$

$$= \frac{-12x - 30}{(x^2 + 5x + 1)^7}$$

9. $\frac{d}{dx}\left[\frac{x^2 - 6x}{x - 2}\right] = \frac{(2x - 6)(x - 2) - (1)(x^2 - 6x)}{(x - 2)^2}$

$$= \frac{2x^2 - 10x + 12 - x^2 + 6x}{(x - 2)^2} = \frac{x^2 - 4x + 12}{(x - 2)^2}$$

10. $\frac{d}{dx}\left[\frac{2x}{2 - 3x}\right] = \frac{2(2 - 3x) - (-3)(2x)}{(2 - 3x)^2} = \frac{4 - 6x + 6x}{(2 - 3x)^2} = \frac{4}{(2 - 3x)^2}$

11. $\frac{d}{dx}\left[\left(\frac{3 - x^2}{x^3}\right)^2\right] = 2\left(\frac{3 - x^2}{x^3}\right)\left(\frac{-2x(x^3) - (3x^2)(3 - x^2)}{x^6}\right)$

$$= 2\left(\frac{3 - x^2}{x^3}\right)\left(\frac{-2x^4 - 9x^2 + 3x^4}{x^6}\right)$$

$$= 2\left(\frac{3 - x^2}{x^3}\right)\left(\frac{x^4 - 9x^2}{x^6}\right) = \left(\frac{3 - x^2}{x^3}\right)\left(\frac{x^2 - 9}{x^4}\right)$$

$$= \frac{2(3 - x^2)(x^2 - 9)}{x^7}$$

12. $\frac{d}{dx}\left[\frac{x^3 + x}{x^2 - x}\right] = \frac{(3x^2 + 1)(x^2 - x) - (2x - 1)(x^3 + x)}{(x^2 - x)^2}$

$$= \frac{3x^4 - 3x^3 + x^2 - x - 2x^4 - 2x^2 + x^3 + x}{(x^2 - x)^2}$$

$$= \frac{x^4 - 2x^3 - x^2}{x^4 - 2x^3 + x^2} = \frac{x^2 - 2x - 1}{x^2 - 2x + 1} = \frac{x^2 - 2x - 1}{(x - 1)^2}$$

13. $f(x) = (3x + 1)^4(3 - x)^5$

$$f'(x) = 4(3x + 1)^3(3)(3 - x)^5 + 5(3 - x)^4(-1)$$

$$= 12(3x + 1)^3(3 - x)^5 - 5(3 - x)^4(3x + 1)^4$$

$$= (3x + 1)^3(3 - x)^4(12(3 - x) - 5(3x + 1))$$

$$= (3x + 1)^3(3 - x)^4(-27x + 31)$$

Let $f'(x) = 0$ and solve for x.

$(3x + 1)^3(3 - x)^4(31 - 27x) = 0$, $x = -\frac{1}{3}$, $x = 3$, $x = \frac{31}{27}$

14. $f(x) = \frac{x^2 + 1}{x^2 + 5}$

$$f'(x) = \frac{2x(x^2 + 5) - (2x)(x^2 + 1)}{(x^2 + 5)^2} = \frac{2x^3 + 10x - 2x^3 - 2x}{(x^2 + 5)^2}$$

$$= \frac{8x}{(x^2 + 5)^2}$$

Let $f'(x) = 0$ and solve for x. $\frac{8x}{(x^2 + 5)^2} = 0$, $x = 0$.

15. $y = (x^3 - 1)(x^2 + 1)^4$

slope $= y' = 3x^2(x^2 + 1)^4 + 4(x^2 + 1)^3(2x)(x^3 - 1)$

$$= 3x^2(x^2 + 1)^4 + 8x(x^2 + 1)^3(x^3 - 1)$$

When $x = -1$, slope $= 3(-1)^2(1 + 1)^4 + 8(-1)(1 + 1)^3(-1 - 1)$

$$= 48 + 128 = 176$$

When $x = -1$, $y = (-1 - 1)(1 + 1)^4 = -32$.

Let $(x_1, y_1) = (-1, -32)$. $y + 32 = 176(x + 1)$

16. $y = \dfrac{x - 3}{\sqrt{4 + x^2}}$

$$\text{slope} = y' = \frac{1(4 + x^2)^{1/2} - \frac{1}{2}(4 + x^2)^{-1/2}(2x)(x - 3)}{4 + x^2}$$

$$= \frac{(4 + x^2)^{1/2}}{(4 + x)^2} - \frac{(x^2 - 3x)}{(4 + x^2)^{3/2}}$$

$$= \frac{4 + x^2 - x^2 + 3x}{(4 + x^2)^{3/2}} = \frac{3x + 4}{(4 + x^2)^{3/2}}$$

When $x = 0$, $y = -\frac{3}{2}$. When $x = 0$, slope $= \frac{4}{8} = \frac{1}{2}$.

Let $(x_1, y_1) = \left(0, -\frac{3}{2}\right)$. $y + \frac{3}{2} = \frac{1}{2}(x - 0)$, $y = \dfrac{x - 3}{2}$

17. The objective is $A = 2y(2) + (x - 4)(2) = 4y + 2x - 8$. The constraint is $(x - 4)(y - 2) = 800$, $y = \dfrac{800}{x - 4} + 2$. Thus,

$$A(x) = 4\left(\frac{800}{x - 4} + 2\right) + 2x - 8 = \frac{3200}{x - 4} + 2x,$$

$$A'(x) = -\frac{3200}{(x - 4)^2} + 2, \quad A'(44) = 0.$$

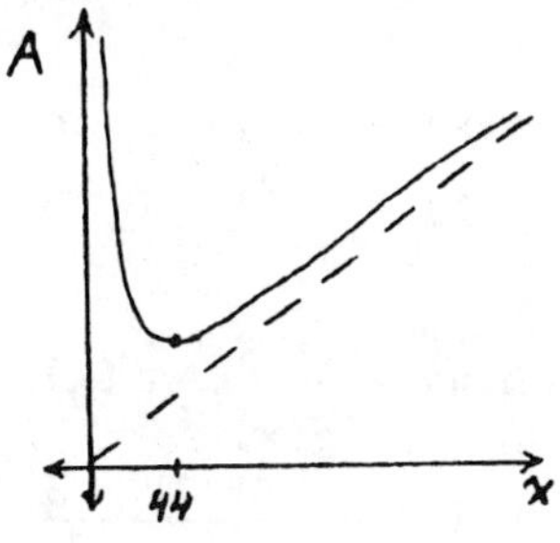

The minimum value of $A(x)$ for $x > 0$ occurs at $x = 44$. The optimal value of x and y are thus,

$$x = 44, \quad y = \frac{800}{44 - 4} + 2 = 22.$$

18. Here the objective is $A = 2(x - 4)(2) + 2(y)(2) = 4x - 16 + 4y$ and the constraint is $(x - 4)(y - 4) = 800$, $y = \frac{800}{x - 4} + 4$.

$A(x) = 4x + 4\left[\frac{800}{x - 4} + 4\right] - 16 = 4x + \frac{3200}{x - 4}$,

$A'(x) = 4 - \frac{3200}{(x - 4)^2}$.

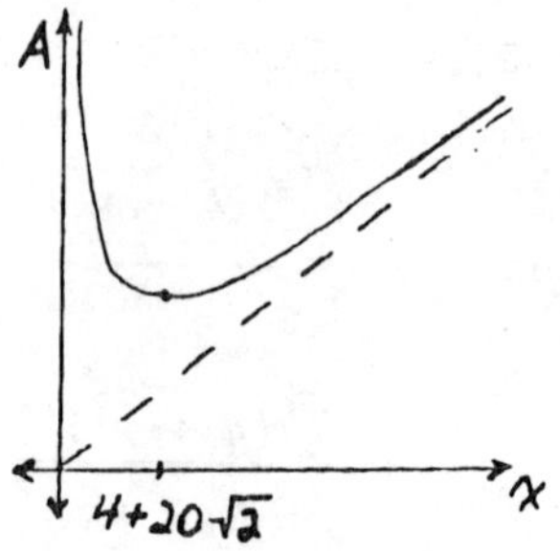

$A'(4 + 20\sqrt{2}) = 0$. The minimum value of $A(x)$ occurs at $x = 4 + 20\sqrt{2}$. The optimal dimensions are $x = y = 4 + 20\sqrt{2}$.

19. We are given that $C(x) = 40x + 30$ and $\frac{dx}{dt} = 3$. By the chain rule, $\frac{dC}{dt} = \frac{dC}{dx}\frac{dx}{dt} = 40 \cdot 3 = 120$.

20. $\frac{dy}{dt} = \frac{dy}{dP}\frac{dP}{dt}$

21. $f(1) = 3$, $f'(1) = 1/2$, $g(1) = 2$, $g'(1) = 3/2$

$h(1) = 2f(1) - 3g(1) = 2(3) - 3(2) = 0$

$h'(1) = 2f'(1) - 3g'(1) = 2(1/2) - 3(3/2) = -7/2$

22. $h(x) = f(x)g(x)$, $h'(x) = f(x)g'(x) + f'(x)g(x)$

$h(1) = 3(2) = 6$; $h'(1) = 3(3/2) + (1/2)(2) = 11/2$

23. $h(x) = \frac{f(x)}{g(x)}$, $h'(x) = \frac{g(x)f'(x) - g'(x)f(x)}{[g(x)]^2}$, $h(1) = 3/2$

$h'(x) = \frac{2(1/2) - 3(3/2)}{2^2} = -7/8$

24. $h(x) = [f(x)]^2$, $h'(x) = 2f(x)f'(x)$, $h(1) = 9$,

$h'(1) = 2(3)(1/2) = 3.$

25. $h(x) = f(g(x)),\ h(x) = f'(g(x))g'(x)$

$g(1) = 2,\ f(2) = 1,\ f'(2) = -1,\ g'(1) = 3/2$

$h(1) = 1,\ h'(1) = f'(2)g'(1) = -1(3/2) = -3/2$

26. $h(x) = g(f(x)),\ h(x) = g'(f(x))f'(x)$

$g'(3) = -1/2,\ g(3) = 1,\ f(1) = 3,\ f'(1) = 1/2$

$h(1) = 1,\ h'(1) = g'(3)f'(1) = (-1/2)(1/2) = -1/4$

27. $g(x) = x^3,\ g'(x) = 3x^2,\ \frac{d}{dx} f(g(x)) = \frac{1}{x^6 + 1}(3x^2) = \frac{3x^2}{x^6 + 1}$

28. $g(x) = \frac{1}{x},\ g(x) = \frac{-1}{x^2}$

$$\frac{d}{dx} f(g(x)) = \frac{1}{\left(\frac{1}{x}\right)^2 + 1}\left(\frac{x - 1}{x^2}\right) = \left(\frac{x^2}{1 + x^2}\right)\left(\frac{-1}{x^2}\right) = \frac{-1}{1 + x^2}$$

29. $g(x) = x^2 + 1,\ g'(x) = 2x$

$$\frac{d}{dx} f(g(x)) = \frac{1}{(x^2 + 1)^2 + 1}(2x) = \frac{2x}{(x^2 + 1)^2 + 1}$$

30. $g(x) = x^2,\ g'(x) = 2x$

$$\frac{d}{dx} f(g(x)) = x^2\sqrt{1 - x^4}(2x) = 2x^3\sqrt{1 - x^4}$$

31. $g(x) = \sqrt{x},\ g'(x) = \frac{1}{2}x^{-1/2}$

$$\frac{d}{dx} f(g(x)) = \sqrt{x}\ \sqrt{1 - x}\left(\frac{1}{2}\cdot\frac{1}{\sqrt{x}}\right) = \frac{\sqrt{1 - x}}{2}$$

32. $g(x) = x^{3/2},\ g'(x) = \frac{3}{2}x^{1/2}$

$$\frac{d}{dx} f(g(x)) = x^{3/2}\sqrt{1 - x^3}\left(\frac{3}{2}x^{1/2}\right) = \frac{3}{2}x^2\sqrt{1 - x^3}$$

33. $\frac{dy}{du} = \frac{u}{u^2 + 1}$, $u = x^{3/2}$, $\frac{du}{dx} = \frac{3}{2}x^{1/2}$, $\frac{dy}{dx} = \frac{dy}{du}\frac{du}{dx}$

$$= \left(\frac{u}{u^2 + 1}\right)\frac{3}{2}x^{1/2}$$

Substitute $x^{3/2}$ for u.

$$\frac{dy}{dx} = \frac{x^{3/2}}{x^3 + 1}\left(\frac{3}{2}x^{1/2}\right) = \frac{3x^2}{2(x^3 + 1)}$$

34. $\frac{dy}{du} = \frac{u}{u^2 + 1}$, $u = x^2 + 1$, $\frac{du}{dx} = 2x$, $\frac{dy}{dx} = \left(\frac{u}{u^2 + 1}\right)(2x)$

Substitute $(x^2 + 1)$ for u.

$$\frac{dy}{dx} = \frac{(x^2 + 1)}{(x^2 + 1)^2 + 1}(2x) = \frac{2x(x^2 + 1)}{(x^2 + 1)^2 + 1}$$

35. $\frac{dy}{du} = \frac{u}{u^2 + 1}$, $u = \frac{5}{x}$, $\frac{du}{dx} = -\frac{5}{x^2}$

$$\frac{dy}{dx} = \frac{\frac{5}{x}}{\frac{25}{x^2} + 1}\left(-\frac{5}{x^2}\right) = \frac{-25}{x(25 + x^2)}$$

36. $\frac{dy}{du} = \frac{u}{\sqrt{1 + u^4}}$, $u = x^2$, $\frac{du}{dx} = 2x$

$$\frac{dy}{dx} = \frac{x^2}{\sqrt{1 + x^8}}(2x) = \frac{2x^3}{\sqrt{1 + x^8}}$$

37. $\frac{dy}{du} = \frac{u}{\sqrt{1 + u^4}}$, $u = \sqrt{x}$, $\frac{du}{dx} = \frac{1}{2}x^{-1/2}$

$$\frac{dy}{dx} = \frac{x^{1/2}}{\sqrt{1 + x^2}}\left(\frac{1}{2}x^{-1/2}\right) = \frac{1}{2\sqrt{1 + x^2}}$$

38. $\frac{dy}{du} = \frac{u}{\sqrt{1 + u^4}}$, $u = \frac{2}{u}$, $\frac{du}{dx} = \frac{-2}{x^2}$

$$\frac{dy}{dx} = \frac{\frac{2}{x}}{\sqrt{1 + \frac{16}{x^4}}}\left(-\frac{2}{x^2}\right) = \frac{-4}{x(x^4 + 16)^{1/2}}$$

39. (a) $\frac{dR}{dA}$, $\frac{dA}{dt}$, $\frac{dx}{dA}$, $\frac{dR}{dx}$ (b) $\frac{dR}{dt} = \frac{dR}{dx}\frac{dx}{dA}\frac{dA}{dt}$

40. (a) $\frac{dP}{dt}$, $\frac{dA}{dP}$, $\frac{dS}{dP}$, $\frac{dA}{dS}$ (b) $\frac{dA}{dt} = \frac{dA}{dS}\frac{dS}{dP}\frac{dP}{dt}$

41. $x^{2/3} + y^{2/3} = 8$

(a) $\frac{2}{3}x^{-1/3} + \frac{2}{3}y^{-1/3}\frac{dy}{dx} = 0$, $\frac{dy}{dx} = -\frac{2}{3}x^{-1/3}\cdot\frac{3}{2}y^{1/3} = -\frac{y^{1/3}}{x^{1/3}}$

(b) slope $= \frac{dy}{dx}$

When x = 8, y = -8, slope = 1.

42. $x^3 + y^3 = 9xy$

(a) $3x^2 + 3y^2\frac{dy}{dx} = 9y + 9x\frac{dy}{dx}$, $\frac{dy}{dx} = \frac{9y - 3x^2}{3y^2 - 9x} = \frac{3y - x^2}{y^2 - 3x}$

(b) slope $= \frac{dy}{dx}$

When x = 2, y = 4, slope $= \frac{3(4) - 4}{16 - 6} = \frac{4}{5}$.

43. $x^2y^2 = 9$

$2xy^2 + 2yx^2\frac{dy}{dx} = 0$, $\frac{dy}{dx} = -\frac{2xy^2}{2yx^2} = -\frac{y}{x}$

When x = 1, y = 3, $\frac{dy}{dx} = -3$.

44. $xy^4 = 48$

$y^4 + 4y^3x\frac{dy}{dx} = 0$, $\frac{dy}{dx} = -\frac{y^4}{4y^3x} = -\frac{y}{4x}$

When $x = 3$, $y = 2$, $\frac{dy}{dx} = \frac{-2}{4(3)} = -\frac{1}{6}$.

45. $x^2 - xy^3 = 20$

$2x - \left(y^3 + 3y^2x\frac{dy}{dx}\right) = 0$, $\frac{dy}{dx} = \frac{2x - y^3}{3y^2x}$

When $x = 5$, $y = 1$, $\frac{dy}{dx} = \frac{10 - 1}{3(1)(5)} = \frac{3}{5}$.

46. $xy^2 - x^3 = 10$

$\left(y^2 + 2yx\frac{dy}{dx}\right) - 3x^2 = 0$, $\frac{dy}{dx} = \frac{3x^2 - y^2}{2yx}$

When $x = 2$, $y = 3$, $\frac{dy}{dx} = \frac{3(4) - 9}{2(3)(2)} = \frac{1}{4}$.

47. $y^2 - 5x^3 = 4$

(a) $2y\frac{dy}{dx} - 15x^2 = 0$, $\frac{dy}{dx} = \frac{15x^2}{2y}$

(b) When $x = 4$, and $y = 18$, $\frac{dy}{dx} = \frac{15(16)}{2(18)} = \frac{20}{3}$ (thousand dollars per thousand unit increase in production)

(c) $\frac{dy}{dt} = \frac{dy}{dx}\frac{dx}{dt} = \frac{15x^2}{2y}\frac{dx}{dt}$

(d) When $x = 4$, $y = 18$, and $\frac{dx}{dt} = .3$, $\frac{dy}{dt} = \frac{15(16)}{2(18)}(.3) = 2$ (thousand dollars per week).

48. $y^3 - 8000x^2 = 0$

(a) $3y^2\frac{dy}{dx} - 1600x = 0$, $\frac{dy}{dx} = \frac{16,000x}{3y^2}$

(b) When $x = 27$, $y = 180$, $\frac{dy}{dx} = \frac{16,000(27)}{3(32,400)} = \frac{40}{9} \sim 4.44$.

(c) $\frac{dy}{dt} = \frac{dy}{dx}\frac{dx}{dt} = \frac{16,000x}{3y^2}\frac{dx}{dt}$

(d) When $x = 27$, $y = 180$, and $\frac{dx}{dt} = 1.8$.

$\frac{dy}{dt} = \frac{16,000(27)}{3(32,400)}(1.8) = 8$ (thousand books per year)

49. $6p + 5x + xp = 50$

$6 + 5\frac{dx}{dp} + p\frac{dx}{dp} + x = 0, \quad \frac{dx}{dp} = \frac{-x - 6}{(5 + p)}$

$\frac{dx}{dt} = \frac{dx}{dp}\frac{dp}{dt} = \left(\frac{-x - 6}{5 + p}\right)\frac{dp}{dt}$

When $x = 4$, $p = 3$, and $\frac{dp}{dt} = -2$, $\frac{dx}{dt} = \left(\frac{-4 - 6}{5 + 3}\right)(-2) = 2.5$

The quantity is increasing at the rate of 2.5 units per unit time.

50. $V = .005\pi r^2$, so $\frac{dV}{dt} = .005\pi(2)r\frac{dr}{dt} = .01\pi r\frac{dr}{dt}$

Now, $\frac{dV}{dt} = 20$, so $20 = .01r\pi\frac{dr}{dt}$, $\frac{dr}{dt} = \frac{200}{\pi r}$.

When $r = 50$, $\frac{dr}{dt} = \frac{200}{50\pi} = \frac{40}{\pi}$ m/hr.

51. $S = .1x^{2/3}$, so $\frac{dS}{dt} = \frac{2}{3}(.1)w^{-1/3}\frac{dw}{dt} = \frac{.2}{3}w^{-1/3}\frac{dw}{dt}$.

When $w = 350$ and $\frac{dw}{dt} = 200$, $\frac{dS}{dt} = \frac{.2}{3\sqrt[3]{350}}(200) = \frac{40}{3\sqrt[3]{350}}$

~ 1.89 m^2/yr.

52. $xy - 6x + 20y = 0$, $\frac{dx}{dt}y + \frac{dy}{dt}x - 6\frac{dx}{dt} + 20\frac{dy}{dt} = 0$

Currently, $x = 10$, $y = 2$, $\frac{dx}{dt} = 1.5$. Thus,

$1.5(2) + \frac{dy}{dt}(10) - 6(1.5) + 20\frac{dy}{dt} = 0$, so $30\frac{dy}{dt} = 6$, $\frac{dy}{dt} = .2$ or 200 dishwashers/month.

Exercises 4.1

1. $4^x = (2^2)^x = 2^{2x}$, $(\sqrt{3})^x = (3^{1/2})^x = 3^{(1/2)x}$,

$\left(\frac{1}{9}\right)^x = (9^{-1})^x = ((3^2)^{-1})^x = 3^{-2x}$

2. $27^x = (3^3)^x = 3^{3x}$, $(\sqrt[3]{2})^x = (2^{1/3})^x = 2^{(1/3)x}$,

$\left(\frac{1}{8}\right)^x = (8^{-1})^x = ((2^3)^{-1})^x = 2^{-3x}$

3. $8^{2x/3} = (2^3)^{2x/3} = 2^{3(2x/3)} = 2^{2x}$,

$9^{3x/2} = (3^2)^{3x/2} = 3^{2(3x/2)} = 3^{3x}$,

$16^{-3x/4} = (2^4)^{-3x/4} = 2^{4(-3x/4)} = 2^{-3x}$

4. $9^{-x/2} = (3^2)^{-x/2} = e^{2(-x/2)} = 3^{-x}$,

$8^{4x/3} = (2^3)^{4x/3} = 2^{3(4x/3)} = 2^{4x}$,

$27^{-2x/3} = (3^3)^{-2x/3} = 3^{3(-2x/3)} = 3^{-2x}$

5. $\left(\frac{1}{4}\right)^{2x} = (4^{-1})^{2x} = ((2^2)^{-1})^{2x} = 2^{-4x}$,

$\left(\frac{1}{8}\right)^{-3x} = (8^{-1})^{-3x} = ((2^3)^{-1})^{-3x} = 2^{9x}$,

$\left(\frac{1}{81}\right)^{x/2} = (81^{-1})^{x/2} = ((3^4)^{-1})^{x/2} = 3^{-2x}$

6. $\left(\frac{1}{9}\right)^{2x} = (9^{-1})^{2x} = ((3^2)^{-1})^{2x} = 3^{-4x}$,

$\left(\frac{1}{27}\right)^{x/3} = (27^{-1})^{x/3} = ((3^3)^{-1})^{x/3} = 3^{-3(x/3)} = 3^{-x}$,

$\left(\frac{1}{16}\right)^{-x/2} = (16^{-1})^{-x/2} = ((2^4)^{-1})^{-x/2} = 2^{-4(-x/2)} = 2^{2x}$

7. $2^{3x} \cdot 2^{-5x/2} = 2^{3x-(5x/2)} = 2^{(6x/2)-(5x/2)} = 2^{x/2}$,

$3^{2x} \cdot \left(\frac{1}{3}\right)^{2x/3} = 3^{2x} \cdot (3^{-1})^{2x/3} = 3^{2x} \cdot 3^{-2x/3} = 3^{(6x/3)-(2x/3)}$

$= 3^{4x/3} = 3^{(4/3)x}$

8. $2^{5x/4}\cdot\left(\frac{1}{2}\right)^{x} = 2^{5x/4}\cdot(2^{-1})^{x} = 2^{5x/4}\cdot 2^{-4x/4} = 2^{(5x/4)-(4x/4)}$

$= 2^{x/4}$,

$3^{-2x}\cdot 3^{5x/2} = 3^{-4x/2}\cdot 3^{5x/2} = 3^{(-4x/2)+(5x/2)} = 3^{x/2}$

9. $(2^{-3x}\cdot 2^{-2x})^{2/5} = (2^{-3x-2x})^{2/5} = (2^{-5x})^{2/5} = 2^{-2x}$,

$(9^{1/2}\cdot 9^{4})^{x/9} = ((3^{2})^{1/2}\cdot(3^{2})^{4})^{x/9} = (3^{1}\cdot 3^{8})^{x/9} = (3^{1+8})^{x/9}$

$= (3^{9})^{x/9} = 3^{x}$

10. $(3^{-x}\cdot 3^{x/5})^{5} = 3^{-5x}\cdot 3^{(x/5)5} = 3^{-5x}\cdot 3^{x} = 3^{-5x+x} = 3^{-4x}$,

$(16^{1/4}\cdot 16^{-3/4})^{3x} = ((2^{4})^{1/4}\cdot(2^{4})^{-3/4})^{3x} = (2^{1}\cdot 2^{-3})^{3x}$

$= (2^{1-3})^{3x} = (2^{-2})^{3x} = 2^{-6x}$

11. $\frac{3^{4x}}{3^{2x}} = 3^{4x}\cdot 3^{-2x} = 3^{4x-2x} = 3^{2x}$,

$\frac{2^{5x+1}}{2\cdot 2^{-x}} = \frac{2^{5x+1}}{2^{1-x}} = 2^{5x+1}\cdot 2^{-(1-x)} = 2^{5x+1-(1-x)} = 2^{6x}$,

$\frac{9^{-x}}{27^{-x/3}} = \frac{(3^{2})^{-x}}{(3^{3})^{-x/3}} = \frac{3^{-2x}}{3^{-x}} = 3^{-2x}\cdot 3^{x} = 3^{-2x+x} = 3^{-x}$

12. $\frac{2^{x}}{6^{x}} = \frac{2^{x}}{(2\cdot 3)^{x}} = \frac{2^{x}}{2^{x}\cdot 3^{x}} = \frac{1}{3^{x}} = 3^{-x}$.

$\frac{3^{-5x}}{3^{-2x}} = 3^{-5x}\cdot 3^{2x} = 3^{-5x+2x} = 3^{-3x}$,

$\frac{16^{x}}{8^{-x}} = \frac{(2^{4})^{x}}{(2^{3})^{-x}} = 2^{4x}\cdot 2^{3x} = 2^{4x+3x} = 2^{7x}$

13. $6^{x}\cdot 3^{-x} = (2\cdot 3)^{x}\cdot 3^{-x} = 2^{x}\cdot(3^{x}\cdot 3^{-x}) = 2^{x}\cdot(3^{x-x}) = 2^{x}$,

$\frac{15^{x}}{5^{x}} = \frac{(3\cdot 5)^{x}}{5^{x}} = \frac{3^{x}\cdot 5^{x}}{5^{x}} = 3^{x}$,

$\frac{12^{x}}{2^{2x}} = \frac{(3\cdot 4)^{x}}{2^{2x}} = \frac{3^{x}\cdot 4^{x}}{2^{2x}} = \frac{3^{x}\cdot(2^{2})^{x}}{2^{2x}} = 3^{x}$

14. $7^{-x}\cdot 14^x = 7^{-x}\cdot(7\cdot 2)^x = 7^{-x}\cdot 7^x\cdot 2^x = 7^{-x+x}\cdot 2^x = 2^x$

$\frac{2^x}{6^x} = \frac{2}{(2.3)^x} = \frac{2^x}{2^x.3^x} = \frac{1}{3^x} = 3^{-x}$

$\frac{3^{2x}}{18^x} = \frac{3^{2x}}{9^x.2^x} = \frac{3^{2x}}{3^{2x}.2^x} = \frac{1}{2^x} = 2^{-x}$

15. (a) $2^3 = 8$ (b) $2^{-3} = \frac{1}{2^3} = \frac{1}{8}$

(c) $2^{5/2} = 2^{2+(1/2)} = 2^2\cdot 2^{1/2} \sim 4(1.414) \sim 5.66$

(d) $2^{4.1} = 2^{4+(1/10)} = 2^4\cdot 2^{1/10} \sim 16(1.072) \sim 17.15$

(e) $2^{.2} = 2^{2(1/10)} = (2^{1/10})^2 \sim (1.072)^2 \sim 1.15$

(f) $2^{.9} = 2^{1-.1} = 2\cdot 2^{-1} = \frac{2}{2^{1/10}} \sim \frac{2}{1.072} \sim 1.87$

(g) $2^{-2.5} = 2^{-3+.5} = 2^{-3}\cdot 2^{.5} = \frac{1}{2^3}\cdot 2^{1/2} \sim \frac{1}{8}(1.414) \sim .18$

(h) $2^{-3.9} = 2^{-4+.1} = 2^{-4}\cdot 2^{.1} = \frac{1}{2^4}\cdot 2^{1/10} \sim \frac{1.072}{16} \sim .07$

16. (a) $2.7^{.2} = 2.7^{2(1/10)} = (2.7^{1/10})^2 \sim (1.104)^2 \sim 1.22$

(b) $2.7^{1.5} = 2.7\cdot 2.7^{1/2} \sim 2.7(1.643) \sim 4.44$

(c) $2.7^0 = 1$ (d) $2.7^{-1} = \frac{1}{2.7} \sim 3.7$

(e) $2.7^{1.1} = 2.7\cdot 2.7^{1/10} \sim 2.7(1.104) \sim 2.98$

(f) $2.7^{.6} = 2.7^{.1+.5} = 2.7^{.1}\cdot 2.7^{.5} \sim (1.643)(1.104) \sim 1.81$

17. $5^{2x} = 5^2$; $2x = 2$; $x = 1$

18. $10^{-x} = 10^2$; $-x = 2$; $x = -2$

19. $(2.5)^{2x+1} = (2.5)^5$; $2x + 1 = 5$; $x = \frac{5 - 1}{2} = 2$

20. $(3.2)^{x-3} = (3.2)^5$; $x - 3 = 5$; $x = 8$

21. $10^{1-x} = 100$; $10^{1-x} = 10^2$; $1 - x = 2$; $x = -1$

22. $2^{4-x} = 8$; $2^{4-x}\ 2^3$; $4 - x = 3$; $x = 1$

23. $3(2.7)^{5x} = 8.1$; $8.1^{5x} = 8.1$; $5x = 1$; $x = \frac{1}{5}$

24. $4(2.7)^{2x-1}10.8$; $10.8^{2x-1} = 10.8$; $2x - 1 = 1$; $x = \frac{1 + 1}{2} = 1$

25. $(2^{x+1}\cdot 2^{-3})^2 = 2$; $(2^{x+1-3})^2 = 2$; $(2^{x-2})^2 = 2$; $2^{2x-4} = 2$;

$2x - 4 = 1$; $x = \frac{4 + 1}{2} = \frac{5}{2}$

26. $(3^{2x}\cdot 3^2)^4 = 3$; $(3^{2x+2})^4 = 3$; $3^{8x+8} = 3$; $8x + 8 = 1$;

$x = \frac{1 - 8}{8} = -\frac{7}{8}$

27. $2^{3x} = 4\cdot 2^{5x}$; $2^{3x} = 2^2\cdot 2^{5x}$; $2^{3x} = 2^{2+5x}$; $3x = 2 + 5x$; $2x = -2$;

$x = -1$

28. $3^{5x}\cdot 3^x - 3 = 0$; $3^{5x+x} = 3$; $5x + x = 1$; $6x = 1$; $x = \frac{1}{6}$

29. $(1 + x)2^{-x} - 5\cdot 2^{-x} = 0$; $2^{-x}(1 + x - 5) = 0$; $2^{-x}(x - 4) = 0$

Since $2^{-x} \neq 0$ for every x, then $x = 4$ is the only solution.

30. $(2 - 3x)5^x + 4\cdot 5^x = 0$; $5^x(2 - 3x + 4) = 0$; $5^x(6 - 3x) = 0$

Since $5^x \neq 0$ for every x, then $x = 2$ is the only solution.

31. $2^{3+h} = 2^3\cdot 2^h$ The missing factor is 2^h.

32. $5^{2+h} = 5^2\cdot 5^h = 25\cdot 5^h$ The missing factor is 5^h.

33. $2^{x+h} - 2^x = 2^x\cdot 2^h - 2^x = 2^x(2^h - 1)$

The missing factor is $2^h - 1$.

34. $5^{x+h} + 5^x = 5^x \cdot 5^h + 5^x = 5^x(5^h + 1)$

The missing factor is $5^h + 1$.

35. $3^{x/2} + 3^{-x/2} = 3^{x-(x/2)} + 3^{-x/2} = 3^x \cdot 3^{-x/2} + 3^{-x/2}$

$= 3^{-x/2}(3^x + 1)$ The missing factor is $3^x + 1$.

36. $5^{7x/2} - 5^{x/2} = 5^{x/2} \cdot 5^{6x/2} - 5^{x/2} = 5^{x/2}(5^{6x/2} - 1)$

$= \sqrt{5^x}\,(5^{3x} - 1)$ The missing factor is $5^{3x} - 1$.

37. $3^{10x} - 1 = 1 \cdot (3^{5x})^2 + 0 \cdot (3^{5x}) + (-1)$

Use the quadratic formula with $a = 1$, $b = 0$, $c = -1$,

$x = \pm 1$. Thus $3^{10x} - 1 = (3^{5x} - 1)(3^{5x} + 1)$.

The missing factor is $3^{5x} + 1$.

38. As in ex. 37, we need to find a binomial such that the inner terms of the product cancel. Thus, by inspection, that factor is $2^{3x} + 2^{x/2}$.

Exercises 4.2

1. If $h = .1$, then $\frac{3^h - 1}{h} \sim \frac{1.11612 - 1}{.1} = 1.612$.

If $h = .01$, then $\frac{3^h - 1}{h} \sim \frac{1.01105 - 1}{.01} = 1.105$.

If $h = .001$, then $\frac{3^h - 1}{h} \sim \frac{1.00110 - 1}{.001} = 1.10$.

Therefore $\frac{d}{dx} 3^x \Big|_{x=0} = \lim_{x \to 0} \frac{3^h - 1}{h} \sim 1.1$.

2. If $h = .1$, then $\dfrac{(2.7)^h - 1}{h} \sim \dfrac{1.10443 - 1}{.1} = 1.0443.$

If $h = .01$, then $\dfrac{(2.7)^h - 1}{h} \sim \dfrac{1.00998 - 1}{.01} = .998.$

If $h = .001$, then $\dfrac{(2.7)^h - 1}{h} \sim \dfrac{1.00099 - 1}{.001} = .99.$

Therefore, $\dfrac{d}{dx}(2.7)^x\Big|_{x=0} = \lim\limits_{x \to 0} \dfrac{(2.7)^h - 1}{h} \sim .99.$

3. At $h = .01$, $\dfrac{e^h - 1}{h} \sim \dfrac{1.01005 - 1}{.01} = 1.005.$

At $h = .005$, $\dfrac{e^h - 1}{h} \sim \dfrac{1.00501 - 1}{.005} = 1.002.$

At $h = .001$, $\dfrac{e^h - 1}{h} \sim \dfrac{1.00100 - 1}{.001} = 1.00.$

4. $\dfrac{d}{dx}(5e^x) = 5(e^x) + e^x(0) = 5e^x$

5. $\dfrac{d}{dx}(e^x)^{10} = 10(e^x)^9 e^x = 10e^{10x}$

6. $\dfrac{d}{dx}(e^{2+x}) = \dfrac{d}{dx}(e^2 \cdot e^x) = e^2(e^x) + e^x(0) = e^{2+x}$

7. $\dfrac{d}{dx}(e^{4x}) = \dfrac{d}{dx}(e^x)^4 = 4(e^x)^3 e^x = 4e^{4x}$

8. $\dfrac{d}{dx}(e^x + x^2) = \dfrac{d}{dx}e^x + \dfrac{d}{dx}x^2 = e^x + 2x$

9. $e^{2x}(1 + e^{3x}) = e^{2x} + e^{2x} \cdot e^{3x} = e^{2x} + e^{5x}$

10. $(e^x)^2 = e^{2x}$

11. $e^{1-x} \cdot e^{2x} = e^{1-x+2x} = e^{1+x}$

12. $\dfrac{5e^{3x}}{e^x} = \dfrac{5e^{3x}}{e^x} \cdot \dfrac{e^{-x}}{e^{-x}} = \dfrac{5e^{2x}}{e^0} = 5e^{2x}$

13. $\dfrac{1}{e^{-2x}} = e^{2x}$

14. $e^3 \cdot e^{x+1} = e^{3+(x+1)} = e^{x+4}$

15. $e^2 \sim 7.3891$

16. $e^{-1.30} \sim .27253$ 17. $e^{-.5} \sim .60653$

18. $e^{3/2} = e^{1.5} \sim 4.48169$ 19. $e^{5x} = e^{20}$; $5x = 20$; $x = 4$

20. $e^{1-x} = e^{2}$; $1 - x = 2$; $x = -1$

21. $e^{x^2-2x} = e^{8}$; $x^2 - 2x = 9$; $x^2 - 2x - 8 = (x - 4)(x + 2)$;

$x = 4$ or $x = -2$

22. $e^{-x} = 1$; $e^{-x} = e^{0}$; $-x = 0$; $x = 0$

23. $\frac{d}{dx} xe^{x} = x(e^{x}) + e^{x}(1) = xe^{x} + e^{x}$

24. $\frac{d}{dx}\left[\frac{e^{x}}{x}\right] = \frac{x(e^{x}) - e^{x}(1)}{x^2} = \frac{e^{x}(x - 1)}{x^2}$

25. $\frac{d}{dx}\left[\frac{e^{x}}{1 + e^{x}}\right] = \frac{(1 + e^{x})(e^{x}) - (e^{x})(e^{x})}{(1 + e^{x})^2} = \frac{e^{x} + e^{2x} - e^{2x}}{(1 + e^{x})^2}$

$= \frac{e^{x}}{(1 + e^{x})^2}$

26. $\frac{d}{dx}\left[(1 + x^2)e^{x}\right] = (1 + x^2)e^{x} + e^{x}(2x) = e^{x}(x^2 + 2x + 1)$

27. $\frac{d}{dx}\left[(1 + 5e^{x})^4\right] = 4(1 + 5e^{x})^3(5e^{x}) = 20e^{x}(1 + 5e^{x})^3$

28. $\frac{d}{dx}\left[(xe^{x} - 1)^{-3}\right] = -3(xe^{x} - 1)^{-4}(xe^{x} + e^{x}) = \frac{-3e^{x}(x + 1)}{(xe^{x} - 1)^4}$

Exercises 4.3

1. $\frac{d}{dx}(e^{-x}) = e^{-x}(-1) = -e^{-x}$ 2. $\frac{d}{dx}(e^{10x}) = e^{10x}(10) = 10e^{10x}$

3. $\frac{d}{dx}(5e^{x}) = 5e^{x} + e^{x}(0) = 5e^{x}$

4. $\frac{d}{dx}\left[\frac{e^{x} + e^{-x}}{2}\right] = \frac{1}{2}\frac{d}{dx}\left[e^{x} + e^{-x}\right] = \frac{1}{2}(e^{x} + e^{-x}(-1)) = \frac{e^{x} - e^{-x}}{2}$

5. $\frac{d}{dt}\left[e^{t^2}\right] = e^{t^2}(2t) = 2te^{t^2}$

6. $\frac{d}{dt}\left[e^{-2t}\right] = e^{-2t}(-2) - 2e^{-2t}$

7. $\frac{d}{dx}\left[\frac{e^{x} - e^{-x}}{2}\right] = \frac{1}{2}\frac{d}{dx}(e^{x} - e^{-x}) = \frac{1}{2}(e^{x} - e^{-x}(-1)) = \frac{e^{x} + e^{-x}}{2}$

8. $\frac{d}{dx}(2e^{1-x}) = 2e^{1-x}(-1) + e^{1-x}(0) = -2e^{1-x}$

9. $\frac{d}{dx}(e^{-2x} - 2x) = e^{-2x}(-2) - 2 = -2(e^{-2x} + 1)$

10. $\frac{d}{dx}\left[\frac{1}{10}e^{-x^2/2}\right] = \frac{1}{10}\left[e^{-x^2/2}\left(-\frac{2x}{2}\right)\right] = \frac{1}{10}\left[-xe^{-x^2/2}\right] = -\frac{1}{10}xe^{-x^2/2}$

11. $\frac{d}{dx}\left[(e^{x} + e^{-x})^3\right] = 3(e^{x} + e^{-x})^2(e^{x} - e^{-x})$

12. $\frac{d}{dx}\left[(e^{-x})^2\right] = 2(e^{-x})(e^{-x})(-1) = -2e^{-2x}$

13. $\frac{d}{dx}\left[\frac{1}{3}e^{3-2x}\right] = \frac{1}{3}\left(e^{3-2x}\right)(-2) = -\frac{2}{3}e^{3-2x}$

14. $\frac{d}{dx}(e^{1/x}) = \frac{d}{dx}\left[e^{x^{-1}}\right] = \left(e^{x^{-1}}\right)\left(-1\cdot x^{-2}\right) = -\frac{1}{x^2}e^{1/x}$

15. $\frac{d}{dt}\left[e^{t}(e^{2t} - e^{-t})\right] = \frac{d}{dt}\left[e^{3t} - e^{0}\right] = \frac{d}{dt}\left[e^{3t} - 1\right] = e^{3t}(3)$
$= 3e^{3t}$

16. $\frac{d}{dt}\left[\frac{e^{t} + e^{-t}}{e^{t}}\right] = \frac{d}{dt}\left[1 + \frac{e^{-t}}{e^{t}}\right] = \frac{d}{dt}\left[1 + e^{-2t}\right] = e^{-2t}(-2)$
$= -2e^{-2t}$

17. $\frac{d}{dx}\left[e^{x^3+x-(1/x)}\right] = e^{x^3+x-(1/x)}\left(3x^2 + 1 + \frac{1}{x^2}\right)$

$= \left(3x^2 + 1 + \frac{1}{x^2}\right)e^{x^3+x-(1/x)}$

18. $\frac{d}{dx}\left[\left(e^{x^2} + x^2\right)^5\right] = 5\left(e^{x^2} + x^2\right)^4\left(e^{x^2}(2x) + 2x\right)$

$= 10x\left(e^{x^2} + 1\right)\left(e^{x^2} + x^2\right)^4$

19. $\frac{d}{dx}\left[(2x + 1 - e^{2x+1})^4\right] = 4(2x + 1 - e^{2x+1})^3(2 - e^{2x+1}(2))$

$= 8(2x + 1 - e^{2x+1})^3(1 - e^{2x+1})$

20. $\frac{d}{dx}\left[e^{1/(3x-7)}\right] = \frac{d}{dx}\left[e^{(3x-7)^{-1}}\right] = e^{(3x-7)^{-1}}(-1)(3x - 7)^{-2}(3)$

$= \frac{-3e^{1/(3x-7)}}{(3x - 7)^2}$

21. $\frac{d}{dx}\left[x^3e^{x^2}\right] = x^3\left[e^{x^2}(2x)\right] + e^{x^2}(3x^2) = 2x^4e^{x^2} + 3x^2e^{x^2}$

22. $\frac{d}{dx}\left[x^2e^{-3x}\right] = x^2\left[e^{-3x}(-3)\right] + e^{-3x}(2x) = -3x^2e^{-3x} + 2xe^{-3x}$

$= xe^{-3x}(2 - 3x)$

23. $\frac{d}{dx}\left[\frac{e^{x^3}}{x}\right] = \frac{xe^{x^3}(3x^2) - e^{x^3}(1)}{x^2} = \frac{e^{x^3}(3x^3 - 1)}{x^2}$

or $3xe^{x^3} - x^{-2}e^{x^3}$

24. $\frac{d}{dx}\left[e^{(x-1)/x}\right] = \frac{d}{dx}\left[e^{(x-1)x^{-1}}\right]$

$= e^{(x-1)x^{-1}}\left[(x - 1)(-1)x^{-2} + (x^{-1})(1)\right]$

$= e^{(x-1)x^{-1}}\left[-x^{-1} + x^{-2} + x^{-1}\right] = \frac{e^{(x-1)/x}}{x^2}$

25. $\frac{d}{dx}\left[(x + 1)e^{-x+2}\right] = (x + 1)e^{-x+2}(-1) + e^{-x+2}(1)$

$= e^{-x+2}(-x - 1 + 1) = -xe^{-x+2}$

26. $\frac{d}{dx}\left[x\sqrt{2 + e^x}\right] = \frac{d}{dx}\left[x(2 + e^x)^{1/2}\right]$

$= x\left(\frac{1}{2}\right)(2 + e^x)^{-1/2}e^x + (2 + e^x)^{1/2}(1)$

$= \frac{1}{2}x(2 + e^x)^{-1/2}e^x + (2 + e^x)^{1/2}$

27. $\frac{d}{dx}\left[\left(\frac{1}{x} + 3\right)e^x\right] = \left(\frac{1}{x} + 3\right)e^x + e^x\left(-\frac{1}{x^2}\right) = e^x\left(-\frac{1}{x^2} + \frac{1}{x} + 3\right)$

28. $\frac{d}{dx}\left[\frac{e^{-3x}}{1 - 3x}\right] = \frac{(1 - 3x)e^{-3x}(-3) - e^{-3x}(-3)}{(1 - 3x)^2}$

$= \frac{-3e^{-3x}(1 - 3x - 1)}{(1 - 3x)^2} = \frac{9xe^{-3x}}{(1 - 3x)^2}$

29. $\frac{d}{dx}\left[\frac{e^x - 1}{e^x + 1}\right] = \frac{(e^x + 1)e^x - (e^x - 1)e^x}{(e^x + 1)^2} = \frac{e^{2x} + e^x - e^{2x} + e^x}{(e^x + 1)^2}$

$= \frac{2e^x}{(e^x + 1)^2}$

30. $\frac{d}{dx}\left[\frac{xe^x - 3}{x + 1}\right] = \frac{(x + 1)(xe^x + e^x(1)) - (xe^x - 3)(1)}{(x + 1)^2}$

$= \frac{x^2e^x + xe^x + xe^x + e^x - xe^x + 3}{(x + 1)^2}$

$= \frac{(x^2 + x + 1)e^x + 3}{(x + 1)^2}$

31. $f(x) = (1 + x)e^{-x/2}$

$f'(x) = (1 + x)e^{-x/2}\left(-\frac{1}{2}\right) + e^{-x/2}(1) = -\frac{1}{2}(1 + x)e^{-x/2} + e^{-x/2}$

$= \left[-\frac{1}{2}(1 + x) + 1\right]e^{-x/2} = \left(\frac{1}{2} - \frac{x}{2}\right)e^{-x/2}$

$f''(x) = \left(\frac{1}{2} - \frac{x}{2}\right)e^{-x/2}\left(-\frac{1}{2}\right) + e^{-x/2}\left(-\frac{1}{2}\right) = -\frac{e^{-x/2}}{2}\left(\frac{1}{2} - \frac{x}{2} + 1\right)$

$$= -\frac{e^{-x/2}}{2}\left(\frac{3}{2} - \frac{x}{2}\right)$$

Setting $f'(x) = 0$ we see that the only critical point is at $x = 1$. $f''(1) = -\frac{e^{-1/2}}{2}\left(\frac{3}{2} - \frac{1}{2}\right) = -\frac{e^{-1/2}}{2} < 0$

Since $f''(1) < 0$, we conclude that $x = 1$ is a relative maximum.

32. $f(x) = (1 - x)e^{-x/2}$

$$f'(x) = (1 - x)e^{-x/2}\left(-\frac{1}{2}\right) + e^{-x/2}(-1)$$

$$= -\frac{1}{2}e^{-x/2} + \frac{x}{2}e^{-x/2} - e^{-x/2} = \left(-\frac{3}{2} + \frac{x}{2}\right)e^{-x/2}$$

$$f''(x) = \left(-\frac{3}{2} + \frac{x}{2}\right)e^{-x/2}\left(-\frac{1}{2}\right) + e^{-x/2}\left(\frac{1}{2}\right) = \frac{1}{2}e^{-x/2}\left(1 + \frac{3}{2} - \frac{x}{2}\right)$$

$$= \frac{1}{2}e^{-x/2}\left(\frac{5}{2} - \frac{x}{2}\right)$$

The only critical point is at $x = 3$. Using the second derivative test, $f''(3) > 0$, so $x = 3$ is a relative minimum.

33. $f(x) = \frac{3 - 2x}{e^{x/4}}$

$$f'(x) = \frac{e^{x/4}(-2) - (3 - 2x)e^{x/4}(1/4)}{(e^{x/4})^2}$$

$$= \frac{\left(-\frac{11}{4} + \frac{1}{2}x\right)e^{x/4}}{(e^{x/4})^2} = \frac{-\frac{11}{4} + \frac{1}{2}x}{e^{x/4}}$$

$$f''(x) = \frac{e^{x/4}\left(\frac{1}{2}\right) - \left(-\frac{11}{4} + \frac{1}{2}x\right)e^{x/4}\left(\frac{1}{4}\right)}{(e^{x/4})^2}$$

$$= \frac{1 + \frac{11}{8} - \frac{1}{4}x}{2e^{x/4}}$$

The only critical point is at $x = \frac{11}{2}$ and since $f''\left(\frac{11}{2}\right) > 0$, $x = \frac{11}{2}$ is a relative minimum.

34. $f(x) = \dfrac{4x - 3}{e^{x/2}}$

$$f'(x) = \frac{e^{x/2}(4) - (4x - 3)e^{x/2}(1/2)}{(e^{x/2})^2} = \frac{e^{x/2}(8 - 4x + 3)}{2(e^{x/2})^2}$$

$$= \frac{4 - 4x}{2e^{x/2}}$$

$$f''(x) = \frac{2e^{x/2}(-4) - (11 - 4x)2e^{x/2}(1/2)}{4(e^{x/2})^2}$$

$$= \frac{e^{x/2}(-8 - 11 - 4x)}{4(e^{x/2})^2} = \frac{4x - 19}{4e^{x/2}}$$

The only critical point occurs at $x = \frac{11}{4}$ and since $f''\left(\frac{11}{4}\right) < 0$,

$x = \frac{11}{4}$ is a relative maximum.

35. $f(x) = (8 - 2x)e^{x+5}$

$$f'(x) = (8 - 2x)e^{x+5} + e^{x+5}(-2) = e^{x+5}(6 - 2x)$$

$$f''(x) = e^{x+5}(-2) + (6 - 2x)e^{x+5} = e^{x+5}(4 - 2x)$$

The only critical point occurs at $x = 3$ and since $f''(3) < 0$,

$x = 3$ is a relative maximum.

36. $f(x) = (4x - 1)e^{3x-2}$

$$f'(x) = (4x - 1)e^{3x-2}(3) + e^{3x-2}(4) = e^{3x-2}(12x + 1)$$

$$f''(x) = e^{3x-2}(12) + (12x + 1)e^{3x-2}(3) = e^{3x-2}(15 + 12x)$$

The only critical point is at $x = -\frac{1}{12}$, and since $f''\left(-\frac{1}{12}\right) > 0$,

$x = -\frac{1}{12}$ is a relative minimum.

37. $f(x) = \dfrac{(x - 1)^2}{e^x}$

$$f'(x) = \frac{e^x(2)(x - 1) - (x - 1)^2e^x}{(e^x)^2}$$

$$= \frac{e^x[2(x - 1) - (x^2 - 2x + 1)]}{(e^x)^2} = \frac{-x^2 + 4x - 3}{e^x}$$

$$f''(x) = \frac{e^x(-2x + 4) - (-x^2 + 4x - 3)e^x}{(e^x)^2} = \frac{x^2 - 6x + 7}{e^x}$$

The critical points occur at $x = 1$ and at $x = 3$. For $x = 1$, $f''(1) = 2 > 0$, hence there is a minimum there. For $x = 3$, $f''(3) = -2 < 0$, so a maximum.

38. $f(x) = (x + 3)^2e^x$

$$f'(x) = (x + 3)^2e^x + e^x(2)(x + 3)$$

$$= e^x(x + 3)(x + 5) \text{ or } e^x(x^2 + 8x + 15)$$

$$f''(x) = e^x(2x + 8) + (x^2 + 8x + 15)e^x = e^x(x^2 + 10x + 23)$$

The critical points are at $x = -3, -5$. By the second derivative test,

$f''(-3) > 0$ implies that $x = -3$ is a relative minimum, and

$f''(-5) < 0$ implies that $x = -5$ is a relative maximum.

39. $f(x) = (x + 5)^2e^{2x-1}$

$$f'(x) = (x + 5)^2e^{2x-1}(2) + e^{2x-1}(2)(x + 5)$$

$$= 2e^{2x-1}(x + 5)(x + 6) \text{ or } 2e^{2x-1}(x^2 + 11x + 30)$$

$$f''(x) = 2e^{2x-1}(2x + 11) + (x^2 + 11x + 30)2e^{2x-1}(2)$$

$$= 2e^{2x-1}(2x^2 + 24x + 71)$$

The critical points are at $x = -5, -6$. Using the second derivative test,

$f''(-5) > 0$ implies that $x = -5$ is a relative minimum, and

$f''(-6) < 0$ implies that $x = -6$ is a relative maximum.

40. $f(x) = \dfrac{(x - 3)^2}{e^{2x}}$

$$f'(x) = \frac{e^{2x}(2)(x - 3) - (x - 3)^2e^{2x}(2)}{(e^{2x})^2}$$

$$= \frac{2e^{2x}(x - 3 - (x^2 - 6x + 9))}{(e^{2x})^2} = \frac{2(-x^2 + 7x - 12)}{e^{2x}}$$

$$f''(x) = \frac{e^{2x}(2)(-2x + 7) - 2(-x^2 + 7x - 12)e^{2x}(2)}{(e^{2x})^2}$$

$$= \frac{2e^{2x}(2x^2 - 16x + 31)}{(e^{2x})^2} = \frac{4x^2 - 32x + 62}{e^{2x}}$$

Using the quadratic formula with $a = -1$, $b = 7$, $c = -12$ we determine that the critical points are at $x = 3, 4$. Using the second derivative test

$f''(3) > 0$ implies that $x = 3$ is a relative minimum, and

$f''(4) < 0$ implies that $x = 4$ is a relative maximum.

41. $y = e^{-2e^{-.01x}}$

$$\frac{dy}{dx} = e^{-2e^{-.01x}}(-2e^{-.01x}(-.01)) = .02e^{-2e^{-.01x}}e^{-.01x}$$

42. $\left.\frac{dv}{dx}\right|_{t=4} = \left.-700e^{-.035t}\right|_{t=4} = -172.62$

so it depreciates at a rate of 172.62\$/yr.

43. $\left.\frac{dv}{dx}\right|_{t=5} = \left.20000e^{-.2t}\right|_{t=5} = 54365.64$ \$/yr.

44. $y = e^{-(1/10)e^{-x/2}}$

$$\frac{dy}{dx} = e^{-(1/10)e^{-x/2}}\left(-\frac{1}{10}\right)\left(e^{-x/2}\right)\left(-\frac{1}{2}\right) = \frac{1}{20}e^{-(1/10)e^{-x/2}}e^{-x/2}$$

45. Since $y' = -4y$ is of the form $y' = ky$, then y is an exponential function of the form $y = Ce^{-4x}$.

46. As in ex. 43, we know that y is an exponential function of the form $y = Ce^{(1/3)x}$.

47. $y' = -.5y$ so for some C, $y = Ce^{-.5x}$. Using that $f(0) = 1$, we have $1 = Ce^{-.5(0)} = C\cdot 1 = C$. Thus $y = e^{-.5x}$.

48. $y' = 3y$ so for some C, $y = Ce^{3x}$. Using that $f(0) = \frac{1}{2}$, we have $\frac{1}{2} = Ce^{3(0)} = C \cdot 1 = C$. Thus, $y = \frac{1}{2}e^{3x}$.

49. Following the hint, suppose that $g(x) = f(x)e^{-kx}$. Then $g'(x) = f(x)e^{-kx}(-k) + e^{-kx}(f'(x)) = e^{-kx}(-k \cdot f(x) + f'(x))$.

But as given in (3), $-k \cdot f(x) + f'(x) = 0$, so $g'(x) = 0$, i.e., $g(x)$ is a constant function, $g(x) = C$ for some C. Then $C = f(x)e^{-kx}$, so $f(x) = Ce^{kx}$ and (3) is verified.

50. $g(x) = f(e^x)$; $g'(x) = f'(e^x) \cdot e^x$

Since $f'(x) = \frac{1}{x}$, $g'(x) = e^{-x} \cdot e^x = 1$.

51. $y = e^{-x^2}$; $y' = e^{-x^2}(-2x) = -2xe^{-x^2}$

$y'' = -2xe^{-x^2}(-2x) + e^{-x^2}(-2) = -2e^{-x^2}(-2x^2 + 1)$

We conclude that $x = 0$ is the only critical point and since $y'' < 0$ for $x = 0$, $x = 0$ is a relative maximum.

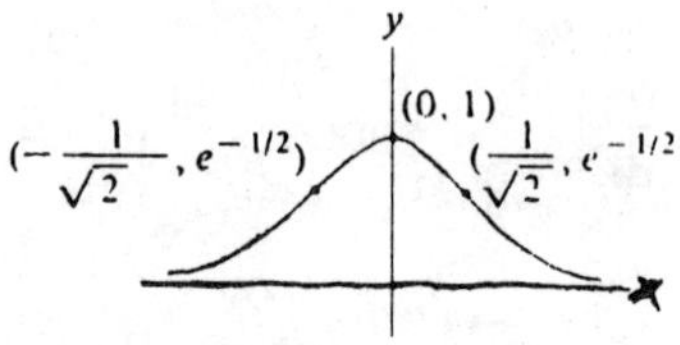

52. $y = xe^{-x}$ for $x \geq 0$; $y' = xe^{-x}(-1) + e^{-x}(1) = e^{-x}(1 - x)$;

$y'' = e^{-x}(-1) + (1 - x)e^{-x}(-1) = -e^{-x}(2 - x)$

We conclude that y has a critical point at $x = 1$, and since $y'' < 0$ for $x = 1$, $x = 1$ is a relative maximum.

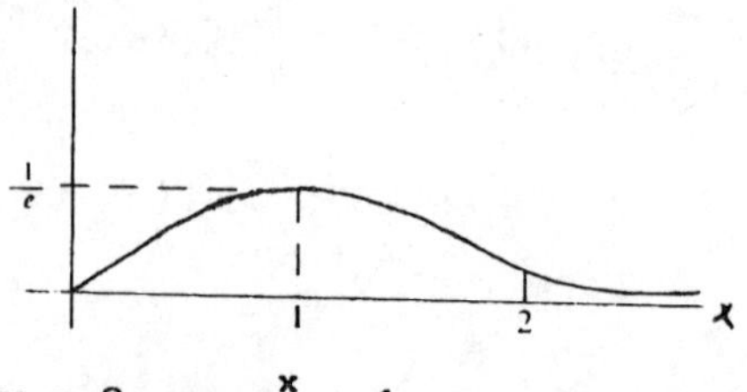

53. $\lim_{h \to 0} \frac{e^{x+h} - e^x}{h} = \frac{d}{dx}(e^x) = e^x$; hence $x = 0$, so $e^x = 1$

54. $\lim_{h\to 0} \dfrac{e^{2(x+h)} - e^{2x}}{h} = \dfrac{d}{dx}(e^{2x}) = 2e^{2x}$; hence if $x = 0$, so $2e^{x} = 2$

Exercises 4.4

1. $\ln(1/e) = \ln e^{-1} = -1$

2. $\ln(\sqrt{e}) = \ln e^{1/2} = \frac{1}{2}$

3. $e^{-x} = 1.7$; $\ln(e^{-x}) = \ln(1.7)$; $-x = \ln(1.7)$; $x = -\ln(1.7)$

4. $e^{x} = 3.5$; $\ln(e^{x}) = \ln(3.5)$; $x = \ln(3.5)$

5. $\ln x = 2.2$; $e^{\ln x} = e^{2.2}$; $x = e^{2.2}$

6. $\ln x = -5.7$; $e^{\ln x} = e^{-5.7}$; $x = e^{-5.7}$

7. $\ln e^{2} = 2$

8. $e^{\ln 1.37} = 1.37$

9. $e^{e^{\ln 1}} = e^{1} = e$

10. $\ln(e^{.73 \ln e}) = .73 \ln e = .73$

11. $e^{5 \ln 1} = e^{5(0)} = 1$

12. $\ln(\ln e) = \ln(1) = 0$

13. $e^{2x} = 5$; $\ln(e^{2x}) = \ln 5$; $2x = \ln 5$; $x = \frac{1}{2}\ln 5$

14. $e^{3x-1} = 4$; $\ln(e^{3x-1}) = \ln 4$; $3x - 1 = \ln 4$; $x = \dfrac{1 + \ln 4}{3}$

15. $\ln(4 - x) = \frac{1}{2}$; $e^{\ln(4 - x)} = e^{1/2}$; $4 - x = e^{1/2}$; $x = 4 - e^{1/2}$

16. $\ln 3x = 2$; $e^{\ln 3x} = e^{2}$; $3x = e^{2}$; $x = \frac{1}{3}e^{2}$

17. $\ln x^{2} = 6$; $e^{\ln x^{2}} = e^{6}$; $x^{2} = e^{6}$; $x = \pm e^{3}$

18. $e^{x^{2}} = 7$; $\ln(e^{x^{2}}) = \ln 7$; $x^{2} = \ln 7$; $x = \pm\sqrt{\ln 7}$

19. $6e^{-.00012x} = 3$; $e^{-.00012x} = \frac{1}{2}$; $\ln(e^{-.00012x}) = \ln\left(\frac{1}{2}\right)$;

$-.00012x = \ln \frac{1}{2}$; $x = \dfrac{\ln \frac{1}{2}}{-.00012}$

20. $2 - \ln x = 0$; $2 = \ln x$; $e^2 = e^{\ln x}$; $x = e^2$

21. $\ln 5x = \ln 3$; $e^{\ln 5x} = e^{\ln 3}$; $5x = 3$; $x = \frac{3}{5}$

22. $\ln(x^2 - 3) = 0$; $e^{\ln(x^2-3)} = e^0$; $x^2 - 3 = 1$; $x = \pm 2$

23. $\ln(\ln 2x) = 0$; $e^{\ln(\ln 2x)} = e^0$; $\ln 2x = 1$; $e^{\ln 2x} = e^1$;
$2x = e$; $x = \frac{e}{2}$

24. $3 \ln x = 8$; $\ln x = \frac{8}{3}$; $e^{\ln x} = e^{8/3}$; $x = e^{8/3}$

25. $2e^{x/3} - 9 = 0$; $e^{x/3} = \frac{9}{2}$; $\ln(e^{x/3}) = \ln\left(\frac{9}{2}\right)$; $x = 3 \ln\left(\frac{9}{2}\right)$

26. $4 - 3e^{x+6} = 0$; $e^{x+6} = \frac{4}{3}$; $\ln(e^{x+6}) = \ln\left(\frac{4}{3}\right)$; $x + 6 = \ln\left(\frac{4}{3}\right)$;
$x = \ln\left(\frac{4}{3}\right) - 6$

27. $300e^{.2x} = 1800$; $e^{.2x} = 6$; $\ln(e^{.2x}) = \ln 6$; $.2x = \ln 6$;
$\frac{2}{10}x = \ln 6$; $x = 5 \ln 6$

28. $750\ e^{-.4x} = 375$; $e^{-.4x} = \frac{1}{2}$; $\ln(e^{-.4x}) = \ln\left(\frac{1}{2}\right)$; $-\frac{4}{10}x = \ln\left(\frac{1}{2}\right)$;
$x = -\frac{5}{2}\ln\left(\frac{1}{2}\right)$

29. $e^{5x}\cdot e^{\ln 5} = 2$; $e^{5x}\cdot 5 = 2$; $\ln(e^{5x}) = \ln\left(\frac{2}{5}\right)$; $5x = \ln\left(\frac{2}{5}\right)$;
$x = \frac{1}{5}\ln\left(\frac{2}{5}\right)$

30. $e^{x^2-5x+6} = 1$; $\ln(e^{x^2-5x+6}) = \ln 1$; $x^2 - 5x + 6 = 0$;
$(x - 3)(x - 2) = 0$; $x = 3$ and $x = 2$

31. $4e^x\cdot e^{-2x} = 6$; $e^x\cdot e^{-2x} = \frac{3}{2}$; $e^{x-2x} = \frac{3}{2}$; $\ln(e^{-x}) = \ln\left(\frac{3}{2}\right)$;

$-x = \ln\left(\frac{3}{2}\right)$; $x = -\ln\left(\frac{3}{2}\right)$

32. $(e^x)^2 \cdot e^{2-3x} = 4$; $e^{2x+2-3x} = 4$; $\ln(e^{2-x}) = \ln 4$; $2 - x = \ln 4$;

$x = 2 - \ln 4$

33. $f(x) = e^{-x} + 3x$; $f'(x) = -e^{-x} + 3$; $f''(x) = e^{-x}$

$f'(x) = 0$ when $-e^{-x} + 3 = 0$; $\ln(e^{-x}) = \ln 3$; $-x = \ln 3$;

$x = -\ln 3$. Thus $f'(-\ln 3) = 0$ and the y-coordinate is

$f(-\ln 3) = e^{\ln 3} - 3\ln 3 = 3 - 3\ln 3$. Using the second derivative test, $f''(-\ln 3) = e^{\ln 3} = 3 > 0$, so

$(-\ln 3, 3 - 3\ln 3)$ is a relative minimum.

34. $f(x) = 5x - 2e^x$; $f'(x) = 5 - 2e^x$; $f''(x) = -2e^x$

$f'(x) = 0$ when $5 - 2e^x = 0$; $2e^x = 5$; $e^x = \frac{5}{2}$; $\ln(e^x) = \ln\left(\frac{5}{2}\right)$;

$x = \ln\left(\frac{5}{2}\right)$. $f\left(\ln\frac{5}{2}\right) = 5\ln\left(\frac{5}{2}\right) - 2e^{\ln(5/2)} = 5\ln\left(\frac{5}{2}\right) - 5$

$= 5\left[\ln\left(\frac{5}{2}\right) - 1\right]$

Using the second derivative test,

$f''\left(\ln\frac{5}{2}\right) = e^{-\ln 5/2} = -\frac{5}{2} < 0$, so $\left(\ln\frac{5}{2}, 5\left[\ln\frac{5}{2} - 1\right]\right)$ is a relative maximum.

35. $f(x) = \frac{1}{3}e^{2x} - x + \frac{1}{2}\ln\frac{3}{2}$; $f'(x) = \frac{2}{3}e^{2x} - 1$; $f''(x) = \frac{4}{3}e^{2x}$

$f'(x) = 0$ when $\frac{2}{3}e^{2x} = 1$; $e^{2x} = \frac{3}{2}$; $\ln(e^{2x}) = \ln\frac{3}{2}$; $2x = \ln\frac{3}{2}$;

$x = \frac{1}{2}\ln\frac{3}{2}$. $f\left(\frac{1}{2}\ln\frac{3}{2}\right) = \frac{1}{3}e^{2\left(\frac{1}{2}\ln\frac{3}{2}\right)} - \frac{1}{2}\ln\frac{3}{2} + \frac{1}{2}\ln\frac{3}{2} = \frac{1}{3}e^{\ln 3/2}$

$= \frac{1}{2}$

$f''\left(\frac{1}{2}\ln\frac{3}{2}\right) = \frac{4}{3}e^{2\left(\frac{1}{2}\ln\frac{3}{2}\right)} = \frac{4}{3}\cdot\frac{3}{2} = 2 > 0$, so $\left(\frac{1}{2}\ln\frac{3}{2}, \frac{1}{2}\right)$ is a relative minimum.

36. $f(x) = 5 - \frac{1}{2}x - e^{-3x}$; $f'(x) = 3e^{-3x} - \frac{1}{2}$; $f''(x) = -9e^{-3x}$

$f'(x) = 0$ when $3e^{-3x} = \frac{1}{2}$; $e^{-3x} = \frac{1}{6}$; $\ln(e^{-3x}) = \ln \frac{1}{6}$;

$-3x = \ln \frac{1}{6}$; $x = -\frac{1}{3}\ln \frac{1}{6}$.

$$f\left(-\frac{1}{3}\ln \frac{1}{6}\right) = 5 - \frac{1}{2}\left(-\frac{1}{3}\ln \frac{1}{6}\right) - e^{-3\left(-\frac{1}{3}\ln \frac{1}{6}\right)} = 5 + \frac{1}{6}\ln \frac{1}{6} - \frac{1}{6}$$

$$= \frac{29}{6} + \frac{1}{6}\ln \frac{1}{6} = \frac{1}{6}\left(29 + \ln \frac{1}{6}\right)$$

$$f''\left(-\frac{1}{3}\ln \frac{1}{6}\right) = -9e^{-3\left(-\frac{1}{3}\ln \frac{1}{6}\right)} = -9\left(\frac{1}{6}\right) = -\frac{9}{6} < 0, \text{ so}$$

$\left(-\frac{1}{3}\ln \frac{1}{6}, \frac{1}{6}\left(29 + \ln \frac{1}{6}\right)\right)$ is a relative maximum.

37. $f(t) = 5(e^{-.06t} - e^{-.51t})$, for $t \geq 0$.

$f'(t) = 5(-.01e^{-.01t} + .51e^{-.51t})$

$f'(t) = 0$ when $5(-.01e^{-.01t} + .51e^{-.51t}) = 0$;

$-e^{.01t} + 51e^{.51t} = 0$; $51e^{-.51t} = e^{-.01t}$;

$51e^{-.51t}\cdot e^{.51t} = e^{-.01t}\cdot e^{.51t}$; $51(1) = e^{.5t}$; $\ln 51 = \ln(e^{.5t})$;

$\ln 51 = .5t$; $t = 2 \ln 51$.

Thus, the maximum must occur at

$t = 2 \ln 51$.

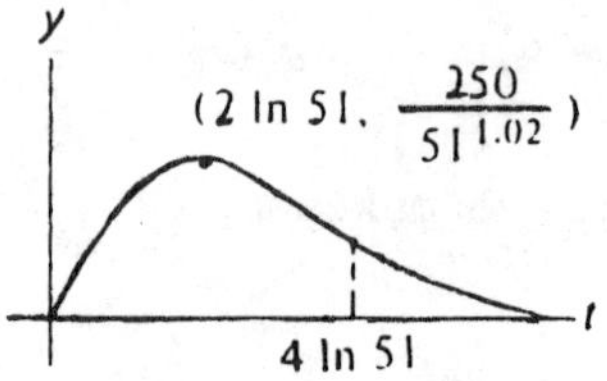

38. (a) $v = K \ln(x/x_0) = 0$; $\ln(x/x_0) = 0$; $e^{\ln(x/x_0)} = e^0 = 1$;

$\frac{x}{x_0} = 1$; $x = x_0 = .7$ cm

(b) $v = K \ln(x/x_0) = 1200$; $\ln(x/x_0) = \frac{1200}{K}$;

$e^{\ln(x/x_0)} = e^{1200/K}$; $\frac{x}{x_0} = e^{1200/K}$; $x = .7e^{1200/300} = .7e^4$ cm

39. Using that $b^x = e^{kx}$ where $k = \ln b$, we have

$1.6^{10} = e^{.47(10)} = e^{4.7} = 109.947$.

40. Using that $b^x = e^{kx}$ where $k = \ln b$, we have $k = \ln 2$.

Exercises 4.5

1. $\frac{d}{dx} \ln 2x = \frac{1}{2x}(2) = \frac{1}{x}$

2. $\frac{d}{dx} \ln x^2 = \frac{1}{x^2}(2x) = \frac{2}{x}$

3. $\frac{d}{dx} \ln(x + 5) = \frac{1}{x + 5}$

4. $\frac{d}{dx} x^2 \ln x = x^2\left(\frac{1}{x}\right) + (\ln x)2x = 2x \ln x + x$

5. $\frac{d}{dx}\left[\frac{1}{x}\ln(x + 1)\right] = \frac{1}{x}\left(\frac{1}{x + 1}\right) + \ln(x + 1)\left(-\frac{1}{x^2}\right)$
$= \frac{1}{x(x + 1)} - \frac{\ln(x + 1)}{x^2}$

6. $\frac{d}{dx} \sqrt{\ln x} = \frac{d}{dx} (\ln x)^{1/2} = \frac{1}{2}(\ln x)^{-1/2}\left(\frac{1}{x}\right) = \frac{1}{2x\sqrt{\ln x}}$

7. $\frac{d}{dx} e^{\ln x + x} = e^{\ln x + x}\left(\frac{1}{x} + 1\right)$

8. $\frac{d}{dx} \ln\left(\frac{x}{x - 3}\right) = \frac{1}{\frac{x}{x - 3}} \cdot \frac{(x - 3)(1) - x(1)}{(x - 3)^2} = \frac{x - 3}{x} \cdot \frac{x - 3 - x}{(x - 3)^2}$
$= \frac{-3}{x(x - 3)}$

9. $\frac{d}{dx}\left[4 + \ln\left(\frac{x}{2}\right)\right] = \frac{1}{\frac{x}{2}}\left(\frac{1}{2}\right) = \frac{1}{x}$

10. $\frac{d}{dx} \ln\sqrt{x} = \frac{1}{\sqrt{x}} \cdot \frac{1}{2}x^{-1/2} = \frac{1}{2x}$

11. $\frac{d}{dx}\left[(\ln x)^2 + \ln x\right] = 2(\ln x)\left(\frac{1}{x}\right) + \frac{1}{x} = \frac{1}{x}(2 \ln x + 1)$

12. $\frac{d}{dx} \ln(x^3 + 2x + 1) = \frac{1}{x^3 + 2x + 1} \cdot 3x^2 + 2 = \frac{3x^2 + 2}{x^3 + 2x + 1}$

13. $\frac{d}{dx} \ln(kx) = \frac{1}{kx} \cdot k = \frac{1}{x}$

14. $\frac{d}{dx} \left[\frac{x}{\ln x}\right] = \frac{(\ln x)(1) - x\left(\frac{1}{x}\right)}{(\ln x)^2} = \frac{\ln x - 1}{(\ln x)^2}$

15. $\frac{d}{dx} \left[\frac{x}{(\ln x)^2}\right] = \frac{(\ln x)^2(1) - x\left[2(\ln x)\left(\frac{1}{x}\right)\right]}{(\ln x)^4} = \frac{(\ln x)^2 - 2 \ln x}{(\ln x)^4}$

$= \frac{\ln x - 2}{(\ln x)^3}$

16. $\frac{d}{dx} \left[(\ln x)e^{-x}\right] = (\ln x)e^{-x}(-1) + e^{-x}\left(\frac{1}{x}\right) = e^{-x}\left(\frac{1}{x} - \ln x\right)$

17. $\frac{d}{dx} \left[e^{2x}\ln x\right] = e^{2x}\left(\frac{1}{x}\right) + (\ln x)e^{2x}(2) = e^{2x}\left(\frac{1}{x} + 2 \ln x\right)$

18. $\frac{d}{dx} \left[(\ln x + 1)^3\right] = 3(\ln x + 1)^2\left(\frac{1}{x}\right) = \frac{3}{x}(\ln x + 1)^2$

19. $\frac{d}{dx} \ln(e^{5x} + 1) = \frac{1}{e^{5x} + 1} \cdot e^{5x}(5) = \frac{5e^{5x}}{e^{5x} + 1}$

20. $\frac{d}{dx} \ln\left(e^{e^x}\right) = \frac{1}{e^{e^x}} \cdot e^{e^x}(e^x) = e^x$

21. $\frac{d}{dt} \left[t^2\ln 4\right] = 2t \ln 4$

22. $\frac{d}{dx} \ln(1 + x^2) = \frac{1}{1 + x^2}(2x) = \frac{2x}{1 + x^2}$

$\frac{d^2}{dx^2} \ln(1 + x^2) = \frac{d}{dx}\left[\frac{2x}{1 + x^2}\right] = \frac{(1 + x^2)(2) - (2x)(2x)}{(1 + x^2)^2}$

$= \frac{2 + 2x^2 - 4x^2}{(1 + x^2)^2} = \frac{2(1 - x^2)}{(1 + x^2)^2}$

23. $\frac{d}{dt}\left[(\ln t)^3\right] = 3(\ln t)^2\left(\frac{1}{t}\right) = \frac{3(\ln t)^2}{t}$

$$\frac{d^2}{dt^2}\left[(\ln t)^3\right] = \frac{d}{dt}\left[\frac{3(\ln t)^2}{t}\right] = \frac{t(6\ln t)\left(\frac{1}{t}\right) - 3(\ln t)^2(1)}{t^2}$$

$$= \frac{6\ln t - 3(\ln t)^2}{t^2}$$

24. $\frac{dy}{dx} = \frac{1}{x}$ Thus at $x = 3$, the slope is $\frac{1}{3}$ and at $x = -3$, the slope $-\frac{1}{3}$.

25. $\frac{dy}{dx} = \frac{1}{x^2 + e}(2x) = \frac{2x}{x^2 + e}$, $\frac{dy}{dx}(0, 0) = \frac{0}{e} = 0$. At $x = 0$,

$y = \ln(0^2 + e) = 1$. Thus the tangent line at $x = 0$ is

$y - 1 = 0(x - 0)$; $y = 1$.

26. $f(x) = \frac{\ln x + 1}{x}$; $f'(x) = \frac{x\left(\frac{1}{x}\right) - (\ln x + 1)(1)}{x^2} = \frac{-\ln x}{x^2}$;

$$f''(x) = \frac{x^2\left(-\frac{1}{x}\right) - (-\ln x)(2x)}{x^4} = \frac{-x + 2x\ln x}{x^4}$$

For $x > 0$, the only critical point is at $x = 1$, $f(1) = 1$.

Using the second derivative test,

$f''(1) = \frac{-1 + 2(1)\ln(1)}{1^4} = -1 < 0$, so $(1, 1)$ is a relative maximum.

27. $f(x) = \frac{\ln x}{\sqrt{x}} = \frac{\ln x}{x^{1/2}}$

$$f'(x) = \frac{(x^{1/2})\left(\frac{1}{x}\right) - (\ln x)\left(\frac{1}{2}\right)x^{-1/2}}{(x^{1/2})^2} = \frac{x^{-1/2}\left[1 - \frac{1}{2}\ln x\right]}{x}$$

$$= \frac{\left[1 - \frac{1}{2}\ln x\right]}{x^{3/2}}$$

$$f''(x) = \frac{(x^{3/2})\left(-\frac{1}{2}\right)\left(\frac{1}{x}\right) - \left(1 - \frac{1}{2}\ln x\right)\left(\frac{3}{2}\right)x^{1/2}}{x^3}$$

$$= \frac{x^{1/2}\left(-\frac{1}{2} - \frac{3}{2} + \frac{3}{4}\ln x\right)}{x^3} = \frac{-2 + \frac{3}{4}\ln x}{x^{5/2}}$$

For $x > 0$, the only critical point occurs at $x = e^2$,

$f(e^2) = \frac{2}{e}$. Since $f''(e^2) = \dfrac{-2 + \frac{3}{4}(2)}{(e^2)^{5/2}} < 0$, then $\left(e^2, \frac{2}{e}\right)$ is a relative maximum.

28. $f(x) = \dfrac{x}{\ln x + x}$

$$f'(x) = \frac{(\ln x + x)(1) - x\left(\frac{1}{x} + 1\right)}{(\ln x + x)^2} = \frac{\ln x - 1}{(\ln x + x)^2}$$

$$f''(x) = \frac{(\ln x + x)^2\left(\frac{1}{x}\right) - (\ln x - 1)2(\ln x + x)\left(\frac{1}{x} + 1\right)}{(\ln x + x)^4}$$

$$= \frac{(\ln x + x)\left(\frac{1}{x}\right) - 2(\ln x - 1)\left(\frac{1}{x} + 1\right)}{(\ln x + x)^3}$$

$$= \frac{\frac{\ln x}{x} + 1 - \frac{2\ln x}{x} - 2\ln x + \frac{1}{x} + 1}{(\ln x + x)^3}$$

$$= \frac{-\frac{\ln x}{x} - 2\ln x + \frac{1}{x} + 2}{(\ln x + x)^3}$$

For $x > 1$, the only critical point is at $x = e$. Since

$f''(e) = \dfrac{-\frac{1}{e} - 2 + \frac{1}{e} + 2}{(1 + e)^3} = 0$, then $\left(e, \frac{e}{1 + e}\right)$ is neither a relative maximum nor a relative minimum.

29.

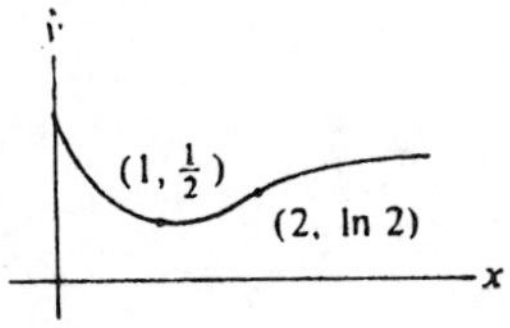

30.

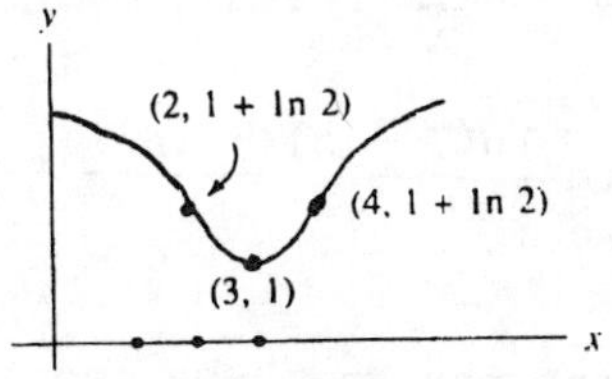

31. $C(x) = \dfrac{\ln x}{40 - 3x}$; $C'(x) = \dfrac{(40 - 3x)\left(\frac{1}{x}\right) - (\ln x)(-3)}{(40 - 3x)^2}$

$$= \frac{\frac{40}{x} - 3 + 3\ln x}{(40 - 3x)^2}$$

At $x = 10$, $C'(x) = \dfrac{\frac{40}{10} - 3 + 3\ln 10}{(40 - 30)^2} = \dfrac{4 - 3 + 3(2.30259)}{10^2}$

$\sim .07908.$

32. The revenue function is $R(x) = x \cdot \dfrac{45}{\ln x} = \dfrac{45x}{\ln x}$. The marginal revenue function is $R'(x) = \dfrac{45\ln x - 45x\left(\frac{1}{x}\right)}{(\ln x)^2} = \dfrac{45(\ln x - 1)}{(\ln x)^2}$.

When $x = 20$, $R'(x) = \dfrac{45(\ln 20 - 1)}{(\ln 20)^2} \sim 10.$

33. We wish to maximize $R(x) - C(x)$.

$$\frac{d}{dx}\Big[R(x) - C(x)\Big] = \frac{d}{dx}\Big[300\ln(x + 1) - 2x\Big] = 300\left(\frac{1}{x + 1}\right) - 2$$

$$= \frac{300}{x + 1} - 2$$

The only critical point occurs when $x = 149$. To show that this is a relative maximum take the second derivative.

$$\frac{d^2}{dx^2}\Big[R(x) - C(x)\Big] = \frac{d}{dx}\left[\frac{300}{x + 1} - 2\right] = \frac{-300}{(x + 1)^2}$$

At $x = 149$, we have $\dfrac{d^2}{dx^2}\Big[R(x) - C(x)\Big] = \dfrac{-300}{150^2} < 0$, so $x = 149$

is a relative maximum.

34. $\lim_{h \to 0} \frac{\ln(7 + h) - \ln 7}{h} = \frac{d}{dx}\left[\ln x\right]\Big|_{x=7} = \frac{1}{x}\Big|_{x=7} = \frac{1}{7}$

35. From the graph we see that

area = A = x(-ln x) = -x ln x

To maximize the area take

the first derivative.

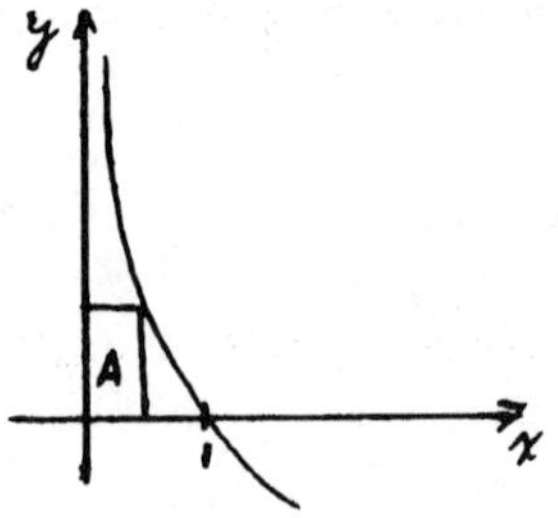

$A' = -x\left(\frac{1}{x}\right) + (\ln x)(-1) = -1 - \ln x.$

Now set A′ = 0; -1 - ln x = 0; ln x = -1; $e^{\ln x} = e^{-1}$;

$x = e^{-1} \sim .36788$. Thus area is maximized when x ~ .36788.

Exercises 4.6

1. $\ln 5 + \ln x = \ln(5x)$

2. $\ln x^5 - \ln x^3 = \ln\left(\frac{x^5}{x^3}\right) = \ln x^2 = 2 \ln x$

3. $\frac{1}{2}\ln 9 = \ln 9^{1/2} = \ln 3$

4. $3 \ln \frac{1}{2} + \ln 16 = \ln\left(\frac{1}{2}\right)^3 + \ln 16 = \ln \frac{1}{8} + \ln 16 = \ln\left(\frac{16}{8}\right) = \ln 2$

5. $\ln 4 + \ln 6 - \ln 12 = \ln\left(\frac{4\cdot 6}{12}\right) = \ln 2$

6. $\ln 2 - \ln x + \ln 3 = \ln\left(\frac{2\cdot 3}{x}\right) = \ln \frac{6}{x}$

7. $e^{2 \ln x} = e^{\ln x^2} = x^2$

8. $\frac{3}{2}\ln 4 - 5\ln 2 = \ln 4^{3/2} - \ln 2^5 = \ln 2^3 - \ln 2^5 = \ln\left(\frac{2^3}{2^5}\right)$

$= \ln 2^{-2} = -\ln 4$

9. $5\ln x - \frac{1}{2}\ln y + 3\ln z = \ln x^5 - \ln y^{1/2} + \ln z^3$

$= \ln\left(\frac{x^5z^3}{y^{1/2}}\right)$

10. $e^{\ln x^2 + 3\ln y} = e^{\ln x^2 + \ln y^3} = e^{\ln x^2y^3} = x^2y^3$

11. $\ln x - \ln x^2 + \ln x^4 = \ln\left(\frac{x\cdot x^4}{x^2}\right) = \ln x^3 = 3\ln x$

12. $\frac{1}{2}\ln xy + \frac{3}{2}\ln\frac{x}{y} = \ln(xy)^{1/2} + \ln\left(\frac{x}{y}\right)^{3/2} = \ln\left[(xy)^{1/2}\cdot\left(\frac{x}{y}\right)^{3/2}\right]$

$= \ln\left(\frac{x^2}{y}\right)$

13. $2\ln 5 = \ln 5^2 = \ln 25$, $3\ln 3 = \ln 3^3 = \ln 27$

Since the natural log function increases as x gets larger,

$3\ln 3 > 2\ln 5$.

14. $\frac{1}{2}\ln 16 = \ln 16^{1/2} = \ln 4$, $\frac{1}{3}\ln 27 = \ln 27^{1/3} = \ln 3$

$\ln 4 > \ln 3$ so $\frac{1}{2}\ln 16 > \frac{1}{3}\ln 27$.

15. (d)$4\ln 2x = \ln(2x)^4 = \ln 16x^4$

16. (d)$\ln 9x - \ln 3x = \ln(9x/3x) = \ln 3$

17. (d)

18. (c)$2\ln 3x = \ln(3x)^2 = \ln 9x^2$

19. $\frac{d}{dx}\left[\ln\left[(x + 5)(2x - 1)(4 - x)\right]\right]$

$= \frac{d}{dx}\left[\ln(x + 5) + \ln(2x - 1) + \ln(4 - x)\right]$

$$= \frac{1}{x+5} + \frac{1}{2x-1}(2) + \frac{1}{4-x}(-1) = \frac{1}{x+5} + \frac{2}{2x-1} - \frac{1}{4-x}$$

20. $$\frac{d}{dx}\left[\ln\left(x^3(x+1)^4\right)\right] = \frac{d}{dx}\left[\ln x^3 + \ln(x+1)^4\right]$$

$$= \frac{1}{x^3}(3x^2) + \frac{1}{(x+1)^4}\cdot 4(x+1)^3$$

$$= \frac{3}{x} + \frac{4}{x+1}$$

21. $$\frac{d}{dx}\left[\ln\left(\frac{(x+1)(3x-2)}{(x+2)}\right)\right]$$

$$= \frac{d}{dx}\left[\ln(x+1) + \ln(3x-2) - \ln(x+2)\right]$$

$$= \frac{1}{x+1} + \frac{1}{3x-2}(3) - \frac{1}{x+2} = \frac{1}{x+1} + \frac{3}{3x-2} - \frac{1}{x+2}$$

22. $$\frac{d}{dx}\left[\ln\left(\frac{x^2}{(3-x)^3}\right)\right] = \frac{d}{dx}\left[\ln x^2 - \ln(3-x)^3\right]$$

$$= \frac{1}{x^2}(2x) - \frac{1}{(3-x)^3}\cdot 3(3-x)^2(-1)$$

$$= \frac{2}{x} + \frac{3}{3-x}$$

23. $$\frac{d}{dx}\left[\ln\left(\frac{\sqrt{x}}{x^2+1}\right)\right] = \frac{d}{dx}\left[\ln x^{1/2} - \ln(x^2+1)\right]$$

$$= \frac{1}{x^{1/2}}\cdot\frac{1}{2}x^{-1/2} - \frac{1}{x^2+1}(2x) = \frac{1}{2x} - \frac{2x}{x^2+1}$$

24. $$\frac{d}{dx}\left[\ln\left(e^{x^2}(x^4+x^2+1)\right)\right] = \frac{d}{dx}\left[\ln e^{x^2} + \ln(x^4+x^2+1)\right]$$

$$= \frac{d}{dx}\left[x^2 + \ln(x^4+x^2+1)\right]$$

$$= 2x + \frac{1}{x^4+x^2+1}(4x^3+2x)$$

$$= 2x + \frac{4x^3+2x}{x^4+x^2+1}$$

$$= \frac{2x(x^4+3x^2+2)}{x^4+x^2+1}$$

25. $\ln f(x) = \ln\left[(x + 3)^2(4x - 1)^2\right] = \ln(x + 1)^3 + \ln(4x - 1)^2$

$= 3\ln(x + 1) + 2\ln(4x - 1)$

Differentiating both sides,

$\dfrac{f'(x)}{f(x)} = \dfrac{3}{x + 1} + \dfrac{8}{4x - 1}$

$f(x) = (x + 1)^3(4x - 1)^2\left[\dfrac{3}{x + 1} + \dfrac{8}{4x - 1}\right].$

26. $\ln f(x) = \ln\left[e^x(x - 4)^8\right] = \ln e^x + 8\ln(x - 4)$

$= x + 8\ln(x - 4)$

Differentiating both sides,

$\dfrac{f'(x)}{f(x)} = 1 + \dfrac{8}{x - 4};\ f'(x) = e^x(x - 4)^8\left[1 + \dfrac{8}{x - 4}\right].$

27. $\ln f(x) = \ln\left[(x - 2)^3(x - 3)^5(x + 2)^{-7}\right]$

$= 3\ln(x - 2) + 5\ln(x - 3) - 7\ln(x + 2)$

Differentiating both sides,

$\dfrac{f'(x)}{f(x)} = \dfrac{3}{x - 2} + \dfrac{5}{x - 3} - \dfrac{7}{x + 2}$

$f'(x) = (x - 2)^3(x - 3)^5(x + 2)^{-7}\left[\dfrac{3}{x - 2} + \dfrac{5}{x - 3} - \dfrac{7}{x + 2}\right].$

28. $\ln f(x) = \ln\left[(x + 1)(2x + 1)(3x + 1)(4x + 1)\right]$

$= \ln(x + 1) + \ln(2x + 1) + \ln(3x + 1) + \ln(4x + 1)$

Differentiating both sides,

$\dfrac{f'(x)}{f(x)} = \dfrac{1}{x + 1} + \dfrac{2}{2x + 1} + \dfrac{3}{3x + 1} + \dfrac{4}{4x + 1}$

$f'(x) = (x + 1)(2x + 1)(3x + 1)(4x + 1)\left[\dfrac{1}{x + 1} + \dfrac{2}{2x + 1}\right.$

$\left. + \dfrac{3}{3x + 1} + \dfrac{4}{4x + 1}\right].$

29. $\ln f(x) = \ln x^x = x \ln x$ Differentiating both sides,

$$\frac{f'(x)}{f(x)} = x\left(\frac{1}{x}\right) + \ln x = 1 + \ln x; \quad f'(x) = x^x\left[1 + \ln x\right].$$

30. $\ln f(x) = \ln x^{1/x} = \frac{1}{x}\ln x$ Differentiating both sides,

$$\frac{f'(x)}{f(x)} = \frac{1}{x}\cdot\frac{1}{x} + (\ln x)\left(-\frac{1}{x^2}\right) = \frac{1}{x^2} - \frac{1}{x^2}\ln x = \frac{1 - \ln x}{x^2}$$

$$f'(x) = x^{1/x}\left[\frac{1 - \ln x}{x^2}\right].$$

31. $\ln f(x) = \ln\left[e^x(x^2 - 1)^{1/2}\right] = \ln e^x + \frac{1}{2}\ln(x^2 - 1)$

$$= x + \frac{1}{2}\ln(x^2 - 1)$$

Differentiating both sides,

$$\frac{f'(x)}{f(x)} = 1 + \frac{1}{2}\cdot\frac{1}{x^2 - 1}(2x) = 1 + \frac{x}{x^2 - 1}$$

$$f'(x) = e^x\sqrt{x^2 - 1}\left[1 + \frac{x}{x^2 - 1}\right].$$

32. $\ln f(x) = \ln 2^x = x \ln 2$ Differentiating both sides,

$$\frac{f'(x)}{f(x)} = \ln 2; \quad f'(x) = 2^x\ln 2.$$

33. $\ln f(x) = \ln(x^{\ln x}) = \ln x \cdot \ln x = (\ln x)^2$

Differentiating both sides,

$$\frac{f'(x)}{f(x)} = 2(\ln x)\left(\frac{1}{x}\right) = \frac{2 \ln x}{x}; \quad f'(x) = \frac{x^{\ln x}2 \ln x}{x}.$$

34. $\ln f(x) = \ln\left[(2x - 1)^{2x}\right] = 2x \ln(2x - 1)$

Differentiating both sides,

$$\frac{f'(x)}{f(x)} = 2x\cdot\frac{1}{2x - 1}(2) + \ln(2x - 1)\cdot 2 = \frac{4x}{2x - 1} + 2\ln(2x - 1)$$

$$f'(x) = (2x - 1)^{2x}\left[\frac{4x}{2x - 1} + 2\ln(2x - 1)\right].$$

35. $\ln f(x) = \ln\left[\dfrac{(x-1)^{1/2}(x-2)}{x^2-3}\right]$

$= \frac{1}{2}\ln(x-1) + \ln(x-2) - \ln(x^2-3)$

Differentiating both sides,

$\dfrac{f'(x)}{f(x)} = \dfrac{1}{2(x-1)} + \dfrac{1}{x-2} - \dfrac{2x}{x^2-3}$

$f'(x) = \dfrac{\sqrt{x-1}\,(x-2)}{x^2-3}\left[\dfrac{1}{2(x-1)} + \dfrac{1}{x-2} - \dfrac{2x}{x^2-3}\right].$

36. $\ln f(x) = \ln\left[\dfrac{xe^x}{(3x^2+1)^{1/2}}\right] = \ln x + \ln e^x - \ln(3x^2+1)^{1/2}$

$= \ln x + x - \ln(3x^2+1)^{1/2}$

Differentiating both sides,

$\dfrac{f'(x)}{f(x)} = \dfrac{1}{x} + 1 - \dfrac{1}{(3x^2+1)^{1/2}}\cdot\dfrac{1}{2}(3x^2+1)^{-1/2}(6x)$

$= \dfrac{1}{x} + 1 - \dfrac{3x}{3x^2+1}$

$f'(x) = \dfrac{xe^x}{\sqrt{3x^2+1}}\left[\dfrac{1}{x} + 1 - \dfrac{3x}{3x^2+1}\right]$

37. $\ln y - k\ln x = \ln x;\ \ln y = \ln c + k\ln x = \ln c + \ln x^k$

$= \ln(cx^k);$

$e^{\ln y} = e^{\ln(cx^k)};\ y = cx^k$

38. $\ln(1-y) - \ln y = C - rt;\ \ln\left(\dfrac{1-y}{y}\right) = C - rt;$

$e^{\ln\left(\frac{1-y}{y}\right)} = e^{C-rt};\ \dfrac{1-y}{y} = e^{C-rt};\ \dfrac{1}{y} - 1 = e^{C-rt};\ y = \dfrac{1}{e^{C-rt}+1}$

39. $y = he^{kx}$

$\left.\begin{array}{l} 6 = he^{k} \\ 48 = he^{4k} \end{array}\right\} \left.\begin{array}{l} \ln\left(\frac{6}{h}\right) = k \\ \ln\left(\frac{48}{h}\right) = 4k \end{array}\right\} 4\ln\left(\frac{6}{h}\right) = \ln\left(\frac{48}{h}\right);\ \ln\left(\frac{6}{h}\right)^4 = \ln\left(\frac{48}{h}\right);$

$\frac{6^4}{h^4} = \frac{48}{h};\ h^3 = \frac{6^4}{48};\ h^3 = 27;\ h = 3$

$6 = 3e^{k};\ e^{k} = 2;\ k = \ln 2$

40. $y = kx^{r}$

$\left.\begin{array}{l} 3 = k2^{r} \\ 15 = k4^{r} \end{array}\right\} \left.\begin{array}{l} k = \frac{3}{2^r} \\ k = \frac{15}{4^r} \end{array}\right\} \frac{3}{2^r} = \frac{15}{4^r};$

$\ln 3 - \ln 2^{r} = \ln 15 - \ln 4^{r};\ r \ln 4 - r \ln 2 = \ln 15 - \ln 3;$

$r \ln\left(\frac{4}{2}\right) = \ln\left(\frac{15}{3}\right);\ r = \frac{\ln 5}{\ln 2} \sim 2.32193$

$k = \frac{3}{2^{2.32193}} \sim .6$

Chapter 4 Supplementary Exercises

1. $27^{4/3} = (3^3)^{4/3} = 3^4 = 81$

2. $4^{1.5} = (2^2)^{3/2} = 2^3 = 8$

3. $5^{-2} = \frac{1}{5^2} = \frac{1}{25}$

4. $16^{-.25} = 16^{-1/4} = \frac{1}{(2^4)^{1/4}} = \frac{1}{2}$

5. $(2^{5/7})^{14/5} = 2^{14/7} = 2^2 = 4$

6. $8^{1/2}\cdot 2^{1/2} = (2^3)^{1/2}\cdot 2^{1/2} = 2^{3/2}\cdot 2^{1/2} = 2^{4/2} = 4$

7. $\frac{9^{5/2}}{9^{3/2}} = \frac{(3^2)^{5/2}}{(3^2)^{3/2}} = \frac{3^5}{3^3} = 3^2 = 9$

8. $4^{.2}\cdot 4^{.3} = 4^{.5} = 4^{1/2} = 2$

9. $\left(e^{x^2}\right)^3 = e^{3x^2}$

10. $e^{5x} \cdot e^{2x} = e^{7x}$

11. $\dfrac{e^{3x}}{e^{x}} = e^{3x-x} = e^{2x}$

12. $2^{x} \cdot 3^{x} = (2 \cdot 3)^{x} = 6^{x}$

13. $(e^{8x} + 7e^{-2x})e^{3x} = e^{11x} + 7e^{x}$

14. $\dfrac{e^{5x/2} - e^{3x}}{\sqrt{e^{x}}} = (e^{5x/2} - e^{3x})e^{(-1/2)x} = e^{4x/2} - e^{5x/2}$

$= e^{2x} - e^{5x/2}$

15. $e^{-3x} = e^{-12}$; $\ln e^{-3x} = \ln e^{-12}$; $-3x = -12$; $x = 4$

16. $e^{x^2 - x} = e^{2}$; $\ln e^{x^2 - x} = \ln e^{2}$; $x^2 - x = 2$; $x^2 - x - 2 = 0$;
$(x - 2)(x + 1) = 0$; $x = 2, -1$

17. $(e^{x} \cdot e^{2})^{3} = e^{-9}$; $e^{3x+6} = e^{-9}$; $\ln e^{3x+6} = \ln e^{-9}$; $3x + 6 = -9$;
$x = -5$

18. $e^{-5x} \cdot e^{4} = e$; $e^{-5x+4} = e$; $\ln e^{-5x+4} = \ln e$; $-5x + 4 = 1$;
$x = \dfrac{3}{5}$

19. $\dfrac{d}{dx} 10e^{7x} = 10e^{7x}(7) = 70e^{7x}$

20. $\dfrac{d}{dx} e^{\sqrt{x}} = \dfrac{d}{dx} e^{x^{1/2}} = e^{x^{1/2}} \cdot \dfrac{1}{2}x^{-1/2} = \dfrac{1}{2\sqrt{x}} e^{\sqrt{x}}$

21. $\dfrac{d}{dx}\left[xe^{x^2}\right] = xe^{x^2}(2x) + e^{x^2}(1) = e^{x^2}(2x^2 + 1)$

22. $\dfrac{d}{dx}\left[\dfrac{e^{x} + 1}{x - 1}\right] = \dfrac{(x - 1)e^{x} - (e^{x} + 1)(1)}{(x - 1)^2} = \dfrac{-2e^{x} + xe^{x} - 1}{(x - 1)^2}$

$= \dfrac{(x - 2)e^{x} - 1}{(x - 1)^2}$

23. $\dfrac{d}{dx}\left[e^{e^{x}}\right] = e^{e^{x}}(e^{x}) = e^{x+e^{x}}$

24. $\dfrac{d}{dx}\left[(\sqrt{x} + 1)e^{-2x}\right] = (\sqrt{x} + 1)e^{-2x}(-2) + e^{-2x}\left(\dfrac{1}{2}x^{-1/2}\right)$

$$= \frac{e^{-2x}}{2\sqrt{x}}(1 - 4\sqrt{x} - 4x)$$

25. $\frac{d}{dx}\left[\frac{x^2 - x + 5}{e^{3x} + 3}\right] = \frac{(e^{3x} + 3)(2x - 1) - (x^2 - x + 5)e^{3x}(3)}{(e^{3x} + 3)^2}$

$$= \frac{(2x - 1)(e^{3x} + 3) - 3e^{3x}(x^2 - x + 5)}{(e^{3x} + 3)^2}$$

26. $\frac{d}{dx} x^e = ex^{e-1}$

27. $y' = -y$; $y = Ce^{-x}$ for some C.

28. $y' = -1.5y$; $f(x) = y = Ce^{-1.5x}$ for some C. Since $f(0) = 2000$, $2000 = Ce^{(-1.5)(0)} = C$. Thus, $y = 2000e^{-1.5x}$.

29. $y' = 1.5y$; $f(x) = y = Ce^{1.5x}$ for some C. Since $f(0) = 2$, $2 = Ce^{(1.5)(0)} = C$. Thus, $y = 2e^{1.5x}$.

30. $y' = \frac{1}{3}y$; $y = Ce^{(1/3)x}$

31.

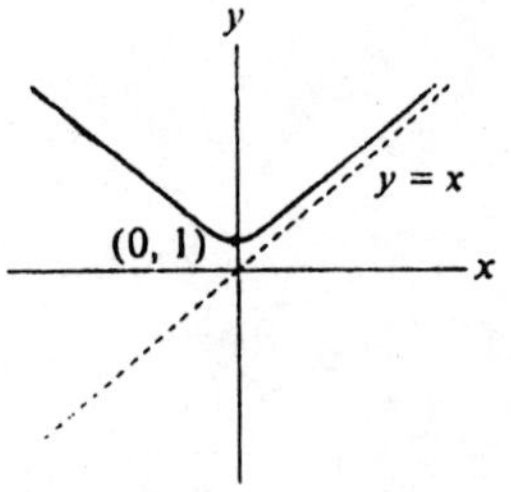

32.

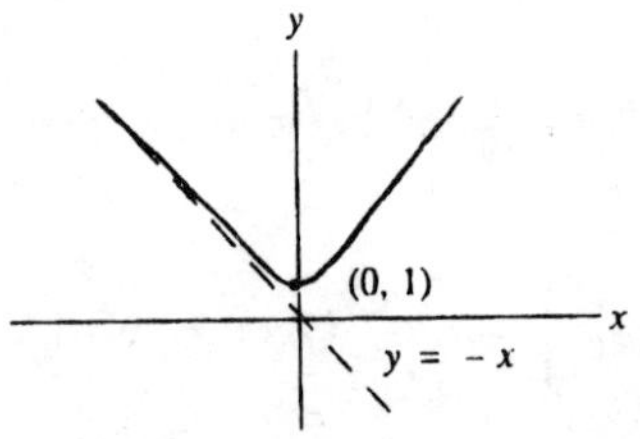

33.

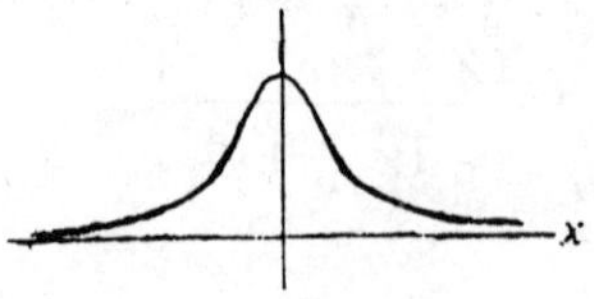

34.

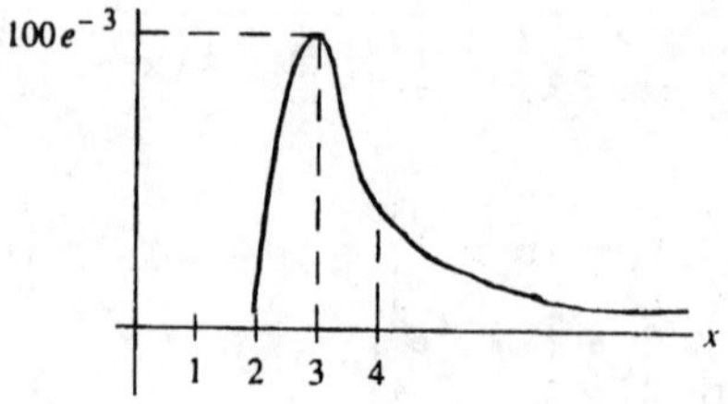

35. $\frac{\ln x^2}{\ln x^3} = \frac{2 \ln x}{3 \ln x} = \frac{2}{3}$

36. $e^{2 \ln 2} = e^{\ln 2^2} = 2^2 = 4$

37. $e^{-5 \ln 1} = e^{-5(0)} = 1$

38. $\left[e^{\ln x}\right]^2 = e^{2 \ln x} = e^{\ln x^2} = x^2$

39. $e^{(\ln 5)/2} = e^{\ln \sqrt{5}} = \sqrt{5}$

40. $e^{\ln(x^2)} = x^2$

41. $3e^{2t} = 15;\ e^{2t} = 5;\ \ln e^{2t} = \ln 5;\ 2t = \ln 5;\ t = \frac{1}{2}\ln 5$

42. $3e^{t/2} - 12 = 0;\ 3(e^{t/2} - 4) = 0;\ e^{t/2} = 4;\ \ln e^{t/2} = \ln 4;$

$t = 2 \ln 4;\ t = \ln 16$

43. $2 \ln t = 5;\ \ln t = \frac{5}{2};\ e^{\ln t} = e^{5/2};\ t = e^{5/2}$

44. $2e^{-.3t} = 1;\ e^{-.3t} = \frac{1}{2};\ \ln e^{-.3t} = \ln \frac{1}{2};\ -.3t = \ln \frac{1}{2};$

$t = -\frac{1}{.3}\ln \frac{1}{2} = \frac{\ln 2}{.3}$

45. $t^{\ln t} = e;\ \ln t^{\ln t} = \ln e;\ \ln t(\ln t) = 1;\ (\ln t)^2 = 1;$

Taking the square root of both sides, $|\ln t| = 1;$

$t = e$ or $t = \frac{1}{e}$.

46. $\ln(\ln 3t) = 0;\ e^{\ln(\ln 3t)} = e^0;\ \ln 3t = 1;\ e^{\ln 3t} = e;\ 3t = e;$

$t = \frac{e}{3}$

47. $\frac{d}{dx} \ln(5x - 7) = \frac{1}{5x - 7}(5) = \frac{5}{5x - 7}$

48. $\frac{d}{dx} \ln(9x) = \frac{1}{9x}(9) = \frac{1}{x}$

49. $\frac{d}{dx} \left[(\ln x)^2\right] = 2(\ln x)\frac{1}{x} = \frac{2 \ln x}{x}$

50. $\frac{d}{dx} \left[(x \ln x)^3\right] = 3(x \ln x)^2\left[x \cdot \frac{1}{x} + \ln x\right] = 3(x \ln x)^2(1 + \ln x)$

51. $\frac{d}{dx} \ln(x^6 + 3x^4 + 1) = \frac{1}{x^6 + 3x^4 + 1}(6x^5 + 12x^3) = \frac{6x^5 + 12x^3}{x^6 + 3x^4 + 1}$

52. $\frac{d}{dx}\left[\frac{x}{\ln x}\right] = \frac{\ln x - x\left(\frac{1}{x}\right)}{(\ln x)^2} = \frac{\ln x - 1}{(\ln x)^2}$

53. $\frac{d}{dx}\ln\left(\frac{xe^x}{\sqrt{1+x}}\right) = \frac{d}{dx}\left(\ln xe^x - \ln\sqrt{1+x}\right)$

$= \frac{1}{xe^x}(xe^x + e^x) - \frac{1}{\sqrt{1+x}}\cdot\frac{1}{2}(1+x)^{-1/2}$

$= 1 + \frac{1}{x} - \frac{1}{2(1+x)}$

54. $\frac{d}{dx}\left[\ln\left((x^2+3)^5(x^3+1)^{-4}\right)\right]$

$= \frac{d}{dx}\left[5\ln(x^2+3) - 4\ln(x^3+1)\right]$

$= 5\frac{1}{x^2+3}(2x) - 4\frac{1}{x^3+1}(3x^2)$

$= \frac{10x}{x^2+3} - \frac{12x^2}{x^3+1}$

55. $\frac{d}{dx}\left[\ln(\ln\sqrt{x})\right] = \frac{1}{\ln\sqrt{x}}\cdot\frac{1}{\sqrt{x}}\cdot\frac{1}{2}x^{-1/2} = \frac{1}{2x\ln\sqrt{x}} = \frac{1}{x\ln x}$

56. $\frac{d}{dx}\left[\frac{1}{\ln x}\right] = \frac{d}{dx}\left[(\ln x)^{-1}\right] = -1(\ln x)^{-2}\left(\frac{1}{x}\right) = -\frac{1}{x(\ln x)^2}$

57. $\frac{d}{dx}\left[x\ln x - x\right] = x\left(\frac{1}{x}\right) + \ln x - 1 = \ln x$

58. $\frac{d}{dx}\left[e^{2\ln(x+1)}\right] = \frac{d}{dx}\left[e^{\ln(x+1)^2}\right] = \frac{d}{dx}\left[(x+1)^2\right] = 2(x+1)$

59. $\frac{d}{dx}\left[e^x\ln x\right] = e^x\left(\frac{1}{x}\right) + e^x\ln x = \frac{e^x}{x} + e^x\ln x$

60. $\frac{d}{dx}\ln(x^2+e^x) = \frac{1}{x^2+e^x}(2x+e^x) = \frac{2x+e^x}{x^2+e^x}$

61. $\ln f(x) = \ln\left[(x^2+5)^6(x^3+7)^8(x^4+9)^{10}\right]$

$= 6\ln(x^2+5) + 8\ln(x^3+7) + 10\ln(x^4+9)$

Differentiating both sides,

$$\frac{f'(x)}{f(x)} = \frac{6}{x^2 + 5}(2x) + \frac{8}{x^3 + 7}(3x^2) + \frac{10}{x^4 + 9}(4x^3)$$

$$f'(x) = (x^2 + 5)^6(x^3 + 7)^8(x^4 + 9)^{10}\left[\frac{12x}{x^2 + 5} + \frac{24x^2}{x^3 + 7} + \frac{40x^3}{x^4 + 9}\right]$$

62. $\ln f(x) = \ln x^{1+x} = (1 + x)\ln x = \ln x + x \ln x$

Differentiating both sides,

$$\frac{f'(x)}{f(x)} = \frac{1}{x} + x\left(\frac{1}{x}\right) + \ln x;\quad f'(x) = x^{1+x}\left(\frac{1}{x} + 1 + \ln x\right).$$

63. $\ln f(x) = \ln 10^x = x \ln 10$ Differentiating both sides,

$$\frac{f'(x)}{f(x)} = \ln 10;\quad f'(x) = 10^x \ln 10.$$

64. $\ln f(x) = \ln\left(\sqrt{x^2 + 5}\, e^{x^2}\right) = \frac{1}{2}\ln(x^2 + 5) + \ln e^{x^2}$

$$= \frac{1}{2}\ln(x^2 + 5) + x^2$$

Differentiating both sides,

$$\frac{f'(x)}{f(x)} = \frac{1}{2}\cdot\frac{1}{x^2 + 5}(2x) + 2x = \frac{x}{x^2 + 5} + 2x$$

$$f'(x) = \sqrt{x^2 + 5}\, e^{x^2}\left[\frac{1}{x^2 + 5} + 2x\right]$$

65.

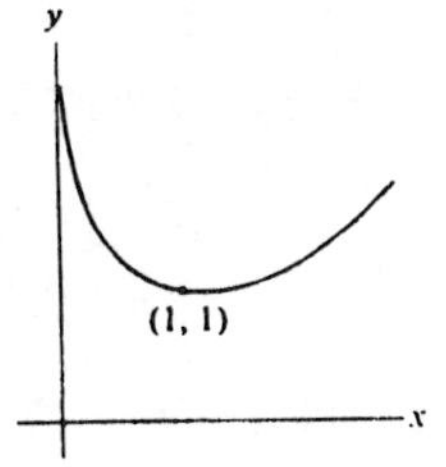

66.

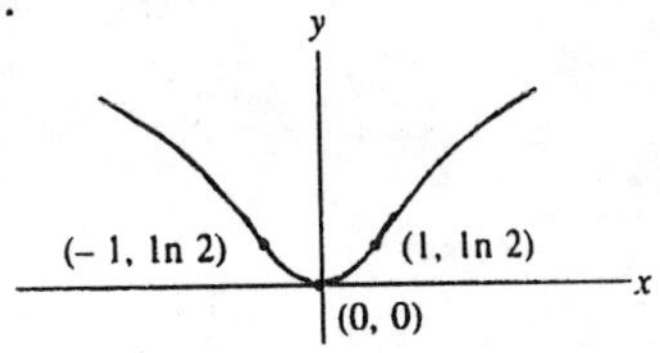

67.

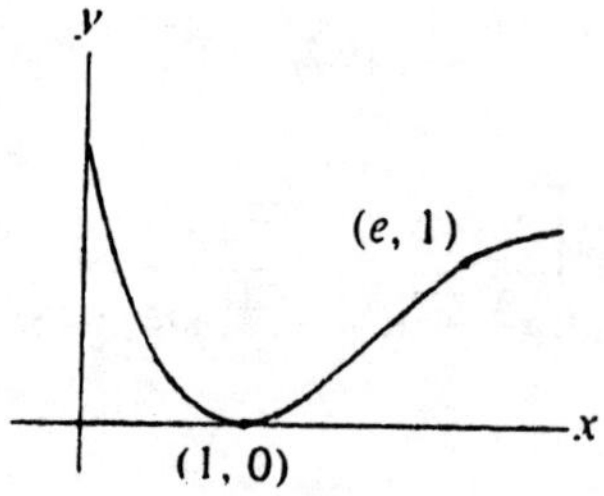

68.

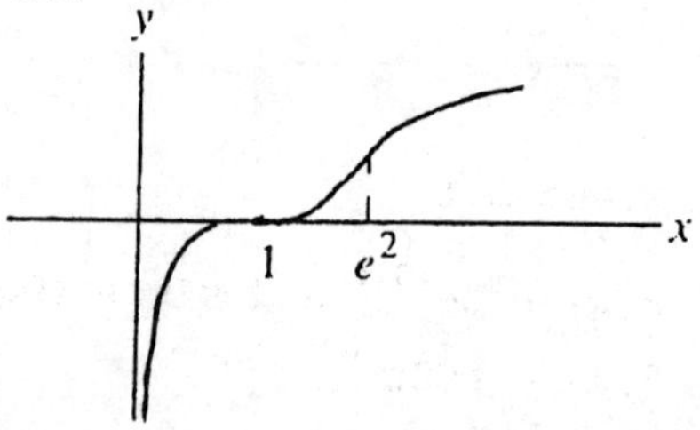

Exercises 5.1

1. $P(t) = Ce^{.07t}$ and since $400 = P(0) = Ce^{(.07)(0)} = C$ we have $P(t) = 400e^{.07t}$.

2. $P(t) = Ce^{.55t}$ and since $10,000 = P(0) = Ce^{(.55)(0)} = C$ we have $P(t) = 10,000e^{.55t}$.

3. (a) $P(0) = 5000e^{.2(0)} = 5000$, so there were 5000 present initially.
   (b) $20,000 = 5000e^{.2t}$; $4 = e^{.2t}$; $\ln 4 = \ln e^{.2t}$; $\ln 4 = .2t$; $t = \dfrac{\ln 4}{2} \sim 6.9$ hours.

4. $600 = 300e^{.01t}$; $2 = e^{.01t}$; $\ln 2 = \ln e^{.01t}$; $\dfrac{\ln 2}{.01} = t$;
   $t \sim 69.3$ days
   $1200 = 300e^{.01t}$; $4 = e^{.01t}$; $\dfrac{\ln 4}{.01} = t$; $t \sim 138.6$ days

5. Let $P(t)$ be the population after t days, $P(t) = P_0e^{kt}$. It is given that $P(40) = 2P(0)$, i.e., $P_0e^{40k} = 2P_0e^{0(k)} = 2P_0$;
   $e^{40k} = 2$; $\ln e^{40k} = \ln 2$; $k = \dfrac{\ln 2}{40} \sim .017$.

6. Let $P(t)$ be the population after t hours, $P(t) = P_0e^{kt}$. It is given that $P(5) = 3P(0)$, i.e., $P_0e^{5k} = 3P_0e^{(0)k} = 3P_0$;
   $e^{5k} = 3$; $\ln e^{5k} = \ln 3$; $k = \dfrac{\ln 3}{5} \sim .22$.

7. (a)Initial population on Jan 1 1991 is 5.4 billion.
   $P(0) = 5.4$; $P(t) = 5.4e^{kt}$; (4 years) later $P(4) = 5.4e^{4k} = 6$
   $e^{4k} = 1.1111 \Rightarrow 4k = \ln(1.1111)$ or $k = .02634$ thus the equation that models population growth is $P(t) = 5.4e^{.0263t}$.

   (b)Jan 1, 2010 is 19 years after Jan 1, 1991
   $P(19) - 5.4e^{.0263(19)} \cong 5.4e^{.5} = 8.9$ billion.

   (c)$P(t) = 7 \Rightarrow 5.4e^{.0263t} = 7$ or $.o263t = \ln(7/5.4) = .262$

$t \cong 9.95$ years. The population reaches 7 billion in 2000.

8. (a) $P(t)$ is the population (in millions) t years after 1980. Then $P(t) = 20.2e^{.032t}$.
$P(2000 - 1970) = P(10) = 20.2e^{.032(10)} = 20.2e^{.32} \sim 8$ (million)
(b) $40.4 = 20.2e^{.032t}$; $\ln 2 = \ln e^{.032t}$; $t = \dfrac{\ln 2}{.032} \sim 21$ so the population will have doubled in $1990 + 21 = 2011$.

9. Let $P(t)$ be the bacteria population after t minutes, $P(t) = P_0e^{kt}$. It is given that
$P(20) = 2P_0$, i.e., $P_0e^{20k} = 2P_0$; $e^{20k} = 2$; $\ln e^{20k} = \ln 2$;
$k = \dfrac{\ln 2}{20} \sim .035$. $12P_0 = P_0e^{.035t}$; $\ln 12 = \ln e^{.035t}$;
$t = \dfrac{\ln 12}{.035} \sim 71$ minutes.

10. Let $P(t)$ be the cell population (in millions) after t hours, $P(t) = P_0e^{kt}$. It is given that $P(0) = 1$ and that $P(10) = 9$.
Thus $9 = 1\cdot e^{k(10)}$; $\ln 9 = \ln e^{10k}$; $k = \dfrac{\ln 9}{10} \sim .22$.
$P(t) = e^{.22t}$ so $P(15) = e^{.22(15)} = e^{3.3} \sim 27$ (million)

11. a - F, b - D, c - A, d - G, e - H, f - C, g - B, h - E

12. (a) $P(t) = 3e^{kt}$ (t in hours, $P(t)$ in millions)
$P(.5) = 2P(0)$; $3e^{.5k} = 2\cdot 3e^{k(0)} = 6$; $e^{.5k} = 2$; $\ln e^{.5k} = \ln 2$;
$k = \dfrac{\ln 2}{.5} \sim 1.39$. $P(t) = 3e^{1.39t}$
$600 = 3e^{1.39t}$; $200 = e^{1.39t}$; $\ln 200 = \ln e^{1.39t}$;
$t = \dfrac{\ln 200}{1.39} \sim 3.8$ hours
(b) $P'(t) = \dfrac{d}{dt}\left[3e^{1.39t}\right] = 3e^{1.39t}(1.39) = 4.17e^{1.39t}$

$600 = 4.17e^{1.39t}$; $\ln\left(\dfrac{600}{4.17}\right) = 1.39t$; $t = \dfrac{\ln\left(\dfrac{600}{4.17}\right)}{1.39} \sim 3.6$ hours.

13. (a) $P(t) = P_0e^{-.08t}$ and since $P(0) = 30$, $P(t) = 30e^{-.08t}$.

(b) $P(10) = 30e^{-.08(10)} = 30e^{-.8} \sim 13.48$ g.

14. (a) $P(t) = 5e^{-.09t}$ (P(t) is milligrams present after t hours)

(b) $P(20) = 5e^{-.09(20)} = 5e^{-1.8} \sim .83$ mg.

15. (a) $P(t) = 100e^{.01t}$ (P(t) is grams present after t years)

(b) $P(30) = 100e^{(-.01)30} = 100e^{-.3} \sim 74$ grams

(c) $50 = 100e^{-.01t}$; $\frac{1}{2} = e^{-.01t}$; $\frac{\ln \frac{1}{2}}{-.01} = t$; $t \sim 69$ years.

16. $\frac{1}{2} = e^{-.023t}$; $\frac{\ln \frac{1}{2}}{-.023} = t$; $t \sim 30$ years

17. (a) $\frac{1}{2} = e^{-\lambda(5.3)}$; $\frac{\ln \frac{1}{2}}{-5.3} = \lambda$; $\lambda \sim .13$

(b) $P(t) = 10e^{-.13t}$; $P(2) = 10e^{-.13(2)} = 10e^{-.26} = 7.7105$ g

18. $P(t) = 5e^{-\lambda t}$; $P(1) = 3 = 5e^{-\lambda(1)} = 5e^{-\lambda}$; $\ln \frac{3}{5} = -\lambda$; $\lambda = .51$

Thus, $P(t) = 5e^{-.51t}$. $1 = 5e^{-.51t}$; $\ln \frac{1}{5} = \ln e^{-.51t}$;

$t = \frac{\ln \frac{1}{5}}{-.51} \sim 3.16$ years

19. Let the original amount be 1. Then $P(t) = 1 \cdot e^{-.00012t}$.

$P(4500) = e^{-.00012t(4500)} = e^{-.54} \sim .583$. Thus about 58.3% remains.

20. Let the original amount be 1, $P(t) = e^{-.00012t}$. Then for some t_0, $P(t_0) = .20 = e^{-.00012t_0}$. $\frac{\ln .20}{-.00012} = t_0$; $t_0 \sim 13{,}400$ years old.

21. Let the original amount be 1, $P(t) = e^{-.00012t}$.

$P(t_0) = .34 = e^{-.00012t_0}$; $\frac{\ln .34}{-.00012} = t_0$; $t_0 \sim 8990$ years old.

22. $P(t) = e^{-.00012t}$; $P(t_0) = .27 = e^{-.00012t_0}$; $\frac{\ln .27}{-.00012} = t_0$;

$t_0 \sim 10,900$ years ago.

23. $f(t) = 1.00e^{-\lambda t}$; Since the dollar lost 15% of its purchasing power in two years then $f(2) = .85 = 1.00e^{-\lambda(2)}$.

$\frac{\ln .85}{2} = -\lambda$; $\lambda \sim .081$. Thus the formula is $f(t) = e^{-.081t}$.

24. $f'(t) = .12f(t)$; $f(t) = Ce^{.12t}$; $f(0) = 100 = Ce^{.12(0)} = C$.

Thus $f(t) = 100e^{.12t}$; Solve $200 = 100e^{.12t}$; $\ln 2 = \ln e^{.12t}$;

$t = \frac{\ln 2}{.12} \sim 5.78$ years.

25. $\frac{1}{2} = e^{-\lambda(28)}$; $\frac{\ln \frac{1}{2}}{28} = -\lambda$; $\lambda \sim .025$. Thus letting the initial level be 1, the decay equation is $P(t) = e^{-.025t}$. Now solve

$\frac{1}{100} = e^{-.025t}$ for t. $\frac{\ln .01}{-.025} = t$; $t \sim 184$ years.

26. For 10^8 bacteria to remain, there must be $10^8 \cdot 10$ present prior to emptying the bladder, i.e., 10^9.

$P(t) = 10^8e^{kt}$ and given that $P(20) = 2P(0)$ we have

$10^8e^{(20)k} = 2 \cdot 10^8e^{(0)k} = 2 \cdot 10^8$; $e^{(20)k} = 2$; $\ln e^{20k} = \ln 2$;

$k = \frac{\ln 2}{20} \sim .035$; $P(t) = 10^8e^{.035t}$. Now solve $10^9 = 10^8e^{.035T}$

for T. $10 = e^{.035T}$; $\frac{\ln 10}{.035} = T$; $T \sim 66$ minutes.

27. First find the decay constant. $\frac{1}{2} = e^{-\lambda(1500)}$; $\frac{\ln \frac{1}{2}}{1500} = -\lambda$;

$\lambda \sim .00046$. Thus, $P(t) = e^{-.00046t}$. If the initial amount is 1, then we wish to solve $.0001 = e^{-.00046t}$ for t.

$\frac{\ln .0001}{-.00046} = t$; $t \sim 20,022$ years.

28. (a) $A(t) = Ce^{kt}$. Letting $A(t)$ be the amount of land (in 10^9 hectares) needed after t years, $A(0) = 1 \cdot e^{kt}$.

$2 = A(30) = e^{(k)30}$; $\ln 2 = \ln e^{30k}$; $k = \frac{\ln 2}{30} \sim .023$. Thus

$A(t) = e^{.023t}$.

(b) $3.2 = e^{.023t}$; $\frac{\ln 3.2}{.032} = t$; $t \sim 50$ years, so approximately

the year 2000.

29. (a) $P(0) = 500$; $P(t) = 500e^{-\lambda t}$; $P(4) = 200 = 500e^{-\lambda(4)}$;

$\ln \frac{2}{5} = \ln e^{-\lambda(4)}$; $\frac{\ln \frac{2}{5}}{4} = -\lambda$; $\lambda \sim .23$. Thus, $P(t) = 500e^{-.23t}$.

(b) $50 = 500e^{-.23t}$; $.1 = e^{-.23t}$; $\frac{\ln .1}{-.23} = t$; $t \sim 10$ months.

30. $f(t) = 8e^{-\lambda t}$; $f(50) = 4 = 8e^{-\lambda(50)}$; $\frac{\ln \frac{1}{2}}{50} = -\lambda$; $\lambda \sim .014$;

$f(t) = 8e^{-.014t}$

31. Following the hint,

(i) In fig. 5 the slope of the tangent line is $-\frac{C}{T}$.

(ii) $\frac{dy}{dt} = Ce^{-\lambda t}(-\lambda) = -\lambda Ce^{-\lambda t}$ so at $t = 0$, the slope is

$-\lambda Ce^{-\lambda(0)} = -\lambda C$.

Since (i) and (ii) both express the slope at $t = 0$, we may write $-\lambda C = -\frac{C}{T}$ or $T = \frac{1}{\lambda}$.

32. $f(t) = 300e^{-(2/3)t}$; $f'(t) = 300e^{-(2/3)t}\left(-\frac{2}{3}\right) = -200e^{-(2/3)t}$

(a) At $t = 0$, $f'(t) = -200$ mg/hr.

(b) Since $T = \frac{1}{\lambda}$ and $\lambda = \frac{2}{3}$, we have $T = \frac{3}{2}$.

33. $M(t) = 80000e^{.05t}$;2000 corresponds to $t = 14$,

$M(14) = 80000e^{.05(14)} = \$161100$.

34. $P(t) = P_o e^{-\lambda t}$ for carbon-14, $\lambda = .00012$.

$P(t) = .91P_o$, so $.91 = e^{-.00012t} \Rightarrow$ (taking ln of both sides)

$-.09431 = -.00012t$, or $t \cong 786$. So the table dates from approximately 1190, and hence could not have belonged to King Arthur.

35.a. $f(5) = 25$ gm
b. $f(t) = 40$ has solution $t = 3$ years.
c. $f'(1) = -15$. The material is disintegrating at a rate of 15 gm/yr
d. $f'(t) = -5$ has solution $t = 6$ yrs.

36.a. $f(7.5) = \$700$
b. $f(r) = 600$ has solution $r = 6\%$.
c. $f(3) - f(2) \approx f'(2.5) = \$50$

Exercises 5.2

1. $A = 1000(1 + .1)^2 = \$1210$

2. $A = 5000\left(1 + \frac{.06}{12}\right)^{4(12)} = 5000(1.005)^{48}$

3. $A = 10,000\left(1 + \frac{.08}{4}\right)^{3(4)} = 10,000(1.02)^{12}$

4. $A = P\left(1 + \frac{.08}{2}\right)^{1(2)} = P(1.0816)$ The effective annual rate is 8.16%.

5. $A = 1000e^{.14(6)} = 1000e^{.84} = \$2316.37$

6. $A = 100,000e^{.12(7)} = 10,000e^{.84} = \$231,636.70$

7. $A = 500\left(1 + \frac{.07}{360}\right)^{(360)3} = \$616.83.$

8. $A = 10,000e^{(.18)(.5)} = 10,000e^{.09} = \$10,941.74$, so \$941.74 interest will be earned.

9. $25 = 10e^{k(8)}$; $\frac{\ln 2.5}{8} = k$; $k \sim .1145 = 11.45\%$

10. Solve $2500 = 1000e^{.06t}$ for t. $\frac{\ln 2.5}{.06} = t$; $t \sim 15.27$ years

11. $A_1 = 10{,}000\left(1 + \frac{.40}{2}\right)^2 = 10{,}000(1.44) = \$14{,}400$

$A_2 = 10{,}000e^{.39} = 14{,}769.81$. Thus, the 39% interest compounded continuously is preferred.

12. $A(t) = 5000e^{kt}$; $A(1985 - 1966) = A(19) = 60{,}000 = 5000e^{k(19)}$;

$12 = e^{19k}$; $\frac{\ln 12}{19} = k$; $k \sim .13$. Thus $A(t) = 5000e^{.13t}$.

$100{,}000 = 5000e^{.13t}$; $20 = e^{.13t}$; $\frac{\ln 20}{.13}$; $t \sim 23$, so in 1989.

13. $38{,}000 = 10{,}000e^{.15t}$; $\frac{\ln 3.8}{.15} = t$; $t \sim 8.9$ years.

14. $2 = 1\cdot e^{.13t}$; $\frac{\ln 2}{.13} = t$; $t \sim 5.33$ years.

15. $A(t) = 1\cdot e^{kt}$ ($A(t)$ in millions)

$A(10) = 3 = e^{k(10)}$; $\frac{\ln 3}{10} = k$; $k \sim .11$. Thus, $A(t) = e^{.11t}$.

Solve $10 = e^{.11t}$ for t. $\frac{\ln 10}{.11} = t$; $t \sim 21$ years so in 1996.

16. $1000 = Pe^{(.14)(3)} = Pe^{.42}$; $\frac{1000}{e^{.42}} = P$; $P \sim \$657.05$

17. $2000 = Pe^{(.15)(10)} = Pe^{1.5}$; $P = \frac{2000}{e^{1.5}} \sim \$446.26$

18. $A(t) = 10{,}000e^{kt}$; $A(5) = 16{,}000 = 10{,}000e^{k(5)}$; $\frac{\ln 1.6}{5} = k$;

$k \sim .094$. Thus $A(t) = 10{,}000e^{.094t}$. Solve

$45{,}000 = 10{,}000e^{.094t}$ for t. $\frac{\ln 4.5}{.094} = t$; $t \sim 16$ years so in 1996.

19. $A = 100e^{.07} = 107.25$ so the effective annual rate of return is 7.25%.

20. $A(t) = 1\cdot e^{kt}$; $A(87 - 76) = A(11) = 3 = e^{11k}$; $\frac{\ln 3}{11} = k$;

$k \sim .1$. Thus, $A(t) = e^{.1t}$. Solve $5 = e^{.1t}$ for t.

$\frac{\ln 5}{.1} = t$; $t \sim 16$ years so in 1992.

21. $10,000 = Pe^{.12(5)} = Pe^{.6}$; $P = \frac{10,000}{e^{.6}} \sim \$5488.12$

22. $A = 70,200e^{.13t}$, $B = 60,000e^{.14t}$. Set A = B and solve for t.

    $70,200e^{.13t} = 60,000e^{.14t}$; $\frac{7.02}{6} = \frac{e^{.14t}}{e^{.13t}} = e^{.01t}$; $\frac{\ln \frac{7.02}{6}}{.01} = t$;

    $t \sim 15.7$ years.

23. $1000 = 559.90e^{2k}$; $\frac{\ln \frac{1000}{559.9}}{2} = k$; $k \sim .29$ so 29%.

24. $f(t) = 2000e^{.05t}$; $f'(t) = 2000e^{.05t}(.05) = 100e^{.05t}$;

    $f'(2) = 100e^{.05(2)} = 100e^{.1} = 110.5$. Thus, at the end of two years interest will be earned at the rate of \$100.52.

25. a - B, b - D, c - G, d - A, e - F, f - E, g - H, h - C

26. a. $f(6) = \$1,800$
    b. $f(r) = 2600$ has solution $r = 10\%$.
    c. $f(9) - f(8) \approx f'(8.5) = \$200$.

Exercises 5.3

1. $f(t) = t^2$; $\frac{f'(t)}{f(t)} = \frac{2t}{t^2} = \frac{2}{t}$; $\frac{f'(10)}{f(10)} = \frac{2}{10} = \frac{1}{5} = 20\%$;

   $\frac{f'(50)}{f(50)} = \frac{2}{50} = \frac{1}{25} = 4\%$

2. $f(t) = t^{10}$; $\frac{f'(t)}{f(t)} = \frac{10t^9}{t^{10}} = \frac{10}{t}$; $\frac{f'(10)}{f(10)} = \frac{10}{10} = 100\%$;

   $\frac{f'(50)}{f(50)} = \frac{10}{50} = 20\%$

3. $f(t) = e^{.3x}$; $\frac{f'(t)}{f(t)} = \frac{.3e^{.3x}}{e^{.3x}} = .3$; Thus for all x, the percentage rate of change is 30%.

4. $f(x) = e^{-.05x}$; $\frac{f'(t)}{f(t)} = \frac{-.05e^{-.05x}}{e^{-.05x}} = -.05$; Thus for all x, the percentage rate of change is -5%.

5. $f(t) = e^{.3t^2}$; $\frac{f'(t)}{f(t)} = \frac{.6te^{.3t^2}}{e^{.3t^2}} = .6t$; $\frac{f'(1)}{f(1)} = .6 = 60\%$;

$\frac{f'(5)}{f(5)} = 3 = 300\%$

6. $G(s) = e^{-.05s^2}$; $\frac{G'(s)}{G(s)} = \frac{-.1se^{-.05s^2}}{e^{-.05s^2}} = -.1s$;

$\frac{G'(1)}{G(1)} = -1(1) = -10\%$; $\frac{G'(10)}{G(10)} = -.1(10) = -100\%$

7. $f(p) = \frac{1}{p+2}$; $\frac{f'(p)}{f(p)} = \frac{\frac{1}{(p+2)^2}}{\frac{1}{p+2}} = -\frac{1}{p+2}$;

$\frac{f'(2)}{f(2)} = -\frac{1}{2+2} = -.25 = -25\%$; $\frac{f'(8)}{f(8)} = -\frac{1}{8+2} = -.1 = -10\%$

8. $g(p) = \frac{5}{2p+3}$; $\frac{f'(p)}{f(p)} = \frac{-\frac{5(2)}{(2p+3)^2}}{\frac{5}{2p+3}} = -\frac{2}{2p+3}$;

$\frac{g'(1)}{g(1)} = -\frac{2}{2+3} = -\frac{2}{5} = -40\%$; $\frac{g'(11)}{g(11)} = -\frac{2}{22+3} = -\frac{2}{25} = -8\%$

9. $$\frac{dS}{dt}\left[\ln\left[50,000\left[e^{t^{1/2}}\right]^{1/2}\right]\right] = \frac{dS}{dt}\left[\ln 50,000 + \frac{1}{2}\ln e^{t^{1/2}}\right]$$
$$= \frac{dS}{dt}\left[\ln 50,000 + \frac{t^{1/2}}{2}\right]$$
$$= 0 + \left(\frac{1}{2}\right)\left(\frac{1}{2}\right)t^{-1/2} = \frac{1}{4\sqrt{t}}$$

At $t = 4$, the percentage rate of growth is $\frac{1}{4\sqrt{4}} = \frac{1}{8} = 12.5\%$.

10. $\frac{f'(t)}{f(t)} = \frac{.001 - (.01)e^{-t}}{4 + .001T + .01e^{-t}}$;
$\frac{f'(0)}{f(0)} = \frac{.001 - .01}{4 + .01} = -.00224 = -.224\%$;
$\frac{f'(1)}{f(1)} = \frac{.001 - (.01)e^{-1}}{4 + .001 + .01e^{-1}} = -.00067 = -.067\%$;
$\frac{f'(2)}{f(2)} = \frac{.001 - (.01)e^{-2}}{4 + .002 + (.01)e^{-2}} = -.000088 = -.0088\%$

11. $\frac{f'(x)}{f(x)} = .12$ so $f(x) = A_0e^{.12t}$ where A_0 is the initial amount.
$2A_0 = A_0e^{.12t}$; $\frac{\ln 2}{.12} = t$; $t \sim 5.8$ years

12. $\frac{f'(t)}{f(t)} = \frac{r}{100}$ so $f(x) = A_0e^{(r/100)t}$
$f(2) = 2A_0 = A_0e^{(r/100)3}$; $\ln 2 = \ln e^{(3/100)r}$; $\frac{100}{3}\ln 2 = r$;
$r \sim 23.1\%$

13. $E(p) = \frac{-p(-5)}{700 - 5p} = \frac{5p}{700 - 5p} = \frac{p}{140 - p}$
$E(80) = \frac{400}{700 - 400} > 1$ so demand is elastic.

14. $E(p) = \frac{-p(-120e^{-.2p})}{600e^{-.2p}} = \frac{pe^{-.2p}}{5e^{-.2p}} = \frac{p}{5}$; $E(10) = \frac{10}{5} > 1$ so demand is elastic.

15. $E(p) = \frac{-p(-800p)}{400(116 - p^2)} = \frac{2p^2}{116 - p^2}$; $E(6) = \frac{2\cdot 36}{116 - 36} = \frac{72}{80} < 1$ so demand is inelastic.

16. $E(p) = \frac{-p\left(-2\frac{77}{p^3}\right)}{\frac{77}{p^2} + 3} = \frac{\frac{2(77)}{p^2}}{\frac{77}{p^2} + 3} = \frac{154}{77 + 3p^2}$; $E(1) = \frac{154}{77 + 3} > 1$ so

demand is elastic.

17. $q' = p^2e^{-(p+3)}(-1) + e^{-(p+3)}2p = pe^{-(p+3)}(2 - p)$

$$E(p) = \frac{-p(pe^{-(p+3)})(2 - p)}{p^2e^{-(p+3)}} = -(2 - p) = p - 2$$

$E(4) = 4 - 2 = 2 > 0$ so demand is elastic.

18. $$E(p) = \frac{-p\left(-\dfrac{700}{(p + 5)^2}\right)}{\dfrac{700}{p + 5}} = \frac{p}{p + 5};\quad E(15) = \frac{15}{15 + 5} < 1$$ so demand is inelastic.

19. (a) $q = 3000 - 600p^{1/2}$; $q' = -300p^{-1/2}$

$$E(p) = \frac{-p(-300p^{-1/2})}{3000 - 600p^{1/2}} = \frac{300p^{1/2}}{3000 - 600p^{1/2}} = \frac{p^{1/2}}{10 - 2p^{1/2}}$$

$$E(4) = \frac{4^{1/2}}{10 - 2\cdot 4^{1/2}} = \frac{2}{6} = \frac{1}{3} < 1$$

(b) Since demand is inelastic, to increase revenue, the ticket price should be increased.

20. (a) $$q' = 9000(-1(p + 60)^{-2}) = -\frac{9000}{(p + 60)^2}$$

$$E(p) = \frac{\dfrac{9000}{(90)^2}(30)}{\dfrac{9000}{90} - 50} = \frac{\dfrac{100(30)}{90}}{100 - 50} = \frac{\dfrac{100}{3}}{50} < 1$$

Thus demand is inelastic.

(b) decrease

21. (a) $$E(p) = \frac{-p\left(-\dfrac{18,000}{p^2}\right)}{\dfrac{18,000}{p} - 1500} = \frac{\dfrac{18,000}{p}}{\dfrac{18,000}{p} - 1500}$$

$$E(6) = \frac{3000}{3000 - 1500} = \frac{3000}{1500} > 1 \quad \text{Thus, demand is elastic.}$$

(b) increase

22. (a) $q = 2000(90 - p)^{1/2}$, $q' = 2000\left(\frac{1}{2}\right)(90 - p)^{-1/2}(-1)$

$$= -\frac{1000}{(90 - p)^{1/2}}$$

$$E(p) = \frac{p\left[\frac{1000}{(90 - p)^{1/2}}\right]}{2000(90 - p)^{1/2}} = \frac{p}{2(90 - p)}$$

$$E(65) = \frac{65}{2(90 - 65)} = \frac{65}{50} > 1 \quad \text{Thus demand is elastic.}$$

(b) lowered

23. (a) $q = \frac{1000}{p^2}$, $q' = -\frac{2000}{p^3}$ $\quad E(p) = \frac{\frac{2000}{p^2}}{\frac{1000}{p^2}} = 2$

(b) yes

24. $q = \frac{a}{p^m}$, $q' = -\frac{ma}{p^{m+1}}$ $\quad E(p) = \frac{-p\left[-\frac{ma}{p^{m+1}}\right]}{\frac{a}{p^m}} = \frac{\frac{ma}{p^m}}{\frac{a}{p^m}} = m$

25. $$E_c(x) = \frac{\frac{d}{dx}\ln C(x)}{\frac{d}{dx}\ln x} = \frac{\frac{C'(x)}{C(x)}}{\frac{1}{x}} = \frac{x\cdot C'(x)}{C(x)}$$

26. The average cost (AC) is $\frac{C(x)}{x}$. The marginal cost (MC) is $C'(x)$. Thus $\frac{MC}{AC} = \frac{C'(x)}{\frac{C(x)}{x}} = \frac{x\cdot C'(x)}{C(x)}$.

27. $C(x) = \frac{1}{10}x^2 + 5x + 300$; $C'(x) = \frac{2}{10}x + 5 = \frac{1}{5}x + 5$

$$E_c(x) = \frac{x\left(\frac{1}{5}x + 5\right)}{\frac{1}{10}x^2 + 5x + 300} = \frac{\frac{1}{5}x^2 + 5x}{\frac{1}{10}x^2 + 5x + 300}$$

$$E_c(50) = \frac{500 + 250}{250 + 250 + 300} = \frac{750}{800} < 1.$$

28. $C(x) = 1000e^{.02x}$, $C'(x) = 20e^{.02x}$ $\quad E_c(x) = \dfrac{20xe^{.02x}}{1000e^{.02x}} = \dfrac{x}{50}$

$E_c(60) = \frac{60}{50} > 1$. An increase in the production level will result in a greater percentage increase in cost.

Exercises 5.4

1. $f(x) = 5(1 - e^{-2x})$, $x \geq 0$

(a) $f'(x) = 5(-e^{-2x}(-2)) = 10e^{-2x}$, $f''(x) = -20e^{-2x}$

Since $f'(x) > 0$ for all $x \geq 0$, $f(x)$ is increasing for all $x \geq 0$. Furthermore, since $f''(x) < 0$ for all $x \geq 0$, $f(x)$ is concave down for all $x \geq 0$.

(b) Since $\lim_{x\to\infty} e^{-2x} = 0$, then for very large values of x, $f(x) \sim 5(1- 0) = 5$.

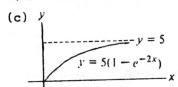

(c) 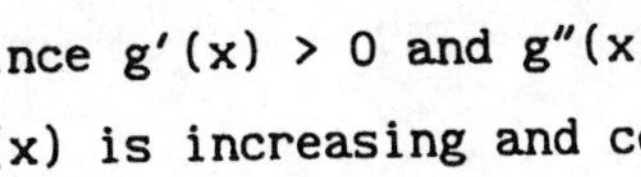

2. $g(x) = 10 - 10e^{-.1x}$, $x \geq 0$.

(a) $g'(x) = e^{-.1x}$, $g''(x) = -.1e^{-.1x}$

Since $g'(x) > 0$ and $g''(x) < 0$ for all $x \geq 0$, it follows that $g(x)$ is increasing and concave down for all $x \geq 0$.

(b) Since $\lim_{x\to\infty} 10e^{-.1x} = 0$, the for very large values of x we have $g(x) \sim 10$.

(c)

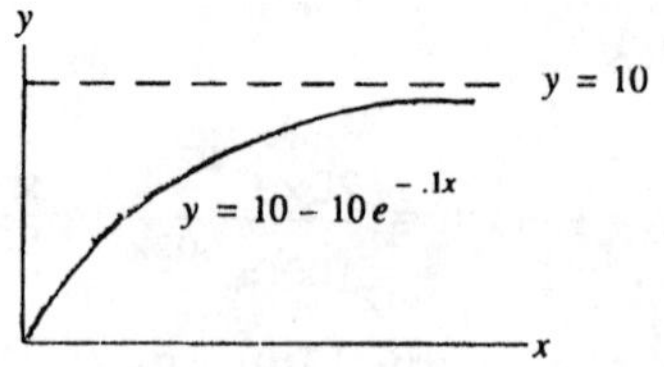

3. $y' = -2e^{-x}(-1) = 2e^{-x}$

$2 - y = 2 - 2(1 - e^{-x}) = 2 - 2 + 2e^{-x} = 2e^{-x} = y'$

4. $y' = -5e^{-2x}(-2) = 10e^{-2x}$

$10 - 2y = 10 - 2(5 - 5e^{-2x}) = 10e^{-2x} = y'$

5. $y' = f'(x) = -3e^{-10x}(-10) = 30e^{-10x}$

$$\begin{aligned} y' - 10(3 - y) &= 30e^{-10x} - 10(3 - (3 - 3e^{-10x})) \\ &= 30e^{-10x} - 10(3e^{-10x}) \\ &= 0 \end{aligned}$$

Also, $f(0) = 3(1 - 1) = 0$.

6.

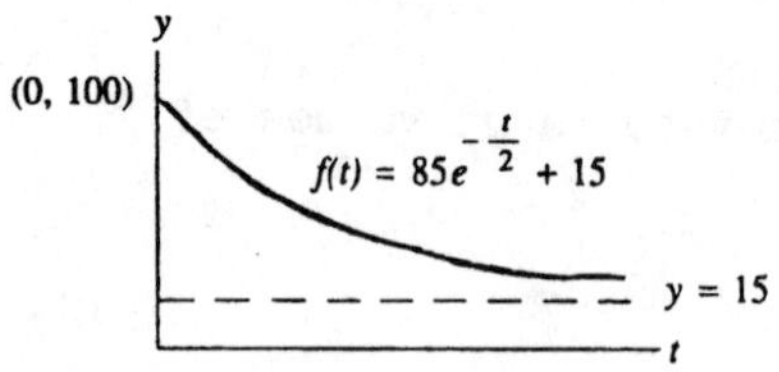

7. The number of people who have heard about the indictment by time t is given by $f(t) = P(1 - e^{-kt})$. It is given that $f(1) = \frac{1}{4}P = P(1 - e^{k(1)})$; $\frac{1}{4} = 1 - e^{-k}$; $e^{-k} = \frac{3}{4}$; $-k = \ln\frac{3}{4}$; $k \sim .29$.

Thus, $f(t) = P(1 - e^{-.29t})$. Solve $\frac{3}{4}P = P(1 - e^{-.29t})$ for t.

$\frac{3}{4} = 1 - e^{-.29t}$; $e^{-.29t} = \frac{1}{4}$; $\frac{\ln .25}{-.29} = t$; $t \sim 4.8$ hours.

8. The amount of glucoses will stabilize at $\frac{2r}{\lambda} = 2M$.

9. Since the body uses up excess glucose at a rate proportional to the amount of excess glucose, we have

$$A(t) = A_0 e^{-\lambda t}$$

which gives the amount of excess glucose above equilibrium at any time, t, after an initial injection of A_0 milligrams. Thus, the doctor should give the injection and (after some know time interval) measure the amount of excess remaining. From this information the velocity constant of elimination, λ, can be calculated from the equation above.

10. $$\frac{dC}{dt} = \frac{A'(t)}{V} = \frac{\lambda(M - A(t))}{V} = \lambda\left(\frac{M}{V} - \frac{A(t)}{V}\right) = \lambda\left(\frac{M}{V} - C\right)$$

11.a. $f(7) = 2,500$
b. $f'(10) = 1,000$ people/day
c. $f(t) = 7000$ has solution $t = 12$. On day 12.
d. $f'(t) = 600$ has solutions $t = 6$, 13.5. On days 6 and 13.5.
e. f' has a maximum at approximately t=9.78 days.

12.a. $f(25) = 64$ ft/sec
b. $f'(10) = 2$ ft/sec^2
c. $f(t) = 56$ has solution $t = 12$ sec.
d. $f'(t) = 4$ has solution $t = 6$ sec.
e. The graph of f has a horizontal asymptote at $t = 65$. Therefore, the terminal velocity is 65 ft/sec.

13.a. $f(1) = 2$ cm
b. $f'(10) = 2$ cm/wk
c. $f(t) = 10$ has solution $t = 5$. In week 5.
d. $f'(t) = 2$ has solutions $t = 3, 10$. In weeks 3 and week 10.
e. $f'(t)$ has a maximum at $t = 6.4$. Answer: At 6.4 weeks, when the weed is 15 cm

14.a. $f(10) = 47,500$
b. $f'(0) = 15,000$ people/day
c. $f(t) = 22500$ has solution $t = 2$. On day 2.
d. $f'(t) = 2500$ has solution $t = 6$. On day 6.

15. a. $f(11) = 400°$

b. $f'(6) = -100$. Decreasing at a rate of 100°/sec

c. $f(t) = 200$ has solution $t = 17$ sec.

d. $f'(t) = -200$ has solution $t = 2$ sec.

Chapter 5 Supplementary Exercises

1. $P(x) = Ce^{-.2x}$; $29.92 = P(0) = Ce^{-.2(0)} = C$

   Thus, $P(x) = 29.92e^{-.2x}$

2. $P(x) = P_0e^{kt}$ (t in years, P_0 = herring gulls in 1990)

   $P(13) = 2P_0 = P_0e^{(k)13}$; $\ln 2 = \ln e^{13k}$; $\frac{\ln 2}{13} = k$; $k \sim .0533$.

   $P(t) = .0533P(t)$

3. $10{,}000 = P_0e^{(.12)5} = P_0e^{.6}$; $P_0 = \frac{10{,}000}{e^{.6}} \sim \$5{,}488.12$

4. Solve $3000 = 1000e^{.1t}$ for t. $\ln 3 = \ln e^{.1t}$; $\frac{\ln 3}{.1} = t$;

   $t \sim 11$ years.

5. $\frac{1}{2} = e^{-\lambda(12)}$; $\frac{\ln .5}{12} = -\lambda$; $\lambda \sim .058$

6. $.63 = e^{-.00012t}$; $\frac{\ln .63}{-.00012} = t$; $t \sim 3{,}850$ years old.

7. (a) $P(t) = 14.2e^{kt}$; $P(10) = 14.2e^{10k} = 17$; $\frac{\ln \frac{17}{14.2}}{10} = k$;
$k \sim .0179$. $P(t) = 141.2e^{.0179t}$ (in millions)
(b) $P(20) = 14.2e^{(.0179)20} \sim 20.346$ (million)
(c) $25 = 14.2e^{.0179t}$; $\frac{\ln \frac{25}{14.2}}{.0179} = t$; $t \sim 31$ years so in 2011.

8. $A(t) = 100{,}000e^{kt}$; $A(2) = 117{,}000 = 100{,}000e^{2k}$; $\frac{\ln \frac{11.7}{10}}{2} = k$;
$k \sim .0785$ so it earned 7.85%.

9. (a) $A(t) = (10{,}000e^{.2(5)})e^{.06(5)} = (10{,}000e)e^{.3} = 10{,}000e^{1.3}$
$\sim \$36{,}693$
(b) $A(t) = 10{,}000e^{.14(10)} = 10{,}000e^{1.4} \sim \$40{,}552$
The second option is better by $3859.

10. $P_1(t) = 1000e^{k_1t}$; $P_1(21) = 2000 = 1000e^{k_1(21)}$; $\frac{\ln 2}{21} = k_1$;
$k_1 \sim .033$. Thus $P_1(t) = 1000e^{.033t}$.
$P_2(t) = 710{,}000e^{k_2t}$; $P_2(33) = 1{,}420{,}000 = 710{,}000e^{k_2(33)}$;
$\frac{\ln 2}{33} = k_2$; $k_2 \sim .021$. Thus $P_2(t) = 710{,}000e^{.021t}$.
Equating P_1 and P_2 and solving for t,
$1000e^{.033t} = 710{,}000e^{.021t}$; $710 = e^{.012t}$; $\frac{\ln 710}{.012} = t$;
$t \sim 547$ minutes.

11. $\frac{f'(t)}{f(t)} = \frac{50e^{.2t^2}(.4t)}{50e^{.2t^2}} = .4t$; $\frac{f'(10)}{f(10)} = .4(10) = 4 = 400\%$

12. $E(p) = \frac{-p(80p)}{4000 - 40p^2} = \frac{80p^2}{4000 - 40p^2} = \frac{2}{\frac{100}{p^2} - 1}$

$$E(5) = \frac{2}{\frac{100}{25} - 1} = \frac{2}{4 - 1} = \frac{2}{3} < 1 \text{ so demand is inelastic.}$$

13. Since a price increase of \$.16 represents a 2% increase in price, the quantity demanded will decrease 1.5(2%) = 3%. Since demand is elastic revenue will decrease.

14. $f(p) = \frac{1}{3p + 1}$, $f'(p) = -\frac{3}{(3p + 1)^2}$

$$\frac{f'(p)}{f(p)} = \frac{-\frac{3}{(3p + 1)^2}}{\frac{1}{3p + 1}} = -\frac{3}{3p + 1}; \quad \frac{f'(1)}{f(1)} = -\frac{3}{3 + 1} = -\frac{3}{4} = -75\%$$

15. $q = 1000p^2e^{-.02(p+5)}$

$$q' = 1000p^2(e^{-.02(p+5)})(-.02) + (e^{-.02(p+5)})2000p$$

$$= -1000pe^{-.02(p+5)}(.02p - 2)$$

$$E(p) = \frac{1000p^2e^{-.02(p+5)}(.02p - 2)}{1000p^2e^{-.02(p+5)}} = .02p - 2$$

$E(200) = .02(200) - 2 = 4 - 2 = 2 > 1$

Thus, demand is elastic so a decrease in price will increase revenue.

16. $E(p) = \frac{-p(ae^{-bp})(-b)}{ae^{-bp}} = pb$. Thus if $p = \frac{1}{b}$, $E(p) = \frac{1}{b}\cdot b = 1$.

17. Since for group A, $f'(t) = k(P - f(t))$, it follows that $f(t) = P(1 - e^{-kt})$.

$f(0) = 0 = 100(1 - e^{-k(0)})$; $f(13) = 66 = 100(1 - e^{-k(13)})$;

$.66 = 1 - e^{-13k}$; $e^{-13k} = .34$; $\frac{\ln .34}{-13} = k$; $k \sim .083$.

Thus, $f(t) = 100(1 - e^{-.083t})$.

18. $f(t) = \frac{M}{1 + Be^{-Mkt}}$ Since 55 is the maximum height for the

weed, M = 55.

$f(9) = 8 = \dfrac{55}{1 + Be^{-55(9)k}}$; $1 + Be^{-55(9)k} = \dfrac{55}{8}$;

$B = \dfrac{47}{8}e^{55(9)k}$

$f(25) = 48 = \dfrac{55}{1 + Be^{-55(25)k}}$; $1 + Be^{-55(25)k} = \dfrac{55}{48}$;

$B = \dfrac{7}{48}e^{55(25)k}$

Thus, $\dfrac{47}{8}e^{55(9)k} = \dfrac{7}{48}e^{55(25)k}$; $\dfrac{47}{8} = \dfrac{7}{48}e^{880k}$; $\dfrac{\ln \dfrac{47 \cdot 48}{8 \cdot 7}}{880} = k$;

$k \sim .0042$, $-Mkt = -.231$.

$8 = \dfrac{55}{1 + Be^{-.231(9)}} = \dfrac{55}{1 + Be^{-2.079}}$; $1 + Be^{-2.079} = \dfrac{55}{8}$;

$B = \dfrac{47}{8}e^{2.079} \sim 46.98$

Thus, $f(t) = \dfrac{55}{1 + 46.98e^{-.231t}}$.

19. a - D, b - G, c - E, d - B, e - H, f - F, g - A, h - C

Exercises 6.1

1. $F(x) = \frac{1}{2}x^2 + C$

2. $F(x) = x^9 + C$

3. $F(x) = \frac{1}{3}e^{3x} + C$

4. $F(x) = -\frac{1}{3}e^{-3x} + C$

5. $F(x) = 3x + C$

6. $F(x) = -2x^2 + C$

7. $k = -\frac{1}{4}$

8. $k = \frac{3}{4}$

9. $k = \frac{2}{3}$

10. $k = -3$

11. $\frac{d}{dt}\left[kt^{-5}\right] = -5kt^{-6} = \frac{10}{t^6}$; $k = -2$

12. $\frac{d}{dt}\left[kt^{1/2}\right] = \frac{1}{2}kt^{-1/2} = 3t^{-1/2}$; $k = 6$

13. $\frac{d}{dt}\left[ke^{-2t}\right] = -2ke^{-2t} = 5e^{-2t}$; $k = -\frac{5}{2}$

14. $\frac{d}{dt}\left[ke^{t/10}\right] = \frac{1}{10}ke^{t/10} = 3e^{t/10}$; $k = 30$

15. $\frac{d}{dx}\left[ke^{4x-1}\right] = 4ke^{4x-1} = 2e^{4x-1}$; $k = \frac{1}{2}$

16. $\frac{d}{dx}\left[\frac{k}{e^{3x+1}}\right] = \frac{d}{dx}\left[ke^{-(3x+1)}\right] = -3ke^{-(3x+1)} = 4e^{-(3x+1)}$; $k = -\frac{4}{3}$

17. $\frac{d}{dx}\left[k(x - 7)^{-1}\right] = -k(x - 7)^{-2} = (x - 7)^{-2}$; $k = -1$

18. $\frac{d}{dx}\left[k(x + 1)^{3/2}\right] = \frac{3}{2}k(x + 1)^{1/2} = (x + 1)^{1/2}$; $k = \frac{2}{3}$

19. $\frac{d}{dx}\left[k \ln(x + 4)\right] = \frac{k}{x + 4} = \frac{1}{x + 4}$; $k = 1$

20. $\frac{d}{dx}\left[k(x - 8)^{-3}\right] = -3k(x - 8)^{-4} = 5(x - 8)^{-4}$; $k = -\frac{5}{3}$

21. $\frac{d}{dx}\left[k(3x + 2)^5\right] = 15k(3x + 2)^4 = (3x + 2)^4$; $k = \frac{1}{15}$

22. $\frac{d}{dx}\left[k(2x - 1)^4\right] = 8k(2x - 1)^3 = (2x - 1)^3;\ k = \frac{1}{8}$

23. $\int (x^2 - x - 1)dx = \frac{1}{3}x^3 - \frac{1}{2}x^2 - x + C$

24. $\int (x^3 + 6x^2 - x)dx = \frac{1}{4}x^4 + 2x^3 - \frac{1}{2}x^2 + C$

25. $\int \left(\frac{2}{\sqrt{x}} - 3\sqrt{x}\right) = \int (2x^{-1/2} - 3x^{1/2})dx = 4x^{1/2} - 2x^{3/2} + C$

26. $\int \left[\frac{\sqrt{t}}{4} - 4(t - 3)^{-2}\right]dt = \int \left[\frac{1}{4}t^{1/2} + 4(t - 3)^{-2}\right]dt$

$= \frac{1}{6}t^{3/2} + 4(t - 3)^{-1} + C$

27. $\int \left(4 - 5e^{-5t} + \frac{1}{3}e^{2t}\right)dt = 4t + e^{-5t} + \frac{1}{6}e^{2t} + C$

28. $\int (e^2 + 3t^2 - 2e^{3t})dt = e^2t + t^3 - \frac{2}{3}e^{3t} + C$

29. $f'(t) = t^{3/2};\ f(t) = \frac{2}{5}t^{5/2} + C$

30. $f'(t) = \frac{4}{6 + t};\ f(t) = 4\ln(6 + t) + C$

31. $f'(t) = 0;\ f(t) = C$

32. $f'(t) = t^2 - 5t - 7;\ f(t) = \frac{1}{3}t^3 - \frac{5}{2}t^2 - 7t + C$

33. $f'(x) = x;\ f(x) = \frac{1}{2}x^2 + C;\ f(0) = 3 = \frac{1}{2}0^2 + C$

Thus, $f(x) = \frac{1}{2}x^2 + 3$.

34. $f'(x) = 8x^{1/3};\ f(x) = 6x^{4/3} + C;\ f(1) = 4 = 6 + C;\ C = -2$

Thus, $f(x) = 6x^{4/3} - 2$.

35. $f'(x) = x^{1/2} + 1;\ f(x) = \frac{2}{3}x^{3/2} + x + C;$

$f(4) = 0 = \frac{2}{3}4^{3/2} + 4 + C = \frac{2}{3}\cdot 8 + 4 + C = \frac{28}{3} + C;\ C = -\frac{28}{3}$

Thus, $f(x) = \frac{2}{3}x^{3/2} + x - \frac{28}{3}$.

36. $f'(x) = x^2 + x^{1/2};\ f(x) = \frac{1}{3}x^3 + \frac{2}{3}x^{3/2} + C;$

$f(1) = 3 = \frac{1}{3} + \frac{2}{3} + C = 1 + C;\ C = 2$

Thus, $f(x) = \frac{1}{3}x^3 + \frac{2}{3}x^{3/2} + 2$.

37. $f(x) = \int \frac{2}{x}\,dx = 2\ln|x| + C$

$f(1) = 2 = 2\ln|1| + C$, $C = 2$, and $f(x) = 2\ln|x| + 2$

38. $f(x) = \int \frac{1}{3}\,dx = \frac{1}{3}x + C$

$f(6) = 3 = \frac{1}{3}(6) + C$, $C = 1$, and $f(x) = \frac{1}{3}x + 1$

39. $\frac{d}{dx}\left[\frac{1}{x} + C\right] \neq \ln x$, $\frac{d}{dx}\left[x\ln x - x + C\right] = \ln x$,

$\frac{d}{dx}\left[\frac{1}{2}(\ln x)^2 + C\right] \neq \ln x$, so the answer is b.

40. $\frac{d}{dx}\left[\frac{2}{5}(x+1)^{5/2} - \frac{2}{3}(x+1)^{3/2} + C\right] = (x+1)^{3/2} - (x+1)^{1/2}$

$= \sqrt{x+1}\,[x + 1 - 1]$

$\frac{d}{dx}\left[\frac{1}{2}x^2\frac{2}{3}(x+1)^{3/2} + C\right] \neq x\sqrt{x+1}$: the answer is a.

41.

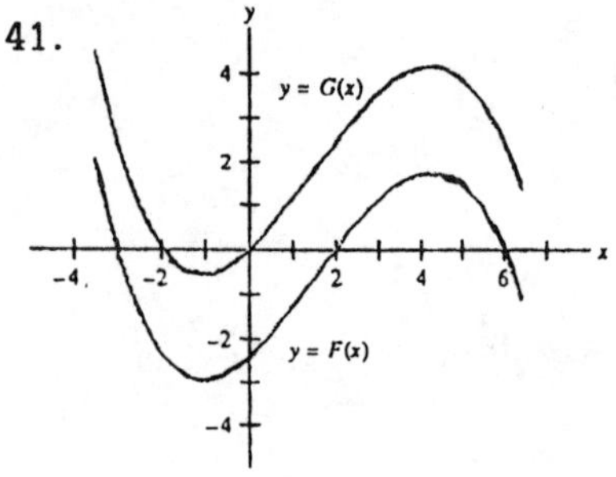

42.

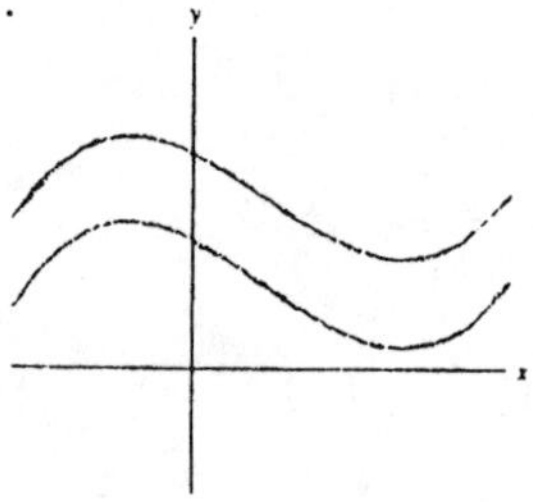

43. $g(x) = f(x) + 3$, $g'(x) = f'(x)$, $g'(5) = 1/4$

44. $h(x) = g(x) - f(x) = f(x) + 2 - f(x) = 2$; $h'(x) = 0$

45. (a) $\int(96 - 32t)dt = 96t - 16t^2 + C$ The initial height is 256 feet so $C = 256$. Thus $s(t) = 96t - 16t^2 + 256$.

(b) Setting $s(t) = 0$, $16t^2 + 96t + 256 = 0$; $t^2 - 6t - 16 = 0$; $(t - 8)(t + 2) = 0$. The only solution that is sensible is $t = 8$ seconds.

(c) Since $s'(t) = 96 - 32t$, $s(t)$ has a maximum when $s'(t) = 0$, i.e., $t = 3$. The ball will reach a maximum height of $s(3) = 400$ ft.

46. (a) $= \int -32t\, dt = -16t^2 + C$ $s(0) = 400 = -16(0) + C$; $C = 400$

Thus $s(t) = -16t^2 + 400$.

(b) The rock hits the ground when $s(t) = -16t^2 + 400 = 0$; $16t^2 = 400$; $t^2 = 25$; $t = 5$ seconds.

(c) $v(5) = -32(5) = -160$ ft/sec

47. $P(t) = \int\left((60 + 2t - \frac{1}{4}t^2\right)dt = 60t + t^2 - \frac{1}{12}t^3 + C$;

$P(0) = 0 = C$; $C = 0$ Thus $P(t) = 60t + t^2 - \frac{1}{12}t^3$.

48. $P(t) = \int\left(40 + 2t - \frac{1}{5}t^2\right)dt = 40t + t^2 - \frac{1}{15}t^3 + C$ Assuming the output is 0 at time $t = 0$, $P(t) = 40t + t^2 - \frac{1}{15}t^3$.

49. $f(t) = \int 10e^{-.4t}dt = -\frac{100}{4}e^{-.4t} + C = -25e^{-.4t} + C$

$f(0) = -5 = -25 + C$; $C = 20$ Thus, $f(t) = -25e^{-.4t} + 20$.

50. $P(t) = \int(120t - 3t^2)dt = 60t^2 - t^3 + C$; $P(0) = 100 = C$

Thus, $P(t) = 60t^2 - t^3 + 100$.

51. $P(x) = \int (1.30 + .06x - .0018x^2)dx$

$= 1.30x + .03x^2 - .0006x^3 + C$

$P(0) = -95 = C$. Thus, $P(x) = 1.30x + .03x^2 - .0006x^3 - 95$.

52. $C(x) = \int (.2x + 1)dx = .1x^2 + x + C$; $C(0) = 200 = C$

Thus, $C(x) = .1x^2 + x + 200$.

53. $f(t) = \int 94e^{.016t}dt = \frac{94}{.016}e^{.016t} + C = 5875e^{.016t} + C$

Since consumption is reckoned from 1980, we have

$f(0) = 0 = 5875(1) + C$; $C = -5875$.

Thus, $f(t) = 5875e^{.016t} - 5875 = 5875(e^{.016t} - 1)$.

54. $T(t) = \int_0^t 17.04e^{.016x}dx = \frac{17.04}{.016}\left[e^{.016t} \Big|_0^t \right]$

$= 1065\left[e^{.016t} - 1\right]$

55. $\int C'(x)dx = C(x) = 1000x + 25x^2 + C_1$

$C(0) = C_1$ = fixed cost; $C(x) = 25x^2 + 1000x + 10000$.

Exercises 6.2

1. $\Delta x = \frac{2 - 0}{4} = .5$. The first midpoint is that of [0,.5] which is .25, so the midpoints are .25, .75, 1.25, 1.75.

2. $\Delta x = \frac{3 - 0}{6} = .5$; The first midpoint is .25, so the midpoints are .25, .75, 1.25, 1.75, 2.25, .2.75.

3. $\Delta x = \frac{4 - 1}{5} = .6$; The first midpoint is that of [1,1.6] which is 1.3, so the midpoints are 1.3, 1.9, 2.5, 3.1, 3.7.

4. $\Delta x = \frac{5 - 3}{5} = .4$; The first midpoint is that of [3,3.4] which is 3.2, so the midpoints are 3.2, 3.6, 4, 4.4, 4.8.

5. $\Delta x = 0.5$; the midpoints are 1.25, 1.75, 2.25, 2.75.

$$\text{Area} = 0.5[f(1.25) + f(1.75) + f(2.25) + f(2.75)]$$
$$= 0.5[(1.25)^2 + (1.75)^2 + (2.25)^2 + (2.75)^2] = 8.625.$$

6. $\Delta x = 1$; the midpoints are -1.5, -.5, .5, 1.5.

$$\text{Area} = 1[(-1.5)^2 + (-.5)^2 + (.5)^2 + (1.5)^2] = 5$$

7. $\Delta x = .4$; the left endpoints are 1, 1.4, 1.8, 2.2, 2.6.

$$\text{Area} = 0.4[1^3 + (1.4)^3 + (1.8)^3 + (2.2)^3 + (2.6)^3] = 15.12$$

8. $\Delta x = .2$; the right endpoints are .2, .4, .6, .8, 1.

$$\text{Area} = 0.2[(.2)^3 + (.4)^3 + (.6)^3 + (.8)^3 + 1^3] = .36.$$

9. $\Delta x = .2$; the right endpoints are 2.2, 2.4, 2.6, 2.8, 3.

$$\text{Area} = 0.2[e^{-2.2} + e^{-2.4} + e^{-2.6} + e^{-2.8} + e^{-3}] = 0.077278$$

10. $\Delta x = .4$; the left endpoints are 2.2, 2.4, 2.8, 3.2, 3.6.

$$\text{Area} = 0.4[\ln 2 + \ln 2.4 + \ln 2.8 + \ln 3.2 + \ln 3.6] = 2.0169$$

7. $\Delta x = .75$; the left endpoints are 1, 1.75, 2.5, 3.25.

$$\text{Area} = 0.75[(4 - 1) + (4 - 1.75) + (4 - 2.5) + (4 - 3.25)]$$
$$= 5.625$$

The midpoints are 1.375, 2.125, 2.875, 3.625

Area = 0.75[(4 - 1.375) + (4 - 2.125) + (4 - 2.875)

+ (4 - 3.625)] = 4.5

7. Δx = .25; the right endpoints are 2.25, 2.5, 2.75, 3.

Area = 0.25[2(2.25) - 4 + 2(2.5) - 4 + 2(2.75) - 4 + 2(3) - 4]

= 1.25

The midpoints are 2.125, 2.375, 2.625, 2.875

Area = 0.25[2(2.125) - 4 + 2(2.375) - 4 + 2(2.625) - 4

+ 2(2.875) - 4] = 1

13. Δx = .4; the midpoints are −.8, −.4, 0, .4, .8.

$$\text{Area} = 0.4\left\{ [1 - (-.8)^2]^{1/2} + [1 - (-.4)^2]^{1/2} + [1]^{1/2} + [1 - (.4)^2]^{1/2} + [1 - (.8)^2]^{1/2} \right\} = 1.61321$$

The error is .04241.

14. Δx = .2; the midpoints are .1, .3, .5, .7, .9.

$$\text{Area} = 0.2\left\{\sqrt{1 - (.1)^2} + \sqrt{1 - (.3)^2} + \sqrt{1 - (.5)^2} + \sqrt{1 - (.7)^2} + \sqrt{1 - (.9)^2}\right\} = .792996$$

To five decimal places = .79300.N.B 2(.79300) = 1.586. The error is .0076, and the error for the semi-circle is .0152 better than in exercise 13.

14. Δx = .4; the midpoints are .2, .6, 1, 1.4, 1.8.

Area = .4[.2 + .7 + .8 + .7 + .3] = 1.08 liters.

16. Area = 40[106 + 101 + 100 + 113] = 16800 ft^2

17. Distance = 10[20 + 44 + 32 + 39 + 65 + 80] = 2800 ft

18. (a) Distance = 1[3 + 5 + 5 + 6 + 6 + 7] = 32 ft

(b) Distance = 1[5 + 5 + 6 + 6 + 7 + 8] = 37 ft

(c) Distance = 2[5 + 6 + 7] = 36 ft

19. Increase in the population (in millions) from 1910 to 1950.

Rate of cigarette consumption t years after 1985.

20 to 50.

20. Total revenue from the sales of the first 10 units of goods.

21. Amount of soil eroded during a five day period.

22. Increase in height form ages 5 to 10.

23. $n = 5, \Delta x = .2$, midpoints are 2.1, 2.3, 2.5, 2.7, 2.9.

Area under $y = x^2$ from $x = 2$ to $x = 3$ by Riemann sum is

$= .2[(2.1)^2 + (2.3)^2 + (2.5)^2 + (2.7)^2 + (2.9)^2]$.

Area under $y = (1/5)x^2$ from $x = 2$ to $x = 3$ by Riemann sum

is $= .2[(1/5)(2.1)^2 + (1/5)(2.3)^2 + (1/5)(2.5)^2$

$+ (1/5)(2.7)^2 + (1/5)(2.9)^2]$.

$= (1/5)\{.2[(2.1)^2 + (2.3)^2 + (2.5)^2 + (2.7)^2 + (2.9)^2]\}$

So the area under $y = x^2$ from $x = 2$ to $x = 3$ is 5 times greater than that under $y = (1/5)x^2$ from $x = 2$ to $x = 3$.

23. $\Delta x = .2$, midpoints are 1.1, 1.3, 1.5, 1.7, 1.9. A = area under $y = x^2$ from $x = 1$ to $x = 2$ by Riemann sum, which is $= .2[(1.1)^2 + (1.3)^2 + (1.5)^2 + (1.7)^2 + (1.9)^2]$. B = area under $y = 2 + x^3$ from $x = 1$ to $x = 2$ by Riemann sum, which is $= .2[2 + (2.1)^3 + 2 + (2.3)^3 + 2 + (2.5)^3 + 2 + (2.7)^3 + 2 + (2.9)^3]$. Finally C = area under $y = x^2 + 2 + x^3$ from $x = 1$ to $x = 2$ is $= .2[(1.1)^2 + 2 + (2.1)^3 + (1.3)^2 + 2 + (2.3)^3$ $(1.5)^2 + 2 + (2.5)^3 + (1.7)^2 + 2 + (2.7)^3 + (1.9)^2 + 2 + (2.9)^3]$

Now clearly A + B = C.

25. Let the midpoints be $x_1, x_2, \cdots, x_n$, and $\Delta x = \frac{b - a}{n}$.

Area under $f(x) = \Delta x[f(x_1) + f(x_2) + \cdots + f(x_n)]$

Area under $g(x) = \Delta x[g(x_1) + g(x_2) + \cdots + g(x_n)]$

Thus the area under

$$f(x) + g(x) = \Delta x[f(x_1) + g(x_1) + f(x_2) + g(x_2) + \cdots + f(x_n) + g(x_n)]$$

= Area under $f(x)$ + Area under $g(x)$.

26. Let the midpoints be $x_1, x_2, \cdots, x_n$, and $\Delta x = \frac{b - a}{n}$.

$$\text{Area under } kf(x) = \Delta x[kf(x_1) + kf(x_2) + \cdots + kf(x_n)]$$

$$= k\Delta x[f(x_1) + f(x_2) + \cdots + f(x_n)]$$

The area under $f(x) = \Delta x[f(x_1) + f(x_2) + \cdots + f(x_n)]$

Hence the area under the graph of $kf(x)$ is k times the the area under the graph of $f(x)$.

Exercises 6.3

1. $\displaystyle\int_{-1}^{1} x\,dx = \frac{1}{2}x^2\Big|_{-1}^{1} = \frac{1}{2} - \frac{1}{2} = 0$

2. $\displaystyle\int_{4}^{5} e^{2x}dx = \frac{1}{2}e^{2x}\Big|_{4}^{5} = \frac{1}{2}(e^{10} - e^{8})$

3. $\displaystyle\int_{1}^{2} 5\,dx = 5x\Big|_{1}^{2} = 10 - 5 = 5$

4. $\displaystyle\int_{-1}^{-1/2} \frac{1}{x^2}\,dx = -\frac{1}{x}\Big|_{-1}^{-1/2} = 2 - 1 = 1$

5. $\displaystyle\int_1^2 8x^3dx = 2x^4\Big|_1^2 = 32 - 2 = 30$

6. $\displaystyle\int_0^1 e^{x/3}dx = 3e^{x/3}\Big|_0^1 = 3e^{1/3} - 3 = 3(e^{1/3} - 1)$

7. $\displaystyle\int_0^1 4e^{-3x}dx = -\frac{4}{3}e^{-3x}\Big|_0^1 = -\frac{4}{3}e^{-3} - \left(-\frac{4}{3}\right) = \frac{4}{3}\left(1 - e^{-3}\right)$

8. $\displaystyle\int_1^3 \frac{5}{x}\,dx = 5\ \ln x\Big|_1^3 = 5\ \ln 3 - 5\ \ln 1 = 5\ \ln 3$

9. $\displaystyle\int_1^4 3\sqrt{x}\,dx = 2x^{3/2}\Big|_1^4 = 2\cdot 8 - 2 = 14$

10. $\displaystyle\int_1^8 2x^{1/3} = \frac{3}{2}x^{4/3}\Big|_1^8 = \frac{3\cdot 16}{2} - \frac{3}{2} = \frac{45}{2}$

11. $\displaystyle\int_0^5 e^{.2t}dt = 5e^{.2t}\Big|_0^5 = 5e - 5 = 5(e - 1)$

12. $\displaystyle\int_0^1 \frac{5}{e^{3t}}\,dt = -\frac{5}{3}e^{-3t}\Big|_0^1 = -\frac{5}{3}e^{-3} - \left(-\frac{5}{3}\right) = \frac{5}{3}(1 - e^{-3})$

13. $\displaystyle\int_3^6 x^{-1}dx = \ln x\Big|_3^6 = \ln 6 - \ln 3 = \ln \frac{6}{3} = \ln 2$

14. $\displaystyle\int_1^3 (5t - 1)^3dt = \frac{1}{20}(5t - 1)^4\Big|_1^3 = \frac{38,416}{20} - \frac{256}{20} = \frac{38,160}{20} = 1908$

15. $\displaystyle\int_{-1}^1 \frac{4}{(t + 2)^3}\,dt = -2(t + 2)^{-2}\Big|_{-1}^1 = -\frac{2}{9} - (-2) = \frac{16}{9}$

16. $\displaystyle\int_{-3}^0 \sqrt{25 + 3t}\,dt = \frac{2}{9}(25 + 3t)^{3/2}\Big|_{-3}^0 = \frac{2\cdot 125}{9} - \frac{2\cdot 64}{9} = \frac{122}{9}$

17. $\displaystyle\int_2^3 (5 - 2t)^4dt = -\frac{1}{10}(5 - 2t)^5\Big|_2^3 = \frac{1}{10} - \left(-\frac{1}{10}\right) = \frac{1}{5}$

18. $\displaystyle\int_4^9 \frac{3}{t-2}\,dt = 3\ln(t-2)\Big|_4^9 = 3\ln 7 - 3\ln 2 = 3\ln\frac{7}{2}$

19. $\displaystyle\int_0^3 (x^3 + x - 7)dx = \left(\frac{1}{4}x^4 + \frac{1}{2}x^2 - 7x\right)\Big|_0^3 = \frac{81}{4} + \frac{9}{2} - 21 = \frac{15}{4}$

20. $\displaystyle\int_{-5}^5 (e^{x/10} - x^2 - 1)dx = \left[10e^{x/10} - \frac{1}{3}x^3 - x\right]\Big|_{-5}^5$

$\displaystyle = 10e^{1/2} - \frac{125}{3} - 5 - \left[10e^{-1/2} + \frac{125}{3} + 5\right]$

$\displaystyle = 10(e^{1/2} - e^{-1/2}) - \frac{280}{3}$

21. $\displaystyle\int_2^4 \left[x^2 + \frac{2}{x^2} - \frac{1}{x+5}\right]dx = \left[\frac{1}{3}x^3 - \frac{2}{x} - \ln(x+5)\right]\Big|_2^4$

$\displaystyle = \frac{64}{3} - \frac{1}{2} - \ln 9 - \left[\frac{8}{3} - 1 - \ln 7\right]$

$\displaystyle = \frac{128}{6} - \frac{3}{6} - \frac{16}{6} + \frac{6}{6} + \ln\frac{7}{9} = \frac{115}{6} + \ln\frac{7}{9}$

22. $\displaystyle\int_1^2 (4x^3 + 3x^{-4} - 5)dx = (x^4 - x^{-3} - 5x)\Big|_1^2$

$\displaystyle = 16 - \frac{1}{8} - 10 - (1 - 1 - 5) = \frac{87}{8}$

23. $\displaystyle\int_2^3 4x\,dx = 2x^2\Big|_2^3 = 18 - 8 = 10$

24. $\displaystyle\int_{-1}^1 3x^2dx = x^3\Big|_{-1}^1 = 1 - (-1) = 2$

25. $\displaystyle\int_0^1 e^{x/2}dx = 2e^{x/2}\Big|_0^1 = 2e^{1/2} - 2 = 2(e^{1/2} - 1)$

26. $\displaystyle\int_0^4 x^{1/x}dx = \frac{2}{3}x^{3/2}\Big|_0^4 = \frac{16}{3}$

27. $\int_1^4 (x - 3)^4 dx = \frac{1}{5}(x - 3)^5 \Big|_1^4 = \frac{1}{5} - \left(-\frac{32}{5}\right) = \frac{33}{5}$

28. $\int_{-1/3}^{0} e^{3x} dx = \frac{1}{3}e^{3x} \Big|_{-1/3}^{0} = \frac{1}{3} - \frac{1}{3e} - \frac{1}{3}\left(1 - \frac{1}{e}\right)$

29. $\int_{-1}^{2} 5\ dx = 5x \Big|_{-1}^{2} = 10 - (-5) = 15$

Area = $5 \cdot 3 = 15$

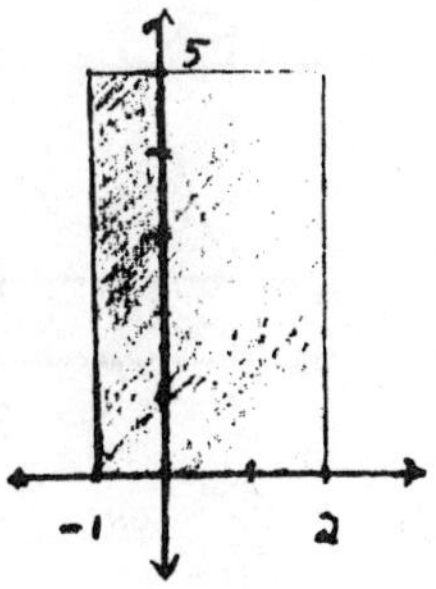

30. $\int_2^4 (x + 1) dx = \frac{1}{2}(x + 1)^2 \Big|_2^4 = \frac{25}{2} - \frac{9}{2} = 8$

Area = $(2 \cdot 3) + \frac{1}{2}(2 \cdot 2) = 8$

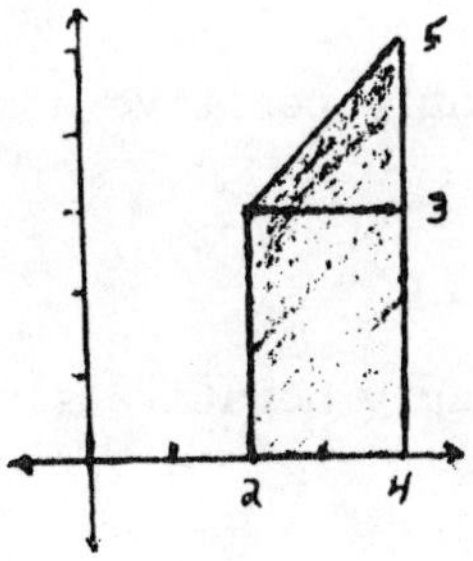

31. $\int_0^3 2x\ dx = x^2 \Big|_0^3 = 9$

Area = $\frac{1}{2}(3 \cdot 6) = 9$

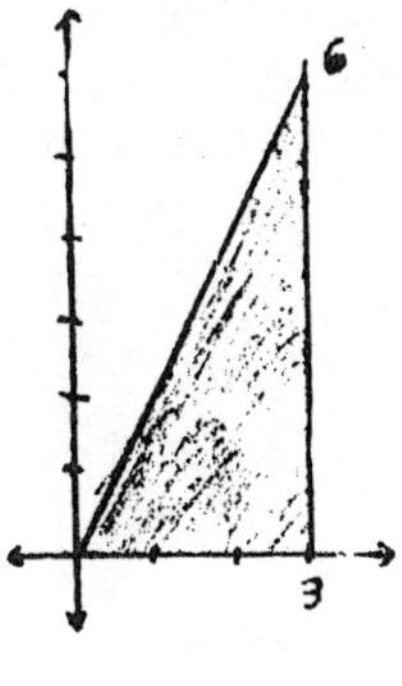

32. $\int_0^3 (2x + 1)dx = \frac{1}{4}(2x + 1)^2\Big|_0^3 = \frac{49}{4} - \frac{1}{4} = \frac{48}{4} = 12$

Area $= (3\cdot 1) + \frac{1}{2}(3\cdot 6) = 12$

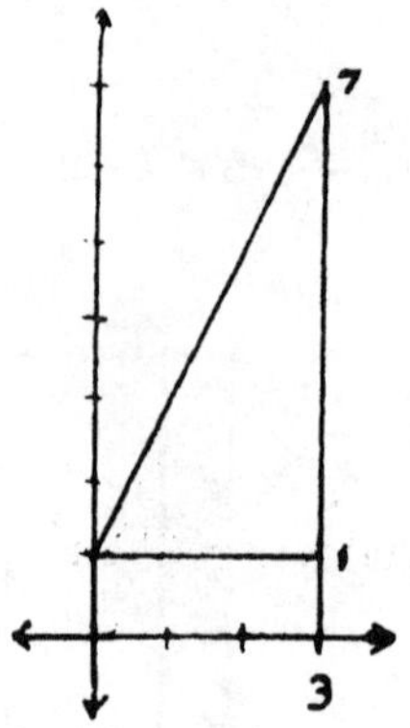

33. There are 2 triangles, so the integral $= \frac{1}{2}2(4) - \frac{1}{2}1(2) = 3$

34. Area $= \frac{1}{2}2(2) + 2 + \frac{1}{2}\pi 3^2 - \frac{1}{2}\pi 3^2 - \frac{1}{2}5(3) = -\frac{7}{2}$

35. $\int_0^7 f(x)dx$ is clearly positive since there is more area above the x-axis.

36. $\int_0^7 g(x)dx$ is clearly negative since there is more area below the x-axis.

37. $\int_{20}^{30} (.1t + 2.4)dt = \frac{.1t^2}{2} + 2.4t\Big|_{20}^{30} = 49$ trillion.

38. The number of tons of pollutants discharged from 1990 to 1992.

39. (a) $\int_2^5 (2t + 1)dt = \frac{1}{4}(2t + 1)^2\Big|_0^5 = \frac{121}{4} - \frac{1}{4} = 30$ feet.

(b)

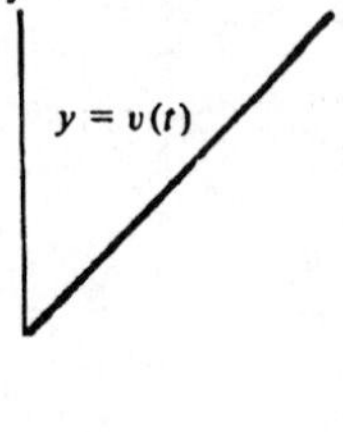

40. (a) $\int_2^5 \left(21 - \frac{4}{5}t\right)dt = -\frac{5}{8}\left(21 - \frac{4}{5}t\right)^2\Big|_2^5 = -\frac{1445}{8} - \left(-\frac{5}{8}\left(\frac{97}{5}\right)^2\right) = 54.6$

(b)

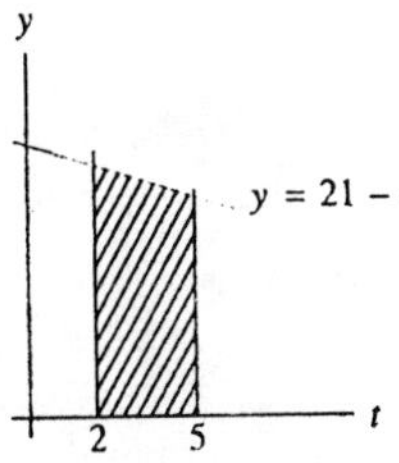

41. (a) $\int_2^8 \left(\frac{3}{32}x^2 - x + 200\right)dx = \left(\frac{1}{32}x^3 - \frac{1}{2}x^2 + 200x\right)\Big|_2^8$

$= 16 - 32 + 1600 - \left(\frac{8}{32} - 2 + 400\right)$

$= \$1185.75$

(b) It is the area under the marginal cost curve from x = 2 to x = 8.

42. (a) $\int_5^8 (100 + 50x - 3x^2)dx = (100x + 25x^2 - x^3)\Big|_5^8$

$= 800 + 1600 - 512 - (500 + 625 - 125)$

$= \$888$

(b) It is the area under the marginal profit function from x = 5 to x = 8.

43. The increase in profits resulting from increasing the production level from 44 to 48 units.

44. The total variable costs of producing 100 units of goods.

45. (a) $\int_0^2 \left[12 + \frac{4}{(t + 3)^2}\right]dt = \left[12t - \frac{4}{t + 3}\right]\Big|_0^2 = 24 - \frac{4}{5} - \left(-\frac{4}{3}\right)$

$= \frac{360}{15} - \frac{12}{15} + \frac{20}{15} = \frac{368}{15} \sim 24.53$

(b) It is the amount the temperature falls during the first 2 hours.

46. (a) $\displaystyle\int_1^9 \left[40 + \frac{8}{(t + 1)^2}\right]dt = \left[40t - \frac{8}{t + 1}\right]\Bigg|_1^9$

$= 360 - \frac{8}{10} - (40 - 4)$

$= \frac{3232}{10} = 323.2$

(b) It is the distance traveled during the time from $t = 1$ to $t = 9$.

47. $\displaystyle\int_0^{20} 76.2e^{.03t}dt = 2540e^{.03}\Big|_0^{20} \sim 2088$ (million cubic meters)

48. (a) $\displaystyle\int_1^b \frac{1}{t}\, dt = \ln t\Big|_1^b = \ln b$

(b) As shown by (a), the area under the curve $y = \frac{1}{t}$ from $t = 1$ to $t = b$ is simply $\ln b$.

49. $A(x) = \displaystyle\int_0^x (x^2 + 1)dx = \left(\frac{1}{3}x^3 + x\right)\Bigg|_0^x = \frac{1}{3}x^3 + x$

$A'(x) = x^2 + 1$; $A'(3) = 10$

50. $A(x) = \displaystyle\int_2^x x^3dx = \frac{1}{4}x^4\Big|_2^x = \frac{1}{4}x^4 - 4$; $A'(x) = x^3$; $A'(6) = 216$

Exercises 6.4

1. $\displaystyle\int_1^2 f(x)dx + \int_3^4 -f(x)dx$

2. $\displaystyle\int_2^3 [f(x) - g(x)]dx$

3.

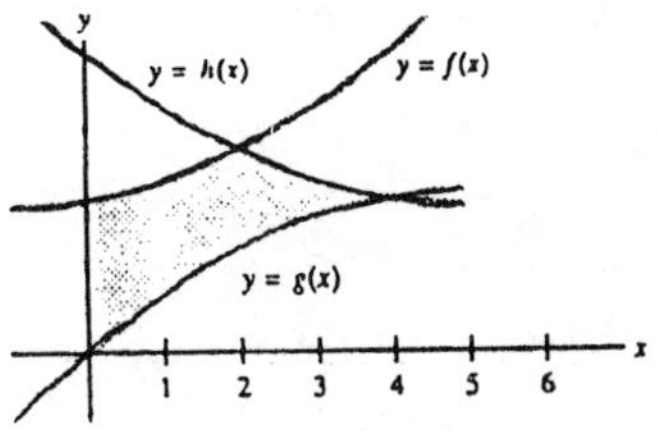

4.

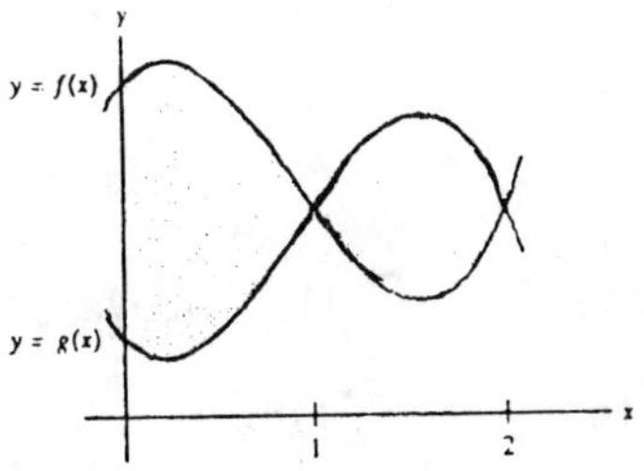

5. Since y = 8 lies above y = 2x on [-2, 2], we should calculate

$$\int_{-2}^{2}(8 - 2x^2)dx = \left(8x - \frac{2}{3}x^3\right)\Bigg|_{-2}^{2} = 16 - \frac{16}{3} - \left(-16 + \frac{16}{3}\right)$$

$$= 32 - \frac{32}{3} = \frac{96}{3} - \frac{32}{3} = \frac{64}{3}$$

6. Since $y = 13 - 3x^2$ lies above y = 1 on [-2, 2], we should calculate $\int_{-2}^{2}(13 - 3x^2 - 1)dx = \int_{-2}^{2}(12 - 3x^2)dx$

$$= (12x - x^3)\Big|_{-2}^{2}$$

$$= 24 - 8 - (-24 + 8) = 32$$

7. Since $y = x^2 - 6x + 12$ lies above y = 1 on [0, 4], we should calculate $\int_{0}^{4}(x^2 - 6x + 11)dx = \left(\frac{1}{3}x^3 - 3x^2 + 11x\right)\Bigg|_{0}^{4}$

$$= \frac{64}{3} - 48 + 44 - 0 = \frac{52}{3}$$

8. Since y = 4 lies above y = x(2 - x) on [0, 2], we should calculate $\int_{0}^{2}(4 - x(2 - x))dx = \int_{0}^{2}(x^2 - 2x + 4)dx$

$$= \left(\frac{1}{3}x^3 - x^2 + 4x\right)\Bigg|_{0}^{2} = \frac{8}{3} - 4 + 8$$

$$= \frac{20}{3}$$

9. Since $y = 3x^2$ lies above $y = -3x^2$ we should calculate

$$\int_{-1}^{2} (3x^2 - (-3x^2))dx = \int_{-1}^{2} 6x^2 dx = 2x^3 \Big|_{-1}^{2} = 16 - (-2) = 18$$

10. Since $y = e^{2x}$ lies above $y = e^{-2x}$ we should calculate

$$\int_{-1}^{1} (e^{2x} - (-e^{2x}))dx = \int_{-1}^{1} 2e^{2x} dx = e^{2x} \Big|_{-1}^{1} = e^2 - e^{-2}$$

11. To find the points of intersection solve $x^2 + x - (3 - x) = 0$; $x^2 + 2x - 3 = 0$; $(x + 3)(x - 1) = 0$. Thus, we want to integrate from $x = -3$ to $x = 1$ with $y = 3 - x$ above $y = x^2 + x$, so $\int_{-3}^{1} (3 - x - (x^2 + x))dx = \int_{-3}^{1} (-x^2 - 2x + 3)dx$

$$= \left(-\frac{1}{3}x^3 - x^2 + 3x\right)\Big|_{-3}^{1}$$

$$= -\frac{1}{3} - 1 + 3 - \left(\frac{27}{3} - 9 - 9\right)$$

$$= -\frac{1}{3} + 11 = \frac{32}{3}$$

12. Solve: $3x - x^2 - (4 - 2x) = 0$; $-x^2 + 5x - 4 = 0$; $x^2 - 5x + 4 = 0$; $(x - 4)(x - 1) = 0$ Thus, the interval is $[1, 4]$ and since $y = 3x - x^2$ lies above $y = 4 - 2x$,

$$\int_{1}^{4} (3x - x^2 - (4 - 2x))dx = \int_{1}^{4} (-x^2 + 5x - 4)dx$$

$$= \left(-\frac{1}{3}x^3 + \frac{5}{2}x^2 - 4x\right)\Big|_{1}^{4}$$

$$= -\frac{64}{3} + 40 - 16 - \left(-\frac{1}{3} + \frac{5}{2} - 4\right)$$

$$= 7 - \frac{5}{2} = \frac{9}{2}$$

13. Solve: $2x - 5 - (-x^2 + 6x - 5) = 0$; $x^2 - 4x = 0$; $x(x - 4) = 0$ Thus, we should integrate over $[0, 4]$ and since $y = -x^2 + 6x - 5$ lies above $y = 2x - 5$ on $[0, 4]$, we have

$$\int_{0}^{4} (-x^2 + 6x - 5 - (2x - 5))dx = \int_{0}^{4} (-x^2 + 4x)dx$$

$$= \left(-\frac{1}{3}x^3 + 2x^2\right)\Bigg|_0^4$$

$$= -\frac{64}{3} + 32 = \frac{32}{2}$$

14. Solve: $2x^2 + x - 7 - (x + 1) = 0$; $2x^2 - 8 = 0$; $x^2 - 4 = 0$

Thus, we should integrate over [-2, 2] and since $y = x + 1$ lies above $y = 2x^2 + x - 7$ on [-2, 2], we have

$$\int_{-2}^{2} (x + 1 - (2x^2 + x - 7))dx = \int_{-2}^{2} (-2x^2 + 8)dx$$

$$= \left(-\frac{2}{3}x^3 + 8x\right)\Bigg|_{-2}^{2}$$

$$= -\frac{16}{3} + 16 - \left(\frac{16}{3} - 16\right)$$

$$= -\frac{32}{3} + 32 = \frac{64}{3}$$

15. $x^2 = 18 - x^2 \Rightarrow 2x^2 = 18 \Rightarrow x = \pm 3$

$$\text{Area} = \int_{-3}^{3} \left[(18 - x^2) - x^2 \right]dx = \int_{-3}^{3} (18 - 2x^2)dx$$

$$= 18x - \frac{2}{3}x^3\Bigg|_{-3}^{3} = 72.$$

16. $2 - 2x^2 = 4x^2 - 24x + 20 \Rightarrow 6x^2 - 24x + 18 = 0$ or

$x^2 - 4x + 3 = 0 \Rightarrow x = 1, 3$

$$\text{Area} = \int_{1}^{3} (2 - 2x^2 - (4x^2 - 24x + 20))dx$$

$$= \int_{1}^{3} (-6x^2 + 24x - 18)dx = -2x^3 + 12x^2 - 18x\Bigg|_{1}^{3} = 8$$

17. $-2x^2 + 6x + 8 = x^2 - 6x - 7 \Rightarrow 3x^2 - 12x - 15 = 0$ or

$x^2 - 4x - 5 = 0 \Rightarrow x = -1, 5$

$$\text{Area} = \int_{-1}^{5} (-2x^2 + 6x + 8 - (x^2 - 6x - 7))dx$$

$$= \int_{-1}^{5} (-3x^2 + 12x + 15)dx = -x^3 + 6x^2 - 15x\Bigg|_{-1}^{5} = 108.$$

18. $-x^2 + 7x + 8 = 2x^2 - 8x + 8 \Rightarrow 3x^2 - 15x = 0$ or

$3x(x - 5) = 0 \Rightarrow x = 0, 5$

$$\text{Area} = \int_0^5 (-x^2 + 7x + 8 - (2x^2 - 8x + 8))dx$$

$$= \int_0^5 (-3x^2 + 15x)dx = -x^3 + \frac{15}{2}x^2 \Big|_0^5 = 62.5$$

19. First solve $x^2 - 3x = 0$ to find any x-intercepts.

$x(x - 3) = 0$ so the x-intercepts are at $x = 0, 3$.

(a) $y = x^2 - 3x$ lies under the x-axis on the interval, so

calculate $\int_0^3 (0 - (x^2 - 3x))dx = \int_0^3 (-x^2 + 3x)dx$

$$= \left(-\frac{1}{3}x^3 + \frac{3}{2}x^2\right)\Big|_0^3 = -\frac{27}{3} + \frac{27}{2}$$

$$= \frac{27}{6} = \frac{9}{2}$$

(b) On [3, 4] $y = x^2 - 3x$ lies above the x-axis so borrowing from part (a), we wish to calculate

$$\int_0^3 -(x^2 - 3x)dx + \int_3^4 (x^2 - 3x)dx = \frac{9}{2} + \left(\frac{1}{3}x^3 - \frac{3}{2}x^2\right)\Big|_3^4$$

$$= \frac{9}{2} + \left(\frac{64}{3} - 24 - \left(\frac{27}{3} - \frac{27}{2}\right)\right)$$

$$= \frac{9}{2} + \left(\frac{64}{3} - 24 + \frac{9}{2}\right) = \frac{19}{3}$$

(c) On [-2, 0], $y = x^2 - 3x$ lies above the x-axis so using results from (a) and (b)

$$\int_{-2}^0 (x^2 - 3x)dx + \int_0^3 -(x^2 - 3x)dx = \left(\frac{1}{3}x^3 - \frac{3}{2}x^2\right)\Big|_{-2}^0 + \frac{9}{2}$$

$$= -\left(-\frac{8}{3} - \frac{12}{2}\right) + \frac{9}{2}$$

$$= \frac{26}{3} + \frac{9}{2} = \frac{79}{6}$$

20. (a) On [1, 4], $y = \frac{1}{x^2}$ lies below $y = x^2$ so calculate

$$\int_1^4 \left(x^2 - \frac{1}{x^2}\right)dx = \left(\frac{1}{3}x^3 + \frac{1}{x}\right)\Bigg|_1^4 = \frac{64}{3} + \frac{1}{4} - \left(\frac{1}{3} + 1\right) = 21 - \frac{3}{4} = \frac{81}{4}$$

(b) On $\left[\frac{1}{2}, 1\right]$, $y = \frac{1}{x^2}$ lies above $y = x^2$, below on [1, 4] so

$$\int_{1/2}^1 \left(\frac{1}{x^2} - x^2\right)dx + \int_1^4 \left(x^2 - \frac{1}{x^2}\right)dx = \left(-\frac{1}{x} - \frac{1}{3}x^3\right)\Bigg|_{1/2}^1 + \frac{81}{4}$$

$$= -1 - \frac{1}{3} - \left(-2 - \frac{1}{24}\right) + \frac{81}{4}$$

$$= \frac{24}{24} - \frac{8}{24} + \frac{1}{24} + \frac{486}{24} = \frac{503}{24}$$

21. First solve: $\frac{1}{x^2} = 8x$. This has a solution at $x = \frac{1}{2}$.

Next solve: $\frac{1}{x^2} = x$. This has a solution at $x = 1$. Thus, the area should be calculated by the following sum of integrals,

$$\int_0^{1/2} (8x - x)dx + \int_{1/2}^1 \left(\frac{1}{x^2} - x\right)dx$$

$$= \left(4x^2 - \frac{1}{2}x^2\right)\Bigg|_0^{1/2} + \left(-\frac{1}{x} - \frac{1}{2}x^2\right)\Bigg|_{1/2}^1$$

$$= 1 - \frac{1}{8} + \left(-1 - \frac{1}{2} - \left(-2 - \frac{1}{8}\right)\right) = \frac{3}{2}$$

22. First solve: $4x = \frac{1}{x}$. This has a solution at $x = \frac{1}{2}$.

Next solve: $\frac{x}{2} = \frac{1}{x}$. This has a solution at $x = \sqrt{x}$.

Thus, the area should be calculated by

$$\int_0^{1/2} \left(4x - \frac{x}{2}\right)dx + \int_{1/2}^{\sqrt{2}} \left(\frac{1}{x} - \frac{x}{2}\right)dx$$

$$= \left(2x^2 - \frac{1}{4}\right)\Bigg|_0^{1/2} + \left(\ln x - \frac{1}{4}x^2\right)\Bigg|_{1/2}^{\sqrt{2}}$$

$$= \frac{1}{2} - \frac{1}{16} + \ln 2^{1/2} - \frac{1}{2} - \left[\ln \frac{1}{2} - \frac{1}{16}\right]$$

$$= \frac{1}{2}\ln 2 - (\ln 1 - \ln 2) = \frac{3}{2}\ln 2.$$

23. First solve: $\frac{12}{x} = \frac{3}{2}\sqrt{x}$; $\frac{24}{3} = x^{3/2}$; $\sqrt{x^3} = 8$; $x^3 = 64$; $x = 4$

Next solve: $\frac{x}{3} = \frac{12}{x}$; $x^2 = 36$; $x = 6$

Thus, we should calculate the area as follows,

$$\int_0^4 \left[\frac{3}{2}x^{1/2} - \frac{x}{3}\right]dx + \int_4^6 \left[\frac{12}{x} - \frac{x}{3}\right]dx$$

$$= \left[x^{3/2} - \frac{1}{6}x^2\right]\Big|_0^4 + \left[12 \ln x - \frac{1}{6}x^2\right]\Big|_4^6$$

$$= 8 - \frac{16}{6} + \ln 6 - \frac{36}{6} - \left[12 \ln 4 - \frac{16}{6}\right] = 2 + 12 \ln \frac{3}{2}$$

24. First solve: $12 - x^2 = 4x$; $x^2 + 4x - 12 = 0$;

$(x + 6)(x - 2) = 0$; only $x = 2$ interests us.

Next solve: $12 - x^2 = x$; $x^2 + x - 12 = 0$; $(x + 4)(x - 3) = 0$;

again we want only the positive solution, $x = 3$. Thus,

$$\int_0^2 3x\,dx + \int_2^3 (12 - x^2 - x)dx = \frac{3}{2}x^2\Big|_0^2 + \left[12x - \frac{1}{3}x^3 - \frac{1}{2}x^2\right]\Big|_2^3$$

$$= 6 + 36 - \frac{27}{3} - \frac{9}{2} - \left[24 - \frac{8}{3} - 2\right]$$

$$= 11 - \frac{9}{2} + \frac{8}{3} = \frac{55}{6}$$

25. $g(t) = 50 - 6.03e^{.09t}$; $C(t) = 76.2e^{.03t}$;

$$\int_0^{20} (76.2e^{.03t} - 50 + 6.03e^{.09t})dt$$

26. $$\int_0^4 (16.2e^{.07t} - 16.1e^{.04t})dt = \left[230e^{.07t} - 402.5e^{.04t}\right]\Big|_0^4$$

$$\sim 4.48$$

So about 4.48 billion barrels would have been saved.

27. $$\int_6^8 \left[(-x^2 + 14x - 24) - (-x^2 + 12x - 20)\right]dx = \int_6^8 (2x - 4)dx$$
$$= \left(x^2 - 4x\right)\Big|_6^8 = 20$$

The company should <u>not</u> adopt the new plan. The area is 20, and it represents the additional profit from using the original plan.

28. A is the difference between the two heights after 10 seconds.

29. (a) The distance between the two cars after 1 hour.
(b) After 2 hours.

Exercises 6.5

1. $$\text{Average} = \frac{1}{b-a}\int_a^b f(x)dx = \frac{1}{3}\int_0^3 x^2 dx = \frac{x^3}{9}\Big|_0^3 = 3$$

2. $$\frac{1}{3}\int_0^3 e^{x/3}dx = \frac{1}{3}\cdot 3e^{x/3}\Big|_0^3 = e - 1$$

3. $$\frac{1}{2}\int_{-1}^1 x^3 dx = \frac{x^4}{8}\Big|_{-1}^1 = 0$$

4. $$\frac{1}{9}\int_1^{10} x dx = \frac{5x}{9}\Big|_1^{10} = \frac{45}{9} = 5$$

5. $$\frac{1}{\frac{1}{2} - \frac{1}{4}}\int_{1/4}^{1/2} 1/x^2 dx = -4/x\Big|_{1/4}^{1/2} = 8$$

6. $$\frac{1}{2}\int_2^4 (2x - 6)dx = \frac{1}{2}x^2 - 3x\Big|_2^4 = 0$$

7. $\frac{1}{12}\int_0^{12}(47 + 4t - \frac{1}{3}t^2)dt = \frac{1}{12}(47t + 2t^2 - \frac{1}{9}t^3)\Big|_0^{12} = 55^\circ$

8. $\frac{1}{50}\int_0^{50} 3e^{.02t}dt = \frac{1}{50}\cdot\frac{3}{(.02)}\cdot e^{.02t}\Big|_0^{50} = 5.5155$

9. $P(t) = P(0)e^{kt}$. Find k: $1 = 2e^{k(1690)} \Rightarrow \frac{1}{2} = e^{1690k}$ or

$1690k = \ln\frac{1}{2} \Rightarrow k = -.00041.$

The average value $= \frac{1}{1000}\int_0^{1000} 100e^{-.00041t}dt = \frac{1}{10}\cdot\frac{e^{-.00041t}}{-.00041}\Big|_0^{1000}$

$\cong$ 82 grammes

10. $\frac{1}{20}\int_0^{20} 100e^{.05t}dt = \frac{5e^{.05t}}{.05}\Big|_0^{20} = \$171.83$

11. $p(20) = 3 - \frac{20}{10} = 1$

$$\int_0^{20}\left(3 - \frac{x}{10} - 1\right)dx = \int_0^{20}\left(2 - \frac{x}{10}\right)dx = \left(2x - \frac{x^2}{20}\right)\Big|_0^{20}$$

$= 40 - 20 = \$20$

12. $p(20) = 2 - 20 + 50 = 32$

$$\int_0^{20}\left(\frac{x^2}{200} - x + 50 - 32\right)dx = \int_0^{20}\left(\frac{x^2}{200} - x + 18\right)dx$$

$$= \left(\frac{x^3}{600} - \frac{1}{2}x^2 + 18x\right)\Big|_0^{20}$$

$$= \frac{8000}{600} - 200 + 360 \sim \$173.33$$

13. $p(40) = \frac{500}{40 + 10} - 3 = 7$

$$\int_0^{40}\left(\frac{500}{x + 10} - 3 - 7\right)dx = \int_0^{40}\left(\frac{500}{x + 10} - 10\right)dx$$

$$= (500 \ln(x + 10) - 10x)\Big|_0^{40}$$

$$= 500 \ln 50 - 400 - 500 \ln 10 \sim \$404.72$$

14. $p(350) = \sqrt{16 - (.02)350} = 3$

$$\int_0^{350} ((16 - .02x)^{1/2} - 3)dx = \left(\frac{2}{3}\left(-\frac{1}{.02}\right)\left(16 - .02x\right)^{3/2} - 3x\right)\Big|_0^{350}$$

$$= -\frac{2}{3}\cdot\frac{9^{3/2}}{.02} - 1050 + \frac{2}{3}\cdot\frac{16^{3/2}}{.02}$$

$$= \$183.33$$

15. $p(200) = .01(200) + 3 = 5$

$$\int_0^{200} (5 - (.01x + 3))dx = \int_0^{200} (2 - .01x)dx = \left(2x - \frac{.01}{2}x^2\right)\Big|_0^{200}$$

$$= 400 - .005(40{,}000) = \$200$$

16. $p(3) = \frac{9}{9} + 1 = 2$

$$\int_0^3 \left(2 - \left(\frac{x^2}{9} + 1\right)\right)dx = \int_0^3 \left(1 - \frac{x^2}{9}\right)dx = \left(x - \frac{x^3}{27}\right)\Big|_0^3$$

$$= 3 - 1 = \$2$$

17. $p(10) = \frac{10}{2} + 7 = 12$

$$\int_0^{10} \left(12 - \left(\frac{x}{2} + 7\right)\right)dx = \int_0^{10} \left(5 - \frac{x}{2}\right) = \left(5x - \frac{x^2}{4}\right)\Big|_0^{10} = 50 - 25 = \$25$$

18. $p(36) = 1 + \frac{1}{2}\sqrt{36} = 4$

$$\int_0^{36} \left(4 - \left(1 + \frac{1}{2}x^{1/2}\right)\right)dx = \int_0^{36} \left(3 - \frac{1}{2}x^{1/2}\right)dx = \left(3x - \frac{1}{3}x^{3/2}\right)\Big|_0^{36}$$

$$= 108 - \frac{216}{3} = \$36$$

19. $12 - \frac{x}{50} = \frac{x}{20} + 5$; $7 = \frac{x}{50} + \frac{x}{20}$; $7 = \frac{2x + 5x}{100}$; $700 = 7x$; $x = 100$;

$p(100) = 12 - \frac{100}{50} = 10$ Thus, they intersect at (100, 10).

$$\text{C.S.} = \int_0^{100} \left[12 - \frac{x}{50} - 10\right]dx = \int_0^{100} \left[2 - \frac{x}{50}\right]dx = \left[2x - \frac{x^2}{100}\right]\Big|_0^{100}$$

$$= 200 - 100 = \$100$$

$$\text{P.S.} = \int_0^{100} \left[10 - \left[\frac{x}{20} + 5\right]\right]dx = \int_0^{100} \left[5 - \frac{x}{20}\right]dx = \left[5x - \frac{x^2}{40}\right]\Big|_0^{100}$$

$$= 500 - 250 = \$250$$

20. $\sqrt{25 - .1x} = \sqrt{.1x + 9} - 2$; $25 - .1x = .1x + 9 + 4 - 4\sqrt{.1x + 9}$;

$4\sqrt{.1x + 9} = .2x - 12$; $16(.1x + 9) = .04x^2 + 144 - 4.8x$;

$.04x^2 - 6.4 = 0$; $x^2 - 160x = 0$; $x(x - 160) = 0$; so

intersection is at $x = 160$, $p(160) = 3$; (160, 3).

$$\text{C.S.} = \int_0^{160} ((25 - .1x)^{1/2} - 3)dx = \left[\frac{-2}{.3}\left[25 - .1x\right]^{3/2} - 3x\right]\Big|_0^{160}$$

$$= \frac{-2}{.3}(27) - 480 + \frac{2}{.3}(125)$$

$$= \$173.33$$

$$\text{P.S.} = \int_0^{160} (3 - ((.1x + 9)^{1/2} - 2))dx = \int_0^{160} (5 - (.1x + 9)^{1/2})dx$$

$$= \left[5x - \frac{2}{.3}(.1x + 9)^{3/2}\right]\Big|_0^{160}$$

$$= 800 - \frac{2}{.3}(125) + \frac{2}{.3}(27)$$

$$= \$146.67$$

21. $A = 1000e^{.05(3-t_1)}\Delta t + 1000e^{.05(3-t_2)}\Delta t + \ldots$

Thus, A is approximately

$$\int_0^3 1000e^{.05(3-t)}dt = -20{,}000e^{.05(3-t)}\Big|_0^3 = \$3236.68.$$

22. $A = 2000e^{.06(2-t_1)}\Delta t + 2000e^{.06(2-t_2)}\Delta t + \ . \ . \ .$

Thus, A is approximately

$$\int_0^2 2000e^{.06(2-t)}dt = -33,333e^{.06(2-t)}\Big|_0^2 = \$4249.85.$$

23. $A = 16,000e^{.08(4-t_1)}\Delta t + 16,000e^{.08(4-t)}\Delta t + \ . \ . \ .$

Thus, A is approximately

$$\int_0^4 16,000e^{.08(4-t)}dt = -200,000e^{.08(4-t)}\Big|_0^4 = \$75,426.$$

24. $A = 14,000e^{.07(6-t_1)}\Delta t + 14,000e^{.07(6-t_2)}\Delta t + \ . \ . \ .$

Thus, A is approximately

$$\int_0^6 14,000e^{.07(6-t)}dt = -200,000e^{.07(6-t)}\Big|_0^6 = \$104,392.$$

25. Solve, $140,000 = \int_0^x 5000e^{.1(x-t)}dt$ for x.

$140,000 = -50,000e^{.1(x-t)}\Big|_0^x = -50,000(1 - e^{.1x});$

$2.8 = e^{.1x} - 1;\ e^{.1x} = 3.8;\ .1x = \ln 3.8;\ x = 13.35$ years

26. Solve $100,000 = \int_0^{10} xe^{.075(10-t)}dt$ for x.

$100,000 = -\frac{x}{.075}e^{.075(10-t)}\Big|_0^{10} = -\frac{x}{.075}\left(1 - e^{.75}\right);$

$\frac{7500}{x} = e^{.75} - 1;\ x = \frac{7500}{e^{.75} - 1} = \$6714.41$ per year

27. (a)Present value $= 1000e^{-.06t}$. The present value at time

$t_i = \Delta t(1000e^{-.06t_i})$

(b)Riemann sum $= \Delta t(1000e^{-.06t_1} + 1000e^{-.06t_2}$

$+ \cdot \cdot \cdot + 1000e^{-.06t_n})$

(c) $f(t) = 1000e^{-.06t};\ 0 \le t \le 5$

(d) $\int_0^5 1000e^{-.06t}dt$

(e) $\int_0^5 1000e^{-.06t}dt = \left.\frac{1000e^{-.06t}}{-.06}\right|_0^5 = \$4319.70$

28. $\int_0^{10} 5000e^{-.05t}dt = \left.\frac{5000e^{-.05t}}{-.05}\right|_0^{10} = \$39346.93$

29. $\int_{-r}^{r} \pi(r^2 - x^2)dx = \left(\pi r^2 x - \frac{\pi x^3}{3}\right)\Big|_{-r}^{r} = \pi r^3 - \frac{\pi r^3}{3} - \left(-\pi r^3 + \frac{\pi r^3}{3}\right)$

$= 2\pi r^3 - \frac{2\pi r^3}{3} = \frac{4}{3}\pi r^3$

30. $\int_0^h \pi r^2 dx = \pi r^2 x\Big|_0^h = \pi r^2 h$

31. $\int_1^2 \pi x^4 dx = \frac{\pi}{5}x^5\Big|_1^2 = \frac{32\pi}{5} - \frac{\pi}{5} = \frac{31\pi}{5}$

32. $\int_1^{100} \frac{\pi}{x^2}\, dx = -\frac{\pi}{x}\Big|_1^{100} = -\frac{\pi}{100} + \pi = \frac{99\pi}{100}$

33. $\int_0^4 \pi x\ dx = \frac{\pi}{2}x^2\Big|_0^4 = 8\pi$

34. $\int_0^2 \pi(2x - x^2)^2 dx = \int_0^2 \pi(x^4 - 4x^3 + 4x^2)dx = \pi\left(\frac{x^5}{5} - x^4 + \frac{4}{3}x^3\right)\Big|_0^2$

$= \pi\left(\frac{32}{5} - 16 + \frac{32}{3}\right) = \pi\left(\frac{16}{15}\right) = \frac{16\pi}{15}$

35. $\int_0^r \pi e^{-2x}dx = -\frac{\pi}{2}e^{-2x}\Big|_0^r = -\frac{\pi}{2}e^{-2r} + \frac{\pi}{2} = \frac{\pi}{2}\left(1 - \frac{1}{e^{2r}}\right)$

36. $\int_0^1 \pi(2x + 1)^2 dx = \int_0^1 \pi(4x^2 + 4x + 1)dx = \pi\left(\frac{4}{3}x^3 + 2x^2 + x\right)\Big|_0^1$

$= \pi\left(\frac{4}{3} + 2 + 1\right) = \frac{13\pi}{3}$

37. $n = 4;\ \frac{b - 8}{4} = .5;\ b = 10;\ f(x) = x^3$

38. $n = 6;\ \frac{b - 1}{6} = .5;\ b = 4;\ f(x) = \frac{3}{x}$

39. $n = 3;\ \frac{b - 4}{3} = 1;\ b = 7;\ f(x) = x + e^x$

40. $n = 5;\ \frac{b - 0}{5} = .6;\ b = 3;\ f(x) = 3x^2$

41. The sum is approximately

$$\int_0^3 (3 - x)^2 dx = \int_0^3 (x^2 - 6x + 9)dx = \left(\frac{1}{3}x^3 - 3x^2 + 9x\right)\Big|_0^3$$
$$= \frac{27}{3} - 27 + 27 = 9.$$

42. The sum is approximately

$$\int_0^1 (2x + x^3)dx = \left(x^2 + \frac{1}{4}x^4\right)\Big|_0^1 = 1 + \frac{1}{4} = \frac{5}{4}.$$

Chapter 6 Supplementary Exercises

1. $\int e^{-x/2}dx = -2e^{-x/2} + C$

2. $\int \frac{5}{\sqrt{x - 7}}\, dx = 10\sqrt{x - 7} + C$

3. $\int (3x^4 - 4x^2)dx = \frac{3}{5}x^5 - x^4 + C$

4. $\int (2x + 3)^7 dx = \frac{1}{16}(2x + 3)^8 + C$

5. $\int\sqrt{4 - x}\, dx = -\frac{2}{3}(4 - x)^{3/2} + C$

6. $\int\left(\frac{5}{x} - \frac{x}{5}\right)dx = 5\ \ln|x| - \frac{x^2}{10} + C$

7. $\int_1^4 \frac{1}{x^2}\, dx = -\frac{1}{x}\Big|_1^4 = -\frac{1}{4} + 1 = \frac{3}{4}$

8. $\int_3^6 e^{2-(x/3)}dx = -3e^{2-(x/3)}\Big|_3^6 = -3 + 3e = 3(e - 1)$

9. $\int_0^5 (5 + 3x)^{-1}dx = \frac{1}{3}\ln(5 + 3x)\Big|_0^5 = \frac{1}{3}\ln\ 20 - \frac{1}{3}\ln\ 5 = \frac{1}{3}\ln\ 4$

10. $\int_1^9 (1 + \sqrt{x})\ dx = \left(x + \frac{2}{3}x^{3/2}\right)\Big|_1^9 = 9 + 18 - \left(1 + \frac{2}{3}\right)$

$= 26 + \frac{2}{3} = \frac{76}{3}$

11. $\int_1^2 (3x - 2)^{-3}dx = \left(-\frac{1}{6}(3x - 2)^{-2}dx\right)\Big|_1^2 = -\frac{1}{6}\cdot\frac{1}{16} + \frac{1}{6} = \frac{15}{96} = \frac{5}{32}$

12. $16 - x^2 = 10 - x$; $x^2 - x - 6 = 0$; $(x - 3)(x - 2) = 0$ Thus, we should calculate

$$\int_{-2}^3 (16 - x^2 - (10 - x))dx = \int_2^3 (-x^2 + x + 6)dx$$

$$= \left(-\frac{1}{3}x^3 + \frac{1}{2}x^2 + 6x\right)\Big|_{-2}^3$$

$$= -9 + \frac{9}{2} + 18 - \left(\frac{8}{3} + 2 - 12\right)$$

$$= 19 + \frac{9}{2} - \frac{8}{3} = \frac{125}{6}$$

13. $x^3 - 3x + 1 = x + 1$; $x^3 - 4x = 0$; $x(x^2 - 4) = 0$ Thus, the graphs intersect at $x = 0, \pm 2$. On $[-2, 0]$, $y = x^3 - 3x + 1$ lies above $y = x + 1$, and below on $[0, 2]$. Thus we should calculate

$$\int_{-2}^{0}((x^3 - 3x + 1) - (x + 1))dx + \int_{0}^{2}((x + 1) - (x^3 - 3x + 1))dx$$

$$= \int_{-2}^{0}(x^3 - 4x)dx + \int_{0}^{2}(-x^2 + 4x)dx$$

$$= \left(\frac{1}{4}x^4 - 2x^2\right)\Big|_{-2}^{0} + \left(-\frac{1}{4}x^4 + 2x^2\right)\Big|_{0}^{2}$$

$$= 0 - (4 - 8) + (-4 + 8) = 8$$

14. $2x^2 + x = x^2 + 2;\ x^2 + x - 2 = 0;\ (x + 2)(x - 1) = 0$ Thus, on the interval [0, 2], the graphs intersect at $x = 1$. On [0, 1], $y = x^2 + 2$ lies above $y = 2x^2 + x$ and below on [1, 2]. Thus, we should calculate

$$\int_{0}^{1}(x^2 + 2 - (2x^2 + x))dx + \int_{1}^{2}(2x^2 + x - (x^2 + 2))dx$$

$$= \int_{0}^{1}(-x^2 - x + 2)dx + \int_{1}^{2}(x^2 + x - 2)dx$$

$$= \left(-\frac{1}{3}x^3 - \frac{1}{2}x^2 + 2x\right)\Big|_{0}^{1} + \left(\frac{1}{3}x^3 + \frac{1}{2}x^2 - 2x\right)\Big|_{1}^{2}$$

$$= -\frac{1}{3} - \frac{1}{2} + 2 + \frac{8}{3} + 2 - 4 - \left(\frac{1}{3} + \frac{1}{2} - 2\right) = \frac{6}{3} - \frac{2}{2} + 2 = 3$$

15. $\int(x - 5)^2dx = \frac{1}{3}(x - 5)^3 + C;\quad f(8) = 2 = \frac{1}{3}(3)^3 + C = 9 + C;$

$C = -7;\ f(x) = \frac{1}{3}(x - 5)^3 - 7$

16. $\int e^{-5x}dx = -\frac{1}{5}e^{-5x} + C;\ f(0) = 1 = -\frac{1}{5} + C;\ C = \frac{6}{5};$

$f(x) = \frac{6}{5} - \frac{1}{5}e^{-5x}$

17. Following the hint:

$$\frac{d}{dt}\left[f(t)e^{-kt^2/2}\right] = f(t)(-kt)e^{-kt^2/2} + f'(x)e^{-kt^2/2}$$

$$= -e^{-kt^2/2}(-f(t)kt + f'(t))$$

$$= e^{-kt^2/2}(0) = 0, \text{ since } f'(t) = ktf(t).$$

Since only constant functions have a zero derivative, $f(t)e^{-kt^2/2} = C$ for some C. Thus, $f(t) = Ce^{kt^2/2}$.

18. (a) $y' = 4t$, $y = 2t^2 + C$ (b) $y' = 4y$, $y = Ce^{4t}$

(c) $y' = e^{4t}$, $y = \frac{1}{4}e^{4t} + C$

19. $C(x) = \int(.04x + 150)dx = .02x^2 + 150x + C$

$C(0) = 500 = 0 + 0 + C = C$. Thus, $C(x) = .02x^2 + 150x + 500$ (dollars).

20. $\int_{10}^{20}(400 - 3x^2)dx = (400x - x^3)\Big|_{10}^{20} = -3000$

Thus, a loss of \$3000 would result.

21. It represents the total quantity of drug (in cubic centimeters) injected during the first 4 minutes.

22. $v(t) = -9.8t + 20$

(a) $\int_0^2(-9.8t + 20)dt = (-4.9t^2 + 20t)\Big|_0^2 = -19.6 + 40$

$= 20.4$ meters

(b)

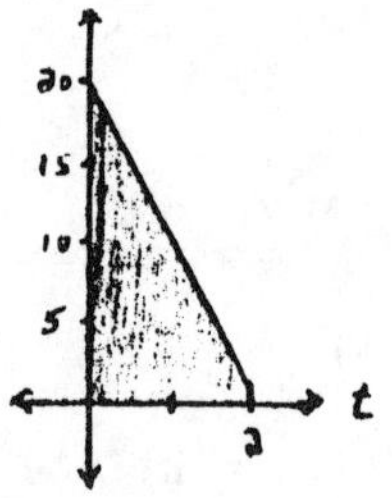

23. $\int_0^1 4500e^{.09(1-t)}dt = -50,000e^{.09(1-t)}\Big|_0^1 = \$4708.71$

24. $p = \int(-.02x + 5)dx = -.01x^2 + 5x$

25. $\Delta x = 1$: the midpoints are .5, 1.5

$$\text{Area} = \left[\frac{1}{.5 + 2} + \frac{1}{1.5 + 2}\right] = .68571$$

$$\int_0^2 \frac{1}{x + 2}\,dx = \ln(x + 2)\Big|_0^2 = \ln 4 - \ln 2 = \ln\frac{4}{2} = .69315$$

26. $\Delta x = .2$: the midpoints are .1, .3, .5, .7, .9.

$$\text{Area} = \left[e^{2(.1)} + e^{2(.3)} + e^{2(.5)} + e^{2(.7)} + e^{2(.9)}\right](.2) = 3.17333$$

27. $p(400) = \sqrt{25 - .04(400)} = 3$

$$\text{C.S.} = \int_0^{400} (\sqrt{25 - .04x} - 3)dx = \left(\frac{2}{-.12}(25 - .04x)^{3/2} - 3x\right)\Big|_0^{400}$$

$$= \frac{2}{-.12}(27) - 1200 - \frac{2}{-.12}(125)$$

$$= \$433.33$$

28. $\frac{1}{10}\int_0^{10} 3000e^{.06t}dt = \int_0^{10} 300e^{.06t}dt = 5000e^{.06t}\Big|_0^{10}$

$$= 5000(e^{.6} - e^0)$$

$$= \$4110.59$$

29. $\frac{1}{\frac{1}{2} - \frac{1}{3}}\int_{1/3}^{1/2} \frac{1}{x^3}\,dx = 6\int_{1/3}^{1/2} \frac{1}{x^3}\,dx = 6\left[-\frac{1}{2x^2}\Big|_{1/3}^{1/2}\right] = 6\left(-2 + \frac{9}{2}\right) = 15$

30. The sum is approximated by

$$\int_0^1 3e^{-x}dx \quad -3e^{-x}\Big|_0^1 = -3e^{-1} + 3 = 3(1 - e^{-1})$$

31. $\int_2^3 \left(e^x - \frac{3}{x}\right)dx = \left(e^x - 3\ln x\right)\Big|_2^3 = e^3 - 3\ln 3 - e^2 + 3\ln 2$

$= 11.48$

32. $\int_0^1 \pi(1 - x^2)^2 dx = \int_0^1 \pi(x^4 - 2x^2 + 1)dx = \pi\left(\frac{x^5}{5} - \frac{2}{3}x^3 + x\right)\Big|_0^1$

$= \pi\left(\frac{1}{5} - \frac{2}{3} + 1\right) = \frac{8\pi}{15}$

33. (a) Since inventory is decreasing, the slope is $-\frac{Q}{A}$. From the graph we can see that

$f(t) = Q - \frac{Q}{A}t.$

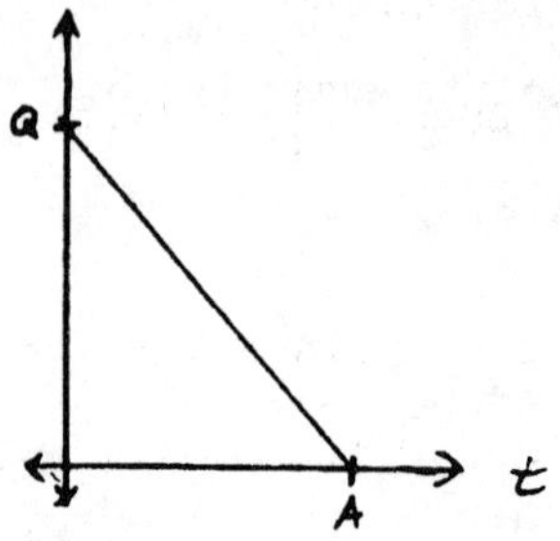

(b) $\frac{1}{A}\int_0^A \left(Q - \frac{Q}{A}t\right)dt = \frac{1}{A}\left(Qt - \frac{Q}{2A}t^2\right)\Big|_0^A = \frac{Q}{A}A - \frac{QA^2}{2A^2} = \frac{Q}{2}$

34. (a) $f(t) = Q - \int_0^t rt\ dt = Q - \frac{rt^2}{2}$

(b) $0 = Q - \frac{rA^2}{2}$; $r = \frac{2Q}{A^2}$

(c) $f(t) = Q - \frac{\frac{2Q}{A^2}t^2}{2} = Q - \frac{Qt^2}{A^2}$

$\frac{1}{A}\int_0^A \left(Q - \frac{Qt^2}{A^2}\right)dt = \frac{1}{A}\left(Qt - \frac{Qt^3}{3A^2}\right)\Big|_0^A = Q - \frac{Q}{3} = \frac{2}{3}Q$

35. (a) g(3) is the area under the curve $y = \dfrac{1}{1 + t^2}$ from t = 0 to t = 3.

(b) $g'(x) = \dfrac{1}{1 + x^2}$

36. (a) h(0) is the area under one-quarter of the unit circle.
h(1) is the area under one-half of the unit circle.

(b) $h'(x) = \sqrt{1 - x^2}$

37. The sum is approximated by

$$\int_0^3 5000e^{.1t}dt = -50{,}000e^{-.1t}\Big|_0^3 \sim 12{,}595 \sim 13{,}000.$$

38. $\Delta x = \frac{1}{n}$: with left endpoints $t_i = i\Delta x$

$$\text{Sum} = \Delta xe^{0} + \Delta xe^{\Delta x} + \Delta xe^{2\Delta x} + \cdots + \Delta xe^{(n-1)\Delta x}$$

$$= \Delta x\left[e^{t0} + e^{t1} + e^{t2} + \cdots + e^{tn-1}\right] \cong \int_0^1 e^x dx = e - 1$$

39. $\Delta x = \frac{1}{n}$: with left endpoints $t_i = i\Delta x$

$$\text{Sum} = \Delta x\left[1 + (1 + \Delta x)^3 + \cdots + (1 + (n - 1)\Delta x)^3\right]$$

$$= \Delta x\left[1 + (1 + t_1)^3 + \cdots + (1 + t_{n-1})^3\right]$$

$$\cong \int_0^1 (1 + x)^3 dx = \frac{(1 + x)^4}{4}\Big|_0^1 = \frac{15}{4}$$

40. by figure 1(a), the average value of F(x) = 4 or $2 \le x \le 6$

41. $3 \le f(x) \le 4$

$$\int_0^5 3dx \le \int_0^5 f(x)dx \le \int_0^5 4dx$$

$15 \le \int_0^5 f(x)dx \le 20$, so $3 \le \frac{1}{5}\int_0^5 f(x)dx \le 4$

42. (a) $\int_{t_1}^{t_2} (20 - 4t_1)dt = (20 - 4t_1)t\Big|_{t_1}^{t_2}$

$= (20 - 4t_1)t_2 - (20 - 4t_1)t_1$

$= (20 - 4t_1)\Delta t$

(b) $\int_0^5 r(t)\,dt$ gives the amount of water added to the tank in the first 5 hours, because it is the area under the curve r(t) (the rate at which water is flowing into the tank at time t).

43. $\int_0^{40} 860e^{.04t}dt = 21{,}500e^{.04t}\Big|_0^{40} = 21{,}500(e^{1.6} - 1)$

$\sim 84{,}990$ cubic kilometers

Exercises 6.6 (Brief Edition)

1. $\int 2x(x^2 + 4)^5 dx$. Let $u = x^2 + 4$; then $du = 2x\ dx$.

$\int 2x(x^2 + 4)^5 dx = \int u^5 du = \frac{1}{6}u^6 + C = \frac{1}{6}(x^2 + 4)^6 + C$

2. $\int 3x^2(x^3 + 1)^2 dx$. Let $u = x^3 + 1$; then $du = 3x^2 dx$.

$\int 3x^2(x^3 + 1)^2 dx = \int u^2 du = \frac{1}{3}u^3 + C = \frac{1}{3}(x^3 + 1)^3 + C$

3. $\int (x^2 - 5x)^3(2x - 5)dx$. Let $u = x^2 - 5x$; then $du = (2x - 5)dx$.

$\int (x^2 - 5x)^3(2x - 5)dx = \int u^3 du = \frac{1}{4}u^4 + C = \frac{1}{4}(x^2 - 5x)^4 + C$

4. $\int 2x\sqrt{x^2 + 3}\ dx$. Let $u = x^2 + 3$; then $du = 2x\ dx$.

$\int 2x\sqrt{x^2 + 3}\ dx = \int u^{1/2} du = \frac{2}{3}u^{3/2} + C = \frac{2}{3}(x^2 + 3)^{3/2} + C$

5. $\int 5e^{5x-3} dx$. Let $u = 5x - 3$, $du = 5\ dx$. Then

$\int 5e^{5x-3} dx = \int e^u du = e^u + C = e^{5x-3} + C.$

6. $\int 2xe^{-x^2} dx$. Let $u = -x^2$; then $du = -2x\ dx$.

$\int 2xe^{-x^2} dx = -\int -2xe^{-x^2} dx = -\int e^u du = -e^u + C = -e^{-x^2} + C.$

7. $\int \frac{3x^2}{x^3 - 1} dx$. Let $u = x^3 - 1$, $du = 3x^2 dx$. Then,

$\int \frac{3x^2}{x^3 - 1} dx = \int \frac{du}{u} = \ln|u| + C = \ln|x^3 - 1| + C$

8. $\int \frac{2x + 1}{(x^2 + x + 3)^6} dx$. Let $u = x^2 + x + 3$, $du = (2x + 1)dx$. Then,

$\int \frac{2x + 1}{(x^2 + x + 3)^6} dx = \int \frac{du}{u} = \ln|u| + C = \ln|x^2 + x + 3| + C.$

9. $\int \frac{x^2}{\sqrt{x^3 - 1}}\, dx$. Let $u = x^3 - 1$, $du = 3x^2\, dx$. Then,

$$\int \frac{x^2}{\sqrt{x^3 - 1}}\, dx = \frac{1}{3}\int \frac{du}{u^{1/2}} = \left(\frac{1}{3}\right)2u^{1/2} + C = \frac{2}{3}(x^3 - 1)^{1/2} + C.$$

10. $\int \frac{1}{\sqrt{2x + 1}}\, dx$. Let $u = 2x + 1$, $du = 2\, dx$. Then,

$$\int \frac{1}{\sqrt{2x + 1}}\, dx = \frac{1}{2}\int \frac{du}{u^{1/2}} = \left(\frac{1}{2}\right)2u^{1/2} + C = (2x + 1)^{1/2} + C.$$

11. $\int \frac{e^{1/x}}{x^2}\, dx$. Let $u = \frac{1}{x}$; then $du = -\frac{1}{x^2}dx$.

$$\int \frac{e^{1/x}}{x^2}\, dx = -\int -\frac{e^{1/x}}{x^2}\, dx = -\int e^u du = -e^u + C = -e^{1/x} + C.$$

12. $\int \frac{e^{3x}}{e^{3x} + 1}\, dx$. Let $u = e^{3x} + 1$, $du = 3e^{3x}dx$. Then,

$$\int \frac{e^{3x}}{e^{3x} + 1}\, dx = \frac{1}{3}\int \frac{du}{u} = \frac{1}{3}\ln|u| + C = \frac{1}{3}\ln(e^{3x} + 1) + C.$$

13. $\int \frac{x^2 + 1}{x^3 + 3x + 2}\, dx$. Let $u = x^3 + 3x + 2$, $du = (3x^2 + 3)dx$. Then,

$$\int \frac{x^2 + 1}{x^3 + 3x + 2}\, dx = \frac{1}{3}\int \frac{du}{u} = \frac{1}{3}\ln|u| + C = \frac{1}{3}\ln|x^3 + 3x + 2| + C.$$

14. $\int \frac{\ln x}{x}\, dx$. Let $u = \ln x$, $du = \frac{1}{x}\, dx$. Then,

$$\int \frac{\ln x}{x}\, dx = \int u\, du = \frac{u^2}{2} + C = \frac{1}{2}(\ln x)^2 + C.$$

15. $\int (x^5 - 2x + 1)^{10}(5x^4 - 2)dx$. Let $u = x^5 - 2x + 1$,

$du = (5x^4 - 2)dx$. Then,

$$\int (x^5 - 2x + 1)^{10}(5x^4 - 2)dx = \int u^{10}du = \frac{1}{11}u^{11} + C$$

$$= \frac{1}{11}(x^5 - 2x + 1)^{11} + C.$$

16. $\int (x + 1)e^{x^2+2x+4}dx$. Let $u = x^2 + 2x + 4$, $du = (2x + 2)dx$.

Then, $\int (x + 1)e^{x^2+2x+4}dx = \frac{1}{2}\int e^u du = \frac{1}{2}e^u + C = \frac{1}{2}e^{x^2+2x+4} + C.$

17. $\int \frac{3}{2x - 4}\,dx$. Let $u = x - 2$, $du = dx$. Then,

$$\int \frac{3}{2x - 4}\,dx = \frac{3}{2}\int \frac{du}{u} = \frac{3}{2}\ln|u| + C = \frac{3}{2}\ln|x - 2| + C.$$

18. $\int \frac{e^{\sqrt{x}}}{\sqrt{x}}\,dx$. Let $u = \sqrt{x}$, $du = \frac{1}{2}x^{-1/2} = \frac{dx}{2\sqrt{x}}$. Then,

$$\int \frac{e^{\sqrt{x}}}{\sqrt{x}}\,dx = 2\int e^u du = 2e^u + C = 2e^{\sqrt{x}} + C.$$

19. $\int \frac{x}{e^{x^2}}\,dx$. Let $u = -x^2$; then $du = -2x\,dx$.

$$\int \frac{x}{e^{x^2}}\,dx = -\frac{1}{2}\int e^{-x^2}(-2x)dx = -\frac{1}{2}\int e^u du = -\frac{1}{2}e^u + C = -\frac{1}{2}e^{-x^2} + C.$$

20. $\int \frac{3x - x^3}{x^4 - 6x^2 + 5}\,dx$. Let $u = x^4 - 6x^2 + 5$, $du = (4x^3 - 12x)dx$.

Then, $\int \frac{3x - x^3}{x^4 - 6x^2 + 5}\,dx = -\frac{1}{4}\int \frac{du}{u} = -\frac{1}{4}\ln|u| + C$

$$= -\frac{1}{4}\ln|x^4 - 6x^2 + 5| + C.$$

21. $\int xe^{5x}dx$. Let $f(x) = x$, $g(x) = e^{5x}$.

Then $f'(x) = 1$ and $G(x) = \frac{1}{5}e^{5x}$.

$$\int xe^{5x}dx = x\left[\frac{1}{5}e^{5x}\right] - \int (1)\frac{1}{5}e^{5x} = \frac{1}{5}xe^{5x} - \frac{1}{25}e^{5x} + C.$$

22. $\int xe^{-x/2}dx$. Let $f(x) = x$, $g(x) = e^{-x/2}$.

Then $f'(x) = 1$ and $G(x) = -2e^{-x/2}$.

$$\int xe^{-x/2}dx = x\left[-2e^{-x/2}\right] - \int (1)\left[-2e^{-x/2}\right]dx$$

$$= -2xe^{-x/2} - 4e^{-x/2} + C.$$

23. $\int x(2x + 1)^4dx$. Let $f(x) = x$, $g(x) = (2x + 1)^4$.

Then $f'(x) = 1$ and $G(x) = \frac{1}{10}(2x + 1)^5$.

$$\int x(2x + 1)^4dx = x\left[\frac{1}{10}(2x + 1)^5\right] - \int (1)\frac{1}{10}(2x + 1)^5dx$$

$$= \frac{1}{10}x(2x + 1)^5 - \frac{1}{120}(2x + 1)^6 + C.$$

24. $\int (x + 1)e^xdx$. Let $f(x) = x + 1$, $g(x) = e^x$.

Then $f'(x) = 1$ and $G(x) = e^x$.

$$\int (x + 1)e^xdx = (x + 1)e^x - \int (1)e^xdx = e^x(x + 1) - e^x + C$$

$$= e^x\left[(x + 1) - 1\right] + C = xe^x + C.$$

25. $\int x\sqrt{x + 1}\,dx$. Let $f(x) = x$, $g(x) = (x + 1)^{1/2}$.

Then $f'(x) = 1$ and $G(x) = \frac{2}{3}(x + 1)^{3/2}$.

$$\int x\sqrt{x + 1}\,dx = x\left[\frac{2}{3}(x + 1)^{3/2}\right] - \int (1)\frac{2}{3}(x + 1)^{3/2}dx$$

$$= \frac{2}{3}x(x + 1)^{3/2} - \frac{4}{15}(x + 1)^{5/2} + C.$$

26. $\int x(x + 5)^{-3}dx$. Let $f(x) = x$, $g(x) = (x + 5)^{-3}$.

Then $f'(x) = 1$ and $G(x) = -\frac{1}{2}(x + 5)^{-2}$.

32. $\int x\sqrt{x^2 - 1}\,dx$. Let $u = x^2 - 1$, $du = 2x\,dx$. Then,

$$\int x\sqrt{x^2 - 1}\,dx = \frac{1}{2}\int \sqrt{u}\,du = \frac{1}{2}\left(\frac{2}{3}\right)u^{3/2} + C = \frac{1}{3}(x^2 - 1)^{3/2} + C.$$

33. $\int xe^{x^2}dx$. Let $u = x^2$, $du = 2x\,dx$. Then,

$$\int xe^{x^2}dx = \frac{1}{2}\int e^u du = \frac{1}{2}e^u + C = \frac{1}{2}e^{x^2} + C.$$

34. $\int x\sqrt{x - 1}\,dx$. Let $f(x) = x$, $g(x) = \sqrt{x - 1}$. Then $f'(x) = 1$,

$G(x) = \frac{2}{3}(x - 1)^{3/2}$ and

$$\int x\sqrt{x - 1} = \frac{2}{3}x(x - 1)^{3/2} - \int \frac{2}{3}(x - 1)^{3/2}dx$$

$$= \frac{2}{3}x(x - 1)^{3/2} - \frac{2}{3}\left(\frac{2}{5}\right)(x - 1)^{5/2} + C$$

$$= \frac{2}{3}x(x - 1)^{3/2} - \frac{4}{15}(x - 1)^{5/2} + C.$$

35. $\int \frac{2x - 1}{\sqrt{3x - 3}}\,dx$. Let $f(x) = 2x - 1$, $g(x) = (3x - 3)^{-1/2}$. Then

$f'(x) = 2$, $G(x) = \frac{2}{3}(3x - 3)^{1/2}$ and

$$\int \frac{2x - 1}{\sqrt{3x - 3}}\,dx = \frac{2}{3}(2x - 1)(3x - 3)^{1/2} - \int \frac{4}{3}(3x - 3)^{1/2}dx$$

$$= \frac{2}{3}(2x - 1)(3x - 3)^{1/2} - \frac{8}{27}(3x - 3)^{3/2}$$

36. $\int \frac{2x - 1}{3x^2 - 3x + 1}\,dx$. Let $u = 3x^2 - 3x + 1$, $du = (6x - 3)dx$.

Then $$\int \frac{2x - 1}{3x^2 - 3x + 1}\,dx = \frac{1}{3}\int \frac{du}{u} = \frac{1}{3}\ln|u| + C$$

$$= \frac{1}{3}\ln|3x^2 - 3x + 1| + C.$$

37. $\int(4x + 3)(6x^2 + 9x)^{-7}dx$. Let $u = 6x^2 + 9x$, $du = (12x + 9)dx$.

Then $\int(4x + 3)(6x^2 + 9x)^{-7}dx = \frac{1}{3}\int u^{-7}du = -\frac{1}{18}u^{-6} + C$

$$= -\frac{1}{18}(6x^2 + 9x)^{-7} + C.$$

38. $\int(4x + 3)(6x + 9)^{-7}dx$. Let $f(x) = 4x + 3$, $g(x) = (6x + 9)^{-7}$.

Then $f'(x) = 4$, $G(x) = -\frac{1}{36}(6x + 9)^{-6}$ and

$\int(4x + 3)(6x + 9)^{-7}dx$

$$= -\frac{1}{36}(4x + 3)(6x + 9)^{-6} - \int -\frac{1}{9}(6x + 9)^{-6}dx$$

$$= -\frac{1}{36}(4x + 3)(6x + 9)^{-6} - \frac{1}{270}(6x + 9)^{-5} + C.$$

39. $\int\frac{\ln x}{\sqrt{x}}\,dx$. Let $u = \sqrt{x}$, $du = \frac{1}{2\sqrt{x}}$.

Then $\int\frac{\ln x}{\sqrt{x}}\,dx = \int\frac{2\ln\sqrt{x}}{\sqrt{x}} = 4\int\ln u\,du = 4(u\ln u - u) + C$

$$= 4(\sqrt{x}\ln\sqrt{x} - \sqrt{x}) + C.$$

40. $\int\frac{x + 4}{e^{4x}}\,dx$. Let $f(x) = x + 4$, $g(x) = e^{-4x}$. Then $f'(x) = 1$,

$G(x) = -\frac{1}{4}e^{-4x}$ and

$$\int\frac{x + 4}{e^{4x}}\,dx = (x + 4)\left(-\frac{1}{4}e^{-4x}\right) - \int-\frac{1}{4}e^{-4x}dx$$

$$= -\frac{1}{4}e^{-4x}(x + 4) - \frac{1}{16}e^{-4x} + C.$$

41. $\int\frac{1}{x\ln 5x}\,dx$. Let $u = \ln 5x$, $du = \frac{1}{x}\,dx$. Then

$$\int\frac{1}{x\ln 5x}\,dx = \int\frac{du}{u} = \ln|u| + C = \ln|\ln 5x| + C.$$

42. $\int x \ln 5x\,dx$. Let $f(x) = \ln 5x$, $g(x) = x$. Then $f'(x) = \frac{1}{x}$,

$G(x) = \frac{1}{2}x^2$ and $\int x \ln 5x\,dx = \frac{1}{2}x^2 \ln 5x - \int \frac{x}{2}\,dx$

$= \frac{1}{2}x^2 \ln 5x - \frac{1}{4}x^2 + C.$

43. The present value is $\int_0^5 (300t - 500)e^{-.1t}dt.$

Let $f(t) = 300t - 500$, $g(t) = e^{-.1t}$. Then $f'(t) = 300$,

$G(t) = -10e^{-.1t}$ and

$\int_0^5 (300t - 500)e^{-.1t}dt$

$= (300t - 500)(-10)e^{-.1t}\Big|_0^5 - \int_0^5 -3000e^{-.1t}dt$

$= \left[(5000 - 3000t)e^{-.1t} - 30000e^{-.1t}\right]\Big|_0^5 \sim \$10739$

44. $\int_0^5 (300t - 500)e^{-.05t}dt$

$= (300t - 500)(-20)e^{-.05t}\Big|_0^5 - \int -6000e^{-.05t}dt$

$= \left[(10{,}000 - 6000t)e^{-.05t} - 120{,}000e^{-.05t}\right]\Big|^5 \sim 20{,}967.89.$

Exercises 6.7 (Brief Edition)

See solutions to Section 9.6.

Exercises 6.8 (Brief Edition)

1. (a) $\int_{1/4}^{1/2} (6x - 6x^2)dx = (3x^2 - 2x^3)\Big|_{1/4}^{1/2} = \frac{11}{32}$

(b) $\int_{0}^{1/3} (6x - 6x^2)dx = (3x^2 - 2x^3)\Big|_{0}^{1/3} = \frac{7}{27}$

(c) $(3x^2 - 2x^3)\Big|_{1/4}^{1} = \frac{17}{32}$

(d) $3x^2 - 2x^3\Big|_{0}^{3/4} = \frac{27}{32}$

2. (a) $\int_{0}^{1} \frac{1}{8}x\ dx = \frac{1}{16}x^2\Big|_{0}^{1} = \frac{1}{16}$

(b) $\frac{1}{16}x^2\Big|_{2}^{2.5} = \frac{1}{16}(6.25 - 4) = \frac{2.25}{16} = \frac{9}{64}$

(c) $\frac{1}{16}x^2\Big|_{3.5}^{4} = \frac{1}{16}(16 - (3.5)^2) = \frac{15}{64}$

3. $\int_{0}^{2} kx^2 dx = \frac{k}{3}x^3\Big|_{0}^{2} = \frac{8}{3}k = 1;$ so $k = \frac{3}{8}$.

4. $\int_{1}^{4} kx^{-1/2}dx = 2kx^{1/2}\Big|_{1}^{4} = 4k - 2k = 2k = 1;$ so $k = \frac{1}{2}$.

5. (a) $\int_{0}^{.1} 2e^{-2x}dx = -e^{-2x}\Big|_{0}^{.1} = 1 - e^{-.2} \sim .1813$

(b) $\int_{.1}^{.5} 2e^{-2x}dx = -e^{-2x}\Big|_{.1}^{.5} = e^{-.2} - e^{-1} \sim .4509$

(c) $\Pr(1 \le X) = 1 - P(X < 1) = 1 - [1 - e^{-2}] \sim .1353$

(d) $E(x) = \frac{1}{2}$

6. (a) $\int_1^2 .25e^{-.25x}dx = -e^{-.25x}\Big|_1^2 = e^{-.25} - e^{-.5} \sim .1723$

(b) $-e^{-.25}\Big|_0^3 = 1 - e^{-.75} \sim .5276$

(c) $1 - \Pr(X < 4) = 1 - [1 - e^{-1}] \sim .3679$

(d) $E(x) = 4$

7. (a) $f(x) = \frac{1}{3}$ (b) $\int_1^3 \frac{1}{3}\,dx = \frac{2}{3}$ (c) $\int_0^1 \frac{1}{3}\,dx = \frac{1}{3}$

8. (a) $\int_5^{25} kx = 1$, so $\frac{1}{2}kx^2\Big|_5^{25} = 1$, $\frac{k}{2}\Big[625 - 25\Big] = 1$, $300k = 1$,

$k = \frac{1}{300}$

(b) $\int_5^{10} \frac{1}{300}x\,dx = \frac{1}{600}x^2\Big|_5^{10} = \frac{1}{600}\Big[100 - 25\Big] = \frac{75}{600} = \frac{1}{8}$

(c) $\int_{20}^{25} \frac{1}{300}x\,dx = \frac{1}{600}x^2\Big|_{20}^{25} = \frac{1}{600}\Big[625 - 400\Big] = \frac{225}{600} = \frac{3}{8}$

9. (a) $\int_{30}^{35} \frac{x - 30}{50}\,dx = \frac{(x - 30)^2}{100}\Big|_{30}^{35} = .25$

(b) From (a), 25% (or 500 acres) produced less than 35 bushels.

10. (a) $\int_0^{.4} (12x^2 - 12x^3)dx = (4x^3 - 3x^4)\Big|_0^{.4} = 4(.4)^3 - 3(.4)^4$

$\sim .1792$

(b) $(4x^3 - 3x^4)\Big|_{.5}^{1} = 1 - \Big[4(.5)^3 - 3(.5)^4\Big] = .6875$

11. $\int_0^4 \frac{11}{10}(x + 1)^{-2}dx = -\frac{11}{10}(x + 1)^{-1}\Big|_0^4 = -\frac{11}{50} + \frac{11}{10} = \frac{44}{50} = \frac{22}{25}$

12. (a) $\int_0^5 2ke^{-kx}dx = -2e^{-kx}\Big|_0^5 = -2e^{-5k} + 2 = 2 - 2e^{\ln(2)(-1/2)}$

$= 2 - \sqrt{2} \sim .58579.$

(b) $\int_{10-M}^{10} 2ke^{-5kx}dx = -2e^{-kx}\Big|_{10-M}^{10}$

$= -2e^{-10k} + 2e^{-10k+Mk}$ $\qquad k = \frac{\ln 2}{10}$

$= -1 + 2^{M/10} = .1$

$2^{M/10} = 1.1;\quad \frac{M}{10}\ln 2 = \ln 1.1;\quad M = 10\frac{\ln 1.1}{\ln 2} \sim 1.37504$ days.

13. (a) $\int_0^2 .25e^{-.25t}dt = -e^{-.25t}\Big|_0^2 = 1 - e^{-.5} \sim .3935$

(b) $1 - \left[1 - e^{-1}\right] \sim .3679$

14. The density function is $f(x) = .02e^{-.02x}$

$\int_0^{12} .02e^{-.02x}dx = -e^{-.02x}\Big|_0^{12} = 1 - e^{-.24} \sim .21337$

15. The density function is $f(x) = .5e^{-.5x}$.

$\Pr(X \geq 3) = 1 - \Pr(X < 3) = 1 - \int_0^3 .5e^{-.5x}dx = 1$

$= 1 - \left[1 - e^{-1.5}\right] \sim .2231$

16. $\Pr(X \leq 4) = \int_0^4 ke^{-kx}dx = -e^{-kx}\Big|_0^4 = 1 - e^{-4k} = .75$, so

$.25 = e^{-4k},\ \ln(.25) = -4k,\ k = -\frac{\ln(.25)}{4} \sim .3466$

Exercises 6.9 (Brief Edition)

See solutions for Section 9.7.

Chapter 6 Supplementary Exercises (Brief Edition)

1. $\int e^{-x/2}dx = -2e^{-x/2} + C$

2. $\int \frac{5}{\sqrt{x-7}}\,dx = 10\sqrt{x-7} + C$

3. $\int \frac{x^2 - 1}{(x^3 - 3x + 2)^2}\,dx$. Let $u = x^3 - 3x + 2$, $du = (3x^2 - 3)dx$.

$$\int \frac{x^2 - 1}{(x^3 - 3x + 2)^2}\,dx = \frac{1}{3}\int \frac{du}{u^2} = -\frac{1}{3}u^{-1} + C = -\frac{1}{3(x^3 - 3x + 2)} + C$$

4. $\int x^4 e^{-x^5}\,dx$. Let $u = x^5$, $du = 5x^4dx$.

$$\int x^4 e^{-x^5}\,dx = \frac{1}{5}\int e^{-u}du = -\frac{1}{5}e^{-u} + C = -\frac{1}{5}e^{-x^5} + C$$

5. $\int (2x + 3)^7dx = \frac{1}{16}(2x + 3)^8 + C$

6. $\int \left(9 - 4e^x + \frac{1}{x^4}\right)dx = 9x - 4e^x - \frac{1}{3}x^{-3} + C$

7. $\int (e^x + 4)^3e^xdx$. Let $u = e^x$, $du = e^xdx$.

$$\int (e^x + 4)^3e^xdx = \int (u + 4)^3du = \frac{1}{4}(u + 4)^4 + C = \frac{1}{4}(e^x + 4)^4 + C$$

8. $\int \frac{x - e^{-2x}}{x^2 + e^{-2x}}\,dx$. Let $u = x^2 + e^{-2x}$, $du = 2x - 2e^{-2x}$.

$$\int \frac{x - e^{-2x}}{x^2 + e^{-2x}}\,dx = \frac{1}{2}\int \frac{du}{u} = \frac{1}{2}\ln|u| + C = \frac{1}{2}\ln|x^2 + e^{-2x}| + C$$

9. $\int (2x + 1)e^{-x/2}dx$. Let $f(x) = 2x + 1$, $g(x) = e^{-x/2}$.

$f'(x) = 2$, $G(x) = -2e^{-x/2}$.

$$\int(2x + 1)e^{-x/2}dx = -2(2x + 1)e^{-x/2} - \int -4e^{-x/2}dx$$

$$= -2(2x + 1)e^{-x/2} - 8e^{-x/2} + C.$$

10. $\int\frac{5x}{\sqrt{x - 7}}\,dx$. Let $f(x) = 5x$, $g(x) = (x - 7)^{-1/2}$. $f'(x) = 5$,

$G(x) = 2(x - 7)^{1/2}$

$$\int\frac{5x}{\sqrt{x - 7}}\,dx = 10x(x - 7)^{1/2} - \int 10(x - 7)^{1/2}dx$$

$$= 10x(x - 7)^{1/2} - \frac{20}{3}(x - 7)^{3/2} + C$$

11. $\int\frac{x^2 - 1}{x^3 - 3x + 2}\,dx$. Let $u = x^3 - 3x + 2$, $du = (3x^2 - 3)dx$.

$$\int\frac{x^2 - 1}{x^3 - 3x + 2}\,dx = \frac{1}{3}\int\frac{du}{u} = \frac{1}{3}\ln|u| + C = \frac{1}{3}\ln|x^3 - 3x + 2| + C$$

12. $\int x^3e^{x^2}\,dx$. Let $u = x^2$, $du = 2x$.

$$\int x^3e^{x^2}\,dx = \frac{1}{2}\int ue^u du = \frac{1}{2}\left[ue^u - \int e^u du\right] = \frac{1}{2}\left[(u - 1)e^u\right] + C$$

$$= \frac{1}{2}\left[(x^2 - 1)e^{x^2}\right] + C$$

13. $\int x(2x + 3)^7dx = \frac{1}{16}x(2x + 3)^8 - \int\frac{1}{16}(2x + 3)^8dx$

$$= \frac{1}{16}x(2x + 3)^8 - \frac{1}{288}(2x + 3)^9 + C$$

14. $\int x^{-2}\ln x\,dx$. Let $f(x) = \ln x$, $g(x) = x^{-2}$. $f'(x) = \frac{1}{x}$,

$G(x) = -x^{-1}$

$$\int x^{-2}\ln x\,dx = -\frac{\ln x}{x} - \int -\frac{1}{x^2}\,dx = -\frac{\ln x}{x} - \frac{1}{x} + C$$

15. $\int(x - 5)^2dx = \frac{1}{3}(x - 5)^3 + C$; $f(8) = 2 = \frac{1}{3}(3)^3 + C = 9 + C$;

$C = -7$; $f(x) = \frac{1}{3}(x - 5)^3 - 7$

16. $\int e^{-5x}dx = -\frac{1}{5}e^{-5x} + C;\ f(0) = 1 = -\frac{1}{5} + C;\ C = \frac{6}{5};$

$f(x) = \frac{6}{5} - \frac{1}{5}e^{-5x}$

17. $\int_1^4 \frac{1}{x^2}\,dx = -\frac{1}{x}\Big|_1^4 = -\frac{1}{4} + 1 = \frac{3}{4}$

18. $\int_3^6 e^{2-(x/3)}dx = -3e^{2-(x/3)}\Big|_3^6 = -3 + 3e = 3(e - 1)$

19. $\int_{-1}^1 x^4e^{x^5-1}dx$: Let $u = x^5 - 1$, $du = 5x^4dx$

$\int_{-1}^1 x^4e^{x^5-1}dx = \int_{-2}^0 \frac{1}{5}e^u du = \frac{1}{5}e^u\Big|_{-2}^0 = \frac{1}{5}(e^0 - e^{-2}) = \frac{1}{5}(1 - e^{-2})$

20. $\int_0^5 \left(\frac{5}{x} - \frac{x}{5}\right)dx = 5\ln|x| - \frac{x^2}{10}\Big|_0^5 = 5(\ln 5) - 5\lim_{x\to 0+} x - 2.5$: undefined

21. $\int_0^1 \frac{e^x - e^{-x}}{e^x + e^{-x}}\,dx = \ln|e^x + e^{-x}|\Big|_0^1 = \ln\left(e + \frac{1}{e}\right) - \ln 2$

22. $\int_2^3 \frac{x - 1}{(x^2 - 2x + 1)^2}\,dx = \int_2^3 \frac{1}{(x - 1)^3}\,dx = -\frac{1}{2}(x - 1)^{-2}\Big|_2^3$

$= -\frac{1}{2}\left[\frac{1}{4} - 1\right] = \frac{3}{8}$

23. $\int_1^9 (1 + \sqrt{x})\,dx = \left(x + \frac{2}{3}x^{3/2}\right)\Big|_1^9 = 9 + 18 - \left(1 + \frac{2}{3}\right)$

$= 26 + \frac{2}{3} = \frac{76}{3}$

24. $\int_1^2 (3x - 2)^{-3}dx = \left(-\frac{1}{6}(3x - 2)^{-2}dx\right)\Big|_1^2 = -\frac{1}{6}\cdot\frac{1}{16} + \frac{1}{6} = \frac{15}{96} = \frac{5}{32}$

25. $16 - x^2 = 10 - x$; $x^2 - x - 6 = 0$; $(x - 3)(x - 2) = 0$ Thus, we should calculate

$$\int_{-2}^{3} (16 - x^2 - (10 - x))dx = \int_{2}^{3} (-x^2 + x + 6)dx$$

$$= \left[-\frac{1}{3}x^3 + \frac{1}{2}x^2 + 6x\right]\Big|_{-2}^{3}$$

$$= -9 + \frac{9}{2} + 18 - \left(\frac{8}{3} + 2 - 12\right)$$

$$= 19 + \frac{9}{2} - \frac{8}{3} = \frac{125}{6}$$

26. $x^3 - 3x + 1 = x + 1$; $x^3 - 4x = 0$; $x(x^2 - 4) = 0$ Thus, the graphs intersect at $x = 0, \pm 2$. On $[-2, 0]$, $y = x^3 - 3x + 1$ lies above $y = x + 1$, and below on $[0, 2]$. Thus we should calculate

$$\int_{-2}^{0} ((x^3 - 3x + 1) - (x + 1))dx + \int_{0}^{2} ((x + 1) - (x^3 - 3x + 1))dx$$

$$= \int_{-2}^{0} (x^3 - 4x)dx + \int_{0}^{2} (-x^2 + 4x)dx$$

$$= \left(\frac{1}{4}x^4 - 2x^2\right)\Big|_{-2}^{0} + \left(-\frac{1}{4}x^4 + 2x^2\right)\Big|_{0}^{2}$$

$$= 0 - (4 - 8) + (-4 + 8) = 8$$

27. $$\int_{2}^{3} \left[\left(5x + \frac{1}{x}\right) - \left(2x + \frac{1}{x}\right)\right] dx = \int_{2}^{3} 3x\,dx = \frac{3}{2}x^2\Big|_{2}^{3} = \frac{3}{2}[9 - 4] = \frac{15}{2}$$

28. $2x^2 + x = x^2 + 2$; $x^2 + x - 2 = 0$; $(x + 2)(x - 1) = 0$ Thus, on the interval $[0, 2]$, the graphs intersect at $x = 1$. On $[0, 1]$, $y = x^2 + 2$ lies above $y = 2x^2 + x$ and below on $[1, 2]$. Thus, we should calculate

$$\int_{0}^{1} (x^2 + 2 - (2x^2 + x))dx + \int_{1}^{2} (2x^2 + x - (x^2 + 2))dx$$

$$= \int_0^1 (-x^2 - x + 2)dx + \int_1^2 (x^2 + x - 2)dx$$

$$= \left(-\frac{1}{3}x^3 - \frac{1}{2}x^2 + 2x\right)\Big|_0^1 + \left(\frac{1}{3}x^3 + \frac{1}{2}x^2 - 2x\right)\Big|_1^2$$

$$= -\frac{1}{3} - \frac{1}{2} + 2 + \frac{8}{3} + 2 - 4 - \left(\frac{1}{3} + \frac{1}{2} - 2\right) = \frac{6}{3} - \frac{2}{2} + 2 = 3$$

29. $\underline{n = 2}$: $\Delta x = \frac{2 - 0}{2} = 1$ $\quad \left[\frac{1}{.5 + 2} + \frac{1}{1.5 + 2}\right](1) = .68571$

exact value:

$$\int_0^2 \frac{1}{x + 2}\,dx = \ln(x + 2)\Big|_0^2 = \ln 4 - \ln 2 = \ln \frac{4}{2} = \ln 2 = .69315$$

30. $\Delta x = \frac{1 - 0}{5} = .2$

$$\left[e^{.1^2} + e^{.3^2} + e^{.5^2} + e^{.7^2} + e^{.9^2}\right](.2) = 1.14537$$

31. $p(400) = \sqrt{25 - .04(400)} = 3$

$$\text{C.S.} = \int_0^{400} (\sqrt{25 - .04x} - 3)dx = \left(\frac{2}{-.12}(25 - .04x)^{3/2} - 3x\right)\Big|_0^{400}$$

$$= \frac{2}{-.12}(27) - 1200 - \frac{2}{-.12}(125)$$

$$= \$433.33$$

32. $\frac{1}{10}\int_0^{10} 3000e^{.06t}dt = \int_0^{10} 300e^{.06t}dt = 5000e^{.06t}\Big|_0^{10}$

$$= 5000(e^{.6} - e^0)$$

$$= \$4110.59$$

33. $\frac{1}{\frac{1}{2} - \frac{1}{3}}\int_{1/3}^{1/2} \frac{1}{x^3}\,dx = 6\int_{1/3}^{1/2} \frac{1}{x^3}\,dx = 6\left[-\frac{1}{2x^2}\Big|_{1/3}^{1/2}\right] = 6\left(-2 + \frac{9}{2}\right) = 15$

34. The sum is approximated by

$$\int_0^1 3e^{-x}dx \quad -3e^{-x}\Big|_0^1 = -3e^{-1} + 3 = 3(1 - e^{-1})$$

35. $C(x) = \int(.04x + 150)dx = .02x^2 + 150x + C$

$C(0) = 500 = 0 + 0 + C = C$. Thus, $C(x) = .02x^2 + 150x + 500$ (dollars).

36. $\int_{10}^{20}(400 - 3x^2)dx = (400x - x^3)\Big|_{10}^{20} = -3000$

Thus, a loss of \$3000 would result.

37. $\int_0^1 \pi(1 - x^2)^2 dx = \int_0^1 \pi(x^4 - 2x^2 + 1)dx = \pi\left[\frac{x^5}{5} - \frac{2}{3}x^3 + x\right]\Big|_0^1$

$= \pi\left[\frac{1}{5} - \frac{2}{3} + 1\right] = \frac{8\pi}{15}$

38. $v(t) = -9.8t + 20$

(a) $\int_0^2(-9.8t + 20)dt = (-4.9t^2 + 20t)\Big|_0^2 = -19.6 + 40$

$= 20.4$ meters

(b)

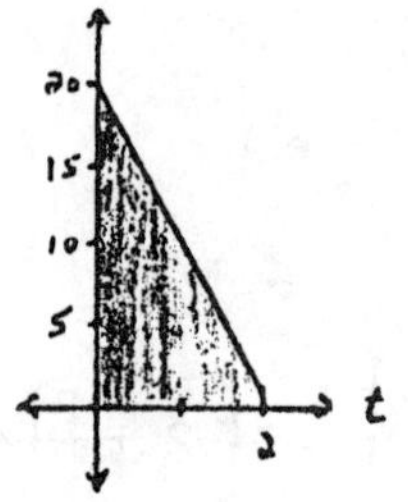

39. $\int_0^1 4500e^{.09(1-t)}dt = -50{,}000e^{.09(1-t)}\Big|_0^1 = \$4708.71$

40. $R = \int(-.02x + 5)dx = -.01x^2 + 5x$: thus the demand equation equation is $p = -.01x + 5$.

41. It is the volume injected during the first four minutes.

42. $\int_0^\infty \frac{e^x}{(1 + e^x)^2}\,dx$. First evaluate the indefinite integral.

Let $u = e^x$, $du = e^x dx$. $\int \frac{e^x}{(1 + e^x)^2}\,dx = \int \frac{du}{u^2} = -u^{-1} + C$

$= -(1 + e^x)^{-1} + C.$

Thus, $\int_0^\infty \frac{e^x}{(1 + e^x)^2}\,dx = \lim_{b\to\infty} \int_0^b \frac{e^x}{(1 + e^x)^2}\,dx$

$= \lim_{b\to\infty} \left[-(1 + e^x)^{-1}\Big|_0^b\right]$

$= \lim_{b\to\infty} \left[\frac{1}{2} - \frac{1}{1 + e^b}\right] = \frac{1}{2}.$

43. $\int_0^\infty e^{6-3x}dx = \lim_{b\to\infty} \int_0^b e^{6-3x}dx.$

Let $u = 6 - 3x$, $du = -3dx$. When $x = 0$, $u = 6$; when $x = b$,

$u = 6 - 3b$.

$\int_0^b e^{6-3x}dx = -\frac{1}{3}\int_6^{6-3b} e^u du = -\frac{1}{3}e^u\Big|_6^{6-3b} = -\frac{1}{3}e^{6-3b} + \frac{1}{3}e^6.$

Thus $\int_0^\infty e^{6-3x}dx = \lim_{b\to\infty} \left[-\frac{1}{3}e^{6-3b} + \frac{e^6}{3}\right] = \frac{e^6}{3}.$

44. $\int_1^{\infty} x^{-2/3}dx = \lim_{b\to\infty} \int_1^b x^{-2/3}dx = \lim_{b\to\infty} \left[3x^{1/3}\Big|_1^b\right] = \lim_{b\to\infty} \left[3b^{1/3} - 3\right].$

As $b \to \infty$, $3b^{1/3}$ increases without bound. Thus, $\int_1^{\infty} x^{-2/3}dx$ diverges.

45. $\int_{-\infty}^{0} \frac{8}{(5 - 2x)^3} dx = \lim_{b\to-\infty} \int_b^0 \frac{8}{(5 - 2x)^3} dx$

$= \lim_{b\to-\infty} \left[2(5 - 2x)^{-2}\Big|_b^0\right]$

$= \lim_{b\to-\infty} \left[\frac{2}{25} - 2(5 - 2b)^{-2}\right] = \frac{2}{25}.$

46. $\int_0^{\infty} x^2e^{-x^3} dx = \lim_{b\to\infty} \int_0^b x^2e^{-x^3} dx = -\frac{1}{3}\left[e^{-b^3} - 1\right].$

Thus $\int_0^{\infty} x^2e^{-x^3} dx = \lim_{b\to\infty} \left[\frac{1}{3} - \frac{e^{-b^3}}{3}\right] = \frac{1}{3}.$

47. $\int_1^{\infty} x^{-1/4}dx = \lim_{b\to\infty} x^{3/4}\Big|_1^b = \lim_{b\to\infty} b^{3/4} - 1 = \infty$

48. $\int_1^{\infty} \frac{1}{x} dx = \lim_{b\to\infty} \ln x\Big|_1^b = \lim_{b\to\infty} \ln b = \infty$, so the integral diverges and the area is unbounded. Ans: no.

49. The probability density function is $f(x) = \frac{1}{72}e^{-(1/72)x}$.

(a) $\Pr(X > 24) = 1 - \Pr(X \le 24) = 1 - \int_0^{24} \frac{1}{72}e^{-(1/72)x}dx$

$$= 1 + e^{-(1/72)x}\Big|_0^{24} = e^{-(1/3)}.$$

(b) $r(t) = \Pr(X > t) = 1 - \Pr(X \le t) = 1 - \int_0^t \frac{1}{72}e^{-(1/72)x}dx$

$$= 1 + e^{-(1/72)x}\Big|_0^{t} = e^{-t/72}.$$

50. I. $e^{A-x} \ge 0$ for all x.

II. $\int_A^\infty e^{A-x}dx = \lim_{b\to\infty}\left[-e^{A-x}\Big|_A^b\right] = \lim_{b\to\infty}\left[1 - e^{A-b}\right] = 1.$ Thus

$f(x) = e^{A-x}$, $x \ge A$ is a density function.

$$F(x) = \int_A^x e^{A-t}dt = -e^{A-t}\Big|_A^x = 1 - e^{A-x}.$$

Exercises 7.1

1. $f(x, y) = x^2 + 8y$; $f(1, 0) = 1^2 + 8(0) = 1$

   $f(0, 1) = 0^2 + 8(1) = 8$; $f(3, 2) = 3^2 + 8(2) = 25$

2. $g(x, y) = 3xe^y$; $g(2,1) = 3(2)e^1 = 6e$; $g(1,0) = 3(1)e^0 = 3$

   $g(0, 0) = 3(0)e^0 = 0$

3. $f(L, K) = 3\sqrt{LK}$; $f(0, 1) = 3\sqrt{0\cdot 1} = 0$; $f(3, 12) = 3\sqrt{3\cdot 12} = 18$

   $f(a, b) = 3\sqrt{a\cdot b} = 3\sqrt{ab}$

4. $f(p, q) = pe^{q/p}$; $f(1, 0) = 1e^{0/1} = 1$; $f(3, 12) = 3e^{12/3} = 3e^4$

   $f(a, b) = ae^{b/a}$

5. $f(x, y, z) = \dfrac{x}{(y - z)}$; $f(2, 3, 4) = \dfrac{2}{3 - 4} = -2$

   $f(7, 46, 44) = \dfrac{7}{46 - 44} = \dfrac{7}{2}$

6. $f(x, y, z) = x^2e^{\sqrt{y^2 + z^2}}$; $f(1, 0, 1) = 1^2e^{\sqrt{0^2 + 1^2}} = e$

   $f(5, 2, 3) = 5^2e^{\sqrt{2^2 + 3^2}} = 25e^{\sqrt{13}}$

7. $f(x, y) = xy$; $f(2, 3 + k) = 2(3 + k) = 6 + 2k$

   $f(2, 3) = 2(3) = 6$; $f(2 + k, 3) - f(2, 3) = 6 + 3k - 6 = 3k$

8. $f(x, y) = xy$; $f(2, 3 + k) = 2(3 + k) = 6 + 2k$

   $f(2, 3) = 2(3) = 6$; $f(2, 3 + k) - f(2, 3) = 6 + 2k - 6 = 2k$

9. $f(x, y) = \dfrac{x^2 + 3xy + 3y^2}{x + y}$

   $$f(2a, 2b) = \frac{(2a)^2 + 3(2a)(2b) + 3(2b)^2}{2a + 2b} = \frac{4a^2 + 12ab + 12b^2}{2a + 2b}$$

   $$= \frac{2(a^2 + 3ab + 3b^2)}{(a + b)}$$

   $2f(a, b) = 2\dfrac{(a^2 + 3ab + 3b^2)}{(a + b)}$ Therefore, $f(2a, 2b)$ is equal to $2f(a, b)$.

10. $f(x, y) = 75x^A y^{1-A}$, $0 < A < 1$

$f(2a, 2b) = 75(2a)^A(2b)^{1-A} = 75(2^A a^A)(2^{1-A}b^{1-A})$

$= 75(2^{A+(1-A)}a^A b^{1-A}) = 150a^A b^{1-A}$

$2f(a, b) = 2(75a^A b^{1-A}) = 150a^A b^{1-A}$ Therefore, $f(2a, 2b)$ is equal to $2f(a, b)$.

11. $P(A, t) = Ae^{-.05t}$; $P(100, 13.8) = 100e^{-.05(13.8)} = 100e^{-.69}$

~ 50.16

\$50 invested at 5% continuously compounded interest will yield \$100 in 13.8 years.

12. $C(x, y)$ is the cost of utilizing x units of labor and y units of capital. $C(x, y) = 100x + 200y$

13. a) $v = 200000$, $x = 5000$, $r = 2.5$;

$T = r(.4v - x)/100 = 2.5(.4(200000) - 5000)/100$

$T = \$1875$

b) If $v = 200000$, $x = 5000$, $r = 3$;

$T = r(.4v - x)/100 = 3(.4(200000) - 5000)/100$

$T = \$2250$

The tax due also increases by 20% (1/5) since

$1875 + (1/5)(1875) = \$2250$

14. a) $v = 100000$, $x = 5000$, $r = 2.2$;

$T = r(.4v - x)/100 = 2.2(.4(100000) - 5000)/100$

$T = \$770$

b) If $v = 120000$, $x = 5000$, $r = 2.2$;

$T = r(.4v - x)/100 = 2.2(.4(1200000) - 5000)/100$

$T = \$946$

20% of \$770 is \$154, so tax due does not increase by 20%.

15. $C = 2x + y$, so $y = 2x + C$

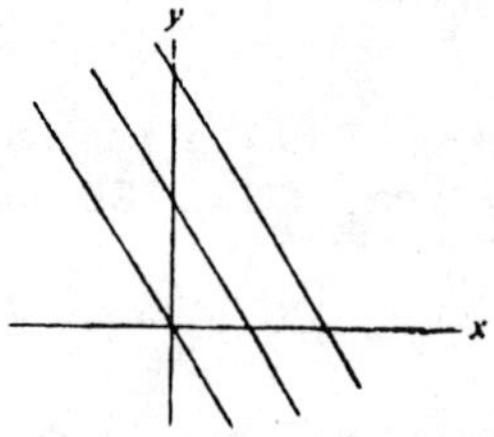

16. $C = -x^2 + y$, so $y = x^2 + C$

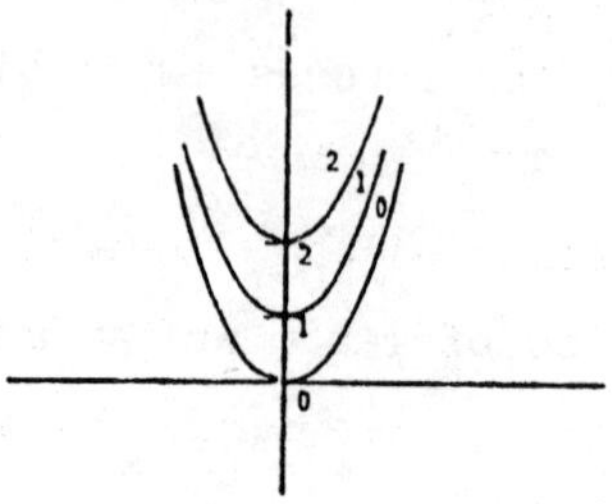

17. $C = x + y$, $y = x + C$

But $3 = 5 + C \Rightarrow C = -2$

so $y = x - 2$

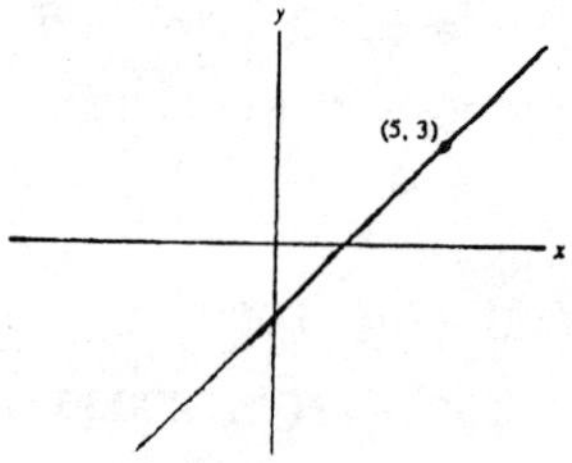

18. $C = xy$, $y = C/x$

But $\frac{1}{2} = \frac{C}{2} \Rightarrow C = 1$

thus $y = 1/x$

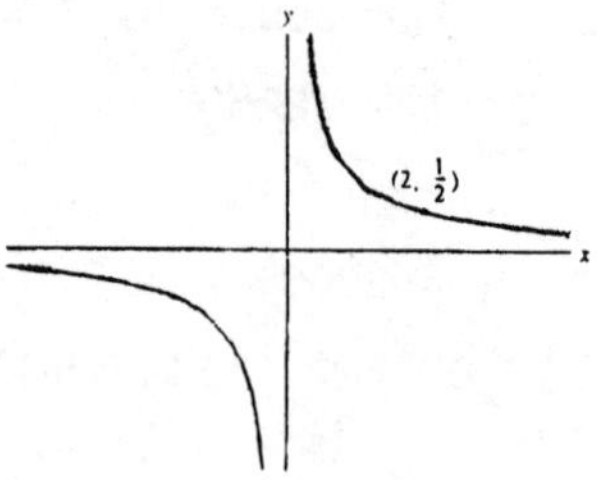

19. $y = 3x - 4$, $y - 3x = -4$: so $y - 3x = C$, some constant

$f(x,y) = y - 3x$.

20. $y = 3/x^2$, $yx^2 = 3$. Thus $yx^2 = C$, some constant $\Rightarrow f(x,y) = x^2y$

21. They correspond to the points having the same altitude above sea level.

22. $C(x, y) = 100x + 200y$ is the cost of using x units of labor and y units of capital. If $C(x, y) = 600$, then $100x + 200y = 600$; $y = 3 - \frac{1}{2}x$. If $C(x, y) = 800$, then $y = 4 - \frac{1}{2}x$. If $C(x, y) = 1000$, then $y = 5 - \frac{1}{2}x$.

Points on the same level curve correspond to production amounts that have the same total cost.

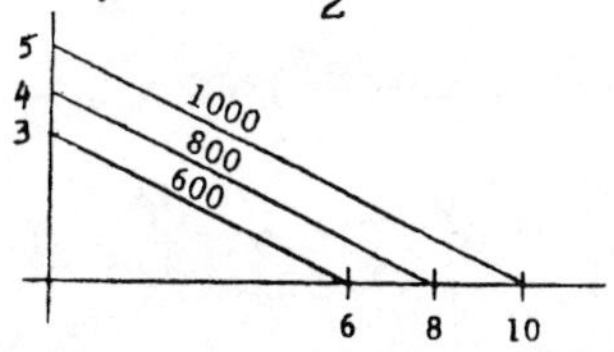

23,24,25,26. 25 is matched with (c). Imagine slicing "near the top" of the "4 humps"; we get a cross-section of 4 circular-like figures As we move further down the larger these figures become. Similarly 23) is matched with (d)
24) is matched with (b)
26) is matched with (a)

Exercises 7.2

1. $f(x, y) = 5xy$; $\frac{\partial f}{\partial x} = 5(1)y = 5y$; $\frac{\partial f}{\partial y} = 5(1)x = 5x$

2. $f(x, y) = 3x^2 + 2y + 1$; $\frac{\partial f}{\partial x} = 3(2x) = 6x$; $\frac{\partial f}{\partial y} = 2(1) = 2$

3. $f(x, y) = 2x^2e^y$; $\frac{\partial f}{\partial x} = 2(2x)e^y = 4xe^y$; $\frac{\partial f}{\partial y} = 2x^2e^y(1) = 2x^2e^y$

4. $f(x, y) = x + e^{xy}$; $\frac{\partial f}{\partial x} = 1 + e^{xy}(y) = 1 + ye^{xy}$

$\frac{\partial f}{\partial y} = e^{xy}(x) = xe^{xy}$

5. $f(x, y) = \frac{y^2}{x} = y^2x^{-1}$; $\frac{\partial f}{\partial x} = y^2(-x^{-2}) = -\frac{y^2}{x^2}$; $\frac{\partial f}{\partial y} = 2y(x^{-1}) = \frac{2y}{x}$

6. $f(x, y) = \frac{x}{1 + e^y}$; $\frac{\partial f}{\partial x} = \frac{(1 + e^y) - 0}{(1 + e^y)^2} = \frac{1}{1 + e^y}$

$\frac{\partial f}{\partial y} = \frac{0 - xe^y}{(1 + e^y)^2} = -\frac{xe^y}{(1 + e^y)^2}$

7. $f(x, y) = (2x - y + 5)^2$; $\dfrac{\partial f}{\partial x} = 2(2xy - y + 5)(2)$

$= 4(2x - y + 5$

$\dfrac{\partial f}{\partial y} = 2(2x - y + 5)(-1) = -2(2x - y + 5)$

8. $f(x, y) = (9x^2y + 3x)^{12}$; $\dfrac{\partial f}{\partial x} = 12(9x^2y + 3x)^{11}(18xy + 3)$

$= 12(18xy + 3)(9x^2y + 3x)^{11}$

$\dfrac{\partial f}{\partial y} = 12(9x^2y + 3x)^{11}(9x^2) = 12(9x^2)(9x^2y + 3x)^{11}$

9. $f(x, y) = x^2e^{3x}\ln y$; $\dfrac{\partial f}{\partial x} = (2xe^{3x} + 3e^{3x}x^2)\ln y$

$= (2xe^{3x} + 3x^2e^{3x})\ln y$

$\dfrac{\partial f}{\partial y} = x^2e^{3x}\left(\dfrac{1}{y}\right) = \dfrac{x^2e^{3x}}{y}$

10. $f(x, y) = (x - \ln y)e^{xy}$; $\dfrac{\partial f}{\partial x} = (1)e^{xy} + (y)e^{xy}(x - \ln y)$

$= e^{xy} + ye^{xy}(x - \ln y)$

$= e^{xy}(1 + xy - y \ln y)$

$\dfrac{\partial f}{\partial y} = -\dfrac{1}{y}(e^{xy}) + (x - \ln y)(x)e^{xy} = -\dfrac{1}{y}e^{xy} + x^2e^{xy} - xe^{xy}\ln y$

$= e^{xy}\left(x^2 - x \ln y - \dfrac{1}{y}\right)$

11. $f(x, y) = \dfrac{x - y}{x + y}$; $\dfrac{\partial f}{\partial x} = \dfrac{1(x + y) - (x - y)(1)}{(x + y)^2} = \dfrac{2y}{(x + y)^2}$

$\dfrac{\partial f}{\partial y} = \dfrac{(-1)(x + y) - (x - y)(1)}{(x + y)^2} = -\dfrac{2x}{(x + y)^2}$

12. $f(x, y) = \dfrac{2xy}{e^x}$; $\dfrac{\partial f}{\partial x} = \dfrac{2y(e^x) - (2xy)(1)e^x}{(e^x)^2} = \dfrac{2ye^x - 2xye^x}{(e^x)^2}$

$= \dfrac{2y(1 - x)}{e^x}$

$\dfrac{\partial f}{\partial y} = \dfrac{2x(e^x) - (2xy)(0)}{(e^x)^2} = \dfrac{2x}{e^x}$

13. $f(L, K) = 3\sqrt{LK}$; $\dfrac{\partial f}{\partial L} = 3\left(\dfrac{1}{2}\right)(KL)^{-1/2}(K) = \dfrac{3\sqrt{K}}{2\sqrt{L}}$

14. $f(p, q) = e^{q/p} = e^{qp^{-1}}$; $\dfrac{\partial f}{\partial p} = e^{qp^{-1}}(-qp^{-2}) = \dfrac{-qe^{q/p}}{p^2}$

$$\frac{\partial f}{\partial q} = e^{qp^{-1}}(p^{-1}) = \frac{e^{q/p}}{p}$$

15. $f(x, y, z) = \dfrac{(1 + x^2y)}{y} = z^{-1} + x^2yz^{-1}$

$$\frac{\partial f}{\partial x} = 0 + 2xyz^{-1} = \frac{2xy}{z};\ \frac{\partial f}{\partial y} = 0 + x^2z^{-1} = \frac{x^2}{z}$$

$$\frac{\partial f}{\partial z} = -z^2 + (-z^{-2}x^2y) = -\frac{1}{z^2} - \frac{x^2y}{zy^2} = -\frac{(1 + x^2y)}{z^2}$$

16. $f(x, y, z) = x^2y + 3yz - z^2;\ \dfrac{\partial f}{\partial x} = 2xy + 0 - 0 = 2xy$

$$\frac{\partial f}{\partial y} = x^2 + 3z - 0 = x^2 + 3z;\ \frac{\partial f}{\partial z} = 0 + 3y - 2z$$

17. $f(x, y, z) = xze^{yz};\ \dfrac{\partial f}{\partial x} = (1)ze^{yz} = ze^{yz}$

$$\frac{\partial f}{\partial y} = e^{yz}(z)xz = xz^2e^{yz}$$

$$\frac{\partial f}{\partial z} = (1)xe^{yz} + e^{yz}(y)xz = xe^{yz}(1 + yz)$$

18. $f(x, y, z) = ze^{z/xy} = ze^{zx^{-1}y^{-1}}$

$$\frac{\partial f}{\partial x} = ze^{zx^{-1}y^{-1}}(-zy^{-1}x^{-2}) = -\frac{z^2e^{z/xy}}{x^2y}$$

$$\frac{\partial f}{\partial y} = ze^{zx^{-1}y^{-1}}(-zx^{-1}y^{-2}) = -\frac{z^2e^{z/xy}}{xy^2}$$

$$\frac{\partial f}{\partial z} = (1)e^{zx^{-1}y^{-1}} + ze^{zx^{-1}y^{-1}}(x^{-1}y^{-1}) = e^{z/xy} + \frac{ze^{z/xy}}{xy}$$

$$= e^{z/xy}\left[1 + \frac{z}{xy}\right]$$

19. $f(x, y) = x^2 + 2xy + y^2 + 3x + 5y$

$$\frac{\partial f}{\partial x} = 2x + 2y(1) + 0 + 3 + 0 = 2x + 2y + 3$$

$$\frac{\partial f}{\partial x}(2, -3) = 2(2) + 2(-3) + 3 = 1$$

$$\frac{\partial f}{\partial y} = 0 + 2x(1) + 2y + 0 + 5 = 2x + 2y + 5$$

$$\frac{\partial f}{\partial y}(2, -3) = 2(2) + 2(-3) + 5 = 3$$

20. $f(x, y) = xye^{2x-y};\ \dfrac{\partial f}{\partial x} = y(1)(e^{2x-y}) + xye^{2x-y}(2)$

$$= ye^{2x-y} + 2xye^{2x-y}$$

$$= ye^{2x-y}(1 + 2x)$$

$\frac{\partial f}{\partial x}(1, 2) = 2e^{2(1)-2}(1 + 2(1)) = 2e^0(3) = 6$

$\frac{\partial f}{\partial y} = x(1)e^{2x-y} + xye^{2x-y}(-1) = xe^{2x-y} - xye^{2x-y}$

$= xe^{2x-y}(1 - y)$

$\frac{\partial f}{\partial y}(1, 2) = (1)e^{2(1)-2}(1 - 2) = e^0(-1) = -1$

21. $f(x, y, z) = xy^2z + 5$; $\frac{\partial f}{\partial y} = 2xyz$; $\frac{\partial f}{\partial y}(2, -1, 3) = 2(2)(-1)(3)$

$= -12$

22. $f(x, y, z) = \frac{x}{y - z} = x(y - z)^{-1}$; $\frac{\partial f}{\partial y} = -x(y - z)^{-2}(1)$

$= -\frac{x}{(y - z)^2}$

$\frac{\partial f}{\partial y}(2, -1, 3) = -\frac{2}{(-1 - 3)^2} = -\frac{1}{8}$

23. $f(x, y) = x^3y + 2xy^2$; $\frac{\partial f}{\partial x} = 3x^2y + 2y^2$; $\frac{\partial^2 f}{\partial x^2} = 6xy$

$\frac{\partial^2 f}{\partial y \partial x} = 3x^2 + 4y$; $\frac{\partial f}{\partial y} = x^3 + 4xy$; $\frac{\partial^2 f}{\partial y^2} = 4x$; $\frac{\partial^2 f}{\partial x \partial y} = 3x^2 + 4y$

24. $f(x,y) = xe^y + x^4y + y^3$; $\frac{\partial f}{\partial x} = e^y + 4x^3y + 0 = e^y + 4x^3y$

$\frac{\partial^2 f}{\partial x^2} = 0 + 12x^2y = 12x^2y$; $\frac{\partial^2 f}{\partial y \partial x} = e^y(1) + 4x^3 = e^y + 4x^3$

$\frac{\partial f}{\partial y} = xe^y(1) + x^4 + 3y^2 = xe^y + x^4 + 3y^2$

$\frac{\partial^2 f}{\partial y^2} = xe^y(1) + 0 + 6y = xe^y + 6y$

$\frac{\partial^2 f}{\partial x \partial y} = e^y + 4x^3 + 0 = e^y + 4x^3$

25. $f(x, y) = 200\sqrt{6x^2 + y^2}$

(a) $\frac{\partial f}{\partial x}$ is the marginal productivity of labor.

$\frac{\partial f}{\partial x} = 200\left(\frac{1}{2}\right)(6x^2 + y^2)^{-1/2}(12x) = 1200x(6x^2 + y^2)^{-1/2}$

$= \frac{1200x}{\sqrt{6x^2 + y^2}}$

When x = 100 and y = 5, $\frac{\partial f}{\partial x}(10, 5) = \frac{1200(10)}{\sqrt{6(10)^2 + 5^2}} = 480.$

$\frac{\partial f}{\partial y}$ is the marginal productivity of capital.

$$\frac{\partial f}{\partial y} = 200\left(\frac{1}{2}\right)(6x^2 + y^2)^{-1/2}(2y) = 200y(6x^2 + y^2)^{-1/2}$$

$$= \frac{200y}{\sqrt{6x^2 + y^2}}$$

When x = 10 and y = 5, $\frac{\partial f}{\partial y}(10, 5) = \frac{200(5)}{\sqrt{6(10)^2 + 5^2}} = 40.$

(b) If capital is fixed at 5 units and labor is decreased by one unit from 10 to 9, then the quantity of goods produced will decrease by approximately 480 units. So if capital is fixed at 5 units and labor decreased by 1/2 unit from 10 to 9.5 units, the number of goods produced will decrease by approximately 240 units.

26. $f(x, y) = 300x^{2/3}y^{1/3}$ is the productivity of a country, where x and y are the amounts of labor and capital.

(a) $\frac{\partial f}{\partial x}$ is the marginal productivity of labor.

$\frac{\partial f}{\partial x} = 300\left(\frac{2}{3}\right)x^{-1/3}y^{1/3} = \frac{200\sqrt[3]{y}}{\sqrt[3]{x}}.$ When x = 125 and y = 64,

$$\frac{\partial f}{\partial x}(125, 64) = \frac{200\sqrt[3]{64}}{\sqrt[3]{125}} = 160.$$

$\frac{\partial f}{\partial y}$ is the marginal productivity of capital.

$\frac{\partial f}{\partial y} = 300^{2/3}\left(\frac{1}{3}\right)y^{-2/3} = \frac{100\sqrt[3]{x^2}}{\sqrt[3]{y^2}}.$ When x = 125 and y = 64,

$$\frac{\partial f}{\partial y}(125, 64) = \frac{100\sqrt[3]{125^2}}{\sqrt[3]{64^2}} = 156.25.$$

(b) If the labor is fixed and capital is decreased by one unit, then the productivity will decrease by 156.25. Thus, if the labor is fixed and capital is decreased by two units, the productivity will decrease by 312.5.

27. As the price of a bus ride increases fewer people will ride the bus if the train fare remains constant. An increase in train ticket prices, coupled with constant bus fare should cause more people to ride the bus.

28. $g(p_1, p_2)$ is the number of people who will take the train when when p_1 is the price of the bus ride and p_2 is the price of the price of the train ride.

$\frac{\partial g}{\partial p_1}$ is positive (an increase in bus fare would mean more people would take the train).

$\frac{\partial g}{\partial p_2}$ is negative (an increase in the train fare would mean less people taking the train).

29. If the price of the average video tape increases and the average price of the VCR remains constant, people will purchase fewer VCR's. An increase in the average price of the VCR's, coupled with constant video tape prices, should cause a decline in the number of video tapes purchased.

30. When the gasoline price is constant, an increase in the price of the car will decrease the demand for the car. If the price of the car is constant and the price of the gasoline increases the demand for the car will decrease.

31. $V = .08\left(\frac{T}{P}\right)$; $\frac{\partial V}{\partial T} = \frac{.08}{P}$; When P = 20, T = 300, $\frac{\partial V}{\partial T} = \frac{.08}{20} = .004$. At this level, increasing the temperature by one unit will increase the volume by approximately .004 units.
$\frac{\partial V}{\partial P} = \frac{-.08T}{P^2}$. When P = 20, T = 300, $\frac{\partial V}{\partial P} = \frac{-.08(300)}{400} = -.06$. At this level, increasing the pressure by one unit will decrease the volume by approximately .06 units.

32. Assuming m, p, r, s > 0, all first partial derivatives are

positive except $\frac{\partial f}{\partial p} = -.727m^{.136}r^{.914}s^{.816}p^{-1.727} < 0$. Thus increases in aggregate income, retail prices of the other goods or the strength of the beer (holding the other quantities constant) should cause an increase in the amount of beer consumed; while an increase in the price of beer itself should cause the amount consumed to decrease.

33. Assuming m, p, r > 0, $\frac{\partial f}{\partial m} > 0$, $\frac{\partial f}{\partial r} > 0$ and $\frac{\partial f}{\partial p} = -.543m^{.595}r^{.922}p^{-1.543} < 0$. Thus increases in aggregate income or retail prices of other goods (holding the other quantities constant) should cause an increase in the amount of food consumed; while an increase in the price of food itself should cause the amount consumed to decrease.

34. $f(x, y) = 60x^{3/4}y^{1/4}$; $\frac{\partial f}{\partial x} = 45y^{1/4}x^{-1/4}$; $\frac{\partial f}{\partial y} = 15x^{3/4}y^{-3/4}$.

$$f(a, b) = 60a^{3/4}b^{1/4} = a\left[45b^{1/4}a^{-1/4}\right] + b\left[15a^{3/4}b^{-3/4}\right]$$

$$= a\left[\frac{\partial f}{\partial x}(a, b)\right] + b\left[\frac{\partial f}{\partial y}(a, b)\right].$$

35. $f(x, y) = 60x^{3/4}y^{1/4}$. $\frac{\partial f}{\partial x} = 45y^{1/4}x^{-1/4}$; $\frac{\partial^2 f}{\partial x^2} = -\frac{45}{4}y^{1/4}x^{-5/4} < 0$ for all x, y > 0. The fact that $\frac{\partial^2 f}{\partial x^2} < 0$ confirms the <u>law of diminishing returns</u>, which says that as additional units of a given productive input are added (holding other factors constant) production increases at a decreasing rate.

36. $f(x, y) = 60x^{3/4}y^{1/4}$; $\frac{\partial f}{\partial y} = 15x^{3/4}y^{-3/4}$; $\frac{\partial^2 f}{\partial y^2} = -\frac{45}{4}x^{3/4}y^{-7/4} < 0$ for all x, y > 0. The fact that $\frac{\partial^2 f}{\partial x^2} < 0$ confirms the law of diminishing returns, which says that as additional units of a given productive input are added (holding other factors constant) production increases at a decreasing rate.

37. $f(x, y) = 3x^2 + 2xy + 5y$

$$f(1 + h, 4) - f(1, 4) = \left[3(1 + h)^2 + 2(1 + h)(4) + 5(4)\right] - \left[3(1)^2 + 2(1)(4) + 5(4)\right] = 3h^2 + 14h$$

38. $A = (.007)W^{.425}H^{.725}$; $\frac{\partial A}{\partial W} = (.002975)W^{-.575}H^{.725}$. When $W = 54$, $H = 165$, $\frac{\partial A}{\partial W} = .002975(54)^{-.575}(165)^{.725} \sim .01216$. If a person weighing 54 kg who is 165 cm tall increases his weight by 1 kg, the surface area of his body will increase by about .0125 cm^2. $\frac{\partial A}{\partial H} = (.005075)W^{.425}H^{-.275}$. When $W = 54$, $H = 165$, $\frac{\partial A}{\partial H} \sim .0068512$. If a person as above increases his height by 1 cm, his body surface will increase by approximately .0068512 m^2.

Exercises 7.3

1. $f(x, y) = x^2 - 3y^2 + 4x + 6y + 8$; $\frac{\partial f}{\partial x} = 2x + 4$; $\frac{\partial f}{\partial y} = -6y + 6$.

$\left.\begin{matrix} 2x + 4 = 0 \\ -6y + 6 = 0 \end{matrix}\right\}$ $\begin{matrix} x = -2 \\ y = 1 \end{matrix}$ The only possible extreme point is $(-2, 1)$.

2. $f(x, y) = \frac{1}{2}x^2 + y^2 - 3x + 2y - 5$, $\frac{\partial f}{\partial x} = x - 3$; $\frac{\partial f}{\partial y} = 2y + 2$.

$\left.\begin{matrix} x - 3 = 0 \\ 2y + 2 = 0 \end{matrix}\right\}$ $\begin{matrix} x = 3 \\ y = -1 \end{matrix}$ The only possible extreme point is $(3, -1)$.

3. $f(x, y) = x^2 - 5xy + 6y^2 + 3x - 2y + 4$; $\frac{\partial f}{\partial x} = 2x - 5y + 3$

$\frac{\partial f}{\partial y} = -5x + 12y - 2$

$\left.\begin{matrix} 2x - 5y + 3 = 0 \\ -5x + 12y - 2 = 0 \end{matrix}\right\}$ $\begin{matrix} x = 26 \\ y = 11 \end{matrix}$ The only possible extreme point is $(26, 11)$.

4. $f(x, y) = -3x^2 + 7xy - 4y^2 + x + y$; $\frac{\partial f}{\partial x} = -6x + 7y + 1$

$\frac{\partial f}{\partial y} = 7x - 8y + 1$

$\left.\begin{array}{r} -6x + 7y + 1 = 0 \\ 7x - 8y + 1 = 0 \end{array}\right\}$ $\begin{array}{l} x = -15 \\ y = -13 \end{array}$ The only possible extreme point is (-15, -13).

5. $f(x, y) = x^3 + y^2 - 3x + 6y;\ \frac{\partial f}{\partial x} = 3x^2 - 3;\ \frac{\partial f}{\partial y} = 2y + 6$

$\left.\begin{array}{r} x^2 - 3 = 0 \\ 2y + 6 = 0 \end{array}\right\}$ $\begin{array}{l} x = \pm 1 \\ y = -3 \end{array}$ The only possible extreme points are (1, -3) and (-1, -3).

6. $f(x, y) = x^2 - y^3 + 5x + 12y + 1;\ \frac{\partial f}{\partial x} = 2x + 5;\ \frac{\partial f}{\partial y} = -3y^2 + 12$

$\left.\begin{array}{r} 2x + 5 = 0 \\ -3y^2 + 12 = 0 \end{array}\right\}$ $\begin{array}{l} x = -\frac{5}{2} \\ y = \pm 2 \end{array}$ The only possible extreme points are $\left(-\frac{5}{2}, 2\right)$ and $\left(-\frac{5}{2}, -2\right)$.

7. $f(x, y) = \frac{1}{3}x^3 - 2y^3 - 5x + 6y - 5;\ \frac{\partial f}{\partial x} = x^2 - 5;\ \frac{\partial f}{\partial y} = -6y^2 + 6$

$\left.\begin{array}{r} x^2 - 5 = 0 \\ -6y2 + 16 = 0 \end{array}\right\}$ $\begin{array}{l} x = \pm\sqrt{5} \\ y = \pm 1 \end{array}$ There are four possible extreme points: $(\pm\sqrt{5}, \pm 1)$.

8. $f(x, y) = x^4 - 8xy + 2y^2 - 3;\ \frac{\partial f}{\partial x} = 4x^3 - 8y;\ \frac{\partial f}{\partial y} = -8x + 4y$

$\left.\begin{array}{r} 4x^3 - 8y = 0 \\ -8x + 4y = 0 \end{array}\right\}$ $\left.\begin{array}{l} y = \frac{1}{2}x^3 \\ -8x + 2x^3 = 0 \end{array}\right\}$ Solutions: (0, 0), (2, 4) and (-2, -4).

9. $f(x, y) = 2x + 3y + 9 - y^2;\ \frac{\partial f}{\partial x} = 2 - 2x - y;\ \frac{\partial f}{\partial y} = 3 - x - 2y$

$\left.\begin{array}{r} 2 - 2x - y = 0 \\ 3 - x - 2y = 0 \end{array}\right\}$ $\left.\begin{array}{l} y = 2 - 2x \\ x = \frac{1}{3} \end{array}\right\}$ $\begin{array}{l} x = \frac{1}{3} \\ y = \frac{4}{3} \end{array}$

Thus $\left(\frac{1}{3}, \frac{4}{3}\right)$ is the only point at which f(x, y) can have a maximum, so the maximum value must occur at this point.

10. $f(x, y) = \frac{1}{2}x^2 + 2xy + 3y^2 - x + 2y;\ \frac{\partial f}{\partial x} = x + 2y - 1$

$\frac{\partial f}{\partial y} = 2x + 6y + 2$

$\left.\begin{array}{r} x + 2y - 1 = 0 \\ 2x + 6y + 2 = 0 \end{array}\right\}$ $\left.\begin{array}{l} x = 1 - 2y \\ y = -2 \end{array}\right\}$ $\begin{array}{l} x = 5 \\ y = -2 \end{array}$

Thus (5, -2) is the only point at which f(x, y) can have a minimum, so the minimum value must occur at this point.

11. $f(x, y) = 3x^2 - 6xy + y^3 - 9y;\ \frac{\partial f}{\partial x} = 6x - 6y;$

$\frac{\partial f}{\partial y} = -6x + 3y^2 - 9;\ \frac{\partial^2 f}{\partial x^2} = 6;\ \frac{\partial^2 f}{\partial y^2} = 6y;\ \frac{\partial^2 f}{\partial x \partial y} = -6$

$D(x,y) = 36y - 36$; $D(3, 3) = 36\cdot 3 - 36 > 0$, $\frac{\partial^2 f}{\partial x^2}(3,3) > 0$

so (3,3) is a relative minimum of f(x,y).

$D(-1, -1) < 0$, $\frac{\partial^2 f}{\partial x^2}(-1,-1) > 0$, so f(x,y) has neither a

relative maximum or a relative minimum at (-1,-1).

12. $f(x, y) = 6xy^2 - 2x^3 - 3y^4$; $\frac{\partial f}{\partial x} = 6y^2 - 6x^2$;

$\frac{\partial f}{\partial y} = 12xy - 12y^3$; $\frac{\partial^2 f}{\partial x^2} = 12x$; $\frac{\partial^2 f}{\partial y^2} = 12x - 36y^2$; $\frac{\partial^2 f}{\partial x \partial y} = 12y$

$D(x,y) = -144x^2 + 432xy^2 - 144y^2$;

$D(0, 0) = 0$, so the test is inconclusive.

$D(1, 1) > 0$, $\frac{\partial^2 f}{\partial x^2}(1,1) < 0$, so f(x,y) has a relative maximum

at (1,1).

$D(1, -1) > 0$, $\frac{\partial^2 f}{\partial x^2}(1,-1) < 0$, so f(x,y) has a relative maximum

at (1,-1).

13. $f(x, y) = 2x^2 - x^4 - y^2$; $\frac{\partial f}{\partial x} = 4x - 4x^3$;

$\frac{\partial f}{\partial y} = -2y$; $\frac{\partial^2 f}{\partial x^2} = 4 - 12x^2$; $\frac{\partial^2 f}{\partial y^2} = -2$; $\frac{\partial^2 f}{\partial x \partial y} = 0$

$D(x,y) = -2(4 - 12x^2)$

$D(-1, 0) > 0$, $\frac{\partial^2 f}{\partial x^2}(-1,0) < 0$, so f(x,y) has a relative maximum

at (-1,0).

$D(0,0) < 0$, so f(x,y) has neither a relative maximum or a

relative minimum at (0,0).

$D(1,0) > 0$, $\frac{\partial^2 f}{\partial x^2}(1,0) < 0$, so f(x,y) has a relative maximum

at (1,0).

14. $f(x, y) = x^4 - 4xy + y^4$; $\frac{\partial f}{\partial x} = 4x^3 - 4y$;

$\frac{\partial f}{\partial y} = -4x + 4y^3$; $\frac{\partial^2 f}{\partial x^2} = 12x^2$; $\frac{\partial^2 f}{\partial y^2} = 12y^2$; $\frac{\partial^2 f}{\partial x \partial y} = -4$

$D(x,y) = -144x^2y^2 - 16$

$D(0, 0) < 0$, so $f(x,y)$ has neither a relative maximum or a relative minimum at $(0,0)$.

$D(1, 1) > 0$, $\frac{\partial^2 f}{\partial x^2}(1,1) > 0$, so $f(x,y)$ has a relative minimum at $(1,1)$.

$D(-1,-1) > 0$, $\frac{\partial^2 f}{\partial x^2}(-1,-1) > 0$, so $f(x,y)$ has a relative minimum at $(1,-1)$.

15. $f(x, y) = ye^x - 3x - y + 5$; $\frac{\partial f}{\partial x} = ye^x - 3$

$\frac{\partial f}{\partial y} = e^x - 1$; $\frac{\partial^2 f}{\partial x^2} = ye^x$; $\frac{\partial^2 f}{\partial y^2} = 0$; $\frac{\partial^2 f}{\partial x \partial y} = e^x$

$D(x,y) = -e^{2x}$, $D(0,3) < 0$, thus $f(x,y)$ has neither a maximum or a minimum at $(0,3)$.

16. $f(x, y) = \frac{1}{x} + \frac{1}{y} + xy$; $\frac{\partial f}{\partial x} = \frac{-1}{x^2}$; $\frac{\partial f}{\partial y} = \frac{-1}{y^2}$

$\frac{\partial^2 f}{\partial x^2} = \frac{2}{x^3}$; $\frac{\partial^2 f}{\partial y^2} = \frac{2}{y^3}$; $\frac{\partial^2 f}{\partial x \partial y} = 1$

$D(x,y) = \frac{4}{x^3y^3} - 1$; $D(1,1) > 0$, $\frac{\partial^2 f}{\partial x^2}(1,1) > 0$, so $f(x,y)$ has a a relative at $(1,1)$.

17. $f(x, y) = x^2 - 2xy + 4y^2$; $\frac{\partial f}{\partial x} = 2x - 2y$; $\frac{\partial f}{\partial y} = -2x + 8y$

$\frac{\partial^2 f}{\partial x^2} = 2$; $\frac{\partial^2 f}{\partial y^2} = 8$; $\frac{\partial^2 f}{\partial x \partial y} = -2$

$$\left.\begin{aligned} 2x - 2y &= 0 \\ -2x + 8y &= 0 \end{aligned}\right\} \quad \begin{aligned} x &= 0 \\ y &= 0 \end{aligned}$$

$D(0, 0) = 2 \cdot 8 - (-2)^2 > 0$, $\frac{\partial^2 f}{\partial x^2}(0, 0) > 0$, so $f(x, y)$ has a relative maximum at (0, 0).

18. $f(x, y) = 2x^2 + 3xy + 5y^2$; $\frac{\partial f}{\partial x} = 4x + 3y$; $\frac{\partial f}{\partial y} = 3x + 10y$

$\frac{\partial^2 f}{\partial x^2} = 4$; $\frac{\partial^2 f}{\partial y^2} = 10$; $\frac{\partial^2 f}{\partial x \partial y} = 3$.

$$\left.\begin{aligned} 4x + 3y &= 0 \\ 3x + 10y &= 0 \end{aligned}\right\} \begin{aligned} x &= 0 \\ y &= 0 \end{aligned}$$

$D(0,0) = 4 \cdot 10 - 3^2 > 0$, $\frac{\partial^2 f}{\partial x^2}(0, 0) > 0$ so $f(x, y)$ has a relative minimum at (0, 0).

19. $f(x, y) = -2x^2 + 2xy - y^2 + 4x - 6y + 5$; $\frac{\partial f}{\partial x} = -4x + 2y + 4$

$\frac{\partial f}{\partial y} = 2x - 2y - 6$; $\frac{\partial^2 f}{\partial x^2} = -4$; $\frac{\partial^2 f}{\partial y^2} = -2$; $\frac{\partial^2 f}{\partial x \partial y} = 2$.

$$\left.\begin{aligned} -4x + 2y + 4 &= 0 \\ 2x - 2y - 6 &= 0 \end{aligned}\right\} \begin{aligned} x &= -1 \\ y &= -4 \end{aligned}$$

$D(-1, -4) = (-4)(-2) - 2^2 > 0$, $\frac{\partial^2 f}{\partial x^2}(-1, -4) < 0$, so $f(x, y)$ has a relative maximum at (-1, -4).

20. $f(x, y) = -x^2 - 8xy - y^2$; $\frac{\partial f}{\partial x} = -2x - 8y$; $\frac{\partial f}{\partial y} = -8x - 2y$

$\frac{\partial^2 f}{\partial x^2} = -2$; $\frac{\partial^2 f}{\partial y^2} = -2$; $\frac{\partial^2 f}{\partial x \partial y} = -8$

$$\left.\begin{aligned} -2x - 8y &= 0 \\ -8x - 2y &= 0 \end{aligned}\right\} \begin{aligned} x &= 0 \\ y &= 0 \end{aligned}$$

$D(0, 0) = (-2)(-2) - (-8)^2 < 0$, so $f(x, y)$ has neither a maximum nor a minimum at (0, 0).

21. $f(x, y) = x^2 + 2xy + 5y^2 + 2x + 10y - 3$; $\frac{\partial f}{\partial x} = 2x + 2y + 2$

$\frac{\partial f}{\partial y} = 2x + 10y + 10$; $\frac{\partial^2 f}{\partial x^2} = 2$; $\frac{\partial^2 f}{\partial y^2} = 6$; $\frac{\partial^2 f}{\partial x \partial y} = -2$.

$$\left.\begin{aligned} 2x - 2y + 4 &= 0 \\ -2x + 6y - 16 &= 0 \end{aligned}\right\} \begin{aligned} x &= 0 \\ y &= -1 \end{aligned}$$

$D(0, -1) = (2)(10) - 2^2 > 0$, $\frac{\partial^2 f}{\partial x^2} > 0$, so $f(x, y)$ has a relative minimum at (0, -1).

22. $f(x, y) = x^2 - 2xy + 3y^2 + 4x - 16y + 22$; $\frac{\partial f}{\partial x} = 2x - 2y + 4$

$\frac{\partial f}{\partial y} = 2x + 6y - 16$; $\frac{\partial^2 f}{\partial x^2} = 2$; $\frac{\partial^2 f}{\partial y^2} = 6$; $\frac{\partial^2 f}{\partial x \partial y} = -2$.

$$\left.\begin{aligned} 2x - 2y + 4 &= 0 \\ -2x + 6y - 16 &= 0 \end{aligned}\right\} \begin{aligned} x &= 1 \\ y &= 3 \end{aligned}$$

$D(1, 3) = 2\cdot 6 - (-2)^2 > 0$, $\frac{\partial^2 f}{\partial x \partial y} > 0$, so $f(x, y)$ has a relative minimum at $(1, 3)$.

23. $f(x, y) = x^3 - y^2 - 3s + 4y$; $\frac{\partial f}{\partial x} = 3x^2 - 3$; $\frac{\partial f}{\partial y} = -2y + 4$

$\frac{\partial^2 f}{\partial x^2} = 6x$; $\frac{\partial^2 f}{\partial y^2} = -2$; $\frac{\partial^2 f}{\partial x \partial y} = 0$.

$$\left.\begin{aligned} 3x^2 - 3 &= 0 \\ -2y + 4 &= 0 \end{aligned}\right\} \begin{aligned} x &= \pm 1 \\ y &= 2 \end{aligned}$$

$D(1, 2) = (6)(-2) - 0 < 0$, so $f(x, y)$ has neither a maximum nor a minimum at $(1, 2)$. $D(-1, 2) = (-6)(-2) > 0$,

$\frac{\partial^2 f}{\partial x^2}(-1, 2) = -6 < 0$, so $f(x, y)$ has a relative maximum at $(-1, 2)$.

24. $f(x, y) = x^3 - 2xy + 4y$; $\frac{\partial f}{\partial x} = 3x^2 - 2y$; $\frac{\partial f}{\partial y} = -2x + 4$

$\frac{\partial^2 f}{\partial x^2} = 6x$; $\frac{\partial^2 f}{\partial y^2} = 0$; $\frac{\partial^2 f}{\partial x \partial y} = -2$.

$$\left.\begin{aligned} 3x^2 - 2y &= 0 \\ -2x + 4 &= 0 \end{aligned}\right\} \begin{aligned} x &= 2 \\ y &= 6 \end{aligned}$$

$D(2, 6) = 6\cdot 2\cdot 0 - (-2)^2 < 0$, so $f(x, y)$ has neither a maximum nor a minimum at $(2, 6)$.

25. $f(x, y) = 2x^2 + y^3 - x - 12y + 7$; $\frac{\partial f}{\partial x} = 4x - 1$; $\frac{\partial f}{\partial y} = 3y^2 - 12$

$\frac{\partial^2 f}{\partial x^2} = 4$; $\frac{\partial^2 f}{\partial y^2} = 6y$; $\frac{\partial^2 f}{\partial x \partial y} = 0$

$$\left.\begin{aligned} 4x - 1 &= 0 \\ 3y^2 - 12 &= 0 \end{aligned}\right\} \begin{aligned} x &= \frac{1}{4} \\ y &= \pm 2 \end{aligned}$$

$D\left(\frac{1}{4}, 2\right) = 4\cdot 6\cdot 2 - 0 > 0$, $\frac{\partial^2 f}{\partial x^2} > 0$ so $f(x, y)$ has a relative minimum at $\left(\frac{1}{4}, 2\right)$. $D\left(\frac{1}{4}, -2\right) = 4\cdot 6\cdot(-2) - 0 < 0$, so $f(x, y)$

has neither a maximum nor a minimum at $\left(\frac{1}{4}, -2\right)$.

26. $f(x, y) = x^2 + 4xy + 2y^4$; $\frac{\partial f}{\partial x} = 2x + 4y$; $\frac{\partial f}{\partial y} = 4x + 8y^3$

$\frac{\partial^2 f}{\partial x^2} = 2$; $\frac{\partial^2 f}{\partial y^2} = 24y^2$; $\frac{\partial^2 f}{\partial x \partial y} = 0$

$\left.\begin{array}{r} 2x + 4y = 0 \\ 4x + 8y^3 = 0 \end{array}\right\}$ $\left.\begin{array}{l} x = -2y \\ 8y^3 - 8y = 0 \end{array}\right\}$ Solutions: (0, 0), (-2, 1), (2, -1).

$D(0, 0) = 2 \cdot 0 - 0 = 0$, so the second derivative test is inconclusive at (0, 0). $D(-2, 1) = 2 \cdot 24(1)^2 - 0 > 0$, $\frac{\partial^2 f}{\partial x^2} > 0$, so f(x, y) has a relative minimum at (-2, 1).

$D(2, -1) = 2 \cdot 24(-1)^2 > 0$, $\frac{\partial^2 f}{\partial x^2} > 0$, so f(x, y) has a relative minimum at (2, -1).

27. $f(x, y, z) = 2x^2 + 3y^2 + z^2 - 2x - y - z$

$\frac{\partial f}{\partial x} = 4x - 2$; $\frac{\partial f}{\partial y} = 6y - 1$; $\frac{\partial f}{\partial z} = 2z - 1$.

$\left.\begin{array}{r} 4x - 2 = 0 \\ 6y - 1 = 0 \\ 2z - 1 = 0 \end{array}\right\}$ $x = \frac{1}{2}$; $y = \frac{1}{6}$; $z = \frac{1}{2}$

$\left(\frac{1}{2}, \frac{1}{6}, \frac{1}{2}\right)$ is the only point at which f(x, y, z) can have a relative minimum.

28. $f(x, y, z) = 5 + 8x - 4y + x^2 + y^2 + z^2$; $\frac{\partial f}{\partial x} = 8 + 2x$

$\frac{\partial f}{\partial y} = 2y - 4$; $\frac{\partial f}{\partial z} = 2z$

$\left.\begin{array}{r} 8 + 2x = 0 \\ 2y - 4 = 0 \\ 2z = 0 \end{array}\right\}$ $\begin{array}{l} x = -4 \\ y = 2 \\ z = 0 \end{array}$

(-4, 2, 0) must give the minimum value of f(x, y, z).

29. Let x, y and ℓ be as shown in fig. 4. Since $\ell = 84 - 2x - 2y$, the volume of the box may be written as

$V(x, y) = xy(84 - 2x - 2y) = 84xy - 2x^2y - 2xy^2$.

$\frac{\partial V}{\partial x} = 84y - 4xy - 2y^2$; $\frac{\partial V}{\partial y} = 84x - 2x^2 - 4xy$; $\frac{\partial^2 V}{\partial x^2} = -4y$;

$\frac{\partial^2 V}{\partial y^2} = -4x$; $\frac{\partial^2 V}{\partial x \partial y} = 84 - 4x - 4y$

$\left.\begin{array}{l} 84y - 4xy - 2y^2 = 0 \\ 84x - 2x^2 - 4xy = 0 \end{array}\right\}$ $\begin{array}{l} x = 0 \\ y = 0 \end{array}$ or $\left.\begin{array}{l} 84 - 4x - 2y = 0 \\ 84 - 4y - 2x = 0 \end{array}\right\}$ $\begin{array}{l} x = 14 \\ y = 14 \end{array}$

Obviously, (0, 0) does not give the maximum value of V(x, y).

To verify that (14, 14) is the maximum, check

$D(14, 14) = -4(14)(-4)(14) - (84 - 4(14) - 4(14))^2 > 0$,

$\frac{\partial^2 V}{\partial x^2} = -4(14) < 0$. Thus the dimensions that give the maximum volume are $x = 14$, $y = 14$, $\ell = 84 - 56 = 28$; or 14 x 14 x 28 inches.

30. Let x, y and z be the dimensions of the box. Since the volume of the box is 1000 in^3, $x > 0$, $y > 0$ and $z = \frac{1000}{xy}$. The surface area is $S(x, y) = 2xy + 2x\left(\frac{1000}{xy}\right) + 2y\left(\frac{1000}{xy}\right)$

$$= 2xy + \frac{2000}{y} + \frac{2000}{x}$$

$\frac{\partial f}{\partial x} = 2y - \frac{2000}{x^2}$; $\frac{\partial f}{\partial y} = 2x - \frac{2000}{y^2}$; $\frac{\partial^2 f}{\partial x^2} = \frac{4000}{x^3}$; $\frac{\partial^2 f}{\partial y^2} = \frac{4000}{y^3}$;

$\frac{\partial^2 f}{\partial x \partial y} = 2$.

$\left.\begin{array}{l} 2y - \frac{2000}{x^2} = 0 \\ 2x - \frac{2000}{y^2} = 0 \end{array}\right\}$ $x = y = 10$.

To verify that (10, 10) is a minimum, check

$D(10, 10) = \left(\frac{4000}{1000}\right)^2 - 2^2 > 0$, $\frac{\partial^2 f}{\partial x^2}(10, 10) > 0$. Thus the dimensions giving the smallest surface area are 10 x 10 x 10 inches.

31. The revenue is $10x + 9y$, so the profit function is

$P(x, y) = 10x + 9y - \left[400 + 2x + 3y + .01(3x^2 + xy + 3y^2)\right]$

$= 8x + 6y - .03x^2 - .01xy - .03y^2 - 400.$

$\frac{\partial P}{\partial x} = 8 - .06x - .01y;\ \frac{\partial P}{\partial y} = 6 - .01x - .06y;\ \frac{\partial^2 P}{\partial x^2} = -.06$

$\frac{\partial^2 P}{\partial y^2} = -.06;\ \frac{\partial^2 P}{\partial x \partial y} = -.01.$

$\left.\begin{array}{l} 8 - .06x - .01y = 0 \\ 6 - .01x - .06y = 0 \end{array}\right\} \begin{array}{l} x = 120 \\ y = 80 \end{array}$

(120, 80) is a maximum, $D(120, 80) = (-.06)^2 - (-.01)^2 > 0$,

$\frac{\partial^2 P}{\partial x^2}(120, 80) = -.06 < 0$. Thus profit is maximized by producing 120 units of product I and 80 units of product II.

32. The cost is $30x + 20y$, so the profit function is

$P(x, y) = 98x + 112y - .04xy - .1x^2 - .2y^2 - 30x - 20y$

$= 68x + 92y - .04xy - .1x^2 - .2y^2$

$\frac{\partial P}{\partial x} = 68 - .04y - .2x;\ \frac{\partial P}{\partial y} = 92 - .04x - .4y;\ \frac{\partial^2 P}{\partial x^2} = -.2$

$\frac{\partial^2 P}{\partial y^2} = -.4;\ \frac{\partial^2 P}{\partial x \partial y} = -.04$

$\left.\begin{array}{l} 68 - .04y - .2x = 0 \\ 92 - .04x - .4y = 0 \end{array}\right\} \begin{array}{l} x = 300 \\ y = 200 \end{array}$

To verify that (300, 200) is a maximum, check $D(300, 200) =$

$(-.2)(-.4) - (.04)^2 > 0,\ \frac{\partial^2 P}{\partial x^2} < 0$. Thus the profit is maximized by producing 300 units of product I and 200 units of product II.

33. Let $P(x, y)$ denote the company's profit from producing x units of product I and y units of product II. Then

$P(x, y) = P_1x + P_2y - C(x, y)$. If (a, b) is the profit maximizing output combination, then

$\frac{\partial P}{\partial x}(a, b) = \frac{\partial P}{\partial y}(a, b) = 0$; so $P_1 - \frac{\partial C}{\partial x} = 0$ and $P_2 - \frac{\partial C}{\partial y} = 0$ or

$\frac{\partial C}{\partial x} = P_1$ and $\frac{\partial C}{\partial y} = P_2$.

34. Let $P(x, y)$ denote the company's profit from producing x units

of product I and y units of product II. Then

$P(x, y) = R(x, y) - P_1x - P_2y$. If (a, b) is the profit maximizing output combination, then

$\frac{\partial P}{\partial x}(a, b) = \frac{\partial P}{\partial y}(a, b) = 0$; so $\frac{\partial R}{\partial x} - P_1 = 0$ and $\frac{\partial R}{\partial y} - P_2 = 0$ or

$\frac{\partial R}{\partial x} = P_1$ and $\frac{\partial R}{\partial y} = P_2$.

45.a. $f(18) = 69°$

b. $f'(20) = 1.6$. Increasing 1.6°/wk

c. $f(t) = 39$ has solutions $t = 6, 44$. In weeks 6 and 44.

d. $f'(t) = -1$ has solutions $t = 28, 48$. In weeks 28 and 48.

e. $f(t)$ has maximum at $t = 25$ and minimum at $t = 51$. So the average weekly temperature is most in week 25, and least in week 51.

f. $f'(t)$ has a maximum at $t = 12$ and a minimum at $t = 38$. So the average weekly temperature is increasing fastest in week 12 and decreasing fastest in week 38.

46.a. 11.5 hrs

b. -.25 hrs/wk

c. weeks 18 and 32

d. weeks 6 and 18

e. longest week 25, shortest week 51

f. increasing fastest – week 12, decreasing fastest – week 38

Exercises 7.4

1. $F(x, y, \lambda) = x^2 + 3y^2 + 10 + \lambda(8 - x - y)$

$$\left.\begin{array}{l}\frac{\partial F}{\partial x} = 2x - \lambda = 0 \\ \frac{\partial F}{\partial y} = 6y - \lambda = 0 \\ \frac{\partial F}{\partial \lambda} = 8 - x - y = 0\end{array}\right\} \begin{array}{l}\lambda = 2x \\ \lambda = 6y \\ 8 - x - y = 0\end{array}\Bigg\} \begin{array}{l}\left.\begin{array}{l}2x = 6y;\ x = 3y \\ \\ \end{array}\right\} \\ 8 - 3y - y = 0\end{array} \begin{array}{l}x = 6 \\ y = 2\end{array}$$

The minimum value is $6^2 + 3\cdot 2^2 + 10 = 58$.

2. $F(x, y, \lambda) = x^2 - y^2 + \lambda(2x + y - 3)$

$$\left.\begin{array}{l}\frac{\partial F}{\partial x} = 2x + 2\lambda = 0 \\ \frac{\partial F}{\partial y} = -2y + \lambda = 0 \\ \frac{\partial F}{\partial \lambda} = 2x + y - 3 = 0\end{array}\right\} \left.\begin{array}{l}\lambda = -x \\ \lambda = 2y \\ 2x + y - 3 = 0\end{array}\right\} \left.\begin{array}{l}x = -2y \\ -4y + y - 3 = 0\end{array}\right\} \begin{array}{l}x = 2 \\ y = -1\end{array}$$

The maximum value is $2^2 - (-1)^2 = 3$.

3. $F(x, y, \lambda) = x^2 + xy - 3y^2 + \lambda(2 - x - 2y)$

$$\left.\begin{array}{l}\frac{\partial F}{\partial x} = 2x + y - \lambda = 0 \\ \frac{\partial F}{\partial y} = x - 6y - 2\lambda = 0 \\ \frac{\partial F}{\partial \lambda} = 2 - x - 2y = 0\end{array}\right\} \left.\begin{array}{l}\lambda = 2x + y \\ \lambda = \frac{1}{2}x - 3y \\ 2 - x - 2y = 0\end{array}\right\} \left.\begin{array}{l}\frac{3}{2}x + 4y = 0 \\ x + 2y = 2\end{array}\right\} \begin{array}{l}x = 8 \\ y = -3\end{array}$$

The maximum value is $8^2 + 8(-3) - 3(-3)^2 = 13$.

4. $F(x, y, \lambda) = \frac{1}{2}x^2 - 3xy + y^2 + \frac{1}{2} + \lambda(3x - y - 1)$

$$\left.\begin{aligned}\frac{\partial F}{\partial x} &= x - 3y + 3\lambda = 0\\ \frac{\partial F}{\partial y} &= -3x + 2y - \lambda = 0\\ \frac{\partial F}{\partial \lambda} &= 3x - y - 1 = 0\end{aligned}\right\} \left.\begin{aligned}\lambda &= -\frac{1}{3}x + y\\ \lambda &= -3x + 2y\\ 3x - y - 1 &= 0\end{aligned}\right\} \left.\begin{aligned}\frac{8}{3}x - y &= 0\\ 3x - y &= 1\end{aligned}\right\} \begin{aligned}x &= 3\\ y &= 8\end{aligned}$$

The minimum value is $\frac{1}{2}3^2 - 3(3)(8) + 8^2 + \frac{1}{2} = -3$.

5. $F(x, y, \lambda) = -2x^2 - 2xy - \frac{3}{2}y^2 + x + 2y + \lambda\left(x + y - \frac{5}{2}\right)$

$$\left.\begin{aligned}\frac{\partial F}{\partial x} &= -4x - 2y + 1 + \lambda = 0\\ \frac{\partial F}{\partial y} &= -2x - 3y + 2 + \lambda = 0\\ \frac{\partial F}{\partial \lambda} &= x + y - \frac{5}{2} = 0\end{aligned}\right\} \left.\begin{aligned}\lambda &= 4x + 2y - 1\\ \lambda &= 2x + 3y - 2\\ x + y - \frac{5}{2} &= 0\end{aligned}\right\} \left.\begin{aligned}2x - y &= -1\\ x + y &= \frac{5}{2}\end{aligned}\right\} \begin{aligned}x &= \frac{1}{2}\\ y &= 2\end{aligned}$$

6. $F(x, y, \lambda) = x^2 + xy + y^2 - 2x - 5y + \lambda(1 - x + y)$

$$\left.\begin{aligned}\frac{\partial F}{\partial x} &= 2x + y - 2 - \lambda = 0\\ \frac{\partial F}{\partial y} &= x + 2y - 5 + \lambda = 0\\ \frac{\partial F}{\partial \lambda} &= 1 - x + y = 0\end{aligned}\right\} \left.\begin{aligned}\lambda &= 2x + y - 2\\ \lambda &= -x - 2y + 5\\ 1 - x + y &= 0\end{aligned}\right\} \left.\begin{aligned}3x + 3y &= 7\\ -x + y &= -1\end{aligned}\right\} \begin{aligned}x &= \frac{5}{3}\\ y &= \frac{2}{3}\end{aligned}$$

7. We want to minimize the function $x + y$ subject to the constraint $xy = 25$ or $xy - 25 = 0$.

$F(x, y, \lambda) = x + y + \lambda(xy - 25)$

$$\left.\begin{aligned}\frac{\partial F}{\partial x} &= 1 + \lambda y = 0\\ \frac{\partial F}{\partial y} &= 1 + \lambda x = 0\\ \frac{\partial F}{\partial \lambda} &= xy - 25 = 0\end{aligned}\right\} \left.\begin{aligned}\lambda &= \frac{-1}{y}\\ \lambda &= \frac{-1}{x}\\ xy - 25 &= 0\end{aligned}\right\} \left.\begin{aligned}x - y &= 0\\ xy &= 25\end{aligned}\right\} \begin{aligned}&x^2 = 25 \text{ or } x = \pm 5\\ &\text{so } x = 5,\ y = 5\\ &\text{i.e., positve}\end{aligned}$$

8. Let x = length of the north side.

Let y = length of the west side.

Cost $= 2x(10) + 2y(15) = 20x + 30y = 480$

We want to maximize xy subject to $20x + 30y - 480 = 0$.

$F(x, y, \lambda) = xy + \lambda(20x + 30y - 480)$

$$\left.\begin{array}{l}\frac{\partial F}{\partial x} = y + 20\lambda = 0\\ \frac{\partial F}{\partial y} = x + 30\lambda = 0\\ \frac{\partial F}{\partial \lambda} = 20x + 30y - 480 = 0\end{array}\right\} \left.\begin{array}{l}\lambda = \frac{-y}{20}\\ \lambda = \frac{-x}{30}\\ 2x + 3y = 48\end{array}\right\} \left.\begin{array}{l}2x - 3y = 0\\ 2x + 3y = 48\end{array}\right\} \begin{array}{l}x = 12\\ y = 8\end{array}$$

9. Let x = length of a side of the base.

Let y = height of the box.

Area = $x^2 + 4xy = 300$

Maximize the volume = x^2y subject to $x^2 + 4xy - 300 = 0$

$F(x, y, \lambda) = x^2y + \lambda(x^2 + 4xy - 300)$

$$\left.\begin{array}{l}\frac{\partial F}{\partial x} = 2xy + 2x\lambda + 4y\lambda = 0\\ \frac{\partial F}{\partial y} = x^2 + 4x\lambda = 0\\ \frac{\partial F}{\partial \lambda} = x^2 + 4xy - 300 = 0\end{array}\right\} \left.\begin{array}{l}\lambda = \frac{-2xy}{2(x + 2y)}\\ \lambda = \frac{-x^2}{4x} = \frac{-x}{4}\\ x^2 + 4xy = 300\end{array}\right\} \left.\begin{array}{l}x - 2y = 0\\ x^2 + 4xy = 300\end{array}\right\} \begin{array}{l}x = 10\\ y = 5\end{array}$$

10. The problem is the minimize $1000\sqrt{6x^2 + y^2}$ subject to $5000 - 480x - 40y$.

$F(x, y, \lambda) = 1000\sqrt{6x^2 + y^2} + \lambda(5000 - 480x - 40y)$

$$\left.\begin{array}{l}\frac{\partial F}{\partial x} = \frac{6000x}{\sqrt{6x^2 + y^2}} + 480\lambda = 0\\ \frac{\partial F}{\partial y} = \frac{1000y}{\sqrt{6x^2 + y^2}} + 40\lambda = 0\\ \frac{\partial F}{\partial \lambda} = 5000 - 480x - 40y = 0\end{array}\right\} \left.\begin{array}{l}\lambda = \frac{12.5x}{\sqrt{6x^2 + y^2}}\\ \lambda = \frac{25y}{\sqrt{6x^2 + y^2}}\\ 480x + 40y = 500\end{array}\right\} \left.\begin{array}{l}x = 2y\\ 1000y = 5000\end{array}\right\}$$

$x = 10,\ y = 5$

11. The problem is to maximize xy subject to $x^2 + y^2 - 1 = 0$.

$F(x, y, \lambda) = xy + \lambda(x^2 + y^2 - 1)$.

$$\left.\begin{array}{l}\frac{\partial F}{\partial x} = y + 2\lambda x = 0\\ \frac{\partial F}{\partial y} = x + 2\lambda y = 0\\ \frac{\partial F}{\partial \lambda} = x^2 + y^2 - 1 = 0\end{array}\right\} \left.\begin{array}{l}x = 0 \text{ or } \lambda = \frac{-y}{2x}\\ y = 0 \text{ or } \lambda = \frac{-x}{2y}\\ x^2 + y^2 = 1\end{array}\right\} \left.\begin{array}{l}\text{Assuming } x \neq 0,\ y \neq 0\\ \frac{-y}{2x} = \frac{-x}{2y}\\ x^2 + y^2 = 1\end{array}\right\}$$

$$\left.\begin{aligned} x^2 &= y^2 \\ 2x^2 &= 1 \end{aligned}\right\} \text{ Assuming } x > 0,\ y > 0,\ x = \frac{\sqrt{2}}{2} \text{ and } y = \frac{\sqrt{2}}{2}.$$

12. Following the hint, the problem is to minimize

$(x - 16)^2 + \left(y - \frac{1}{2}\right)^2$ subject to $y - x^2 = 0$.

$F(x, y, \lambda) = (x - 16)^2 + \left(y - \frac{1}{2}\right)^2 + \lambda(y - x^2)$

$$\left.\begin{aligned} \frac{\partial F}{\partial x} &= 2x - 32 - 2\lambda x = 0 \\ \frac{\partial F}{\partial y} &= 2y - 1 + \lambda = 0 \\ \frac{\partial F}{\partial \lambda} &= y - x^2 = 0 \end{aligned}\right\} \left.\begin{aligned} &x = 0 \text{ or } \lambda = 1 - \frac{16}{x} \\ &\lambda = 1 - 2y \\ &y = x^2 \end{aligned}\right\} \left.\begin{aligned} &x = 0,\ y = 0 \text{ or} \\ &\frac{16}{x} = 2y \\ &y = x^2 \end{aligned}\right\}$$

$x = 0$, $y = 0$ or $x = 2$, $y = 4$. To decide which of (0, 0) or (2, 4) is the closest point, check that

$(0 - 16)^2 + \left(0 - \frac{1}{2}\right)^2 > (2 - 16)^2 + \left(4 - \frac{1}{2}\right)^2$. Thus (2, 4) is the desired point.

13. The problem is to maximize $3x + 4y$ subject to $18{,}000 - 9x^2 - 4y^2 = 0$, $x \geq 0$, $y \geq 0$.

$F(x, y, \lambda) = 3x + 4y + \lambda(18{,}000 - 9x^2 - 4y^2)$

$$\left.\begin{aligned} \frac{\partial F}{\partial x} &= 3 - 18\lambda x = 0 \\ \frac{\partial F}{\partial y} &= 4 - 8\lambda y = 0 \\ \frac{\partial F}{\partial \lambda} &= 18{,}000 - 9x^2 - 4y^2 = 0 \end{aligned}\right\} \left.\begin{aligned} &x = 0 \text{ or } \lambda = \frac{1}{6x} \\ &y = 0 \text{ or } \lambda = \frac{1}{2y} \\ &9x^2 + 4y^2 = 18{,}000 \end{aligned}\right\} \text{If } x \neq 0 \text{ and } y \neq 0 \text{ then}$$

$$\left.\begin{aligned} y &= 3x \\ 9x^2 + 36x^2 &= 18{,}000 \end{aligned}\right\} \begin{aligned} x &= 20 \\ y &= 60 \end{aligned}$$

Technically, we should also check the solution $x = 0$, $y = \sqrt{\frac{18{,}000}{9}} \sim 44.7$ and $y = 0$, $x = \sqrt{\frac{18{,}000}{36}} \sim 22.4$. These both give smaller values in the objective function $3x + 4y$ than does (20, 60).

14. We want to minimize the function $P = 2x + 10y$ subject to the constraint $4x^2 + 25y^2 = 50000$.

$F(x, y, \lambda) = 2x + 10y + \lambda(4x^2 + 25y^2 - 50000)$

$$\left.\begin{array}{l} \frac{\partial F}{\partial x} = 2 + 8\lambda x = 0 \\ \frac{\partial F}{\partial y} = 10 + 50\lambda y = 0 \\ \frac{\partial F}{\partial \lambda} = 4x^2 + 25y^2 - 50000 = 0 \end{array}\right\} \left.\begin{array}{l} \lambda = \frac{-1}{4x} \\ \lambda = \frac{-1}{5y} \\ 4x^2 + 25y^2 = 50000 \end{array}\right\} \begin{array}{l} x = \frac{5}{4}y \\ 4x^2 + 25y^2 = 50000 \end{array}$$

Solving gives $\frac{125}{4}y^2 = 50000$ or $y = 40$, hence $x = 50$.

15. (a) $F(x, y, \lambda) = 96x + 16y + \lambda(3456 - 64x^{3/4}y^{1/4})$

Note that $3456 = x^{3/4}y^{1/4}$ implies $x \neq 0$, $y \neq 0$.

$$\left.\begin{array}{l} \frac{\partial F}{\partial x} = 96 - 48\lambda x^{-1/4}y^{1/4} \\ \frac{\partial F}{\partial y} = 162 - 16\lambda x^{3/4}y^{-3/4} \\ \frac{\partial F}{\partial \lambda} = 3456 - 64x^{3/4}y^{1/4} \end{array}\right\} \left.\begin{array}{l} \lambda = 2x^{1/4}y^{-1/4} \\ \lambda = \frac{81}{8}x^{-3/4}y^{3/4} \\ 3456 = 64x^{3/4}y^{1/4} \end{array}\right\} \text{Dividing 1st equation by 2nd gives}$$

$$\left.\begin{array}{l} \frac{16}{81}xy^{-1} = 1 \\ y = \frac{16}{81}x \\ 3456 = 64\left(\frac{16}{81}\right)^{1/4} \end{array}\right\} \begin{array}{l} x = 81 \\ y = 16 \end{array}$$

(b) $\lambda = 2(81)^{1/4}(16)^{-1/4} = 3$

(c) The production function is $f(x, y) = 64x^{3/4}y^{1/4}$.

Thus $\dfrac{\frac{\partial f}{\partial x}}{\frac{\partial f}{\partial y}} = \dfrac{48x^{-1/4}y^{1/4}}{16x^{3/4}y^{-3/4}} = \dfrac{48y}{16x}$. When $x = 81$ and $y = 16$,

$\dfrac{48y}{16x} = \dfrac{48 \cdot 16}{16 \cdot 81} = \dfrac{96}{162}$, which is the ratio of the unit cost of labor and capital.

16. $F(x, y, \lambda) = 94x - \frac{x^2}{10} + 80y - \frac{y^2}{20} - 20{,}000 + \lambda\left[14 - \frac{x}{10} + \frac{y}{20}\right]$

$$\left.\begin{array}{l} \frac{\partial F}{\partial x} = 94 - \frac{x}{5} - \frac{\lambda}{10} = 0 \\ \frac{\partial F}{\partial y} = 80 - \frac{y}{10} + \frac{\lambda}{20} = 0 \\ \frac{\partial F}{\partial \lambda} = 14 - \frac{x}{10} + \frac{y}{20} = 0 \end{array}\right\} \left.\begin{array}{l} \lambda = 940 - 2x \\ \lambda = -1600 + 2y \\ \frac{x}{10} - \frac{y}{20} = 14 \end{array}\right\} \left.\begin{array}{l} 2x + 2y = 2540 \\ 2x - y = 280 \end{array}\right\} \begin{array}{l} x = \frac{1550}{3} \\ y = \frac{2260}{3} \end{array}$$

17. $F(x, y, z, \lambda) = xyz + \lambda(36 - x - 6y - 3z)$

$$\left.\begin{aligned} \frac{\partial F}{\partial x} &= yz + \lambda = 0 \\ \frac{\partial F}{\partial y} &= xz + 6\lambda = 0 \\ \frac{\partial F}{\partial z} &= xy + 3\lambda = 0 \\ \frac{\partial F}{\partial \lambda} &= 36 - x - 6y - 3z = 0 \end{aligned}\right\} \left.\begin{aligned} \lambda &= -yz \\ \lambda &= \frac{-xz}{6} \\ \lambda &= \frac{-xy}{3} \\ 2x + y + z &= 20 \end{aligned}\right\} \left.\begin{aligned} x - 2y &= -\frac{7}{2} \\ 2y - 2z &= 4 \\ 2x + y + z &= 20 \end{aligned}\right\}$$

$$\left.\begin{aligned} x &= 2y - \frac{7}{2} \\ z &= -2 + y \\ 2\left(2y - \frac{7}{2}\right) &+ y + y - 2 = 20 \end{aligned}\right\} \begin{aligned} x &= \frac{37}{6} \\ y &= \frac{29}{6} \\ z &= \frac{17}{6} \end{aligned}$$

18. $F(x, y, z, \lambda) = xy + 3xy + 3yz + \lambda(9 - xyz)$

$$\left.\begin{aligned} \frac{\partial F}{\partial x} &= y + 3z - \lambda yz = 0 \\ \frac{\partial F}{\partial y} &= x + 3z - \lambda xz = 0 \\ \frac{\partial F}{\partial z} &= 3x + 3y - \lambda xy = 0 \\ \frac{\partial F}{\partial \lambda} &= 9 - xyz = 0 \end{aligned}\right\} \left.\begin{aligned} \lambda &= \frac{1}{z} - \frac{3}{y} \\ \lambda &= \frac{1}{z} - \frac{3}{x} \\ \lambda &= \frac{3}{y} - \frac{3}{x} \\ xyz &= 9 \end{aligned}\right\} \left.\begin{aligned} \frac{3}{y} &= \frac{3}{x} \\ \frac{1}{z} &= \frac{3}{y} \\ xyz &= 9 \end{aligned}\right\}$$

$$\left.\begin{aligned} x &= y \\ z &= y/3 \\ yy(y/3) &= 9 \text{ or } y^3 = 27 \text{ or } y = 3 \end{aligned}\right\} \begin{aligned} x &= 3 \\ y &= 3 \\ z &= 1 \end{aligned}$$

19. $F(x, y, z, \lambda) = 3x + 5y + z - x^2 - y^2 - z^2 + \lambda(6 - x - y - z)$

$$\left.\begin{aligned} \frac{\partial F}{\partial x} &= 3 - 2x - \lambda = 0 \\ \frac{\partial F}{\partial y} &= 5 - 2y - \lambda = 0 \\ \frac{\partial F}{\partial z} &= 1 - 2z - \lambda = 0 \\ \frac{\partial F}{\partial \lambda} &= 6 - x - y - z = 0 \end{aligned}\right\} \left.\begin{aligned} \lambda &= 3 - 2x \\ \lambda &= 5 - 2y \\ \lambda &= 1 - 2z \\ x + y + z &= 6 \end{aligned}\right\} \left.\begin{aligned} -2x + 2y &= 2 \\ -2y + 2z &= -4 \\ x + y + z &= 6 \end{aligned}\right\}$$

$$\left.\begin{aligned} x &= -1 + y \\ z &= -2 + y \\ (-1 + y) + y &+ (-2 + y) = 6 \end{aligned}\right\} \begin{aligned} x &= 2 \\ y &= 3 \\ z &= 1 \end{aligned}$$

20. $F(x, y, z, \lambda) = x^2 + y^2 + z^2 - 3x - 5y - z + \lambda(20 - 2x - y - z)$

$$\left.\begin{aligned} \frac{\partial F}{\partial x} &= 2x - 3 - 2\lambda = 0 \\ \frac{\partial F}{\partial y} &= 2y - 5 - \lambda = 0 \\ \frac{\partial F}{\partial z} &= 2z - 1 - \lambda = 0 \\ \frac{\partial F}{\partial \lambda} &= 20 - 2x - y - z = 0 \end{aligned}\right\} \left.\begin{aligned} \lambda &= x - \frac{3}{2} \\ \lambda &= 2y - 5 \\ \lambda &= 2z - 1 \\ 2x + y &+ z = 20 \end{aligned}\right\} \left.\begin{aligned} x - 2y &= -\frac{7}{2} \\ 2y - 2z &= 4 \\ 2x + y + z &= 20 \end{aligned}\right\}$$

$$\left.\begin{aligned} x &= 2y - \frac{7}{2} \\ z &= -2 + y \\ 2\left(2y - \frac{7}{2}\right) &+ y + y - 2 = 20 \end{aligned}\right\} \begin{aligned} x &= \frac{37}{6} \\ y &= \frac{29}{6} \\ z &= \frac{17}{6} \end{aligned}$$

21. The problem is to minimize $3xy + 2xz + 2yz$ subject to the constraint $xyz = 12$ or $xyz - 12 = 0$.

$F(x, y, z, \lambda) = 3xy + 2xz + 2yz + \lambda(xyz - 12)$.

(Note that $xyz = 12$ implies that $x \neq 0$, $y \neq 0$, $z \neq 0$.)

$$\left.\begin{aligned} \frac{\partial F}{\partial x} &= 3y + 2z + \lambda yz = 0 \\ \frac{\partial F}{\partial y} &= 3x + 2z + \lambda xz = 0 \\ \frac{\partial F}{\partial z} &= 2x + 2y + \lambda xy = 0 \\ \frac{\partial F}{\partial \lambda} &= xyz - 12 = 0 \end{aligned}\right\} \left.\begin{aligned} \lambda &= -\frac{3}{z} - \frac{2}{y} \\ \lambda &= -\frac{3}{z} - \frac{2}{z} \\ \lambda &= -\frac{2}{y} - \frac{2}{x} \\ xyz &= 12 \end{aligned}\right\} \left.\begin{aligned} x &= y \\ z &= \frac{3}{2}y \\ y^2\frac{3}{2}y &= 12 \end{aligned}\right\} \begin{aligned} x &= 2 \\ y &= 2 \\ z &= 3 \end{aligned}$$

22. The problem is to maximize xyz subject to $x + y + z - 15 = 0$, $x > 0$, $y > 0$, $z > 0$.

$F(x, y, z, \lambda) = xyz + \lambda(x + y + z - 15)$

$$\left.\begin{aligned}\frac{\partial F}{\partial x} &= yz + \lambda = 0\\ \frac{\partial F}{\partial y} &= xz + \lambda = 0\\ \frac{\partial F}{\partial z} &= xy + \lambda = 0\\ \frac{\partial F}{\partial \lambda} &= x + y + z - 15 = 0\end{aligned}\right\} \begin{aligned}\lambda &= -yz\\ \lambda &= -xz\\ \lambda &= -xy\\ x + y + z &= 15\end{aligned} \quad \left.\begin{aligned}x &= y\\ z &= y\\ 3y &= 15\end{aligned}\right\} \begin{aligned}x &= 5\\ y &= 5\\ z &= 5\end{aligned}$$

23. Let x, y, z be as shown in fig. 2(a). The problem is to minimize $xy + 2xz + 2yz$ subject to $xyz = 32$.

$F(x, y, z, \lambda) = xy + 2xz + 2yz + \lambda(xyz - 32)$.

(Note that $xyz = 32$ implies $x \neq 0$, $y \neq 0$, $z \neq 0$.)

$$\left.\begin{aligned}\frac{\partial F}{\partial x} &= y + 2z + \lambda yz = 0\\ \frac{\partial F}{\partial y} &= x + 2z + \lambda xz = 0\\ \frac{\partial F}{\partial z} &= 2x + 2y + \lambda xy = 0\\ \frac{\partial F}{\partial \lambda} &= xyz - 32 = 0\end{aligned}\right\} \left.\begin{aligned}\lambda &= -\frac{1}{z} - \frac{2}{y}\\ \lambda &= -\frac{1}{z} - \frac{2}{x}\\ \lambda &= -\frac{2}{y} - \frac{2}{x}\\ xyz &= 32\end{aligned}\right\} \left.\begin{aligned}x &= y\\ z &= \frac{1}{2}y\\ \frac{1}{2}y^3 &= 32\end{aligned}\right\} \begin{aligned}x &= 4 \text{ ft.}\\ y &= 4 \text{ ft.}\\ z &= 2 \text{ ft.}\end{aligned}$$

24. Let x, y, z be as shown in fig. 2(b). The problem is to maximize xyz subject to $xy + xz + 2yz = 96$, $x > 0$, $y > 0$, $z > 0$.

$F(x, y, z, \lambda) = xyz + \lambda(xy + xz + 2yz - 96)$

$$\left.\begin{aligned}\frac{\partial F}{\partial x} &= yz + \lambda(y + z) = 0\\ \frac{\partial F}{\partial y} &= xz + \lambda(x + 2z) = 0\\ \frac{\partial F}{\partial z} &= xy + \lambda(x + 2y) = 0\\ \frac{\partial F}{\partial \lambda} &= xy + xz + 2yz - 96 = 0\end{aligned}\right\} \left.\begin{aligned}\lambda &= -\frac{yz}{y + z}\\ \lambda &= -\frac{xz}{x + 2z}\\ \lambda &= -\frac{xy}{x + 2y}\\ xy + xz + 2yz &= 96\end{aligned}\right\} \text{Equating expressions for } \lambda \text{ gives}$$

$$\left.\begin{aligned}x &= \frac{2y^2}{z}\\ z &= y \text{ (so } x = 2y)\\ 2y^2 + 2y^2 + 2y^2 &= 96\end{aligned}\right\} \begin{aligned}x &= 8 \text{ ft.}\\ y &= 4 \text{ ft.}\\ z &= 4 \text{ ft.}\end{aligned}$$

25. $F(x, y, \lambda) = f(x, y) + \lambda(c - ax - by)$. The values of x and y that minimize production subject to the cost constraint satisfy

$$\left.\begin{aligned} \frac{\partial F}{\partial x}(x, y) &= \frac{\partial f}{\partial x}(x, y) - \lambda a = 0 \\ \frac{\partial F}{\partial y}(x, y) &= \frac{\partial f}{\partial y}(x, y) - \lambda b = 0 \\ \frac{\partial F}{\partial \lambda}(x, y) &= c - ax - by = 0 \end{aligned}\right\} \quad \left.\begin{aligned} \frac{\partial f}{\partial x}(x, y) &= \lambda a \\ \frac{\partial f}{\partial y}(x, y) &= \lambda b \end{aligned}\right\}$$

Dividing the 1st equation by the second gives

$$\frac{\frac{\partial f}{\partial x}(x, y)}{\frac{\partial f}{\partial y}(x, y)} = \frac{a}{b}.$$

26. Given $f(x, y) = kx^{\alpha}y^{\beta}$, $\frac{\partial f}{\partial x} = \alpha kx^{\alpha-1}y^{\beta}$ and $\frac{\partial f}{\partial y} = \beta kx^{\alpha}y^{\beta-1}$. Applying the result of ex. 19, at the optimal values (x, y), $\frac{\alpha kx^{\alpha-1}y^{\beta}}{\beta kx^{\alpha}y^{\beta-1}} = \frac{a}{b}$; so $\frac{\alpha y}{\beta x} = \frac{a}{b}$; or $\frac{y}{x} = \frac{\beta a}{\alpha b}$.

Exercises 7.5

1. Given $y = Ax + B$, the error function is

$f(A, B) = (A + B - 0)^2 + (2A + B + 1)^2 + (3A + B + 4)^2.$

$$\frac{\partial f}{\partial A} = 2(A + B) + 4(2A + B + 1) + 6(3A + B + 4)$$
$$= 28A + 12B + 28 = 0$$
$$\frac{\partial f}{\partial B} = 2(A + B) + 2(2A + B + 1) + 2(3A + B + 4)$$
$$= 12A + 6B + 11 = 0$$

$$\left.\begin{aligned} 28A + 12B &= -28 \\ 12A + 6B &= -11 \end{aligned}\right\} \quad \begin{aligned} A &= -2 \\ B &= \frac{7}{3} \end{aligned}$$

2. Given $y = Ax + B$, the error function is

$f(A, B) = (2A + B - 2)^2 + (3A + B)^2 + (7A + B + 1)^2.$

$$\frac{\partial f}{\partial A} = 4(2A + B - 2) + 6(3A + B) + 14(7A + B + 1)$$
$$= 124A + 24B = 0$$
$$\frac{\partial f}{\partial B} = 2(2A + B - 2) + 2(3A + B) + 2(7A + B + 1)$$
$$= 24A + 6B - 2 = 0$$

$$\left.\begin{aligned} 124A + 24B &= -6 \\ 24A + 6B &= 2 \end{aligned}\right\} \quad \begin{aligned} A &= -\frac{1}{2} \\ B &= \frac{7}{3} \end{aligned}$$

The equation of the line is $y = -\frac{1}{2}x + \frac{7}{3}.$

3. Given $y = Ax + B$, the error function is

$f(A, B) = (A + B - 6)^2 + (3A + B)^2 + (6A + B - 10)^2.$

$$\frac{\partial f}{\partial A} = 2(A + B - 6) + 6(3A + B) + 12(6A + B - 10)$$
$$= 92A + 20B - 132 = 0$$
$$\frac{\partial f}{\partial B} = 2(A + B - 6) + 2(3A + B) + 2(6A + B - 10)$$
$$= 20A + 6B - 32 = 0$$

$$\left.\begin{aligned} 92A + 20B &= 132 \\ 20A + 6B &= 32 \end{aligned}\right\} \begin{aligned} A &= 1 \\ B &= 2 \end{aligned}$$ The equation of the line is $y = x + 2$.

4. $f(A, B) = (A + B - 1)^2 + (3A + B - 2)^2 + (5A + B - 4)^2$

$$\frac{\partial f}{\partial A} = 2(A + B - 1) + 6(3A + B - 2) + 10(5A + B - 4)$$
$$= 70A + 18B - 54 = 0$$
$$\frac{\partial f}{\partial B} = 2(A + B - 1) + 2(3A + B - 2) + 2(5A + B - 4)$$
$$= 18A + 6B - 14 = 0$$

$$\left.\begin{aligned} 70A + 18B &= 54 \\ 18A + 6B &= 14 \end{aligned}\right\} \begin{aligned} A &= .75 \\ B &= .0833 \end{aligned}$$ The equation of the line is $y = .75x + .0833$

5. $f(A, B) = (B - 1)^2 + (A + B - 1)^2 + (2A + B - 2)^2 + (3A + B - 2)^2$

$$\frac{\partial f}{\partial A} = 2(A + B - 1) + 4(2A + B - 2) + 6(3A + B - 2)^2$$
$$= 28A + 12B - 22 = 0$$
$$\frac{\partial f}{\partial B} = 2(B - 1) + 2(A + B - 1) + 2(2A + B - 2) + 2(3A + B - 2)$$
$$= 12A + 8B - 12 = 0$$

$$\left.\begin{aligned} 28A + 12B &= 22 \\ 12A + 8B &= 12 \end{aligned}\right\} \begin{aligned} A &= .4 \\ B &= .9 \end{aligned}$$ The equation of the line is $y = .4x + .9$.

6. $f(A, B) = (B - 2)^2 + (A + B - 2)^2 + (2A + B - 2)^2 + (3A + B - 3)^2$

$$\frac{\partial f}{\partial A} = 2(A + B - 2) + 4(2A + B - 2) + 6(3A + B - 3)$$
$$= 28A + 12B - 30 = 0$$
$$\frac{\partial f}{\partial B} = 2(B - 2) + 2(A + B - 2) + 2(24 + B - 2) + 2(3A + B - 3)$$
$$= 12A + 8B - 18 = 0$$

$$\left.\begin{aligned} 28A + 12B &= 30 \\ 12A + 8B &= 18 \end{aligned}\right\} \begin{aligned} A &= .3 \\ B &= 1.8 \end{aligned}$$ The equation of the line is $y = .3x + 1.8$.

7. $f(A, B) = (A + B)^2 + (2A + B - 1)^2 + (4A + B - 2)^2 + (5A + B - 3)^2$

$$\frac{\partial f}{\partial A} = 2(A + B) + 4(2A + B - 1) + 8(4A + B - 2) + 10(5A + B - 3)$$
$$= 92A + 24B - 50 = 0$$
$$\frac{\partial f}{\partial B} = 2(A + B) + 2(2A + B - 1) + 2(4A + B - 2) + 2(5A + B - 3)$$

$$= 24A + 8B - 12 = 0$$

$$\left.\begin{aligned} 92A + 24B &= 50 \\ 24A + 8B &= 12 \end{aligned}\right\} \begin{aligned} A &= .7 \\ B &= -.6 \end{aligned}$$

The equation of the line is $y = .7x - .6$.

8. $f(A, B) = (3A + B - 2)^2 + (5A + B - 1)^2 + (6A + B)^2$

$$\frac{\partial f}{\partial A} = 4(2A + B - 3) + 6(3A + B - 2) + 10(5A + B - 2)$$
$$+ 12(6A + B) = 148A + 28B - 34 = 0$$

$$\frac{\partial f}{\partial B} = 2(2A + B - 3) + 2(3A + B - 2) + 2(5A + B - 1) + 2(6A + B)$$
$$= 32A + 8B - 12 = 0$$

$$\left.\begin{aligned} 480A + 40B &= 356 \\ 32A + 8B &= 12 \end{aligned}\right\} \begin{aligned} A &= -.7 \\ B &= 4.3 \end{aligned}$$

The equation of the line is $y = -.7x + 4.3$.

9. $f(A, B) = (A + B)^2 + (2A + B - 1)^2 + 3A + B - 5)^2$
$+ (15A + B - 11)^2 + (-A + B - 4)^2$

$$\frac{\partial f}{\partial A} = 2(A + B) + 4(2A + B - 1) + 6(3A + B - 5)$$
$$+ 30(15A + B - 11) - 2(-A + B - 4)$$
$$= 480A + 40B - 356 = 0$$

$$\frac{\partial f}{\partial B} = 2(A + B) + 2(2A + B - 1) + 2(3A + B - 5)$$
$$+ 2(15A + B - 11) + 2(-A + B - 4)$$
$$= 40A + 10B - 42 = 0$$

$$\left.\begin{aligned} 480A + 40B &= 356 \\ 40A + 10B &= 42 \end{aligned}\right\} \begin{aligned} A &= .5875 \\ B &= 1.85 \end{aligned}$$

The equation of the line is $y = .5875x + 1.85$.

10. $f(A, B) = (B - 7)^2 + (-A + B - 5)^2 + (A + B - 9)^2$
$+ (3A + B - 14)^2 + (-2A + B)^2$

$$\frac{\partial f}{\partial A} = -2(-A + B - 5) + 2(A + B - 9) + 6(3A + B - 14)$$
$$+ 4(-2A + B)$$
$$= 30A + 2B - 92 = 0$$

$$\frac{\partial f}{\partial B} = 2(B - 7) + 2(-A + B - 5) + 2(A + B - 9) + 2(3A + B - 14)$$
$$+ 2(-2A + B)$$
$$= 2A + 10B - 70$$

$$\left.\begin{aligned} 30A + 2B &= 92 \\ 2A + 10B &= 70 \end{aligned}\right\} \begin{aligned} A &= 2.635 \\ B &= 6.473 \end{aligned}$$

The equation of the line is $y = 2.635x + 6.473$.

11. $\left(\frac{11}{3} - 4\right)^2 + \left(5 - \frac{17}{3}\right)^2 + \left(\frac{23}{3} - 8\right)^2 = \frac{2}{3}$

12. $\left(95 - 106\right)^2 + \left(120 - 123\right)^2 + \left(140 - 165\right)^2 + \left(170 - 181\right)^2 = 876.$

13. Letting the week number and number of cars sold correspond to x- and y-coordinates respectively, the least-squares function is $f(A, B) = (A + B - 38)^2 + (2A + B - 40)^2 + (3A + B - 41)^2 + (4A + B - 39)^2 + (5A + B - 45)^2$

$$\frac{\partial f}{\partial A} = 2(A + B - 38) + 4(2A + B - 40) + 6(3A + B - 41) + 8(4A + B - 39) + 10(5A + B - 45)$$
$$= 110A + 30B - 1244 = 0$$
$$\frac{\partial f}{\partial B} = 2(A + B - 38) + 2(2A + B - 40) + 2(3A + B - 41) + 2(4A + B - 39) + 2(5A + B - 45)$$
$$= 30A + 10B - 406 = 0$$

$$\left.\begin{array}{l} 110A + 30B = 1244 \\ 30A + 10B = 406 \end{array}\right\} \begin{array}{l} A = 1.3 \\ B = 36.7 \end{array}$$

The equation of the line is $y = 1.3x + 36.7$.

Thus, to estimate the number of cars sold during the seventh week is $1.3(7) + 36.7 = 45.8$; so approximately 46 cars will be sold.

14. Measuring profits in thousands, if $y = Ax + B$, the error function is

$$f(A, B) = (78A + B - 52)^2 + (83A + B - 55)^2 + (100A + B - 68)^2 + (110A + B - 77)^2 + (129A + B - 97)^2$$

$$\frac{\partial f}{\partial A} = 156(78A + B - 52) + 166(83A + B - 55) + 200(100A + B - 68) + 220(110A + B - 77) + 258(129A + B - 97)$$
$$= 103{,}428A + 1000B - 72{,}808 = 0.$$
$$\frac{\partial f}{\partial B} = 2(78A + B - 52) + 2(83A + B - 55) + 2(100A + B - 68) + 2(110A + B - 77) + 2(129A + B - 97)$$
$$= 1000A + 10B - 698 = 0$$

$$\left.\begin{array}{l}103{,}428A + 1000B = 72{,}808 \\ 1000A + 10B = 698\end{array}\right\} \begin{array}{l}A = .87748 \\ B = -17.94787\end{array}$$

If production is increased to 150 (thousand units), then we estimate the profit should be $y = .87748(150) - 17.94787$

$\sim 113.6741.$

Since we measured profit in thousands, this represents about $113,674.

15. (a) $f(A, B) = (2.7A + B - 11.2)^2 + (2.8A + B - 10)^2 + (3A + B - 8.5)^2 + (3.5A + B - 7.5)^2$

$$\frac{\partial f}{\partial A} = 5.4(2.7A + B - 11.2) + 5.6(2.8A + B - 10) + 6(3a + B - 8.5) + 7(3.5A + B - 7.5)$$

$$= 72.76A + 24B - 219.98 = 0$$

$$\frac{\partial f}{\partial B} = 2(2.7A + B - 11.2) + 2(2.8A + B - 10) + 2(3A + B - 8.5) + 2(3.5A + B - 7.5)$$

$$= 24A + 8B - 74.4 = 0$$

$$\left.\begin{array}{l}72.76A + 24B = 219.98 \\ 24A + 8B = 74.4\end{array}\right\} \begin{array}{l}A = -4.233686 \sim -4.2 \\ B = 22.01058 \sim 22\end{array}$$

The equation of the line is $y = -4.2x + 22.$

(b) $y = -4.2(3.2) + 22 = 8.6^\circ C.$

16. (a) $f(A, B) = (30A + B - 120)^2 + (35A + B - 100)^2 + (40A + B - 90)^2$

$$\frac{\partial f}{\partial A} = 60(30A + B - 120) + 70(35A + B - 100) + 80(40A + B - 90)$$

$$= 7450A + 210B - 21{,}400 = 0$$

$$\frac{\partial f}{\partial B} = 2(30A + B - 120) + 2(35A + B - 100) + 2(40A + B - 90)$$

$$= 210A + 6B - 620 = 0$$

$$\left.\begin{array}{l}7450A + 210B = 21{,}400 \\ 210A + 6B = 620\end{array}\right\} \begin{array}{l}A = -3 \\ B = 208.33\end{array}$$

The equation of the line is $y = -3x + 208.33.$

17. The points are (1.00,1400),(2.20,620),(2.80,700),(3.10,580) and (3.85,420).

$$E = \left[A + B - 1400\right]^2 + \left[2.2A + B - 620\right]^2 + \left[2.8A + B - 700\right]^2$$

$$+ \left[3.1A + B - 580\right]^2 + \left[3.85A + B - 420\right]^2$$

$$\frac{\partial E}{\partial A} = 2\Big[\left[A + B - 1400\right]1 + \left[2.2A + B - 620\right]2.2$$

$$+ \left[2.8A + B - 700\right]2.8 + \left[3.1A + B - 580\right]3.1$$

$$+ \left[3.85A + B - 420\right]3.85\Big] = 0$$

$$\frac{\partial E}{\partial A} = 2\Big[A + B - 1400 + 2.2A + B - 620 + 2.8A + B - 700 + 3.1A$$

$$+ B - 580 + 3.85A + B - 420\Big] = 0$$

We get the following equations:

$$38.1125A + 12.95B = 8139$$

$$12.95A + 5B = 3720$$

Solving: we find A = -327.2, B = 1591.4.

So the line is $y = -327.2x + 1591.4$

(b) Set $x = 3.0$ and substitute into the line equation.

$$y = -327.2(3) + 1591.4 = 610$$

Exercises 7.6

1. $\int_0^1 \left(\int_0^1 e^{x+y} dy \right) dx = \int_0^1 \left(e^{x+y} \Big|_0^1 \right) dx = \int_0^1 \left(e^{x+1} - e^x \right) dx$

$= e^{x+1} - e^x \Big|_{y=0}^1$

$= e^2 - e - e + 1 = e^2 - 2e + 1.$

2. $\int_{-1}^1 \left(\int_0^2 (3x^3 + y^2) dy \right) dx = \int_{-1}^1 \left(\left[3x^3 + \frac{1}{3}y^3 \right] \Big|_{y=0}^2 \right) dx$

$= \int_{-1}^1 \left(6x^3 + \frac{8}{3} - 0 \right) dx = \frac{3}{2}x^4 + \frac{8}{3}x \Big|_{-1}^1$

$= \frac{3}{2} + \frac{8}{3} - \frac{3}{2} + \frac{8}{3} = \frac{16}{3}.$

3. $\int_{-2}^0 \left(\int_{-1}^1 xe^{xy} dy \right) dx = \int_{-2}^0 \left(e^{xy} \Big|_{y=-1}^1 \right) dx = \int_{-2}^0 (e^x - e^{-x}) dx$

$= e^x + e^{-x} \Big|_{-2}^0 = 1 + 1 - e^{-2} - e^{-2}$

$= 2 - e^{-2} - e^2.$

4. $\int_1^3 \left(\int_2^5 (2y - 3x) dy \right) dx = \int_1^3 \left((y^2 - 3xy) \Big|_{y=2}^5 \right) dx$

$= \int_1^3 (25 - 15x - 14 + 6x) dx$

$= \int_1^3 (21 - 9x) dx = 21x - \frac{9}{2}x^2 \Big|_1^3$

$= 63 - \frac{81}{2} - 21 + \frac{9}{2} = 6$

5. $\int_1^4 \left(\int_x^{x^2} xy \, dy \right) dx = \int_1^4 \left(\frac{1}{2}xy^2 \Big|_{y=x}^{x^2} \right) dx = \int_1^4 \left(\frac{1}{2}x^5 - \frac{1}{2}x^3 \right) dx$

$= \frac{1}{12}x^6 - \frac{1}{8}x^4 \Big|_1^4 = \frac{1}{12}4^6 - \frac{1}{8}4^4 - \frac{1}{12} + \frac{1}{8} = 309\frac{3}{8}.$

6. $$\int_0^2\left[\int_x^{5x} y\,dy\right]dx = \int_0^2\left[\frac{1}{2}y^2\Big|_{y=x}^{5x}\right]dx = \int_0^2\left[\frac{25}{2}x^2 - \frac{1}{2}x^2\right]dx$$

$$= \int_0^2 12x^2dx = 4x^3\Big|_0^2 = 32.$$

7. $$\int_{-1}^1\left[\int_x^{2x}(x + y)dy\right]dx = \int_{-1}^1\left[\left[xy + \frac{1}{2}y^2\right]\Big|_{y=x}^{2x}\right]dx$$

$$= \int_{-1}^1\left[2x^2 + 2x^2 - x^2 - \frac{1}{2}x^2\right]dx = \int_{-1}^1 \frac{5}{2}x^2dx$$

$$= \frac{5}{6}x^3\Big|_{-1}^1 = \frac{5}{6} + \frac{5}{6} = \frac{5}{3}.$$

8. $$\int_0^1\left[\int_0^x(x + e^y)dy\right]dx = \int_0^1\left[(xy + e^y)\Big|_{y=0}^x\right]dx = \int_0^1(x^2 + e^x - 1)dx$$

$$= \frac{1}{3}x^3 + e^x - x\Big|_0^1 = \frac{1}{3} + e - 1 - 1 = e - \frac{5}{3}.$$

9. $$\int_0^2\left[\int_2^3 xy^2dy\right]dx = \int_0^2\left[\frac{1}{3}xy^3\Big|_{y=2}^3\right]dx = \int_0^2\left[9x - \frac{8}{3}x\right]dx = \int_0^2\frac{19}{3}x\,dx$$

$$= \frac{19}{6}x^2\Big|_0^2 = \frac{38}{3}.$$

10. $$\int_0^2\left[\int_2^3(xy + y^2)dy\right]dx = \int_0^2\left[\left[\frac{1}{2}xy^2 + \frac{1}{3}y^3\right]\Big|_{y=2}^3\right]dx$$

$$= \int_0^2\left[\frac{9}{2}x + 9 - 2x - \frac{8}{3}\right]dx = \int_0^2\left[\frac{5}{2}x + \frac{19}{3}\right]dx$$

$$= \frac{5}{4}x^2 + \frac{19}{3}x\Big|_0^2 = 5 + \frac{38}{3} = \frac{53}{3}.$$

11. $$\int_0^2\left[\int_2^3 e^{-x-y}dy\right]dx = \int_0^2\left[-e^{-x-y}\Big|_{y=2}^3\right]dx = \int_0^2(e^{-x-2} - e^{-x-3})dx$$

$$= -e^{-x-2} + e^{-x-3}\Big|_0^2 = -e^{-4} + e^{-5} + e^{-2} - e^{-3}.$$

12. $$\int_0^2\left[\int_2^3 e^{y-x}dy\right]dx = \int_0^2\left[e^{y-x}\Big|_{y=2}^3\right]dx = \int_0^2(e^{3-x} - e^{2-x})dx$$

7. $f(x, y) = e^{x/y}$; $\frac{\partial f}{\partial x} = \frac{1}{y}e^{x/y}$; $\frac{\partial f}{\partial y} = -\frac{x}{y^2}e^{x/y}$

8. $f(x, y) = \frac{x}{(x - 2y)}$; $\frac{\partial f}{\partial x} = \frac{(1)(x - 2y) - (x)(1)}{(x - 2y)^2} = \frac{-2y}{(x - 2y)^2}$

$\frac{\partial f}{\partial y} = \frac{(0)(x - 2y) - (-2)(x)}{(x - 2y)^2} = \frac{2x}{(x - 2y)^2}$

9. $f(x, y, z) = x^3 - yz^2$; $\frac{\partial f}{\partial x} = 3x^2$; $\frac{\partial f}{\partial y} = -z^2$; $\frac{\partial f}{\partial z} = -2yz$

10. $f(x, y, \lambda) = xy + \lambda(5 - x - y) = xy + 5\lambda - x\lambda - y\lambda$

$\frac{\partial f}{\partial x} = y - \lambda$; $\frac{\partial f}{\partial y} = x - \lambda$; $\frac{\partial f}{\partial \lambda} = 5 - x - y$

11. $f(x, y) = x^3y + 8$; $\frac{\partial f}{\partial x} = 3x^2y$; $\frac{\partial f}{\partial x}(1, 2) = 3(1)^2(2) = 6$

$\frac{\partial f}{\partial y} = x^3$; $\frac{\partial f}{\partial y}(1, 2) = 1$

12. $f(x, y, z) = (x + y)z = xz + yz$; $\frac{\partial f}{\partial y} = y$; $\frac{\partial f}{\partial y}(2, 3, 4) = 4$

13. $f(x, y) = x^5 - 2x^3y + \frac{y^4}{2}$; $\frac{\partial f}{\partial x} = 5x^4 - 6x^2y$; $\frac{\partial^2 f}{\partial x^2} = 6y^2$

$\frac{\partial f}{\partial y} = -2x^3 + 2y^3$; $\frac{\partial^2 f}{\partial y^2} = 6y^2$; $\frac{\partial^2 f}{\partial x \partial y} = -6x^2$; $\frac{\partial^2 f}{\partial y \partial x} = -6x^2$

14. $f(x, y) = 2x^3 + x^2y - y^2$; $\frac{\partial f}{\partial x} = 6x^2 + 2xy$; $\frac{\partial^2 f}{\partial x^2} = 12x + 2y$

$\frac{\partial^2 f}{\partial x^2}(1, 2) = 16$; $\frac{\partial f}{\partial y} = x^2 - 2y$; $\frac{\partial^2 f}{\partial y^2} = -2$; $\frac{\partial^2 f}{\partial y2}(1, 2) = -2$

$\frac{\partial^2 f}{\partial x \partial y} = 2x$; $\frac{\partial^2 f}{\partial x \partial y}(1, 2) = 2$

15. $f(p, t) = -p + 6t - .02pt$; $\frac{\partial f}{\partial p} = -1 - .02t$

$\frac{\partial f}{\partial p}(25, 10{,}000) = -201$; $\frac{\partial f}{\partial t} = 6 - .02p$; $\frac{\partial f}{\partial t}(25, 10{,}000) = 5.5$

At the level p = 25, t = 10,000, an increase in price of $1 will result in a loss in sales of approximately 201 calculators, and an increase in advertising of $1 will result in the sale of approximately 5.5 additional calculators.

16. The crime rate increases with increased unemployment and decreases with increased social services and police force size.

17. $f(x, y) = -x^2 + 2y^2 + 6x - 8y + 5$; $\frac{\partial f}{\partial x} = -2x + 6$; $\frac{\partial f}{\partial y} = 4y - 8$

$-2x + 6 = 0$; $x = 3$; $4y - 8 = 0$; $y = 2$

The only possibility is $(x, y) = (3, 2)$.

18. $f(x, y) = x^2 + 3xy - y^2 - x - 8y + 4$; $\frac{\partial f}{\partial x} = 2x + 3y - 1$

$\frac{\partial f}{\partial y} = 3x - 2y - 8$

$\left.\begin{array}{l} 2x + 3y = 1 \\ 3x - 2y = 8 \end{array}\right\} \begin{array}{l} x = 2 \\ y = -1 \end{array}$ The only possibility is $(x, y) = (2, -1)$.

19. $f(x, y) = x^3 + 3x^2 + 3y^2 - 6y + 7$; $\frac{\partial f}{\partial x} = 3x^2 + 6x$; $\frac{\partial f}{\partial y} = 6y - 6$

$3x^2 + 6x = 0$; $x(3x + 6) = 0$; $x = 0$; $3x + 6 = 0$; $x = -2$

$6y - 6 = 0$; $y = 1$; $(x, y) = (0, 1), (-2, 1)$

20. $f(x, y) = \frac{1}{2}x^2 + 4xy + y^3 + 8y^2 + 3x + 2$; $\frac{\partial f}{\partial x} = x + 4y + 3$

$\frac{\partial f}{\partial y} = 4x + 3y^2 + 16y$; $x + 4y + 3 = 0$; $y = \frac{-3 - x}{4}$

$4x + 3y^2 + 16y = 0$; $y = \frac{-3y^2 - 4x}{16}$; $\frac{-3 - x}{4} = \frac{-3y^2 - 4x}{16}$

$-12 - 4x = -3y^2 - 4x$; $y = \pm 2$; $\frac{-3 - x}{4} = \pm 2$; $x = -11, 5$

$(x, y) = (-11, 2), (5, -2)$

21. $f(x, y) = x^2 + 3xy + 4y^2 - 13x - 30y + 12$

$\frac{\partial f}{\partial x} = 2x + 3y - 13$; $\frac{\partial^2 f}{\partial x^2} = 2$; $\frac{\partial f}{\partial y} = 3x + 8y - 30$; $\frac{\partial^2 f}{\partial y^2} = 8$

$\frac{\partial^2 f}{\partial x \partial y} = 3$

$\left.\begin{array}{l} 2x + 3y = 13 \\ 3x + 8y = 30 \end{array}\right\} \begin{array}{l} x = 2 \\ y = 3 \end{array}$

$D(2, 3) = 2 \cdot 8 - 32 > 0$, $\frac{\partial^2 f}{\partial x^2} > 0$, so $f(x, y)$ has a relative minimum at $(2, 3)$.

22. $f(x, y) = 7x^2 - 5xy + y^2 + x - y + 6$

$\frac{\partial f}{\partial x} = 14x - 5y + 1$; $\frac{\partial^2 f}{\partial x^2} = 14$; $\frac{\partial f}{\partial y} = -5x + 2y - 1$; $\frac{\partial^2 f}{\partial y^2} = 2$

$\frac{\partial^2 f}{\partial x \partial y} = -5$

$$\left.\begin{array}{l} 14x - 5y = -1 \\ -5x + 2y = 1 \end{array}\right\} \begin{array}{l} x = 1 \\ y = 3 \end{array}$$

$D(1, 3) = 14 \cdot 2 - (-5)^2 > 0$, $\frac{\partial^2 f}{\partial x^2} > 0$, so $f(x, y)$ has a relative minimum at $(1, 3)$.

23. $f(x, y) = x^3 + y^2 - 3x - 8y + 12$

$\frac{\partial f}{\partial x} = 3x^2 - 3$; $\frac{\partial^2 f}{\partial x^2} = 6x$; $\frac{\partial f}{\partial y} = 2y - 8$; $\frac{\partial^2 f}{\partial y^2} = 2$; $\frac{\partial^2 f}{\partial x \partial y} = 0$

$3x^2 = 3$ $\quad$ $2y = 8$

$x = \pm 1$ $\quad$ $y = 4$

$D(1, 4) = 6(1)(4) - 0^2 > 0$; $\frac{\partial^2 f}{\partial x^2}(1, 4) > 0$, so $f(x, y)$ has a relative minimum at $(1, 4)$.

$D(-1, 4) = 6(-1)(2) - 0^2 < 0$, so $f(x, y)$ has neither a maximum nor a minimum at $(-1, 4)$.

24. $f(x, y, z) = x^2 + 4y^2 + 5z^2 - 6x + 8y + 3$

$\frac{\partial f}{\partial x} = 2x - 6 = 0$; $x = 3$; $\frac{\partial f}{\partial y} = 8y + 8 = 0$; $y = -1$

$\frac{\partial f}{\partial z} = 10z = 0$; $z = 0$

$f(x, y, z)$ must assume its minimum value at $(3, -1, 0)$.

25. $F(x, y, \lambda) = 3x^2 + 2xy - y^2 + \lambda(5 - 2x - y)$

$$\left.\begin{array}{l} \frac{\partial F}{\partial x} = 6x + 2y - 2\lambda = 0 \\ \frac{\partial F}{\partial y} = 2x - 2y - \lambda = 0 \\ \frac{\partial F}{\partial \lambda} = 5 - 2x - y = 0 \end{array}\right\} \left.\begin{array}{l} \lambda = 3x + y \\ \lambda = 2x - 2y \\ 2x + y = 5 \end{array}\right\} \left.\begin{array}{l} x + 3y = 0 \\ 2x + y = 5 \end{array}\right\} \begin{array}{l} x = 3 \\ y = -1 \end{array}$$

The maximum value of $f(x, y)$ subject to the constraint is $3(3^2) + 2(3)(-1) - (-1)^2 = 20$ occurring at $(x, y) = (3, -1)$.

26. $F(x, y, \lambda) = -x^2 - 3xy - \frac{1}{2}y^2 + y + 10 + \lambda(10 - x - y)$

$$\left.\begin{array}{l} \frac{\partial F}{\partial x} = -2x - 3y - \lambda = 0 \\ \frac{\partial F}{\partial y} = -3x - y + 1 - \lambda = 0 \\ \frac{\partial F}{\partial \lambda} = 10 - x - y = 0 \end{array}\right\} \left.\begin{array}{l} \lambda = -2x - 3y \\ \lambda = -3x - y + 1 \\ x + y = 10 \end{array}\right\} \left.\begin{array}{l} x - 2y = 1 \\ x + y = 10 \end{array}\right\} \begin{array}{l} x = 7 \\ y = 3 \end{array}$$

27. $F(x, y, z, \lambda) = 3x^2 + 2y^2 + z^2 + 4x + y + 3z + \lambda(4 - x - y - z)$

$$\left.\begin{array}{l} \frac{\partial F}{\partial x} = 6x + 4 - \lambda = 0 \\ \frac{\partial F}{\partial y} = 4y + 1 - \lambda = 0 \\ \frac{\partial F}{\partial z} = 2z + 3 - \lambda = 0 \\ \frac{\partial F}{\partial \lambda} = 4 - x - y - z = 0 \end{array}\right\} \left.\begin{array}{l} 6x + 4 = 4y + 1 \\ 4y + 1 = 2z + 3 \\ x + y + z = 4 \end{array}\right\} \left.\begin{array}{l} x = \frac{2}{3}y - \frac{1}{2} \\ z = 2y - 1 \\ \frac{2}{3}y - \frac{3}{2} + y + 2y = 4 \end{array}\right\}$$

$x = \frac{1}{2};\ y = \frac{3}{2};\ z = 2$

28. The problem is to minimize $x + y + z$ subject to $xyz = 1000$.

$(x > 0,\ y > 0,\ z > 0)$

$F(x, y, z, \lambda) = x + y + z + \lambda(1000 - xyz)$

(Assuming $x \neq 0,\ y \neq 0,\ z \neq 0$)

$$\left.\begin{array}{l} \frac{\partial F}{\partial x} = 1 - \lambda yz \\ \frac{\partial F}{\partial y} = 1 - \lambda xz \\ \frac{\partial F}{\partial z} = 1 - \lambda xy \\ \frac{\partial F}{\partial \lambda} = 1000 - xyz \end{array}\right\} \left.\begin{array}{l} \lambda = \frac{1}{yz} \\ \lambda = \frac{1}{xz} \\ \lambda = \frac{1}{xy} \\ xyz = 1000 \end{array}\right\} \left.\begin{array}{l} yz = xz \\ xz = xy \\ xyz = 1000 \end{array}\right\} \left.\begin{array}{l} x = y = z \\ xyz = 1000 \end{array}\right\} \begin{array}{l} x = 10 \\ y = 10 \\ z = 10 \end{array}$$

The optimal dimensions are 10 in. x 10 in. x 10 in.

29. The problem is to maximize xy subject to $2x + y = 40$.

$F(x, y, \lambda) = xy + \lambda(40 - 2x - y)$

$$\left.\begin{array}{l} \frac{\partial F}{\partial x} = y - 2\lambda = 0 \\ \frac{\partial F}{\partial y} = x - \lambda = 0 \\ \frac{\partial F}{\partial \lambda} = 40 - 2x - y = 0 \end{array}\right\} \left.\begin{array}{l} \lambda = \frac{1}{2}y \\ \lambda = x \\ 2x + y = 40 \end{array}\right\} \left.\begin{array}{l} x = \frac{1}{2}y \\ 2y = 40 \end{array}\right\} \begin{array}{l} x = 10 \text{ ft.} \\ y = 20 \text{ ft.} \end{array}$$

30. If one more foot of fencing becomes available, the dimensions

of the garden can be increased by k and ℓ, where $2k + \ell = 1$. Let $f(x, y) = xy$. The resulting increase in area is $f(10 + k, 20 + \ell) - f(10, 20) \sim \frac{\partial f}{\partial x}(10, 20)k + \frac{\partial f}{\partial y}(10, 20)\ell$. In the solution to ex. 29, we see that $\frac{\partial f}{\partial x}(10, 20) = 2\lambda$ (from L-1) and $\frac{\partial f}{\partial y}(10, 20) = \lambda$ (from L-2). Thus the increase in area is approximately $2\lambda k + \lambda\ell = \lambda(2k + \ell) = \lambda$.

31. $f(x, y) = x\sqrt{y}$; $\frac{\partial f}{\partial x} = \sqrt{y}$, $\frac{\partial f}{\partial y} = \frac{x}{2\sqrt{y}}$

$f(12 + .1, 4 - .2) \sim 12\sqrt{4} + \sqrt{4}(.1) + \frac{12}{2\sqrt{4}}(-.2) = 23.6$

32. $f(x, y) = x^5 + \frac{y}{x}$; $\frac{\partial f}{\partial x} = 5x^4 - \frac{y}{x^2}$; $\frac{\partial f}{\partial y} = \frac{1}{x}$

$f(1 - .02, 3 - .01) \sim 1^5 + \frac{3}{1} + (5 - 3)(-.02) + 1(-.01) = 3.95$

33. $f(x, y, z) = \frac{x^2 + y}{z - 1}$; $\frac{\partial f}{\partial x} = \frac{2x}{z - 1}$; $\frac{\partial f}{\partial y} = \frac{1}{z - 1}$; $\frac{\partial f}{\partial z} = \frac{-x^2 - y}{(z - 1)^2}$

$f(2 - .02, 2 - .01, 2 + .01)$

$\sim \frac{2^2 + 2}{2 - 1} + \frac{2(2)}{2 - 1}(-.02) + \frac{1}{2 - 1}(-.01) - \frac{2^2 + 2}{(2 - 1)^2}(.01)$

$= 5.85$

34. Let $f(x, y) = 2x^2 + 4xy$. If the measurements of x and y are off by k and ℓ respectively, then the resulting error in $f(x, y)$ is $f(30, 50) - f(30 + k, 50 + \ell) \sim \frac{\partial f}{\partial x}(30, 50)k + \frac{\partial f}{\partial y}(30, 50)\ell$. Now $\frac{\partial f}{\partial x} = 4x + 4y$; $\frac{\partial f}{\partial y} = 4x$, so the error estimate is $320k + 120\ell$. If $|k| < 1$ and $|\ell| < 1$, then $|\text{error}| \le |320(1) + 120(1)| = 440 \text{ mm}^2$.

35. Let $y = Ax + B$ be the line. The error function is

$F(A, B) = (A + B - 1)^2 + (2A + B - 3)^2 + (3A + B - 6)^2$.

$\frac{\partial F}{\partial A} = 2(A + B - 1) + 4(2A + B - 3) + 6(3A + B - 6)$

$= 28A + 12B - 50$

$\frac{\partial F}{\partial B} = 2(A + B - 1) + 2(2A + B - 3) + 2(3A + B - 6)$

$= 12A + 6B - 20$

$$\left.\begin{aligned} 28A + 12B &= 50 \\ 12A + 6B &= 20 \end{aligned}\right\} \quad A = \frac{5}{2}, \quad B = -\frac{5}{3}$$

The equation of the line is $y = \frac{5}{2}x - \frac{5}{3}$.

36. The error function is

$F(A, B) = (A + B - 1)^2 + (3A + B - 4)^2 + (5A + B - 7)^2.$

$\frac{\partial F}{\partial A} = 2(A + B - 1) + 6(3A + B - 4) + 10(5A + B - 7)$

$= 70A + 18B - 96$

$\frac{\partial F}{\partial B} = 2(A + B - 1) + 2(3A + B - 4) + 2(5A + B - 7)$

$= 18A + 6B - 24$

$$\left.\begin{aligned} 70A + 18B &= 96 \\ 18A + 6B &= 24 \end{aligned}\right\} \quad A = \frac{3}{2}, \quad B = -\frac{1}{2}$$

The equation of the line is $y = \frac{3}{2}x - \frac{1}{2}$.

37. The error function is

$F(A, B) = (B - 1)^2 + (A + B + 1)^2 + (2A + B + 3)^2 + (3A + B + 5)^2.$

$\frac{\partial F}{\partial A} = 2(A + B + 1) + 4(2A + B + 3) + 6(3A + B + 5)$

$= 28A + 12B + 44$

$\frac{\partial F}{\partial B} = 2(B - 1) + 2(A + B + 1) + 2(2A + B + 3) + 2(3A + B + 5)$

$= 12A + 8B + 16$

$$\left.\begin{aligned} 28A + 12B &= -44 \\ 12A + 8B &= -16 \end{aligned}\right\} \quad A = -2, \quad B = 1$$

The equation of the line is $y = -2x + 1$.

38. $$\int_0^4\left[\int_0^1 (x\sqrt{y} + y)dx\right]dy = \int_0^4\left[\frac{1}{2}x^2\sqrt{y} + xy\Big|_{x=0}^{1}\right]dy = \int_0^4\left[\frac{1}{2}\sqrt{y} + y\right]dy$$

$$= \frac{1}{3}y^{3/2} + \frac{1}{2}y^2\Big|_0^4 = \frac{8}{3} + 8 = \frac{32}{3}.$$

39. $$\int_1^4\left[\int_0^5 (2xy^4 + 3)dx\right]dy = \int_1^4\left[x^2y^4 + 3x\Big|_{x=0}^{5}\right]dy = \int_1^4 (25y^4 + 15)dy$$

$$= 5y^5 + 15y\Big|_1^4 = 5(4^5) + 15(4) - 20$$

$$= 5160.$$

40. $\int_0^4\left[\int_1^3(2x + 3y)dy\right]dx = \int_0^4\left[2xy + \frac{3}{2}y^2\Big|_{y=1}^{3}\right]dx$

$= \int_0^4\left[6x + \frac{27}{2} - 2x - \frac{3}{2}\right]dx = \int_0^4(4x + 12)dx$

$= 2x^2 + 12x\Big|_0^4 = 80.$

41. $\iint_R 5\ dx\ dy$ represents the volume of a box with dimensions

$(4 - 0) \times (3 - 1) \times 5$. So $\iint_R 5\ dx\ dy = 4\cdot 2\cdot 5 = 40.$

42.

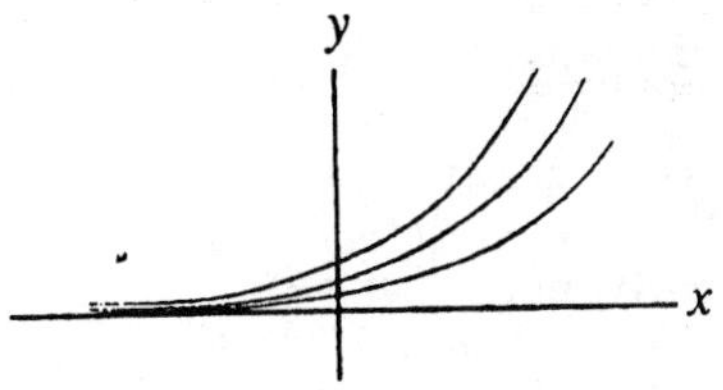

Exercises 8.1

1. $30^\circ = 30 \cdot \frac{\pi}{180}$ radians $= \frac{\pi}{6}$ radians.

   $120^\circ = 120 \cdot \frac{\pi}{180}$ radians $= \frac{2\pi}{3}$ radians.

   $315^\circ = 315 \cdot \frac{\pi}{180}$ radians $= \frac{7\pi}{4}$ radians.

2. $18^\circ = 18 \cdot \frac{\pi}{180}$ radians $= \frac{\pi}{10}$ radians.

   $72^\circ = 72 \cdot \frac{\pi}{180}$ radians $= \frac{2\pi}{5}$ radians.

   $150^\circ = 150 \cdot \frac{\pi}{180}$ radians $= \frac{5\pi}{6}$ radians.

3. $450^\circ = 450 \cdot \frac{\pi}{180}$ radians $= \frac{5\pi}{2}$ radians.

   $-210^\circ = -210 \cdot \frac{\pi}{180}$ radians $= \frac{-7\pi}{6}$ radians.

   $-90^\circ = -90 \cdot \frac{\pi}{180}$ radians $= -\frac{\pi}{2}$ radians.

4. $990^\circ = 990 \cdot \frac{\pi}{180}$ radians $= \frac{11\pi}{2}$ radians.

   $-270^\circ = 270 \cdot \frac{\pi}{180}$ radians $= -\frac{3\pi}{2}$ radians.

   $-540^\circ = -540 \cdot \frac{\pi}{180}$ radians $= -3\pi$ radians.

5. $t = 8 \cdot \frac{\pi}{2} = 4\pi$ radians

6. $t = -3 \cdot \frac{\pi}{2} = -\frac{3\pi}{2}$ radians

7. $t = 7 \cdot \frac{\pi}{2} = \frac{7\pi}{2}$ radians

8. $t = 9 \cdot \frac{\pi}{2} = \frac{9\pi}{2}$ radians

9. $t = -6 \cdot \frac{\pi}{2} = -3\pi$ radians

10. $t = -5 \cdot \frac{\pi}{2} = -\frac{5\pi}{2}$ radians

11. $t = 2 \cdot \frac{\pi}{3} = \frac{2\pi}{3}$ radians

12. $t = 5 \cdot \frac{\pi}{2} = \frac{5\pi}{4}$ radians

13.

14.

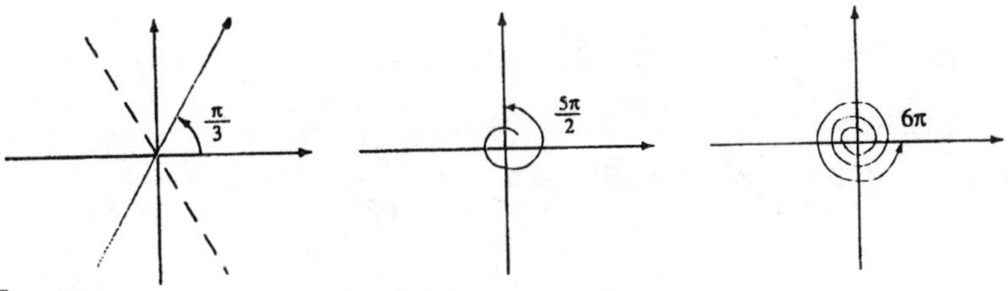
$\frac{\pi}{3}$
$\frac{5\pi}{2}$
6π

15.

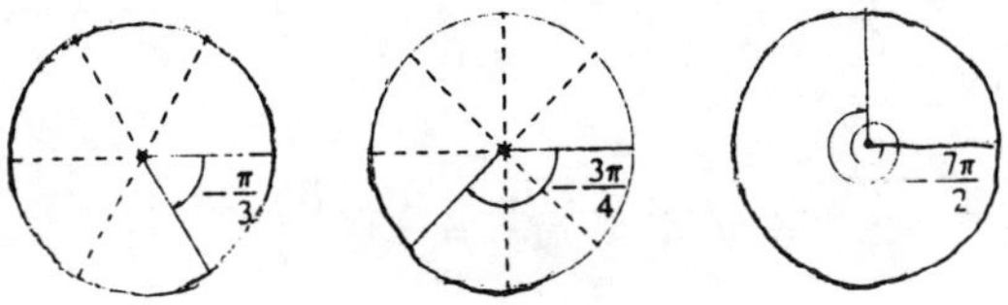
$-\frac{\pi}{3}$
$-\frac{3\pi}{4}$
$-\frac{7\pi}{2}$

16.

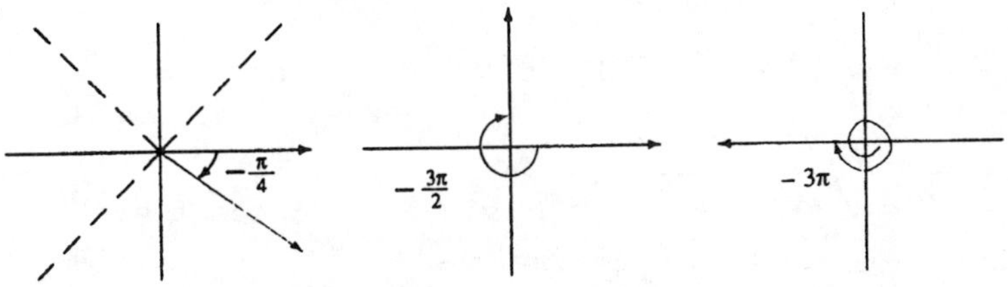
$-\frac{\pi}{4}$
$-\frac{3\pi}{2}$
-3π

17.

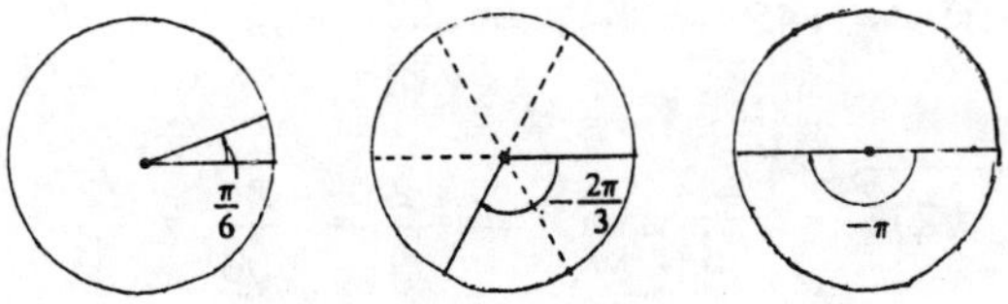
$\frac{\pi}{6}$
$-\frac{2\pi}{3}$
$-\pi$

18.

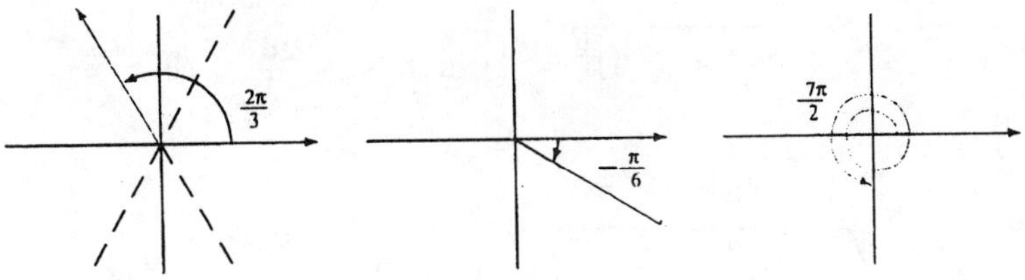
$\frac{2\pi}{3}$
$-\frac{\pi}{6}$
$\frac{7\pi}{2}$

Exercises 8.2

1. $\sin t = \frac{1}{2}$; $\cos t = \frac{\sqrt{3}}{2}$

2. $\sin t = \frac{2}{3}$; $\cos t = \frac{\sqrt{5}}{3}$

3. $r = \sqrt{x^2 + y^2} = \sqrt{3^2 + 2^2} = \sqrt{13}$; $\sin t = \frac{2}{\sqrt{13}}$; $\cos t = \frac{3}{\sqrt{13}}$

4. $r = \sqrt{x^2 + y^2}$, so $x = \sqrt{r^2 - y^2} = \sqrt{13^2 - 12^2} = 5$
   $\sin t = \frac{12}{13}$; $\cos t = \frac{5}{13}$

5. $r = \sqrt{x^2 + y^2}$, so $x = \sqrt{r^2 - y^2} = \sqrt{4^2 - 1^2} = \sqrt{15}$
   $\sin t = \frac{1}{4}$; $\cos t = \frac{\sqrt{15}}{4}$

6. $r = \sqrt{x^2 + y^2} = \sqrt{3^2 + 5^2} = \sqrt{34}$; $\sin t = \frac{5}{\sqrt{34}}$; $\cos t = \frac{3}{\sqrt{34}}$

7. $r = \sqrt{x^2 + y^2} = \sqrt{(-2)^2 + 1^2} = \sqrt{5}$; $\sin t = \frac{1}{\sqrt{5}}$; $\cos t = \frac{-2}{\sqrt{5}}$

8. $r = \sqrt{x^2 + y^2} = \sqrt{2^2 + (-3)^2} = \sqrt{13}$; $\sin t = \frac{-3}{\sqrt{13}}$; $\cos t = \frac{2}{\sqrt{13}}$

9. $r = \sqrt{x^2 + y^2} = \sqrt{(-2)^2 + (2)^2} = \sqrt{8}$; $\sin t = \frac{2}{\sqrt{8}} = \frac{2}{2\sqrt{2}} = \frac{\sqrt{2}}{2}$
   $\cos t = \frac{-2}{\sqrt{8}} = -\frac{2}{2\sqrt{2}} = -\frac{\sqrt{2}}{2}$

10. $r = \sqrt{x^2 + y^2} = \sqrt{(.6)^2 + (.8)^2} = 1$; $\sin t = \frac{.8}{1} = .8$
    $\cos t = \frac{.6}{1} = .6$

11. $r = \sqrt{x^2 + y^2} = \sqrt{(-.6)^2 + (-.8)^2} = 1$; $\sin t = \frac{-.8}{1} = -.8$
    $\cos t = \frac{-.6}{1} = -.6$

12. $r = \sqrt{x^2 + y^2} = \sqrt{(.8)^2 + (-.6)^2} = 1$; $\sin t = \frac{-.6}{1} = -.6$
    $\cos t = \frac{.8}{1} = .8$

13. $a = 12$, $b = 5$ and $c = 13$; $\sin t = \frac{b}{c} = \frac{5}{13} = .3846$

$t = .4$ radians

14. $t = 1.1$ and $c = 100$; $\sin t = \frac{b}{c}$; $\sin 1.1 = \frac{6}{10.0}$

$b = .8912(10.0) = 8.91$

15. $t = 1.1$ and $b = 3.2$; $\sin t = \frac{b}{c}$; $\sin 1.1 = \frac{3.2}{c}$

$c = \frac{3.2}{.8912} = 3.59$

16. $t = .4$ and $c = 5.0$; $\cos t = \frac{a}{c}$; $\cos .4 = \frac{a}{5.0}$

$a = (.92106)(5.0) = 4.61$

17. $t = .4$, $a = 10.0$; $\cos .4 = \frac{10.0}{c}$; $c = \frac{10.0}{.92106} = 10.86$

18. $t = .9$, $c = 20.0$; $\sin t = \frac{b}{c}$; $\sin .9 = \frac{b}{20.0}$

$b = (.78333)(20.0) = 15.67$; $\cos t = \frac{a}{c}$; $\cos .9 = \frac{a}{20.0}$

$a = (.62161)(20) = 12.43$

19. $t = .5$, $a = 2.4$; $\cos t = \frac{a}{c}$; $\cos .5 = \frac{2.4}{c}$; $c = \frac{2.4}{.87758} = 2.73$

$\sin t = \frac{b}{c}$; $\sin .5 = \frac{b}{2.73}$; $b = (.47943)(2.73) = 1.31$

20. $t = 1.1$, $b = 3.5$; $\sin t = \frac{b}{c}$; $\sin 1.1 = \frac{3.5}{c}$; $c = \frac{3.5}{.89121} = 3.93$

$\cos t = \frac{a}{c}$; $\cos 1.1 = \frac{a}{3.93}$; $a = (.45360)(3.93) = 1.78$

21. $\cos t = \cos\left(-\frac{\pi}{6}\right)$; $t = \frac{\pi}{6}$

22. $\cos t = \cos\left(\frac{3\pi}{2}\right)$; $t = \frac{\pi}{2}$

23. $\cos t = \cos\left(\frac{5\pi}{4}\right)$; $t = \frac{3\pi}{4}$

24. $\cos t = \cos\left(-\frac{4\pi}{6}\right)$; $t = \frac{2\pi}{3}$

25. $\cos t = \cos\left(-\frac{5\pi}{8}\right)$; $t = \frac{5\pi}{8}$

26. $\cos t = \cos\left(-\frac{3\pi}{4}\right)$; $t= \frac{3\pi}{4}$

27. $\sin t = \sin\left(\frac{3\pi}{4}\right)$; $t = \frac{\pi}{4}$

28. $\sin t = \sin\left(\frac{7\pi}{6}\right)$; $t = -\frac{\pi}{6}$

29. $\sin t = \sin\left(-\frac{4\pi}{3}\right)$; $t = \frac{\pi}{3}$

30. $\sin t = -\sin\left(\frac{3\pi}{8}\right)$; $t = -\frac{3\pi}{8}$

31. $\sin t = -\sin\left(\frac{\pi}{6}\right)$; $t = -\frac{\pi}{6}$

32. $\sin t = -\sin\left(-\frac{\pi}{3}\right)$; $t = \frac{\pi}{3}$

33. $\sin t = \cos t$; $t = \frac{\pi}{4}$

34. $\sin t = -\cos t$; $t = -\frac{\pi}{4}$

35. cos t decreases from 1 to -1

36. As t increases from π to $\frac{3\pi}{2}$, sin t decreases from 0 to -1. From $\frac{3\pi}{2}$ to 2π, sin t increases from -1 to 0.

37. $\sin 5t = 0$; $\sin(-2\pi) = 0$; $\sin\left(\frac{17\pi}{2}\right) = 1$; $\sin\left(\frac{-13\pi}{2}\right) = -1$

38. $\cos 5\pi = -1$; $\cos(-2\pi) = 1$; $\cos\left(\frac{17\pi}{2}\right) = 0$; $\cos\left(\frac{-13\pi}{2}\right) = 0$

39. $\sin(.19) = \sqrt{1 - \cos^2 .19} = .2$

$\cos(.19 - 4\pi) = \cos(.19) = .98$

$\cos(-.19) = \cos(.19) = .98$

$\sin(-.19) = -\sin(.19) = -.2$

40. $\sin(.42) = .41$

$\sin(6\pi - .42) = \sin(-.19) = -\sin(.42) = -.41$

$\cos(.42) = \sqrt{1 - \sin^2 .42} = .91$

$\sin(-.42) = -\sin(.42) = -.41$

Exercises 8.3

1. $\frac{d}{dt} \sin 4t = \cos 4t(4) = 4\cos 4t$

2. $\frac{d}{dt} -3\cos t = -3(-\sin t) = 3\sin t$

3. $\frac{d}{dt} 4\sin t = 4(\cos t) = 4\cos t$

4. $\frac{d}{dt} \cos(-3t) = -\sin(-3t)(-3) = 3\sin(-3t)$

5. $\frac{d}{dt} 2\cos 3t = 2(-\sin 3t)(3) = -6\sin 3t$

6. $\frac{d}{dt} 2\sin \pi t = 2(\cos \pi t)(\pi) = 2\pi \cos \pi t$

7. $\frac{d}{dt} (t + \cos \pi t) = 1 + (-\sin \pi t)(\pi) = 1 - \pi \sin \pi t$

8. $\frac{d}{dt} (t^2 - 2\sin 4t) = 2t - 2(\cos 4t)(4) = 2t - 8\cos 4t$

9. $\frac{d}{dt} \sin(\pi - t) = -\cos(\pi - t)$

10. $\frac{d}{dt} 2\cos(t + \pi) = -2\sin(t + \pi)$

11. $\frac{d}{dt} \cos^3 t = 3\cos^2 t(-\sin t) = -3\cos^2 t \sin t$

12. $\frac{d}{dt} \sin t^3 = \cos t^3(3t^2) = 3t^2 \cos t^3$

13. $\frac{d}{dx} \sin \sqrt{x - 1} = \cos(x - 1)^{1/2} \cdot \frac{1}{2}(x - 1)^{-1/2}(1) = \frac{\cos\sqrt{x - 1}}{2\sqrt{x - 1}}$

14. $\frac{d}{dx} \cos e^x = -\sin e^x(e^x) = -e^x \sin e^x$

15. $\frac{d}{dx} \sqrt{\sin(x - 1)} = \frac{1}{2}(\sin(x - 1))^{-1/2}(\cos(x - 1))(1)$

$= \frac{\cos(x - 1)}{2\sqrt{\sin(x - 1)}}$

16. $\frac{d}{dx} e^{\cos x} = e^{\cos x}(-\sin x) = -(\sin x)e^{\cos x}$

17. $\frac{d}{dt} (1 + \cos t)^8 = 8(1 + \cos t)^7(-\sin t) = -8\sin t\ (1 + \cos t)^7$

18. $\frac{d}{dt} \sqrt[3]{\sin \pi t} = \frac{1}{3}(\sin \pi t)^{-2/3}(\cos \pi t)(\pi) = \frac{1}{3}\pi \cos \pi t\ (\sin \pi t)^{-2/3}$

$= \frac{\pi \cos \pi t}{3\sqrt[3]{(\sin \pi t)^2}}$

19. $\frac{d}{dx} \cos^2 x^3 = 2(\cos x^3)(-\sin x^3)(3x^2) = -6x^2 \sin x^3 \cos x^3$

20. $\frac{d}{dx} (\sin^3 x + 4\sin^2 x) = 3(\sin x)^2(\cos x) + 4(2)(\sin x)(\cos x)$

$= 3\sin^2 x \cos x + 8\sin x \cos x$

$= \sin x \cos x(3\sin x + 8)$

21. $\frac{d}{dx} e^x \sin x = e^x(1)\sin x + e^x \cos x\ (1) = e^x \sin x + e^x \cos x$
$= e^x(\sin x + \cos x)$

22. $\frac{d}{dx} (x\sqrt{\cos x}) = (1)(\cos x)^{1/2} + x(-\sin x)(1)\left(\frac{1}{2}\right)(\cos x)^{-1/2}$
$= \sqrt{\cos x} - \frac{x \sin x}{2\sqrt{\cos x}}$

23. $\frac{d}{dx} \sin 2x \cos 3x = \cos 2x\ (2) \cos 3x + \sin 2x\ (-\sin 3x)(3)$
$= 2\cos 2x \cos 3x - 3\sin 2x \sin 3x$

24. $\frac{d}{dx} \sin^3 x \cos x = 3(\sin x)^2(\cos x)(\cos x) + (\sin x)^3(-\sin x)$
$= 3\sin^2 x \cos^2 x - \sin^4 x$
$= \sin^2 x\ (3\cos^2 x - \sin^2 x)$

25. $\frac{d}{dt} \frac{\sin t}{\cos t} = \cos t\ (\cos t)^{-1} + \sin t \cdot -1(\cos t)^{-2}(-\sin t)$
$= \frac{\cos t}{\cos t} + \frac{\sin^2 t}{\cos^2 t} = \frac{\cos^2 t + \sin^2 t}{\cos^2 t} = \frac{1}{\cos^2 t} = \cos^{-2} t$

26. $\frac{d}{dt} \left[\frac{e^t}{\cos 2t}\right] = e^t(1)(\cos 2t)^{-1} + e^t \cdot -1(\cos 2t)^{-2}(-\sin 2t)(2)$
$= \frac{e^t}{\cos 2t} + \frac{2e^t \sin 2t}{\cos^2 2t} = \frac{e^t(\cos 2t + 2\sin 2t)}{\cos^2 2t}$

27. $\frac{d}{dt} \ln(\cos t) = \frac{1}{\cos t} (-\sin t) = -\frac{\sin t}{\cos t}$

28. $\frac{d}{dt} \ln(\sin 2t) = \frac{1}{\sin 2t}(\cos 2t)(2) = \frac{2\cos 2t}{\sin 2t}$

29. $\frac{d}{dt} \sin(\ln t) = \cos(\ln t)\cdot\left(\frac{1}{t}\right) = \frac{\cos(\ln t)}{t}$

30. $\frac{d}{dt} (\cos t)\ln t = -\sin t(\ln t) + \cos t \cdot \frac{1}{t}$
$= -\sin t \ln t + \frac{\cos t}{t}$

31. $y = \cos 3x$; slope $= \frac{dy}{dx} = -\sin 3x(3) = -3\sin 3x$ When $x = \frac{13\pi}{6}$,
slope $= -3\sin 3\left(\frac{13\pi}{6}\right) = -3(1) = -3.$

32. $y = \sin 2x$; slope $= \frac{dy}{dx} = \cos 2x\ (2) = 2\cos 2x$ When $x = \frac{5\pi}{4}$,
slope $= 2\cos\left(2 \cdot \frac{5\pi}{4}\right) = 2(0) = 0.$

33. $y = 3\sin x + \cos 2x$ When $x = \frac{\pi}{2}$, $y = 3\sin\frac{\pi}{2} + \cos 2\left(\frac{\pi}{2}\right)$
$= 3 + (-1) = 2.$

slope $= \frac{dy}{dx} = 3\cos x + (-\sin 2x)2 = 3\cos x - 2\sin 2x$

When $x = \frac{\pi}{2}$, slope $= 3\cos\frac{\pi}{2} - 2\sin 2\cdot\frac{\pi}{2} = 0 - 0 = 0.$

The equation of the tangent line is $y - 2 = 0\left(x - \frac{\pi}{2}\right)$ or

$y = 2.$

34. $y = 3\sin 2x - \cos 2x$ When $x = \frac{3\pi}{4}$, $y = 3\sin 2\left(\frac{3\pi}{4}\right) - \cos 2\left(\frac{3\pi}{4}\right)$
$= -3 - 0 = -3.$

slope $= \frac{dy}{dx} = 3(\cos 2x)2 - (-\sin 2x)2 = 6\cos 2x + 2\sin 2x$

When $x = \frac{3\pi}{4}$, slope $= 6\cos 2\left(\frac{3\pi}{4}\right) + 2\sin 2\left(\frac{3\pi}{4}\right) = 0 - 2 = -2.$

The equation of the tangent line is $y - (-3) = -2\left(x - \frac{3\pi}{4}\right)$ or

$y = -2x + \frac{3\pi}{2} - 3.$

35. $\int \cos 2x\,dx = \frac{1}{2}\sin 2x + C$

36. $\int \sin\frac{x}{3}\,dx = -3\cos\frac{x}{3} + C$

37. $\int \sin(yx + 1)dx = -\frac{1}{4}\cos(4x + 1) + C = -\frac{\cos(4x + 1)}{4} + C$

38. $\int \cos(5 - x)dx = -\sin(5 - x) + C$

39. (a) $P = 100 + 20\cos 6t$; $\frac{dP}{dt} = (-20\sin 6t)6 = -120\sin 6t$

$\frac{d^2P}{dt^2} = (-120\cos 6t)6 = -720\cos 6t$

Setting $\frac{dP}{dt} = 0$ gives $-120\sin 6t = 0$; $\sin 6t = 0$

$6t = 0, \pi, 2\pi, 3\pi, \ldots$; $t = 0, \frac{\pi}{6}, \frac{\pi}{3}, \frac{\pi}{2}, \ldots$

When $t = 0$, $\frac{d^2P}{dt^2} = -720\cos 6(0) = -720.$

When $t = \frac{\pi}{6}$, $\frac{d^2P}{dt^2} = -720\cos 6\left(\frac{\pi}{6}\right) = -720(-1) = 720.$

When $t = \frac{\pi}{3}$, $\frac{d^2P}{dt^2} = -720\cos 6\left(\frac{\pi}{3}\right) = -720(1) = -720.$

When $t = \frac{\pi}{2}$, $\frac{d^2P}{dt^2} = -720\cos 6\left(\frac{\pi}{2}\right) = -720(-1) = 720$.

Thus $t = 0$ and $t = \frac{\pi}{3}$ give relative maximum values for P. The maximum value is $P = 100 + 20\cos 6\left(\frac{\pi}{3}\right) = 100 + 20(1) = 120$.

The minimum value is $P = 100 + 20\cos 6\left(\frac{\pi}{6}\right) = 100 - 20 = 80$.

(b) The length of time between two maximum values of P is $\frac{\pi}{3}$ seconds. The heart beats every $\frac{\pi}{3}$ seconds. Therefore the heart rate is $\frac{180}{\pi} \sim 57$ beats per minute.

40. $BMR(t) = .4 + .2\sin\left(\frac{\pi t}{12}\right)$; $BM = \int BMR(t)dt$

$$\int\left[.4 + .2\sin\left(\frac{\pi t}{12}\right)\right]dt = .4t - 2.4\pi\cos\left(\frac{\pi t}{12}\right) + C$$

$$\int_0^{24}\left[.4 + .2\sin\left(\frac{\pi t}{12}\right)\right]dt = \left[.4t - 2.4\pi\cos\left(\frac{\pi t}{12}\right)\right]\Bigg|_0^{24}$$

$$= \left[.4(24) - 2.4\pi\cos 2\pi\right] - \left[.4(0) - 2.4\pi(1)\right]$$

$$= 9.6 \text{ kcal}$$

41. $T = 59 + 12\cos\left[\frac{(t - 208)\pi}{183}\right]$, $0 \le t \le 365$

$$T' = -14\sin\left[\frac{(t - 208)\pi}{183}\right]\cdot\frac{\pi}{183} = -\frac{14\pi}{183}\sin\left[\frac{(t - 208)\pi}{183}\right]$$

$$T'' = -\frac{14\pi}{183}\cos\left[\frac{(t - 208)}{183}\pi\right]\cdot\frac{\pi}{183} = -\frac{14\pi^2}{183^2}\cos\left[\frac{(t - 208)}{183}\pi\right]$$

$$= .004\cos\left[\frac{\pi t - 208\pi}{183}\right]$$

Setting $T' = 0$ gives $-\frac{14\pi}{183}\sin\left[\frac{\pi t - 208\pi}{183}\right] = 0$; $t = 25, 208$.

When $t = 25$, $T'' = .004\cos\left[\frac{25\pi - 208\pi}{183}\right] = -.004$.

When $t = 208$, $T'' = .004\cos\left[\frac{208\pi - 208\pi}{183}\right] = .004$.

When $t = 208$, $T = 59 + 14\cos\left[\frac{(208 - 208)\pi}{183}\right] = 59 + 14 = 73^\circ F$.

When $t = 25$, $T = 59 + 14\cos\left[\frac{(25 - 208)\pi}{183}\right] = 59 - 14 = 45^\circ F$.

The maximum temperature 73°F occurs at t = 208 and the minimum temperature 45°F occurs at t = 25.

42. $D = 720 + 200\sin\left[\frac{(t - 79.5)\pi}{183}\right]$, $0 \le t \le 365$.

$$D' = 200\cos\left[\frac{(t - 79.5)\pi}{183}\right]\cdot\frac{\pi}{183} = \frac{200\pi}{183}\cos\left[\frac{(t - 79.5)\pi}{183}\right]$$

$$D'' = -\frac{200\pi}{183}\sin\left[\frac{(t - 79.5)\pi}{183}\right]\cdot\frac{\pi}{183} = -\frac{200\pi^2}{183^2}\sin\left[\frac{(t - 79.5)\pi}{183}\right]$$

Set $D' = 0$. $D' = \frac{200\pi}{183}\cos\left[\frac{(t - 79.5)\pi}{183}\right] = 0$; $t = 171.0, 354$

When $t = 171$, $D'' = -\frac{200\pi^2}{183^2}\sin\left[\frac{(171 - 79.5)\pi}{183}\right]$

$$= -\frac{200\pi^2}{183^2}\sin\left[\frac{\pi}{2}\right] = -\frac{200\pi^2}{183^2}.$$

When $t = 354$, $D'' = -\frac{200\pi^2}{183^2}\sin\left[\frac{(354 - 79.5)\pi}{183}\right] = -\frac{200\pi^2}{183^2}\sin\left[\frac{3\pi}{2}\right]$

$$= \frac{200\pi^2}{183^2}.$$

When $t = 171$, $D = 720 + 200\sin\left[\frac{(171 - 79.5)\pi}{183}\right]$

$$= 720 + 200\sin\left[\frac{\pi}{2}\right] = 920 \text{ minutes.}$$

When $t = 354$, $D = 720 + 200\sin\left[\frac{(354 - 79.5)\pi}{183}\right]$

$$= 720 - 200 = 520 \text{ minutes.}$$

The maximum of 920 minutes occurs at t = 171 and the minimum of 520 minutes occurs at t = 354 days.

43. Find $\lim_{h\to 0}\frac{\sin(\pi/2 + h) - 1}{h} = \lim_{h\to 0}\frac{\sin(\pi/2 + h) - \sin(\pi/2)}{h}$

Since $f'(a) \cong \frac{f(a + \Delta x) - f(a)}{\Delta x}$ we see that $f(x) = \sin x$ and $\Delta x = h$, $a = \pi/2$, so

$$(\sin(a))' = \frac{\sin(a + h) - \sin(a)}{h} = \frac{\sin(\pi/2 + h) - \sin(\pi/2)}{h}$$

but $(\sin(a))' = (\sin(\pi/2))' = \cos(\pi/2) = 0$

44. $f(x) = \cos(x)$, $\Delta x = h$, $a = \pi$

$$\text{So, } (\cos(a))' = \lim_{h\to 0} \frac{\cos(\pi + h) - \cos(\pi)}{h}$$
$$= \lim_{h\to 0} \frac{\cos(\pi + h) + 1}{h}$$
$$(\cos(\pi))' = -\sin\pi = 0.$$

Exercises 8.4

1. $\sec t = \dfrac{1}{\cos t} = \dfrac{\text{hyp.}}{\text{adj.}}$

2. $\cot t = \dfrac{\cos t}{\sin t} = \dfrac{\text{adj.}}{\text{opp.}}$

3. $(\text{adj.})^2 + 5^2 = 13^2$; adj. $= 12$; $\tan t = \dfrac{\sin t}{\cos t} = \dfrac{\text{opp.}}{\text{adj.}} = \dfrac{5}{12}$

$\sec t = \dfrac{1}{\cos t} = \dfrac{\text{hyp.}}{\text{adj.}} = \dfrac{13}{12}$

4. $3^2 + (\text{opp.})^2 = 4^2$; opp. $= \sqrt{7}$; $\tan t = \dfrac{\text{opp.}}{\text{adj.}} = \dfrac{\sqrt{7}}{3}$

$\sec t = \dfrac{\text{hyp.}}{\text{adj.}} = \dfrac{4}{3}$

5. $(-2)^2 + (1)^2 = (\text{hyp.})^2$; hyp. $= \sqrt{5}$; $\tan t = \dfrac{\sin t}{\cos t} = -\dfrac{1}{2}$

$\sec t = \dfrac{1}{\cos t} = -\dfrac{\sqrt{5}}{2}$

6. $2^2 + (-3)^2 = (\text{hyp.})^2$; hyp. $= \sqrt{13}$; $\tan t = \dfrac{\sin t}{\cos t} = \dfrac{\text{opp.}}{\text{adj.}} = \dfrac{-3}{2}$

$\sec t = \dfrac{\text{hyp.}}{\text{adj.}} = \dfrac{\sqrt{13}}{2}$

7. $(-2)^2 + 2^2 = (\text{hyp.})^2$; hyp. $= \sqrt{8}$

$\tan t = \dfrac{\sin t}{\cos t} = \dfrac{\text{opp.}}{\text{adj.}} = \dfrac{2}{\sqrt{2}} = -1$; $\sec t = \dfrac{1}{\cos t} = \dfrac{\text{hyp.}}{\text{adj.}} = \dfrac{\sqrt{8}}{-2}$

$= -\sqrt{2}$

8. $(.6)^2 + (.8)^2 = (\text{hyp.})^2$; hyp. $= 1$

$\tan t = \dfrac{\sin t}{\cos t} = \dfrac{\text{opp.}}{\text{adj.}} = \dfrac{.8}{.6} = \dfrac{4}{3}$; $\sec t = \dfrac{1}{\cos t} = \dfrac{\text{hyp.}}{\text{adj.}} = \dfrac{1}{.6} = \dfrac{5}{3}$

9. $(-.6)^2 + (-.8)^2 = (\text{hyp.})^2$; hyp. = 1

$$\tan t = \frac{\sin t}{\cos t} = \frac{\text{opp.}}{\text{adj.}} = \frac{-.8}{-.6} = \frac{4}{3};\ \sec t = \frac{1}{\cos t} = \frac{\text{hyp.}}{\text{adj.}} = \frac{1}{-.6} = -\frac{5}{3}$$

10. $(.8)^2 + (-.6)^2 = (\text{hyp.})^2$; hyp. = 1

$$\tan t = \frac{\sin t}{\cos t} = \frac{\text{opp.}}{\text{adj.}} = \frac{-.6}{.8} = -\frac{3}{4};\ \sec t = \frac{1}{\cos t} = \frac{\text{hyp.}}{\text{adj.}} = \frac{1}{.8} = \frac{5}{4}$$

11. $40 = 40 \times \frac{\pi}{180} = \frac{2\pi}{9} \sim .7;\ \tan(.7) = \frac{\sin t}{\cos t} = \frac{\text{opp.}}{\text{adj.}} = \frac{AB}{75}$

$AB = 75\tan(.7) \sim 63$ feet.

12. Let AC be the distance between the ground and the top of the spire, and AB be the distance between the ground and the top of the church.

$\tan(.4) = \frac{AC}{70}$; $AC = 70\tan(.4)$; $\tan(.3) = \frac{AB}{70}$; $AB = 70\tan(.3)$

The height of the spire is $70\tan(.4) - \tan(.3) \sim 7.94$ meters.

13. $$\frac{d}{dt}\sec t = \frac{d}{dt}\left(\frac{1}{\cos t}\right) = \frac{0 - (-\sin t)(1)}{\cos^2 t} = \tan t \sec t$$

14. $$\frac{d}{dt}\left(\frac{1}{\sin t}\right) = \frac{0 - (\cos t)(1)}{\sin^2 t} = -\cot t \csc t$$

15. $$\frac{d}{dt}\left(\frac{\cos t}{\sin t}\right) = \frac{-\sin t\ (\sin t) - \cos t\ (\cos t)}{\sin^2 t} = \frac{-\sin^2 t - \cos^2 t}{\sin^2 t} = \frac{-1}{\sin^2 t} = -\csc^2 t$$

16. $f(t) = \cot 3t = \frac{\cos 3t}{\sin 3t}$

$$\frac{d}{dt}\left(\frac{\cos 3t}{\sin 3t}\right) = \frac{(-\sin 3t)(3)\sin 3t - \cos 3t\ (\cos 3t)(3)}{\sin^2 3t}$$

$$= \frac{-3(\sin^2 3t + \cos^2 3t)}{\sin^2 3t} = \frac{-(3)(1)}{\sin^2 3t} = -3\csc^2 3t$$

17. $f(t) = \tan 4t = \frac{\sin 4t}{\cos 4t}$

$$\frac{d}{dt}\left(\frac{\sin 4t}{\cos 4t}\right) = \frac{(\cos 4t)(4)\cos 4t - \sin 4t\,(-\sin 4t)(4)}{\cos^2 4t}$$

$$= \frac{4(\cos^2 4t + \sin^2 4t)}{\cos^2 4t} = \frac{4(1)}{\cos^2 4t} = 4\sec^2 4t$$

18. $f(t) = \tan \pi t = \dfrac{\sin \pi t}{\cos \pi t}$

$$\frac{d}{dt}\left(\frac{\sin \pi t}{\cos \pi t}\right) = \frac{(\cos \pi t)(\pi)\cos \pi t - (-\sin \pi t)(\pi)\sin \pi t}{\cos^2 \pi t}$$

$$= \frac{\pi(\cos^2 \pi t + \sin^2 \pi t)}{\cos^2 \pi t} = \frac{\pi(1)}{\cos^2 \pi t} = \pi \sec^2 \pi t$$

19. $\dfrac{d}{dx}\left[3\tan(\pi - x)\right] = 3\dfrac{d}{dx}\left[\dfrac{\sin(\pi - x)}{\cos(\pi - x)}\right]$

$$= 3\left[\frac{\cos(\pi - x)(-1)\cos(\pi - x) - (-\sin(\pi - x)(-1)\sin(\pi - x)}{\cos^2(\pi - x)}\right]$$

$$= 3\left[\frac{-\cos^2(\pi - x) - \sin^2(\pi - x)}{\cos^2(\pi - x)}\right] = -3\frac{1}{\cos^2(\pi - x)}$$

$$= -3\sec^2(\pi - x)$$

20. $f(x) = 5\tan(2x + 1) = 5\left(\dfrac{\sin(2x + 1)}{\cos(2x + 1)}\right)$

$$\frac{d}{dx}\left[5\left(\frac{\sin(2x + 1)}{\cos(2x + 1)}\right)\right]$$

$$= 5\left[\frac{\cos(2x + 1)(2)\cos(2x + 1) - (-\sin(2x + 1)(2)\sin(2x + 1)}{\cos^2(2x + 1)}\right]$$

$$= 10\left[\frac{\cos^2(2x + 1) + \sin^2(2x + 1)}{\cos^2(2x + 1)}\right] = 10\left[\frac{1}{\cos^2(2x + 1)}\right]$$

$$= 10\sec^2(2x + 1)$$

21. $f(x) = 4\tan(x^2 + x + 3) = 4\left[\dfrac{\sin(x^2 + x + 3)}{\cos(x^2 + x + 3)}\right]$

$$\frac{d}{dx}\left[4\frac{\sin(x^2 + x + 3)}{\cos(x^2 + x + 3)}\right]$$

$$= 4\left[\frac{\cos(x^2 + x + 3)\cos(x^2 + x + 3)(2x + 1) - (-\sin(x^2 + x + 3)(2x + 1)\sin(x^2 + x + 3))}{\cos^2(x^2 + x + 3)}\right]$$

$$= 4\left[\frac{(2x + 1)\cos^2(x^2 + x + 3) + (2x + 1)\sin^2(x^2 + x + 3)}{\cos^2(x^2 + x + 3)}\right]$$

$$= 4(2x + 1)\left[\frac{1}{\cos^2(x^2 + x + 3)}\right] = 4(2x + 1)\sec^2(x^2 + x + 3)$$

22. $f(x) = 3\tan(1 - x^2) = \dfrac{3\sin(1 - x^2)}{\cos(1 - x^2)}$

$$\frac{d}{dx}\left[\frac{3\sin(1 - x^2)}{\cos(1 - x^2)}\right]$$

$$= 3\left[\frac{\cos(1 - x^2)(-2x)\cos(1 - x^2) - (-\sin(1 - x^2)(-2x)\sin(1 - x^2))}{\cos^2(1 - x^2)}\right]$$

$$3(-2x)\left[\frac{\cos^2(1 - x^2) + \sin^2(1 - x^2)}{\cos^2(1 - x^2)}\right] = -6x\left[\frac{1}{\cos^2(1 - x^2)}\right]$$

$$= -6x\sec^2(1 - x^2)$$

23. $y = \tan\sqrt{x} = \tan x^{1/2}$

$$\frac{d}{dx}\tan x^{1/2} = \frac{d}{dx}\left(\frac{\sin x^{1/2}}{\cos x^{1/2}}\right)$$

$$= \frac{\cos x^{1/2}\left(\frac{1}{2}x^{-1/2}\right)\cos x^{1/2} - (-\sin x^{1/2})\left(\frac{1}{2}x^{-1/2}\right)\sin x^{1/2}}{\cos^2 x^{1/2}}$$

$$= \frac{1}{2}x^{-1/2}\left(\frac{\cos^2 x^{1/2} + \sin^2 x^{1/2}}{\cos^2 x^{1/2}}\right) = \frac{1}{2\sqrt{x}}\left(\frac{1}{\cos^2 x^{1/2}}\right) = \frac{\sec^2\sqrt{x}}{\sqrt{x}}$$

24. $y = 2\tan\sqrt{x^2 - 4} = 2\tan(x^2 - 4)^{1/2} = \dfrac{2\sin(x^2 - 4)^{1/2}}{\cos(x^2 - 4)^{1/2}}$

$$\frac{d}{dx}\left[\frac{2\sin(x^2 - 4)^{1/2}}{\cos(x^2 - 4)^{1/2}}\right]$$

$$= 2\left[\frac{\cos(x^2 - 4)^{1/2}\left(\frac{1}{2}\right)(x^2 - 4)^{-1/2}(2x)\cos(x^2 - 4)^{1/2}}{\cos^2(x^2 - 4)^{1/2}}\right.$$

$$\left.\frac{-(-\sin(x^2 - 4)^{1/2})\left(\frac{1}{2}\right)(x^2 - 4)^{-1/2}(2x)\sin(x^2 - 4)^{1/2}}{}\right]$$

$$= 2x(x^2 - 4)^{-1/2}\left[\frac{\cos^2(x^2 - 4)^{1/2} + \sin^2(x^2 - 4)^{1/2}}{\cos^2(x^2 - 4)^{1/2}}\right]$$

$$= \frac{2x}{\sqrt{x^2 - 4}}\left[\frac{1}{\cos^2(x^2 - 4)^{1/2}}\right] = \frac{2x\sec^2\sqrt{x^2 - 4}}{\sqrt{x^2 - 4}}$$

25. $y = x\tan x = x\left(\frac{\sin x}{\cos x}\right)$

$$\frac{d}{dx}\left[x\left(\frac{\sin x}{\cos x}\right)\right] = (1)\left(\frac{\sin x}{\cos x}\right) + x\left(\frac{\cos x\cos x - (-\sin x)\sin x}{\cos^2 x}\right)$$

$$= \frac{\sin x}{\cos x} + x\left(\frac{\cos^2 x + \sin^2 x}{\cos^2 x}\right)$$

$$= \frac{\sin x}{\cos x} + \frac{x}{\cos^2 x}$$

$$= \tan x + x\sec^2 x$$

26. $y = e^{3x}\tan 2x = e^{3x}\left(\frac{\sin 2x}{\cos 2x}\right)$

$$\frac{d}{dx}\left[e^{3x}\left(\frac{\sin 2x}{\cos 2x}\right)\right] = e^{3x}(3)\left(\frac{\sin 2x}{\cos 2x}\right) +$$

$$e^{3x}\left(\frac{(\cos 2x)(2)\cos 2x - (-\sin 2x)(2)\sin 2x}{\cos^2 2x}\right)$$

$$= 3e^{3x}\tan 2x + 2e^{3x}\left(\frac{\cos^2 2x + \sin^2 2x}{\cos^2 2x}\right)$$

$$= 3e^{3x}\tan 2x + 2e^{3x}\left[\frac{1}{\cos^2 2x}\right]$$

$$= 3e^{3x}\tan 2x + 2e^{3x}\sec^2 2x = e^{3x}(2\sec^2 2x + 3\tan 2x)$$

27. $y = \tan^2 x = (\tan x)^2 = \left[\frac{\sin x}{\cos x}\right]^2$

$$\frac{d}{dx}\left[\left[\frac{\sin x}{\cos x}\right]^2\right] = 2\left[\frac{\sin x}{\cos x}\right]\left[\frac{\cos x \cos x - (-\sin x) \sin x}{\cos^2 x}\right]$$

$$= 2\tan x\left[\frac{\cos^2 x + \sin^2 x}{\cos^2 x}\right]$$

$$= 2\tan x\left[\frac{1}{\cos^2 x}\right] = 2\tan x \sec^2 x$$

28. $y = \sqrt{\tan x} = \left[\frac{\sin x}{\cos x}\right]^{1/2}$

$$\frac{d}{dx}\left[\left[\frac{\sin x}{\cos x}\right]^{1/2}\right] = \frac{1}{2}\left[\frac{\sin x}{\cos x}\right]^{-1/2}\left[\frac{\cos x \cos x - (-\sin x)\sin x}{\cos^2 x}\right]$$

$$= \frac{1}{2\sqrt{\tan x}}\left[\frac{\cos^2 x + \sin^2 x}{\cos^2 x}\right]$$

$$= \frac{\sec^2 x}{2\sqrt{\tan x}}$$

29. $y = (1 + \tan 2t)^3 = \left[1 + \frac{\sin 2t}{\cos 2t}\right]^3$

$$\frac{d}{dt}\left[\left[1 + \frac{\sin 2t}{\cos 2t}\right]^3\right]$$

$$= 3\left[1 + \frac{\sin 2t}{\cos 2t}\right]^2\left[\frac{\cos 2t\ (\cos 2t)(2) - (-\sin 2t)(2) \sin 2t}{\cos^2 2t}\right]$$

$$= 3(1 + \tan 2t)^2(2)\left[\frac{\cos^2 2t + \sin^2 2t}{\cos^2 2t}\right]$$

$$= 6(1 + \tan 2t)^2 \sec^2 2t = 6\sec^2 2t\ (1 + \tan 2t)^2$$

30. $y = \tan^4 3t = (\tan 3t)^4 = \left(\dfrac{\sin 3t}{\cos 3t}\right)^4$

$$\frac{d}{dt}\left[\left(\frac{\sin 3t}{\cos 3t}\right)^4\right]$$

$$= 4\left(\frac{\sin 3t}{\cos 3t}\right)^3\left(\frac{(\cos 3t)(3)\cos 3t - (-\sin 3t)(3)\sin 3t}{\cos^2 3t}\right)$$

$$= 4\tan^3 3t\ (3)\left(\frac{\cos^2 3t + \sin^2 3t}{\cos^2 3t}\right)$$

$$= 12\tan^3 3t\ \sec^2 3t$$

31. $y = \ln(\tan t + \sec t) = \ln\left(\dfrac{\sin t}{\cos t} + \dfrac{1}{\cos t}\right)$

$$\frac{d}{dt}\left[\ln\left(\frac{\sin t}{\cos t} + \frac{1}{\cos t}\right)\right]$$

$$= \frac{1}{\left(\dfrac{\sin t}{\cos t} + \dfrac{1}{\cos t}\right)}\left[\left(\frac{\cos t \cos t - (-\sin t)\sin t}{\cos^2 t}\right) + \left(\frac{0 - (-\sin t)(1)}{\cos^2 t}\right)\right]$$

$$= \frac{\cos t}{1 + \sin t}\left(\frac{1}{\cos^2 t} + \frac{\sin t}{\cos^2 t}\right) = \frac{\cos t}{1 + \sin t}\left(\frac{1 + \sin t}{\cos^2 t}\right)$$

$$= \frac{1}{\cos t} = \sec t$$

32. $y = \ln(\tan t)$; $\dfrac{d}{dx}\left[\ln(\tan t)\right] = \dfrac{1}{\tan t}\sec^2 t = \dfrac{\cos t}{\sin t}\cdot\dfrac{1}{\cos^2 t}$

$$= \frac{1}{\sin t \cos t} = \csc t \sec t$$

Chapter 8 Supplementary Exercises

1. $t = \frac{3\pi}{2}$ 2. $t = -\frac{7\pi}{2}$ 3. $t = -\frac{3\pi}{4}$

4.

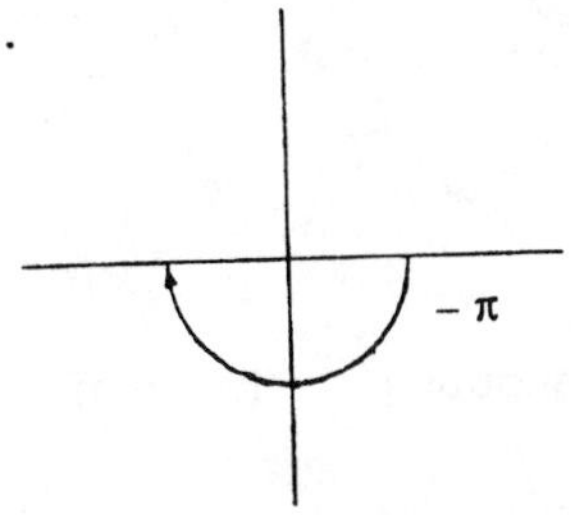

5.

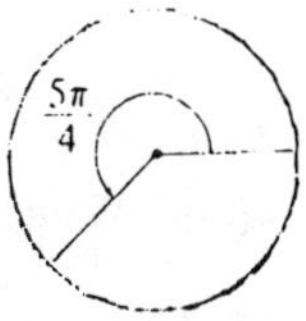

6.

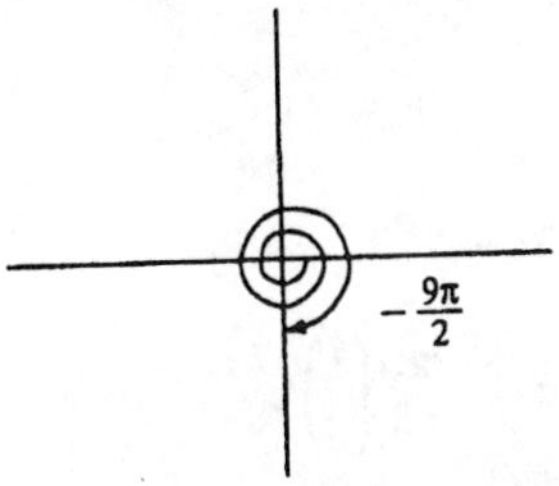

7. $3^2 + 4^2 = (\text{hyp.})^2$; hyp. = 5; $\sin t = \frac{\text{opp.}}{\text{hyp.}} = \frac{4}{5}$

$\cos t = \frac{\text{adj.}}{\text{hyp.}} = \frac{3}{5}$; $\tan t = \frac{\text{opp.}}{\text{adj.}} = \frac{4}{3}$

8. $(-.6)^2 + (.8)^2 = (\text{hyp.})^2$; hyp = 1; $\sin t = \frac{\text{opp.}}{\text{hyp.}} = .8$

$\cos t = \frac{\text{adj.}}{\text{hyp.}} = -.6$; $\tan t = \frac{\text{opp.}}{\text{adj.}} = \frac{.8}{-.6} = -\frac{4}{3}$

9. $(-.6)^2 + (-.8)^2 = (\text{hyp.})^2$; hyp. = 1; $\sin t = \frac{\text{opp.}}{\text{hyp.}} = -.8$

$\cos t = \frac{\text{adj.}}{\text{hyp.}} = -.6$; $\tan t = \frac{\text{opp.}}{\text{adj.}} = \frac{4}{3}$

10. $3^2 + (-4)^2 = (\text{hyp.})^2$; hyp. = 5; $\sin t = \frac{\text{opp.}}{\text{hyp.}} = -\frac{4}{5}$

$\cos t = \frac{\text{adj.}}{\text{hyp.}} = \frac{3}{5}$; $\tan t = \frac{\text{opp.}}{\text{adj.}} = -\frac{4}{3}$

11. $\sin t = \frac{1}{5}$; $(\text{opp.})^2 + (\text{adj.})^2 = (\text{hyp.})^2$; $1 + (\text{adj.})^2 = 25$

adj. $= \pm\sqrt{24}$; $\cos t = \frac{\text{adj.}}{\text{hyp.}} = \frac{\pm 2\sqrt{6}}{5}$

12. $\cos t = -\frac{2}{5}$; $(\text{opp.})^2 + (\text{adj.})^2 = (\text{hyp.})^2$; $(\text{opp.})^2 + 4 = 9$

opp. $= \pm\sqrt{5}$; $\sin t = \frac{\pm\sqrt{5}}{3}$

13. $\frac{\pi}{4}, \frac{5\pi}{4}, -\frac{3\pi}{4}, -\frac{7\pi}{4}$

14. $\frac{3\pi}{4}, \frac{7\pi}{4}, -\frac{\pi}{4}, -\frac{5\pi}{4}$

15. negative

16. positive

17. Let r be the length of the rafter needed to support the roof. $r^2 = (15)^2 + [15(\tan 23)]^2$; $r \sim 16.3$ feet.

18. Let t be the height of the tree. $t = 60(\tan 53) \sim 79.63$ feet.

19. $f(t) = 3\sin t$; $\frac{d}{dt} 3\sin t = 3\cos t$

20. $f(t) = \sin 3t$; $\frac{d}{dt} \sin 3t = (\cos 3t)(3) = 3\cos 3t$

21. $f(t) = \sin\sqrt{t} = \sin t^{1/2}$; $\frac{d}{dt} \sin t^{1/2} = (\cos t^{1/2})\left(\frac{1}{2}\right)t^{-1/2}$

$$= \frac{\cos\sqrt{t}}{2\sqrt{t}}$$

22. $f(t) = \cos t^3$; $\frac{d}{dt} \cos t^3 = (-\sin t^3)(3t^2) = -3t^2\sin t^3$

23. $g(x) = x^3\sin x$; $\frac{d}{dx} x^3\sin x = 3x^2\sin x + x^3\cos x$

24. $g(x) = \sin(-2x) \cos 5x$

$$\frac{d}{dx}\left[\sin(-2x) \cos 5x\right]$$

$$= \cos(-2x)(-2) \cos 5x + (-\sin 5x)(5)\sin(-2x)$$

$$= -2\cos(-2x) \cos 5x - 5\sin 5x \sin(-2x)$$

25. $f(x) = \frac{\cos 2x}{\sin 3x}$

$$\frac{d}{dx}\left(\frac{\cos 2x}{\sin 3x}\right) = \frac{(-\sin 2x)(2)\sin 3x - \cos 2x (\cos 3x)(3)}{\sin^2 3x}$$

$$= \frac{-2\sin 2x \sin 3x - 3\cos 2x \cos 3x}{\sin^2 3x}$$

26. $f(x) = \dfrac{\cos x - 1}{x^3}$

$$\frac{d}{dx}\left[\frac{\cos x - 1}{x^3}\right] = \frac{(-\sin x)(x^3) - 3x^2(\cos x - 1)}{x^6}$$
$$= \frac{-x^3\sin x - 3x^2(\cos x - 1)}{x^6}$$

27. $f(x) = \cos^3 4x$; $\dfrac{d}{dx}\cos^3 4x = 3\cos^2 4x\ (-\sin 4x)(4)$
$= -12\cos^2 4x \sin 4x$

28. $f(x) = \tan^3 2x$; $\dfrac{d}{dx}\tan^3 2x = 3\tan^2 x\ (\sec^2 2x)(2) = 6\tan^2 2x\ \sec^2 2x$

29. $y = \tan(x^4 + x^2)$; $\dfrac{d}{dx}\tan(x^4 + x^2) = (\sec^2(x^4 + x^2))(4x^3 + 2x)$
$= (4x^3 + 2x)\ \sec^2(x^4 + x^2)$

30. $y = \tan e^{-2x}$; $\dfrac{d}{dx}\tan e^{-2x} = (\sec^2 e^{-2x})e^{-2x}(-2)$
$= -2e^{-2x}\sec^2 e^{-2x}$

31. $y = \sin(\tan x)$; $\dfrac{d}{dx}\sin(\tan x) = \cos(\tan x)\ \sec^2 x$

32. $y = \tan(\sin x)$; $\dfrac{d}{dx}\tan(\sin x) = \sec^2(\sin x)\ \cos x$

33. $y = \sin x \tan x$; $\dfrac{d}{dx}\left[\sin x \tan x\right] = \cos x \tan x + \sec^2 x \sin x$
$= \sin x + \sec^2 x \sin x$

34. $y = \ln x \cos x$; $\dfrac{d}{dx}\left[\ln x \cos x\right] = \dfrac{1}{x}\cos x + (-\sin x)\ln x$
$= \dfrac{\cos x}{x} - \ln x \sin x$

35. $y = \ln(\sin x)$; $\dfrac{d}{dx}\ln(\sin x) = \dfrac{1}{\sin x}(\cos x) = \cot x$

36. $y = \ln(\cos x)$; $\dfrac{d}{dx}\ln(\cos x) = \dfrac{1}{\cos x}(-\sin x) = -\tan x$

37. $y = e^{3x}\sin^4 x$; $\dfrac{d}{dx}\left[e^{3x}\sin^4 x\right] = e^{3x}(3)\sin^4 x + 4\sin^3 x\ e^{3x}\cos x$
$= e^{3x}\sin^3 x(3\sin x + 4\cos x)$

38. $y = \sin^4 e^{3x}$; $\dfrac{d}{dx}\sin^4 e^{3x} = 4\sin^3 e^{3x}\ (\cos e^{3x})(e^{3x})(3)$
$= 12e^{3x}(\cos e^{3x})(\sin^3 e^{3x})$

39. $f(t) = \dfrac{\sin t}{\tan 3t}$; $\dfrac{d}{dt}\left(\dfrac{\sin t}{\tan 3t}\right) = \dfrac{\cos t \tan 3t - (\sec^2 3t)(3)\sin t}{\tan^2 3t}$

$= \dfrac{\cos t \tan 3t - 3\sin t \sec^2 3t}{\tan^2 3t}$

40. $f(t) = \dfrac{\tan 2t}{\cos t}$; $\dfrac{d}{dt}\left[\dfrac{\tan 2t}{\cos t}\right] = \dfrac{(\sec^2 2t)(2)\cos t - (-\sin t)\tan 2t}{\cos^2 t}$

$= \dfrac{2\cos t \sec^2 2t + \sin t \tan 2t}{\cos^2 t}$

41. $f(t) = e^{\tan t}$; $\dfrac{d}{dt} e^{\tan t} = e^{\tan t}(\sec^2 t)$

42. $f(t) = e^t \tan t$; $\dfrac{d}{dt}\left[e^t \tan t\right] = e^t(1)\tan t + \sec^2 t\ e^t$

$= e^t \tan t + e^t \sec^2 t$

$= e^t(\tan t + \sec^2 t)$

43. $f(t) = -\sin^2 t$; $f'(t) = 2\sin t \cos t$

$f''(t) = 2\left[\cos t \cos t + \sin t\ (-\sin t)\right] = 2(\cos^2 t - \sin^2 t)$

44. $y = 3\sin 2t + \cos 2t$; $y' = 3\left[(\cos 2t)(2) + (-\sin 2t)(2)\right]$

$= 3(2\cos 2t - 2\sin 2t)$

$y'' = 6\left[(-\sin 2t)(2) - (\cos 2t)(2)\right] = -12(\sin 2t + \cos 2t)$

$-4y = -12(\sin 2t + \cos 2t)$ Therefore y'' and $-4y$ are equal.

45. $f(s, t) = \sin s \cos 2t$; $\dfrac{\partial f}{\partial s} = \cos s \cos 2t$

$\dfrac{\partial f}{\partial t} = \sin s(-\sin 2t)(2) = -2\sin s \sin 2t$

46. $z = \sin wt$; $\dfrac{\partial z}{\partial w} = t \cos wt$; $\dfrac{\partial z}{\partial t} = w \cos wt$

47. $f(s, t) = t \sin st$; $\dfrac{\partial f}{\partial s} = t((\cos st)(t)) = t^2 \cos st$

$\dfrac{\partial f}{\partial t} = (1)\sin st + (\cos st)(s)(t) = \sin st + st \cos st$

48. $\sin(s + t) = \sin s \cos t + \cos s \sin t$

$\dfrac{d}{dt} \sin(s + t) = \cos(s + t)$

$$\frac{d}{dt}\left[\sin s \cos t + \cos s \sin t\right] = \sin s(-\sin t) + \cos s \cos t$$
$$= \cos s \cos t - \sin s \sin t$$

Thus, $\cos(s + t) = \cos s \cos t - \sin s \sin t$

49. $y = \tan t = \dfrac{\sin t}{\cos t}$ When $t = \frac{\pi}{4}$, $y = \dfrac{\sin \frac{\pi}{4}}{\cos \frac{\pi}{4}} = 1$

slope $= y' = \sec^2 t = \sec^2\left(\frac{\pi}{4}\right) = 2$

The tangent line is $y - 1 = 2\left(t - \frac{\pi}{4}\right)$; or $y = 2t - \frac{\pi}{2} + 1$.

50.

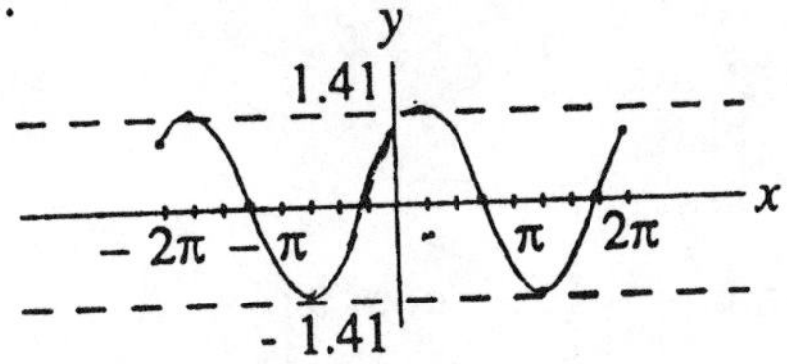

51.

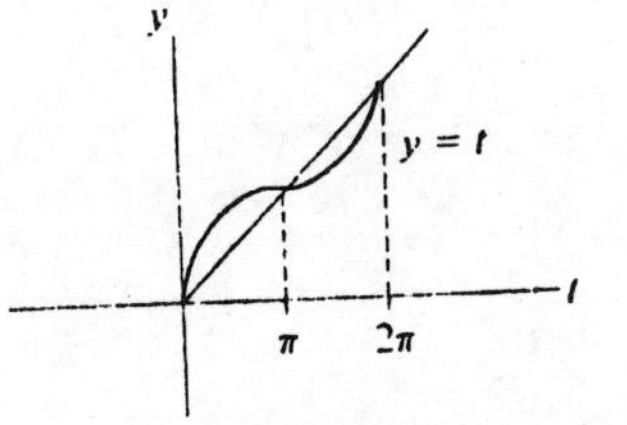

52. $y = 2 + \sin 3t$; Area under the curve is $\int_0^{\pi/2} 2 + \sin 3t\, dt$.

$$\int_0^{\pi/2} 2 + \sin 3t\, dt = 2t - (\cos 3t)\left(\frac{1}{3}\right)\Bigg|_0^{\pi/2} = 2t - \frac{1}{3}\cos 3t\Bigg|_0^{\pi/2}$$
$$= \pi - \frac{1}{3}(0) - \left(0 - \frac{1}{3}\right) = \frac{1}{3} + \pi$$

53. The desired area is

$$\int_0^{\pi} \sin t\, dt + \int_{\pi}^{2\pi} -\sin t\, dt = -\cos t\Big|_0^{\pi} + \cos t\Big|_{\pi}^{2\pi} = 2 + 2 = 4.$$

54. The desired area is

$$\int_0^{\pi/2} \cos t\, dt + \int_{\pi/2}^{3\pi/2} -\cos t\, dt = \sin t\Big|_0^{\pi/2} + -\sin t\Big|_{\pi/2}^{3\pi/2}$$

$$= 1 + 2 = 3$$

55. It is easy to check that the line $y = x$ is tangent to the graph of $y = \sin x$ at $x = 0$. From the graph of $y = \sin x$, it is clear that $y = \sin x$ lies below $y = x$ for $x \geq 0$. So the area between these two curves form $x = 0$ to $x = \pi$ is given by

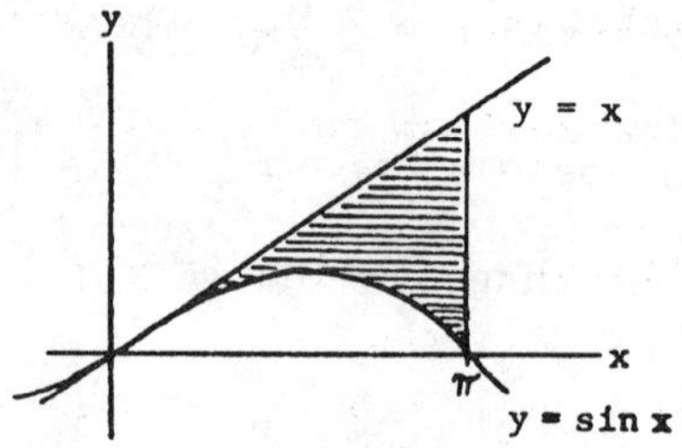

$$\text{Area} = \int_0^{\pi} (x - \sin x)dx = \left[\frac{x^2}{2} \cos x\right]\Bigg|_0^{\pi}$$

$$= \left[\frac{\pi^2}{2} + \cos \pi\right] - \left[0 - \cos 0\right]$$

$$= \frac{\pi^2}{2} + (-1) - [0 + 1] = \frac{\pi^2}{2} - 2.$$

56. (a) $V(0) = 2.95$, $V\left(\frac{1}{320}\right) = 3$, $V\left(\frac{1}{160}\right) = 3.05$, $V\left(\frac{1}{80}\right) = 2.95$

(b) $V'(t) = .05 \cos\left(160\pi t - \frac{\pi}{2}\right)(160\pi)$. Setting $V'(t) = 0$ gives

$\cos\left(160\pi t - \frac{\pi}{2}\right) = 0$; $160\pi t - \frac{\pi}{2} = \frac{\pi}{2}, \frac{3\pi}{2}, \frac{5\pi}{2}, \ldots$

$160t = \pi, 2\pi, 3\pi, \ldots$; $t = \frac{1}{160}, \frac{2}{160}, \frac{3}{160}, \ldots$

$V''(t) = -8\pi \sin\left(160\pi t - \frac{\pi}{2}\right)(160\pi) = -1280\pi^2 \sin\left(160\pi t - \frac{\pi}{2}\right)$

At $t = \frac{1}{160}$, $V''(t) < 0$, so this value of t gives a relative maximum and the maximum lung volume is $V\left(\frac{1}{160}\right) = 3.05$ liters.

57. (a) $V'(t) = 8\pi \cos\left(160\pi t - \frac{\pi}{2}\right)$

(b) Inspiration (in the first cycle) occurs from $t = 0$ to $t = \frac{1}{160}$. To find the maximum rate of inspiration, we need to find the maximum of $V'(t)$ on $\left[0, \frac{1}{160}\right]$.

$V''(t) = -1280\pi^2 \sin\left(160\pi t - \frac{\pi}{2}\right)$. Setting $V''(t) = 0$ gives

$160\pi t - \frac{\pi}{2} = 0, \pi, 2\pi, \ldots;\ 160\pi t = \frac{\pi}{2}, \frac{3\pi}{2}, \frac{5\pi}{2}, \ldots$

$t = \frac{1}{320}, \frac{3}{320}, \frac{5}{320}, \ldots$ Among these values, only $\frac{1}{320}$ is within $\left[0, \frac{1}{160}\right]$. Since $V'(t) = 0$ at the end points of the interval, $t = \frac{1}{320}$ must give the maximum value of $V'(t)$. Thus the maximum rate of air flow is $V'\left(\frac{1}{320}\right) = 8\pi$ liters/min.

(c) The average value of $V'(t)$ on $\left[0, \frac{1}{160}\right]$ is

$$\frac{1}{\frac{1}{160}}\int_0^{1/160} V'(t)dt = 160\int_0^{1/160} 8\pi\cos\left(160\pi t - \frac{\pi}{2}\right)dt$$

$$= 1280\pi\left[\frac{1}{160\pi}\sin\left(160\pi t - \frac{\pi}{2}\right)\right]\Bigg|_0^{1/160}$$

$$= 8(1 + 1) = 16 \text{ (liters/min)}$$

58. During one minute, there will be 80 inspirations. Each inspiration represents $V\left(\frac{1}{160}\right) - V(0) = 3 + .05 - (3 - .05)$ $= .1$ liters. Therefore, the minute volume is $.1(80)$ $= 8$ liters. In ex. 57(b), we found that the peak respiratory flow was 8π liters/min. Thus the first statement is verified. In ex. 57(c), we found the mean inspiratory flow to be $16 = 8\cdot 2$ liters/min, verifying the second statement.

Exercises 9.1

1. $\int 2x(x^2 + 4)^5 dx$. Let $u = x^2 + 4$; then $du = 2x\,dx$.

$\int 2x(x^2 + 4)^5 dx = \int u^5 du = \frac{1}{6}u^6 + C = \frac{1}{6}(x^2 + 4)^6 + C$

2. $\int 3x^2(x^3 + 1)^2 dx$. Let $u = x^3 + 1$; then $du = 3x^2 dx$.

$\int 3x^2(x^3 + 1)^2 dx = \int u^2 du = \frac{1}{3}u^3 + C = \frac{1}{3}(x^3 + 1)^3 + C$

3. $\int (x^2 - 5x)^3(2x - 5)dx$. Let $u = x^2 - 5x$; then $du = (2x - 5)dx$.

$\int (x^2 - 5x)^3(2x - 5)dx = \int u^3 du = \frac{1}{4}u^4 + C = \frac{1}{4}(x^2 - 5x)^4 + C$

4. $\int 2x\sqrt{x^2 + 3}\,dx$. Let $u = x^2 + 3$; then $du = 2x\,dx$.

$\int 2x\sqrt{x^2 + 3}\,dx = \int u^{1/2} du = \frac{2}{3}u^{3/2} + C = \frac{2}{3}(x^2 + 3)^{3/2} + C$

5. $\int 3x^2 e^{(x^3 - 1)} dx$. Let $u = x^3 - 1$; then $du = 3x^2 dx$.

$\int 3x^2 e^{(x^3 - 1)} dx = \int e^u du = e^u + C = e^{x^3 - 1} + C.$

6. $\int 2xe^{-x^2} dx$. Let $u = -x^2$; then $du = -2x\,dx$.

$\int 2xe^{-x^2} dx = -\int -2xe^{-x^2} dx = -\int e^u du = -e^u + C = -e^{-x^2} + C.$

7. $\int x\sqrt{4 - x^2}\,dx$. Let $u = 4 - x^2$; then $du = -2x\,dx$.

$\int x\sqrt{4 - x^2}\,dx = -\frac{1}{2}\int u^{1/2}\,du = -\frac{1}{3}u^{3/2} + C$

$= -\frac{1}{3}(4 - x^2)^{3/2} + C.$

8. $\int \frac{1}{x}(\ln x)^3 dx$. Let $u = \ln x$; then $du = \frac{1}{x}dx$.

$\int \frac{1}{x}(\ln x)^3 dx = \int u^3 du = \frac{1}{4}u^4 + C = \frac{1}{4}(\ln x)^4 + C.$

9. $\displaystyle\int \frac{1}{\sqrt{2x+1}}\,dx$. Let $u = 2x + 1$; then $du = 2dx$.

$$\int \frac{1}{\sqrt{2x+1}}\,dx = \frac{1}{2}\int \frac{2}{\sqrt{2x+1}}\,dx = \frac{1}{2}\int \frac{du}{\sqrt{u}} = \frac{1}{2}\int u^{-1/2}du$$

$$= \left(\frac{1}{2}\right)\left(2u^{1/2}\right) + C = u^{1/2} + C = \sqrt{2x+1} + C.$$

10. $\int (x^3 - 6x)^7(x^2 - 2)dx$. Let $u = x^3 - 6x$;

then $du = (3x^2 - 6)dx$.

$$\int (x^3 - 6x)^7(x^2 - 2)dx = \frac{1}{3}\int u^7du = \frac{1}{24}u^8 + C = \frac{1}{24}(x^3 - 6x)^8 + C.$$

11. $\int xe^{3x^2}dx$. Let $u = 3x^2$; $du = 6x\,dx$.

$$\int xe^{3x^2}dx = \frac{1}{6}\int e^u du = \frac{e^u}{6} + C = \frac{1}{6}e^{3x^2} + C.$$

12. $\displaystyle\int \frac{e^{\sqrt{x}}}{\sqrt{x}}\,dx$. Let $u = \sqrt{x}$; then $du = \dfrac{1}{2\sqrt{x}}\,dx$.

$$\int \frac{e^{\sqrt{x}}}{\sqrt{x}}\,dx = 2\int \frac{e^{\sqrt{x}}}{2\sqrt{x}}\,dx = 2\int e^u du = 2e^u + C = 2e^{\sqrt{x}} + C.$$

13. $\displaystyle\int \frac{\ln(2x)}{x}\,dx$. Let $u = \ln(2x)$; then $du = \dfrac{1}{2x}(2)dx = \dfrac{1}{x}dx$.

$$\int \frac{\ln(2x)}{x}\,dx = \int u\,du = \frac{u^2}{2} + C = \frac{(\ln(2x))^2}{2} + C.$$

14. $\displaystyle\int \frac{\sqrt{\ln x}}{x}\,dx$. Let $u = \ln x$; $du = \dfrac{1}{x}dx$.

$$\int \frac{\sqrt{\ln x}}{x}\,dx = \int u^{1/2}du = \frac{2}{3}u^{3/2} + C = \frac{2}{3}(\ln x)^{3/2} + C.$$

15. $\displaystyle\int \frac{x^4}{x^5 + 1}\,dx$. Let $u = x^5 + 1$; $du = 5x^4dx$.

$$\int \frac{x^4}{x^5+1}\,dx = \frac{1}{5}\int \frac{du}{u} = \frac{1}{5}\ln u + C = \frac{1}{5}\ln\left|x^5 + 1\right| + C$$

16. $\displaystyle\int \frac{x}{\sqrt{x^2+1}}\,dx$. Let $u = x^2 + 1$; then $du = 2x\,dx$.

$$\int \frac{x}{\sqrt{x^2+1}}\,dx = \frac{1}{2}\int \frac{2x\,dx}{\sqrt{x^2+1}}\,dx = \frac{1}{2}\int u^{-1/2}du = \left(\frac{1}{2}\right)2u^{1/2} + C$$

$$= u^{1/2} + C = (x^2+1)^{1/2} + C.$$

17. $\displaystyle\int \frac{x^3}{(x^4+4)^4}\,dx$. Let $u = x^4 + 4$; $du = 4x^3dx$.

$$\int \frac{x^3}{(x^4+4)^4}\,dx = \frac{1}{4}\int \frac{4x^3}{(x^4+4)^4}\,dx = \frac{1}{4}\int \frac{du}{u^4} = \left(\frac{1}{4}\right)\left(-\frac{1}{3}u^{-3}\right) + C$$

$$= -\frac{1}{12}u^{-3} + C = -\frac{1}{12(x^4+4)^3} + C.$$

18. $\displaystyle\int \frac{(x^{-1}+2)^5}{x}\,dx$. Let $u = x^{-1} + 5$; $du = -x^{-2}dx$.

$$\int \frac{(x^{-1}+2)^5}{x}\,dx = -\int u^5\,du = -\frac{u^6}{6} + C = -\frac{1}{6}(x^{-1}+2)^6 + C$$

19. $\displaystyle\int \frac{\ln\sqrt{x}}{x}\,dx$. Let $u = \ln x$; $du = \frac{1}{x}dx$.

$$\int \frac{\ln\sqrt{x}}{x}\,dx = \int \frac{\ln(x^{1/2})}{x}\,dx = \int \frac{\left(\frac{1}{2}\right)\ln x}{x}\,dx = \frac{1}{2}\int \frac{\ln x}{x}\,dx$$

$$= \frac{1}{2}\int u\,du = \left(\frac{1}{2}\right)\frac{1}{2}u^2 + C = \frac{1}{4}u^2 + C$$

$$= \frac{1}{4}(\ln x)^2 + C \quad\text{or}\quad (\ln\sqrt{x})^2 + C$$

20. $\displaystyle\int \frac{x^2}{3-x^3}\,dx$. Let $u = 3 - x^3$; $du = -3x^2dx$.

$$\int \frac{x^2}{3-x^3}\,dx = -\frac{1}{3}\int \frac{du}{u} = -\frac{1}{3}\ln|u| + C = -\frac{1}{3}\ln\left|3 - x^3\right| + C$$

21. $\displaystyle\int \frac{x^2-2x}{x^3-3x^2+1}\,dx$. Let $u = x^3 - 3x^2 + 1$; $du = (3x^2 - 6x)dx$

$$= 3(x^2 - 2x)dx.$$

$$\int \frac{x^2 - 2x}{x^3 - 3x^2 + 1}\,dx = \frac{1}{3}\int \frac{3(x^2 - 2x)}{x^3 - 3x^2 + 1}\,dx = \frac{1}{3}\int \frac{du}{u} = \frac{1}{3}\ln|u| + C$$

$$= \frac{1}{3}\ln|x^3 - 3x^2 + 1| + C.$$

22. $\displaystyle\int \frac{\ln(3x)}{3x}\,dx.$ this is similar to question 13.

$$\int \frac{\ln(3x)}{3x}\,dx = \frac{1}{3}\left[\frac{(\ln(3x))^2}{2}\right] + C = \frac{(\ln(3x))^2}{6} + C.$$

23. $\displaystyle\int \frac{8x}{e^{x^2}}\,dx.$ Let $u = x^2$; then $du = 2x\,dx$.

$$\int \frac{8x}{e^{x^2}}\,dx = 4\int e^{-u}du = -4e^{-u} + C = -4e^{-x^2} + C.$$

24. $\displaystyle\int \frac{x + 4}{(1 - 8x - x^2)^3}dx.$ Let $u = 1 - 8x - x^2$; $du = (-8 - 2x)dx$

$= -2(x + 4)dx.$

$$\int \frac{x + 4}{(1 - 8x - x^2)^3}dx = -\frac{1}{2}\int u^{-3}du = \frac{1}{4}u^{-2} + C$$

$$= \frac{1}{4(1 - 8x - x^2)^2} + C.$$

25. $\displaystyle\int \frac{1}{x\ln x^2}\,dx.$ Let $u = x^2$; $du = 2xdx$

$$\int \frac{1}{x\ln x^2}\,dx = \frac{1}{2}\int \frac{1}{u\ln u}\,du = \frac{1}{2}\ln\left|\ln u\right| + C = \frac{1}{2}\ln\left|\ln x^2\right| + C.$$

26. $\displaystyle\int \frac{1}{x\ln x}\,dx.$ Let $u = \ln x$; $du = \frac{1}{x}dx.$

$$\int \frac{1}{x\ln x}\,dx = \int \frac{du}{u} = \ln|u| + C = \ln|\ln x| + C.$$

27. $\displaystyle\int (3 - x)(x^2 - 6x)^4dx.$ Let $u = x^2 - 6x$; then $du = -2(3 - x)dx.$

$$\int (3 - x)(x^2 - 6x)^4dx = -\frac{1}{2}\int u^4du = -\frac{1}{10}u^5 + C = -\frac{1}{10}(x^2 - 6x)^5 + C$$

28. $\displaystyle\int \frac{e^x}{1 + 3e^x}\,dx.$ Let $u = 1 + 3e^x$; $du = 3e^xdx$

$$\int \frac{e^x}{1 + 3e^x}\,dx = \frac{1}{3}\int \frac{du}{u} = \frac{1}{3}\ln|u| + C = \frac{1}{3}\ln|1 + 3e^x| + C.$$

29. $\displaystyle\int \frac{1}{x(\ln x)^3}\,dx.$ Let $u = \ln x$; then $du = \frac{1}{x}dx.$

$$\int \frac{1}{x(\ln x)^3}\,dx = \int u^{-3}du = -\frac{1}{2}u^{-2} + C = -\frac{1}{2}(\ln x)^{-2} + C.$$

30. $\displaystyle\int \frac{e^{2x} - e^{-2x}}{e^{2x} + e^{-2x}}\,dx.$ Let $u = e^{2x} + e^{-2x}$; $du = (2e^{2x} - 2e^{-2x})dx$

$$= 2(e^{2x} - e^{-2x})dx.$$

$$\int \frac{e^{2x} - e^{-2x}}{e^{2x} + e^{-2x}}\,dx = \frac{1}{2}\int \frac{2(e^{2x} - e^{-2x})}{e^{2x} + e^{-2x}}\,dx = \frac{1}{2}\int \frac{du}{u} = \frac{1}{2}\ln|u| + C$$

$$= \frac{1}{2}\ln|e^{2x} + e^{-2x}| + C.$$

31. $f'(x) = \dfrac{x}{\sqrt{x^2 + 9}}$, so $f(x) = \displaystyle\int \frac{x}{\sqrt{x^2 + 9}}\,dx$

Let $u = x^2 + 9$; then $du = 2x\,dx$.

$$\int \frac{x}{\sqrt{x^2 + 9}}\,dx = \frac{1}{2}\int \frac{2xdx}{\sqrt{x^2 + 9}}\,dx = \frac{1}{2}\int u^{-1/2}du = \left(\frac{1}{2}\right)2u^{1/2} + C$$

$$= u^{1/2} + C = (x^2 + 9)^{1/2} + C.$$

But $f(4) = 8$, so $8 = (4^2 + 9)^{1/2} + C$, hence $C = 3$, and

$f(x) = (x^2 + 9)^{1/2} + 3.$

32. $f'(x) = \dfrac{2\sqrt{x} + 1}{\sqrt{x}}$, so $f(x) = \displaystyle\int \frac{2\sqrt{x} + 1}{\sqrt{x}}\,dx = 2\int dx + \int \frac{dx}{\sqrt{x}}$

But $f(4) = 15$, so $15 = 8 + 4 + C$, hence $C = 3$, and

$f(x) = 2x + 2\sqrt{x} + 3.$

33. $$\int (x + 5)^{-1/2}e^{(x+5)^{1/2}}\,dx = 2\int \frac{1}{2}(x + 5)^{-1/2}e^{(x+5)^{1/2}}\,dx$$

$$= 2\int e^u du = 2e^u + C = 2e^{(x+5)^{1/2}} + C.$$

34. $\int \frac{x^4}{x^5 - 7} \ln(x^5 - 7)dx$. Let $u = \ln(x^5 - 7)$; $du = \frac{1}{x^5 - 7}(5x^4)dx$

$= 5\frac{x^4}{x^5 - 7}dx$

$\int \frac{x^4}{x^5 - 7} \ln(x^5 - 7)dx = \frac{1}{5}\int \frac{5x^4}{x^5 - 7} \ln(x^5 - 7)dx = \frac{1}{5}\int u\, du$

$= \left(\frac{1}{5}\right)\frac{1}{2}u^2 + C = \frac{1}{10}(\ln(x^5 - 7))^2 + C.$

35. $\int x\sec^2(x^2)\, dx$. Let $u = x^2$; then $du = 2x\, dx$.

$\int x\sec^2(x^2)\, dx = \frac{1}{2}\int \sec^2 u\, du = \frac{1}{2}\tan u + C = \frac{1}{2}\tan x^2 + C$

36. $\int (1 + \ln x)\sin(x\ln x)dx$. Let $u = x \ln x$;

then $du = \left[\ln x + x\left(\frac{1}{x}\right)\right]dx = (\ln x + 1)dx.$

$\int (1 + \ln x)\sin(x \ln x)dx = \int \sin u\, du = -\cos u + C$

$= -\cos(x \ln x) + C.$

37. $\int \sin x \cos x\, dx$. Let $u = \sin x$; then $du = \cos x\, dx$.

$\int \sin x \cos x\, dx = \int u\, du = \frac{u^2}{2} + C = \frac{\sin^2 x}{2} + C.$

38. $\int 2x \cos^2 x\, dx$. Let $u = x^2$; then $du = 2x\, dx$.

$\int 2x \cos^2 x\, dx = \int \cos u\, du = \sin u + C = \sin(x^2) + C.$

39. $\int \frac{\cos \sqrt{x}}{\sqrt{x}}\, dx$. Let $u = \sqrt{x}$; $du = \frac{1}{2}x^{-1/2}dx = \frac{1}{2\sqrt{x}}\, dx.$

$\int \frac{\cos \sqrt{x}}{\sqrt{x}}\, dx = 2\int \frac{\cos \sqrt{x}}{2\sqrt{x}}\, dx = 2\int \cos u\, du = 2\sin u + C$

$= 2\sin\sqrt{x} + C$

40. $\int \frac{\cos x}{(2 + \sin x)^3}\, dx$. Let $u = 2 + \sin x$; $du = \cos x\, dx$.

$\int \frac{\cos x}{(2 + \sin x)^3}\, dx = \int u^{-3}du = -\frac{1}{2}u^{-2} + C = \frac{-1}{2(2 + \sin x)^2} + C.$

41. $\int \cos^3 x \sin x\, dx$. Let $u = \cos x$; then $du = -\sin x\, dx$.

$\int \cos^3 x \sin x\, dx = -\int \cos^3 x(-\sin x)dx = -\int u^3 du = -\frac{1}{4}u^4 + C$

$= -\frac{1}{4}\cos^4 x + C.$

42. $\int (\sin 2x)e^{\cos 2x}dx$. Let $u = \cos 2x$; then $du = -2\sin x\, dx$.

$\int (\sin 2x)e^{\cos 2x}dx = -\frac{1}{2}\int e^u du = -\frac{1}{2}e^u + C = -\frac{1}{2}e^{\cos 2x} + C.$

43. $\int \cos 5x \sqrt{1 - \sin 5x}\, dx$. Let $u = 1 - \sin 5x$; then

$du = -5\cos 5x\, dx$.

$\int \cos 5x \sqrt{1 - \sin 5x}\, dx = -\frac{1}{5}\int u^{1/2}du = -\frac{2}{15}u^{3/2} + C$

$= -\frac{2}{15}(1 - \sin 5x)^{3/2} + C.$

44. $\int \frac{\sin 2x}{\sqrt{4 - \cos 2x}} dx$. Let $u = 4 - \cos 2x$; then $du = 2\sin 2x\, dx$.

$\int \frac{\sin 2x}{\sqrt{4 - \cos 2x}} dx = \frac{1}{2}\int u^{-1/2}du = \left(\frac{1}{2}\right)2u^{1/2} + C$

$= u^{1/2} + C = \sqrt{4 - \cos 2x} + C$

45. $\int (2x - 1)\sin(x^2 - x + 1)\, dx$. Let $u = x^2 - x + 1$; then

$du = (2x - 1)\, dx$.

$\int (2x - 1)\sin(x^2 - x + 1)\, dx = \int \sin u\, du = -\cos u + C$

$= -\cos(x^2 - x + 1) + C$

46. $\int e^{\sin(x^2)} x \cos(x^2)dx$. Let $u = \sin(x^2)$; then $du = 2x \cos(x^2)dx$

$\int e^{\sin(x^2)} x \cos(x^2)dx = \frac{1}{2}\int e^{\sin(x^2)} 2x \cos(x^2)dx$

$= \frac{1}{2}\int e^u du = \frac{1}{2}e^u + C = \frac{1}{2}e^{\sin(x^2)} + C.$

47. $\int 2x(x^2 + 5)\, dx$. Let $u = x^2 + 5$; then $du = 2xdx$.

$$\int 2x(x^2 + 5)\,dx = \int u\,du = \frac{1}{2}u^2 + C = \frac{1}{2}(x^2 + 5)^2 + C.$$

$$= \frac{1}{2}x^4 + 5x^2 + \frac{25}{2} + C.$$

Also, $\int 2x(x^2 + 5)\,dx = \int (2x^3 + 10x)\,dx = \frac{1}{2}x^4 + 5x + C_1.$

Exercises 9.2

1. $\int xe^{5x}dx.$ Let $f(x) = x$, $g(x) = e^{5x}$.

   Then $f'(x) = 1$ and $G(x) = \frac{1}{5}e^{5x}$.

   $$\int xe^{5x}dx = x\left[\frac{1}{5}e^{5x}\right] - \int (1)\frac{1}{5}e^{5x} = \frac{1}{5}xe^{5x} - \frac{1}{25}e^{5x} + C.$$

2. $\int xe^{-x/2}dx.$ Let $f(x) = x$, $g(x) = e^{-x/2}$.

   Then $f'(x) = 1$ and $G(x) = -2e^{-x/2}$.

   $$\int xe^{-x/2}dx = x\left[-2e^{-x/2}\right] - \int (1)\left[-2e^{-x/2}\right]dx$$

   $$= -2xe^{-x/2} - 4e^{-x/2} + C.$$

3. $\int x(2x + 1)^4dx.$ Let $f(x) = x$, $g(x) = (2x + 1)^4$.

   Then $f'(x) = 1$ and $G(x) = \frac{1}{10}(2x + 1)^5$.

   $$\int x(2x + 1)^4dx = x\left[\frac{1}{10}(2x + 1)^5\right] - \int (1)\frac{1}{10}(2x + 1)^5dx$$

   $$= \frac{1}{10}x(2x + 1)^5 - \frac{1}{120}(2x + 1)^6 + C.$$

4. $\int (x + 1)e^xdx.$ Let $f(x) = x + 1$, $g(x) = e^x$.

   Then $f'(x) = 1$ and $G(x) = e^x$.

   $$\int (x + 1)e^xdx = (x + 1)e^x - \int (1)e^xdx = e^x(x + 1) - e^x + C$$

   $$= e^x\left[(x + 1) - 1\right] + C = xe^x + C.$$

5. $\int xe^{-x}dx$. Let $f(x) = x$, $g(x) = e^{-x}$.

Then $f'x = 1$ and $G(x) = -e^{-x}$.

$$\int xe^{-x}dx = x(-e^{-x}) - \int(1)\left[-e^{-x}\right]dx = -xe^{-x} - e^{-x} + C.$$

6. $\int x(x + 5)^{-3}dx$. Let $f(x) = x$, $g(x) = (x + 5)^{-3}$.

Then $f'(x) = 1$ and $G(x) = -\frac{1}{2}(x + 5)^{-2}$.

$$\int x(x + 5)^{-3}dx = x\left[-\frac{1}{2}(x + 5)^{-2}\right] - \int(1)\left[-\frac{1}{2}(x + 5)^{-2}\right]dx$$
$$= -\frac{1}{2}x(x + 5)^{-2} - \frac{1}{2}(x + 5)^{-1} + C.$$

7. $\int \frac{x}{\sqrt{x + 1}}dx$. Let $f(x) = x$, $g(x) = (x + 1)^{-1/2}$.

Then $f'(x) = 1$ and $G(x) = 2(x + 1)^{1/2}$.

$$\int \frac{x}{\sqrt{x + 1}}dx = x\left[2(x + 1)^{1/2}\right] - \int(1)2(x + 1)^{1/2}dx$$
$$= 2x(x + 1)^{1/2} - \frac{4}{3}(x + 1)^{3/2} + C.$$

8. $\int \frac{x}{\sqrt{3 + 2x}}dx$. Let $f(x) = x$, $g(x) = (3 + 2x)^{-1/2}$.

Then $f'(x) = 1$ and $G(x) = (3 + 2x)^{1/2}$.

$$\int \frac{x}{\sqrt{3 + 2x}}dx = x(3 + 2x)^{1/2} - \int(1)(3 + 2x)^{1/2}dx$$
$$= x(3 + 2x)^{1/2} - \frac{1}{3}(3 + 2x)^{3/2} + C.$$

9. $\int x(1 - 3x)^4dx$. Let $f(x) = x$, $g(x) = (1 - 3x)^4$.

Then $f'(x) = 1$ and $G(x) = -\frac{1}{15}(1 - 3x)^5$.

$$\int x(1 - 3x)^4dx = x\left[-\frac{1}{15}(1 - 3x)^5\right] - \int(1)\left[-\frac{1}{15}(1 - 3x)^5\right]dx$$
$$= -\frac{1}{15}x(1 - 3x)^5 - \frac{1}{270}(1 - 3x)^6 + C.$$

10. $\int x(1 + x)^{10}dx$. Let $f(x) = x$, $g(x) = (1 + x)^{10}$.

Then $f'(x) = 1$ and $G(x) = \frac{1}{11}(1 + x)^{11}$.

$$\int x(1 + x)^{10}dx = x\left[\frac{1}{11}(1 + x)^{11}\right] - \int(1)\frac{1}{11}(1 + x)^{11}dx$$
$$= \frac{1}{11}x(1 + x)^{11} - \frac{1}{132}(1 + x)^{12} + C.$$

11. $\int \frac{3x}{e^x}\,dx$. Let $f(x) = 3x$, $g(x) = e^{-x}$.

Then $f'(x) = 3$ and $G(x) = -e^{-x}$.

$$\int \frac{3x}{e^x}\,dx = 3x(-e^{-x}) - \int 3(-e^{-x})dx = -3xe^{-x} - 3e^{-x} + C.$$

12. $\int \frac{x + 1}{e^x}\,dx$. Let $f(x) = x + 1$, $g(x) = e^{-x}$.

Then $f'(x) = 1$ and $G(x) = -e^{-x}$.

$$\int \frac{x + 1}{e^x}\,dx = (x + 1)(-e^{-x}) - \int(1)\left[-e^{-x}\right]dx$$
$$= -(x + 1)e^{-x} - e^{-x} + C$$
$$= e^{-x}(-(x + 1) - 1) + C = -e^{-x}(x + 2) + C.$$

13. $\int x\sqrt{x + 1}\,dx$. Let $f(x) = x$, $g(x) = (x + 1)^{1/2}$.

Then $f'(x) = 1$ and $G(x) = \frac{2}{3}(x + 1)^{3/2}$.

$$\int x\sqrt{x + 1}\,dx = x\left[\frac{2}{3}(x + 1)^{3/2}\right] - \int(1)\frac{2}{3}(x + 1)^{3/2}dx$$
$$= \frac{2}{3}x(x + 1)^{3/2} - \frac{4}{15}(x + 1)^{5/2} + C.$$

14. $\int x\sqrt{2 - x}\,dx$. Let $f(x) = x$, $g(x) = (2 - x)^{1/2}$.

Then $f'(x) = 1$ and $G(x) = -\frac{2}{3}(2 - x)^{3/2}$.

$$\int x\sqrt{2 - x}\,dx = x\left[-\frac{2}{3}(2 - x)^{3/2}\right] - \int(1)\left[-\frac{2}{3}(2 - x)^{3/2}\right]dx$$
$$= -\frac{2}{3}x(2 - x)^{3/2} - \frac{4}{15}(2 - x)^{5/2}.$$

15. $\int \sqrt{x}\,\ln\sqrt{x}\,dx$. Let $f(x) = \ln\sqrt{x}$, $g(x) = \sqrt{x}$.

Then $f'(x) = \frac{1}{\sqrt{x}}\left(\frac{1}{2}x^{-1/2}\right) = \frac{1}{2x}$ and $G(x) = \frac{2}{3}x^{3/2}$.

$$\int \sqrt{x}\ \ln\sqrt{x}\ dx = \ln\sqrt{x}\left[\frac{2}{3}x^{3/2}\right] - \int\left(\frac{1}{2x}\right)\frac{2}{3}x^{3/2}dx$$

$$= \frac{2}{3}x^{3/2}\ln\sqrt{x} - \int\frac{1}{3}x^{1/2}dx = \frac{2}{3}x^{3/2}\ln\sqrt{x} - \frac{2}{9}x^{3/2} + C$$

$$= \frac{1}{3}x^{3/2}\ln x - \frac{2}{9}x^{3/2} + C.$$

16. $\int x^5 \ln x\ dx$. Let $f(x) = \ln x$, $g(x) = x^5$.

Then $f'(x) = \frac{1}{x}$ and $G(x) = \frac{1}{6}x^6$.

$$\int x^5 \ln x\ dx = \ln x\left(\frac{1}{6}x^6\right) - \int\left(\frac{1}{x}\right)\frac{1}{6}x^6 dx$$

$$= \frac{1}{6}x^6 \ln x - \frac{1}{36}x^6 + C \quad \text{or} \quad \frac{1}{6}x^6\left(\ln x - \frac{1}{6}\right) + C$$

17. $\int x^2 \cos x\ dx$. Let $f(x) = x^2$, $g(x) = \cos x$.

Then $f'(x) = 2x$ and $G(x) = \sin x$.

$$\int x^2 \cos x\ dx = x^2 \sin x - \int 2x \sin x\ dx$$

To evaluate $\int 2x \sin x\ dx$, use integration by parts again:

Let $f(x) = 2x$, $g(x) = \sin x$.

Then $f'(x) = 2$ and $G(x) = -\cos x$.

$$\int 2x \sin x\ dx = 2x(-\cos x) - \int 2(-\cos x)dx$$

$$= -2x \cos x + 2\sin x + C.$$

Therefore, $\int x^2 \cos x\ dx = x^2 \sin x - \left[-2x \cos x + 2\sin x\right] + C$

$$= (x^2 - 2)\sin x + 2x \cos x + C.$$

18. $\int x \sin 8x\ dx$. Let $f(x) = x$, $g(x) = \sin 8x$.

Then $f'(x) = 1$ and $G(x) = -\frac{1}{8}\cos 8x$.

$$\int x \sin 8x\ dx = x\left(-\frac{1}{8}\cos 8x\right) - \int(1)\left(-\frac{1}{8}\cos 8x\right)dx$$

$$= -\frac{1}{8}x \cos 8x + \frac{1}{64}\sin 8x + C.$$

19. $\int x \ln 5x\ dx$. Let $f(x) = \ln 5x$, $g(x) = x$.

Then $f'(x) = \frac{1}{5x}(5) = \frac{1}{x}$ and $G(x) = \frac{x^2}{2}$.

$\int x \ln 5x\, dx = \ln 5x\left(\frac{x^2}{2}\right) - \int \frac{1}{x}\left(\frac{x^2}{2}\right)dx = \frac{1}{2}x^2 \ln 5x - \frac{1}{4}x^2 + C.$

20. $\int x^{-3} \ln x\, dx.$ Let $f(x) = \ln x$, $g(x) = x^{-3}$.

Then $f'(x) = \frac{1}{x}$ and $G(x) = -\frac{1}{2}x^{-2}$.

$\int x^{-3} \ln x\, dx = \ln x\left(-\frac{1}{2}x^{-2}\right) - \int \frac{1}{x}\left(-\frac{1}{2}x^{-2}\right)dx$

$= -\frac{1}{2}x^{-2} \ln x - \frac{1}{4}x^{-2} + C.$

21. $\int \ln x^4\, dx.$ Let $f(x) = \ln x^4$, $g(x) = 1$.

Then $f'(x) = \frac{1}{x^4}(4x^3) = \frac{4}{x}$ and $G(x) = x$.

$\int \ln x^4\, dx = \ln x^4\ (x) - \int \frac{4}{x}(x)dx$

$= x \ln x^4 - 4x + C$ or $4x \ln x - 4x + C.$

22. $\int x^2 \sin 3x\, dx.$ Let $f(x) = x^2$, $g(x) = \sin 3x$.

Then $f'(x) = 2x$ and $G(x) = -\frac{1}{3}\cos 3x$.

$\int x^2 \sin 3x\, dx = x^2\left(-\frac{1}{3}\cos 3x\right) - \int 2x\left(-\frac{1}{3}\cos 3x\right)dx$

$= -\frac{1}{3}x^2 \cos 3x + \int \frac{2}{3}x \cos 3x\, dx.$

To evaluate $\int \frac{2}{3}x \cos 3x\, dx$, use parts again:

Let $f(x) = \frac{2}{3}x$, $g(x) = \cos 3x$.

Then $f'(x) = \frac{2}{3}$ and $G(x) = \frac{1}{3}\sin 3x$.

$\int \frac{2}{3}x \cos 3x\, dx = \frac{2}{3}x\left(\frac{1}{3}\sin 3x\right) - \int \left(\frac{2}{3}\right)\frac{1}{3}\sin 3x\, dx$

$= \frac{2}{9}x \sin 3x + \frac{2}{27}\cos 3x + C.$

Thus, $\int x^2 \sin 3x\, dx = -\frac{1}{3}x^2 \cos 3x + \frac{2}{9}x \sin 3x + \frac{2}{27}\cos 3x + C.$

23. $\int x^2 e^{-x} dx.$ Let $f(x) = x^2$, $g(x) = e^{-x}$.

Then $f'(x) = 2x$ and $G(x) = -e^{-x}$.

$\int x^2 e^{-x} dx = x^2(-e^{-x}) - \int 2x(-e^{-x})dx = -x^2 e^{-x} + \int 2xe^{-x} dx.$

To evaluate $\int 2xe^{-x} dx$, use parts again:

Let $f(x) = 2x$, $g(x) = e^{-x}$. Then $f'(x) = 2$ and $G(x) = -e^{-x}$.

$\int 2xe^{-x}dx = 2x(-e^{-x}) - \int 2(-e^{-x})dx = -2xe^{-x} - 2e^{-x} + C.$

Therefore, $\int x^2e^{-x}dx = -x^2e^{-x} - 2xe^{-x} - 2e^{-x} + C$

or $-e^{-x}(x^2 + 2x + 2) + C.$

24. $\int \frac{x^2}{\sqrt{x + 4}} dx.$ Let $f(x) = x^2$, $g(x) = (x + 4)^{-1/2}$.

Then $f'(x) = 2x$ and $G(x) = 2(x + 4)^{1/2}$.

$$\int \frac{x^2}{\sqrt{x + 4}} dx = x^2\left[2(x + 4)^{1/2}\right] - \int (2x)2(x + 4)^{1/2}dx$$

$$= 2x^2(x + 4)^{1/2} - \int 4x(x + 4)^{1/2}dx.$$

To evaluate $\int 4x(x + 4)^{1/2}dx$, use parts again:

Let $f(x) = 4x$, $g(x) = (x + 4)^{1/2}$.

Then $f'(x) = 4$ and $G(x) = \frac{2}{3}(x + 4)^{3/2}$.

$$\int 4x(x + 4)^{1/2}dx = (4x)\frac{2}{3}(x + 4)^{3/2} - \int (4)\frac{2}{3}(x + 4)^{3/2}dx$$

$$= \frac{8}{3}x(x + 4)^{3/2} - \frac{16}{15}(x + 4)^{5/2} + C.$$

Therefore, $\int \frac{x^2}{\sqrt{x + 4}} dx = 2x^2(x + 4)^{1/2} - \frac{8}{3}x(x + 4)^{3/2}$

$+ \frac{16}{15}(x + 4)^{5/2} + C.$

25. $\int x(x + 5)^4dx.$ Use integration by parts: let $f(x) = x$,

$g(x) = (x + 5)^4$. Then $f'(x) = 1$ and $G(x) = \frac{1}{5}(x + 5)^5$.

$$\int x(x + 5)^4dx = x\left[\frac{1}{5}(x + 5)^5\right] - \int (1)\frac{1}{5}(x + 5)^5dx$$

$$= \frac{1}{5}x(x + 5)^5 - \frac{1}{30}(x + 5)^6 + C.$$

26. $\int 4x\cos(x^2 + 1)dx.$ Use the substitution, $u = x^2 + 1$;

then $du = 2x\,dx$.

$$\int 4x\cos(x^2 + 1)dx = 2\int \cos(x^2 + 1)2x\,dx = 2\int \cos u\,du$$

$$= 2\sin u + C = 2\sin(x^2 + 1) + C.$$

27. $\int x(x^2 + 5)^4 dx$. Use the substitution, $u = x^2 + 5$;

then $du = 2x\ dx$.

$$\int x(x^2 + 5)^4 dx = \frac{1}{2}\int (x^2 + 5)^4 2x\ dx = \frac{1}{2}\int u^4 du = \left(\frac{1}{2}\right)\frac{1}{5}u^5 + C$$
$$= \frac{1}{10}u^5 + C = \frac{1}{10}(x^2 + 5)^5 + C.$$

28. $\int 4x \cos(x + 1)dx$. Use integration by parts with

$f(x) = 4x$, $g(x) = \cos(x + 1)$.

Then $f'(x) = 4$ and $G(x) = \sin(x + 1)$.

$$\int 4x \cos(x + 1)dx = 4x \sin(x + 1) - \int 4\sin(x + 1)dx$$
$$= 4x \sin(x + 1) + 4\cos(x + 1) + C.$$

29. $\int (3x + 1)e^{x/3}dx$. Use integration by parts with

$f(x) = 3x + 1$, $g(x) = e^{x/3}$.

Then $f'(x) = 3$ and $G(x) = 3e^{x/3}$.

$$\int (3x + 1)e^{x/3}dx = (3x + 1)(3e^{x/3}) - \int (3)3e^{x/3}dx$$
$$= (9x + 3)e^{x/3} - 27e^{x/3} + C$$
$$= e^{x/3}(9x - 24) + C.$$

30. $\int \frac{(\ln x)^5}{x} dx$. Use the substitution, $u = \ln x$; $du = \frac{1}{x}dx$.

$$\int \frac{(\ln x)^5}{x} dx = \int u^5 du = \frac{1}{6}u^6 + C = \frac{1}{6}(\ln x)^6 + C.$$

31. $\int x \sec^2(x^2 + 1)dx$. Use the substitution, $u = x^2 + 1$;

then $du = 2x\ dx$.

$$\int x \sec^2(x^2 + 1)dx = \frac{1}{2}\int \sec^2(x^2 + 1)2x\ dx = \frac{1}{2}\int \sec^2 u\ du$$
$$= \frac{1}{2}\tan u + C = \frac{1}{2}\tan(x^2 + 1) + C.$$

32. $\int x^{3/2}\ln 2x\,dx$. Use integration by parts, with

$f(x) = \ln 2x,\ g(x) = x^{3/2}$.

Then $f'(x) = \frac{1}{2x}(2) = \frac{1}{x}$ and $G(x) = \frac{2}{5}x^{5/2}$.

$$\int x^{3/2}\ln 2x\,dx = \ln 2x\left(\frac{2}{5}x^{5/2}\right) - \int\left(\frac{1}{x}\right)\frac{2}{5}x^{5/2}dx$$

$$= \frac{2}{5}x^{5/2}\ln 2x - \frac{4}{25}x^{5/2} + C$$

$$= \frac{2}{5}x^{5/2}\left(\ln 2x - \frac{2}{5}\right) + C.$$

33. $\int(xe^{2x} + x^2)dx = \int xe^{2x}dx + \int x^2dx$.

$\int x^2dx = \frac{1}{3}x^3 + C$. To evaluate $\int xe^{2x}dx$, use integration by parts with $f(x) = x,\ g(x) = e^{2x}$.

Then $f'(x) = 1$ and $G(x) = \frac{1}{2}e^{2x}$.

$$\int xe^{2x}dx = x\left(\frac{1}{2}e^{2x}\right) - \int\frac{1}{2}e^{2x}dx = \frac{1}{2}xe^{2x} - \frac{1}{4}e^{2x} + C$$

$$= \frac{1}{2}e^{2x}\left(x - \frac{1}{2}\right) + C.$$

Therefore, $\int(xe^{2x} + x^2)dx = \frac{1}{3}x^3 + \frac{1}{2}e^{2x}\left(x - \frac{1}{2}\right) + C$.

34. $\int(x^{3/2} + \ln 2x)dx = \int x^{3/2}dx + \int\ln 2x\,dx$.

$\int x^{3/2}dx = \frac{2}{5}x^{5/2} + C$. To evaluate $\int\ln 2x\,dx$, use integration by parts with $f(x) = \ln 2x,\ g(x) = 1$.

Then $f'(x) = \frac{1}{2x}(2) = \frac{1}{x}$ and $G(x) = x$.

$$\int\ln 2x\,dx = \ln 2x\,(x) - \int\frac{1}{x}(x)dx = x\ln 2x - x + C.$$

Therefore, $\int(x^{3/2} + \ln 2x)dx = \frac{2}{5}x^{5/2} + x\ln 2x - x + C$.

35. $\int(xe^{x^2} - 2x)dx = \int xe^{x^2}dx - \int 2x\,dx$.

$\int 2x\,dx = x^2 + C$. To evaluate $\int xe^{x^2}dx$, use the substitution, $u = x^2$; $du = 2x\,dx$. $\int xe^{x^2}dx = \frac{1}{2}\int e^{x^2}2x\,dx = \frac{1}{2}\int e^u du$

$$= \frac{1}{2}e^u + C = \frac{1}{2}e^{x^2} + C.$$

Therefore, $\int (xe^{x^2} - 2x)dx = \frac{1}{2}e^{x^2} - x^2 + C.$

36. $\int (x^2 - x \sin 2x)dx = \int x^2 dx - \int x \sin 2x\, dx.$

$\int x^2 dx = \frac{1}{3}x^3 + C.$ To evaluate $\int x \sin 2x\, dx$, use integration by parts with $f(x) = x$, $g(x) = \sin 2x$.

Then $f'(x) = 1$ and $G(x) = -\frac{1}{2}\cos 2x.$

$$\int x \sin 2x\, dx = x\left[-\frac{1}{2}\cos 2x\right] - \int -\frac{1}{2}\cos 2x\, dx$$

$$= -\frac{1}{2}x \cos 2x + \frac{1}{4}\sin 2x + C.$$

Therefore, $\int (x^2 - x \sin 2x)dx = \frac{1}{3}x^3 + \frac{1}{2}x \cos 2x - \frac{1}{4}\sin 2x + C.$

31. The slope is $\frac{x}{\sqrt{x + 9}}dx$, so $f'(x) = \int \frac{x}{\sqrt{x + 9}} dx$ and thus

$f(x) = \int \frac{x}{\sqrt{x + 9}} dx$ Let $u = x + 9$, so $x = u - 9$; $du = dx$.

$$\int \frac{x}{\sqrt{x + 9}} dx = \int \frac{u - 9}{\sqrt{u}} dx = \int (u^{1/2} - 9u^{-1/2})\, du$$

$$= \frac{2}{3}u^{3/2} - 18u^{1/2} + C = \frac{2}{3}(x + 9)^{3/2} - 18(x + 9)^{1/2} + C$$

(0,2) on the graph, so at $x = 0$, $f(x) = 2$ which implies

$2 = \frac{2}{3}(0 + 9)^{3/2} - 18(0 + 9)^{1/2} + C$, or $C = 38$, thus

$$f(x) = \frac{2}{3}(x + 9)^{3/2} - 18(x + 9)^{1/2} + 38$$

$$= \sqrt{x + 9}[(2/3)x + 6 - 18] + 38$$

$$= \sqrt{x + 9}[2x - (4/3)x - 12] + 38$$

$$= \sqrt{x + 9}[2x - (4/3)\{x + 9\}] + 38$$

$$= 2x\sqrt{x + 9} - (4/3)(x + 9)^{3/2} + 38$$

38. Slope is $\frac{x}{e^x}$, so $f'(x) = \frac{x}{e^x}$, $f(x) = \int xe^{-x/3}dx$

by parts: $u = x$ $\qquad dv = e^{-x/3}dx$

$$du = dx \qquad\qquad v = -3e^{-x/3}$$

$$f(x) = \int xe^{-x/3}dx = -3xe^{-x/3} + 3\int e^{-x/3}dx$$

$$= -3xe^{-x/3} - 9e^{-x/3} + C$$

$f(0) = 6$, $6 = -9 + C$, so $f(x) = -3xe^{-x/3} - 9e^{-x/3} + 15$.

Exercises 9.3

1. $\int_{3/2}^{2} 2(2x - 3)^{17}dx$. Let $u = 2x - 3$; $du = 2x\ dx$. When $x = \frac{3}{2}$,

$u = 2\left(\frac{3}{2}\right) - 3 = 0$. When $x = 2$, $u = 2(2) - 3 = 1$.

$$\int_{3/2}^{2} 2(2x - 3)^{17}dx = \int_{0}^{1} u^{17}du = \frac{1}{18}u^{18}\Big|_0^1 = \frac{1}{18}(1)^{18} - 0 = \frac{1}{18}.$$

2. $\int_{\sqrt{3}}^{2} ((x^2 - 3)^{17} - (x^2 - 3))x\ dx$. Let $u = x^2 - 3$; $du = 2x\ dx$.

When $x = \sqrt{3}$, $u = (\sqrt{3})^2 - 3 = 0$. When $x = 2$, $u = 2^2 - 3 = 1$.

$$\int_{\sqrt{3}}^{2} ((x^2 - 3)^{17} - (x^2 - 3))x\ dx$$

$$= \frac{1}{2}\int_{\sqrt{3}}^{2} ((x^2 - 3)^{17} - (x^2 - 3))2x\ dx$$

$$= \frac{1}{2}\int_{0}^{1} (u^{17} - u)du = \left(\frac{1}{18}u^{18} - \frac{1}{2}u^2\right)\Big|_0^1$$

$$= \frac{1}{2}\left[\left(\frac{1}{18}(1)^{18} - \frac{1}{2}(1)^2\right) - (0 - 0)\right] = \frac{1}{2}\left(\frac{1}{18} - \frac{1}{2}\right) = \frac{1}{2}\left(-\frac{8}{18}\right) = -\frac{2}{9}.$$

3. $\int_1^3 ((4 - 2x)\sin(4x - x^2))dx$. Let $u = 4x - x^2$; $du = (4 - 2x)dx$.

When $x = 1$, $u = 4(1) - 1^2 = 3$. When $x = 3$, $u = 4(3) - 3^2 = 3$.

$$\int_1^3 ((4 - 2x)\sin(4x - x^2))dx = \int_3^3 \sin u \, du = 0.$$

4. $\int_5^{13} x\sqrt{x^2 - 25}\, dx$. Let $u = x^2 - 25$; $du = 2x\, dx$.

When $x = 5$, $u = 5^2 - 25 = 0$. When $x = 13$, $u = 13^2 - 25 = 144$.

$$\int_5^{13} x\sqrt{x^2 - 25}\, dx = \frac{1}{2}\int_5^{13} \sqrt{x^2 - 25}\, 2x\, dx = \frac{1}{2}\int_0^{144} u^{1/2}du$$

$$= \left(\frac{1}{2}\right)\frac{2}{3}u^{3/2}\Big|_0^{144} = \frac{1}{3}144^{3/2} - 0$$

$$= \frac{1}{3}12^3 = 576.$$

5. $\int_1^e \frac{\ln x}{x}\, dx$. Let $u = \ln x$; $du = \frac{1}{x}dx$.

When $x = 1$, $u = \ln(1) = 0$. When $x = e$, $u = \ln(e) = 1$.

$$\int_1^e \frac{\ln x}{x}\, dx = \int_0^1 u\, du = \frac{u^2}{2}\Big|_0^1 = \frac{1^2}{2} - 0 = \frac{1}{2}.$$

6. $\int_0^{\pi} e^{(\sin x)^2} \cos x \sin x\, dx$. Let $u = (\sin x)^2$;

$du = 2\sin x \cos x\, dx$. When $x = 0$, $u = (\sin(0))^2 = 0$.

When $x = \pi$, $u = (\sin(\pi))^2 = 0$.

$$\int_0^{\pi} e^{(\sin x)^2} \cos x \sin x\, dx = \frac{1}{2}\int_0^{\pi} e^{(\sin x)^2} 2\sin x \cos x\, dx$$

$$= \frac{1}{2}\int_0^0 e^u du = 0.$$

7. $\displaystyle\int_2^5 x\sqrt{x^2 - 9}\,dx.$ Let $u = x^2 - 9$, $du = 2x\,dx$.

When $x = 3$, $u = 3^2 - 9 = 0$. When $x = 5$, $u = 5^2 - 9 = 16$.

$$\int_2^5 x\sqrt{x^2 - 9}\,dx = \frac{1}{2}\int_2^5 \sqrt{x^2 - 9}\,2x\,dx = \frac{1}{2}\int_0^{16} u^{1/2}du$$

$$= \left(\frac{1}{2}\right)\frac{2}{3}u^{3/2}\bigg|_0^{16} = \frac{1}{3}16^{3/2} - 0 = \frac{64}{3}.$$

8. $\displaystyle\int_1^{e^{\pi/2}} \frac{\sin(\ln x)}{x}\,dx.$ Let $u = \ln x$; $du = \frac{1}{x}dx$.

When $x = 1$, $u = \ln(1) = 0$. When $x = e^{\pi/2}$, $u = \ln(e^{\pi/2}) = \frac{\pi}{2}$.

$$\int_1^{e^{\pi/2}} \frac{\sin(\ln x)}{x}\,dx = \int_0^{\pi/2} \sin u\,du = -\cos u\bigg|_0^{\pi/2}$$

$$= -\cos\left(\frac{\pi}{2}\right) + \cos(0) = 0 + 1 = 1.$$

9. $\displaystyle\int_{\ln 1}^{\ln 2} \left[(e^x)^2 + e^x\right]e^x\,dx.$ Let $u = e^x$; $du = e^x dx$.

When $x = \ln 1$, $u = e^{\ln 1} = 1$. When $x = \ln 2$, $u = e^{\ln 2} = 2$.

$$\int_{\ln 1}^{\ln 2} \left[(e^x)^2 + e^x\right]e^x\,dx = \int_1^2 (u^2 + u)du = \left(\frac{1}{3}u^3 + \frac{1}{2}u^2\right)\bigg|_1^2$$

$$= \left(\frac{1}{3}2^3 + \frac{1}{2}2^2\right) - \left(\frac{1}{3}1^3 + \frac{1}{3}1^2\right)$$

$$= \frac{8}{3} + 2 - \frac{1}{3} - \frac{1}{2} = \frac{23}{6}.$$

10. $\int_{\pi/4}^{\pi/2} \ln(\sin x) \cos x \, dx$. Use integration by parts with

$f(x) = \ln(\sin x)$, $g(x) = \cos x$.

Then $f'(x) = \frac{1}{\sin x}(\cos x)$ and $G(x) = \sin x$.

$$\int_{\pi/4}^{\pi/2} \ln(\sin x) \cos x \, dx$$

$$= \ln(\sin x)(\sin x)\Big|_{\pi/4}^{\pi/2} - \int_{\pi/2}^{\pi/2} \frac{\cos x}{\sin x}\sin x \, dx$$

$$= \left[\ln\left(\sin \frac{\pi}{2}\right)\left(\sin \frac{\pi}{2}\right) - \ln\left(\sin \frac{\pi}{4}\right)\left(\sin \frac{\pi}{4}\right)\right] - \sin x\Big|_{\pi/4}^{\pi/2}$$

$$= \left[0 - \frac{\sqrt{2}}{2}\ln \frac{\sqrt{2}}{2}\right] - \left[\sin \frac{\pi}{2} - \sin \frac{\pi}{4}\right]$$

$$= -\frac{\sqrt{2}}{2}\ln \frac{\sqrt{2}}{2} - \left[1 - \frac{\sqrt{2}}{2}\right] = -\frac{\sqrt{2}}{2}\ln \frac{\sqrt{2}}{2} - 1 + \frac{\sqrt{2}}{2}.$$

11. $\int_{-1}^{2} (x^3 - 2x - 1)^2(3x^2 - 2)dx$. Let $u = x^3 - 2x - 1$;

$du = (3x^2 - 2)dx$. When $x = -1$, $u = (-1)^3 - 2(-1) - 1 = 0$.

When $x = 2$, $u = 2^3 - 2(2) - 1 = 3$.

$$\int_{-1}^{2} (x^3 - 2x - 1)^2(3x^2 - 2)dx = \int_{0}^{3} u^2 du = \frac{1}{3}u^3\Big|_{0}^{3} = \frac{1}{3}3^3 - 0 = 9.$$

12. $\int_{-1}^{1} x \sin x^4 \, dx$. Let $u = x^2$; $du = 2x \, dx$.

When $x = -1$, $u = (-1)^2 = 1$. When $x = 1$, $u = 1^2 = 1$.

$$\int_{-1}^{1} x \sin x^4 \, dx = \frac{1}{2}\int_{-1}^{1} \sin((x^2)^2) \, 2x \, dx = \frac{1}{2}\int_{-1}^{1} \sin u^2 \, du = 0.$$

13. $\int_1^3 x^2e^{x^3}dx$. Let $u = x^3$; $du = 3x^2dx$.

When $x = 1$, $u = 1^3 = 1$. When $x = 3$, $u = 3^3 = 27$.

$$\int_1^3 x^2e^{x^3}dx = \frac{1}{3}\int_1^3 e^{x^3}3x^2dx = \frac{1}{3}\int_1^{27} e^u du = \frac{1}{3}e^u\Big|_1^{27} = \frac{1}{3}e^{27} - \frac{1}{3}e.$$

14. $\int_0^1 (2x - 1)(x^2 - x)^{10}dx$. Let $u = x^2 - x$; $du = (2x - 1)dx$.

When $x = 0$, $u = 0^2 - 0 = 0$. When $x = 1$, $u = 1^2 - 1 = 0$.

$$\int_0^1 (2x - 1)(x^2 - x)^{10}dx = \int_0^0 u^{10}du = 0.$$

15. $\int_{-\pi}^{2\pi} \sin(8x - \pi)dx$. Let $u = 8x - \pi$; $du = 8dx$.

When $x = -\pi$, $u = 8(-\pi) - \pi = -9\pi$.

When $x = 2\pi$, $u = 8(2\pi) - \pi = 15\pi$.

$$\int_{-\pi}^{2\pi} \sin(8x - \pi)dx = \frac{1}{8}\int_{-\pi}^{2\pi} \sin(8x - \pi)8dx = \frac{1}{8}\int_{-9\pi}^{15\pi} \sin u\, du$$

$$= -\frac{1}{8}\cos u\Big|_{-9\pi}^{15\pi} = -\frac{1}{8}\cos(15\pi) + \frac{1}{8}\cos(-9\pi)$$

$$= \frac{1}{8} - \frac{1}{8} = 0.$$

16. $\int_{-2}^{2} 2x \sin x^2dx$. Let $u = x^2$; $du = 2x\, dx$.

When $x = -2$, $u = (-2)^2 = 4$. When $x = 2$, $u = 2^2 = 4$.

$$\int_{-2}^{2} 2x \sin x^2dx = \int_4^4 \sin u\, du = 0.$$

17. $\int_0^2 xe^{x/2}dx$. Use integration by parts with

$f(x) = x$, $g(x) = e^{x/2}$. Then $f'(x) = 1$ and $G(x) = 2e^{x/2}$.

$$\int_0^2 xe^{x/2}dx = x\left[2e^{x/2}\right]\Big|_0^2 = 4e - \left[4e^1 - 4e^0\right] = 4.$$

18. $\int_0^4 8x(x + 4)^{-3}dx$. Use integration by parts with $f(x) = 8x$,

$g(x) = (x + 4)^{-3}$. Then $f'(x) = 8$ and $G(x) = -\frac{1}{2}(x + 4)^{-2}$.

$$\int_0^4 8x(x + 4)^{-3}dx = 8x\left[-\frac{1}{2}(x + 4)^{-2}\right]\Big|_0^4 - \int_0^4 8\left(-\frac{1}{2}\right)(x + 4)^{-2}dx$$

$$= \left[-4(4)(8)^{-2} - 0\right] - 4(x + 4)^{-1}\Big|_0^4$$

$$= -\frac{1}{4} - \left[\frac{1}{2} - 1\right] = \frac{1}{4}.$$

19. $\int_0^1 x \sin \pi x \, dx$. Use integration by parts, with $f(x) = x$,

$g(x) = \sin \pi x$. Then $f'(x) = 1$ and $G(x) = -\frac{1}{\pi}\cos \pi x$.

$$\int_0^1 x \sin \pi x \, dx = x\left[-\frac{1}{\pi}\cos \pi x\right]\Big|_0^1 - \int_0^1 -\frac{1}{\pi}\cos \pi x \, dx$$

$$= \left[-\frac{1}{\pi}\cos \pi - 0\right] + \frac{1}{\pi^2}\sin \pi x\Big|_0^1$$

$$= \frac{1}{\pi} + \frac{1}{\pi^2}\sin \pi - \frac{1}{\pi^2}\sin 0 = \frac{1}{\pi}.$$

20. $\int_1^4 \ln x \, dx$. Use integration by parts with $f(x) = \ln x$,

$g(x) = 1$. Then $f'(x) = \frac{1}{x}$ and $G(x) = x$.

$$\int_1^4 \ln x\, dx = x \ln x\Big|_1^4 - \int_1^4 \frac{1}{x}(x)\, dx = \Big[4\ln 4 - \ln(1)\Big] - \Big[4 - 1\Big]$$

$$= 4\ln 4 - 3.$$

21. $\displaystyle\int_{-\pi/2}^{\pi/2} \sqrt{1 - \sin^2 x}\, \cos x\, dx.$ Let $u = \sin x$; $du = \cos x\, dx$.

When $x = -\frac{\pi}{2}$, $u = \sin\left(-\frac{\pi}{2}\right) = -1$. When $x = \frac{\pi}{2}$, $u = \sin\left(\frac{\pi}{2}\right) = 1$.

$$\int_{-\pi/2}^{\pi/2} \sqrt{1 - \sin^2 x}\, \cos x\, dx = \int_{-1}^{1} \sqrt{1 - u^2}\, du$$

= the area of the top half of the circle $u^2 + v^2 + 1$

$= \frac{1}{2}\pi(1)^2 = \frac{\pi}{2}$.

22. $\displaystyle\int_0^{\sqrt{2}} \sqrt{4 - x^4}\, 2x\, dx.$ Let $u = x^2$; $du = 2x\, dx$.

When $x = 0$, $u = 0^2 = 0$. When $x = \sqrt{2}$, $u = \sqrt{2}^2 = 2$.

$$\int_0^{\sqrt{2}} \sqrt{4 - x^4}\, 2x\, dx = \int_0^2 \sqrt{4 - u^2}\, du$$

= the area of top right-hand quarter of the circle $u^2 + v^2 = 4$

$= \frac{1}{4}\pi(2)^2 = \pi$.

23. $\displaystyle\int_{-6}^{0} \sqrt{-x^2 - 6x}\, dx = \int_{-6}^{0} \sqrt{9 - (x + 3)^2}\, dx.$

Let $u = (x + 3)$; $du = dx$. When $x = -6$, $u = -6 + 3 = -3$.

When $x = 0$, $u = 0 + 3 = 3$.

$$\int_{-6}^{0} \sqrt{9 - (x + 3)^2}\, dx = \int_{-3}^{3} \sqrt{9 - u^2}\, du$$

= the area of the top half of the circle, $u^2 + v^2 = 9$,

$= \frac{1}{2}\pi(3)^2 = \frac{9}{2}\pi$.

24. $y = -x\sqrt{9 - x^2}$. If $y = 0$ then $9 - x^2 = 0$ or $x = 0$

$\Rightarrow x = -3, 0, 3$ are the intersection points with the x-axis.

Area of the portion from $x = -3$ to $x = 0$ is given by

$\int_{-3}^{0} -x\sqrt{9 - x^2}\, dx$. Let $u = 9 - x^2$, $du = -2xdx$.

$$\int -x\sqrt{9 - x^2}\, dx = \frac{1}{2}\int u^{1/2}du = \frac{1}{3}u^{3/2} + C = \frac{1}{3}(9 - x^2)^{3/2} + C$$

$$\int_{-3}^{0} -x\sqrt{9 - x^2}\, dx = \frac{1}{3}(9 - x^2)^{3/2}\Big|_{-3}^{0} = 9$$

By symmetry area from $x = 0$ to $x = 3$ is also 9, so the total area is 18.

25. $y = x\sqrt{4 - x^2}$. If $y = 0$ then $4 - x^2 = 0$ or $x = 0$

$\Rightarrow x = -2, 0, 2$ are the intersection points with the x-axis.

Area of the portion from $x = -2$ to $x = 0$ is given by

$\int_{3} x\sqrt{4 - x^2}\, dx$. Let $u = 4 - x^2$, $du = -2xdx$.

$$\int x\sqrt{4 - x^2}\, dx = -\frac{1}{2}\int u^{1/2}du = -\frac{1}{3}u^{3/2} + C = -\frac{1}{3}(4 - x^2)^{3/2} + C$$

$$\int_{-3}^{0} x\sqrt{4 - x^2}\, dx = \frac{1}{3}(4 - x^2)^{3/2}\Big|_{-2}^{0} = -\frac{8}{3}$$

By symmetry area from $x = 0$ to $x = 2$ is also $\frac{8}{3}$, so the total area is $\frac{16}{3}$.

Exercises 9.4

1. $\Delta x = (5 - 3)/5 = \frac{2}{5} = .4$

   $a_0 = 3,\ a_1 = 3.4,\ a_2 = 3.8,\ a_3 = 4.2,\ a_4 = 4.6,\ a_5 = 5$

2. $\Delta x = (2 - 1)/5 = \frac{3}{5} = .6$

   $a_0 = -1,\ a_1 = -.4,\ a_2 = .2,\ a_3 = .8,\ a_4 = 1.4,\ a_5 = 2$

3. $\Delta x = (2 - 0)/4 = .5$

   $x_1 = \frac{.5}{2} = .25,\ x_2 = .75,\ x_3 = 1.25,\ x_4 = 1.75$

4. $\Delta x = (3 - 0)/6 = .5 \qquad x_1 = 0 + \frac{.5}{2} = .25,\ x_2 = .75,$

   $x_3 = 1.25,\ x_4 = 1.75,\ x_5 = 2.25,\ x_6 = 2.75$

5. $\Delta x = (4 - 1)/5 = .6$

   $x_1 = 1 + \frac{.6}{2} = 1.3,\ x_2 = 1.9,\ x_3 = 2.5,\ x_4 = 3.1,\ x_5 = 3.7$

6. $\Delta x = (5 - 3)/10 = .2 \qquad x_1 = 3 + \frac{.2}{2} = 3.1,\ x_2 = 3.3,$

   $x_3 = 3.5,\ x_4 = 3.7,\ x_5 = 3.9,\ x_6 = 4.1,\ x_7 = 4.3,\ x_8 = 4.5,$

   $x_9 = 4.7,\ x_{10} = 4.9$

7. If $n = 2$, then $\Delta x = (4 - 0)/2 = 2$. The first midpoint is $x_1 = 0 + \frac{2}{2} = 1$. The second midpoint is $x_2 = 1 + 2 = 3$. Using the midpoint rule we have

   $$\int_0^4 (x^2 + 5)dx \sim [f(1) + f(3)]\cdot 2 = (6 + 14) = 40.$$

   If $n = 4$, then $\Delta x = (4 - 0)/4 = 1$, $x_1 = 0 + 1/2 = .5$, $x_2 = 1.5$, $x_3 = 2.5$, $x_4 = 3.5$. Hence,

   $$\int_0^4 (x^2 + 5)dx \sim [f(x_1) + f(x_2) + f(x_3) + f(x_4)]\Delta x$$

   $$= \Big[[(.5)^2 + 5] + [(1.5)^2 + 5] + [(2.5)^2 + 5] + [(3.5)^2 + 5]\Big](1) = 41$$

   Evaluating the integral directly, we have

$$\int_0^4 (x^2 + 5)dx = \left(\frac{1}{3}x^3 + 5x\right)\Big|_0^4 = \frac{64}{3} + 20 = 41\frac{1}{3}.$$

8. $\int_0^5 (x - 1)^2dx$. Using the midpoint rule with n = 2, we have

$\Delta x = (5 - 1)/2 = 2$. The midpoints are $x_1 = 1 + \frac{2}{2} = 2$ and $x_2 = 4$.

$$\int_0^5 (x - 1)^2dx \sim 2\left[(2 - 1)^2 + (4 - 1)^2\right] = 20.$$

With n = 4, $\Delta x = (5 - 1)/4 = 1$. The midpoints are $x_1 = 1 + \frac{1}{2} = \frac{3}{2}$, $x_2 = \frac{5}{2}$, $x_3 = \frac{7}{2}$, $x_4 = \frac{9}{2}$.

$$\int_0^5 (x - 1)^2dx \sim (1)\left[\left(\frac{3}{2} - 1\right)^2 + \left(\frac{5}{2} - 1\right)^2 + \left(\frac{7}{2} - 1\right)^2 + \left(\frac{9}{2} - 1\right)^2\right]$$

$$= 21$$

The exact value is

$$\int_0^5 (x - 1)^2dx = \frac{1}{3}(x - 1)^3\Big|_9^5 = \frac{1}{3}\left(4^3 - 0\right) = \frac{64}{3}.$$

9. $\int_0^1 e^{-x}dx$. Using the midpoint rule with n = 5, we have

$\Delta x = (1 - 0)/5 = .2$ and the midpoints are $x_1 = 0 + \frac{.2}{2} = .1$, $x_2 = .3$, $x_3 = .5$, $x_4 = .7$, $x_5 = .9$

$$\int_0^1 e^{-x}dx \sim .2\left[e^{-.1} + e^{-.3} + e^{-.5} + e^{-.7} + e^{-.9}\right] = .63107.$$

The exact value is

$$\int_0^1 e^{-x}dx = -e^{-x}\Big|_0^1 = -\frac{1}{e} + 1 \sim .63212.$$

10. $\int_1^2 \frac{1}{x + 1}\,dx$. Using the midpoint rule with n = 5, we have

$\Delta x = (2 - 1)/5 = .2$ and the midpoints are $x_1 = 1 + \frac{.2}{2} = 1.1$, $x_2 = 1.3$, $x_3 = 1.5$, $x_4 = 1.7$, and $x_5 = 1.9$.

$$\int_1^2 \frac{1}{x + 1}\,dx \sim$$

$$.2\left[\frac{1}{1.1 + 1} + \frac{1}{1.3 + 1} + \frac{1}{1.5 + 1} + \frac{1}{1.7 + 1} + \frac{1}{1.9 + 1}\right] \sim .40523$$

The exact value is

$$\int_1^2 \frac{1}{x + 1}\,dx = \ln|x + 1|\Big|_1^2 = \ln 3 - \ln 2 \sim .40547.$$

11. $\int_0^1 \left(x - \frac{1}{2}\right)^2 dx.$ Using the trapezoidal rule with n = 4, we have $\Delta x = (1 - 0)/4 = .25$ and the endpoints are $a_0 = 0$, $a_1 = .25$, $a_2 = .5$, $a_3 = .75$ and $a_4 = 1$.

$$\int_0^1 \left(x - \frac{1}{2}\right)^2 dx \sim \left[\left(0 - \frac{1}{2}\right)^2 + 2\left(.25 - \frac{1}{2}\right)^2 + 2\left(.5 - \frac{1}{2}\right)^2 + \right.$$

$$\left. + 2\left(.75 - \frac{1}{2}\right)^2 + \left(1 - \frac{1}{2}\right)^2\right]\frac{.25}{2} = .09375$$

The exact value is

$$\int_0^1 \left(x - \frac{1}{2}\right)^2 dx = \frac{1}{3}\left(x - \frac{1}{2}\right)^3\Big|_0^1 = \frac{1}{3}\left[.5^3 - (-.5)^3\right] \sim .08333.$$

12. $\int_4^9 \frac{1}{x - 3}\,dx.$ Using the trapezoidal rule with n = 5, we have $\Delta x = (9 - 4)/5 = 1$. The endpoints are $a_0 = 4$, $a_1 = 5$, $a_2 = 6$, $a_3 = 7$, $a_4 = 8$ and $a_5 = 9$.

$$\int_4^9 \frac{1}{x - 3}\,dx \sim \left[\frac{1}{4 - 3} + (2)\frac{1}{5 - 3} + (2)\frac{1}{6 - 3} + (2)\frac{1}{7 - 3}\right.$$

$$\left. + (2)\frac{1}{8 - 3} + \frac{1}{9 - 3}\right]\frac{1}{2} \sim 1.86667$$

The exact value is

$$\int_4^9 \frac{1}{x - 3}\, dx = \ln|x - 3|\Big|_4^9 \ \ln 6 - \ln 1 \sim 1.79176.$$

13. $\int_1^5 \frac{1}{x^2}\, dx$. Using the trapezoidal rule with n = 3, we have

$\Delta x = (5 - 1)/3 \sim 1.33333$. The endpoints are $a_0 = 1$, $a_1 = 2.33333$, $a_2 = 3.66667$ and $a_3 = 5$.

$$\int_1^5 \frac{1}{x^2}\, dx \sim \left[\frac{1}{1^2} + (2)\frac{1}{2.33333^2} + (2)\frac{1}{3.66667^2} + \frac{1}{5^2}\right] \sim 1.03741$$

The exact value is

$$\int_1^5 \frac{1}{x^2}\, dx = -\frac{1}{x}\Big|_1^5 = -\frac{1}{5} + 1 = .8$$

14. $\int_{-1}^1 e^{2x}\, dx$. Using the trapezoidal rule with n = 2, we have

$\Delta x = (1 - (-1))/2 = 1$ and the endpoints are $a_0 = -1$, $a_1 = 0$ and $a_2 = 1$.

$$\int_{-1}^1 e^{2x}\, dx \sim \left[e^{2(-1)} + 2e^{2(0)} + e^{2(1)}\right]\frac{1}{3} \sim 4.76222$$

With n = 4, $\Delta x = (1 - (-1))/4 = .5$, $a_0 = -1$, $a_1 = -.5$, $a_2 = 0$, $a_3 = .5$ and $a_4 = 1$.

$$\int_{-1}^1 e^{2x}\, dx \sim \left[e^{2(-1)} + e^{2(-.5)} + 2e^{2(0)} + 2e^{2(.5)} + e^{2(1)}\right]\frac{.5}{2}$$

$$\sim 3.92418.$$

The exact value is

$$\int_{-1}^1 e^{2x}\, dx = \frac{1}{2}e^{2x}\Big|_{-1}^1 = \frac{1}{2}\left[e^2 - e^{-2}\right] \sim 3.62688.$$

15. $\int_1^4 (2x - 3)^3 dx$; n = 3. $\Delta x = (4 - 1)/3 = 1$

<u>Midpoint rule</u>: $x_1 = 1 + \frac{1}{2} = 1.5$, $x_2 = 2.5$, $x_3 = 3.5$

$$\int_1^4 (2x - 3)^3 dx \sim \left[(2(1.5) - 3)^3 + (2(2.5) - 3)^3 + (2(3.5) - 3)^3\right] = 72$$

Trapezoidal rule: $a_0 = 1,\ a_1 = 2,\ a_2 = 3,\ a_3 = 4$

$$\int_1^4 (2x - 3)^3 dx \sim \left[(2(1) - 3)^3 + 2(2(2) - 3)^3 + 2(2(3) - 3)^3 + (2(4) - 3)^3\right]\frac{1}{2} = 90.$$

Simpson's rule:

$$\int_1^4 (2x - 3)^3 dx \sim S = \frac{2R + T}{3} = \frac{2(72) + 90}{3} = 78.$$

Exact value:

$$\int_1^4 (2x - 3)^3 dx = \frac{1}{8}(2x - 3)^4\Big|_1^4$$

$$= \frac{1}{8}\left[(2(4) - 3)^4 - (2(1) - 3)^4\right] = 78.$$

16. $\displaystyle\int_{10}^{20} \frac{\ln x}{x} dx$; $n = 5$. $\Delta x = (20 - 10)/5 = 2.$

Midpoint rule: $x_1 = 10 + \frac{2}{2} = 11,\ x_2 = 13,\ x_3 = 15,\ x_4 = 17,$ $x_5 = 19.$

$$\int_{10}^{20} \frac{\ln x}{x} dx \sim 2\left[\frac{\ln 11}{11} + \frac{\ln 13}{13} + \frac{\ln 15}{15} + \frac{\ln 17}{17} + \frac{\ln 19}{19}\right] \sim 1.83492.$$

Trapezoidal rule: $a_0 = 10,\ a_1 = 12,\ a_2 = 14,\ a_3 = 16,$

$a_4 = 18,\ a_5 = 20.$

$$\int_{10}^{20} \frac{\ln x}{x}\,dx \sim \left[\frac{\ln 10}{10} + 2\frac{\ln 12}{12} + 2\frac{\ln 14}{14} + 2\frac{\ln 16}{16} + 2\frac{\ln 18}{18} + \frac{\ln 20}{20}\right]\frac{2}{2} \sim 1.83893.$$

Simpson's rule:

$$\int_{10}^{20} \frac{\ln x}{x}\,dx \sim S = \frac{2R + T}{3} = \frac{2(1.83492) + 1.83893}{3} = 1.83626$$

Exact value:

$$\int_{10}^{20} \frac{\ln x}{x}\,dx. \quad \text{Let } u = \ln x,\ du = \frac{1}{x}dx.$$

When $x = 10$, $u = \ln 10$ and when $x = 20$, $u = \ln 20$.

$$\int_{10}^{20} \frac{\ln x}{x}\,dx = \int_{\ln 10}^{\ln 20} u\,du = \frac{1}{2}u^2\Big|_{\ln 10}^{\ln 20} = \frac{1}{2}\left[(\ln 20)^2 - \ln(10)^2\right]$$

$$\sim 1.83626.$$

17. $\int_0^2 2xe^{x^2}dx$; $n = 4$. $\Delta x = (2 - 0)/4 = .5$

Midpoint rule: $x_1 = 0 + \frac{.5}{2} = .25$, $x_2 = .75$, $x_3 = 1.25$, $x_4 = 1.75$.

$$\int_0^2 2xe^{x^2}dx \sim .5\left[2(.25)e^{(.25)^2} + 2(.75)e^{(.75)^2} + 2(1.25)e^{(1.25)^2} + 2(1.75)e^{(1.75)^2}\right] \sim 44.96248$$

Trapezoidal rule: $a_0 = 0$, $a_1 = .5$, $a_2 = 1$, $a_3 = 1.5$, $a_4 = 2$.

$$\int_0^2 2xe^{x^2}dx \sim \left[2(0)e^{0^2} + 2(2)(.5)e^{(.5)^2} + 2(2)(1)e^{1^2} + 2(2)(1.5)e^{(1.5)^2} + 2(2)e^{2^2}\right. \sim 72.19005$$

Simpson's rule:

$$\int_0^2 2xe^{x^2}\,dx \sim S = \frac{2R + T}{3} = \frac{2(44.96248) + 72.19005}{3} = 54.03834.$$

Exact value:

$$\int_0^2 2xe^{x^2}\,dx = e^{x^2}\Big|_0^2 = e^4 - 1 \sim 53.59815.$$

18. $\int_0^3 x\sqrt{4 - x}\,dx$; $n = 5$. $\Delta x = (3 - 0)/.5 = .6$

Midpoint rule: $x_1 = 0 + \frac{.6}{2} = .3$, $x_2 = .9$, $x_3 = 1.5$, $x_4 = 2.1$, $x_5 = 2.7$

$$\int_0^3 x\sqrt{4 - x}\,dx \sim \Big[(.3)\sqrt{4 - .3} + (.9)\sqrt{4 - .9} + (1.5)\sqrt{4 - 1.5} + (2.1)\sqrt{4 - 2.1} + (2.7)\sqrt{4 - 2.7}\Big] \sim 6.30390.$$

Trapezoidal rule: $a_0 = 0$, $a_1 = .6$, $a_2 = 1.2$, $a_3 = 1.8$, $a_4 = 2.4$, $a_5 = 3.0$

$$\int_0^3 x\sqrt{4 - x}\,dx \sim \Big[(0)\sqrt{4 - 0} + 2(.6)\sqrt{4 - .6} + 2(1.2)\sqrt{4 - 1.2} + 2(1.8)\sqrt{4 - 1.8} + 2(2.4)\sqrt{4 - 2.4} + (3.0)\sqrt{4 - 3}\Big]\frac{.6}{2} \sim 6.19197.$$

Simpson's rule:

$$\int_0^3 x\sqrt{4 - x}\,dx \sim \frac{2(6.3039) + 6.19197}{3} = 6.26659.$$

Exact value:

$\int_0^3 x\sqrt{4 - x}\,dx$. Use integration by parts with $f(x) = x$, $g(x) = \sqrt{4 - x}$. Then $f'(x) = 1$ and $G(x) = -\frac{2}{3}(4 - x)^{3/2}$ and

$$\int_0^3 x\sqrt{4-x}\,dx = -\frac{2}{3}x(4-x)^{3/2}\Big|_0^3 + \int_0^3 -\frac{2}{3}(4-x)^{3/2}dx$$

$$= \left[-\frac{2}{3}(3)1^{3/2} - 0\right] - \frac{4}{15}(4-x)^{5/2}\Big|_0^3 \sim 6.26667.$$

19. $\int_2^5 xe^x dx$; $n = 5$. $\Delta x = (5 - 2)/5 = .6$

Midpoint rule: $x_1 = 2 + \frac{.6}{2} = 2.3$, $x_2 = 2.9$, $x_3 = 3.5$, $x_4 = 4.1$, $x_5 = 4.7$.

$$\int_2^5 xe^x dx \sim .6\Big[(2.3)e^{2.3} + (2.9)e^{2.9} + (3.5)e^{3.5} + (4.1)e^{4.1}$$
$$+ (4.7)e^{4.7}\Big] \sim 573.41792$$

Trapezoidal rule: $a_0 = 2$, $a_1 = 2.6$, $a_2 = 3.2$, $a_3 = 3.8$, $a_4 = 4.4$, $a_5 = 5$.

$$\int_2^5 xe^x dx \sim \Big[2e^2 + 2(2.6)e^{2.6} + 2(3.2)e^{3.2} + 2(3.8)e^{3.8}$$
$$+ 2(4.4)e^{4.4} + 5e^5\Big]\frac{.6}{2} \sim 612.10802$$

Simpson's rule:

$$\int_2^5 xe^x dx \sim \frac{2(573.41792) + 612.10802}{3} \sim 586.31461$$

Exact value:

$\int_2^5 xe^x dx$. Use integration by parts with $f(x) = x$, $g(x) = e^x$.

Then $f'(x) = 1$ and $G(x) = e^x$.

$$\int_2^5 xe^x dx = xe^x\Big|_2^5 - \int_2^5 e^x dx = xe^x - e^x\Big|_2^5 = e^x(x-1)\Big|_2^5$$

$$= 4e^5 - e^2 \sim 586.26358$$

20. $\int_1^5 (4x^3 - 3x^2)dx$; $n = 2$. $\Delta x = (5 - 1)/2 = 2$

Midpoint rule: $x_1 = 1 + \frac{2}{2} = 2$, $x_2 = 4$

$$\int_1^5 (4x^3 - 3x^2)dx \sim 2\left[(4(2)^3 - 3(2)^2) + (4(4)^3 - 3(4)^2)\right] = 456$$

Trapezoid rule: $a_0 = 1$, $a_1 = 3$, $a_3 = 5$.

$$\int_1^5 (4x^3 - 3x^2)dx \sim \left[(4(1^3) - 3(1^2)) + 2(4(3^3) - 3(3^2)) + (4(5^3) - (5^2))\right]\frac{2}{2} = 588$$

Simpson's rule:

$$\int_1^5 (4x^3 - 3x^2)dx \sim \frac{2(456) + 588}{3} = 500$$

Exact value:

$$\int_1^5 (4x^3 - 3x^2)dx = (x^4 - x^3)\Big|_1^5 = 5^4 - 5^3 = 500.$$

21. $\int_0^2 \sqrt{1 + x^3}\,dx$; $n = 4$. $\Delta x = (2 - 0)/4 = .5$

$a_0 = 0$, $a_1 = .5$, $a_2 = 1$, $a_3 = 1.5$, $a_4 = 2$,
$x_1 = .25$, $x_2 = .75$, $x_3 = 1.25$, $x_4 = 1.75$

$$\int_0^2 \sqrt{1 + x^3}\,dx \sim \left[\sqrt{1 + 0^3} + 4\sqrt{1 + .25^3} + 2\sqrt{1 + .5^3} + 4\sqrt{1 + .75^3} + 2\sqrt{1 + 1^3} + 4\sqrt{1 + 1.25^3} + 2\sqrt{1 + 1.5^3} + 4\sqrt{1 + 1.75^3} + \sqrt{1 + 2^3}\right]\frac{.5}{6}$$
$$\sim 3.24124$$

22. $\int_0^1 \frac{1}{x^3 + 1}\,dx$; $n = 2$. $\Delta x = (1 - 0)/2 = .5$

$a_0 = 0$, $a_1 = .5$, $a_2 = 1$, $x_1 = .25$, $x_2 = .75$

$$\int_0^1 \frac{1}{x^3+1}\,dx \approx \left[\frac{1}{0^3+1} + (4)\frac{1}{.25^3+1} + (2)\frac{1}{.5^3+1} + (4)\frac{1}{.75^3+1} + \frac{1}{1+1^2}\right]\frac{.5}{6} \approx .83579.$$

23. $\int_0^2 \sqrt{\sin x}\,dx$; $n = 5$. $\Delta x = (2 - 0)/5 = .4$

$a_0 = 0$, $a_1 = .4$, $a_2 = .8$, $a_3 = 1.2$, $a_4 = 1.6$, $a_5 = 2$,
$x_1 = .2$, $x_2 = .6$, $x_3 = 1$, $x_4 = 1.4$, $x_5 = 1.8$.

$$\int_0^2 \sqrt{\sin x}\,dx \approx \left[\sqrt{\sin 0} + 4\sqrt{\sin(.2)} + 2\sqrt{\sin(.4)} + 4\sqrt{\sin(.6)} + 2\sqrt{\sin(.8)} + 4\sqrt{\sin 1} + 2\sqrt{\sin(1.2)} + 4\sqrt{\sin(1.4)} + 2\sqrt{\sin(1.6)} + 4\sqrt{\sin(1.8)} + \sqrt{\sin(1.8)}\right]\frac{.4}{2}$$

≈ 1.61347

24. $\int_{-1}^1 \sqrt{1+x^4}\,dx$; $n = 4$. $\Delta x = (1 - (-1))/4 = .5$

$$\int_{-1}^1 \sqrt{1+x^4}\,dx \approx \left[\sqrt{1+(-1)^4} + 4\sqrt{1+(-.75)^4} + 2\sqrt{1+(-.5)^4} + 4\sqrt{1+(-.25)^4} + 2\sqrt{1+0^4} + 4\sqrt{1+.25^4} + 2\sqrt{1+.5^4} + 4\sqrt{1+.75^4} + \sqrt{1+1^4}\right]\frac{.5}{2}$$

≈ 2.17883

25. View the distance to the water as a function f of the position x of a point on the side. If a corresponds to the top of the diagram and b to the bottom, the area we wish to compute is $\int_a^b f(x)dx$. Since the measurements are taken 50 feet apart, Δx

= 50. Thus the trapezoidal rule with n = 4 gives

$$\int_a^b f(x)dx \sim \left[100 + 2(90) + 2(25) + 2(150) + 200\right]\frac{50}{2}$$

$$= 25{,}750 \text{ sq. ft.}$$

26. View the depth of the river as a function f of the position x of a point between the banks. If a corresponds to the bank on which the first reading was taken and b corresponds to the last reading, then the cross-sectional area we wish to compute is $\int_a^b f(x)dx$. Using the trapezoidal rule with n = 5 and $\Delta x = 20$ (distance between readings) we have, after converting readings to yards,

$$\int_a^b f(x)dx \sim \left[0 + 2(.2) + 2(.4) + 2(.6) + 2(.8) + 0\right]\frac{20}{2}$$

$$= 280 \text{ sq. yd.}$$

27. (See hint.) Using the trapezoidal rule with $\Delta t = 10$, n = 10,

$$S(10) = \int_0^{10} v(t)dt \sim \left[0 + 2(30) + 2(75) + 2(115) + 2(155)\right.$$
$$+ 2(200) + 2(250) + 2(300) + 2(360)$$
$$\left. + 2(420) + 490\right]\frac{1}{2} = 2150 \text{ ft.}$$

28. (See hint for Ex. 27)

$S(5) = \int_0^5 v(t)dt$ with v(t) velocity at time t measured in miles per minute.

$$\int_0^5 v(t)dt \sim \left[33 + 2(32) + 2(28) + 2(30) + 2(32) + 35\right]\left(\frac{1}{2}\right)\frac{1}{60}$$

$$= 2.6 \text{ miles.}$$

29. Let $f(x) = x^4 + 3x^2$. Then $f'(x) = 4x^3 + 6x$ and $|f''(x)| = |12x^2 + 6| \le 12(2^2) + 6 = 54$ for all x satisfying

$0 \le x \le 2$. Thus the error in the estimate is at most

$$\frac{54(2-0)^3}{24(100)^2} = .0018.$$

30. Let $f(x) = 3 \ln x$, $f'(x) = \frac{3}{x}$, $f''(x) = \frac{-3}{x^2}$, $f'''(x) = \frac{6}{x^3}$,

$|f^{(4)}(x)| = \left|\frac{-18}{x^4}\right| = \frac{18}{x^4} \le \frac{18}{1^4} = 18$ for all x such that

$1 \le x \le 2$. Thus the error in the estimate is at most

$$\frac{18(2-1)^5}{2880(5)^4} = .00001$$

31. (See figure 8)

(a) The area of the triangle on top is $\frac{1}{2}(k - h)\ell$ and the area of the rectangle on the bottom is ℓh. Thus the area of the trapezoid is $\frac{1}{2}(k - h)\ell + \ell h = \frac{1}{2}(h + k)\ell$.

(b) From part (a) with $f(a_0) = h$, $\Delta x = \ell$ and $f(a_1) = k$, we have area $= \frac{1}{2}\left[f(a_0) + f(a_1)\right]\Delta x$.

(c) Applying the result in (a) as in part (b), we see that the sum of the areas of the 4 trapezoids in fig. 8(b) is

$$\frac{1}{2}\left[f(a_0) + f(a_1)\right]\Delta x + \frac{1}{2}\left[f(a_1) + f(a_2)\right]\Delta x + \frac{1}{2}\left[f(a_2) + f(a_3)\right]\Delta x$$

$$+ \left[f(a_3) + f(a_4)\right]\Delta x$$

$$= \left[f(a_0) + 2f(a_1) + 2f(a_2) + 2f(a_3) + a_4\right]\frac{\Delta x}{2}.$$

32. (See fig. 9)

Using the approximation described in the problem,

$$\int_a^b f(x)dx \sim \frac{\Delta x}{2}f(a_0) + \Delta x f(a_1) + \Delta f(a_2) + \Delta x f(a_3) + \frac{\Delta x}{2}f(a_4)$$

$$= \left[f(a_0) + 2f(a_1) + 2f(a_2) + 2f(a_3) + f(a_4)\right]\frac{\Delta x}{2}$$

= estimate using the trapezoidal rule with n = 4.

33. (a) Let T_1 be the top triangle in fig. 10(b) and let T_2 be the bottom one. Both are right triangles, their bases have the same length ($\Delta x/2$) and the angle between the base and

hypotenuse is the same in each. Therefore the area, $A(T_1) = A(T_2)$, the area of T_2. The shaded area in fig. 10(a) = area shaded in 10(c) + $A(T_1) - A(T_2)$.

(b) $\int_a^b f(x)dx \le R$ follows from (a), using the fact that

$$\int_a^b f(x)dx = \int_a^{a_1} f(x)dx + \int_{a_1}^{a_2} f(x)dx + \ . \ . \ . + \int_{a_{n-1}}^b f(x)dx \ .$$

Decomposing $\int_a^b f(x)dx$ as above, $T \le \int_a^b f(x)dx$ follows from the fact that if f is concave downward then each of the trapezoids whose area is summed to obtain T is entirely <u>below</u> the graph of f. (See, for example, fig. 8(b), where the first two trapezoids are drawn over a region on which f is concave downward).

34. (a) R is defined to be the rate at which the heart pumps blood, measured in liters per minute. Thus R/60 measures the same quantity in liters per sec. So in Δt_i seconds (the length of the ith time interval) approximately $(R/60)\Delta t_i$ liters of blood will flow past the monitoring point.

(b) If $(R/60)\Delta t_i$ liters of blood flows past the monitoring point during the ith interval and the concentration of dye in the blood is approximately $c(t_i)$ mg/liter (here we are using $c(t_i)$ to approximate $c(t)$ for all t in the ith interval), then approximately $c(t_i)(R/60)\Delta t_i$ mg of dye will flow past the monitoring point during this time.

(c) If essentially all of the dye flows past the monitoring point during the first 22 seconds, then

$$D \sim R(60)\Big[c(t_1) + \ . \ . \ . + c(t_n)\Big]\Delta t$$

follows directly from (b), since D is the total quantity of dye. The approximation improves as n gets large, since as Δt_i gets smaller, $c(t_i)$ becomes a better approximation for $c(t)$ on the ith interval.

(d) $\left[c(t_1) + c(t_2) + . . . + c(t_n)\right]\Delta t$ is a Riemann sum of c(t) on [0,22]. The result in (c) says that

$$\lim_{\Delta t \to 0}\left[c(t_1) + c(t_2) + . . . + c(t_n)\right]\Delta t = \frac{60}{R}D .$$

Thus $\int_0^{22} c(t)dt = \frac{60}{R}D$ or $D = \int_0^{22} \frac{R}{60}c(t)dt$.

Therefore, $R = \dfrac{60D}{\int_0^{22} c(t)dt}$.

Exercises 9.5

1. The present value is

$\int_0^6 (8000 - 200t)e^{-.1t}dt$. Using integration by parts with

$f(t) = 8000 - 200t$ and $g(t) = e^{-.1t}$, we have

$f'(t) = -200$ and $G(t) = -10e^{-.1t}$ and

$$\int_0^6 (8000 - 200t)e^{-.1t}dt$$

$$= (8000 - 200t)(-10)e^{-.1t}\Big|_0^6 - \int_0^6 2000e^{-.1t}dt$$

$$= e^{-.6}(12,000 - 80,000) - (-80,000) + 20,000e^{-.1t}\Big|_0^6$$

$\sim$ \$33,656.

1. Present value = $\int_{T_1}^{T_2} k(t)e^{-rt}dt$; $k(t) = 40000$, $T_1 = 0$, $T_2 = 5$

$r = 0.8$. So P. V. = $\int_0^5 40,000e^{-.08t}dt = -\frac{100}{8}(40,000)e^{-08t}\Big|_0^5$

$$= -500,000\left[e^{-.08(5)} - e^{-0}\right]$$

$$= \$164,840.$$

2. $k(t) = 60000,\ T_1 = 2,\ T_2 = 5,\ r = 0.8.$

So P. V. $= \displaystyle\int_2^5 60,000e^{-.12t}dt = -\frac{100}{12}(60,000)e^{-12t}\Big|_2^5$

$$= \$118,908.$$

2. $\displaystyle\int_0^6 (8000 - 200t)e^{-.15t}dt.$ Use integration by parts with

$f(t) = 8000 - 200t,\ g(t) = e^{-.15t}.$

Then $f'(t) = -200$ and $G(t) = \dfrac{-100}{15}e^{-.15t} = \dfrac{-20}{3}e^{-.15t}.$

$$\int_0^6 (8000 - 200t)e^{-.15t}dt$$

$$= (8000 - 200t)\left(\frac{-20}{3}\right)e^{-.15t}\Big|_0^6 - \int_0^6 \frac{4000}{3}e^{-.15t}dt$$

$$= e^{-.9}\left(8000 - \frac{16,000}{3}\right) + \frac{16,000}{3} + \frac{80,000}{9}e^{-.15t}\Big|_0^6$$

$\sim$ \$29,627.

3. $k(t) = 80e^{-.09t}, T_1 = 0,\ T_2 = 3,\ r = .12.$ The present value is

$$\int_0^3 (80e^{-.09t})e^{-.11t}dt = \int_0^3 80e^{-.2t}dt = -\frac{100}{2}(80)e^{-.2t}\Big|_0^3$$

$$= -\frac{8000}{2}\left[e^{-.6} - 1\right]$$

$$= \$180,475.$$

4. $k(t) = 25e^{-.02t}, T_1 = 0,\ T_2 = 4,\ r = .08.$ The present value is

$$\int_0^4 (80e^{-.02t})e^{-.08t}dt = \int_0^4 80e^{-.t}dt = -\frac{100}{1}(80)e^{-.t}\Big|_0^4$$

$$= -\frac{8000}{1}\left[e^{-.4} - 1\right]$$

$$= \$82,420.$$

5. $k(t) = 30 + 5t, T_1 = 0,\ T_2 = 2,\ r = .1.$ The present value is

$\int_0^2 (30 + 5t)e^{-.t}dt.$ Use integration by parts with

$f(x) = 30 + 5t,\ g(x) = e^{-.t}.$

Then $f'(x) = 5$ and $G(x) = \frac{-10}{1}e^{-.t} = -10e^{-.t}.$

$$\int_0^2 (30 + 5t)e^{-.15t}dt$$

$$= (30 + 5t)(-10)e^{-.t}\Big|_0^2 - \int_0^2 (5)\left(-10\right)e^{-.t}dt$$

$$= 300 - e^{-.2}\left(300 + 100\right) - 500e^{-.t}\Big|_0^2$$

$$= \$63,142,321.$$

6. $\int_0^2 (50 + 7t)e^{-.1t}dt.$ Use integration by parts with

$f(t) = 50 + 7t,\ g(t) = e^{-.1t}.$

Then $f'(x) = 7$ and $G(t) = -10e^{-.1t}.$

$$\int_0^2 (50 + 7t)e^{-.1t}dt = (50 + 7t)(-10)e^{-.1t}\Big|_0^2 - \int_0^2 (7)(-10)e^{-.1t}dt$$

$$= 500 - e^{-.2}(500 + 140) - 700e^{-.1t}\Big|_0^2$$

$$\sim 102.9 \text{ (thousand dollars)}$$

7. (See example 3)

$$\int_0^5 (2\pi t)120e^{-.65t}dt = 240\pi\int_0^5 te^{-.65t}dt.$$

Use integration by parts with $f(t) = t,\ g(t) = e^{-.65t}.$

Then $f'(t) = 1$ and $G(t) = \frac{-100}{65}e^{-.65t} = \frac{-20}{13}e^{-.65t}.$

$$\int_0^5 te^{-.65t}dt = t\left(\frac{-20}{3}\right)e^{-.65t}\Big|_0^5 - \int_0^5 \frac{-20}{13}e^{-.65t}dt$$

$$P(10) = h(10)P(0) + \int_0^{10} h(10 - t)r(t)\,dt.$$

Given that $P(0) = 250{,}000$, $h(t) = e^{-t/50}$ and $r(t) = 10{,}000 - 100t$, this yields

$$P(10) = e^{-1/5}(250{,}000) + \int_0^{10} e^{(t-10)/50}(10{,}000 - 100t)dt.$$

Using integration by parts with $f(t) = 10{,}000 - 100t$, $g(t) = e^{(t-10)/50}$, $f'(x) = -100$, $G(t) = 50e^{(t-10)/50}$, we have

$$\int_0^{10} e^{(t-10)/50}(10{,}000 - 100t)dt$$

$$= (10{,}000 - 100t)50e^{(t-10)/50}\Big|_0^{10} - \int_0^{10} (-100)50e^{(t-10)/50}dt$$

$$= \left[(10{,}000 - 100t)50e^{(t-10)/50} + 250{,}000e^{(t-10)/50}\right]\Big|_0^{10}$$

$$= \left[750{,}000 - 5000(10)\right] - e^{-1/5}(750{,}000)$$

$$\sim 85{,}952.$$

Thus $P(10) \sim 250{,}000e^{-1/5} + 85{,}952 \sim 290{,}635$.

12. (a) Formula (4) is $P(T) = h(T)P(0) + \int_0^{T} h(T - t)r(t)\,dt.$

If $P(0) = 300{,}000$, $r(t) = 10{,}000$ and $h(t) = e^{-t/40}$, this gives

$$P(T) = 300{,}000e^{-T/40} + \int_0^{T} e^{(t-T)/40}(10{,}000)dt$$

$$= 300{,}000e^{-T/40} + 400{,}000e^{(t-T)/40}\Big|_0^{T}$$

$$= 300{,}000e^{-T/40} + 400{,}000 - 400{,}000e^{-T/40}$$

$$= 400{,}000 - 100{,}000e^{-t/40}$$

(b) As T gets large, $100{,}000^{-T/40}$ approaches 0. Thus P(T) approaches 400,000.

13. (a)The area of the ring is $2\pi t(\Delta t)$. The population density is $40e^{-.5t}$. So the population is $2\pi t(\Delta t)40e^{-.5t}$

$$= 80\pi t(\Delta t)e^{-.5t} \text{ thousand.}$$

(b) $\frac{dP}{dt}$ or $P'(t)$

(c)It represents the number of people who live between $(5 + \Delta t)$ miles from the sity centre and 5 miles from the city centre.

(d)$P(t + \Delta t) - P(t) = 80\pi t(\Delta t)e^{-.5t}$ from (a).

So, $\dfrac{P(t + \Delta t) - P(t)}{\Delta t} \cong P'(t) = 80\pi te^{-.5t}$

$$\int_a^b P'(t)\,dt = P(b) - P(a) = \int_a^b 80\pi te^{-.5t}dt.$$

Exercises 9.6

1. As $b \to \infty$, $\frac{5}{b}$ approaches 0.

2. As $b \to \infty$, b^2 increases without bound.

3. As $b \to \infty$, $-3e^{2b}$ decreases without bound.

4. As $b \to \infty$, $\frac{1}{b} + \frac{1}{3}$ approaches $0 + \frac{1}{3} = \frac{1}{3}$.

5. As $b \to \infty$, $\frac{1}{4} - \frac{1}{b^2}$ approaches $\frac{1}{4} - 0 = \frac{1}{4}$.

6. As $b \to \infty$, $\frac{1}{2}\sqrt{b}$ increases without bound.

7. As $b \to \infty$, $2 - (b + 1)^{-1/2}$ approaches $2 - 0 = 2$.

8. As $b \to \infty$, $\frac{2}{b} - \frac{2}{b^{3/2}}$ approaches $0 - 0 = 0$.

9. As $b \to \infty$, $5 - \frac{1}{b - 1}$ approaches $5 - 0 = 5$.

10. As $b \to \infty$, $5(b^2 + 3)^{-1}$ approaches 0.

11. As $b \to \infty$, $6 - 3b^{-2}$ approaches $6 - 0 = 6$.

12. As $b \to \infty$, $e^{-b/2} + 5$ approaches $0 + 5 = 5$.

13. As $b \to \infty$, $2(1 - e^{-3b})$ approaches 2.

14. As $b \to \infty$, $4(1 - b^{-3/4})$ approaches 4.

15. The given area is $\int_2^\infty \frac{1}{x^2}\,dx$. $\int_2^b \frac{1}{x^2}\,dx = -\frac{1}{x}\Big|_2^b = -\frac{1}{b} + \frac{1}{2}$.

As $b \to \infty$, $-\frac{1}{b} + \frac{1}{2}$ approaches $\frac{1}{2}$.

Thus, $\int_2^\infty \frac{1}{x^2}\,dx = \lim_{b\to\infty} \int_2^b \frac{1}{x^2}\,dx = \frac{1}{2}$.

16. The given area is $\int_0^\infty (x + 1)^{-2}dx$.

$\int_0^b (x + 1)^{-2}dx = -\frac{1}{x + 1}\Big|_0^b = -\frac{1}{b + 1} + 1$.

As $b \to \infty$, $-\frac{1}{b + 1} + 1$ approaches 1.

Thus, $\int_0^\infty (x + 1)^{-2}dx = \lim_{b\to\infty} \int_0^b (x + 1)^{-2}dx = 1$.

17. The given area is $\int_0^\infty e^{-x/2}dx$.

$\int_0^b e^{-x/2}dx = -2e^{-x/2}\Big|_0^b = -2e^{-b/2} + 2$.

As $b \to \infty$, $-2e^{-b/2} + 2$ approaches 2.

Thus, $\int_0^{\infty} e^{-x/2}dx = \lim_{b\to\infty} \int_0^b e^{-x/2}dx = 2.$

18. The given area is $\int_0^{\infty} 4e^{-4x}dx.$

$$\int_0^b 4e^{-4x}dx = -e^{-4x}\Big|_0^b = -e^{-4b} + 1.$$

As $b \to \infty$, $-e^{-4b} + 1$ approaches 1.

Thus, $\int_0^{\infty} 4e^{-4x}dx = \lim_{b\to\infty} \int_0^b 4e^{-4x}dx = 1.$

19. The given area is $\int_3^{\infty} (x + 1)^{-3/2}dx.$

$$\int_3^b (x + 1)^{-3/2}dx = -2(x + 1)^{-1/2}\Big|_3^b = -2(b + 1)^{-1/2} + 1.$$

As $b \to \infty$, $-2(b + 1)^{-1/2} + 1$ approaches 1.

Thus, $\int_3^{\infty} (x + 1)^{-3/2}dx = \lim_{b\to\infty} \int_3^b (x + 1)^{-3/2}dx = 1.$

20. The given area is $\int_1^{\infty} (2x + 6)^{-4/3}dx.$

$$\int_1^b (2x + 6)^{-4/3}dx = -\frac{3}{2}(2x + 6)^{-1/3}\Big|_1^b = -\frac{3}{2}(2b + 6)^{-1/3} + \frac{3}{4}.$$

As $b \to \infty$, $-\frac{3}{2}(2b + 6)^{-1/3} + \frac{3}{4}$ approaches $\frac{3}{4}$.

Thus, $\int_1^{\infty} (2x + 6)^{-4/3}dx = \lim_{b\to\infty} \int_1^b (2x + 6)^{-4/3}dx = \frac{3}{4}.$

21. $\int_1^b (14x + 18)^{-4/5}dx = \frac{5}{14}(14x + 18)^{1/5}\Big|_1^b = \frac{5}{14}(14b + 18)^{1/5} - \frac{5}{7}.$

As $b \to \infty$, $\frac{5}{14}(14b + 18)^{1/5} - \frac{5}{7}$ increases without bound.

Thus, $\int_1^{\infty} (14x + 18)^{-4/5}dx$ diverges and the area under the curve

$y = (14x + 18)^{-4/5}$ for $x \geq 1$ cannot be assigned any finite number

22. $\int_2^b (x - 1)^{-1/3}dx = \frac{3}{2}(x - 1)^{2/3}\Big|_2^b = \frac{3}{2}(b - 1)^{2/3} - \frac{3}{2}.$

As $b \to \infty$, $\frac{3}{2}(b -1)^{2/3} - \frac{3}{2}$ increases without bound.

Thus, $\int_2^{\infty} (x - 1)^{-1/3}dx$ diverges and the area under the curve

$y = (x - 1)^{-1/3}$ for $x \geq 2$ cannot be assigned a finite number.

23. $\int_1^{\infty} \frac{1}{x^3} dx = \lim_{b\to\infty} \int_1^b \frac{1}{x^3} dx = \lim_{b\to\infty} \left[-\frac{1}{2}x^{-2}\Big|_1^b\right] = \lim_{b\to\infty} \left[-\frac{1}{2}b^{-2} + \frac{1}{2}\right] = \frac{1}{2}.$

24. $\int_1^{\infty} \frac{2}{x^{3/2}} dx = \lim_{b\to\infty} \int_1^b \frac{2}{x^{3/2}} dx = \lim_{b\to\infty} \left[-4x^{-1/2}\Big|_1^b\right]$

$= \lim_{b\to\infty} \left[-4b^{-1/2} + 4\right] = 4.$

25. $\int_0^{\infty} e^{2x}dx = \lim_{b\to\infty} \int_0^b e^{2x}dx = \lim_{b\to\infty} \left[\frac{1}{2}e^{2x}\Big|_0^b\right] = \lim_{b\to\infty} \left[\frac{1}{2}e^{2b} - \frac{1}{2}\right].$

As $b \to \infty$, $\frac{1}{2}e^{2b} - \frac{1}{2}$ increases without bound.

Therefore, $\int_0^{\infty} e^{2x}dx$ diverges.

26. $\int_0^{\infty} (x^2 + 1)dx = \lim_{b\to\infty} \int_0^b (x^2 + 1)dx = \lim_{b\to\infty} \left[\frac{x^3}{3} + x\Big|_0^b\right]$

$= \lim_{b\to\infty} \left[\frac{b^3}{3} + b - 0\right].$

As $b \to \infty$, $\frac{b^3}{3} + b$ increases without bound. Therefore,

$\int_0^\infty (x^2 + 1)dx$ diverges.

27. $$\int_0^\infty \frac{1}{(2x + 3)^2}\, dx = \int_0^b \frac{1}{(2x + 3)^2}\, dx = \lim_{b\to\infty} \left[-\frac{1}{2}(2x + 3)^{-1}\Big|_0^b\right]$$
$$= \lim_{b\to\infty} \left[-\frac{1}{2}(2b + 3)^{-1} + \frac{1}{6}\right] = \frac{1}{6}.$$

28. $$\int_0^\infty e^{-3x}dx = \lim_{b\to\infty} \int_0^b e^{-3x}dx = \lim_{b\to\infty} \left[-\frac{1}{3}e^{-3x}\Big|_0^b\right] = \lim_{b\to\infty} \left[-\frac{1}{3}e^{-3b} + \frac{1}{3}\right]$$
$$= \frac{1}{3}.$$

29. $$\int_2^\infty \frac{1}{(x - 1)^{5/2}}\, dx = \lim_{b\to\infty} \int_2^b \frac{1}{(x - 1)^{5/2}}\, dx = \lim_{b\to\infty} \left[-\frac{2}{3}(x - 1)^{-3/2}\Big|_2^b\right]$$
$$= \lim_{b\to\infty} \left[-\frac{2}{3}(b - 1)^{-3/2} + \frac{2}{3}\right] = \frac{2}{3}.$$

30. $$\int_2^\infty e^{2-x}dx = \lim_{b\to\infty} \int_2^b e^{2-x}dx = \lim_{b\to\infty} \left[-e^{2-x}\Big|_2^b\right] = \lim_{b\to\infty} \left[-e^{2-b} + 1\right] = 1.$$

31. $$\int_0^\infty .01e^{-.01x}dx = \lim_{b\to\infty} \int_0^b .01e^{-.01x}dx = \lim_{b\to\infty} \left[-e^{-.01x}\Big|_0^b\right]$$
$$= \lim_{b\to\infty} \left[-e^{.01b} + 1\right] = 1.$$

32. $$\int_0^\infty \frac{4}{(2x + 1)^3}\, dx = \lim_{b\to\infty} \int_0^b \frac{4}{(2x + 1)^3}\, dx = \lim_{b\to\infty} \left[(2x + 1)^{-2}\Big|_0^b\right]$$
$$= \lim_{b\to\infty} \left[-(2b + 1)^{-2} + 1\right] = 1.$$

33. $\displaystyle\int_0^{\infty} 6e^{1-3x}dx = \lim_{b\to\infty}\int_0^b 6e^{1-3x}dx = \lim_{b\to\infty}\left[-2e^{1-3x}\Big|_0^b\right]$

$\displaystyle = \lim_{b\to\infty}\left[-2e^{1-3b} + 2e\right] = 2e.$

34. $\displaystyle\int_1^{\infty} e^{-.2x}dx = \lim_{b\to\infty}\int_1^b e^{-.2x}dx = \lim_{b\to\infty}\left[-5e^{-.2x}\Big|_1^b\right]$

$\displaystyle = \lim_{b\to\infty}\left[-5e^{-.2b} + 5e^{-.2}\right] = 5e^{-.2} \sim 4.09365.$

35. $\displaystyle\int_3^{\infty} \frac{x^2}{\sqrt{x^3 - 1}}\,dx = \lim_{b\to\infty}\int_3^b \frac{x^2}{\sqrt{x^3 - 1}}\,dx.$

To evaluate $\displaystyle\int_3^b \frac{x^2}{\sqrt{x^3 - 1}}\,dx$, use the substitution $u = x^3 - 1$,

$du = 3x^2dx$. When $x = 3$, $u = 3^3 - 1 = 26$; when $x = b$,

$u = b^3 - 1$. Thus,

$$\int_3^b \frac{x^2}{\sqrt{x^3 - 1}}\,dx = \frac{1}{3}\int_{26}^{b^3-1} \frac{1}{\sqrt{u}}\,du = \left(\frac{1}{3}\right)2u^{1/2}\Big|_{26}^{b^3-1}$$

$$= \frac{2}{3}(b^3 - 1) - \frac{2}{3}\sqrt{26}.$$

As $b \to \infty$, $\frac{2}{3}(b^3 - 1)^{1/2} - \frac{2}{3}\sqrt{26}$ increases without bound.

Therefore, $\displaystyle\int_3^{\infty} \frac{x^2}{\sqrt{x^3 - 1}}\,dx$ diverges.

36. $\displaystyle\int_2^{\infty} \frac{1}{x \ln x}\,dx = \lim_{b\to\infty}\int_2^b \frac{1}{x \ln x}\,dx.$

To evaluate $\displaystyle\int_2^b \frac{1}{x \ln x}\,dx$, use the substitution $u = \ln x$,

$du = \frac{1}{x}dx$. When $x = 2$, $u = \ln 2$; when $x = b$, $u = \ln b$.

Thus, $\int_2^b \frac{1}{x \ln x}\,dx = \int_{\ln 2}^{\ln b} \frac{1}{u}\,du = \ln|u|\Big|_{\ln 2}^{\ln b} = \ln(\ln b) - \ln(\ln 2)$

As $b \to \infty$, $\ln(\ln b) - \ln(\ln 2)$ increases without bound.

Therefore, $\int_2^\infty \frac{1}{x \ln x}\,dx$ diverges.

37. $\int_0^\infty xe^{-x^2}\,dx = \lim_{b\to\infty} \int_0^b xe^{-x^2}\,dx$.

To evaluate $\int_0^b xe^{-x^2}\,dx$, use the substitution $u = -x^2$,

$du = -2x\,dx$. When $x = 0$, $u = 0$; when $x = b$, $u = -b^2$.

$$\int_0^b xe^{-x^2}\,dx = -\frac{1}{2}\int_0^{-b^2} e^u\,du = -\frac{1}{2}e^u\Big|_0^{-b^2} = -\frac{1}{2}e^{-b^2} + \frac{1}{2}.$$

Thus, $\int_0^\infty xe^{-x^2}\,dx = \lim_{b\to\infty} \left[-\frac{1}{2}e^{-b^2} + \frac{1}{2}\right] = \frac{1}{2}$.

38. $\int_0^\infty \frac{x}{x^2 + 1}\,dx = \lim_{b\to\infty} \int_0^b \frac{x}{x^2 + 1}\,dx$.

To evaluate $\int_0^b \frac{x}{x^2 + 1}\,dx$, use the substitution $u = x^2 + 1$,

$du = 2x\,dx$. When $x = 0$, $u = 1$; when $x = b$, $u = b^2 + 1$.

$$\int_0^b \frac{x}{x^2 + 1}\,dx = \frac{1}{2}\int_1^{b^2+1} \frac{1}{u}\,du = \frac{1}{2}\ln|u|\Big|_1^{b^2+1} = \frac{1}{2}\ln(b^2 + 1) - 0.$$

As $b \to \infty$, $\frac{1}{2}\ln(b^2 + 1)$ increases without bound. Thus,

$$\int_0^\infty \frac{x}{x^2+1}\,dx \text{ diverges.}$$

39. $$\int_0^\infty 2x(x^2+1)^{-3/2}dx = \lim_{b\to\infty}\int_0^b 2x(x^2+1)^{-3/2}dx.$$

To evaluate $\int_0^b 2x(x^2+1)^{-3/2}dx$, use the substitution

$u = x^2 + 1$, $du = 2x\,dx$. When $x = 0$, $u = 1$; when $x = b$, $u = b^2 + 1$.

$$\int_0^b 2x(x^2+1)^{-3/2}dx = \int_1^{b^2+1} u^{-3/2}du = -2u^{-1/2}\Big|_1^{b^2+1}$$

$$= -2(b^2+1)^{-1/2} + 2.$$

Thus, $$\int_0^\infty 2x(x^2+1)^{-3/2}dx = \lim_{b\to\infty}\left[-2(b^2+1)^{-1/2} + 2\right] = 2.$$

40. $$\int_1^\infty (5x+1)^{-4}dx = \lim_{b\to\infty}\int_1^b (5x+1)^{-4}dx = \lim_{b\to\infty}\left[-\frac{1}{15}(5x+1)^{-3}\Big|_1^b\right]$$

$$= \lim_{b\to\infty}\left[-\frac{1}{15}(5b+1)^{-3} + \frac{1}{3240}\right] = \frac{1}{3240} \sim .00031.$$

41. $$\int_{-\infty}^0 e^{4x}dx = \lim_{b\to-\infty}\int_b^0 e^{4x}dx = \lim_{b\to-\infty}\left[\frac{1}{4}e^{4x}\Big|_b^0\right] = \lim_{b\to-\infty}\left[\frac{1}{4} - \frac{1}{4}e^{4b}\right] = \frac{1}{4}.$$

42. $$\int_{-\infty}^0 \frac{8}{(x-5)^2}dx = \lim_{b\to-\infty}\int_b^0 \frac{8}{(x-5)^2}dx = \lim_{b\to-\infty}\left[\frac{8}{(x-5)}\Big|_b^0\right]$$

$$= \lim_{b\to-\infty}\left[\frac{8}{5} + \frac{8}{b-5}\right] = \frac{8}{5}.$$

43. $$\int_{-\infty}^{0} \frac{6}{(1-3x)^2}\,dx = \lim_{b\to-\infty} \int_{b}^{0} \frac{6}{(1-3x)^2}\,dx = \lim_{b\to-\infty} \left[\frac{2}{(1-3x)}\Bigg|_{b}^{0}\right]$$

$$= \lim_{b\to-\infty} \left[2 + \frac{2}{1-3b}\right] = 2.$$

44. $$\int_{-\infty}^{0} \frac{1}{\sqrt{4-x}}\,dx = \lim_{b\to-\infty} \int_{b}^{0} \frac{1}{\sqrt{4-x}}\,dx = \lim_{b\to-\infty} \left[-2\sqrt{4-x}\,\Big|_{b}^{0}\right]$$

$$= \lim_{b\to-\infty} \left[-4 + 2\sqrt{4-b}\right].$$

As $b \to -\infty$, $-4 + 2\sqrt{4-b}$ increases without bound.

Thus, $\displaystyle\int_{-\infty}^{0} \frac{1}{\sqrt{4-x}}\,dx$ diverges.

45. $$\int_{0}^{\infty} \frac{e^{-x}}{(e^{-x}+2)^2}\,dx = \lim_{b\to\infty} \int_{0}^{b} \frac{e^{-x}}{(e^{-x}+2)^2}\,dx.$$

To evaluate $\displaystyle\int_{0}^{b} \frac{e^{-x}}{(e^{-x}+2)^2}\,dx$, use the substitution $u = e^{-x} + 2$,

$du = -e^{-x}dx$. When $x = 0$, $u = 3$; when $x = b$, $u = e^{-b} + 2$.

$$\int_{0}^{b} \frac{e^{-x}}{(e^{-x}+2)^2}\,dx = -\int_{3}^{e^{-b}+2} \frac{du}{u^2} = u^{-1}\Big|_{3}^{e^{-b}+2} = \frac{1}{e^{-b}+2} - \frac{1}{3}.$$

Thus, $$\int_{0}^{\infty} \frac{e^{-x}}{(e^{-x}+2)^2}\,dx = \lim_{b\to\infty} \left[\frac{1}{e^{-b}+2} - \frac{1}{3}\right] = \frac{1}{2} - \frac{1}{3} = \frac{1}{6}.$$

46. $$\int_{-\infty}^{\infty} \frac{e^{-x}}{(e^{-x}+2)^2}\,dx = \int_{0}^{\infty} \frac{e^{-x}}{(e^{-x}+2)^2}\,dx + \int_{-\infty}^{0} \frac{e^{-x}}{(e^{-x}+2)^2}\,dx.$$

$$\int_{0}^{\infty} \frac{e^{-x}}{(e^{-x}+2)^2}\,dx = \frac{1}{6}$$ (see previous exercise)

$$\int_{-b}^{0} \frac{e^{-x}}{(e^{-x}+2)^2}\,dx = -\int_{0}^{b} \frac{e^{-x}}{(e^{-x}+2)^2}\,dx = \frac{1}{3} - \frac{1}{e^{-b}+2}$$ (see prev. exercise)

Thus, $\displaystyle\int_{-\infty}^{0} \frac{e^{-x}}{(e^{-x}+2)^2}\,dx = \lim_{b\to-\infty}\left[\frac{1}{3} - \frac{1}{e^{-b}+2}\right] = \frac{1}{3} - 0 = \frac{1}{3}.$

Combining these gives $\displaystyle\int_{-\infty}^{\infty} \frac{e^{-x}}{(e^{-x}+2)^2}\,dx = \frac{1}{6} + \frac{1}{3} = \frac{1}{2}.$

47. $\displaystyle\int_0^{\infty} ke^{-kx}dx = \lim_{b\to\infty}\int_0^b ke^{-kx}dx = \lim\left[-e^{-kx}\Big|_0^b\right] = \lim_{b\to\infty}\left[-e^{-kb} + 1\right].$

If $k > 0$, as $b \to \infty$, $-e^{-kb}$ approaches 0.

Thus, in this case $\displaystyle\int_0^{\infty} ke^{-kx}dx = 1.$

48. $\displaystyle\int_1^{\infty} \frac{k}{x^{k+1}}\,dx = \lim_{b\to\infty}\int_1^b \frac{k}{x^{k+1}}\,dx = \lim_{b\to\infty}\left[-x^{-k}\Big|_1^b\right] = \lim_{b\to\infty}\left[-b^{-k} + 1\right].$

If $k > 0$, as $b \to \infty$, $-b^{-k}$ approaches 0.

Thus, in this case $\displaystyle\int_1^{\infty} \frac{k}{x^{k+1}}\,dx = 1.$

49. $\displaystyle\int_0^{\infty} 5000e^{-.1t}dt = \lim_{b\to\infty}\int_0^b 5000e^{-.1t}dt = \lim_{b\to\infty}\left[-50{,}000e^{-.1t}\Big|_0^b\right]$

$\displaystyle= \lim_{b\to\infty}\left[-50{,}000e^{-.1b} + 50{,}000\right] = 50{,}000.$

50. $\displaystyle\int_0^{\infty} 6000e^{.04t}e^{-.16t}dt = \int_0^{\infty} 6000e^{-.12t}dt = \lim_{b\to\infty}\int_0^b 6000e^{-.12t}dt$

$\displaystyle= \lim_{b\to\infty}\left[-\frac{25}{3}(6000)e^{-.12t}\Big|_0^b\right]$

$\displaystyle= \lim_{b\to\infty}\left[-50{,}000e^{-.12b} + 50{,}000\right] = \$50{,}000.$

Chapter 9 Supplementary Exercises

1. $\int x \sin 3x^2\, dx$. Let $u = 3x^2$, $du = 6x\, dx$.

Then $\int x \sin 3x^2\, dx = \frac{1}{6}\int \sin u\, du = -\frac{1}{6}\cos u + C$

$= -\frac{1}{6}\cos 3x^2 + C.$

2. $\int \sqrt{2x + 1}\, dx = \frac{1}{3}(2x + 1)^{3/2} + C.$

3. $\int x(1 - 3x^2)^5 dx$. Let $u = 1 - 3x^2$, $du = -6x\, dx$. Then

$\int x(1 - 3x^2)^5 dx = -\frac{1}{6}\int u^2 du = -\frac{1}{36}u^6 + C = -\frac{1}{36}(1 - 3x^2)^6 + C.$

4. $\int \frac{(\ln x)^5}{x}\, dx$. Let $u = \ln x$, $du = \frac{1}{x}dx$. Then

$\int \frac{(\ln x)^5}{x}\, dx = \int u^5 du = \frac{1}{6}u^6 + C = \frac{1}{6}(\ln x)^6 + C.$

5. $\int \frac{(\ln x)^2}{x}\, dx$. Let $u = \ln x$, $du = \frac{1}{x}dx$. Then

$\int \frac{(\ln x)^2}{x}\, dx = \int u^2 du = \frac{u^3}{3} + C = \frac{(\ln x)^3}{3} + C.$

6. $\int \frac{1}{\sqrt{4x + 3}}\, dx = \frac{1}{2}(4x + 3)^{1/2} + C.$

7. $\int x\sqrt{4 - x^2}\, dx$. Let $u = 4 - x^2$, $du = -2x\, dx$. Then

$\int x\sqrt{4 - x^2}\, dx = -\frac{1}{2}\int \sqrt{u}\, du = -\frac{1}{3}u^{3/2} + C = -\frac{1}{3}(4x^2)^{3/2} + C$

8. $\int x \sin 3x\, dx$. Use integration by parts with $f(x) = x$,

$g(x) = \sin 3x$. Then $f'(x) = 1$, $G(x) = -\frac{1}{3}\cos 3x$ and

$$\int x \sin 3x\,dx = -\frac{1}{3}x \cos 3x + \frac{1}{3}\int \cos 3x\,dx$$

$$= -\frac{1}{3}x \cos 3x + \frac{1}{9}\sin 3x + C.$$

9. $\int x^2 e^{-x^3}dx$. Let $u = -x^3$, $du = -3x^2dx$.

Then $\int x^2 e^{-x^3}dx = -\frac{1}{3}\int e^u du = -\frac{1}{3}e^u + C = -\frac{1}{3}e^{-x^3} + C.$

10. $\int \frac{x \ln(x^2 + 1)}{x^2 + 1}dx$. Let $u = \ln(x^2 + 1)$, $du = \frac{2x}{x^2 + 1}dx$.

Then $\int \frac{x \ln(x^2 + 1)}{x^2 + 1}dx = \frac{1}{2}\int u\,du = \frac{1}{4}u^2 + C = \frac{1}{4}u^2 + C$

$$= \frac{1}{4}(\ln(x^2 + 1))^2 + C.$$

11. $\int x^2 \cos 3x\,dx$. Use integration by parts with $f(x) = x^2$,

$g(x) = \cos 3x$. Then $f'(x) = 2x$, $G(x) = \frac{1}{3}\sin 3x$ and

$$\int x^2 \cos 3x\,dx = \frac{1}{3}x^2\sin 3x - \frac{2}{3}\int x \sin 3x\,dx.$$

To evaluate $\int x \sin 3x\,dx$ integrate by parts again:

$$\int x \sin 3x\,dx = -\frac{1}{3}x \cos 3x + \frac{1}{3}\int \cos 3x\,dx$$

$$= -\frac{1}{3}x \cos 3x + \frac{1}{9}\sin 3x + C.$$

Thus, $\int x^2 \cos 3x\,dx = \frac{1}{3}x^2\sin 3x + \frac{2}{9}x \cos 3x - \frac{2}{27}\sin 3x + C.$

12. $\int \frac{\ln(\ln x)}{x \ln x}dx$. Let $u = \ln(\ln x)$, $du = \frac{1}{x \ln x}dx$.

Then $\int \frac{\ln(\ln x)}{x \ln x}dx = \int u\,du = \frac{u^2}{2} + C = \frac{(\ln(\ln x))^2}{2} + C.$

13. $\int \ln x^2 dx = \int 2 \ln x\,dx = 2\int \ln x\,dx.$

To evaluate $\int \ln x\,dx$, use integration by parts with

$f(x) = \ln x$, $g(x) = 1$. Then $f'(x) = \frac{1}{x}$, $G(x) = x$ and

$$\int \ln x\,dx = x \ln x - \int dx = x \ln - x + C.$$

Thus $\displaystyle\int \ln x^2 dx = 2x \ln x - 2x + C.$

14. $\displaystyle\int x\sqrt{x+1}\,dx$. Use integration by parts with $f(x) = x$,

$g(x) = \sqrt{x+1}$. Then $f'(x) = 1$, $G(x) = \frac{2}{3}(x+1)^{3/2}$ and

$$\int x\sqrt{x+1}\,dx = \frac{2}{3}x(x+1)^{3/2} - \frac{2}{3}\int (x+1)^{3/2}dx$$

$$= \frac{2}{3}x(x+1)^{3/2} - \frac{4}{15}(x+1)^{5/2} + C.$$

15. $\displaystyle\int \frac{x}{\sqrt{3x-1}}\,dx$. Use integration by parts with $f(x) = x$,

$g(x) = (3x-1)^{-1/2}$. Then $f'(x) = 1$, $G(x) = \frac{2}{3}(3x-1)^{1/2}$ and

$$\int \frac{x}{\sqrt{3x-1}}\,dx = \frac{2}{3}x(3x-1)^{1/2} - \frac{2}{3}\int (3x-1)^{1/2}dx$$

$$= \frac{2}{3}x(3x-1)^{1/2} - \frac{4}{27}(3x-1)^{3/2} + C.$$

16. $\displaystyle\int x^2 \ln x^2\,dx$. Use integration by parts with $f(x) = \ln x^2$,

$g(x) = x^2$. Then $f'(x) = \frac{2x}{x^2} = \frac{2}{x}$, $G(x) = \frac{x^3}{3}$ and

$$\int x^2 \ln x^2\,dx = \frac{x^3}{3}\ln x^2 - \int \frac{2x^2}{3}\,dx = \frac{x^3}{3}\ln x^2 - \frac{2}{9}x^3 + C.$$

17. $\displaystyle\int \frac{x}{(1-x)^5}\,dx$. Use integration by parts with $f(x) = x$,

$g(x) = (1-x)^{-5}$. Then $f'(x) = 1$, $G(x) = \frac{1}{4}(1-x)^{-4}$ and

$$\int \frac{x}{(1-x)^5}\,dx = \frac{1}{4}x(1-x)^{-4} - \frac{1}{4}\int (1-x)^{-4}dx$$

$$= \frac{1}{4}x(1-x)^{-4} - \frac{1}{12}(1-x)^{-3} + C.$$

18. $\displaystyle\int x(\ln x)^2 dx$. Use integration by parts with $f(x) = (\ln x)^2$,

$g(x) = x$. Then $f'(x) = \frac{2 \ln x}{x}$, $G(x) = \frac{x^2}{2}$ and

$$\int x(\ln x)^2 dx = \frac{x^2(\ln x)^2}{2} - \int x \ln x\, dx$$

$$= \frac{x^2(\ln x)^2}{2} - \left[\frac{x^2}{2}\ln x - \int \frac{x}{2} dx\right] \quad \text{(using parts again)}$$

$$= \frac{x^2(\ln x)^2}{2} - \frac{x^2}{2}\ln x - \frac{x^2}{4} + C$$

$$= \frac{x^2}{2}\left[(\ln x)^2 - \ln x - \frac{1}{2}\right] + C.$$

19. Integration by parts: $f(x) = x$, $g(x) = e^{2x}$.

20. Integration by parts: $f(x) = x - 3$, $g(x) = e^{-x}$.

21. Substitution: $u = \sqrt{x + 1}$.

22. Substitution: $u = x^3 - 1$.

23. Substitution: $u = x^4 - x^2 + 4$.

24. Integration by parts: $f(x) = \ln\sqrt{5 - x}$, $g(x) = 1$.

25. Repeated integration by parts, starting with $f(x) = (3x - 1)^2$, $g(x) = e^{-x}$.

26. Substitution: $u = 3 - x^2$.

27. Integration by parts: $f(x) = 500 - 4x$, $g(x) = e^{-x/2}$.

28. Integration by parts: $f(x) = \ln x$, $g(x) = x^{5/2}$.

29. Integration by parts: $f(x) = \ln(x + 2)$, $g(x) = \sqrt{x + 2}$.

30. Repeated integration by parts, starting with $f(x) = (x + 1)^2$, $g(x) = e^{3x}$.

31. Substitution: $u = x^2 + 6x$.

32. Substitution: $u = \sin x$.

33. Substitution: $u = x^2 - 9$.

34. Integration by parts: $f(x) = 3 - x$, $g(x) = \sin x$.

35. Substitution: $u = x^3 - 6x$.

36. Substitution: $u = \ln x$.

37. $\displaystyle\int_0^1 \frac{2x}{(x^2+1)^3}\,dx$. Let $u = x^2 + 1$, $du = 2x\,dx$.

When $x = 0$, $u = 1$; when $x = 1$, $u = 2$.

$$\int_0^1 \frac{2x}{(x^2+1)^3}\,dx = \int_1^2 \frac{du}{u^3} = -\frac{1}{2}u^{-2}\Big|_1^2 = -\frac{1}{8} + \frac{1}{2} = \frac{3}{8}.$$

38. $\displaystyle\int_0^{\pi/2} x \sin 8x\,dx$. Using integration by parts with $f(x) = x$,

$g(x) = \sin 8x$, we have $f'(x) = 1$, $G(x) = -\frac{1}{8}\cos 8x$ and

$$\int_0^{\pi/2} x\sin 8x\,dx = -\frac{1}{8}x\cos 8x\Big|_0^{\pi/2} + \frac{1}{8}\int_0^{\pi/2}\cos 8x\,dx$$

$$= \frac{1}{8}\left[-x\cos 8x + \frac{1}{8}\sin 8x\right]\Bigg|_0^{\pi/2} = \frac{1}{8}\left[-\frac{\pi}{2} - 0\right] = -\frac{\pi}{16}.$$

39. $\displaystyle\int_0^2 xe^{-(1/2)x^2}\,dx$. Let $u = -\frac{1}{2}x^2$, $du = -x\,dx$. When $x = 0$, $u = 0$;

when $x = 2$, $u = -2$.

$$\int_0^2 xe^{-(1/2)x^2}\,dx = -\int_0^{-2} e^u\,du = -e^u\Big|_0^{-2} = 1 - e^{-2}.$$

40. $\displaystyle\int_{1/2}^1 \frac{\ln(2x+3)}{2x+3}\,dx$. Let $u = \ln(2x+3)$, $du = \frac{2}{2x+3}\,dx$.

When $x = \frac{1}{2}$, $u = \ln(4)$, when $x = 1$, $u = \ln 5$.

$$\int_{1/2}^1 \frac{\ln(2x+3)}{2x+3}\,dx = \frac{1}{2}\int_{\ln 4}^{\ln 5} u\,du = \frac{u^2}{4}\Big|_{\ln 4}^{\ln 5} = \frac{1}{4}\left[(\ln 5)^2 - (\ln 4)^2\right].$$

41. $\int_1^2 xe^{-2x}dx$. Use integration by parts with $f(x) = x$, $g(x) = e^{-2x}$. Then $f'(x) = 1$, $G(x) = -\frac{1}{2}e^{-2x}$ and

$$\int_1^2 xe^{-2x}dx = -\frac{1}{2}xe^{-2x}\Big|_1^2 + \frac{1}{2}\int_1^2 e^{-2x}dx = \left(-\frac{1}{2}xe^{-2x} - \frac{1}{4}e^{-2x}\right)\Big|_1^2$$

$$= -\frac{1}{2}\left[e^{-2x}\left(x + \frac{1}{2}\right)\Big|_1^2\right] = -\frac{1}{2}\left[\frac{5}{2}e^{-4} - \frac{3}{2}e^{-2}\right]$$

$$= \frac{3}{4}e^{-2} - \frac{5}{4}e^{-4}.$$

42. $\int_1^2 x^{-3/2}\ln x\, dx$. Use integration by parts with $f(x) = \ln x$, $g(x) = x^{-3/2}$. Then $f'(x) = \frac{1}{x}$, $G(x) = -2x^{-1/2}$ and

$$\int_1^2 x^{-3/2}\ln x\, dx = -2x^{-1/2}\ln x\Big|_1^2 + \int_1^2 2x^{-3/2}dx$$

$$= \left(-2x^{-1/2}\ln x - 4x^{-1/2}\right)\Big|_1^2$$

$$= -2\left[x^{-1/2}(\ln x + 2)\Big|_1^2\right] = -2\left[\frac{\ln 2 + 2}{\sqrt{2}} - 2\right]$$

$$= -\sqrt{2}\ln 2 - 2\sqrt{2} + 4.$$

43. $\int_1^9 \frac{1}{\sqrt{x}}dx$; $n = 4$, $\Delta x = (9 - 1)/4 = 2$.

Midpoint rule: $x_1 = 1 + \frac{2}{2} = 2$, $x_2 = 4$, $x_3 = 6$, $x_4 = 8$.

$$\int_1^9 \frac{1}{\sqrt{x}}dx \sim \left[\frac{1}{\sqrt{2}} + \frac{1}{\sqrt{4}} + \frac{1}{\sqrt{6}} + \frac{1}{\sqrt{8}}\right](2) \sim 3.93782.$$

Trapezoidal rule: $a_0 = 1$, $a_1 = 3$, $a_2 = 5$, $a_3 = 7$, $a_4 = 9$

$$\int_1^9 \frac{1}{\sqrt{x}}dx \sim \left[\frac{1}{\sqrt{1}} + \frac{2}{\sqrt{3}} + \frac{2}{\sqrt{5}} + \frac{2}{\sqrt{7}} + \frac{1}{\sqrt{9}}\right]\frac{2}{2} \sim 4.13839$$

Simpson's rule:

$$\int_1^9 \frac{1}{\sqrt{x}}\,dx \sim \frac{2(3.93782) + 4.13839}{3} \sim 4.00468.$$

44. $\int_0^{10} e^{\sqrt{x}}dx$; $n = 5$, $\Delta x = (10 - 0)/5 = 2$.

Midpoint rule: $x_1 = 1$, $x_2 = 3$, $x_3 = 5$, $x_4 = 7$, $x_5 = 9$.

$$\int_0^{10} e^{\sqrt{x}}dx \sim \left[e^{\sqrt{1}} + e^{\sqrt{3}} + e^{\sqrt{7}} + e^{\sqrt{9}}\right]2 \sim 103.81310.$$

Trapezoidal rule: $a_0 = 0$, $a_1 = 2$, $a_2 = 4$, $a_3 = 6$, $a_4 = 8$,

$a_5 = 10$.

$$\int_0^{10} e^{\sqrt{x}}dx \sim \left[e^{\sqrt{0}} + 2e^{\sqrt{2}} + 2e^{\sqrt{4}} + 2e^{\sqrt{6}} + 2e^{\sqrt{8}} + e^{\sqrt{10}}\right]\frac{2}{2}$$

$\sim 104.63148.$

Simpson's rule:

$$\int_0^{10} e^{\sqrt{x}}dx \sim \frac{2(103.8130) + 104.63148}{3} \sim 104.08589$$

45. $\int_1^4 \frac{e^x}{x + 1}\,dx$; $n = 5$, $\Delta x = 3/5 = .6$

Midpoint rule: $x_1 = 1.3$, $x_2 = 1.9$, $x_3 = 2.5$, $x_4 = 3.1$,

$x_5 = 3.7$

$$\int_1^4 \frac{e^x}{x + 1}\,dx \sim \left[\frac{e^{1.3}}{2.3} + \frac{e^{1.9}}{2.9} + \frac{e^{2.5}}{3.5} + \frac{e^{3.1}}{4.1} + \frac{e^{3.7}}{4.7}\right](.6)$$

~ 12.84089

Trapezoidal rule: $a_0 = 1$, $a_1 = 1.6$, $a_2 = 2.2$, $a_3 = 2.8$,

$a_4 = 3.4$, $a_5 = 4$

$$\int_1^4 \frac{e^x}{x+1}\,dx \sim \left[\frac{e}{2} + \frac{2e^{1.6}}{2.6} + \frac{2e^{2.2}}{3.2} + \frac{2e^{2.8}}{3.8} + \frac{2e^{3.4}}{4.4} + \frac{e^4}{5}\right](.3)$$

$$\sim 13.20137.$$

Simpson's rule:

$$\int_1^4 \frac{e^x}{x+1}\,dx \sim \frac{2(12.84089) + 13.20137}{3} \sim 12.96105.$$

46. $\displaystyle\int_{-1}^{-1} \frac{1}{1+x^2}\,dx$; $n = 5$, $\Delta x = 2/5 = .4$

Midpoint rule: $x_1 = -.8$, $x_2 = -.4$, $x_3 = 0$, $x_4 = .4$, $x_5 = .8$

$$\int_{-1}^{-1} \frac{1}{1+x^2}\,dx \sim \left[\frac{1}{1+(-.8)^2} + \frac{1}{1+(-.4)^2} + \frac{1}{1+0^2} + \frac{1}{1+(.4)^2} + \frac{1}{1+(.8)^2}\right].8 \sim 1.57746.$$

Trapezoidal rule: $a_0 = -1$, $a_1 = -.6$, $a_2 = -.2$, $a_3 = .2$,

$a_4 = .6$, $a_5 = 1$

$$\int_{-1}^{-1} \frac{1}{1+x^2}\,dx \sim \left[\frac{1}{1+(-1)^2} + \frac{2}{1+(-.6)^2} + \frac{2}{1+(-.2)^2} + \frac{2}{1+(.2)^2} + \frac{2}{1+(.6)^2} + \frac{1}{1+1^2}\right](.2)$$

$$\sim 1.55747.$$

Simpson's rule:

$$\int_{-1}^{-1} \frac{1}{1+x^2}\,dx \sim \frac{2(1.57746) + 1.55747}{3} \sim 1.57080.$$

47. $\displaystyle\int_0^\infty e^{6-3x}dx = \lim_{b\to\infty} \int_0^b e^{6-3x}dx.$

Let $u = 6 - 3x$, $du = -3dx$. When $x = 0$, $u = 6$; when $x = b$,

$u = 6 - 3b.$

$$\int_0^b e^{6-3x}dx = -\frac{1}{3}\int_6^{6-3b} e^u du = -\frac{1}{3}e^u\Big|_6^{6-3b} = -\frac{1}{3}e^{6-3b} + \frac{1}{3}e^6.$$

Thus $\displaystyle\int_0^{\infty} e^{6-3x}dx = \lim_{b\to\infty}\left[-\frac{1}{3}e^{6-3b} + \frac{e^6}{3}\right] = \frac{e^6}{3}.$

48. $$\int_1^{\infty} x^{-2/3}dx = \lim_{b\to\infty}\int_1^b x^{-2/3}dx = \lim_{b\to\infty}\left[3x^{1/3}\Big|_1^b\right] = \lim_{b\to\infty}\left[3b^{1/3} - 3\right].$$

As $b \to \infty$, $3b^{1/3}$ increases without bound. Thus, $\displaystyle\int_1^{\infty} x^{-2/3}dx$ diverges.

49. $$\int_1^{\infty} \frac{x + 2}{x^2 + 4x - 2}dx = \lim_{b\to\infty}\int_1^b \frac{x + 2}{x^2 + 4x - 2}dx.$$

Let $u = x^2 + 4x - 2$, $du = (2x + 4)dx = 2(x + 2)dx.$

$$\int_1^b \frac{x + 2}{x^2 + 4x - 2}dx = \frac{1}{2}\int_3^{b^2+4b-2} \frac{du}{u} = \frac{1}{2}\ln|u|\Big|_3^{b^2+4b-2}$$

$$= \frac{1}{2}\left[\ln|b^2 + 4b - 2| - \ln 3\right].$$

Thus $\displaystyle\int_1^{\infty} \frac{x + 2}{x^2 + 4x - 2}dx$ diverges.

50. $$\int_0^{\infty} x^2e^{-x^3}dx = \lim_{b\to\infty}\int_0^b x^2e^{-x^3}dx = -\frac{1}{3}\left[e^{-b^3} - 1\right].$$

Thus $\displaystyle\int_0^{\infty} x^2e^{-x^3}dx = \lim_{b\to\infty}\left[\frac{1}{3} - \frac{e^{-b^3}}{3}\right] = \frac{1}{3}.$

51. $$\int_{-1}^{\infty} (x + 3)^{-5/4}dx = \lim_{b\to\infty}\int_{-1}^b (x + 3)^{-5/4}dx$$

$$= \lim_{b\to\infty}\left[-4(x+3)^{-1/4}\Big|_{-1}^{b}\right]$$

$$= \lim_{b\to\infty}\left[2^{7/4} - 4(b+3)^{-1/4}\right] = 2^{7/4}.$$

52. $$\int_{-\infty}^{0} \frac{8}{(5-2x)^3}\,dx = \lim_{b\to-\infty}\int_{b}^{0}\frac{8}{(5-2x)^3}\,dx$$

$$= \lim_{b\to-\infty}\left[2(5-2x)^{-2}\Big|_{b}^{0}\right]$$

$$= \lim_{b\to-\infty}\left[\frac{2}{25} - 2(5-2b)^{-2}\right] = \frac{2}{25}.$$

53. $\int_{1}^{\infty} xe^{-x}dx = \lim_{b\to\infty}\int_{1}^{b} xe^{-x}dx$. Using integration by parts with

$f(x) = x$, $g(x) = e^{-x}$; $f'(x) = 1$, $G(x) = -e^{-x}$ gives

$$\int_{1}^{b} xe^{-x}dx = -xe^{-x}\Big|_{1}^{b} + \int_{1}^{b} e^{-x}dx = (-xe^{-x} - e^{-x})\Big|_{1}^{b}$$

$$= 2e^{-1} - be^{-b} + e^{-b}.$$

Thus $\int_{1}^{\infty} xe^{-x}dx = \lim_{b\to\infty}\left[\frac{2}{e} - be^{-b} + e^{-b}\right] = \frac{2}{e}$ (since $\lim_{b\to\infty} be^{-b} = 0$)

54. $\int_{0}^{\infty} xe^{-kx}dx = \lim_{b\to\infty}\int_{0}^{b} xe^{-kx}dx$. Using integration by parts with

$f(x) = x$, $g(x) = e^{-kx}$; $f'(x) = 1$, $G(x) = -\frac{1}{k}e^{-kx}$, we have

$$\int_{0}^{b} xe^{-kx}dx = \frac{-x}{k}e^{-kx}\Big|_{0}^{b} + \frac{1}{k}\int_{0}^{b} e^{-kx}dx = \left(-\frac{x}{k}e^{-kx} - \frac{1}{k^2}e^{-kx}\right)\Big|_{0}^{b}$$

$$= \frac{1}{k^2} - \frac{b}{k}e^{-kb} - \frac{1}{k^2}e^{-kb}.$$

Thus $\int_{0}^{\infty} xe^{-kx}dx = \lim_{b\to\infty}\left[\frac{1}{k}\left(\frac{1}{k} - be^{-kb} - \frac{1}{k}e^{-kb}\right)\right]$

$$= \frac{1}{k^2} \quad (\text{since } \lim_{b\to\infty} be^{-kb} = 0)$$

55. The present value is $\displaystyle\int_0^4 50e^{-.08t}e^{-.12t}dt$

$$= \int_0^4 50e^{-.2t}dt = -250e^{-.2t}\Big|_0^4 = 250 - 250e^{-.8} \sim \$137{,}668.$$

56. Using the method of example 3, sect 9.5, the total tax revenue is $\displaystyle\int_0^{10} (2\pi t)50e^{-t/20}dt = 100\pi\int_0^{10} te^{-t/20}dt$

$$= 100\pi\left[-20te^{-t/20}\Big|_0^{10} + \int_0^{10} 20e^{-t/20}dt\right]$$

$$= 100\pi\left[-20te^{-t/20} - 400e^{-t/20}\Big|_0^{10}\right]$$

$$= 100\pi\left[400 - e^{-.5}(600)\right]$$

$\sim$ 11,335 (thousand dollars)

57. (a) $M(t_1)\Delta t + \ldots + M(t_n)\Delta t \sim \displaystyle\int_0^t M(t)dt$

(b) $M(t_1)e^{-.1t}\Delta t + \ldots + Mt(t_n)e^{-.1t}\Delta t \sim \displaystyle\int_0^2 M(t)e^{-.1t}dt$

58. $80 + \displaystyle\int_0^\infty 50e^{-rt}dt.$

Exercises 10.1

1. $y = f(t) = \frac{3}{2}e^{t^2} - \frac{1}{2}$; $y' = 3te^{t^2}$

$y' - 2ty = 3te^{t^2} - 2t\left[\frac{3}{2}e^{t^2} - \frac{1}{2}\right] = t.$

2. $y = f(t) = t^2 - \frac{1}{2}$; $y' = 2t$

$(y')^2 - 4y = (2t)^2 - 4\left[t^2 - \frac{1}{2}\right] = 2.$

3. $y = f(t) = (e^{-t} + 1)^{-1}$; $y' = \frac{1}{(e^{-t} + 1)^2} = y$

$y(0) = \frac{1}{1 + 1} = \frac{1}{2}.$

4. $y = f(t) = 5e^{2t}$; $y' = 10e^{2t}$; $y'' = 20e^{2t}$.

$y'' - 3y' + 2y = 20e^{2t} - 3(10e^{2t}) + 2(5e^{2t}) = 0$

$y(0) = 5e^0 = 5$, $y'(0) = 10e^0 = 10.$

5. If $y' = \frac{1}{2}y$, then $y = he^{(1/2)t}$ for some constant h.

Since $y(0) = 1$, $1 = he^{(1/2)0}$, so $h = 1$.

Therefore the solution is $y = f(t) = e^{(1/2)t}$.

6. $y' = e^{t/2}$; $y(0) = 1$. If $y' = e^{t/2}$, then $y = 2e^{t/2} + C$ for some constant C. Since $y(0) = 1$, $1 = 2e^{0/2} + C$, so $C = -1$.

Therefore the solution is $y = f(t) = 2e^{t/2} - 1$.

7. Yes. Given $y = f(t) = 3$, $y' = 0 = 6 - 2(3)$.

8. Yes. Given $y = f(t) = -4$, $y' = 0 = t^2(-4 + 4)$ for all t.

9. $y' = t^2y - 5t^2 = t^2(y - 5)$; $y = 5$ is a constant solution.

$\frac{dy}{dt} = 0$, and $t^2(y - 5) = 0$ at $y = 5$.

10. $y' = 4y(y - 7)$; $y = 0$ or $y = 7$ are constant solutions.

$\frac{dy}{dt} = 0$, and $4y(y - 7) = 0$ at $y = 0$ and $y = 7$.

11. $f(0) = y(0) = 4$. Since $y' = 2y - 3$, we must have in particular $y'(0) = 2y(0) - 3 = 2(4) - 3 = 5$.

12. $f(0) = y(0) = 0$. Since $y' = e^t + y$, we must have in particular $y'(0) = e^0 + y(0) = 1 + 0 = 1$.

13. $y' = k(C - y)$, $k > 0$.

14. $y' = k(20 - y)$, $k > 0$.

Exercises 10.2

1. $\frac{dy}{dt} = \frac{5 - t}{y^2}$; $y^2\frac{dy}{dt} = 5 - t$; $\int y^2 dy = \int (5 - t)dt$;

$\frac{y^3}{3} + C_1 = 5t - \frac{t^2}{2} + C_2$; $y = \sqrt[3]{15t - \frac{3}{2}t^2 + C}$

2. $\frac{dy}{dt} = \frac{e^t}{e^y}$; $\frac{dy}{dt}e^y = e^t$; $\int e^y dy = \int e^t dt$

$e^y + C_1 = e^t + C_2$; $y = \ln|e^t + C|$.

3. $y' = te^{2y}$; $\frac{dy}{dt}e^{-2y} = t$; $\int e^{-2y}dy = \int t\ dt$

$-\frac{1}{2}e^{-2y} + C_1 = \frac{t^2}{2} + C_2$; $e^{-2y} = -t^2 + C$; $y = -\frac{1}{2}\ln(-t^2 + C)$.

4. $\frac{dy}{dt} = \frac{-1}{y^2t^2}$; $y^2\frac{dy}{dt} = \frac{-1}{t^2}$; $\int y^2 dy = \int \frac{-1}{t^2}\ dt$;

$\frac{y^3}{3} + C_1 = \frac{1}{t} + C_2$; $y = \sqrt[3]{\frac{3}{t} + C}$

5. First check for constant solutions; if $y = 0$ then $\frac{dy}{dt} = 0$, and $t^{1/2}y^2 = 0$, otherwise assume $y \neq 0$.

$y' = t^{1/2}y^2$; $y'y^{-2} = t^{1/2}$ (Assuming $y \neq 0$).

$\int y^{-2}dy = \int t^{1/2}\ dt$; $-\frac{1}{y} + C_1 = \frac{2}{3}t^{3/2} + C_2$

$y = \dfrac{1}{C - (2/3)t^{3/2}}$, $y = 0$.

6. $y' = \dfrac{t^2y^2}{t^3 + 8}$; $\dfrac{dy}{dt}y^{-2} = \dfrac{t^2}{t^3 + 8}$ (Assuming $y \neq 0$)

$\int y^{-2}dy = \int \dfrac{t^2}{t^3 + 8}\,dt$; $-y^{-1} + C_1 = \frac{1}{3}\ln|t^3 + 8| + C_2$

$y = -\dfrac{3}{\ln|t^3 + 8| + C}$; $y = 0$ is also a solution.

7. $y' = \dfrac{e^{t^3}t^2}{y^2}$; $y^2\dfrac{dy}{dt} = t^2e^{t^3}$; $\int y^2\,dy = \int t^2e^{t^3}dt$

$\dfrac{y^3}{3} + C_1 = \dfrac{e^{t^3}}{3} + C_2$; $y = \sqrt[3]{e^{t^3} + C}$.

8. $y' = e^{4y}t^3 - e^{4y} = e^{4y}(t^3 - 1)$; $\dfrac{dy}{dt}e^{-4y} = t^3 - 1$

$\int e^{-4y}dy = \int(t^3 - 1)dt$; $-\frac{1}{4}e^{-4y} + C_1 = \dfrac{t^4}{4} - t + C_2$

$e^{-4y} = -t^4 + 4t + C$; $y = -\frac{1}{4}\ln(-t^4 + 4t + C)$.

9. $\dfrac{dy}{dt} = \sqrt{\dfrac{y}{t}} = \dfrac{\sqrt{y}}{\sqrt{t}}$; $y^{-1/2}\dfrac{dy}{dt} = t^{-1/2}$ (Assuming $y \neq 0$)

$\int y^{-1/2}dy = \int t^{-1/2}dt$; $2y^{1/2} + C_1 = 2t^{1/2} + C_2$; $y = (\sqrt{t} + C)^2$

$y = 0$ is also a solution.

10. $\dfrac{dy}{dt} = \left(\dfrac{e^t}{y}\right)^2$; $\dfrac{dy}{dt}y^2 = e^{2t}$; $\int y^2dy = \int e^{2t}dt$

$\dfrac{y^3}{3} + C_1 = \frac{1}{2}e^{2t} + C_2$; $y = \sqrt[3]{\frac{3}{2}e^{2t} + C}$.

11. $y' = 4(y - 3)$; $\dfrac{dy}{dt}(y - 3)^{-1} = 4$; $\int \dfrac{1}{y - 3}\,dy = \int 4\,dt$

$\ln|y - 3| + C_1 = 4t + C_2$; $|y - 3| = e^{4t+C} = e^Ce^{4t}$

$y = 3 + Ae^{4t}$.

12. $y' = -\frac{1}{2}(y - 4)$; $(y - 4)^{-1}\dfrac{dy}{dt} = -\frac{1}{2}$; $\int(y - 4)^{-1}dy = \int -\frac{1}{2}dt$

$\ln|y - 4| + C_1 = -\frac{1}{2}t + C_2;\quad y = 4 + Ae^{-(1/2)t}.$

13. $y' = 2 - y;\quad \frac{dy}{dt}(2 - y)^{-1} = 1;\quad \int(2 - y)^{-1}dy = \int dt$

$-\ln|2 - y| = t + C_2;\quad |2 - y| = e^{-t+C_2};\quad y = 2 + Ae^{-t}$

14. $y' = \frac{1}{ty + y} = \frac{1}{y(t + 1)};\quad \frac{dy}{dt}y = \frac{1}{t + 1};\quad \int y\,dy = \int\frac{1}{t + 1}\,dt$

$\frac{y^2}{2} C_1 = \ln|t + 1| + C_2;\quad y = \pm\sqrt{2\ln|t + 1| + C}.$

15. $y' = \frac{\ln t}{ty};\quad y\frac{dy}{dt} = \frac{\ln t}{t};\quad \int y\,dy = \int\frac{\ln t}{t}\,dt$

$\frac{y^2}{2} + C_1 = \frac{(\ln t)^2}{2} + C_2;\quad y = \pm\sqrt{(\ln t)^2 + C}.$

16. $y' = 4t - ty$, $y = 4$ is a constant solution. Assume $y \neq 4$.

$y' = t\left[4 - y\right];\quad \left[4 - y\right]^{-1}\frac{dy}{dt} = t$

$\int\left[4 - y\right]^{-1}dy = \int t\,dt;\quad -\ln|4 - y| + C_1 = \frac{1}{2}t^2 + C_2$

$y = 4 + Ae^{-t^2/2}.$

17. $y' = (y - 3)^2\ln t;\quad (y - 3)^{-2}\frac{dy}{dt} = \ln t$ (Assuming $y \neq 3$)

$\int(y - 3)^{-2}dy = \int\ln t\,dt;\quad -(y - 3)^{-1} + C_1 = t\ln t - t + C_2$

$y = 3 + \frac{1}{t - t\ln t + C};\quad y = 3$ is also a solution.

18. $yy' = t\cos(t^2 + 1);\quad \int y\,dy = \int t\cos(t^2 + 1)dt$

$\frac{y^2}{2} + C_1 = \frac{1}{2}\sin(t^2 + 1) + C_2;\quad y = \pm\sqrt{\sin(t^2 + 1) + C}.$

19. $y' = 2te^{-2y} - e^{-2y},\ y(0) = 3;\quad e^{2y}\frac{dy}{dt} = 2t - 1$

$\int e^{2y}dy = \int(2t - 1)dt;\quad \frac{1}{2}e^{2y} + C_1 = t^2 - t + C_2$

$e^{2y} = 2t^2 - 2t + C;\quad y = \frac{1}{2}\ln(2t^2 - 2t + C).\quad y(0) = 3 = \frac{1}{2}\ln C,$

so $C = e^6$ and the solution is $y = \frac{1}{2}\ln(2t^2 - 2t + e^6).$

20. $y' = y^2 - e^{3t}y^2 = y^2(1 - e^{3t})$; $y'y^{-2} = (1 - e^{3t})$ (Assuming $y \neq 0$)

$\int y^{-2}dy = \int (1 - e^{3t})dt$; $-\frac{1}{y} + C_1 = t - \frac{1}{3}e^{3t} + C_2$

$y = \dfrac{3}{e^{3t} - 3t + C}$; Given that $y(0) = 1 = \dfrac{3}{1 + C}$; $C = 2$

$y = \dfrac{3}{e^{3t} - 3t + 2}$.

21. $y^2y' = t \sin t$, $y(0) = 2$; $\int y^2 dy = \int t \sin t\, dt$

$\frac{y^3}{3} + C_1 = -t \cos t + \sin t + C_2$; $y = \sqrt[3]{3\sin t - 3t \cos t + C}$

$y(0) = 2 = \sqrt[3]{C}$ so $C = 8$ and the solution is

$y = \sqrt[3]{3\sin t - 3t \cos t + 8}$.

22. $y' = t^2e^{-3y}$, $y(0) = 2$; $e^{3y}\frac{dy}{dt} = t^2$; $\int e^{3y}dy = \int t^2 dt$

$\frac{1}{3}e^{3y} + C_1 = \frac{t^3}{3} + C_2$; $e^{3y} = t^3 + C$; $y = \dfrac{\ln(t^3 + C)}{3}$

$y(0) = 2 = \dfrac{\ln(0^3 + C)}{3}$; So $C = e^6$ and the solution is

$y = \dfrac{\ln(t^3 + e^6)}{3}$.

23. $y' = 5 - 3y$, $y(0) = 1$. $(3y - 5)^{-1}\frac{dy}{dt} = -1$; $\int (3y - 5)^{-1}dy = \int -1\, dt$

$\ln|3y - 5| + C_1 = -t + C_2$; $y = \frac{5}{3} + Ae^{-t}$. $y(0) = 1 = \frac{5}{3} + A$,

so $A = -\frac{2}{3}$ and the solution is $y = \frac{5}{3} - \frac{2}{3}e^{-3t}$.

24. $y' = \frac{1}{2}y - 3$, $y(0) = 4$; $(y - 6)^{-1}\frac{dy}{dt} = \frac{1}{2}$; $\int (y - 6)^{-1}dy = \int \frac{1}{2}dt$

$\ln|y - 6| + C_1 = \frac{1}{2}t + C_2$; $y = 6 + Ae^{(1/2)t}$. $y(0) = 4 = 6 + A$,

so $A = -2$ and the solution is $y = 6 - 2e^{(1/2)t}$.

25. $\dfrac{dy}{dt} = \dfrac{t + 1}{ty}$, $t > 0$, $y(1) = -3$. $y\dfrac{dy}{dt} = \dfrac{t + 1}{t}$

$\int y\, dy = \int \left(1 + \frac{1}{t}\right)dt$; $\frac{y^2}{2} = t + \ln|t| + C$

$y = \pm\sqrt{2t + 2\ln|t| + C}$. Since $y(1) = -3$, "-" solution

applies and $-3 = -\sqrt{2 + C}$, so $C = 7$ and the solution is

$y = -\sqrt{2t + 2\ln t + 7}$.

26. $y' = \left(\frac{1 + t}{1 + y}\right)^2$, $y(0) = 2$. $(1 + y)^2\frac{dy}{dt} = (1 + t)^2$

$\int(1 + y)^2dy = \int(1 + t)^2dt$; $\frac{(1 + y)^3}{3} = \frac{(1 + t)^3}{3} + C$

$y = \sqrt[3]{(1 + t)^3 + C} - 1$. $y(0) = 2 = \sqrt[3]{1 + C} - 1$, so $C = 26$ and

the solution is $y = \sqrt[3]{(1 + t)^3 + 26} - 1$.

27. $y' = 5ty - 2t$, $y(0) = 1$. $y' = 5t\left(y - \frac{2}{5}\right)$; $\left(y - \frac{2}{5}\right)^{-1}\frac{dy}{dt} = 5t$

$\int\left(y - \frac{2}{5}\right)^{-1}dy = \int 5t\ dt$; $\ln\left|y - \frac{2}{5}\right| = \frac{5}{2}t^2 + C$

$y = \frac{2}{5} + Ae^{(5/2)t^2}$. $y(0) = 1 = \frac{2}{5} + A$, so $A = \frac{3}{5}$ and the

solution is $y = \frac{2}{5} + \frac{3}{5}e^{(5/2)t^2}$.

28. $y' = \frac{t^2}{y}$, $y(0) = -5$; $y\frac{dy}{dt} = t^2$; $\int y\ dy = \int t^2dt$

$\frac{y^2}{2} + C_1 = \frac{t^3}{3} + C_2$; $y = \pm\sqrt{\frac{2}{3}t^3 + C}$. Since $y(0) = -1$, "-"

solution and $-5 = -\sqrt{C}$, so $C = 25$ and the solution is

$y = -\sqrt{\frac{2}{3}t^3 + 25}$.

29. $\frac{dy}{dx} = \frac{\ln x}{\sqrt{xy}}$, $y(1) = 4$. $\int\sqrt{y}\ dy = \int\frac{\ln x}{\sqrt{x}}\ dx = \int\frac{2\ln\sqrt{x}}{\sqrt{x}}dx$

$\frac{2}{3}y^{3/2} = 4\sqrt{x}\ln\sqrt{x} - 4\sqrt{x} + C$; $y = (6\sqrt{x}\ln\sqrt{x} - 6\sqrt{x} + C)^{2/3}$

$y(1) = 4 = (-6 + C)^{2/3}$, so $C = 14$ and the solution is

$y = (6\sqrt{x}\ln\sqrt{x} - 6\sqrt{x} + 14)^{2/3} = (3\sqrt{x}\ln x - 6\sqrt{x} + 14)^{2/3}$.

30. $\frac{dN}{dt} = -2tN^2$, $N(0) = 5$. $\frac{dN}{dt}N^{-2} = -2t$; $\int N^{-2}dN = \int -2t\ dt$

$-\frac{1}{N} = -t^2 + C$; $N = \frac{1}{t^2 + C}$. $N(0) = 5 = \frac{1}{C}$, so $C = \frac{1}{5}$ and the solution is $N = \frac{1}{t^2 + (1/5)} = \frac{5}{5t^2 + 1}$.

31. $\frac{dy}{dp} = -\frac{1}{2}\left(\frac{y}{p + 3}\right)$; $y^{-1}\frac{dy}{dp} = -\frac{1}{2(p + 3)}$ (Assuming $y \neq 0$)
$\int y^{-1}dy = -\frac{1}{2}\int\frac{1}{p + 3}\,dp$; $\ln|y| = -\frac{1}{2}\ln|p + 3| + C$; $y = \frac{A}{\sqrt{p + 3}}$.

32. $\frac{dy}{ds} = k\frac{y}{s}$; $\int y^{-1}dy = k\int s^{-1}ds$; $\ln|y| = k \ln|s| + C$
$y = As^k$. $(A > 0,\ s > 0)$

33. $\frac{dp}{dt} = k(1 - p)$ $0 \le p \le 1$, $p(0) = 0$.
$\int(1 - p)^{-1}dp = \int k\,dt$ (Assuming $p \neq 1$); $-\ln|1 - p| = kt + C$
$1 - p = e^{-kt+C}$; $p = 1 - e^{-kt+C} = 1 - Ae^{-kt}$. Since $p(0) = 0$,
$A = 1$, the solution is $p(t) = 1 - e^{-kt}$.

34. $\frac{dy}{dt} = k(M - y)$, $y(0) = 0$. $\int(M - y)^{-1}dy = \int k\,dt$ $(y \neq M)$
$-\ln|M - y| = kt + C$; $y = M - Ae^{-kt}$. $y(0) = 0 = M - A$, so
$A = M$ and the solution is $y = M - Me^{-kt}$.

3s. $\frac{dV}{dt} = kV^{2/3}$, $k > 0$ (since decreases), $V(0) = 27$, $V(4) = 15.625$
Solving the differential equation gives $\int V^{-2/3}dV = \int dt$
$\Rightarrow 3V^{1/3} = kt + C$, or $V = \left(\frac{kt}{3} + C\right)^3$. Using the conditions
$V(0) = 27 \Rightarrow 27 = C^3$ or $C = 3$
$V(4) = 15.625 \Rightarrow 15.625 = \left(\frac{4k}{3} + 3\right)^3$, or $k = -\frac{3}{8}$
Hence $V = \left(3 - \frac{t}{8}\right)^3$, $V = 0$ when $t = 24$ weeks.

36. $y' = ky$; $k > 0$, $f(0) = 100$, $f(2) = 115$,
$\frac{dy}{dt}y^{-1} = k$ (Assuming $y \neq 0$)

$\int\frac{dy}{y} = \int k\ dt;\quad \ln|y| + C_1 = kt + C_2;\quad |y| = Ae^{kt}$

$100 = Ae^0 \Rightarrow A = 100$, and $115 = 100e^{kt} = 100e$

$f(0) = 100 \Rightarrow 100 = Ae^0 \Rightarrow A = 100$

$f(2) = 115 \Rightarrow 115 = 100e^{2k} \Rightarrow k = 0.07$

Hence $y = 100e^{0.07t}$.

$y(t) = 200 = 100e^{0.07t}$, or $2 = e^{0.07t}$. Thus $.07t = \ln 2$

$\Rightarrow t = 9.9$.

The CCI will reach 200 when t = 9.9, in approximately October or November of 1999.

37. $\frac{dy}{dt} = -ay\ln\left(\frac{y}{b}\right);\quad \int\frac{1}{y\ln(y/b)}\,dy = \int -a\ dt\quad (y \neq 0,\ b)$

$\ln\left|\ln\left(\frac{y}{n}\right)\right| = -at + C_0;\quad \ln\left(\frac{y}{b}\right) = Ae^{-at};\quad y = be^{Ae^{-at}}$.

(y = 0 is not a solution and the solution y = b corresponds to A = 0).

38. $y' = ky^2$, $k < 0$. $\int y^{-2}dy = \int k\ dt\quad (y \neq 0);\quad -\frac{1}{y} = kt + C$

$y = -\frac{1}{kt + C} = \frac{1}{C - kt}$.

Exercises 10.3

1. The slope of the graph $= y' = ty - 5 = 2\cdot4 - 5 = 3$.

2. $y' = t^2 - y^2 = 2^2 - 3^2 = -5$.

3. $y' = y^2 + ty - 7 = 3^2 + 0(3) - 7 = 2$.

4. $y' = y^2 + ty - 7$, $y(0) = 2$. $y'(0) = 2^2 + 0(2) - 7 = -3$. Thus f(t) is decreasing at t = 0.

5. $g(t, y) = y' = t^2y$; $y(0) = 2$; $h = .5$. The iterates are

t_i	y_i
0	-2
.5	-2
1	-2.25

So $y(1) \sim 2.25$.

6. $g(t, y) = t - 2y$; $y(2) = 3$, $h = .5$. The iterates are

t_i	y_i
2	3
2.5	1
3	1.25

So $y(3) \sim 1.25$

7. $g(t, y) = 2t - y + 1$; $y(0) = 5$; $h = .5$. The iterates are

t_i	y_i
0	5
.5	3
1	2.5
1.5	2.75
2	3.375

So $y(2) \sim 3.375$.

8. $g(t, y) = y(2t - 1)$; $y(0) = 8$, $h = .25$. The iterates are

t_i	y_i
0	8
.25	6
.5	5.25
.75	5.25
1	5.90625

So $y(1) \sim 5.90625$.

9. <u>Euler's Method</u>: $g(t, y) = -(t + 1)y^2$; $y(0) = 1$; $h = .2$. The

iterates are:

t_i	y_i
0	1
.2	.8
.4	.6464
.6	.52941
.8	.43972
1	.37011

So $y(1) \sim 3.7011$

Exact Solution: $\int y^{-2}dy = \int -(t + 1)dt;\quad -\frac{1}{y} = \frac{-t^2}{2} - t + C$

$y = \dfrac{1}{(t^2/2) + t + C} = \dfrac{2}{t^2 + 2t + C}.\quad y(0) = 1 = \frac{2}{C},$ so $C = 2$ and

the exact solution is $y = \dfrac{2}{t^2 + 2t + 2}.\quad y(1) = \frac{2}{5} = .4.$

Therefore, the error in the estimate is $|.4 - .37011| =$

$.02989.$

10. Euler's Method: $g(t, y) = 10 - y;\quad y(0) = 1,\ h = .2.$ The

iterates are:

t_i	y_i
0	1
.2	2.8
.4	4.24
.6	5.392
.8	6.3136
1	7.05088

So $y(1) \sim 7.05088.$

Exact Solution: $\int (y - 10)^{-1}dy = -\int dt;\quad \ln|y - 10| = -t + C$

$y = 10 + Ae^{-t}.\quad y(0) = 1 = 10 + A,$ so $A = -9$ and the exact

solution is $y = 10 - 9e^{-t}.\quad y(1) = 10 - \frac{9}{e} \sim 6.68909.$

11. (a) $y' = k(1 - y),\ y(0) = 0\quad (k > 0).$

(b) Using Euler's method with $n = 3,\ h = 1$ gives the sequence of iterates:

t_i	y_i
0	0
1	$0 + k(1 - 0)$
2	$k + k(1 - k) = 2k - k^2$
3	$2k - k^2 + k(1 - 2k + k^2) = k^3 - 3k^2 + 3k$

So $y(3) \sim k^3 - 3k^2 + 3k$.

(c) $\int (y - 1)^{-1} dy = -\int k\, dt$; $\ln|y - 1| = -kt + C$

$y = 1 + Ae^{-kt}$; $y(0) = 0 = 1 + A$; so $A = -1$.

$y(3) = 1 - e^{-3k}$

(d) Given $k = .1$, Euler's method gives

$y(3) = (.1)^3 - 3(.1)^2 + 3(.1) = .27100$. The exact value is

$y(3) = 1 - e^{-.3} \sim .25918$.

12. (a) $y' = ky$, $k < 0$, $y(0) = 2$

(b) Euler's method with $n = 2$, $h = \frac{1}{2}$ gives the sequence of

iterates:

t_i	y_i
0	2
1/2	$2 + (2k)(1/2) = 2 + k$
1	$(2 + k) + k(2 + k)(1/2) = 2k + (k^2/2) + 2$

So $y(1) \sim 2k + \frac{k^2}{2} + 2$.

(c) $\int \frac{1}{y}\, dy = \int k\, dt$; $\ln|y| = kt + C$; $y = Ae^{kt}$. $y(0) = 2 = A$,

so the exact solution is $y = 2e^{kt}$, and $y(1)\ 2e^{k}$.

(d) The estimates from Euler's method with $k = -.3$ is

$y(1) \sim 2(-.3) + \frac{(-.3)^2}{2} + 2 = 1.445$. The exact value is

$y(1) = 2e^{-.3} \sim 1.48164$.

13.

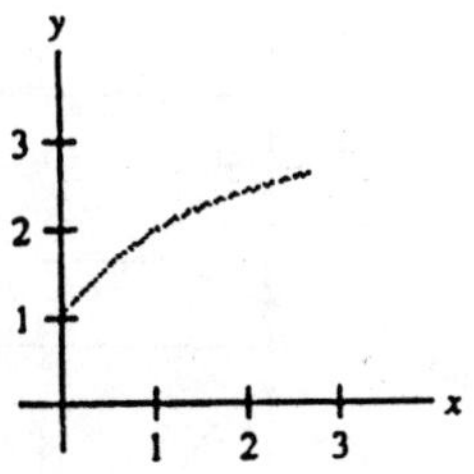

14.

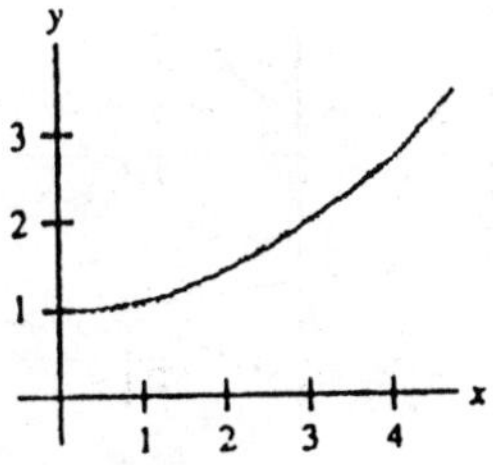

15.

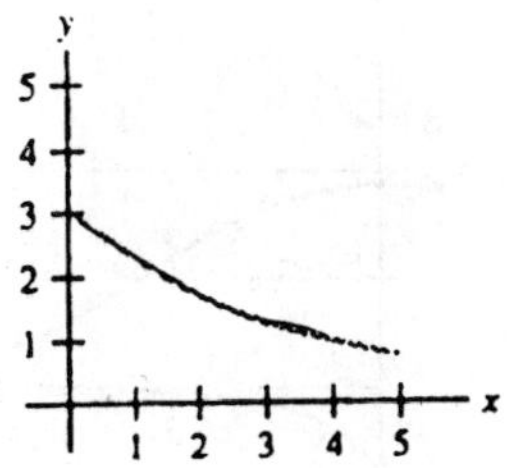

16.

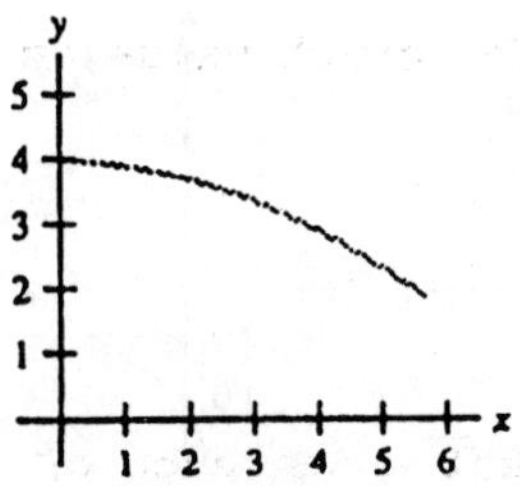

17.

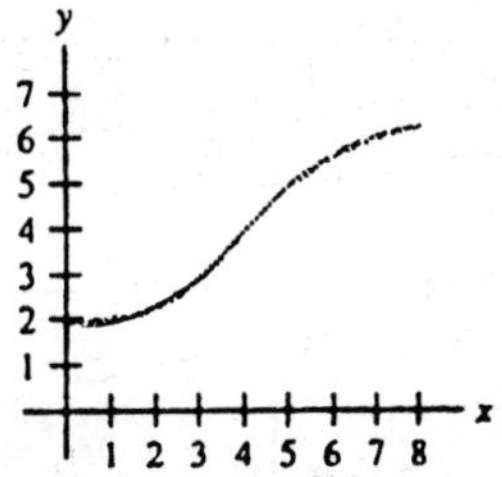

18.

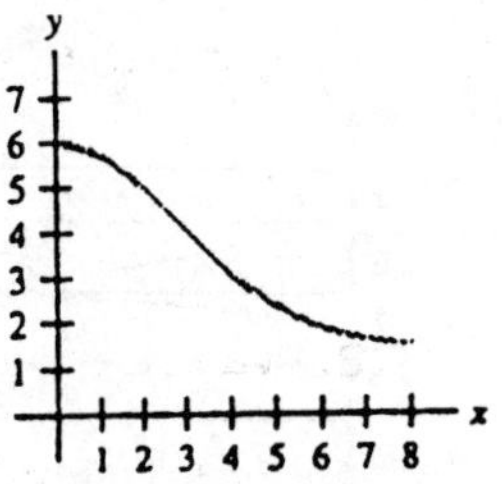

Exercises 10.4

1.

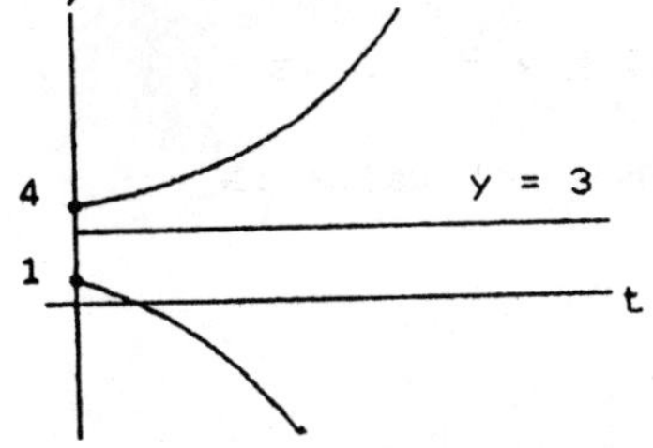

2.

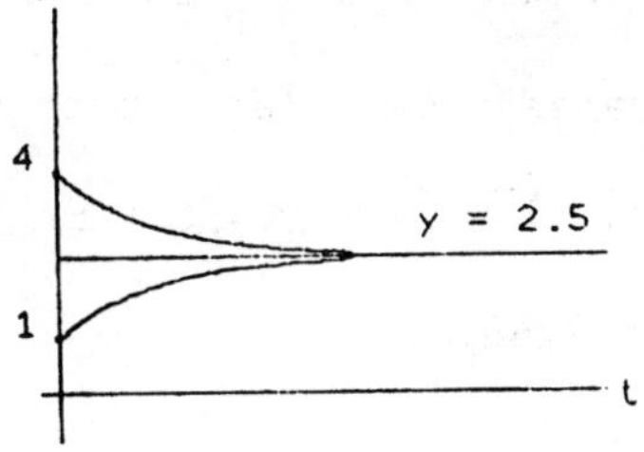

3.

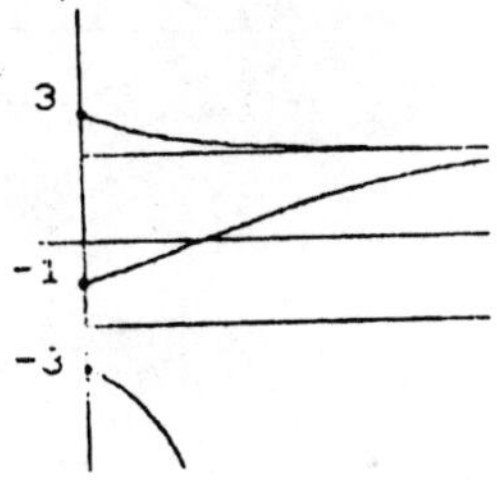

4.

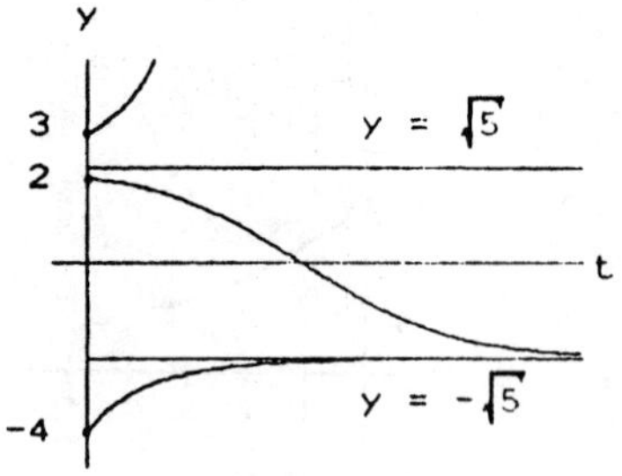

5.

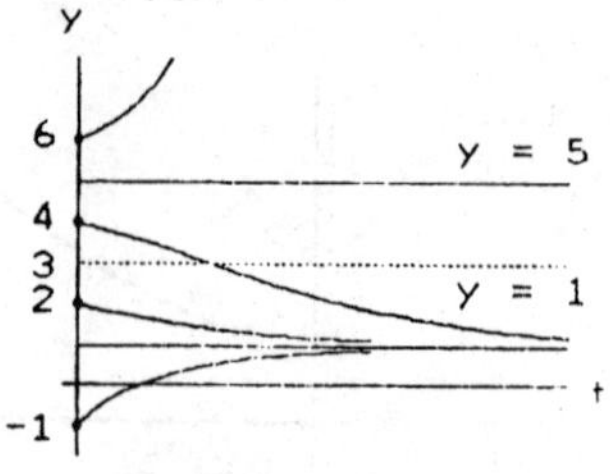

6.

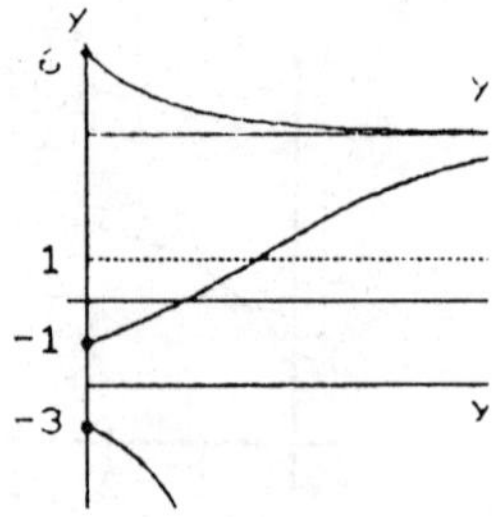

7.

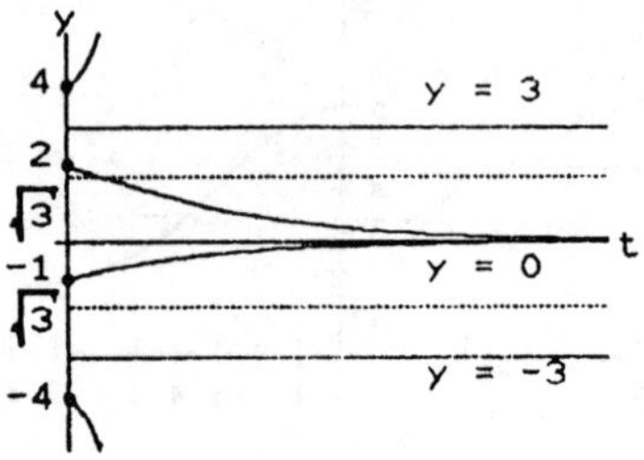

8.

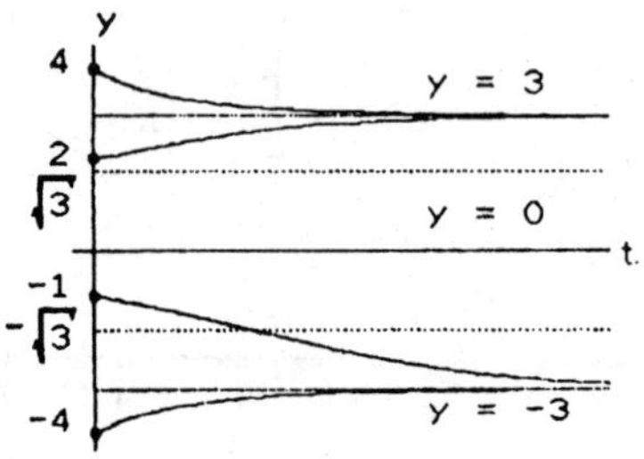

9.

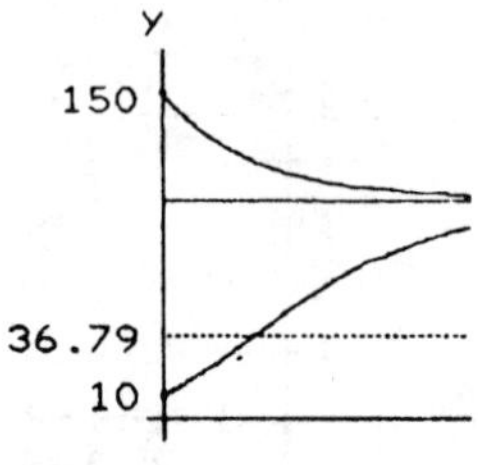

10.

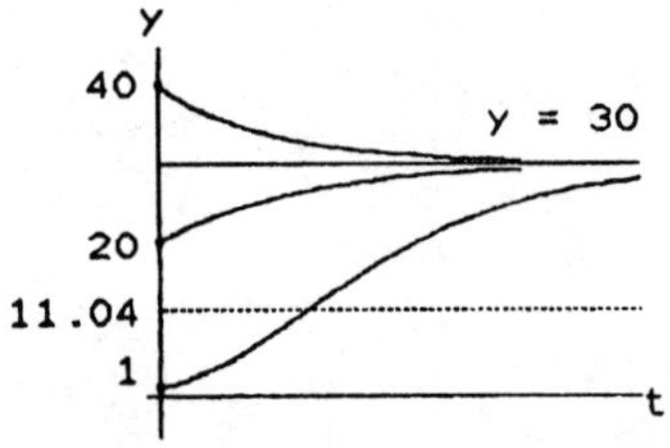

11.

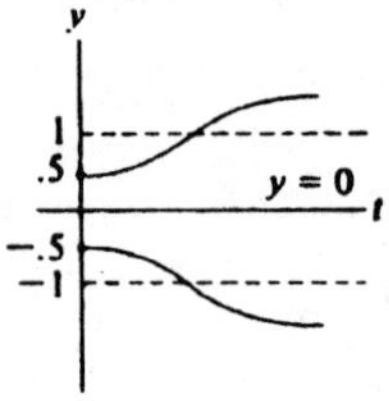

12.

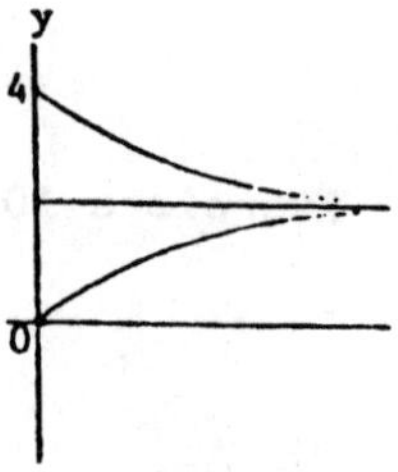

13.

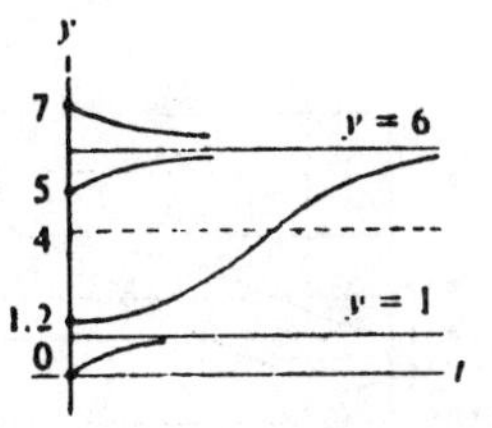

14.

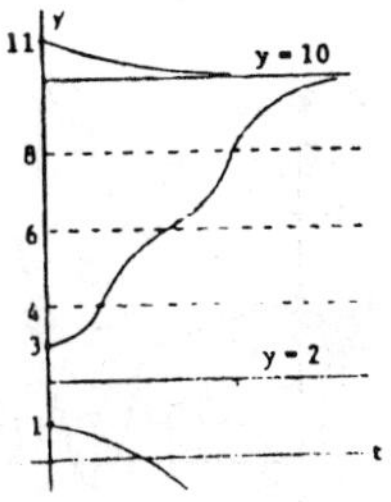

15.

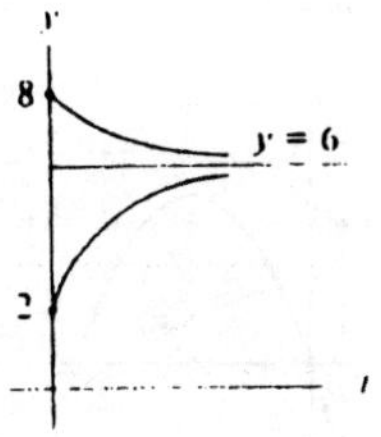

16.

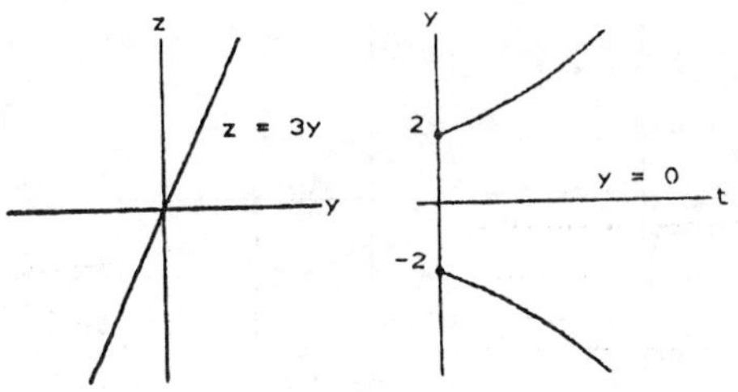

17.

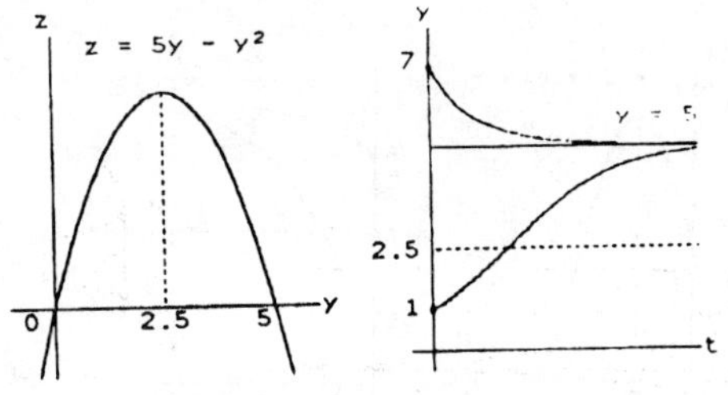

18.

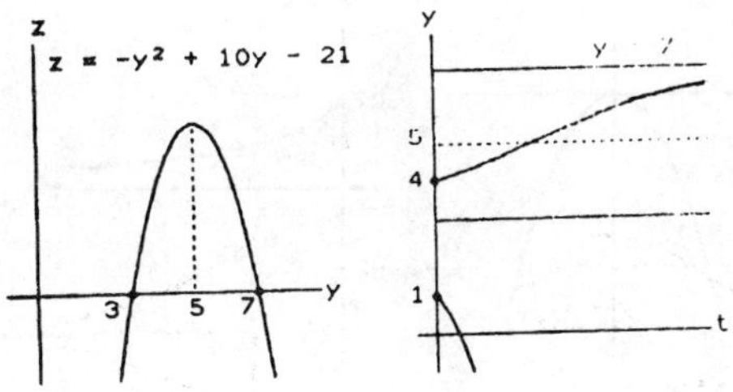

19.

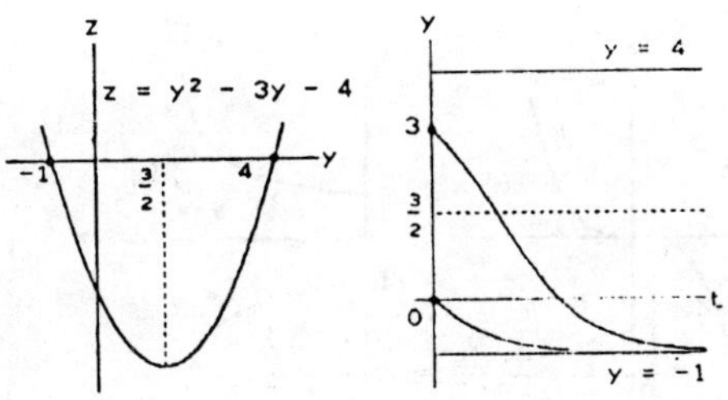

20.

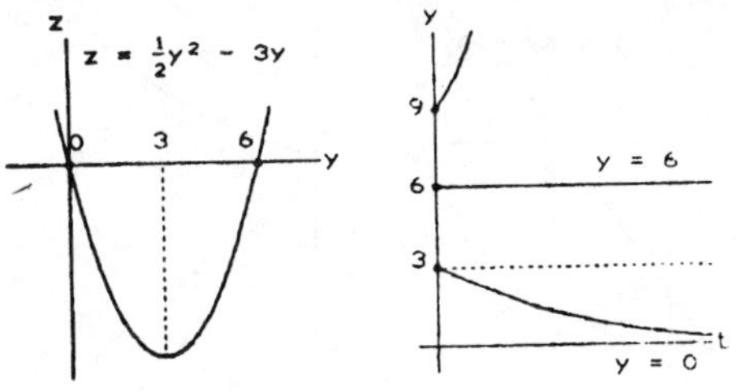

21.

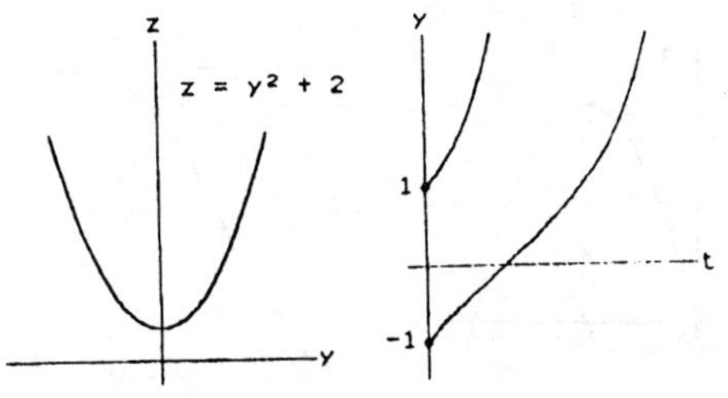

22.

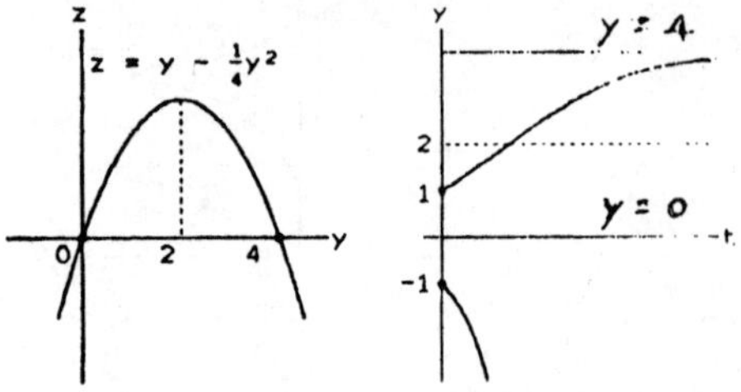

23.

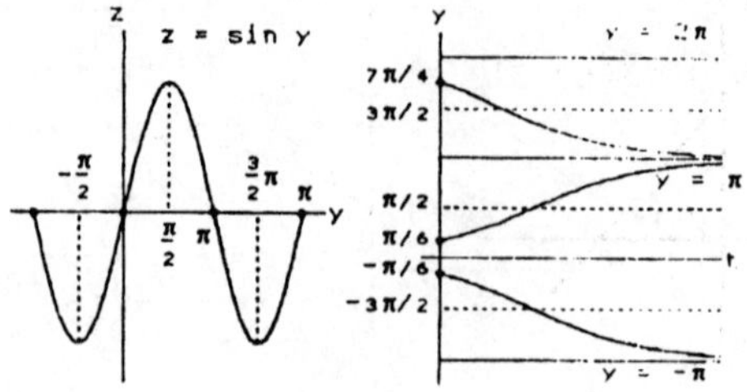

24.

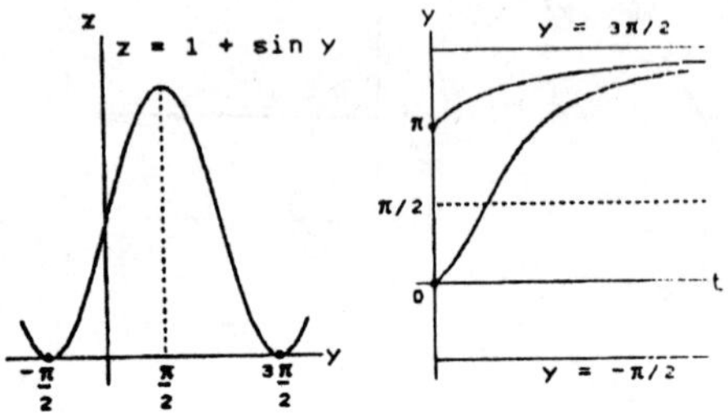

25.

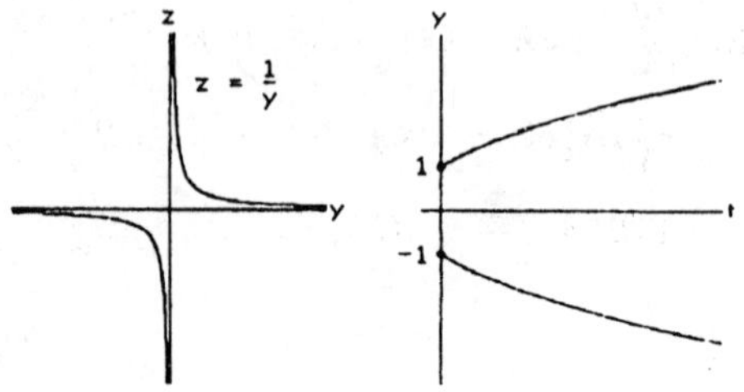

26.

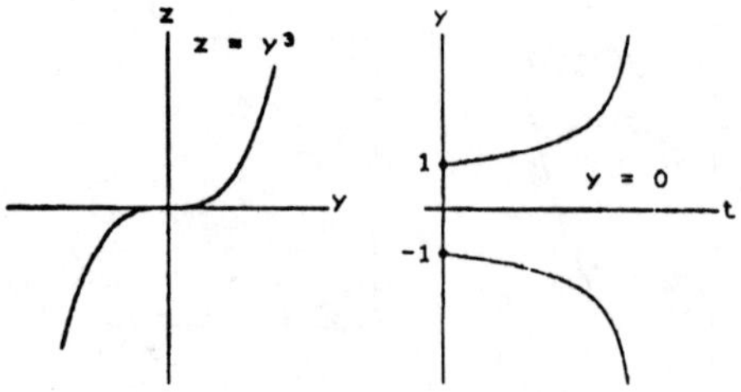

27.

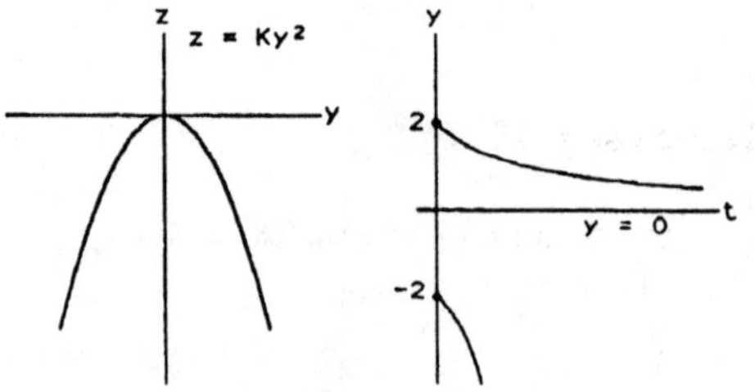

28.

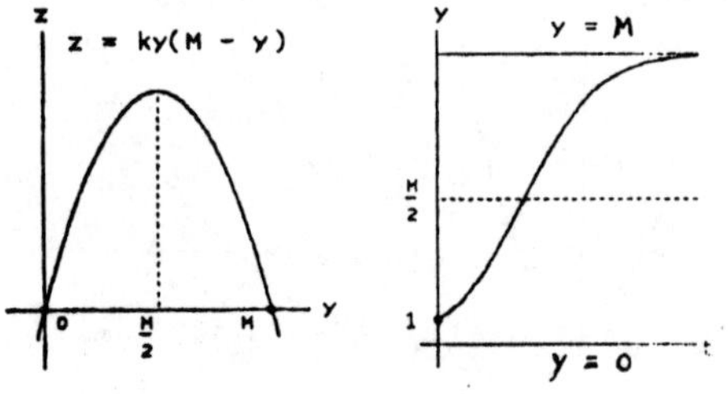

29.

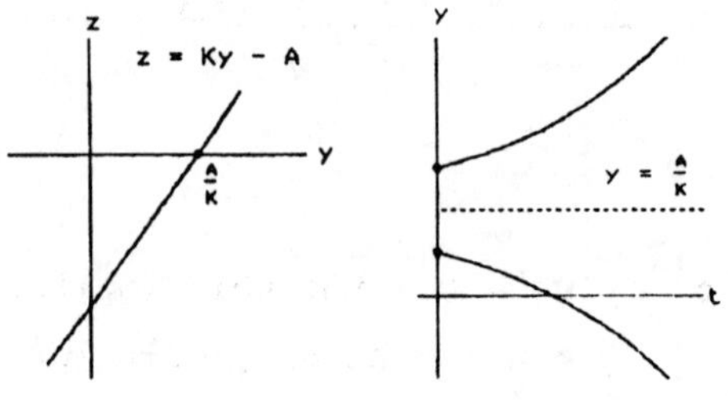

30.

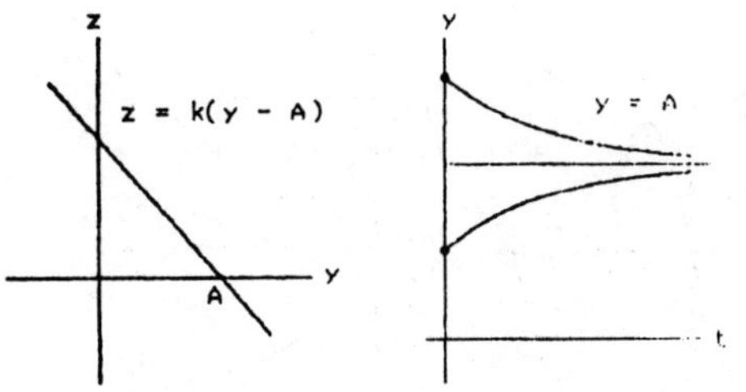

31. $y' = ky(H - y)$, $k > 0$

H = height at maturity

$y = f(t)$

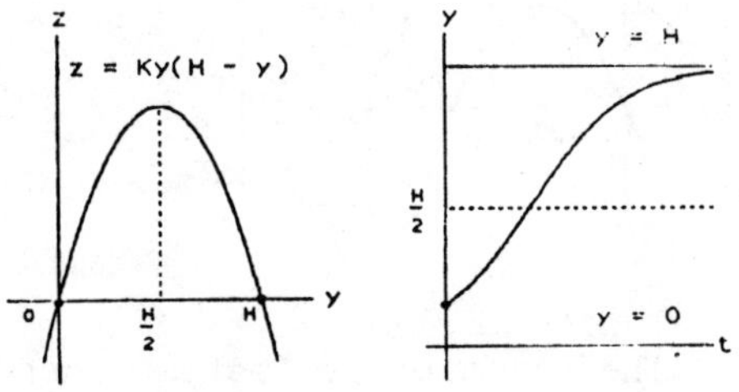

32. $\frac{dV}{dt} = 32 - kV$, $\int (32 - kV)^{-1} dy = \int dt \quad (32 \neq kV)$

$-\frac{1}{k}\ln|32 - kV| = t + C$; $y = \frac{32}{k} + Ae^{-kt}$. As $t \to \infty$ $V = 176$ so

$176 = \frac{32}{k} + A.0$, or $k = \frac{32}{176} = \frac{2}{11}$.

Exercises 10.5

1. $y' = k(100 - y)$, $k > 0$

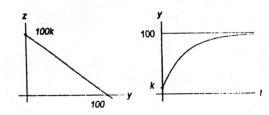

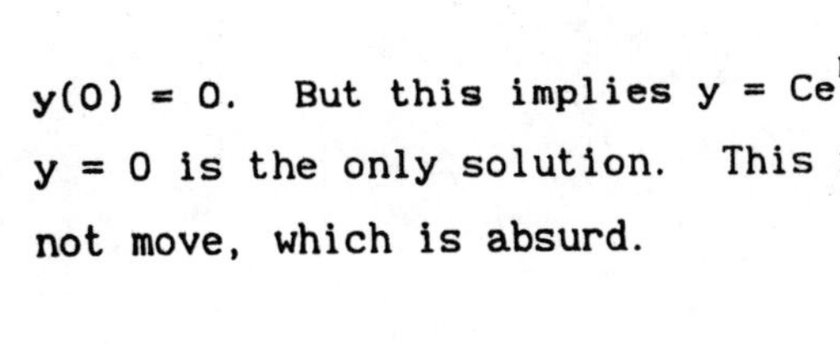

2. $y(0) = 0$. But this implies $y = Ce^{kt}$; $C = y(0) = 0$; so $y = 0$ is the only solution. This means that the object will not move, which is absurd.

3. $y' = ky(M - y)$, $k > 0$

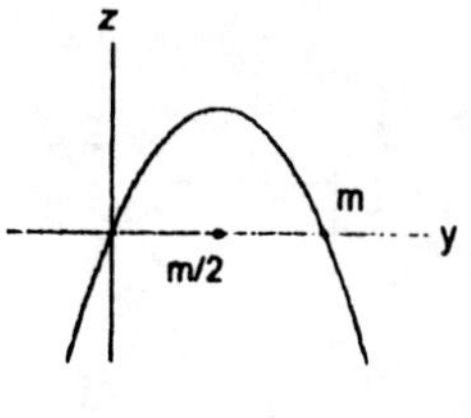

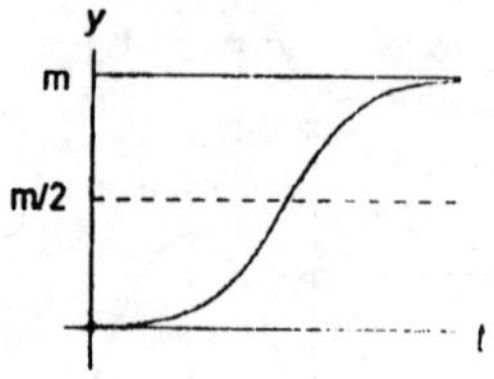

The reaction is fastest when $y = \frac{M}{2}$.

4. $y' = ky$, $k < 0$; so $y = Ce^{kt}$ where $C = y(0)$.

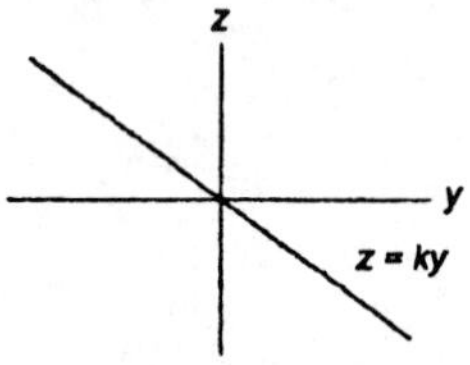

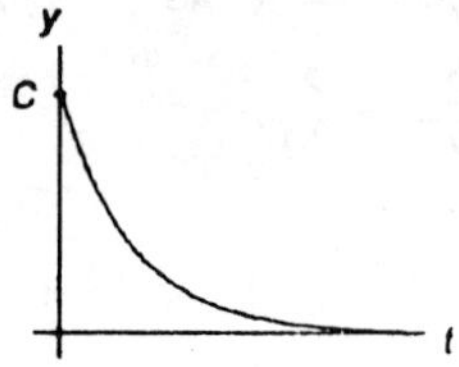

5. $y' = k(c - y)$, $k > 0$, $c > 0$

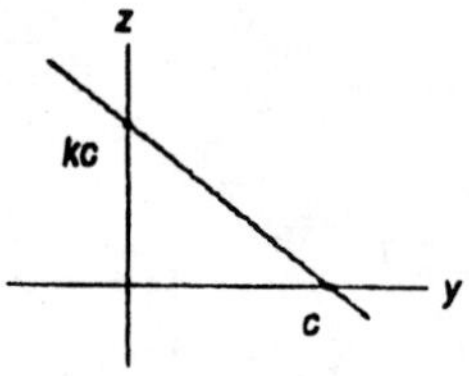

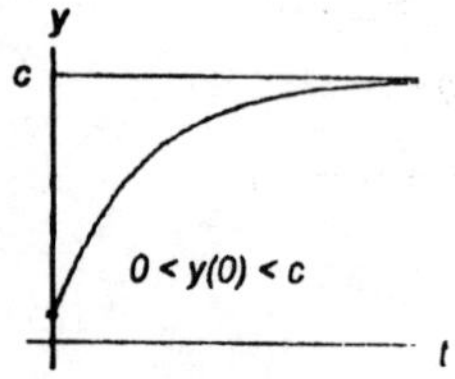

6. $y' = ky^2$, $k > 0$

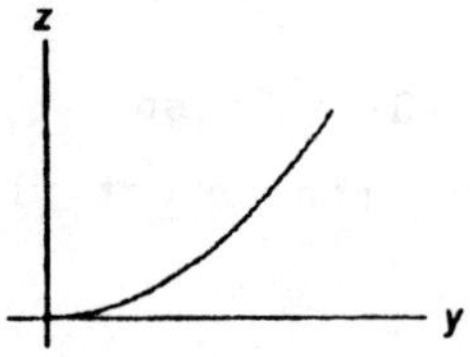

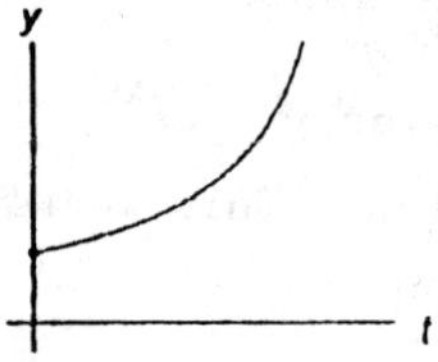

7. $y' = ky^2$, $k < 0$

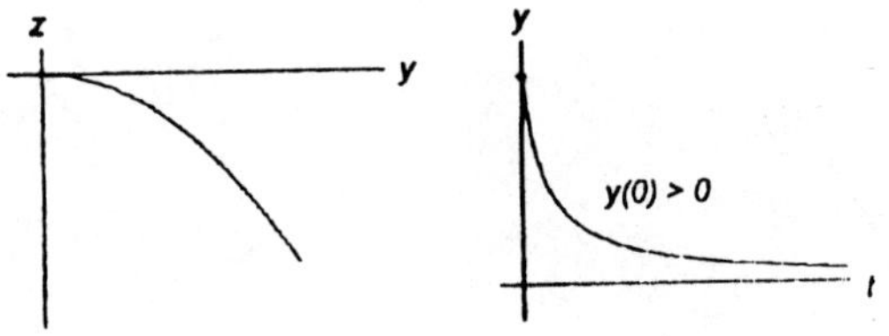

8. $y' = ky(100 - y)$, $k > 0$

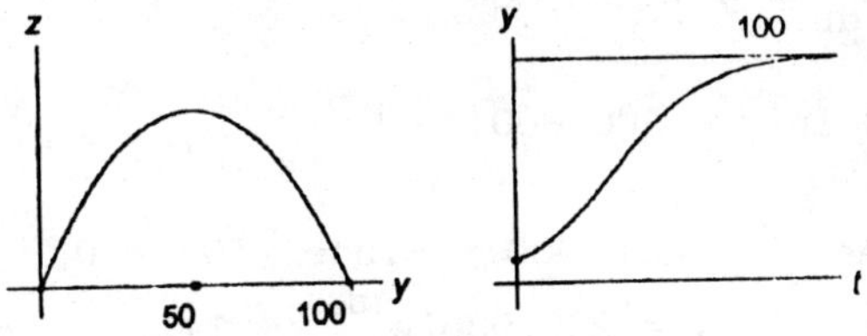

9. $y' = k(E - y)$, $k > 0$

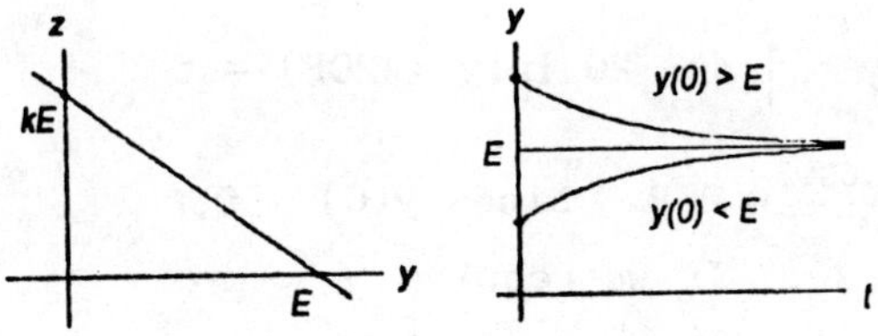

10. $y' = k(P_0 - y)$, $k > 0$

11. $y' = .055y - 300$, $y(0) = 5400$

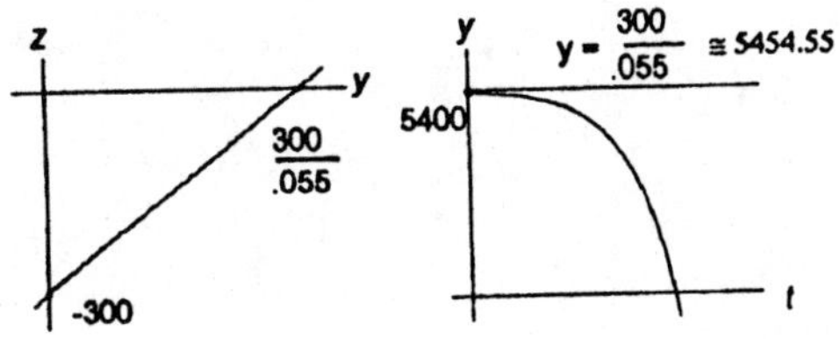

12. $y' = .06y - 600$

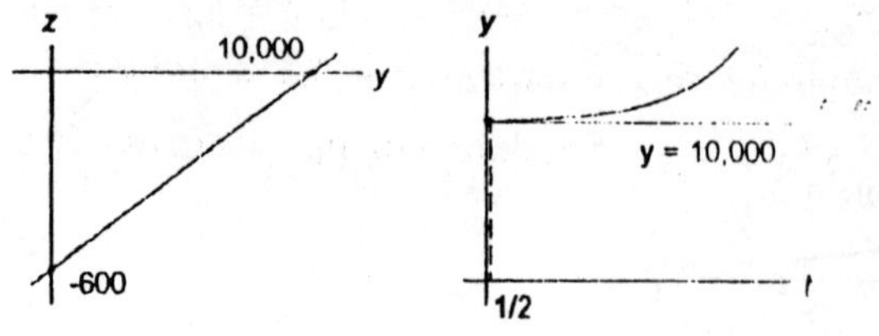

13. (a) $y' = .005y + 10{,}000$, $y(0) = 0$

(b) $\frac{dy}{dt} = .05y + 10{,}000$; $20\frac{dy}{dt} = y + 200{,}000$

$\int \frac{20}{y + 200{,}000}\, dy = \int dt$; $20 \ln(y + 200{,}000) = t + C$

$y + 200{,}000 = Ae^{.05t}$; $y = Ae^{.05t} - 200{,}000$. Since $y(0) = 0$, $A = 200{,}000$; so the solution is $y = 200{,}000(e^{.05t} - 1)$.

(c) $y(5) = 200{,}000(e^{.05(5)} - 1) \sim \$56{,}806$.

14. The balance of the account after t years satisfies the differential equation $y' = .05y + P$, $y(0) = 0$.

$20\frac{dy}{dt} = y + 20P$; $\int \frac{20}{y + 20P}\, dy = \int dt$; $20 \ln(y + 20P) = t$

$y + 20P = e^{.05t} + C$; $y = Ae^{.05t} - 20P$. Since $y(0) = 0$, $A = 20P$; so $y = 20P(e^{.05t} - 1)$. If we require $y(4) = 20P(e^{.2} - 1) = 50{,}000$; we must have $P \sim \$11{,}292$.

15. (a) $y' = .05 - .2y$, $y(0) = 6.25$

(b) $y' = .13 - .2y$, $y(0) = 6.25$

16.

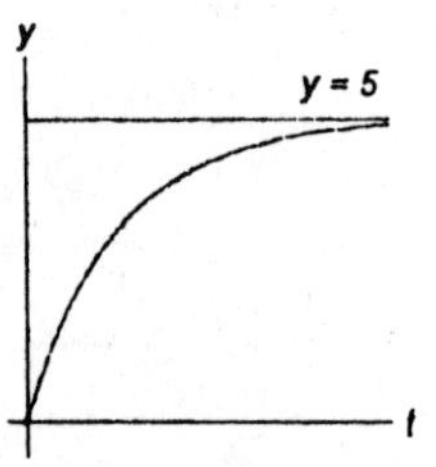

17. $y' = -.14y$

18. $y' = 10 - .8y$ The equilibrium point will be the value of y that makes $y' = 10 - .8y = 0$, i.e. $y = 12.5$. If $y(0) < 12.5$, the solution $y(t)$ will increase approaching $y = 12.5$ asymptotically. If $y(0) > 12.5$, $y(t)$ will be decreasing approaching $y = 12.5$ from above.

19.

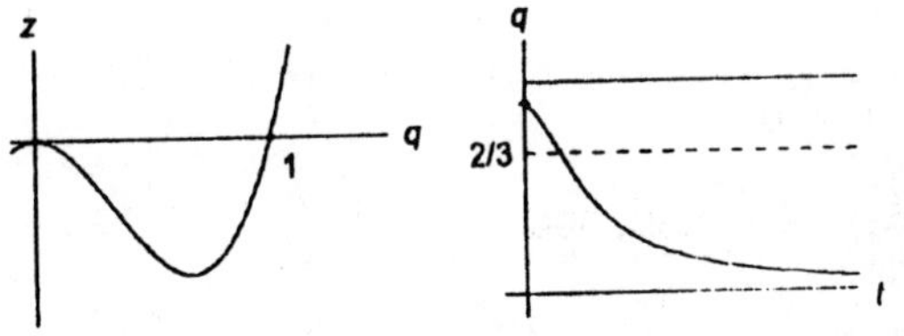

20 (a)

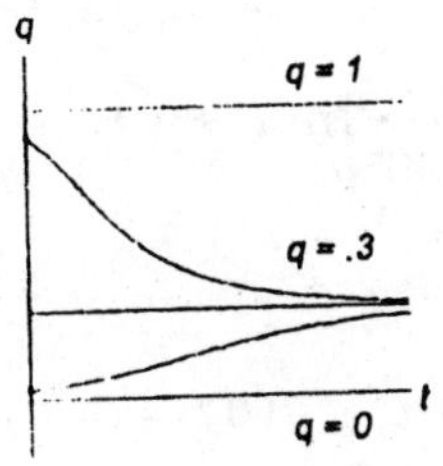

(b)

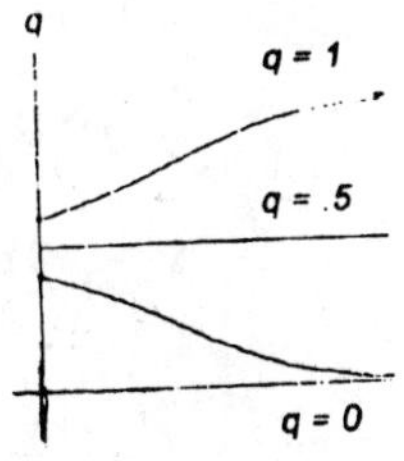

Chapter 10 Supplementary Exercises

1. $y^2y' = 4t^3 - 3t^2 + 2; \quad \int y^2 dy = \int (4t^3 - 3t^2 + 2)dt$

   $\frac{y^3}{3} = t^4 - t^3 + 2t + C; \quad y = (3t^4 - 3t^3 + 6t + C)^{1/3}$

2. $\frac{y'}{t + 1} = y + 1; \quad \int (y + 1)^{-1} dy = \int (t + 1)dt \qquad (y \neq 1)$

   $\ln|y + 1| = \frac{t^2}{2} + t + C; \quad y = -1 + Ae^{t^2/2 + t}.$

3. $y' = \frac{y}{t} - 3y,\ t > 0; \quad \int y^{-1} dy = \int \left(\frac{1}{t} - 3\right) dt \qquad (y \neq 0)$

   $\ln|y| = \ln|t| - 3t + C; \quad y = Ate^{-3t}.$

4. $(y')^2 = t; \quad y' = \pm\sqrt{t}; \quad \int dy = \pm\int t^{1/2} dt; \quad y = \pm\frac{2}{3}e^{3/2} + C.$

5. $y = 7y' + ty'$, $y(0) = 3$. $\int y^{-1}dy = \int (7 + t)^{-1}dt$ $(y \neq 0)$

$\ln|y| = \ln|7 + t|^{-1} + C$; $y = A(7 + t)$; $y(0) = 3 = 7A$ so

$A = \frac{3}{7}$ and the solution is $y = \frac{3}{7}(7 + t) = 3 + \frac{3}{7}t$.

6. $y' = te^{t+y}$, $y(0) = 0$. $\int e^{-y}dy = \int te^{t}dt$; $-e^{-y} = te^{t} - e^{t} + C$

$e^{-y} = -te^{t} + e^{t} + C$; $y = -\ln(-te^{t} + e^{t} + C)$

$y(0) = 0 = -\ln(1 + C)$, so $C = 0$ and the solution is

$y = -\ln(-te^{t} + e^{t})$.

7. $y' + t = 6t^2$, $y(0) = 7$. $\int y\,dy = \int (6t^2 - t)dt$

$\frac{y^2}{2} = 2t^3 - \frac{t^2}{2} + C$; $y = \pm\sqrt{4t^3 - t^2 + C}$; $y(0) = 7 = \sqrt{C}$, so

$C = 49$ and the solution is $y = \sqrt{4t^3 - t^2 + 49}$.

8. $y' = 5 - 8y$, $y(0) = 1$. $y' = -8\left(-\frac{5}{8} + y\right)$; $\int \left(y - \frac{5}{8}\right)^{-1} dy = \int -8\ dt$

$\ln\left|y - \frac{5}{8}\right| = -8t + C$; $y = \frac{5}{8} + Ae^{-8t}$. $y(0) = 1 = \frac{5}{8} + A$, so

$A = \frac{3}{8}$ and the solution is $y = \frac{5}{8} + \frac{3}{8}e^{-8t}$.

9. Euler's Method: $y' = 2e^{2t - y}$, $y(0) = 0$; $n = 4$; $h = .5$. The

iterates are:

t_i	y_i
0	0
.5	1
1	2
1.5	3
2	4

so the estimate is $f(2) \sim 4$.

Exact Solution: $\int e^{y}dy = \int 2e^{2t}dt$; $e^{y} = e^{2t} + C$

$y = \ln(e^{2t} + C)$. $y(0) = 0 = \ln(1 + C)$, so $C = 0$ and the exact

solution is $y = \ln(e^{2t}) = 2t$. Thus $f(2) = 4$ and the estimate above is exact.

10. $y' = (t + 1)/y$, $y(0) = 1$.

Euler's Method: $n = 3$, $h = \frac{1}{3}$. The iterates are:

t_i	y_i
0	1
1/3	4/3
2/3	5/3
1	2

So $y(1) \sim 2$.

Exact Solution: $\int y\,dy = \int (t + 1)dt$; $\frac{y^2}{2} = \frac{t^2}{2} + t + C$

$y = \pm\sqrt{t^2 + 2t + C}$; $y(0) = 1 = \sqrt{C}$, so $C = 1$ and the exact solution is $y = \sqrt{t^2 + 2t + 1} = \sqrt{(t + 1)^2} = |t + 1|$. Thus $y(1) = 2$ and the above estimate is exact.

11. If $f(t) = y(t) = 3$, then $y'(t) = (2 - 3)e^{-3} = -e^{-3} < 0$, so $f(t)$ is decreasing at this point.

12. Note that the constant solution $y = 1$ is a solution to the given equation. Since this solution satisfies the initial condition $y(0) = 1$, it must be the desired particular solution.

13. $y' = .1y(20 - y)$, $y(0) = 2$; $n = 6$, $h = .5$.

t_i	y_i
0	2
.5	3.8
1	6.878
1.5	11.39066
2	16.29396
2.5	19.31326
3	19.97642

$y(3) \sim 19.97642$

14. $y' = \frac{1}{2}y(y - 10)$; $y(0) = 9$; $n = 5$, $h = .2$

t_i	y_i
0	9
.2	8.1
.4	6.561
.6	4.30467
.8	1.85302
1	.34337

$y(1) \sim .34337$

15. $z = 2\cos y$

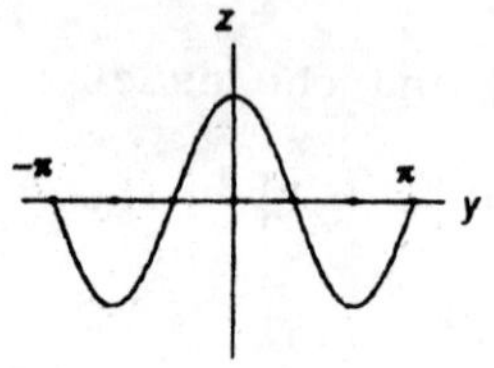

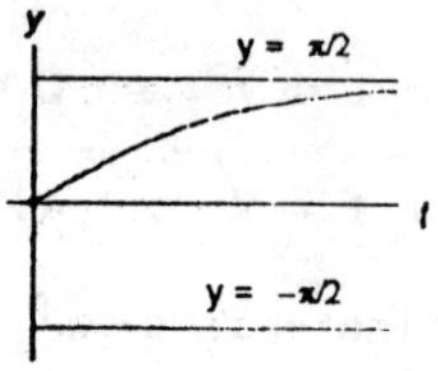

16. $z = 5 + 4y - y^2$

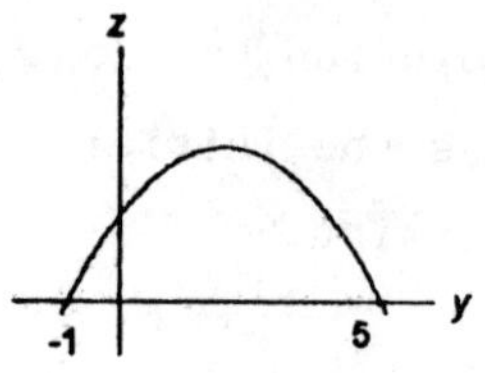

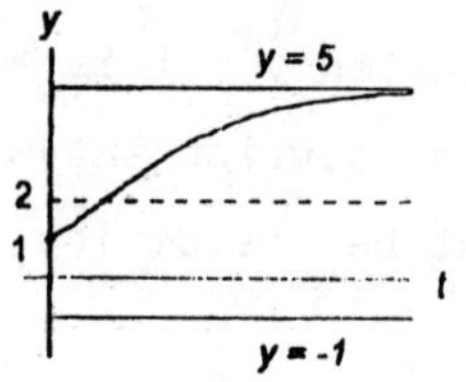

17. $z = y^2 + y$

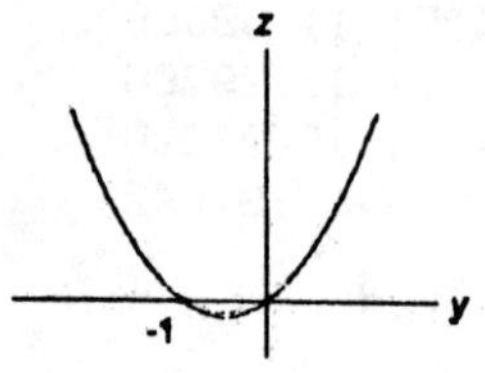

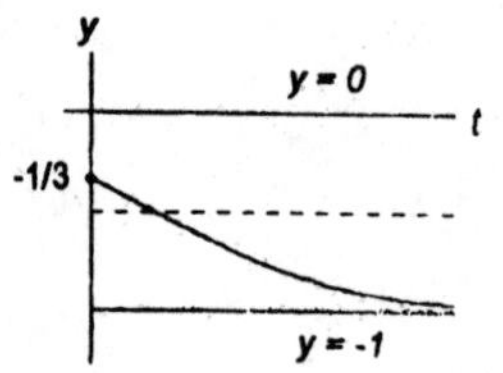

18. $z = y^2 - 2y + 1$

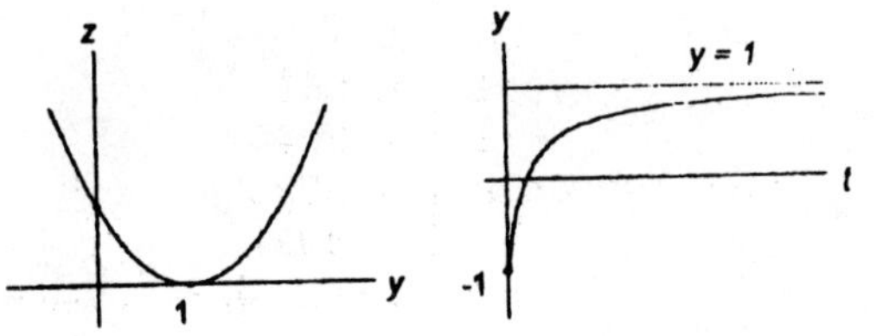

19. $z = \ln y$

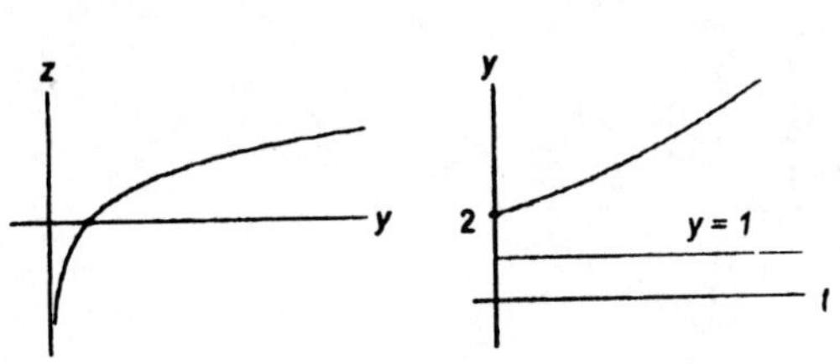

20. $z = \cos y + 1$

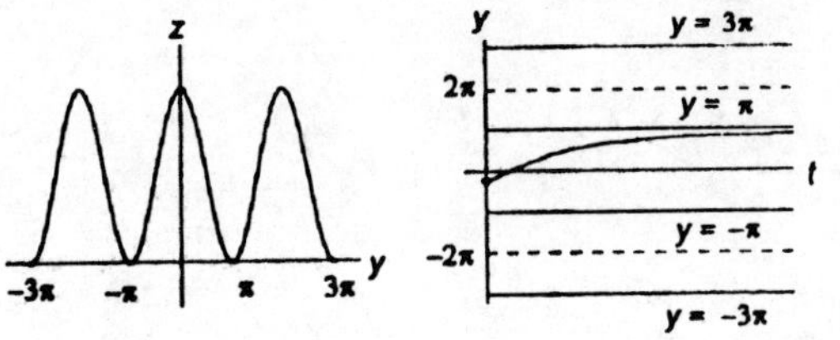

21. $z = \dfrac{1}{y^2 + 1}$

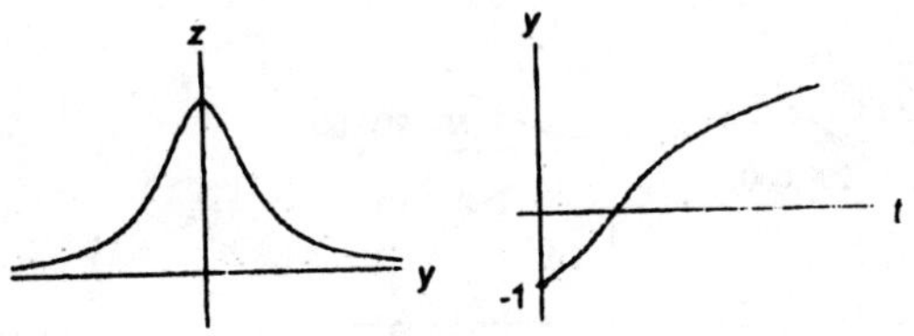

22. $z = \frac{3}{y + 3}$

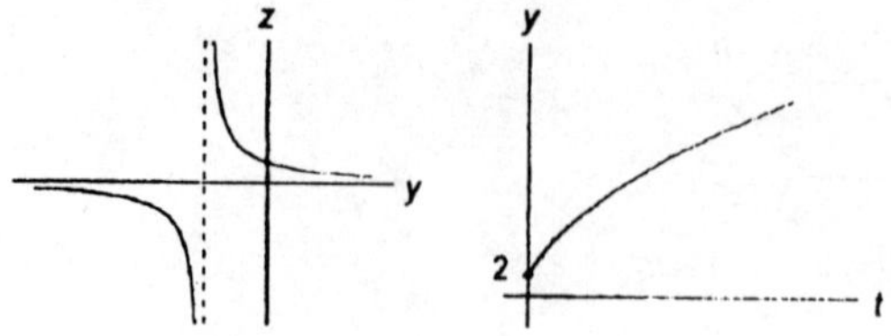

23. $z = .4y^2(1 - y)$

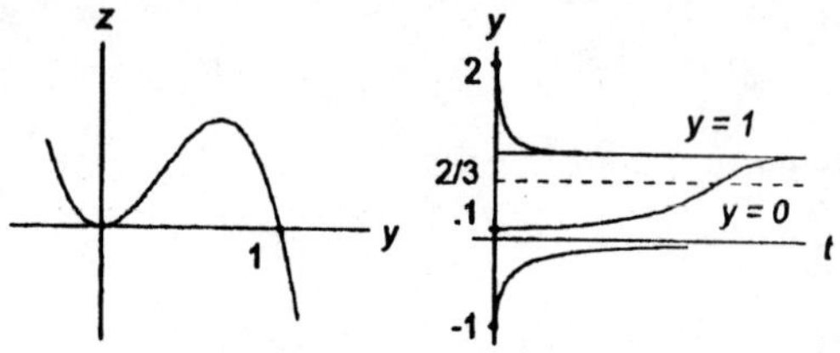

24. $z = y^3 - 6y^2 + 9y$

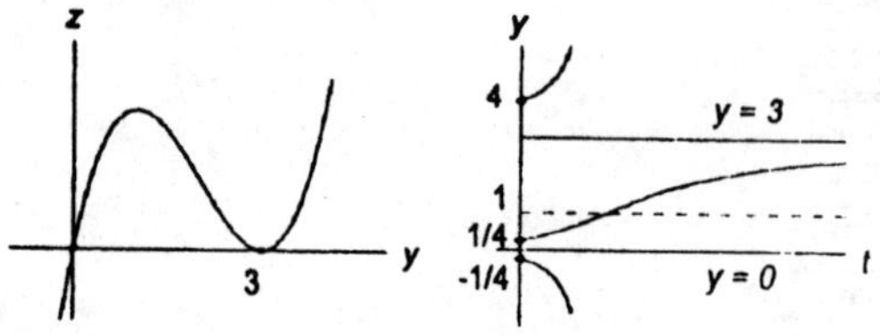

25. (a) $N' = .015N - 3000$

(b) There is a constant solution N = 200,000, but it is unstable. It is unlikely a city would have such a constant population.

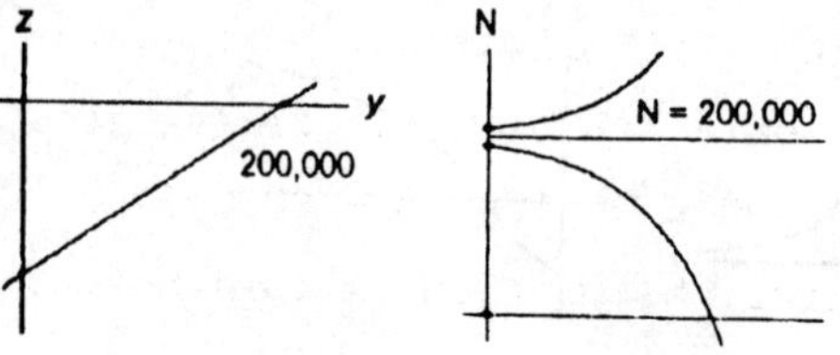

26. (a) $\left(10 - \frac{1}{4}y\right)$ represents the amount of unreacted substance A present and $\left(15 - \frac{3}{4}y\right)$ that of B.

(b) $k > 0$ since the amount of C is increasing.

(c)

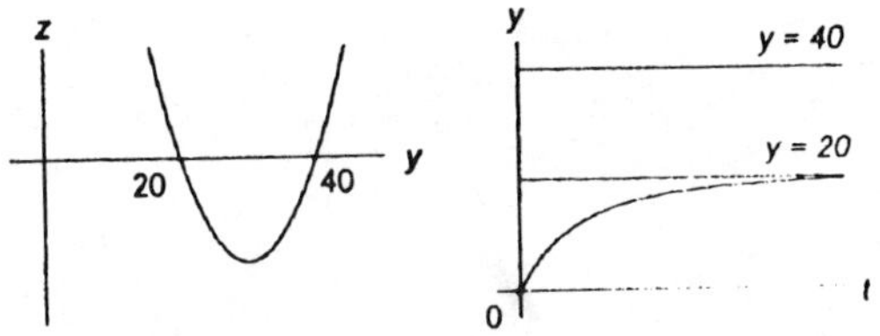

27. Let $f(t)$ be the balance in the account after t years. Then $f(t)$ satisfies the differential equation $y' = .05y - 2000$, $y(0) = 20{,}000$. Thus $y' = .05(y - 40{,}000)$

$\int (y - 40{,}000)^{-1}dy = \int .05\,dt$; $\ln|y - 40{,}000| = .05t + C$

$y = 40{,}000 + Ae^{.05t}$. $y(0) = 20{,}000 = 40{,}000 + A$ so $A = -20{,}000$ and $f(t) = 40{,}000 - 20{,}000e^{.05t}$. We want to find t so that $f(t) = 0$, i.e. $40{,}000 = 20{,}000\ e^{.05t}$, $2 = e^{.05t}$

$\ln 2 = .05t$; $t = 20 \ln 2 \sim 13.86294$ years.

28. Let $f(t)$ be the balance in the savings account after t years and let $M = f(0)$ be the initial amount. Then $f(t)$ satisfies the differential equation $y' = .06y - 12{,}000$

$$= .06(y - 200{,}000);\ y(0) = M.$$

Thus $\int (y - 200{,}000)^{-1}dy = \int .06\,dt$; $\ln|y - 200{,}000| = .06t + C$

$y = 200{,}000 + Ae^{.06t}$. Now $y(0) = M = 200{,}000 + A$, so $A = M - 200{,}000$ and $f(t) = 200{,}000 + (M - 200{,}000)e^{.06t}$.

(a) If the initial amount M is to fund the endowment forever, we must have $f(t) > 0$ for all t. This happens first in case $M - 200{,}000 \geq 0$, i.e. $M \geq \$200{,}000$. (A \$200,000 endowment would give the constant solution $f(t) = 200{,}000$).

(b) Using the expression for f(t) above, setting $f(20) = 0$ gives $200{,}000 = (200{,}000 - M)e^{1.2}$; $200{,}000 \sim 664023 - 3.32012M$ $M \sim \$139{,}761.16$.

Exercises 11.1

1. $E(X) = 0\left(\frac{1}{5}\right) + 1\left(\frac{4}{5}\right) = \frac{4}{5}$. $V(X) = \left(0 - \frac{4}{5}\right)^2\left(\frac{1}{5}\right) + \left(1 - \frac{4}{5}\right)^2\left(\frac{4}{5}\right)$

$= \frac{20}{125} = .16$

Standard deviation $= \sqrt{.16} = .4$.

2. $E(X) = 1\left(\frac{4}{9}\right) + 2\left(\frac{4}{9}\right) + 3\left(\frac{1}{9}\right) = \frac{15}{9} = \frac{5}{3}$.

$V(X) = \left(1 - \frac{5}{3}\right)^2\left(\frac{4}{9}\right) + \left(2 - \frac{5}{3}\right)^2\left(\frac{4}{9}\right) + \left(3 - \frac{5}{3}\right)^2\left(\frac{1}{9}\right) = \frac{36}{81} = \frac{4}{9}$.

Standard deviation $= \sqrt{\frac{36}{81}} = \frac{6}{9} = \frac{2}{3}$.

3. (a) $E(X) = 4(.5) + 6(.5) = 5$

$V(X) = (4 - 5)^2(.5) + (6 - 5)^2(.5) = 1$

(b) $E(X) = 1(.5) + 9(.5) = 5$

$V(X) = (3 - 5)^2(.5) + (7 - 5)^2(.5) = 4$

(c) $E(X) = 1(.5) + 9(.5) = 5$

$V(X) = (1 - 5)^2(.5) + (9 - 5)(.5) = 16$.

As the difference between the maximum and minimum values increases, so does the variance.

4. (a) $E(X) = 2(.1) + 4(.4) + 6(.4) + 8(.1) = 5$

$V(X) = (2 - 5)^2(.1) + (4 - 5)^2(.4) + (6 - 5)^2(.4)$
$+ (8 - 5)^2(.1) = 2.6$

(b) $E(X) = 2(.3) + 4(.2) + 6(.2) + 8(.3) = 5$

$V(X) = (2 - 5)^2(.3) + (4 - 5)^2(.2) + (6 - 5)^2(.2)$
$+ (8 - 5)^2(.3) = 5.8$

Values of (b) farther from E(X) have higher probabilities. Thus the variance is larger.

5. (a)

Outcome	0	1	2	3
Probability	$\frac{11}{52}$	$\frac{26}{52}$	$\frac{13}{52}$	$\frac{2}{52}$

(b) $E(X) = 0\left(\frac{11}{52}\right) + 1\left(\frac{26}{52}\right) + 2\left(\frac{13}{52}\right) + 3\left(\frac{2}{52}\right) = \frac{58}{52} = \frac{29}{26} \sim 1.12$

(c) E(X) is the average number of accidents per week in the given year.

6. (a)

Outcome	0	1	2
Probability	$\frac{30}{60}$	$\frac{20}{60}$	$\frac{10}{60}$

(b) $E(X) = 0\left(\frac{30}{60}\right) + 1\left(\frac{20}{60}\right) + 2\left(\frac{10}{60}\right) = \frac{2}{3}.$

(c) E(X) is the average number of calls coming into the switchboard each minute.

7. (a) $\frac{\text{Area within (1/2) unit of center}}{\text{Total area}} = \frac{\pi(1/2)^2}{\pi(1)^2} = \frac{1}{4} = .25$

Thus 25% of the points in the circle are within $\frac{1}{2}$ units of the center.

(b) $100 \times \frac{\pi c^2}{\pi 1^2} = 100c^2$

8. (a) $\frac{1}{2}$ (b) $\frac{1}{4}$ (c) $\frac{1}{100}$ (d) 0

9. Let X be the profit that the grower makes if he does not protect the fruit. Then
$E(X) = 100{,}000(.75) + 60{,}000(.25) = 90{,}000 < 95{,}000.$
Therefore he <u>should</u> spend the $5000 to protect the fruit.

10. Let X be the random variable defined in ex. 9. In this case,
$E(X) = 100{,}000(.80) + 85{,}000(.1) + 75{,}000(.1)$
$= 96{,}000 > 95{,}000$, so
the grower should <u>not</u> spend the money to protect the fruit.

Exercises 11.2

1. I. $\frac{1}{18}x \geq 0$ for all $0 \leq x \leq 6$

II. $\int_0^6 \frac{1}{18}x\, dx = \frac{1}{18}\frac{x^2}{2}\Big|_0^6 = 1 - 0 = 1.$

2. I. $2(x - 1) = 2x - 2 \geq 0$ for all $1 \leq x \leq 2$

II. $\int_1^2 (2x - 2)\, dx = x^2 - 2x\Big|_1^2 = 0 + 1 = 1.$

3. I. $\frac{1}{4} \geq 0$ II. $\int_1^5 \frac{1}{4}\, dx = \frac{1}{4}x\Big|_1^5 = \frac{5}{4} - \frac{1}{4} = 1$

4. I. $\frac{8}{9}x \geq 0$ for all $0 \leq x \leq \frac{3}{2}$.

II. $\int_0^{3/2} \frac{8}{9}x\, dx = \frac{4}{9}x^2\Big|_0^{3/2} = 1 - 0 = 1.$

5. I. $5x^4 \geq 0$ for all x. II. $\int_0^1 5x^4 dx = x^5\Big|_0^1 = 1 - 0 = 1.$

6. I. $\frac{3}{2}x - \frac{3}{4}x^2 = \frac{3}{2}x\left(1 - \frac{1}{2}x\right) \geq 0$ for all $0 \leq x \leq 2$.

II. $\int_0^2 \left(\frac{3}{2}x - \frac{3}{4}x^2\right)dx = \frac{3}{4}x^2 - \frac{1}{4}x^3\Big|_0^2 = 1 - 0 = 1.$

7. $\int_1^3 kx\, dx = \frac{k}{2}x^2\Big|_1^3 = \frac{9k}{2} - \frac{k}{2} = 4k = 1;$ so $k = \frac{1}{4}$.

8. $\int_0^2 kx^2 dx = \frac{k}{3}x^3\Big|_0^2 = \frac{8}{3}k = 1;$ so $k = \frac{3}{8}$.

9. $\int_5^{20} k\, dx = 15k = 1;$ so $k = \frac{1}{15}$.

10. $\int_1^4 kx^{-1/2}dx = 2kx^{1/2}\Big|_1^4 = 4k - 2k = 2k = 1$; so $k = \frac{1}{2}$.

11. $\int_0^1 kx^2(1 - x)dx = \int_0^1 (kx^2 - kx^3)dx = \frac{k}{3}x^3 - \frac{k}{4}x^4\Big|_0^1 = \frac{k}{12} = 1$;
so $k = 12$.

12. $\int_0^3 k(3x - x^2)dx = \frac{3}{2}kx^2 - \frac{k}{3}x^3\Big|_0^3 = \frac{27}{2}k - 9k = \frac{9}{2}k = 1$; so $k = \frac{2}{9}$.

13.

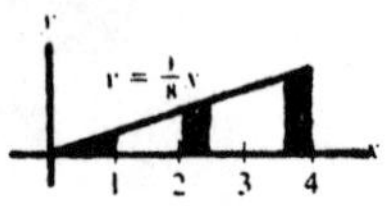

14.

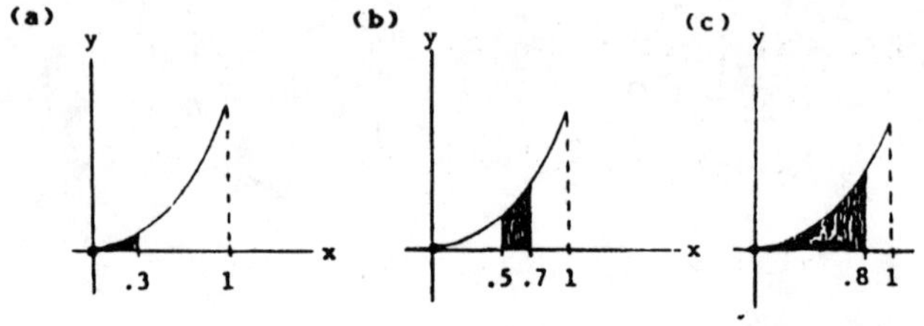

15. $\int_1^2 \frac{1}{18}\, x\, dx = \frac{x^2}{36}\Big|_1^2 = \frac{1}{12}$.

16. $\int_{1.5}^{1.7} (2x - 2)dx = x^2 - 2x\Big|_{1.5}^{1.7} = .24$

17. $\int_1^3 \frac{1}{4}\, dx = \frac{1}{2}$

18. $\int_1^{3/2} \frac{8}{9}x\, dx = \frac{4}{9}x^2\Big|_1^{3/2} = \frac{5}{9}$.

19. $\int_{35}^{50} \frac{1}{20}dx = \frac{15}{20} = \frac{3}{4}$.

20. $\int_0^4 \frac{11}{10}(x + 1)^{-2}dx = -\frac{11}{10}(x + 1)^{-1}\Big|_0^4$
$= -\frac{11}{50} + \frac{11}{10} = \frac{44}{50} = \frac{22}{25}$

21. $f(x) = F'(x) = \frac{8}{3}x^{-3}$.

22. $f(x) = F'(x) = \frac{1}{4}(x - 1)^{-1/2}$

23. $F(x) = \frac{1}{5}x + C$; $F(2) = \frac{1}{5}(2) + C = 0$; so $C = -\frac{2}{5}$ and
$F(x) = -\frac{1}{5}x - \frac{2}{5}$.

24. $F(x) = \frac{3}{2}x - \frac{x^2}{4} + C$; $F(1) = \frac{5}{4} + C = 0$; so $C = -\frac{5}{4}$ and
$F(x) = \frac{3}{2}x - \frac{1}{4}x^2 - \frac{5}{4}$.

25. (a) $\int_2^3 \frac{1}{21}x^2 dx = \frac{x^3}{63}\Big|_2^3 = \frac{19}{63}$.

(b) $F(1) = \frac{x^3}{63} + C$; $F(1) = \frac{1}{63} + C = 0$; $F(x) = \frac{x^3}{63} - \frac{1}{63}$

(c) $F(3) - F(2) = \frac{19}{63}$.

26. (a) $\int_3^4 \left(\frac{4}{9}x - \frac{1}{9}x^2\right)dx = \frac{2}{9}x^2 - \frac{1}{27}x^3\Big|_3^4 = \frac{32}{27} - 1 = \frac{5}{27}$.

(b) $F(x) = \frac{2}{9}x^2 - \frac{1}{27}x^3 + C$; $F(1) = \frac{2}{9} - \frac{1}{27} + C = 0$; $C = \frac{-5}{27}$
$F(x) = \frac{2}{9}x^2 - \frac{1}{27}x^3 - \frac{5}{27}$. (c) $F(4) - F(3) = \frac{5}{27}$.

27. Points whose largest coordinate have value $\le x$ lie within the square with vertices (0, 0), (0, x), (x, x), and (x, 0). The area of this square is x^2. Thus F(x) = the probability that a randomly selected point has maximum coordinate $\le x = \frac{x^2}{4}$.

28. $f(x) = F'(x) = \frac{x}{2}$, $0 \le x \le 2$.

29. Points whose coordinates sum to a value $\le x$ lie within the triangle with vertices (0, 0), (0, x), and (x, 0). The area of this triangle is $\frac{x^2}{2}$. Thus F(x) = the probability that a randomly selected point has coordinates summing to a value $\le x$
$= \frac{(x^2/2)}{2} = \frac{x^2}{4}$.

30. $f(x) = F'(x) = \frac{x}{2}$, $0 \le x \le 2$.

31. $\int_0^5 2ke^{-kx}dx = -2e^{-kx}\Big|_0^5 = -2e^{-5k} + 2 = 2 - 2e^{\ln(2)(-1/2)}$

$= 2 - \sqrt{2} \sim .58579.$

32. $\int_{10-M}^{10} 2ke^{-5kx}dx = -2e^{-kx}\Big|_{10-M}^{10} = -2e^{-10k} + 2e^{-10k+Mk} \qquad k = \frac{\ln 2}{10}$

$= -1 + 2^{M/10} = .1$

$2^{M/10} = 1.1; \quad \frac{M}{10}\ln 2 = \ln 1.1; \quad M = 10\frac{\ln 1.1}{\ln 2} \sim 1.37504$ days.

33. $\int_0^b \frac{1}{3}\, dx = \frac{1}{3}b = .6;$ so $b = 1.8$

34. $\int_a^2 \frac{2}{3}x\, dx = \frac{1}{3}x^2\Big|_a^2 = \frac{4}{3} - \frac{a^2}{3} = \frac{1}{3};$ so $a = \sqrt{3}.$

35. $F(b) = \frac{b^2}{4} = .09;$ so $b = .6$

36. $F(b) = (b - 1)^2 = \frac{1}{4};$ so $b - 1 = \frac{1}{2};\ b = \frac{3}{2}.$

37. (a) I. $4x^{-5} \geq 0$ for all $x \geq 1.$

II. $\int_1^\infty 4x^{-5}dx = \lim_{b\to\infty}\int_1^b 4x^{-5}dx = \lim_{b\to\infty}\left[-x^{-4}\Big|_1^b\right] = \lim_{b\to\infty}\left[1 - \frac{1}{b^4}\right] = 1$

(b) $F(x) = -x^{-4} + C;\ F(1) = -1 + C = 0;$ so $C = 1$ and $F(x) = -x^{-4} + 1.$

(c) $\Pr(1 \leq X \leq 2) = F(2) - F(1) = 1 - \frac{1}{16} = \frac{15}{16}$

$\Pr(2 \leq X) = 1 - \Pr(1 \leq X \leq 2) = \frac{1}{16}.$

38. (a) I. $2(x + 1)^{-3} \geq 0$ for all $x \geq 0.$

II. $\int_0^\infty 2(x + 1)^{-3}dx = \lim_{b\to\infty}\int_0^b 2(x + 1)^{-3}dx = \lim_{b\to\infty}\left[-(x + 1)^{-2}\Big|_0^b\right]$

$= \lim_{b\to\infty}\left[1 - \frac{1}{(b + 1)^2}\right] = 1.$

(b) $F(x) = -\frac{1}{(x + 1)^2} + C,\ F(0) = -1 + C = 0;$ so $C = 1$ and

$F(x) = 1 - \dfrac{1}{(x+1)^2}.$

(c) $\Pr(1 \le X \le 2) = F(2) - F(1) = \frac{8}{9} - \frac{3}{4} = \frac{5}{36}$

$\Pr(3 \le x) = 1 - \Pr(x \le 3) = 1 - F(3) = 1 - \left(1 - \frac{1}{16}\right) = \frac{1}{16}.$

Exercises 11.3

1. $E(X) = \displaystyle\int_0^6 \frac{1}{18}x^2dx = \frac{x^3}{54}\Big|_0^6 = 4;\ V(X) = \int_0^6 \frac{1}{18}x^3dx - 4^2 = \frac{x^4}{72}\Big|_0^6 - 16 = 2$

2. $E(X) = \displaystyle\int_1^2 (2x^2 - 2x)dx = \frac{2}{3}x^3 - x^2\Big|_1^2 = \frac{5}{3}$

$V(X) = \displaystyle\int_1^2 (2x^3 - 2x^2)dx - \left(\frac{5}{3}\right)^2 = \frac{1}{2}x^4 - \frac{2}{3}x^3\Big|_1^2 - \frac{25}{9} = \frac{17}{6} - \frac{25}{9} = \frac{1}{18}$

3. $E(X) = \displaystyle\int_1^5 \frac{1}{4}x\ dx = \frac{x^2}{8}\Big|_1^5 = 3$

$V(X) = \displaystyle\int_1^5 \frac{1}{4}x^2dx - 3^2 = \frac{x^3}{12}\Big|_1^5 - 9 = \frac{31}{3} - 9 = \frac{4}{3}.$

4. $E(X) = \displaystyle\int_0^{3/2} \frac{8}{9}x^2dx = \frac{8}{27}x^3\Big|_0^{3/2} = 1$

$V(X) = \displaystyle\int_0^{3/2} \frac{8}{9}x^3dx - 1^2 = \frac{8}{36}x^4\Big|_0^{3/2} - 1 = \frac{9}{8} - 1 = \frac{1}{8}.$

5. $E(X) = \displaystyle\int_0^1 5x^5dx = \frac{5}{6}x^6\Big|_0^1 = \frac{5}{6}.$

$V(X) = \displaystyle\int_0^1 5x^6dx - \left(\frac{5}{6}\right)^2 = \frac{5}{7}x^7\Big|_0^1 - \left(\frac{5}{6}\right)^2 = \frac{5}{252}$

6. $E(X) = \int_0^2 \left(\frac{3}{2}x^2 - \frac{3}{4}x^3\right)dx = \frac{x^3}{2} - \frac{3}{16}x^4\Big|_0^2 = 1$

$V(X) = \int_0^2 \left(\frac{3}{2}x^3 - \frac{3}{4}x^4\right)dx - 1^2 = \frac{3}{8}x^4 - \frac{3}{20}x^5\Big|_0^2 - 1 = \frac{6}{5} - 1 = \frac{1}{5}$

7. $E(X) = \int_0^1 12x^2(1 - x)^2dx = \int_0^1 (12x^4 - 24x^3 + 12x^2)dx$

$= \frac{12}{5}x^5 - 6x^4 + 4x^3\Big|_0^1 = \frac{2}{5}$

$V(X) = \int_0^1 (12x^5 - 24x^4 + 12x^3)dx - \frac{4}{25} = 2x^6 - \frac{24}{5}x^5 + 3x^4\Big|_0^1 - \frac{4}{25}$

$= \frac{1}{25}$

8. $E(X) = \int_0^4 \frac{3}{16}x^{3/2}dx = \frac{3}{40}x^{5/2}\Big|_0^4 = \frac{12}{5}$

$V(X) = \int_0^4 \frac{3}{16}x^{5/2}dx - \frac{144}{25} = \frac{3}{36}x^{7/2}\Big|_0^4 - \frac{144}{25} = \frac{48}{7} - \frac{144}{25} = \frac{192}{175}$

9. (a) $f(x) = 30x^2(1 - x)^2 = 30x^4 - 60x^3 + 30x^2$, so
$F(x) = 6x^5 - 15x^4 + 10x^3 + C$; $F(0) = C = 0$, so
$F(x) = 6x^5 - 15x^4 + 10x^3$.

(b) $F(.25) = \frac{53}{512}$

(c) $E(X) = \int_0^1 (30x^5 - 60x^4 + 30x^3)dx = \frac{10}{2}x^6 - 12x^5 + \frac{15}{2}x^4\Big|_0^1 = \frac{1}{2}$

On average, the newspaper devotes $\frac{1}{2}$ of its space to advertising.

(d) $V(X) = \int_0^1 (30x^6 - 60x^5 + 30x^4)dx - \frac{1}{4}$

$= \frac{30}{7}x^7 - 10x^6 + 6x^5\Big|_0^1 - \frac{1}{4} = \frac{1}{28}.$

10. $f(x) = 20x^3(1 - x) = -20x^4 + 20x^3$

(a) $E(X) = \int_0^1 -20x^5 + 20x^4 dx = -\frac{10}{3}x^6 + 4x^5\Big|_0^1 = \frac{2}{3}.$

On average, $\frac{2}{3}$ of the new restaurants make a profit during their first year of operation.

(b) $V(X) = \int_0^1 (-20x^6 + 20x^5)dx - \frac{4}{9} = -\frac{20}{7}x^7 + \frac{10}{3}x^6\Big|_0^1 - \frac{4}{9} = \frac{2}{63}.$

11. (a) Since $F(x) = \frac{1}{9}x^2$, $0 \le x \le 3$ $f(x) = \frac{2}{9}x$ and

$E(X) = \int_0^3 \frac{2}{9}x^2 dx = \frac{2}{27}x^3\Big|_0^3 = 2.$ This means that the average useful life of the component is 200 hrs.

(b) $V(X) = \int_0^3 \frac{2}{9}x^3 dx - 4 = \frac{1}{18}x^4\Big|_0^3 - 4 = \frac{1}{2}.$

12. (a) Since $F(x) = \frac{1}{125}x^3$, $0 \le x \le 5$, $f(x) = \frac{3}{125}x^2$ and

$E(X) = \int_0^5 \frac{3}{125}x^3 dx = \frac{3}{4\cdot125}x^4\Big|_0^5 = \frac{15}{4}.$ On average, the assembly takes $\frac{15}{4}$ minutes to complete.

(b) $V(X) = \int_0^5 \frac{3}{125}x^4 dx - \left(\frac{15}{4}\right)^2 = \frac{3}{5^4}x^5\Big|_0^5 - \left(\frac{15}{4}\right)^2 = 15 - \frac{225}{16} = \frac{15}{16}.$

13. $E(X) = \int_0^{12} \frac{1}{72}x^2 dx = \frac{x^3}{72\cdot3}\Big|_0^{12} = 8$ (minutes)

14. $E(X) = \int_0^{10} \frac{-6x^3 + 60x^2}{1000} dx = \frac{1}{1000}\left[-\frac{3}{2}x^4 + 20x^3\Big|_0^{10}\right] = 5.$

15. (a) $f(x) = (6x - x^2)/18$, $3 \le x \le 6$, so $F(x) = \frac{x^2}{6} - \frac{x^3}{54} + C.$

$F(3) = \frac{3}{2} - \frac{1}{2} + C = 0$; so $C = -1$ and $F(x) = \frac{x^2}{6} - \frac{x^3}{54} - 1.$

(b) $F(5) = \frac{25}{6} - \frac{125}{54} - 1 = \frac{23}{27}.$

(c) $E(X) = \frac{1}{18}\int_3^6 (6x^2 - x^3)dx = \frac{1}{18}\left[2x^3 - \frac{x^4}{4}\right]\Big|_3^6 = \frac{1}{18}\left[108 - \frac{135}{4}\right]$

$= 4.125$ Thus the mean completion time is 412.5 hrs.

(d) $V(X) = \frac{1}{18}\int_3^6 (6x^3 - x^4)dx - (4.125)^2$

$= \frac{1}{18}\left[\frac{3}{2}x^4 - \frac{x^5}{5}\right]\Big|_3^6 - (4.125)^2 = .5344$

16. $f(x) = 4(x - 1)^3,\ 1 \le x \le 2.$

(a) $\Pr(X > 1.5) = 1 - \Pr(X \le 1.5) = 1 - \int_1^{1.5} 4(x - 1)^3 dx$

$= 1 - (x - 1)^4\Big|_1^{1.5} = 1 - \frac{1}{16} = \frac{15}{16}.$

(b) $E(X) = \int_1^2 4x(x - 1)^3 dx = \int_1^2 (4x^4 - 12x^3 + 12x^2 - 4x)dx$

$= \frac{4}{5}x^5 - 3x^4 + 4x^3 - 2x^2\Big|_1^2 = \frac{9}{5} = 1.8$ (thousand gal.)

17. $E(X) = \int_1^\infty 4x^{-4}dx = \lim_{b\to\infty}\left[-\frac{4}{3}x^{-3}\Big|_1^b\right] = \lim_{b\to\infty}\left[\frac{4}{3} - \frac{4}{3}b^{-3}\right] = \frac{4}{3}$

$V(X) = \int_1^\infty 4x^{-3}dx - \left(\frac{4}{3}\right)^2 = \lim_{b\to\infty}\left[-2x^{-2}\Big|_1^b\right] - \frac{16}{9}$

$= \lim_{b\to\infty}\left[2 - b^{-4}\right] - \frac{16}{9} = \frac{2}{9}.$

18. $f(x) = 3x^{-4},\ x \ge 1$

$E(X) = \int_1^\infty 3x^{-3}dx = \lim_{b\to\infty}\left[-\frac{3}{2}x^{-2}\Big|_1^b\right] = \lim_{b\to\infty}\left[\frac{3}{2} - \frac{3}{2}b^{-2}\right] = \frac{3}{2}.$

$V(X) = \int_1^\infty 3x^{-2}dx - \left(\frac{3}{2}\right)^2 = \lim_{b\to\infty}\left[-3x^{-1}\Big|_1^b\right] - \frac{9}{4} = \frac{3}{4}.$

19. $\int_0^M \frac{1}{18}x\,dx = \frac{1}{2};\quad \frac{x^2}{36}\Big|_0^M = \frac{1}{2};\quad \frac{M^2}{36} = \frac{1}{2};\quad M = \sqrt{18} = 3\sqrt{2}.$

20. $\int_1^M (2x - 2)dx = \frac{1}{2};\quad x^2 - 2x\Big|_1^M = \frac{1}{2};\quad M^2 - 2M + 1 = \frac{1}{2}$

$(M - 1)^2 = \frac{1}{2}$; $M = 1 + \frac{\sqrt{2}}{2}$.

21. $F(M) = \frac{1}{9}x^2 = \frac{1}{2}$; $x = \frac{3\sqrt{2}}{2}$ (hundred hours)

22. $F(T) = \frac{1}{125}T^3 = \frac{1}{2}$; $T = \frac{5}{\sqrt[3]{2}}$.

23. $\int_0^T \frac{11}{10(x + 1)^2}\,dx = \frac{1}{2}$; $-\frac{11}{10}(x + 1)^{-1}\Big|_0^T = \frac{1}{2}$; $\frac{11}{10} - \frac{11}{10(T + 1)} = \frac{1}{2}$

$T = \frac{5}{6}$ minutes

24. $\int_1^M 4(x - 1)^3 dx = \frac{1}{2}$; $(x - 1)^4\Big|_0^M = \frac{1}{2}$; $(M - 1)^4 = \frac{1}{2}$; $M = \sqrt[4]{\frac{1}{2}} + 1$

25. By definition, $E(X) = \int_A^B xf(x)\,dx$ and $F'(x) = f(x)$.

Using integration by parts,

$$\int_A^B xf(x)\,dx = xF(x)\Big|_A^B - \int_A^B F(x)\,dx = BF(B) - AF(A) - \int_A^B F(x)\,dx$$

$$= B(1) - A(0) - \int_A^B F(x)\,dx = B - \int_A^B F(x)\,dx$$

$$E(X) = B - \int_A^B F(x)\,dx = 5 - \int_0^5 \frac{1}{125}x^3 dx = 5 - \frac{x^4}{500}\Big|_0^5 = 5 - \frac{625}{500} = \frac{15}{4}$$

26. $5 - \int_0^5 \frac{1}{125}x^3\,dx = 5 - \frac{5^3}{4(125)} + 0 = 3.75$

Exercises 11.4

1. $E(X) = \frac{1}{3}$; $V(X) = \frac{1}{9}$

2. $E(X) = 4$; $V(X) = 16$

3. $E(X) = 5$; $V(X) = 25$

4. $E(X) = \frac{2}{3}$; $V(X) = \frac{4}{9}$

5. The probability density function is $f(x) = 2e^{-2x}$. Thus
$$\Pr\left(\frac{1}{2} < X < 1\right) = \int_{1/2}^{1} 2e^{-2x}dx = -e^{-2x}\Big|_{1/2}^{1} = e^{-1} - e^{-2}.$$

6. $\displaystyle\int_{0}^{1/3} 2e^{-2x}dx = -e^{-2x}\Big|_{0}^{1/3} = 1 - e^{-2/3}.$

7. The probability density function is $f(x) = \frac{1}{3}e^{-(1/3)x}$. Thus
$$\Pr(X < 2) = \int_{0}^{2} \frac{1}{3}e^{-(1/3)x}dx = -e^{-(1/3)x}\Big|_{0}^{2} = 1 - e^{-2/3}.$$

8. $$\Pr(X > 5) = 1 - \Pr(X \le 5) = 1 - \int_{0}^{5} \frac{1}{3}e^{-(1/3)x}dx = 1 + e^{-(1/3)x}\Big|_{0}^{5}$$
$$= e^{-5/3}.$$

9. The probability density function is $f(x) = \frac{1}{20}e^{-(1/20)x}$. Thus
$$\Pr(X > 60) = 1 - \Pr(X \le 60) = 1 - \int_{0}^{60} \frac{1}{20}e^{-(1/20)x}dx$$
$$= 1 + e^{-(1/20)x}\Big|_{0}^{60} = e^{-3}.$$

10. $$\Pr(10 < X < 30) = \int_{10}^{30} \frac{1}{20}e^{-(1/20)x}dx = -e^{-(1/20)x}\Big|_{10}^{30}$$
$$= e^{-1/2} - e^{-3/2}.$$

11. The probability density function is $f(x) = \frac{1}{2}e^{-(1/2)x}$. Thus

$$\Pr(X < 4) = \int_0^4 \frac{1}{2}e^{-(1/2)x}dx = -e^{-(1/2)x}\Big|_0^4 = 1 - e^{-2}.$$

12. $\Pr(X \geq 5) = 1 - \Pr(X < 5) = 1 - \int_0^5 \frac{1}{2}e^{-(1/2)x}dx = 1 + e^{-(1/2)x}\Big|_0^5$

$$= e^{-5/2}.$$

13. The probability density function is $f(x) = \frac{1}{72}e^{-(1/72)x}$.

(a) $\Pr(X > 24) = 1 - \Pr(X \leq 24) = 1 - \int_0^{24} \frac{1}{72}e^{-(1/72)x}dx$

$$= 1 + e^{-(1/72)x}\Big|_0^{24} = e^{-(1/3)}.$$

(b) $r(t) = \Pr(X > t) = 1 - \Pr(X \leq t) = 1 - \int_0^t \frac{1}{72}e^{-(1/72)x}dx$

$$= 1 + e^{-(1/72)x}\Big|_0^t = e^{-t/72}.$$

14. (a) $S(x) = \Pr(X \geq x) = 1 - \Pr(X < x) = 1 - F(x)$

$$= 1 - \int_0^x ke^{-kt}dt = 1 + e^{-kt}\Big|_0^x = e^{-kx}.$$

(b) $S(5) = e^{-5k} = .9;\quad -5k = \ln(.9);\quad k = \frac{-\ln(.9)}{5} \sim .02107.$

15. $\mu = 4,\ \sigma = 1$ 16. $\mu = -5,\ \sigma = 1$ 17. $\mu = 0,\ \sigma = 3$

18. $\mu = 3,\ \sigma = 5$

19. $f(x) = e^{-x^2/2}$; $f'(x) = -xe^{-x^2/2}$. Thus $f'(0) = 0$.

$f''(x) = -\left[e^{-x^2/2} - x^2e^{-x^2/2}\right]$; so $f''(0) = -1 < 0$. Therefore

$f(x)$ has a relative maximum at $x = 0$.

20. $f(x) = e^{-1/2\left(\frac{x - \mu}{\sigma}\right)^2}$. $f'(x) = \frac{\mu - x}{\sigma}e^{-\frac{1}{2}\left(\frac{x - \mu}{\sigma}\right)^2}$.

Thus $f'(\mu) = 0$. $f''(x) = -e^{-\frac{1}{2}\left(\frac{x-\mu}{\sigma}\right)^2} + \left(\frac{\mu - x}{\sigma}\right)^2 e^{-\frac{1}{2}\left(\frac{x-\mu}{\sigma}\right)^2}$;

so $f''(\mu) = -1 < 0$. Therefore $f(x)$ has a relative maximum at $x = \mu$.

21. $f(x) = e^{-x^2/2}$. $f''(x) = x^2e^{-x^2/2}$ (see ex. 19); $f''(\pm 1) = 0$, but $f'(\pm 1) = \pm 1e^{-1/2} \neq 0$; so $f(x)$ has inflection points at $x = \pm 1$.

22. $f(x) = e^{-\frac{1}{2}\left(\frac{x-\mu}{\sigma}\right)^2}$. (See ex. 20)

$f''(x) = \left(\frac{\mu - x}{\sigma}\right)^2 e^{-\frac{1}{2}\left(\frac{x-\mu}{\sigma}\right)^2} - e^{-\frac{1}{2}\left(\frac{x-\mu}{\sigma}\right)^2}$. $f''(\mu \pm \sigma) = 0$, but

$f'(\mu \pm \sigma) = \frac{\mu - (\mu \pm \sigma)}{\sigma} e^{-\frac{1}{2}\left(\frac{\mu \pm \sigma - \mu}{\sigma}\right)^2} = \mp e^{\mp 1/2} \neq 0$; so $f(x)$ has inflection points at $x = \mu \pm \sigma$.

23. (a) $\Pr(-1.3 \le Z \le 0) = A(1.3) = .4032$

(b) $\Pr(.25 \le Z) = 1 - \Pr(Z \le .25) = 1 - (.5 + A(.25)) = .4013$

(c) $\Pr(-1 \le Z \le 2.5) = A(1) + A(2.5) = .3413 + .4938 = .8351$

(d) $\Pr(Z \le 2) = .5 + A(2) = .9772$

24. (a) $A(1.5) - A(.5) = .2417$ (b) $2A(.75) = .5468$

(c) $1 - (A(.3) + .5) = .3821$ (d) $A(1) + .5 = .8413$.

25. $\mu = 6$, $\sigma = \frac{1}{2}$. (a) $\Pr(6 \le X \le 7) = A\left(\frac{7-6}{(1/2)}\right) = A(2) = .4772$

So 47.72% of births occur between 6 and 7 months.

(b) same as (a)

26. $\mu = 25{,}000$; $\sigma = 2000$

(a) $\Pr(28{,}000 < X < 30{,}000) = A\left(\frac{30{,}000 - 25{,}000}{2000}\right) - A\left(\frac{28{,}000 - 25{,}000}{2000}\right)$

$= A(2.5) - A(1.5) = .0606$

(b) $Pr(X > 29{,}000) = 1 - Pr(X \le 29{,}000)$

$= 1 - \left[.5 + A\left(\frac{29{,}000 - 25{,}000}{2000}\right)\right]$

$= 1 - [.5 - A(2)] = .0228$

27. $M = 128.2$, $\sigma = .2$. $Pr(X < 128) = 1 - \left[.5 - A\left(\frac{128.2 - 128}{.2}\right)\right]$

$= 1 - [.5 + A(1)] = .1587$

28. $M = 43$, $\sigma = 1.5$. $Pr(X < 40) = 1 - \left[.5 + A\left(\frac{93 - 40}{1.5}\right)\right]$

$= 1 - [.5 + A(2)] = .0228$

29. Let B be the amount of time the Beltway route takes and let L be the time it takes on the local route. Then $\mu_B = 25$, $\sigma_B = 5$ $\mu_L = 28$, $\sigma_L = 3$ and

$Pr(B < 30) = .5 + A\left(\frac{30 - 25}{5}\right) = .5 + A(1) = .8413;$

$Pr(L < 30) = .5 + A\left(\frac{30 - 28}{3}\right) = .5 + A\left(\frac{2}{3}\right) = .7454.$

Therefore the student should take the Beltway route.

30. (See ex. 29)

$Pr(B < 34) = .5 + A\left(\frac{34 - 25}{5}\right) = .5 + A\left(\frac{9}{5}\right)$

$Pr(L < 34) = .5 + A\left(\frac{34 - 28}{3}\right) = .5 + A(2).$

Since $2 > \frac{9}{5}$, in this case $Pr(L < 34) > Pr(B < 34)$ so she should take the local route.

31. Let X be the diameter of a randomly selected bolt.

$Pr(X > 20) = 1 - Pr(X \le 20) = 1 - \left[.5 + A\left(\frac{20 - 18.2}{.8}\right)\right]$

$= 1 - [.5 + A(2.25)] = .0122$

Therefore about 1.22% of the bolts will be discarded.

32. (a) $\Pr(500 < X < 600) = A\left(\frac{600 - 535}{100}\right) + A\left(\frac{535 - 500}{100}\right)$

$= A(.65) + A(.35) = .2422 + .1368 = .379$

Thus approximately 37.9% of the scores were between 500 and 600.

(b) $\Pr(X < t) = .5 + A\left(\frac{t - 535}{100}\right) = .9;\quad A\left(\frac{t - 535}{100}\right) = .4.$

In the table $A(1.28) = .3997 \sim .4$; so $\frac{t - 535}{100} = 1.28$; $t = .663$

Thus the top 10% consists of scores 663 and above.

33. If X has density $f(x) = ke^{-kx}$, then

$$\frac{\Pr(a \le X \le a + b)}{\Pr(a \le X)} = \frac{\int_a^{a+b} ke^{-kx}dx}{1 - \int_0^a ke^{-kx}dx} = \frac{-e^{-kx}\Big|_a^{a+b}}{1 + e^{-kx}\Big|_0^a}$$

$$= \frac{e^{-ak} - e^{-ak-bk}}{e^{-ak}} = 1 - e^{-bk} = -e^{-kx}\Big|_0^b$$

$$= \int_0^b ke^{-kx}dx = \Pr(0 \le X \le b)$$

34. $\Pr(X \le M) = \int_0^M ke^{-kx}dx = -e^{-kx}\Big|_0^M = 1 - e^{-Mk} = .5;\ e^{-Mk} = .5$

$-Mk = \ln\left(\frac{1}{2}\right) = -\ln 2;\ M = \frac{\ln 2}{k}.$

Chapter 11 Supplementary Exercises

1. $f(x) = \frac{3}{8}x^2,\ 0 \le x \le 2.$

(a) $\Pr(X \le 1) = \int_0^1 \frac{3}{8}x^2dx = \frac{1}{8}x^3\Big|_0^1 = \frac{1}{8}.$

$\Pr(1 \le X \le 1.5) = \frac{1}{8}x^3\Big|_1^{1.5} = \frac{19}{64}.$

(b) $E(X) = \int_0^2 \frac{3}{8}x^3 dx = \frac{3}{32}x^4\Big|_0^2 = \frac{3}{2}$

$V(X) = \int_0^2 \frac{3}{8}x^4 dx - \frac{9}{4} = \frac{3}{40}x^5\Big|_0^2 - \frac{9}{4} = \frac{12}{5} - \frac{9}{4} = \frac{3}{20}$

2. $f(x) = 2x - 6,\ 3 \le x \le 4.$

(a) $Pr(3.2 \le X) = \int_{3.2}^{2} (2x - 6)dx = x^2 - 6x\Big|_{3.2}^{2} = -8 + 8.96 = .96$

(b) $E(X) = \int_3^4 (2x^2 - 6x)dx = \frac{2}{3}x^3 - 3x^2\Big|_3^4 = \frac{11}{3}.$

$\int_3^4 (2x^3 - 6x^2)dx = \frac{1}{2}x^4 - 2x^3\Big|_3^4;\quad V(X) = \frac{27}{2} - \left(\frac{11}{3}\right)^2 = \frac{1}{18}.$

3. I. $e^{A-x} \ge 0$ for all x.

II. $\int_A^\infty e^{A-x}dx = \lim_{b\to\infty} \left[-e^{A-x}\Big|_A^b\right] = \lim_{b\to\infty} \left[1 - e^{A-b}\right] = 1.$ Thus

$f(x) = e^{A-x},\ x \ge A$ is a density function.

$F(x) = \int_A^x e^{A-t}dt = -e^{A-t}\Big|_A^x = 1 - e^{A-x}.$

4. $f(x) = kA^k/x^{k+1},\ k > 0,\ A > 0,\ x \ge A.$

I. Since k and A are > 0, $f(x) \ge 0$ for all $x \ge A$.

II. $\int_A^\infty (kA^k/x^{k+1})dx = \lim_{b\to\infty} \left[-A^k/x^k\Big|_A^b\right] = \lim_{b\to\infty} \left[1 - \frac{A^k}{b^k}\right] = 1.$ Thus

f(x) is a density function.

$F(X) = \int_A^x (kA/t^{k+1})dt = -A^k/t^k\Big|_A^x = 1 - \frac{A^k}{x^k}.$

5. For $n \ge 2$, any choice of $c_n > 0$ will ensure $f_n(x) \ge 0$ for all $x \ge 0$. Thus we need only $\int_0^\infty c_n x^{n-2}e^{-x/2}dx = 1.$ If n = 2 this becomes $c_2\int_0^\infty e^{-x/2}dx = 1;\ c_2 \lim_{b\to\infty} \left[-2e^{-x/2}\Big|_0^b\right] = 1;\ 2c_2 = 1$

$c = \frac{1}{2}$. For n = 4, we have $c_4\int_0^{\infty} x^2 e^{-x/2}dx = 1$. Integrating by parts twice gives $\int_0^b x^2 e^{-x/2}dx = -e^{-x/2}[2x^2 + 8x + 16]\Big|_0^b$.

Therefore, $16c_4 = 1$ and $c_4 = \frac{1}{16}$.

6. I. If k > 0, $\frac{1}{2k^3}x^2 e^{-x/k} \geq 0$ for all x.

II. Integrating by parts twice,

$\int_0^b x^2 e^{-x/k}dx = -e^{x/k}[kx^2 + 2k^2x + 2k^3]\Big|_0^b$. Thus

$\int_0^{\infty} \frac{1}{2k^3}x^2 e^{-x/k}dx = \frac{1}{2k^3}(2k^3) = 1.$

7. (a) E(X) = 1(.599) + 11(.401) = 5.01

(b) 200 samples is 20 batches of 10. Thus they can expect to run 20(5.01) ≅ 100 tests.

8. (a) E(X) = 1(.774) + 6(.226) = 2.13

(b) 200 samples is 40 batches of 5. Thus they can expect to run 40(2.13) = 85.20 tests.

9. $F(x) = 1 - \frac{1}{4}(2 - x)^2,\ 0 \leq x \leq 2.$

(a) Pr(X < 1.6) = F(1.6) = .96

(b) $\Pr(X < t) = 1 - \frac{1}{4}(2 - t)^2 = .99$; t = 1.8 (thousand gal.)

(c) $f(x) = F'(x) = \frac{1}{2}(2 - x)$.

10. $E(X) = \frac{1}{625}\int_0^5 x(x - 5)^4 dx = \frac{1}{625}\left[\frac{x}{5}(x - 5)^5 - \frac{1}{30}(x - 5)^6\right]\Big|_0^5$ (Int. by pts.)

$= \frac{(-5)^6}{30 \cdot 5^4} = \frac{5}{6} = .8333$ (hundred dollars)

Thus on average the manufacturer can expect to make

100 - 83.33 = \$16.67 on each service contract sold.

11. (a) $E(X) = \int_{20}^{25} \frac{1}{5}x\,dx = \frac{1}{10}x^2\Big|_{20}^{25} = 22.5$

$V(X) = \int_{20}^{25} \frac{1}{5}x^2 dx - 22.5^2 = \frac{x^3}{15}\Big|_{20}^{25} - 22.5^2 = 2.0833$

(b) $Pr(X \le b) = \int_{20}^{b} \frac{1}{5}\,dx = .3;\ \frac{1}{5}b - 4 = .3;\ b = 21.5$

12. $F(x) = (x^2 - 9)/16,\ 3 \le x \le 5.$

(a) $f(x) = F'(x) = \frac{x}{8},\ 3 \le x \le 5.$

(b) $Pr(a \le X) = \frac{1}{4}$, so $Pr(X \le a) = F(a) = \frac{(a^2 - 9)}{16} = \frac{3}{4};\ a = \sqrt{21}$

13. $f(x) = kx,\ 5 \le x \le 25.$

(a) We need $k > 0$ and $\int_{5}^{25} kx\,dx = 1;\ \frac{k}{2}x^2\Big|_{5}^{25} = 1;\ \frac{k}{2}(600) = 1$

$k = \frac{1}{300}$

(b) $Pr(X \ge 20) = \int_{20}^{25} \frac{1}{300}x\,dx = \frac{1}{600}x^2\Big|_{20}^{25} = .375.$

(c) $E(X) = \int_{5}^{25} \frac{1}{300}x^2\,dx = \frac{1}{900}x^3\Big|_{5}^{25} = 17.222$ (thousand dollars)

14. (a) $F(x) = \int_{0}^{x} k(k + 1)t^{k-1}(1 - t)dt = k(k + 1)\left[\frac{t^k}{k} - \frac{t^{k+1}}{k + 1}\right]\Big|_{0}^{x}$

$= (k + 1)x^k - kx^{k+1}.$

(b) $E(X) = \int_{0}^{1} k(k + 1)x^k(1 - x)dx = k(k + 1)\left[\frac{x^{k+1}}{k + 1} - \frac{x^{k+2}}{k + 2}\right]\Big|_{0}^{1}$

$= k - \frac{k(k + 1)}{k + 2} = \frac{k}{k + 2}.$

15. Points (θ, y) satisfying the given condition are precisely those points under the curve $y = \sin\theta,\ 0 \le \theta \le \pi$. This

region has area $\int_0^{\pi} \sin\theta\, d\theta = -\cos\theta \Big|_0^{\pi} = 2$. The area of the rectangle is π, so the probability that a randomly selected point falls in the region $y \le \sin\theta$ is $\frac{2}{\pi}$.

16. Since the length of the needle is 1 unit, $\sin\theta$ is the difference in the y-coordinates of the base and end of the needle (see fig. 7(a)). The needle will touch a ruled line if and only if this difference exceeds y, the vertical distance from the base to the next ruled line. To compute $\Pr(y \le \sin\theta)$, view dropping the needle as a random choice of a point (θ, y) from the square $0 \le \theta \le \pi$; $0 \le y \le 1$. Then exercise 15 applies.

17. Let X be the lifetime of the TV tube. Then

$$\Pr(Y = 0) = \Pr(X \le 3) = \int_0^3 \frac{1}{5}e^{-(1/5)x}dx = -e^{-(1/5)x}\Big|_0^3$$

$$= 1 - e^{-3/5} \sim .45119.$$

Thus $\Pr(Y = 100) \sim .54881$ and $E(Y) \sim \$54.88$.

18. Let Y be as in the hint and let X be the life span of the motor. Then $\Pr(Y = 300) = \Pr(X \le 1) = \int_0^1 \frac{1}{10}e^{-(1/10)x}dx$

$$= -e^{-(1/10)x}\Big|_0^1 = 1 - e^{-1/10} \sim .09516.$$

Thus $E(Y) \sim 300(.09516) \sim \$28.55$. Since the insurance costs $25 to buy, you should buy it for the first year.

19. $\Pr(X \le 4) = \int_0^4 ke^{-kx}dx = -e^{-kx}\Big|_0^4 = 1 - e^{-4k} = .75$; so $e^{-4k} = .25$

$-4k = \ln\left(\frac{1}{4}\right) = -2\ln 2$; $k = \frac{\ln 2}{2} \sim .34657$.

20. $E(X) = .01\int_0^{\infty} x^2e^{-x/10}dx$. Integrating by parts twice,

$\int_0^b x^2e^{-x/10}dx = -e^{-x/10}(10x^2 + 200x + 2000)\Big|_0^b$. So

$E(X) = 2000(.01) = 20$ (thousand hours) and the expected additional earnings from the machine are $20(5000) = \$100,000$. Since this amount exceeds the price, the machine should be purchased.

21. $f(x)$ is the density of a normal random variable X with $\mu = 50$, $\sigma = 8$. Thus $\Pr(30 \le X \le 50) = A\left(\frac{50 - 30}{8}\right) = A(2.5) = .4938$.

22. Let X be the length of a randomly selected part. Then

$\Pr(79.95 \le X \le 80.05) = A\left(\frac{79.99 - 79.95}{.02}\right) + A\left(\frac{80.05 - 79.99}{.02}\right)$

$= A(2) + A(3) = .4772 + .4987 = .9759.$

Hence out of a lot of 1000 parts, $1000(.9759) = 975.9$ should be within the tolerance limits; leaving 24.1 defective parts.

23. Let X be the height of a randomly selected man in the city. $\Pr(X \ge 69) = .5 + A\left(\frac{70 - 69}{2}\right) = .5 + A(.5) = .6915$. Thus about 69.15% of the men in the city are eligible.

24. Let Y be the height of a randomly selected woman from the city. $\Pr(Y \ge 69) = .5 - A\left(\frac{69 - 65}{1.6}\right) = .5 - A(2.5) = .0062$. So only about .62% of the women are eligible.

25. $\Pr(a \le Z) = .4$, then we must have $a > 0$ and $\Pr(0 \le a \le A) = A(a) = .5 - .4 = .1$. From the table, $A(.25) = .0987 \sim .1$, so $a \sim .25$.

26. Using the result of ex. 25, the cutoff grade t must satisfy $\frac{t - 500}{100} = .25$; $t = 525$.

27. (a) $\Pr(-1 \le Z \le 1) = 24(1) = .6826.$

(b) $\Pr(\mu - \sigma < X < \mu + \sigma) = \Pr(-2 < Z < 2) = .9544.$

28. (a) $\Pr(-2 \le Z \le 2) = 2A(2) = .9544$

(b) $\Pr(\mu - 2\sigma < X < \mu + 2\sigma) = \Pr(-2 < Z < 2) = .9544.$

29. (a) Let X be an exponential random variable with density $f(x) = ke^{-kx}$. Then $E(X) = \mu = \frac{1}{k}$ and $V(X) = \sigma^2 = \frac{1}{k^2}$.

Applying the inequality with n = 2 gives

$$\Pr\left(\frac{1}{k} - \frac{2}{k} \le X \le \frac{1}{k} + \frac{2}{k}\right) = \Pr\left(-\frac{1}{k} \le X \le \frac{3}{k}\right) = \Pr\left(0 \le X \le \frac{3}{k}\right) \ge 1 - \frac{1}{2^2} = \frac{3}{4}$$

(b) $$\Pr\left(0 \le X \le \frac{3}{k}\right) = \int_0^{3/k} ke^{-kx}dx = -e^{-kx}\Big|_0^{3/k} = 1 - e^{-3} \sim .95.$$

30. Let X be a normal random variable with $E(X) = \mu$ and $V(X) = \sigma^2$.

Applying the inequality with n = 2 gives

$\Pr(\mu - 2\sigma \le X \le \mu + 2\sigma) \ge 1 - \frac{1}{2^2} = \frac{3}{4}.$ The exact value is

$\Pr(\mu - 2\sigma \le X \le \mu + 2\sigma) = 2A(2) = .9544.$

Exercises 12.1

1. $p_3(x) = \sin 0 + \frac{-\sin 0}{2!}x^2 + \frac{-\cos 0}{3!}x^3 = x - \frac{1}{6}x^3.$

2. $p_3(x) = e^{-0/2} - \frac{1}{2}e^{-0/2}x + \frac{1}{4}e^{-0/2}\frac{x^2}{2!} - \frac{1}{8}e^{-0/2}\frac{x^3}{3!}$

$= 1 - \frac{1}{2}x + \frac{1}{8}x^2 - \frac{1}{48}x^3.$

3. $p_3(x) = 5e^{2(0)} + 10e^{2(0)}x + 20e^{2(0)}\frac{x^2}{2!} + 40e^{2(0)}\frac{x^3}{3!}$

$= 5 + 10x + 10x^2 + \frac{20}{3}x^3.$

4. $p_3(x) = \cos\pi + (5\sin\pi)x - (25\cos\pi)\frac{x^2}{2!} - (125\sin\pi)\frac{x^3}{3!}$

$= -1 + \frac{25}{2}x^2.$

5. $p_3(x) = 1 + 2x - \frac{4x^2}{2!} + 24\frac{x^3}{3!} = 1 + 2x - 2x^2 + 4x^3.$

6. $p_3(x) = \frac{1}{2} - \frac{1}{2^2}x + \frac{2}{2^3}\frac{x^2}{2!} - \frac{6}{2^4}\frac{x^3}{3!} = \frac{1}{2} - \frac{1}{4}x + \frac{x^2}{8} - \frac{x^3}{16}.$

7. $f(x) = xe^{3x}$; $f'(x) = e^{3x} + 3xe^{3x}$;

$f''(x) = 3e^{3x} + 3e^{3x} + 9xe^{3x} = 6e^{3x} + 9xe^{3x}$

$f'''(x) = 18e^{3x} + 9e^{3x} + 27xe^{3x} = 27e^{3x} + 27xe^{3x}.$

$p_3(x) = 0 + (1 + 0)x + (6 + 0)\frac{x^2}{2!} + (27 + 0)\frac{x^3}{3!} = x + 3x^2 + \frac{9}{2}x^3.$

8. $p_3(x) = 1 - \frac{1}{2}x - \frac{1}{4}\frac{x^2}{2!} - \frac{3}{8}\frac{x^3}{3!} = 1 - \frac{1}{2}x - \frac{1}{8}x^2 - \frac{1}{16}x^3.$

9. $p_4(x) = e^0 + e^0x + e^0\frac{x^2}{2!} + e^0\frac{x^3}{3!} + e^0\frac{x^4}{4!}$

$= 1 + x + \frac{x^2}{2} + \frac{x^3}{6} + \frac{x^4}{24}$

$e^{.01} \sim 1 + .01 + \frac{(.01)^2}{2} + \frac{(.01)^3}{6} + \frac{(.01)^4}{24} = 1.01005$

10. $f(x) = \ln(1 - x)$; $f'(x) = -\dfrac{1}{(1 - x)}$; $f''(x) = -\dfrac{1}{(1 - x)^2}$;

$f'''(x) = -\dfrac{2}{(1 - x)^3}$; $f^4(x) = -\dfrac{6}{(1 - x)^4}$

$p_4(x) = 0 - x - \dfrac{x^2}{2!} - \dfrac{2x^3}{3!} - \dfrac{6x^4}{4!} = -x - \dfrac{1}{2}x^2 - \dfrac{1}{3}x^3 - \dfrac{1}{4}x^4.$

$\ln(.9) - \ln(1 - .1) \sim -.1 - \dfrac{(.1)^2}{2} - \dfrac{(.1)^3}{3} - \dfrac{(.1)^4}{4} = -.10536$

11.

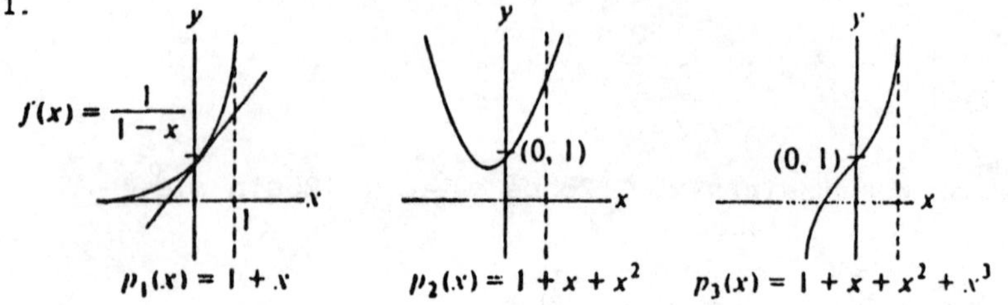

12. Lift graph from original

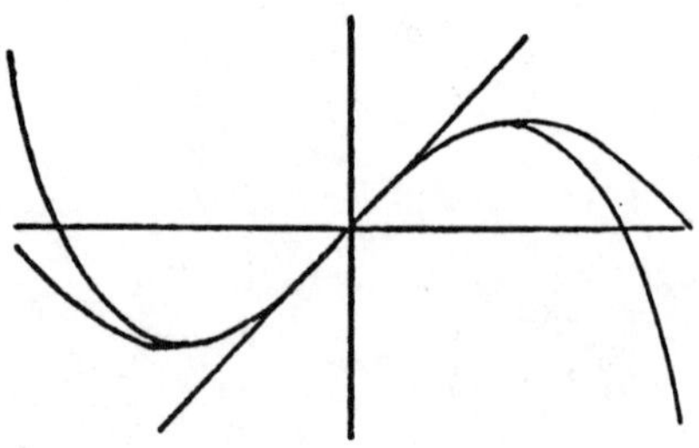

13. $p_n(x) = e^0 + e^0x + e^0\dfrac{x^2}{2!} + \ . \ . \ . + e^0\dfrac{x^n}{n!}$

$= 1 + x + \dfrac{x^2}{2} + \dfrac{x^3}{6} + \ . \ . \ . + \dfrac{x^n}{n!}$

14. $p_0(x) = 0^2 + 2(0) + 1 = 1$; $p_1(x) = 1 + (2(0) + 2)x = 1 + 2x$;

$p_2(x)\ 1 + 2x + 2\dfrac{x^2}{2!} = 1 + 2x + x^2$

Since $f''(x) = 0$ for all $n \geq 2$, $p_n(x) = p_2(x) = f(x)$ for all $n \geq 2$.

15. $f(x) = \ln(1 + x^2)$; $f'(x) = \dfrac{2x}{1 + x^2}$;

$$f''(x) = \frac{2(1 + x^2) - 4x^3}{(1 + x^2)^2} = \frac{-4x^3 + 2x^2 + 2}{(1 + x^2)^2}$$

$$p_2(x) = \ln(1) + (0)x + 2\frac{x^2}{2!} = x^2.$$

$$\int_0^{1/2} \ln(1 + x^2)dx \sim \int_0^{1/2} x^2 dx = \frac{x^3}{3}\Big|_0^{1/2} = \frac{1}{24}.$$

16. $f(x) = \sqrt{\cos x}$; $f'(x) = \dfrac{-\sin x}{2\sqrt{\cos x}}$;

$$f''(x) \frac{-\cos x(2(\sqrt{\cos x}) + \sin^2 x\ (1/2)(\cos x)^{-1/2}}{4\cos x}.$$

$$p_2(x) = 1 + \frac{0}{2(1)}x - \frac{1}{2}\frac{x^2}{2!} = 1 - \frac{1}{4}x^2$$

$$\int_{-1}^{1} \cos x\ dx \sim \int_{-1}^{1} \left(1 - \frac{1}{4}x^2\right)dx = x - \frac{1}{12}x^3\Big|_{-1}^{1} = \frac{11}{6}.$$

17. $f(x) = \dfrac{1}{5 - x}$; $f'(x) = \dfrac{1}{(5 - x)^2}$; $f''(x) = \dfrac{2}{(5 - x)^3}$;

$$f'''(x) = \frac{6}{(5 - x)^4}$$

$$p_3(x) = \frac{1}{5 - 4} + \frac{1}{(5 - 4)^2}(x - 4) + \frac{2}{(5 - 4)^3}\frac{(x - 4)^2}{2!} + \frac{6}{(5 - 4)^4}\frac{(x - 4)^3}{3!}$$

$$= 1 + (x - 4) + (x - 4)^2 + (x - 4)^3$$

18. $f(x) = \ln x$; $f'(x) = \dfrac{1}{x}$; $f''(x) = \dfrac{-1}{x^2}$; $f'''(x) = \dfrac{2}{x^3}$; $f^{(4)}(x) = \dfrac{-6}{x^4}$

$$p_4(x) = \ln(1) + \frac{1}{1}(x - 1) - \frac{1}{1^2}\frac{(x - 1)^2}{2!} + \frac{2}{1^3}\frac{(x - 1)^3}{3!} + \frac{6}{1^4}\frac{(x - 1)^4}{4!}$$

$$= (x - 1) - \frac{1}{2}(x - 1)^2 + \frac{1}{3}(x - 1)^3 - \frac{1}{4}(x - 1)^4.$$

19. $f(x) = \cos x$; $f'(x) = -\sin x$; $f''(x) = -\cos x$; $f^{(4)}(x) = \cos x$

$$p_3(x) = \cos \pi - \sin \pi\,(x - \pi) - \cos \pi \,\frac{(x - \pi)^2}{2!} + \sin \pi \,\frac{(x - \pi)^3}{3!}$$

$$= -1 + \frac{1}{2}(x - \pi)^2$$

$$p_4(x) = p_3(x) + \cos \pi \,\frac{(x - 4)^4}{4!}$$

$$= -1 + \frac{1}{2}(x - \pi)^2 - \frac{1}{24}(\pi - 4)^4.$$

20. $f(x) = x^3 + 3x - 1$; $f'(x) = 3x^2 + 3$; $f''(x) = 6x$; $f'''(x) = 6$; $f^{(n)}(x) = 0$ for all $n > 3$.

$$p_3(x) = (-1)^3 + 3(-1) - 1 + (3(-1)^2 + 3)(x + 1) + 6(-1)\frac{(x + 1)^2}{2!} + 6\frac{(x + 1)^3}{3!}$$

$$= -5 + 6(x + 1) - 3(x + 1)^2 + (x + 1)^3.$$

$p_4(x) = p_3(x)$.

21. $f(x) = \sqrt{x}$; $f'(x) = \dfrac{1}{2\sqrt{x}}$; $f''(x) = -\frac{1}{4}x^{-3/2}$

$$p_2(x) = \sqrt{9} + \frac{1}{2\sqrt{9}}(x - 9) - \frac{1}{4}9^{-3/2}\,\frac{(x - 9)^2}{2!}$$

$$= 3 + \frac{1}{6}(x - 9) - \frac{1}{216}(x - 9)^2$$

$$p_2(9.3) = 3 + \frac{.3}{6} - \frac{(.3)^2}{216} = 3.04958$$

22. $f(x) = \ln x$; $f'(x) = \frac{1}{x}$; $f''(x) = -\dfrac{1}{x^2}$

$$p_2(x) = \ln(1) + \frac{1}{1}(x - 1) - \frac{1}{1^2}\,\frac{(x - 1)^2}{2!}$$

$$= (x - 1) - \frac{1}{2}(x - 1)^2.$$

$$p_2(.8) = -.2 - \frac{1}{2}(-.2)^2 = -.22.$$

23. $f(x) = x^4 + x + 1$; $f'(x) = 4x^3 + 1$; $f''(x) = 12x^2$; $f'''(x) = 24x$; $f^{(4)}(x) = 24$; $f^{(n)}(x) = 0$ all $n > 4$.

$p_0(x) = 2^4 + 2 + 1 = 19$

$p_1(x) = 19 + (4(2)^3 + 1)(x - 2) = 19 - 33(x - 2)$

$p_2(x) = 19 + 33(x - 2) + 12(2)^2 \dfrac{(x - 2)^2}{2!}$

$= 19 + 33(x - 2) + 24(x - 2)^2$

$p_3(x) = 19 + 33(x - 2) + 24(x - 2)^2 + 24(2)\dfrac{(x - 2)^3}{3!}$

$= 19 + 33(x - 2) + 24(x - 2)^2 + 8(x - 2)^3$

$p_4(x) = p_3(x) + 24\dfrac{(x - 2)^4}{4!}$

$= 19 + 33(x - 2) + 24(x - 2)^2 + 8(x - 2)^3 + (x - 2)^4$

$p_n(x) = p_4(x)$ for all $n > 4$.

24. $f(x) = \dfrac{1}{x}$; $f'(x) = -\dfrac{1}{x^2}$; $f''(x) = \dfrac{2}{x^3}$; $f'''(x) = -\dfrac{3\cdot 2}{x^4}$;

$f^{(n)}(x) = \dfrac{(-1)^n n!}{x^n}$.

$p_n(x) = 1 - (x - 1) + \dfrac{2(x - 1)^2}{2!} - \dfrac{3!(x - 1)^3}{3!}$

$+ \ldots + (-1)^n n!\dfrac{(x - 1)^n}{n!}$

$= 1 - (x - 1) + (x - 1)^2 - (x - 1)^3$

$+ \ldots + (-1)^n(x - 1)^n$.

25. $f(z) = \dfrac{1}{\sqrt{2\pi}} e^{-z^2/2}$; $f'(z) = \dfrac{-z}{\sqrt{2\pi}} e^{-z^2/2}$;

$f''(z) = -\dfrac{1}{\sqrt{2\pi}} e^{-z^2/2} + \dfrac{z^2}{\sqrt{2\pi}} e^{-z^2/2} = \dfrac{e^{-z^2}}{\sqrt{2\pi}} \left[z^2 - 1\right]$

$p_2(z) = \dfrac{1}{\sqrt{2\pi}} - \dfrac{1}{2\sqrt{2\pi}} (z - 0)^2$.

26. $\displaystyle\int_0^2 \left[\frac{1}{\sqrt{2\pi}} - \frac{1}{2\sqrt{2\pi}} z^2 \right] dz = \frac{z}{\sqrt{2\pi}} - \frac{z^3}{6\sqrt{2\pi}} \Big|_0^2 = \frac{2}{\sqrt{2\pi}} - \frac{4}{3\sqrt{2\pi}}$

$\sim \dfrac{2}{.3969} - \dfrac{4}{3(.3969)} = .2385$

The value from the table is .2386.

27. The expression on the right must be the Taylor expansion of $f(x)$ at $x = 0$. Therefore, $f''(0) = -5$ and $f'''(0) = 7$.

28. The expression of the right must be the Taylor expansion of $f(x)$ at $x = 1$. Therefore, $f''(1) = 3$ and $f'''(1) = -5$.

29. (a) If $f(x) = \cos x$, then $f^{(4)}(x) = \cos x$ as well, so $|f^{(4)}(c)| \le |\cos(c)| \le 1$ for all c.

(b) $R_3(.12) = \frac{\cos(c)}{4!}(.12)^4$ for some c between 0 and .12. From part (a), it follows that $|R_3(.12)| = |\text{error in the approximation}| \le \frac{1}{4!}(.12)^4 = 8.64 \times 10^{-6}$

30. $|R_4(.1)| = |\text{error in the approximation}| = \left|\frac{f^{(5)}(c)}{5!}(.1)^5\right|$ for some c, $0 \le c \le .1$.

Now $f^{(5)}(c) = e^c \le e^{.1}$ for all $0 \le c \le .1$, since e^x is increasing on [0, .1]. Therefore

$|\text{error in approximation}| \le \frac{e^{.1}}{5!}(.1)^5 \sim 2.5 \times 10^{-7}$.

31. (a) $R_2(x) = \frac{\frac{3}{8}c^{-5/2}}{3!}(x-9)^3 = \frac{c^{-5/2}}{18}(x-9)^3$ for some c between 9 and x.

(b) The function $f^{(3)}(x) = \frac{3}{8}x^{-5/2}$ is positive and decreasing for $x > 0$. Thus for all $x \ge 9$, $|f^{(3)}(x)| \le f^{(3)}(9) = \frac{3}{8 \cdot 3^5}$

$= \frac{1}{648}$.

(c) $|\text{error}| = R_2(9.3) \le \frac{\frac{1}{648}}{3!}(.3)^3$ (using part (b))

$= \frac{1}{144} \times 10^{-3} < 7 \times 10^{-6}$.

32. (a) $f^{(3)}(c) = \frac{2}{c^3} \le \frac{2}{(.8)^3} < 4$ for all $c \ge .8$, since

$f^{(3)}(x) = \frac{2}{c^3}$ is positive and decreasing for $x > 0$.

(b) $R_2(x) = \frac{(2/c)^3}{3!}(x - 1)^3 = \frac{1}{6c^3}(x - 1)^3$ for some c between 1 and x. By part (a), $|R_2(.8)| \le \left|\frac{4}{3!}(.8 - 1)^3\right| = \frac{16}{3} \times 10^{-3}$
$< .0054.$

Exercises 12.2

1. Let $f(x) = x^2 - 5$, $f'(x) = 2x$. $x_0 = 2$; $x_1 = 2 - \frac{2^2 - 5}{2(2)} = 2.25$

$x_2 = 2.25 - \frac{(2.25)^2 - 5}{2(2.25)} = 2.2361$; $x_3 = 2.2361 - \frac{(2.2361)^2 - 5}{2(2.2361)}$
$= 2.23607$

2. Let $f(x) = x^2 - 7$; $f'(x) = 2x$. $x_0 = 3$

$x_1 = 3 - \frac{3^2 - 7}{2(3)} = 2.667$; $x_2 = 2.667 - \frac{2.667^2 - 7}{2(2.667)} = 2.6458$

$x_3 = 2.6458 - \frac{2.6458^2 - 7}{2(2.6458)} = 2.64575.$

3. Let $f(x) = x^3 - 6$; $f'(x) = 3x^2$. $x_0 = 2$

$x_1 = 2 - \frac{2^3 - 6}{3(2^2)} = 1.8333$; $x_2 = 1.8333 - \frac{1.8333^2 - 6}{3(1.8333)^2} = 1.81726$

$x_3 = 1.81726 - \frac{1.81726^2 - 6}{3(1.81726)^2} = 1.81712.$

4. Let $f(x) = x^3 - 11$; $f'(x) = 3x^2$. $x_0 = 2$

$x_1 = 2 - \frac{2^3 - 11}{3(2)^2} = 2.25$; $x_2 = 2.25 - \frac{2.25^3 - 11}{3(2.25)^2} = 2.2243$

$x_3 = 2.6477 - \frac{2.6477^3 - 11}{3(2.6477)^2} = 2.22398.$

5. $f(x) = x^2 - x - 5$; $f'(x) = 2x - 1$. $x_0 = 2$

$x_1 = \frac{2^2 - 2 - 5}{2(2) - 1} = 3$; $x_2 = 3 - \frac{3^2 - 3 - 5}{2(3) - 1} = 2.8$

$x_3 = 2.8 - \frac{2.8^2 - 2.8 - 5}{2(2.8) - 1} = 2.7913$

6. $f(x) = x^2 + 3x - 11$; $f'(x) = 2x + 3$. $x_0 = -5$

$x_1 = -5 - \frac{(-5)^2 + 3(-5) - 11}{2(-5) + 3} = -5.1429$

$$x_2 = -5.1429 - \frac{(-5.1429)^2 - 3(-5.1429) - 11}{2(-5.1429) + 3} = -5.1401$$

$x_3 = -5.14005.$

7. $f(x) = \sin x + x^2 - 1$; $f'(x) = \cos x + 2x$.

   $x_0 = 0$, $x_1 = 1$, $x_2 = .66875$, $x_3 = .63707$

8. $f(x) = e^x + 10x - 3$; $f'(x) = e^x + 10$

   $x_0 = 0$, $x_1 = .18182$, $x_2 = .18025$, $x_3 = 1.8025$

9.

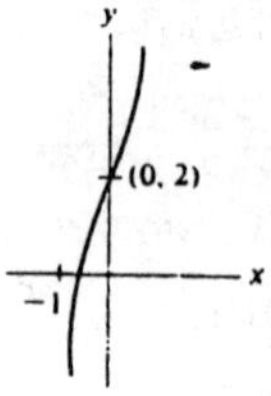

$x_0 = -1$
$x_1 = -.8$
$x_2 = -.77143$
$x_3 = -.77092$

10.

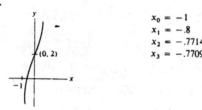

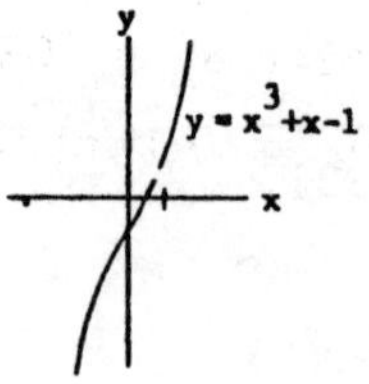

If $x_0 = 0$, then $x_1 = 1$, $x_2 = .75$, $x_3 = .68605$.

11. $f(x) = e^{-x} - x^2$; $f'(x) = -e^{-x} - 2x$

    $x_0 = 1$, $x_1 = .73304$, $x_2 = .70381$, $x_3 = .70347$.

12. $f(x) = e^{5-x} + x - 10$; $f'(x) = -e^{5-x} + 1$. There are 2 roots. Starting with $x_0 = 3$ gives $x_1 = 3.06089$, $x_2 = 3.06315$, $x_3 = 3.06315$. Starting with $x_0 = 9$ gives $x_1 = 10$, $x_2 = 9.932$, $x_3 = 9.99322$.

13. The internal rate of return i satisfies the equation $500(1+i)^3 - 100(1+i)^2 - 200(1+i) - 300 = 0$. Putting $x = 1 + i$, $f(x) = 500x^3 - 100x^2 - 200x - 300$ $f'(x) = 1500x^2 - 200x - 200$; $x_0 = 1.1$, we have $x_1 = 1.08244$, $x_2 = 1.08208$, $x_3 = 1.08208$. Thus i = .0821 or 8.21%.

14. The internal rate of return satisfies the equation $1000(1+i)^2 - 10(1+i) - 1050 = 0$. Putting $x = 1 + i$, $f(x) = 1000x^2 - 10x - 1050$, $f'(x) = 2000x - 10$, $x_0 = 1.1$, we have $x_1 = 1.03196$, $x_2 = 1.02971$, $x_3 = 1.02971$. Thus i = .02971 or 2.971%.

15. The interest rate i satisfies the equation $f(i) = 0$. where $f(i) = 563i + 116((1+i)^{-5} - 1)$; $f'(i) = 563 - 580(1+i)^{-6}$. Starting with $i_0 = .02$ gives $i_1 = .01323$, $i_2 = .01062$, $i_3 = .01003$. Thus the monthly interest rate is about 1%.

16. The interest rate i satisfies $f(i) = 0$, where $f(i) = 100{,}050i + 900((1+i)^{-240} - 1)$ $f'(i) = 100{,}050 - 216{,}000(1+i)^{-241}$. Starting with $i_0 = .02$ gives $i_1 = .00871$, $i_2 = .00757$, $i_3 = .00750$. Thus the monthly interest rate is about .75%.

17.

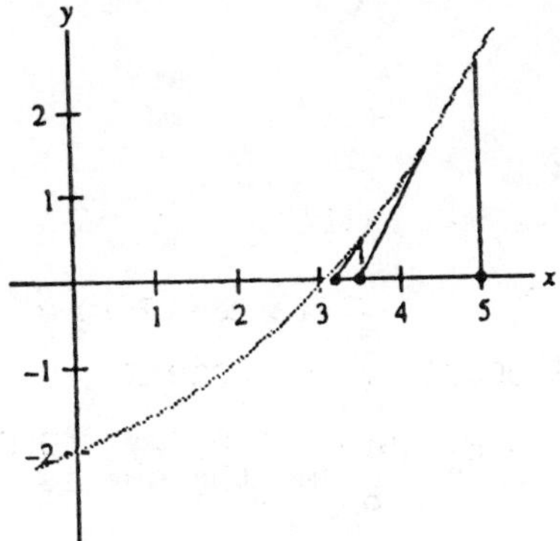

18.

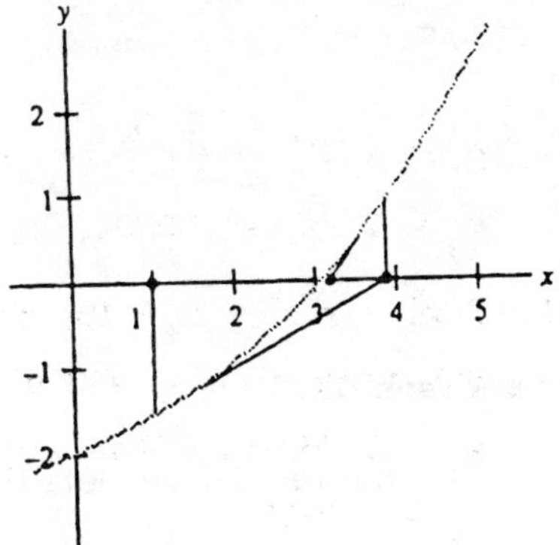

19. $m = 4(\text{slope}) = f'(x)$ at $x = 3$, $f(3) = 17$;

$$x_1 = x_0 - \frac{f(x_0)}{f'(x_0)} = 3 - \frac{1.7}{4} = -\frac{5}{4}.$$

20. $m = -2$ (slope) $= f'(x)$ at $x = 1$, $f(1) = 2$;

$$x_1 = x_0 - \frac{f(x_0)}{f'(x_0)} = 1 - \frac{2}{(-2)} = 2.$$

21. $x_0 > 0$

22. $-1 < x_0 < 1$

23. Given any x_0, x_1 will be a zero of $f(x)$.

$$x_1 = x_0 - \frac{mx_0 + b}{m}, \quad f(x_1) = m\left(x_0 - \frac{mx_0 + b}{m}\right) + b = 0.$$

24. Then $x_1 = x_2 = \ldots = x_0$, since $x_1 = x_0 - \frac{0}{f'(x_0)} = x_0$.

25. $f(x) = x^{1/3}$; $f'(x) = \frac{1}{3}x^{-2/3}$. $x_0 = 1$; $x_1 = 1 - \frac{1}{(1/3)} = -2$

$x_2 = -2 - \dfrac{\sqrt[3]{2}}{\left(\frac{1}{3}\right)2^{-2/3}} = 4$; $x_3 = 4 - 12 = -8$. The iterates diverge.

26. $x_0 = 1$; $x_1 = 1 - \dfrac{1}{\left(\frac{1}{2}\right)(1)} = -1$; $x_2 = -1 - \dfrac{-\sqrt{1}}{\left(\frac{1}{2}\right)} = 1$; $x_3 = -1$, etc.

Exercises 12.3

1. The series is geometric with $a = 1$, $r = \frac{1}{6}$, so the sum is $\dfrac{1}{1 - \frac{1}{6}} = \dfrac{6}{5}$.

2. The series is geometric with a = 1, $r = \frac{3}{4}$, so the sum is

$$\frac{1}{1 - \frac{3}{6}} = 4.$$

3. $a = 1,\ r = -\frac{1}{9};\ \text{sum} = \frac{1}{1 + \frac{1}{9}} = \frac{9}{10}.$

4. $a = 1,\ r = \frac{1}{8};\ \text{sum} = \frac{1}{1 - \frac{1}{8}} = \frac{8}{7}.$

5. $a = 2,\ r = \frac{1}{3};\ \text{sum} = \frac{2}{1 - \frac{1}{3}} = 3.$

6. $a = 3,\ r = \frac{2}{5};\ \text{sum} = \frac{3}{1 - \frac{2}{5}} = 5.$

7. $a = \frac{1}{5},\ r = \frac{\frac{1}{5^4}}{\frac{1}{5}} = \frac{1}{5^3} = \frac{1}{125};\ \text{sum} = \frac{\frac{1}{5}}{1 - \frac{1}{125}} = \frac{25}{124}.$

8. $a = \frac{1}{3^2} = \frac{1}{9};\ r = -\frac{\frac{1}{3^3}}{\frac{1}{3^2}} = -\frac{1}{3};\ \text{sum} = \frac{\frac{1}{9}}{1 + \frac{1}{3}} = \frac{1}{12}.$

9. $a = \frac{1}{3^2} = \frac{1}{9};\ r = \frac{-\frac{3^2}{7}}{3} = -\frac{9}{21};\ \text{sum} = \frac{3}{1 + \frac{9}{21}} = \frac{21}{10}.$

10. $a = 6;\ r = \frac{-1.2}{6} = -.2;\ \text{sum} = \frac{6}{1 + .2} = 5.$

11. $a = \frac{2}{5^4} = \frac{2}{625};\ r = \frac{\frac{-2^4}{5^5}}{\frac{2}{5^4}} = \frac{-2^3}{5} = -\frac{8}{5}.$ Since $|r| > 1$, the series diverges.

12. $a = \frac{3^2}{2^5} = \frac{9}{32}$; $r = \dfrac{\frac{3^4}{2^8}}{\frac{3^2}{2^5}} = \frac{9}{8}$; since $\left|\frac{9}{8}\right| > 1$, the series diverges.

13. $a = 5$; $r = \frac{4}{5}$; sum $= \dfrac{5}{1 - \frac{4}{5}} = 25$.

14. $a = \frac{5^3}{3} = \frac{125}{3}$; $r = \dfrac{-\frac{5^5}{3^4}}{\frac{5^3}{3}} = -\frac{25}{27}$; sum $= \dfrac{\frac{125}{3}}{1 + \frac{25}{27}} = \frac{1125}{52}$.

15. $.2727\underline{27}\ldots = \frac{27}{100} + \frac{27}{100^2} + \frac{27}{100^3} + \ldots$

This is a geometric series with $a = \frac{27}{100}$; $r = \frac{1}{100}$.

Therefore $.2727\underline{27}\ldots = \dfrac{\frac{27}{100}}{1 - \frac{1}{100}} = \frac{3}{11}$.

16. $.173\underline{173}\ldots = \frac{173}{1000} + \frac{173}{1000^2} + \ldots$

This is a geometric series with $a = \frac{173}{1000}$; $r = \frac{1}{1000}$.

Therefore, $.173\underline{173}\ldots = \dfrac{\frac{173}{1000}}{1 - \frac{1}{1000}} = \frac{173}{999}$.

17. $.222\underline{2}\ldots = \frac{2}{10} + \frac{2}{10^2} + \frac{2}{10^3} + \ldots$ This is a geometric series with $a = \frac{1}{5}$; $r = \frac{1}{10}$. Therefore, $.222\underline{2} = \dfrac{\frac{3}{20}}{1 - \frac{1}{10}} = \frac{2}{9}$.

18. $.1515\underline{15}\ldots = \frac{15}{100} + \frac{15}{100^2} + \frac{15}{100^3} + \ldots$ This is a geometric

series with $a = \frac{3}{20}$; $r = \frac{1}{100}$. Therefore,

$$.1515\underline{15} = \frac{\frac{3}{20}}{1 - \frac{1}{100}} = \frac{5}{33}.$$

19. $.011\underline{011}\ldots = \frac{11}{1000} + \frac{11}{1000^2} + \ldots$ This is a geometric series with $a = \frac{11}{1000}$ and $r = \frac{1}{1000}$. Its sum is $\frac{\frac{11}{1000}}{1 - \frac{1}{1000}} = \frac{11}{999}$.

Therefore, $4.011\underline{011}\ldots = 4 + \frac{11}{999} = \frac{4007}{999}$.

20. $.44\underline{4}\ldots = \frac{4}{10} + \frac{4}{10^2} + \frac{4}{10^3} + \ldots$ This is a geometric series with $a = \frac{2}{5}$, $r = \frac{1}{10}$. Its sum is $\frac{\frac{2}{5}}{1 - \frac{1}{10}} = \frac{4}{9}$. Therefore,

$5.44\underline{4}\ldots = 5 + \frac{4}{9} = \frac{49}{9}$.

21. $.99\underline{9}\ldots = \frac{9}{10} + \frac{9}{10^2} + \frac{9}{10^3} + \ldots$ This is a geometric series with $a = \frac{9}{10}$, $r = \frac{1}{10}$. Therefore, $.99\underline{9} = \frac{\frac{9}{10}}{1 - \frac{1}{10}} = 1$.

22. $.1212\underline{1212}\ldots = \frac{.1212}{1 - .0001} = \frac{1212}{9999} = \frac{4}{33}$.

23. The additional spending would be

$10(.95) + 10(.95)^2 + 10(.95)^3 + \ldots = \frac{10(.95)}{1 - .95} = 190$ (billion dollars)

24. The effect on spending would be

$20(.98) + 20(.98)^2 + 20(.98)^3 + \ldots = \frac{20(.98)}{1 - .98} = 980$ (billion dollars)

In this case, the "multiplier" is $\frac{.98}{1 - .98} = 49$.

25. (a) $\sum_{k=1}^{\infty} 100(1.01)^{-k}$ (b) $a = 100$; $r = \frac{1}{1.01}$

$$\text{sum} = \frac{100}{1 - \frac{100}{101}} = \$10,100.$$

26. The capital value is $\sum_{i=1}^{\infty} P(1 + r)^{-i} = \sum_{i=0}^{\infty} P(1 + r)^{-i} - P$

$$= \frac{P}{1 - \frac{1}{1 + r}} - P = \frac{P}{r}.$$

27. Total distance is :

$$1,000,000[1 + (.28) + (.28)^2 + \ . \ . \ . + (.28)^n + \ . \ . \ . \ ].$$

$$= 1.000,000\left[\frac{1}{1 - .28}\right] = \$1,388,889.$$

28. Total distance $= 6 + .7(6) + .7(6) + .7(.7(6)) + \ . \ . \ .$

$$= 6 + 2(6).7[1 + .7 + (.7)^2 + (.7)^3 + \ . \ . \ . \ ]$$

$$= 6 + 8.4\left(\frac{1}{1 - .7}\right) = 34 \text{ feet.}$$

29. $6 + 6(.7) + 6(.7)^2 + 6(.7)^3 + \ . \ . \ . = \frac{6}{1 - .7} = 20$ mg.

30. $2 + 2(.8) + 2(.8)^2 + 2(.8)^3 + \ . \ . \ . = 2\left(\frac{1}{1 - .8}\right) = 2(5) = 10$mg
Since we wish to know before a dose is given,
so, $10 - 2 = 8$mg.

31. $M + \frac{3}{4}M + \left(\frac{3}{4}\right)^2 + \ . \ . = M\left(\frac{1}{1 - .75}\right) = 4M$
$4M = 20$, so $M = 5$mg.

32. $M + M(1 - q) + M(1 - q)^2 + \ . \ . \ . = \frac{M}{1 - (1 - q)} = \frac{M}{q}$ mg.

33. (a) $s_{10} = 3 - \frac{5}{10} = 2.5$

(b) years since $\lim_{n \to \infty}\left(3 - \frac{5}{n}\right) = 3$

34. (a) $s_{10} = 10 - \frac{1}{9} = \frac{89}{9}$

(b) no since $\lim_{n\to\infty}\left(n - \frac{1}{n}\right)$ does not exist.

35. $\dfrac{1}{1 - \frac{5}{6}} = 6$ 36. $\dfrac{7}{1 - \frac{1}{10}} = \frac{70}{9}$ 37. $\dfrac{\frac{1}{5}}{1 - \frac{1}{5}} = \frac{1}{4}$

38. $\dfrac{1}{1 + \frac{1}{3}} = \frac{3}{4}$ 39. $\dfrac{3}{1 + \frac{3}{5}} = \frac{15}{8}$

40. $\displaystyle\sum_{k=1}^{\infty}\left(\frac{1}{3}\right)^{2k} = \sum_{k=1}^{\infty}\left(\frac{1}{9}\right)^{k} = \frac{\frac{1}{9}}{1 - \frac{1}{9}} = \frac{1}{8}$

41. (a) $(1 - r)(a + ar + ar^2 + \ldots + ar^n)$

$= a + ar + ar^2 + \ldots + ar^n - ar - ar^2 - \ldots - ar^n - ar^{n+1}$

$= a - ar^{n+1}$.

Thus $a + ar + ar^2 + \ldots + ar^n = \dfrac{a - ar^{n+1}}{1 - r} = \dfrac{a}{1 - r} - \dfrac{ar^{n+1}}{1 - r}$.

(b) As $n \to \infty$, $\dfrac{ar^{n+1}}{1 - r}$ approaches 0 if $|r| < 1$. Hence, in this case, $\displaystyle\sum_{k=0}^{\infty} ar^k = \lim_{n\to\infty}\sum_{k=0}^{n} ar^k = \lim_{n\to\infty}\left[\frac{a}{1 - r} - \frac{ar^{n+1}}{1 - r}\right] = \frac{a}{1 - r}$.

(c) If $|r| > 1$, $\left|\dfrac{ar^{n+1}}{1 - r}\right|$ increases without bound as $n \to \infty$. Thus in this case, the series diverges.

(d) If $r = 1$, then the series is $a + a + a + \ldots$ which clearly diverges. If $r = -1$, then the expression in part (a) is $\frac{a}{2} - \frac{a(-1)^n}{2}$. Thus the partial sums alternate between 0 and a and $\displaystyle\lim_{n\to\infty}\sum_{k=1}^{\infty} ar^k$ does not exist.

42. From the limit it follows that $\displaystyle\sum_{k=1}^{\infty}\frac{1}{k} > \sum_{k=1}^{\infty}\frac{1}{2}$ which diverges.

Exercises 12.4

1. $\displaystyle\int_1^b \frac{3}{\sqrt{x}}\,dx = 6x^{1/2}\Big|_1^b = 6b^{1/2} - 6.$ Thus $\displaystyle\int_1^\infty \frac{3}{\sqrt{x}}\,dx$ diverges, so $\displaystyle\sum_{k=1}^\infty \frac{3}{\sqrt{k}}$ diverges.

2. $\displaystyle\int_1^b \frac{5}{x^{3/2}}\,dx = -10x^{-1/2}\Big|_1^{10} = 10 - 10b^{-1/2}.$ Thus $\displaystyle\int_1^\infty \frac{5}{x^{3/2}}\,dx = 10$ and $\displaystyle\sum_{k=1}^\infty \frac{5}{k^{3/2}}$ converges.

3. $\displaystyle\int_2^b \frac{1}{(x-1)^3}\,dx = -\frac{1}{2}(x-1)^{-2}\Big|_2^b = \frac{1}{2} - \frac{1}{2}(b-1)^{-2}.$ Thus $\displaystyle\int_2^\infty \frac{1}{(x-1)^3}\,dx = \frac{1}{2}$ and $\displaystyle\sum_{k=2}^\infty \frac{1}{(k-1)^3}$ converges.

4. $\displaystyle\int_0^b \frac{7}{x+100}\,dx = 7\ln(x+100)\Big|_0^b = 7\ln(b+100) - 7\ln 100.$ Thus $\displaystyle\int_0^\infty \frac{7}{x+100}\,dx$ diverges, so $\displaystyle\sum_{k=0}^\infty \frac{7}{k+100}$ diverges.

5. $\displaystyle\int_1^b \frac{2}{5x-1}\,dx = \frac{2}{5}\ln(5x-1)\Big|_1^b = \frac{2}{5}\ln(5b-1) - \frac{2}{5}\ln 4.$ Thus $\displaystyle\int_1^\infty \frac{2}{5x-1}\,dx$ diverges, so $\displaystyle\sum_{k=1}^\infty \frac{2}{5k-1}$ diverges.

6. $\displaystyle\int_2^b \frac{1}{x\sqrt{\ln x}}\,dx = 2\sqrt{\ln x}\Big|_2^b = 2\sqrt{\ln b} - 2\sqrt{\ln 2};$ so $\displaystyle\sum_{k=2}^\infty \frac{1}{k\sqrt{\ln k}}$

diverges.

7. $$\int_2^{\infty} \frac{x}{(x^2+1)^{3/2}}\,dx = -(x^2+1)^{-1/2}\Big|_2^b = \frac{1}{\sqrt{5}} - (b^2+1)^{-1/2};$$

$$\sum_{k=2}^{\infty} \frac{k}{(k^2+1)^{3/2}}$$ converges.

8. $$\int_1^b \frac{1}{(2x+1)^3}\,dx = -\frac{1}{4}(2x+1)^{-2}\Big|_1^b = \frac{3}{4} - \frac{1}{4}(2b+1)^{-2};$$

so $$\sum_{k=1}^{\infty} \frac{1}{(2k+1)^3}$$ converges.

9. $$\int_2^b \frac{1}{x(\ln x)^2}\,dx = -(\ln x)^{-1}\Big|_2^b = \frac{1}{\ln 2} - \frac{1}{\ln b};\text{ so } \sum_{k=2}^{\infty} \frac{1}{k(\ln k)^2}\,dx$$

converges.

10. $$\int_1^b \frac{1}{9x^2}\,dx = -\frac{1}{9}x^{-1}\Big|_1^b = \frac{1}{9} - \frac{1}{9b};\text{ so } \sum_{k=1}^{\infty} \frac{1}{(3k)^2}\text{ converges.}$$

11. $$\int_1^b e^{3-x}dx = -e^{3-x}\Big|_1^b = e^2 - e^{3-b};\text{ so } \sum_{k=1}^{\infty} e^{3-k}\text{ converges.}$$

12. $$\int_1^b e^{-2x-1}dx = -\frac{1}{2}e^{-2x-1}\Big|_1^b = \frac{1}{2}e^{-3} - \frac{1}{2}e^{-2b-1};\text{ so } \sum_{k=1}^{\infty} e^{-2k-1}$$

converges.

13. $$\int_1^b xe^{-x^2}dx = -\frac{1}{2}e^{-x^2}\Big|_1^b = \frac{1}{2}e^{-1} - \frac{1}{2}e^{-b^2};\text{ so } \sum_{k=1}^{\infty} ke^{-k^2}\text{ converges.}$$

14. $$\int_1^b x^{-3/4}dx = 4x^{1/4}\Big|_1^b = 4b^{1/4} - 4;\text{ so } \sum_{k=1}^{\infty} k^{-3/4}\text{ diverges.}$$

15. $\displaystyle\int_1^b \frac{2x + 1}{x^2 + x + 2}\,dx = \ln(x^2 + x + 2)\Big|_1^b = \ln(b^2 + b + 2) - \ln 4;$

so $\displaystyle\sum_{k=1}^{\infty} \frac{2k + 1}{k^2 + k + 2}$ diverges.

16. $\displaystyle\int_2^b \frac{x + 1}{(x^2 + 2x + 2)^2}\,dx = -\frac{1}{2}(x^2 + 2x + 1)^{-1}\Big|_2^b$

$\displaystyle= \frac{1}{18} - \frac{1}{2}(b^2 + 2b + 1)^{-1};$

so $\displaystyle\sum_{k=1}^{\infty} \frac{k + 1}{(k^2 + 2k + 1)^2}$ converges.

17. The series is $\displaystyle\sum_{k=0}^{\infty} \frac{3}{9 + k^2}$. Let $f(x) = \dfrac{3}{9 + x^2}$. Then $f(x) > 0$ for all $x \geq 0$, $f(x)$ is continuous, $f'(x) = -6x(9 + x^2)^{-2} < 0$ for all $x > 0$, so $f(x)$ is decreasing. Therefore the series converges.

18. Let $f(x) = \dfrac{e^{1/x}}{x^2}$. Then $f(x) > 0$ for all $x \geq 1$,

$$f'(x) = \frac{-\frac{1}{x^2}e^{1/x}x^2 - 2xe^{1/x}}{x^4} = \frac{-e^{1/x} - 2xe^{1/x}}{x^4} < 0 \text{ for all}$$

$x > 0$; so $f(x)$ is decreasing, positive and continuous for $x \geq 1$. $\displaystyle\int_1^b \frac{e^{1/x}}{x^2}\,dx = -e^{1/x}\Big|_1^b = 1 - e^{1/b}$; so $\displaystyle\sum_{k=1}^{\infty} \frac{e^{1/k}}{k^2}$ converges.

19. Let $f(x) = \dfrac{x}{e^x} = xe^{-x}$. Then $f(x) > 0$ for all $x \geq 1$, $f(x)$ is continuous and $f'(x) = \dfrac{e^x - xe^x}{e^{2x}} < 0$ for all $x \geq 1$.

Integrating by parts, $\int_1^b xe^{-x}dx = -xe^{-x} - e^{-x}\Big|_1^b$

$= 2e^{-1} - be^{-b} - e^{-b}.$

Thus $\sum_{k=1}^{\infty} ke^{-k}$ converges.

20. The series is convergent, since it is geometric with $r = \frac{3}{4} < 1$. The integral $\int_1^{\infty} \frac{3^x}{4^x}\,dx$ must also converge, since $f(x) = \left(\frac{3}{4}\right)^x = e^{\ln(3/4)x}$ is continuous, positive, decreasing $f'(x) = \ln\left(\frac{3}{4}\right)e^{\ln(3/4)x}$ and $\ln\left(\frac{3}{4}\right) < 0$. Therefore the integral test applies.

21. For all $k \ge 2$, $\frac{1}{k^2 + 5} < \frac{1}{k^2}$. The series $\sum_{k=2}^{\infty} \frac{1}{k^2}$ is shown in the test to be convergent. Thus $\sum_{k=2}^{\infty} \frac{1}{k^2 + 5}$ converges by the comparison test.

22. For $k \ge 2$, $\frac{1}{\sqrt{k^2 - 1}} > \frac{1}{\sqrt{k^2}} = \frac{1}{k}$. $\sum_{k=1}^{\infty} \frac{1}{k}$ diverges, so $\sum_{k=2}^{\infty} \frac{1}{\sqrt{k^2 - 1}}$ diverges.

23. For $k \ge 1$, $\frac{1}{2^k + k} < \frac{1}{2^k}$. $\sum_{k=1}^{\infty} \frac{1}{2^k}$ converges (it is geometric with $r = 1/2$) so $\sum_{k=1}^{\infty} \frac{1}{2^k + k}$ converges.

24. For $k \geq 1$, $\frac{1}{k3^k} < \frac{1}{3^k}$. $\sum_{k=1}^{\infty} \frac{1}{3^k}$ converges (geometric with $r = \frac{1}{3}$)

so $\sum_{k=1}^{\infty} \frac{1}{k3^k}$ converges.

25. For $k \geq 1$, $\frac{1}{5^k} \cos^2\left(\frac{k\pi}{4}\right) \leq \frac{1}{5^k}(1)$. $\sum_{k=1}^{\infty} \frac{1}{5^k}$ converges

(geometric with $r = 1$), so $\sum_{k=1}^{\infty} \frac{1}{5^k} \cos^2\left(\frac{k\pi}{4}\right)$ converges.

26. For $k \geq 0$, $\frac{1}{\left(\frac{3}{4}\right)^k + \left(\frac{5}{4}\right)^k} < \frac{1}{\left(\frac{5}{4}\right)^k} = \left(\frac{4}{5}\right)^k$. $\sum_{k=0}^{\infty} \left(\frac{4}{5}\right)^k$ converges,

so $\sum_{k=0}^{\infty} \frac{1}{\left(\frac{3}{4}\right)^k + \left(\frac{5}{4}\right)^k}$ converges.

27. No. In order for the comparison test to yield any information we would need $\frac{1}{k \ln k} > \frac{1}{k}$ for $k \geq 2$, which is false.

28. Yes. The first series converges, since for $k \geq 3$ $\frac{1}{k^2 \ln k} < \frac{1}{k^2}$. (Actually to apply the test, we need to start the series at $k = 3$, since $\ln 2 < 1$).

29. Area of top set of rectangles: T
Area of bottom set of rect: S
Area of combined set of rect: S + T

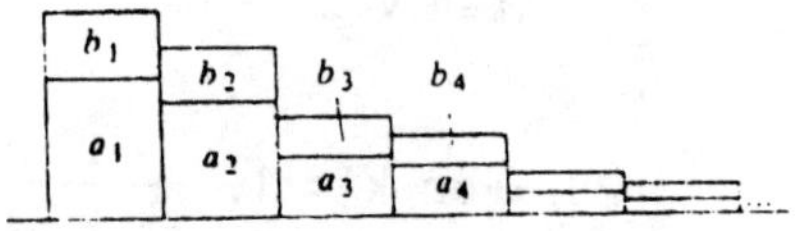

30. When the height of each rectangle is doubled; the area is doubled. Hence the area of the set of all the taller

rectangles is twice the area of the set of all the original rectangles.

31. $$\sum_{k=0}^{\infty} \frac{8^k + 9^k}{10^k} = \sum_{k=0}^{\infty} \left(\frac{8}{10}\right)^k + \sum_{k=0}^{\infty} \left(\frac{9}{10}\right)^k = \frac{1}{1 - \frac{8}{10}} + \frac{1}{1 - \frac{9}{10}}$$

$$= 5 + 10 = 15.$$

32. We know from the text that $\sum_{k=1}^{\infty} \frac{1}{k^2}$ is convergent. By exercise 30, $\sum_{k=1}^{\infty} 3\left(\frac{1}{k^2}\right)$ is also convergent. Since $e^{1/k} \le e < 3$ for all $k \ge 1$, the series $\sum_{k=1}^{\infty} \frac{e^{1/k}}{k^2}$ is convergent, by the comparison test.

Exercises 12.5

1. $f(x) = \frac{1}{2x + 3}$; $f'(x) = -\frac{2}{(2x + 3)^2}$; $f''(x) = \frac{2^2 \cdot 2}{(2x + 3)^3}$

$f'''(x) = \frac{-2^3 \cdot 2 \cdot 3}{(2x + 3)^4}$. The Taylor series at $x = 0$ is

$$f(x) = \frac{1}{3} - \frac{2}{9}x + \frac{2^2 \cdot 2}{3^3 \cdot 2!}x^2 - \frac{2^3 \cdot 3!}{3^4 \cdot 3!}x^3 + \ldots$$

$$= \frac{1}{3} - \frac{2}{9}x + \frac{4}{27}x^2 - \frac{2^3}{3^4}x^3 + \frac{2^4}{3^5}x^4 - \ldots$$

2. $f(x) = \ln(1 - 3x)$; $f'(x) = -\frac{3}{1 - 3x}$; $f''(x) = -\frac{3^2}{(1 - 3x)^2}$;

$f'''(x) = \frac{-3^3 \cdot 2}{(1 - 3x)^3}$; $f^{(4)}(x) = \frac{-3^4 \cdot 3!}{(1 - x)^4}$. The Taylor series at

$x = 0$ is $f(x) = 0 - 3x - \frac{9x^2}{2!} - \frac{3^3 \cdot 2!}{3!}x^3 - \frac{3^4 \cdot 3!}{4!}x^4 + \ldots$

3. $f(x) = (1 + x)^{-5/2}$; $f'(x) = \frac{1}{2}(1 + x)^{-1/2}$

$f''(x) = -\frac{1}{2^2}(1 + x)^{-3/2}$; $f'''(x) = \frac{3}{2^3}(1 + x)^{-5/2}$

$f^{(4)}(x) = \frac{-3.5}{2^4}(1 + x)^{-7/2}$. The Taylor series at $x = 0$ is

$$f(x) = 1 + \frac{1}{2}x - \frac{1}{2^2 \cdot 2!}x^2 + \frac{3}{2^3 \cdot 3!}x^3 - \frac{3 \cdot 5}{2^4 \cdot 4!}x^4 + \ldots$$

4. $f(x) = (1 + x)^3$; $f'(x) = 3(1 + x)^2$; $f''(x) = 6(1 + x)$

$f'''(x) = 6$; $f^{(n)}(x) = 0$ for all $n \geq 4$. the Taylor series at

$x = 0$ is $f(x) = 1 + 3x + \frac{6}{2!}x^2 + \frac{6}{3!}x^3 = 1 + 3x + 3x^2 + x^3$.

5. $\dfrac{1}{1 - 3x} = 1 + 3x + (3x)^2 + (3x)^3 + \ldots$

6. $\dfrac{1}{1 + x} = 1 + (-x) + (-x)^2 + (-x)^3 + (-x)^4 + \ldots$

7. $\dfrac{1}{1 + x^2} = 1 - x^2 + x^4 - x^6 + x^8 - \ldots$ (using ex. 6)

8. $\dfrac{x}{1 + x^2} = x - x^3 + x^5 - x^7 + x^9 - \ldots$ (using ex. 7)

9. $\dfrac{1}{(1 + x)^2} = -\dfrac{d}{dx}\left[\dfrac{1}{1 + x}\right] = 1 - 2x + 3x^2 - 4x^3 + \ldots$

(using ex.6)

10. $\dfrac{x}{(1 - x)^3} = \dfrac{x}{2}\dfrac{d^2}{dx^2}\left[\dfrac{1}{1 - x}\right] = \dfrac{x}{2}(0 + 0 + 2 + 6x + 4\cdot 3x^2 + 5\cdot 4x^3$

$+ 6\cdot 5x^4 + \ldots)$

$= x + \dfrac{3\cdot 2}{2}x^2 + \dfrac{4\cdot 3}{2}x^3 + \dfrac{5\cdot 4}{2}x^4 + \dfrac{6\cdot 5}{2}x^5 + \ldots$

11. $5e^{x/3} = 5 + 5\left(\dfrac{x}{3}\right) + \dfrac{5}{2!}\left(\dfrac{x}{3}\right)^2 + \dfrac{5}{3!}\left(\dfrac{x}{3}\right)^3 + \dfrac{5}{4!}\left(\dfrac{x}{3}\right)^4 + \ldots$

12. $x^3e^{x^2} = x^3\left[1 + x^2 + \dfrac{x^4}{2!} + \dfrac{x^6}{3!} + \dfrac{x^8}{4!} + \ldots\right]$

$= x^3 + x^5 + \dfrac{x^7}{2!} + \dfrac{x^{11}}{4!} + \ldots$

13. $1 - e^{-x} = 1 - \left[1 - x + \dfrac{(-x)^2}{2!} + \dfrac{(-x)^3}{3!} + \dfrac{(-x)^4}{4!} + \ldots\right]$

$= x - \dfrac{x^2}{2!} + \dfrac{x^3}{3!} - \dfrac{x^4}{4!} + \ldots$

14. $3(e^{-2x} - 2) = \left[3 - 6x + \dfrac{3(-2x)^2}{2!} + \dfrac{3(-2x)^3}{3!} + \dfrac{3(-2x)^4}{4!} + \ldots\right] - 6$

$= -3 - 6x + \dfrac{3\cdot 2^2x^2}{2!} - \dfrac{3\cdot 2^3x^3}{3!} + \dfrac{3\cdot 2^4x^4}{4!} - \ldots$

15. $\ln(1 + x) = \displaystyle\int \dfrac{1}{(1 + x)}\,dx + C$

$$= \int(1 - x + x^2 - x^3 + x^4 - \ldots)dx + C$$

$$= C + x - \frac{x^2}{2} + \frac{x^3}{3} - \frac{x^4}{4} + \frac{x^5}{5} - \ldots$$

For $x = 0$, $\ln(1 + x) = 0 = C + 0 + 0 + \ldots$; so $C = 0$ and

$$\ln(1 + x) = x - \frac{x^2}{2} + \frac{x^3}{3} - \frac{x^4}{4} + \frac{x^5}{5} - \ldots$$

16. $\ln(1 + x^2) = x^2 - \frac{x^4}{2} + \frac{x^6}{3} - \frac{x^8}{4} + \frac{x^{10}}{5} - \ldots$ (from ex. 15)

17. $\cos 3x = 1 - \frac{1}{2!}(3x)^2 + \frac{1}{4!}(3x)^4 - \ldots$

18. $\cos x^2 = 1 - \frac{1}{2!}x^4 + \frac{1}{4!}x^8 - \ldots$

19. $\sin 3x = 3x - \frac{1}{3!}(3x)^3 + \frac{1}{5!}(3x)^5 - \ldots$

20. $x \sin x^2 = x\left[x^2 - \frac{1}{3!}x^6 + \frac{1}{5!}x^{10} - \ldots\right]$

$$= x^3 - \frac{1}{3!}x^7 + \frac{1}{5!}x^{11} - \ldots$$

21. $xe^{x^2} = x\left[1 + x^2 + \frac{x^4}{2!} + \frac{x^6}{3!} + \frac{x^8}{4!} + \ldots\right]$

$$= x + x^3 + \frac{x^5}{2!} + \frac{x^7}{3!} + \frac{x^9}{4!} + \ldots$$

22. $\ln\left(\frac{1 + x}{1 - x}\right) = \ln(1 + x) - \ln(1 - x)$

$$= \left[x - \frac{x^2}{2} + \frac{x^3}{3} - \frac{x^4}{4} + \ldots\right] - \left[-x - \frac{x^2}{2} - \frac{x^3}{3} - \frac{x^4}{4} - \ldots\right]$$

$$= 2x + \frac{2x^3}{3} + \frac{2x^5}{5} + \ldots$$

23. (a) $f(x) = \frac{1}{2}(e^x + e^{-x})$; $f'(x) = \frac{1}{2}(e^x - e^{-x})$

$f''(x) = \frac{1}{2}(e^x + e^{-x})$ The Taylor expansion at $x = 0$ is

$$\cosh x = \frac{1}{2}(2) + \frac{1}{2}(2)\frac{x^2}{2!} + \frac{1}{2}(2)\frac{x^4}{4!} + \frac{1}{2}(2)\frac{x^6}{6!} + \ldots$$

$$= 1 + \frac{x^2}{2!} + \frac{x^4}{4!} + \frac{x^6}{6!} + \ldots$$

(b) $\cosh x = \frac{1}{2}\left(\left[1 + x + \frac{x^2}{2!} + \frac{x^3}{3!} + \frac{x^4}{4!} + \frac{x^5}{5!} + \cdots\right] + \left[1 - x + \frac{x^2}{2!} - \frac{x^3}{3!} + \frac{x^4}{4!} - \frac{x^5}{5!} + \cdots\right]\right)$

$= 1 + \frac{x^2}{2!} + \frac{x^4}{4!} + \frac{x^6}{6!} + \cdots$

24. $f(x) = \frac{1}{2}(e^x - e^{-x})$; $f'(x) = \frac{1}{2}(e^x + e^{-x})$; $f''(x) = f(x)$

The Taylor expansion of $f(x)$ at $x = 0$ is

$\sinh x = \frac{1}{2}(0) + \frac{1}{2}(2)x + \frac{1}{2}(0)\frac{x^2}{2!} + \frac{1}{2}(2)\frac{x^3}{3!} + \frac{1}{2}(0)\frac{x^4}{4!} + \frac{1}{2}(2)\frac{x^5}{5!} \cdots$

$= x + \frac{x^3}{3!} + \frac{x^5}{5!} + \frac{x^7}{7!} + \cdots$

(b) $\sinh x = \frac{1}{2}\left(\left[1 + x + \frac{x^2}{2!} + \frac{x^3}{3!} + \frac{x^4}{4!} + \cdots\right] - \left[1 - x + \frac{x^2}{2!} - \frac{x^3}{3!} + \frac{x^4}{4!} - \cdots\right]\right)$

$= x + \frac{x^3}{3!} + \frac{x^5}{5!} + \frac{x^7}{7!} + \cdots$

25. Substituting $-x$ for x in the given series yields

$\frac{1}{\sqrt{1 - x}} = 1 + \frac{1}{2}x + \frac{1}{2}\cdot\frac{3}{4}x^2 + \frac{1\cdot3\cdot5}{2\cdot4\cdot6}x^3 + \frac{1\cdot3\cdot5\cdot7}{2\cdot4\cdot6\cdot8}x^4 + \cdots$ at $x = 0$

26. Substituting x^2 for x in the given series of ex. 25 gives

$\frac{1}{\sqrt{1 - x^2}} = 1 + \frac{1}{2}x^2 + \frac{1\cdot3}{2\cdot4}x^4 + \frac{1\cdot3\cdot5}{2\cdot4\cdot6}x^6 + \frac{1\cdot3\cdot5\cdot7}{2\cdot4\cdot6\cdot8}x^8 + \cdots$

27. Substituting $-x^2$ for x in the series of ex. 25 gives

$\frac{1}{\sqrt{1 + x^2}} = 1 - \frac{1}{2}x^2 + \frac{1\cdot3}{2\cdot4}x^4 - \frac{1\cdot3\cdot5}{2\cdot4\cdot6}x^6 + \frac{1\cdot3\cdot5\cdot7}{2\cdot4\cdot6\cdot8}x^8 - \cdots$

Since $\ln\left(x + \sqrt{1 + x^2}\right) = \int \frac{1}{\sqrt{1 + x^2}}\,dx + C$, it follows that

$\ln\left(x + \sqrt{1 + x^2}\right) = \left[x - \frac{1}{2\cdot3}x^3 + \frac{1\cdot3}{2\cdot4\cdot5}x^5 - \frac{1\cdot3\cdot5}{2\cdot4\cdot6\cdot7}x^7 + \frac{1\cdot3\cdot5\cdot7}{2\cdot4\cdot6\cdot8\cdot9}x^9 - \cdots\right] + C.$

Since $\ln(0 + 1) = 0 = \left[0 + 0 + . . .\right] + C$, $C = 0$.

28. $\dfrac{x}{(1 - x^2)} = x\dfrac{d}{dx}\left[\dfrac{1}{1 - x}\right] = x + 2x^2 + 3x^3 + 4x^4 + 5x^5 + . . .$

Thus $\dfrac{d}{dx}\left[\dfrac{x}{(1 - x)^2}\right] = \dfrac{(1 - x)^2 + 2x(1 - x)}{(1 - x)^4} = \dfrac{1 - x^2}{(1 - x)^4}$

$= 1 + 4x + 9x^2 + 16x^3 + 25x^4 + . . .$

29. $e^x = 1 + x + \dfrac{x^2}{2!} + \dfrac{x^3}{3!} + . . .$

$\dfrac{d}{dx}\left[e^x\right] = 0 + 1 + \dfrac{2x}{2!} + \dfrac{3x^2}{3!} + . . .$

$= 1 + x + \dfrac{x^2}{2!} + \dfrac{x^3}{3!} + . . . = e^x$

30. $\cos x = 1 - \dfrac{1}{2!}x^2 + \dfrac{1}{4!}x^4 - \dfrac{1}{6!}x^6 + . . .$ Substituting $-x$ for x,

$\cos(-x) = 1 - \dfrac{1}{2!}(-x)^2 + \dfrac{1}{4!}(-x)^4 - \dfrac{1}{6!}(-x)^6 + . . .$

$= 1 - \dfrac{1}{2!}x^2 + \dfrac{1}{4!} - \dfrac{1}{6!}x^6 + . . . = \cos x.$

31. The coefficient of x^5 in the series must equal $\dfrac{f^{(5)}(0)}{5!}$.

Thus $f^{(5)}(0) = 5!\left(\dfrac{2}{5}\right) = 48.$

32. The coefficient of x^4 in the series must equal $\dfrac{f^{(4)}(0)}{4!}$.

Thus $f^{(4)}(0) = 4!\left(\dfrac{5}{24}\right) = 5.$

33. The coefficient x^4 in the series must equal $\dfrac{f^{(4)}(0)}{4!}$.

Thus $f^{(4)}(0) = 4!(0) = 0.$

34. The coefficient of x^2 in the given series will be the coefficient of x^4 in the expansion of $f(x)$. Thus

$\dfrac{f^{(4)}(0)}{4!} = 2$; $f^{(4)}(0) = 48.$

35. $\int e^{-x^2}dx = \int\left[1 - x^2 + \frac{x^4}{2!} - \frac{x^6}{3!} + \frac{x^8}{4!} - \ldots\right]dx$

$= \left[x - \frac{x^3}{3} + \frac{x^5}{5\cdot 2!} - \frac{x^7}{7\cdot 3!} + \frac{x^9}{9\cdot 4!} - \ldots\right] + C$

36. $\int xe^{x^3}dx = \int\left[x + x^4 + \frac{x^7}{2!} + \frac{x^{10}}{3!} + \ldots\right]dx$

$= \left[\frac{x^2}{2} + \frac{x^5}{5} + \frac{x^8}{8\cdot 2!} + \frac{x^{11}}{11\cdot 3!} + \ldots\right] + C.$

37. $\int\frac{1}{1 + x^3} = \int\left[1 - x^3 + x^6 - x^9 + \ldots\right]dx$

$= \left[x - \frac{x^4}{4} + \frac{x^7}{7} - \frac{x^{10}}{10} + \ldots\right] + C$

38. $\int_0^1 \sin x^2\, dx = \int_0^1\left[x^2 - \frac{1}{3!}x^6 + \frac{1}{5!}x^{10} - \ldots\right]dx$

$= \left[\frac{x^3}{3} - \frac{1}{7\cdot 3!}x^7 + \frac{1}{11\cdot 5!}x^{11} - \ldots\right]\Big|_0^1$

$= \frac{1}{3} - \frac{1}{7\cdot 3!} + \frac{1}{11\cdot 5!} - \ldots$

39. $\int_0^1 e^{-x^2}dx = \int_0^1\left[1 - x^2 + \frac{x^4}{2!} - \frac{x^6}{3!} + \frac{x^8}{4!} - \ldots\right]dx$

$= \left[x - \frac{x^3}{3} + \frac{x^5}{5\cdot 2!} - \frac{x^7}{7\cdot 3!} + \frac{x^9}{9\cdot 4!} - \ldots\right]\Big|_0^1$

$= 1 - \frac{1}{3} + \frac{1}{5\cdot 2!} - \frac{1}{7\cdot 3!} + \frac{1}{9\cdot 4!} - \ldots$

40. $\int_0^1 xe^{x^3}dx = \int_0^1\left[x + x^4 + \frac{x^7}{2!} + \frac{x^{10}}{3!} + \ldots\right]dx$

$= \left[\frac{x^2}{2} + \frac{x^5}{5} + \frac{x^8}{8\cdot 2!} + \frac{x^{11}}{11\cdot 3!} + \ldots\right]\Big|_0^1$

$= \frac{1}{2} + \frac{1}{5} + \frac{1}{8\cdot 2!} + \frac{1}{11\cdot 3!} + \ldots$

41. (a) $e^x = 1 + x + \frac{x^2}{2!} + \frac{x^3}{3!} + \ldots$

Since the above expansion is valid for all x and all of the

terms are positive for $x > 0$, it follows that

$e^x > 1 + x + \frac{x^2}{2!} > \frac{x^2}{2}$.

(b) For $x > 0$, e^x and $\frac{x^2}{2}$ are both positive. Thus from $\frac{x^2}{2} < e^x$, it follows that $\frac{1}{(x^2/2)} > \frac{1}{e^x}$; or $\frac{2}{x^2} > e^{-x}$.

(c) For $x > 0$, from (b) we have $xe^{-x} < \frac{2x}{x^2} = \frac{2}{x}$. Now $xe^{-x} > 0$ for all $x > 0$. Thus $0 < xe^{-x} < \frac{2}{x}$ for all $x > 0$. Since $\frac{2}{x} \to 0$ as $x \to \infty$, it follows that $\lim_{x\to\infty} xe^{-x} = 0$.

42. (a) The expansion $e^{kx} = 1 + kx + \frac{k^2x^2}{2!} + \frac{k^3x^3}{3!} + \ . \ . \ .$ is valid for all x. Assuming k and x are both positive, all terms of the series are positive. Hence for $x > 0$, $e^{kx} > 1 + kx + \frac{k^2x^2}{2!} > \frac{k^2x^2}{2}$.

(b) For $x > 0$, e^{kx} and $\frac{k^2x^2}{2}$ are both positive. Thus from $e^{kx} > \frac{k^2x^2}{2}$, we may conclude $\frac{1}{e^{kx}} < \frac{1}{\left(\frac{k^2x^2}{2}\right)}$; or $e^{-kx} < \frac{2}{k^2x^2}$.

(c) For $x > 0$, from (b) we have $xe^{-kx} < \frac{2x}{k^2x^2} = \frac{2}{k^2x}$. Now $xe^{-kx} > 0$ for all $x > 0$. Thus $0 < xe^{-kx} < \frac{2}{k^2x}$. Since $\frac{2}{k^2x} \to 0$ as $x \to \infty$, it follows that $\lim_{x\to\infty} xe^{-kx} = 0$.

43. The expression $e^x = 1 + x + \frac{x^2}{2!} + \frac{x^3}{3!} + \frac{x^4}{4!} + \ . \ . \ .$ is valid for all x. For $x > 0$, all terms in the series are positive. Therefore, for $x > 0$, $e^x > 1 + x + \frac{x^2}{2} + \frac{x^3}{6} > \frac{x^3}{6}$. Thus for $x > 0$, $\frac{1}{e^x} < \frac{1}{\left(\frac{x^3}{6}\right)}$; or $e^{-x} < \frac{6}{x^3}$, which implies

$x^2e^{-x} < \frac{6x^2}{x^3} = \frac{6}{x}$. Therefore, for $x > 0$, $0 < x^2e^{-x} < \frac{6}{x}$. Since $\lim_{x\to\infty} \frac{6}{x} = 0$, it follows that $\lim_{x\to\infty} x^2e^{-x} = 0$.

44. Replace x by kx in the solution to ex. 43.

Exercises 12.6

1. $\left(\frac{1}{2}\right)^4 = \frac{1}{16}$ 2. $\left(\frac{1}{2}\right)^4 + \left(\frac{1}{2}\right)^6 = \frac{5}{64}$

3. $1 - \left[\frac{1}{2} + \left(\frac{1}{2}\right)^2 + \left(\frac{1}{2}\right)^3\right] = \frac{1}{8}$

4. $\frac{1}{2} + \left(\frac{1}{2}\right)^3 + \left(\frac{1}{2}\right)^5 + . . . = \dfrac{\frac{1}{2}}{1 - \frac{1}{4}} = \frac{2}{3}$

5. $\left(\frac{1}{2}\right)^4 + \left(\frac{1}{2}\right)^6 + \left(\frac{1}{2}\right)^8 + . . . \dfrac{\left(\frac{1}{4}\right)^4}{1 - \frac{1}{4}} = \frac{1}{12}$

6. $\frac{1}{2} + \left(\frac{1}{2}\right)^6 + \left(\frac{1}{2}\right)^{11} + . . . = \dfrac{\frac{1}{6}}{1 - \left(\frac{1}{2}\right)^5} = \frac{16}{31}$ (we consider 0 to be divisible by 5)

7. $\left(\frac{2}{3}\right)^n\left(\frac{1}{3}\right)$

8. $0\left(\frac{1}{3}\right) + 1\left(\frac{2}{3}\right)\left(\frac{1}{3}\right) + 2\left(\frac{2}{3}\right)^2\left(\frac{1}{3}\right) + n\left(\frac{2}{3}\right)^n\left(\frac{1}{3}\right) + . . .$

$= \left(\frac{2}{3}\right)\frac{1}{3}\left[1 + 2\left(\frac{2}{3}\right) + 3\left(\frac{2}{3}\right)^2 + . . . + n\left(\frac{2}{3}\right)^{n-1} + . . .\right]$

$= \frac{2}{9}\left[\dfrac{1}{\left(1 - \frac{2}{3}\right)^2}\right] = 2$ (using the power series for $\dfrac{1}{1 - x^2}$)

9. $\frac{1}{3} + \left(\frac{2}{3}\right)^3\left(\frac{1}{3}\right) + \left(\frac{2}{3}\right)^6\left(\frac{1}{3}\right) + . . . = \dfrac{\frac{1}{3}}{1 - \left(\frac{2}{3}\right)^3} = \frac{9}{19}$

10. $p_0 = e^{-5} \sim .00674$ 11. $p_2 = \frac{5^2}{1 \cdot 2}e^{-5} \sim .08524$

12. $\left[\frac{5^2}{1 \cdot 2} + \frac{5^3}{1 \cdot 2 \cdot 3}\right]e^{-5} \sim .22467$

13. $p_0 + p_1 + (p_2 + p_3) + p_4 \sim .44063$ 14. 5

15. $1 - (p_0 + p_1) \sim .95956$ 16. $p_0 = e^{-10} = .0000453$

17. $1 - \left[p_0 + p_1 + p_2\right] = 1 - \left[e^{-10} + 10e^{-10} + \frac{100}{2}e^{-10}\right] \sim .99723$

18. 10 19. $p_1 + p_2 + p_3 = 10e^{-10} + \frac{100}{2!}e^{-10} + \frac{10^3}{3!}e^{-10} \sim .01029$

20. $\frac{dp}{dx} = 5x^4 - 6x^5 = x^4(5 - 6x)$; $\frac{dp}{dx} = 0$ when $x = 0$ and when $x = \frac{5}{6}$.

$\frac{d^2p}{dx^2} = 20x^3 - 30x^2 = x^2(20x - 30)$. The graph of $p(x)$ is concave downward for $0 < x < 1$. Thus $x = \frac{5}{6}$ is a relative maximum and since $p(5/6) > p(0)$ it follows that $x = \frac{5}{6}$ gives the maximum value of p for $0 \le x \le 1$.

21. $p = \frac{\lambda^2}{2}e^{-\lambda}$; $\frac{dp}{d\lambda} = \lambda e^{-\lambda} - \frac{\lambda^2}{2}e^{-\lambda} = \lambda e^{-\lambda}\left(1 - \frac{\lambda}{2}\right)$; $\frac{dp}{d\lambda} = 0$

when $\lambda = 0, 2$. $\frac{d^2p}{d\lambda^2} = e^{-\lambda} - \lambda e^{-\lambda} - \lambda e^{-\lambda} + \frac{\lambda^2}{2}e^{-\lambda}$

$$= e^{-\lambda}\left(1 - 2\lambda + \frac{\lambda^2}{2}\right).$$

At $\lambda = 2$, $\frac{d^2p}{d\lambda^2} < 0$, so $\lambda = 2$ is a relative maximum value.

Since $p(2) > p(0) = 0$ and $\frac{dp}{d\lambda} < 0$ for $\lambda > 2$, it follows that $\lambda = 2$ gives the maximum value of p for $\lambda > 0$.

22. The variance of the random variable in example 1 is given by the series $\frac{1}{2} + 1^2 \cdot \frac{1}{2^3} + 2^2 \cdot \frac{1}{2^4} + 3^2 \cdot \frac{1}{2^5} + \ldots$ which is the series in ex. 28, section 12.6 with $x = \frac{1}{2}$. In that exercise, we saw that the given series was the Taylor series for $f(x) = \frac{1 + x}{(1 - x)^2}$. Therefore, the above series sums

to $f(1/2) = \dfrac{1 + \frac{1}{2}}{\left(1 - \frac{1}{2}\right)^2} = 2$.

23. If r is the number of raisins used in the batter and X is the number of raisins in a particular cookie, then $E(X) = \frac{r}{100}$ and $p_0 = e^{-r/100} = .01$. Thus $-\frac{r}{100} = \ln(.01)$; $r \sim 461$.

24. The probability that X is even is given by the series

$$\left[e^{-\lambda} + \frac{\lambda^2}{2!}e^{-\lambda} + \frac{\lambda^4}{4!}e^{-\lambda} + \ldots\right] = e^{-\lambda}\left[1 + \frac{\lambda^2}{2!} + \frac{\lambda^4}{4!} + \ldots\right].$$

The second series is the power series for $\cosh\lambda$ obtained in sect. 12.6, ex. 23.

25. The probability that X is odd is given by the series

$$\left[\lambda e^{-\lambda} + \frac{\lambda^3}{3!}e^{-\lambda} + \frac{\lambda^5}{5!}e^{-\lambda} + \ldots\right] = e^{-\lambda}\left[\lambda + \frac{\lambda^3}{3!} + \frac{\lambda^5}{5!} + \ldots\right].$$

The second series is the power series for $\sinh\lambda$ obtained in sect. 12.6, ex. 24.

Chapter 12 Supplementary Exercises

1. $f(x) = x(x + 1)^{3/2}$; $f'(x) = (x + 1)^{3/2} + \frac{3}{2}x(x + 1)^{1/2}$

$$f''(x) = \frac{3}{2}(x + 1)^{1/2} + \frac{3}{2}(x + 1)^{1/2} + \frac{3}{4}x(x + 1)^{-1/2}$$
$$= 3(x + 1)^{1/2} + \frac{3}{4}x(x + 1)^{-1/2}$$
$$p_2(x) = 0 + x + \frac{3}{2!}x^2 = x + \frac{3}{2}x^2$$

2. $f(x) = (2x + 1)^{3/2}$; $f'(x) = 3(2x + 1)^{1/2}$

$f''(x) = 3(2x + 1)^{-1/2}$; $f'''(x) = -3(2x + 1)^{-3/2}$

$f^{(4)}(x) = 9(2x + 1)^{-5/2}$.

$$p_4(x) = 1 + 3x + \frac{3x^2}{2!} - \frac{3x^3}{3!} + \frac{9x^4}{4!} = 1 + 3x + \frac{3}{2}x^2 - \frac{1}{2}x^3 + \frac{3}{8}x^4.$$

3. For all $n \geq 3$, $p_n(x) = x^3 - 7x^2 + 8$.

4. $f(x) = \frac{2}{2 - x} = \frac{1}{1 - \frac{x}{2}}$

$$= 1 + \frac{x}{2} + \left(\frac{x}{2}\right)^2 + \left(\frac{x}{2}\right)^3 + \ldots + \left(\frac{x}{2}\right)^n + \ldots$$

So $p_n(x) = 1 + \frac{x}{2} + \frac{1}{2^2}x^2 + \frac{1}{2^3}x^3 + \ldots + \frac{1}{2^n}x^n$

5. $f(x) = x^2$, $f'(x) = 2x$, $f''(x) = 2$, $f^{(n)}(x) = 0$ for $n \geq 3$.

$$p_3(x) = 3^2 + 2(3)(x - 3) + \frac{2}{2!}(x - 3)^2 + 0$$
$$= 9 + 6(x - 3) + (x - 3)^2.$$

6. $f(x) = e^x = f^{(n)}(x)$ for all $n \geq 1$.

$$p_3(x) = e^2 + e^2(x - 2) + \frac{e^2}{2!}(x - 2)^2 + \frac{e^2}{3!}(x - 2)^3.$$

7. $f(t) = -\ln(\cos 2t)$, $f'(t) = \frac{2\sin 2t}{\cos 2t}$; $f''(t) = 4 + \frac{4\sin^2 2t}{\cos^2 2t}$

$$p_2(t) = 0 + \frac{0}{1!}t + \frac{4}{2!}t^2 = 2t^2.$$
$$\int_0^{1/2} f(t)dt \sim \int_0^{1/2} 2t^2 dt = \frac{2}{3}\left(\frac{3}{8}\right) = \frac{1}{12}.$$

8. $f(x) = \tan x$; $f'(x) = \sec^2 x$; $f''(x) = (2\sec x)\sec x \tan x$
$$= 2\sec^2 x \tan x$$

$p_2(x) = 0 + x + 0 = x$. $\tan(.1) \sim p_2(.1) = .1$.

9. (a) $f(x) = x^{1/2}$; $f'(x) = \frac{1}{2}x^{-1/2}$; $f''(x) = -\frac{1}{4}x^{-3/2}$.

$$p_2(x) = 3 + \frac{1}{6}(x - 9) - \frac{1}{216}(x - 9)^2$$

(b) $p_2(8.7) = 3 + \frac{1}{6}(-.3) - \frac{1}{216}(-.3)^2 = 2.989583$

(c) $f(x) = x^2 - 8.7$; $f'(x) = 2x$. $x_0 = 3$, $x_1 = 3 - \frac{.3}{6} = 2.95$

$$x_2 = 2.95 - \frac{(2.95)^2 - 8.7}{2(2.95)} = 2.949576$$

16. The series is geometric with $a = \frac{4}{7}$, $r = -\frac{8}{7}$, so it diverges.

17. The series is geometric with $a = \frac{1}{m + 1}$, $r = \frac{m}{m + 1}$, so (since $m > 0$) it converges to $\dfrac{\frac{1}{m+1}}{1 - \frac{m}{m+1}} = \frac{1}{m+1}\left(\frac{m+1}{1}\right) = 1$.

18. The series is geometric with $a = \frac{1}{m}$, $r = -\frac{1}{m}$, so it converges iff $m > 1$. In this case, the sum is $\dfrac{\frac{1}{m}}{1 + \frac{1}{m}} = \frac{1}{m + 1}$.

19. This is the Taylor series for e^x with $x = 2$. Thus the sum is e^2.

20. This is the Taylor series for e^x with $x = \frac{1}{3}$. Thus the sum is $e^{1/3}$.

21. Since $\sum_{k=0}^{\infty} \frac{1}{3^k}$ converges to $\dfrac{1}{1 - \frac{1}{3}} = \frac{3}{2}$ and $\sum_{k=0}^{\infty} \left(\frac{2}{3}\right)^k$ converges to $\dfrac{1}{1 - \frac{2}{3}} = 3$; $\sum_{k=0}^{\infty} \frac{1}{3^k} + \left(\frac{2}{3}\right)^k = \sum_{k=0}^{\infty} \frac{1 + 2^k}{3^k} = \frac{3}{2} + 3 = \frac{9}{2}$.

22. $$\sum_{k=0}^{\infty} \frac{3^k + 5^k}{7^k} = \sum_{k=0}^{\infty} \left(\frac{5}{7}\right)^k = \frac{1}{1 - \frac{3}{7}} + \frac{1}{1 - \frac{5}{7}} = \frac{7}{4} + \frac{7}{2} = \frac{21}{4}.$$

23. Area of blocks $= \frac{1}{2} + \frac{1}{3} + \frac{1}{4} + \ldots + \frac{1}{n}$

Area under the curve from 1 to n $= \int_1^n \frac{1}{x}\,dx = \ln|x|\Big|_1^n = \ln n$

and so, $\frac{1}{2} + \frac{1}{3} + \frac{1}{4} + \ldots + \frac{1}{n} < \ln n$

or $\sum_{k=1}^{n} \frac{1}{k} < 1 + \ln n$

24. Area of blocks $= 1 + \frac{1}{2} + \frac{1}{3} + \frac{1}{4} + \ldots + \frac{1}{n}$

Area under the curve from 1 to n + 1 $= \int_1^{n+1} \frac{1}{x}\,dx = \ln|x|\Big|_1^{n+1}$

$= \ln(n + 1)$, and so, $1 + \frac{1}{2} + \frac{1}{3} + \frac{1}{4} + \ldots + \frac{1}{n} < \ln(n + 1)$

or $\ln(n + 1) < \sum_{k=1}^{n} \frac{1}{k}$

25. Since $\int_0^{\infty} \frac{1}{x^3}\,dx = \lim_{b\to\infty} \left[-\frac{1}{2}x^{-2}\Big|_1^b\right] = \lim_{b\to\infty} \left[\frac{1}{2} - \frac{1}{2b^2}\right] = \frac{1}{2}$, the given series converges by the integral test.

26. The series is geometric with $r = \frac{1}{3}$, so it converges.

27. $\int_1^{\infty} \frac{\ln x}{x}\,dx = \lim_{b\to\infty} \left[\frac{(\ln x)^2}{2}\Big|_1^b\right] = \lim_{b\to\infty} \left[\frac{(\ln b)^2}{2}\right] = \infty$. Thus the series diverges by the integral test.

28. $\frac{k^3}{(k^4 + 1)^2} = \frac{k^3}{k^8 + 2k^4 + 1} \le \frac{k^3}{k^8} = \frac{1}{k^5}$ for $k \ge 0$. Thus since $\sum_{k=0}^{\infty} \frac{1}{k^5}$ converges by the integral test, $\sum_{k=0}^{\infty} \frac{k^3}{(k^4 + 1)^2}$ converges by the comparison test.

29. The series converges iff $\int_1^{\infty} \frac{1}{x^p}\,dx$ converges (by the integral test).

$\int_1^{\infty} \frac{1}{x^p}\,dx = \lim_{b\to\infty} \left[\frac{1}{-p + 1}x^{-p+1}\Big|_1^b\right] = \lim_{b\to\infty} \left[\frac{1}{1 - p}\left(b^{1-p} - 1\right)\right]$. This limit is finite iff $p > 1$.

30. $\dfrac{\frac{1}{p^{k+1}}}{\frac{1}{p^k}} = \dfrac{1}{p}$, so if $|p| > 1$ the series converges, and it $|p| < 1$ the series diverges by the ratio test. If $p = 1$, the series is $1 + 1 + 1 + . . .$ which diverges. Therefore, the series converges iff $|p| > 1$.

31. Replacing x by $-x^3$ in the series for $\dfrac{1}{1 - x}$ gives

$$\frac{1}{1 + x^3} = 1 - x^3 + x^6 - x^9 + x^{12} - . . .$$

32. $\dfrac{d}{dx}\left[\ln(1 + x^3)\right] = \dfrac{3x^2}{1 + x^3}$; so

$$\ln(1 + x^3) = \int\left[3x^2 - 3x^5 + 3x^8 - 3x^{11} + 3x^{14} - . . .\right]dx + C$$

(using the expansion of $\dfrac{1}{1 + x^3}$ in ex. 31]

$$= \left[x^3 - \frac{3}{6}x^6 + \frac{3}{9}x^9 - \frac{3}{12}x^{12} + . . .\right] + C$$

$\ln(1) = 0 = 0 + 0 + . . . + C$; so $C = 0$.

33. $$\frac{1}{(1 - 3x)^2} = \frac{1}{3}\frac{d}{dx}\left[\frac{1}{1 - 3x}\right] = \frac{1}{3}\frac{d}{dx}\left[1 + 3x + 3^2x^2 + 3^3x^3 + . . .\right]$$

$$= 1 + 6x + 27x^2 + 4\cdot 3^3x^3 + . . .]$$

34. $$\frac{e^x - 1}{x} = \frac{1}{x}\left[x + \frac{x^2}{2!} + \frac{x^3}{3!} + . . .\right] = 1 + \frac{1}{2!}x + \frac{1}{3!}x^2 + \frac{1}{4!}x^3 + . . .$$

35. (a) $$\cos 2x = 1 - \frac{1}{2!}(2x)^2 + \frac{1}{4!}(2x)^4 - \frac{1}{6!}(2x)^6 + . . .$$

$$= 1 - \frac{2^2}{2!}x^2 + \frac{2^4}{4!}x^4 - \frac{2^6}{6!}x^6 + . . .$$

(b) $\sin^2 x = \frac{1}{2}(1 - \cos 2x) = \frac{1}{2} - \frac{1}{2}\cos 2x$

$$= \frac{1}{2} - \frac{1}{2}\left[1 - \frac{2^2}{2!}x^2 + \frac{2^4}{4!}x^4 - \frac{2^6}{6!}x^6 + \ldots\right]$$

$$= \frac{2}{2!}x^2 - \frac{2^3}{4!}x^4 + \frac{2^5}{6!}x^6 - \ldots$$

36. (a) $\cos 3x = 1 - \frac{1}{2!}(3x)^2 + \frac{1}{4!}(3x)^4 - \frac{1}{6!}(3x)^6 + \ldots$

$$= 1 - \frac{3^2}{2!}x^2 + \frac{3^4}{4!}x^4 - \frac{3^6}{6!}x^6 + \ldots$$

(b) Adding the first four terms above to the corresponding terms in the expansion of 3cos x and multiplying by $\frac{1}{4}$ gives

$$p_4(x) = \frac{1}{4}\left[(1 + 3) - \left(\frac{3^2}{2!} + \frac{3}{2!}\right)x^2 + \left(\frac{3^4}{4!} + \frac{3}{4!}\right)x^4\right]$$

$$= 1 - \frac{3}{2}x^2 + \frac{7}{8}x^4.$$

37. $\frac{1 + x}{1 - x} = \frac{1}{1 - x} + \frac{x}{1 - x} = \left[1 + x + x^2 + x^3 + \ldots\right]$

$$+ \left[x + x^2 + x^3 + x^4 + \ldots\right]$$

$$= 1 + 2x + 2x^2 + 2x^3 + \ldots$$

38. Using ex. 34,

$$\int_0^{1/2} \frac{e^x - 1}{x}\,dx = \int_0^{1/2} \left[1 + \frac{1}{2}x + \frac{1}{3!}x^2 + \frac{1}{4!}x^3 + \ldots\right]$$

$$= \left[x + \frac{1}{4}x^2 + \frac{1}{3!\cdot 3}x^3 + \frac{1}{4!\cdot 4}x^4 + \ldots\right]\Bigg|_0^{1/2}$$

$$= \frac{1}{2} + \frac{1}{4\cdot 2^2} + \frac{1}{3!\cdot 3\cdot 2^3} + \frac{1}{4!\cdot 4\cdot 2^4} + \ldots$$

39. (a) x^2 (b) 0

(c) $\int_0^1 \sin x^2\,dx \sim \int_0^1 (x^2 - \frac{1}{6}x^6)dx = \frac{x^3}{3} - \frac{1}{42}x^7\Big|_0^1 = \frac{1}{3} - \frac{1}{42}$

$\sim .3095$ (exact value: .3103)

40. $p_4(x) = x + \frac{1}{3!}x^3$

41. (a) $f'(x) = 2x + 4x^3 + 6x^5 + \ldots$

(b) The series given for f(x) is the Taylor series of $\dfrac{1}{1 - x^2}$.

Thus $f(x) = \dfrac{1}{1 - x^2}$ and $f'(x) = \dfrac{2x}{(1 - x^2)^2}$.

42. (a) $\int f(x)\,dx = \int\left[x - 2x^3 + 4x^5 - 8x^7 + 16x^9 - \ldots\right]dx$

$= \left[\frac{1}{2}x^2 - \frac{1}{2}x^4 + \frac{2}{3}x^6 - x^8 + \frac{8}{5}x^{10} - \ldots\right] + C$

(b) The series given for

$f(x) = x\left[1 - 2x^2 + 2^2x^4 - 2^3x^6 + x^4x^8 - \ldots\right]$ is the Taylor

expansion of $\dfrac{x}{1 + 2x^2}$. Thus $f(x) = \dfrac{x}{1 + 2x^2}$ and

$\int f(x)\,dx = \frac{1}{4}\ln(1 + 2x^2) + C$

43. $100 + 100(.85) + 100(.85)^2 + 100(.85)^3 + \ldots = \dfrac{100}{1 - .85}$

$= 666.6\underline{6}$

Thus $566,666,667 new dollars are "created" beyond the original $100 million.

44. $100 + (.85)100 + (.80)(.85)^2 100 + (.80)^2(.85)^3 100 + \ldots$

$= 1000 + \dfrac{85}{1 - (.80)(.85)} = 365.625$ (million dollars), or

$265,625,000 beyond the original $100 million.

45. $\sum_{k=1}^{\infty} 10,000\, e^{-.08k} = \sum_{k=1}^{\infty} 10,000\,(e^{-.08})^k = \dfrac{10,000(e^{-.08})}{1 - e^{-.08}}$

$\sim \$120,066.70$

46. $\sum_{k=1}^{\infty} 10,000(.9)^k e^{-.08k} = \sum_{k=1}^{\infty} 10,000(.9e^{-.08})^k$

$= \dfrac{10,000(.9)e^{-.08}}{1 - .9e^{-.08}} \sim \$49,\ 103$

47. $\sum_{k=1}^{\infty} 10,000(1.08)^k e^{-.08k} = \sum_{k=1}^{\infty} 10,000(1.08e^{-.08})^k$

$$= \frac{10,000(1.08)e^{-.08}}{1 - 1.08e^{-.08}} \sim \$3,285,603$$

48. Yes, by comparison with $\sum_{k=1}^{\infty} M(e^{-.08})^k$.

49. $p_n = \left(\frac{7}{9}\right)^n \left(\frac{2}{9}\right)$

50. $p_1 + p_3 + p_5 + \cdots = \frac{7}{9}\left(\frac{2}{9}\right) + \left(\frac{7}{9}\right)^3\left(\frac{2}{9}\right) + \left(\frac{7}{9}\right)^5\left(\frac{2}{9}\right) + \cdots$

$$= \frac{\frac{7}{9}\left(\frac{2}{9}\right)}{1 - \left(\frac{7}{9}\right)^2} = \frac{7}{16}$$

51. $1\cdot p_1 + 2\cdot p_2 + 3\cdot p_3 + \cdots = \frac{7}{9}\left(\frac{2}{9}\right) + 2\cdot\left(\frac{7}{9}\right)^2\left(\frac{2}{9}\right) + 3\cdot\left(\frac{7}{9}\right)^3\left(\frac{2}{9}\right) + \cdots$

$$= \left(\frac{7}{9}\right)\left(\frac{2}{9}\right)\left[\frac{1}{\left(1 - \frac{7}{9}\right)^2}\right]$$ (using expansion of $\frac{1}{(1 - x)^2}$)

$$= \frac{7}{2}$$

52. $p_4 = \frac{4^4}{4!}e^{-4} \sim .19537$

53. $p_8 = 1 - \left[p_0 + p_1 + p_2 + p_3 + p_4 + p_5 + p_6 + p_7\right] \sim .05113$

54. $E(X) = \lambda = 4.$

NOTES

NOTES

NOTES

NOTES

NOTES

NOTES

NOTES

$\left(\frac{1}{2}\right)^n = \frac{1}{10}$

$2^n = 10$